ELECTROCHEMICAL REMEDIATION TECHNOLOGIES FOR POLLUTED SOILS, SEDIMENTS AND GROUNDWATER

ELECTROCHEMICAL REMEDIATION TECHNOLOGIES FOR POLLUTED SOILS, SEDIMENTS AND GROUNDWATER

Edited by

KRISHNA R. REDDY
University of Illinois at Chicago

CLAUDIO CAMESELLE
University of Vigo

A JOHN WILEY & SONS, INC., PUBLICATION

Copyright © 2009 by John Wiley & Sons, Inc. All rights reserved

Published by John Wiley & Sons, Inc., Hoboken, New Jersey
Published simultaneously in Canada

No part of this publication may be reproduced, stored in a retrieval system, or transmitted in any form or by any means, electronic, mechanical, photocopying, recording, scanning, or otherwise, except as permitted under Section 107 or 108 of the 1976 United States Copyright Act, without either the prior written permission of the Publisher, or authorization through payment of the appropriate per-copy fee to the Copyright Clearance Center, Inc., 222 Rosewood Drive, Danvers, MA 01923, (978) 750-8400, fax (978) 750-4470, or on the web at www.copyright.com. Requests to the Publisher for permission should be addressed to the Permissions Department, John Wiley & Sons, Inc., 111 River Street, Hoboken, NJ 07030, (201) 748-6011, fax (201) 748-6008, or online at http://www.wiley.com/go/permission.

Limit of Liability/Disclaimer of Warranty: While the publisher and author have used their best efforts in preparing this book, they make no representations or warranties with respect to the accuracy or completeness of the contents of this book and specifically disclaim any implied warranties of merchantability or fitness for a particular purpose. No warranty may be created or extended by sales representatives or written sales materials. The advice and strategies contained herein may not be suitable for your situation. You should consult with a professional where appropriate. Neither the publisher nor author shall be liable for any loss of profit or any other commercial damages, including but not limited to special, incidental, consequential, or other damages.

For general information on our other products and services or for technical support, please contact our Customer Care Department within the United States at (800) 762-2974, outside the United States at (317) 572-3993 or fax (317) 572-4002.

Wiley also publishes its books in a variety of electronic formats. Some content that appears in print may not be available in electronic formats. For more information about Wiley products, visit our web site at www.wiley.com.

Library of Congress Cataloging-in-Publication Data:

Reddy, Krishna R.
  Electrochemical remediation technologies for polluted soils, sediments and groundwater / Krishna R. Reddy, Claudio Cameselle.
    p. cm.
  Includes index.
  ISBN 978-0-470-38343-8 (cloth)
  1. Soil remediation. 2. Soils–Electric properties. 3. Contaminated sediments. 4. Groundwater–Pollution. 5. Electrokinetics. 6. Electrolysis. I. Cameselle, Claudio, 1967– II. Title.
  TD878.47.R43 2009
  628.5′5–dc22

2009009333

Printed in the United States of America

10 9 8 7 6 5 4 3 2 1

CONTENTS

PREFACE   xv

CONTRIBUTORS   xix

PART I   Introduction and Basic Principles   1

1   Overview of Electrochemical Remediation Technologies   3
Krishna R. Reddy and Claudio Cameselle

    1.1. Introduction, 3
    1.2. Electrochemical Technologies for Site Remediation, 4
    1.3. Electrochemical Transport, Transfer, and Transformation Processes, 6
    1.4. Electrochemical Removal of Inorganic Pollutants, 11
    1.5. Electrochemical Removal of Organic Pollutants, 13
    1.6. Electrochemical Removal of Contaminant Mixtures, 15
    1.7. Special Considerations in Remediating Polluted Sediments, 17
    1.8. Electrokinetic Barriers for Pollution Containment, 17
    1.9. Coupled (or Integrated) Electrochemical Remediation Technologies, 18
    1.10. Mathematical Modeling of Electrochemical Remediation, 23
    1.11. Economic and Regulatory Considerations, 24
    1.12. Field Applications and Lessons Learned, 25
    1.13. Future Directions, 27
    References, 28

2   Electrochemical Transport and Transformations   29
Sibel Pamukcu

    2.1. Introduction, 29
    2.2. Overview, 29

2.3. Electrochemical Transport in Bulk Fluid, 30
2.4. Electrochemical Transport in Clays in the Direction of Applied Electric Field, 31
2.5. Electrochemical Transformations, 50
2.6. Summary, 60
References, 61

3 Geochemical Processes Affecting Electrochemical Remediation   65

Albert T. Yeung

3.1. Introduction, 65
3.2. Soil–Fluid–Chemical System as Active Electrochemical System, 67
3.3. Generation of pH Gradient, 70
3.4. Change of Zeta Potential of Soil Particle Surfaces, 72
3.5. Change in Direction of Electroosmotic Flow, 76
3.6. Sorption and Desorption of Contaminants onto/from Soil Particle Surfaces, 79
3.7. Buffer Capacity of Soil, 82
3.8. Complexation, 83
3.9. Oxidation–Reduction (Redox) Reactions, 87
3.10. Interactions of Geochemical Processes, 89
3.11. Summary, 90
References, 91

PART II   Remediation of Heavy Metals and Other Inorganic Pollutants   95

4 Electrokinetic Removal of Heavy Metals   97

Lisbeth M. Ottosen, Henrik K. Hansen, and Pernille E. Jensen

4.1. Introduction, 97
4.2. Principle of EK Removal of Heavy Metals from Soils, 98
4.3. Heavy Metal and Soil Type, 99
4.4. Enhancement Methods, 111
4.5. Remediation of Mine Tailings, Ashes, Sediments, and Sludge, 116
4.6. Summary, 119
References, 120

5 Electrokinetic Removal of Radionuclides   127

Vladimir A. Korolev

5.1. Introduction, 127
5.2. Electrokinetic Localization of Radioactive Nuclide Pollution, 129
5.3. Electrokinetic Cleaning of Ground from Radioactive Nuclides, 132
5.4. Summary, 138
References, 138

6 Electrokinetic Removal of Nitrate and Fluoride 141
Kitae Baek and Jung-Seok Yang

    6.1. Introduction, 141
    6.2. Pollution and Health Effects of Anionic Pollutants, 142
    6.3. Removal of Anionic Pollutants by Electrokinetics, 143
    6.4. Summary, 146
    References, 147

7 Electrokinetic Treatment of Contaminated Marine Sediments 149
Giorgia De Gioannis, Aldo Muntoni, Alessandra Polettini, and Raffaella Pomi

    7.1. Introduction, 149
    7.2. Contaminated Sediment Treatment Options, 151
    7.3. Electrokinetic Treatment of Sediments, 152
    7.4. Case Study: Tests on Electrokinetic Remediation of Sea Harbor Sediments, 155
    7.5. Summary, 172
    References, 174

8 Electrokinetic Stabilization of Chromium (VI)-Contaminated Soils 179
Laurence Hopkinson, Andrew Cundy, David Faulkner, Anne Hansen, and Ross Pollock

    8.1. Introduction, 179
    8.2. Materials and Methods, 181
    8.3. Experimental Results, 182
    8.4. Discussion, 187
    8.5. Summary, 191
    Acknowledgments, 191
    References, 192

PART III Remediation of Organic Pollutants 195

9 Electrokinetic Removal of PAHs 197
Ji-Won Yang and You-Jin Lee

    9.1. Introduction, 197
    9.2. Backgrounds, 198
    9.3. Electrokinetic Removal of PAHs Using Facilitating Agents, 204
    9.4. Summary, 213
    References, 214

10 Electrokinetic Removal of Chlorinated Organic Compounds 219
Xiaohua Lu and Songhu Yuan

    10.1. Introduction, 219
    10.2. Electrokinetic Removal of Chlorinated Aliphatic Hydrocarbons, 219

10.3. Electrokinetic Removal of Chlorophenols, 223
10.4. Electrokinetic Removal of Chlorobenzenes, 227
10.5. Summary, 232
References, 232

11 Electrokinetic Transport of Chlorinated Organic Pesticides 235
Ahmet Karagunduz

11.1. Introduction, 235
11.2. Electrokinetic Removal of Chlorinated Pesticides, 236
11.3. Surfactant-Enhanced Electrokinetic Remediation of Chlorinated Pesticides, 239
11.4. Cosolvent-Enhanced Electrokinetic Remediation of Chlorinated Pesticides, 246
11.5. Summary, 246
References, 247

12 Electrokinetic Removal of Herbicides from Soils 249
Alexandra B. Ribeiro and Eduardo P. Mateus

12.1. Introduction, 249
12.2. Herbicides, 250
12.3. Case Study, 252
12.4. Summary, 261
Acknowledgments, 261
References, 262

13 Electrokinetic Removal of Energetic Compounds 265
David A. Kessler, Charles P. Marsh and Sean Morefield

13.1. Introduction, 265
13.2. Chemistry of Clay–Energetic Compound Complexes, 266
13.3. Remediation Strategies, 268
13.4. Electrokinetics to Enhance Remediation Strategies, 273
13.5. Summary, 279
References, 279

PART IV Remediation of Mixed Contaminants 285

14 Electrokinetic Remediation of Mixed Metal Contaminants 287
Kyoung-Woong Kim, Keun-Young Lee and Soon-Oh Kim

14.1. Introduction, 287
14.2. General Principle for Mixed Metal Contaminants, 288
14.3. Representative Studies on Electrokinetic Remediation of Mixed Heavy Metals, 298
14.4. Specific Insight for Removal of Mixed Heavy Metals, Including Cr, As, and Hg, 306

14.5. Summary, 310
References, 310

15 Electrokinetic Remediation of Mixed Metals and Organic Contaminants   315
Maria Elektorowicz

15.1. Challenge in Remediation of Mixed Contaminated Soils, 315
15.2. Application of Electrokinetic Phenomena to the Removal of Organic and Inorganic Contaminants from Soils, 318
15.3. Summary, 328
References, 329

PART V Electrokinetic Barriers   333

16 Electrokinetic Barriers for Preventing Groundwater Pollution   335
Rod Lynch

16.1. Introduction, 335
16.2. History of Electrokinetic Barrier Development, 339
16.3. Recent Studies, 340
16.4. Use With Other Technologies, 351
16.5. Summary, 353
Acknowledgments, 354
References, 354

17 Electrokinetic Biofences   357
Reinout Lageman and Wiebe Pool

17.1. Introduction, 357
17.2. Application in the Field, 358
17.3. Case Study, 359
17.4. Summary, 365
Reference, 366

PART VI Integrated (Coupled) Technologies   367

18 Coupling Electrokinetics to the Bioremediation of Organic Contaminants: Principles and Fundamental Interactions   369
Lukas Y. Wick

18.1. Introduction, 369
18.2. Principles and Fundamental Interactions of Electrobioremediation, 371
18.3. Research Needs, 379
Acknowledgments, 380
References, 380

19 Coupled Electrokinetic–Bioremediation: Applied Aspects 389
Svenja T. Lohner, Andreas Tiehm, Simon A. Jackman, and Penny Carter

19.1. Bioremediation of Soils, 389
19.2. Combination of Electrokinetics and Bioremediation, 395
19.3. Practical Considerations and Limitations for Coupled Bio-Electro Processes, 406
19.4. Summary, 409
Acknowledgments, 409
References, 410

20 Influence of Coupled Electrokinetic–Phytoremediation on Soil Remediation 417
M.C. Lobo Bedmar, A. Pérez-Sanz, M.J. Martínez-Iñigo, and A. Plaza Benito

20.1. Soil Contamination: Legislation, 417
20.2. What is the Limit of the Remediation? Soil Recovery, 419
20.3. Influence of the Electrokinetic Technology on Soil Properties, 421
20.4. Phytoremediation, 424
20.5. Use of the Electrokinetic Process to Improve Phytoremediation, 428
20.6. Phytoremediation after Electrokinetic Process, 430
20.7. Summary, 432
References, 433

21 Electrokinetic–Chemical Oxidation/Reduction 439
Gordon C. C. Yang

21.1. Introduction, 439
21.2. General Principles, 439
21.3. Representative Studies, 447
21.4. Electrokinetic Treatment Coupled with Injection of Nanomaterials, 459
21.5. Prospective, 465
References, 465

22 Electrosynthesis of Oxidants and Their Electrokinetic Distribution 473
W. Wesner, Andrea Diamant, B. Schrammel, and M. Unterberger

22.1. Oxidants for Soil Remediation, 473
22.2. Production of Oxidants, 479
22.3. Distribution of Oxidants, 481
References, 482

23 Coupled Electrokinetic–Permeable Reactive Barriers 483
Chih-Huang Weng

23.1. Introduction, 483
23.2. Design of Reactive Barrier in the EK–PRB Process, 486
23.3. Implementation of EK–PRB to Polluted Soil, 490

23.4. Perspectives, 499
References, 500

24 Coupled Electrokinetic–Thermal Desorption 505
Gregory J. Smith

24.1. Fundamental Principles, 505
24.2. Thermal Principles, 505
24.3. Physical and Chemical Principles, 515
24.4. Fluid and Energy Transport, 517
24.5. Hydraulic Principles, 521
24.6. Biological Processes at Elevated Temperatures, 522
24.7. Summary, 531
References, 532

PART VII Mathematical Modeling 537

25 Electrokinetic Modeling of Heavy Metals 539
José Miguel Rodríguez-Maroto and Carlos Vereda-Alonso

25.1. Introduction, 539
25.2. One-Dimensional EKR Simple Model, 541
25.3. Two-Dimensional Model, 551
Notation, 559
References, 561

26 Electrokinetic Barriers: Modeling and Validation 563
R. Sri Ranjan

26.1. Introduction, 563
26.2. Electrokinetic Phenomena, 565
26.3. Direct and Coupled Flow and Transport of Ions, 565
26.4. Model Development, 569
26.5. Model Validation, 571
26.6. Field Application Scenarios, 572
26.7. Summary, 577
Acknowledgments, 578
References, 578

PART VIII Economic and Regulatory Considerations 581

27 Cost Estimates for Electrokinetic Remediation 583
Christopher J. Athmer

27.1. Introduction, 583
27.2. Cost Factors, 584
27.3. Cost Breakdown, 586

27.4. Summary, 586
References, 587

28 Regulatory Aspects of Implementing Electrokinetic Remediation — 589
Randy A. Parker

28.1. Introduction, 589
28.2. Overview of Environmental Regulation in the USA, 590
28.3. Regulatory Considerations for Implementing Electrokinetic Remediation, 595
28.4. Summary, 605
References, 605

PART IX Field Applications and Performance Assessment — 607

29 Field Applications of Electrokinetic Remediation of Soils Contaminated with Heavy Metals — 609
Anshy Oonnittan, Mika Sillanpaa, Claudio Cameselle, and Krishna R. Reddy

29.1. Introduction, 609
29.2. Description of Processes Involved in Field Applications, 610
29.3. Electrokinetic Remediation Setup in Field Applications, 615
29.4. Outcome of Field-Scale Experiments, 616
29.5. Factors that Limit the Applicability of Electrokinetic Technology, 619
29.6. Prerequisites and Site Information Needed, 621
29.7. Advantages and Disadvantages of the Technology, 622
29.8. Summary, 622
References, 623

30 Field Studies: Organic-Contaminated Soil Remediation with Lasagna Technology — 625
Christopher J. Athmer and Sa V. Ho

30.1. Introduction, 625
30.2. Field Implementation Considerations, 627
30.3. Case Studies, 632
30.4. Summary and Future Activities, 643
References, 644

31 Coupled Electrokinetic PRB for Remediation of Metals in Groundwater — 647
Ha Ik Chung and MyungHo Lee

31.1. Introduction, 647
31.2. Electrokinetic (EK) Extraction System, 647
31.3. Permeable Reactive Barrier (PRB) System, 648

31.4. Combined System of Electrokinetics and Permeable Reactive Barrier, 650
  31.5. Field Application, 651
  31.6. Summary, 657
  References, 658

32 Field Studies on Sediment Remediation   **661**
J. Kenneth Wittle, Sibel Pamukcu, Dave Bowman, Lawrence M. Zanko and Falk Doering

  32.1. Introduction, 661
  32.2. Background on the Need for Remediation and the Duluth Project, 663
  32.3. What is ECGO Technology? 667
  32.4. The Remediation of Minnesota Slip Sediments at the Erie Pier CDF in Duluth, Minnesota, 672
  32.5. Summary, 694
  References, 695

33 Experiences With Field Applications of Electrokinetic Remediation   **697**
Reinout Lageman and Wiebe Pool

  33.1. Introduction, 697
  33.2. ER, 697
  33.3. Investigation and Design of ER, 705
  33.4. Some Project Results, 717
  33.5. Summary, 717

INDEX   **719**

PREFACE

Industrial activities had released to the environment many toxic chemicals, that is, heavy metals and persistent organic pollutants, due to accidental spills or improper management. It resulted in many contaminated sites all over the world. Soil, sediment, and groundwater contamination has been a major problem at these polluted sites, which need urgent remediation to protect public health and the environment. Unfortunately, many conventional *in situ* remediation technologies are found to be ineffective and/or expensive to remediate sites with low permeability and heterogeneous subsurface conditions and contaminant mixtures. There is an urgent need to develop new technologies that can overcome these challenges and that are cost-effective. Recently, electrochemical remediation technologies are shown to have great potential to remediate such complex sites.

The electrokinetic technology for the remediation of soils, sediments, and groundwater relies on the application of a low-intensity electric field directly to the soil in the polluted site. The effect of the electric field mobilizes ionic species that are removed from the soil and collected at the electrodes. At the same time, the electric field provokes the mobilization of the interstitial fluid in the soil, generating an electroosmotic flow toward the cathode. The electroosmotic flow permit the removal of soluble contaminants. The success of the electrokinetic process rely on the effective extraction and solubilization of the contaminant and on their transportation toward the electrodes, where they can be collected, pumped out, and treated. Many studies have been carried out to determine the influence of the operating conditions and the effect of the soil and contaminant nature in order to improve the applicability and effectiveness of the electrokinetic treatment.

Numerous bench-scale studies that use ideal soils such as kaolin spiked with a selected single contaminant (e.g. lead or phenanthrene) to understand the contaminant transport processes have been reported. However, only a limited number of studies have been reported on real-world soils contaminated with a wide range of aged contaminants, and these studies have been helpful in recognizing complex geochemical interactions under induced electric potential. All of the bench-scale studies have clearly documented that nonuniform pH conditions are induced by applying a low direct current or electric potential, complicating the electrochemical remediation process. Low removal of contaminants was observed in these studies, and detailed geochemical assessments were made to understand the hindering mechanisms leading to low contaminant removal. For example, in low acid-buffering soils, high-pH conditions near the cathode cause adsorption and precipitation of cationic metal contaminants, whereas low-pH conditions near the anode cause adsorption of anionic metal contaminants. In high acid-buffering soils, high-pH conditions prevail throughout the soil, causing the immobilization of cationic contaminants without any migration, and anionic contaminants to exist in soluble form and migrate toward the anode. The removal of organic contaminants is dependent on the electroosmotic flow, which varies spatially under applied electric potential. Initially, flow occurs toward the cathode, but it gradually decreases as electric current decreases due to depletion of ions in pore water. If the soil pH reduces to less than the point of zero charge (PZC), electroosmotic flow direction can reverse and flow could cease.

Several studies have investigated strategies to enhance contaminant removal by using different electrode-conditioning solutions, changing the magnitude and mode of electric potential application, or both. The electrode-conditioning solutions aim to increase the solubility of the contaminants and/or increase electroosmotic flow. When dealing with metal contaminants (including radionuclides), organic acids (e.g. acetic acid) are introduced in the cathode to neutralize alkaline conditions, thereby preventing high-pH conditions in the soil. This allows the cationic metal contaminants to be transported and removed at the cathode. Alkaline solutions are introduced in the anode to increase pH near the anode. This allows anionic contaminants to exist in soluble form and be transported and removed at the anode. Instead of using acids, complexing agents (e.g. ethylenediaminetetraacetic acid (EDTA)) can be used in the cathode. When these agents enter the soil, they form negative metal complexes that can be transported and removed at the anode. When addressing organic contaminants (including energetic compounds), solubilizing agents such as surfactants, cosolvents, and cyclodextrins are introduced in the anode. When transported into the soil by electroosmosis, these agents solubilize the contaminants. Alkaline solutions are also induced to maintain soil pH greater than the PZC to enhance electroosmotic flow.

Mixed contaminants (combinations of cationic and anionic metals and organic contaminants) are commonly encountered at contaminated sites. In general, the presence of multiple contaminants is shown to retard contaminant migration and removal. Synergistic effects of multiple contaminants should be assessed prior to the selection of an enhancement strategy for the remediation process. It is found that the removal of multiple contaminants in a single-step process is difficult. Therefore, sequential conditioning systems have been developed to enhance removal of a mixture of cationic and anionic metal contaminants and/or a mixture of metal and organic contaminants. In addition to the use of electrode-conditioning solutions, the magnitude and mode of electric potential application can be altered. An increase in the magnitude of electric potential increases the electromigration rate and initial electroosmotic flow rate.

Although excellent removal efficiencies can be achieved at bench scale by the use of different enhanced electrochemical remediation strategies, several practical problems arise in using them at actual field sites. These problems include high cost of electrode-conditioning solutions, regulatory concerns over injecting conditioning solutions into subsurface, high energy requirements and costs, longer treatment time, potential adverse effects on soil fertility, and costs for treatment of effluents collected at the electrodes. As a result of all these problems, the full-scale field applications of electrochemical remediation are very limited.

Despite the challenges, *in situ* electrochemical remediation holds promise to remediate difficult subsurface conditions, particularly low permeability and heterogeneous subsurface environments, where most of other conventional technologies fail. The electrochemical remediation technology can also be applied to remediate diverse and mixed contaminants even when they are nonuniformly distributed in the subsurface. Standard electrochemical remediation method is essentially an electrokinetically enhanced flushing process. However, the electrochemical remediation can be made efficient and practical, as well as less expensive, by integrating or coupling it with other proven remediation technologies. Such integrated technologies include electrokinetic–chemical oxidation/reduction, electrokinetic–bioremediation, electrokinetic–phytoremediation, electrokinetic–thermal desorption, electrokinetic–permeable reactive barriers, electrokinetic–stabilization, electrokinetic–barriers (fences), and others. These integrated technologies have potential for the simultaneous remediation of mixed contaminants in any subsurface environment. Several successful bench-scale and demonstration projects have been reported recently on integrated electrochemical remediation technologies. Such integrated electrochemical projects are expected to grow in the near future and becoming important technologies for the remediation of actual contaminated sites.

This book compiles the various studies ranging from fundamental processes to field implementation of different electrochemical remediation technologies in a format that can best serve as a valuable resource to all environmental engineers, scientists, regulators, and policy makers to consider electrochemical technologies as potential candidate technologies to remediate contaminated sites. The objectives of this book are as follows: (a) to provide the state of the knowledge on electrochemical remediation technologies for polluted soils, sediments, and groundwater; (b) to present the difficulties in implementing the standard electrochemical remediation process; (c) to outline the opportunities and challenges in developing and implementing promising integrated electrochemical remediation technologies; and (d) to describe field applications and highlight economic and regulatory considerations in implementing these technologies at contaminated sites. It is hoped that this book will stimulate researchers, practicing professionals, and students to better understand electrochemical remediation technologies and also make further advances in fundamental processes and innovative field applications.

Each chapter in this book is prepared by individuals who possess extensive experience in the field of electrochemical remediation technologies. We are grateful to all of these authors for their time and effort in preparing their contribution. Each chapter was peer reviewed by two reviewers, and we are thankful to these reviewers for their constructive comments. Finally, the support of the University of Illinois at Chicago and the University of Vigo during this endeavor is highly appreciated.

KRISHNA R. REDDY
CLAUDIO CAMESELLE

February 2009

CONTRIBUTORS

Christopher J. Athmer, Terran Corporation, Beavercreek, OH, USA

Kitae Baek, Department of Environmental Engineering, Kumoh National Institute of Technology, Gyeongbuk, Korea

Dave Bowman, US Army Corps of Engineers, Detroit, MI, USA

Claudio Cameselle, Department of Chemical Engineering, University of Vigo, Vigo, Spain

Penny Carter, Department of Earth Sciences, University of Oxford, Oxford, UK

Ha Ik Chung, Geotechnical Engineering Research Department, Korea Institute of Construction Technology (KICT), Gyeonggi-do, Korea

Andrew Cundy, School of Environment and Technology, University of Brighton, Brighton, UK

Giorgia De Gioannis, Department of Geoengineering and Environmental Technologies, University of Cagliari, Cagliari, Italy

Andrea Diamant, Echem, Kompetenzzentrum für Angewandte Elektrochemie, Wiener Neustadt, Austria

Falk Doering, Electrochemical Processes, llc., Stuttgart Research Center, Stuttgart, Germany

Maria Elektorowicz, Department of Building, Civil and Environmental Engineering (BCEE), Concordia University, Montreal, Canada

David Faulkner, School of Environment and Technology, University of Brighton, Brighton, UK

Anne Hansen, Churngold Remediation Limited, Bristol, UK

Henrik K. Hansen, Departamento de Procesos Químicos, Biotecnológicos y Ambientales, Universidad Técnica Federico Santa María, Valparaíso, Chile

Sa V. Ho, Pfizer Corporation, Chesterfield, MO, USA

Laurence Hopkinson, School of Environment and Technology, University of Brighton, Brighton, UK

Simon A. Jackman, Department of Earth Sciences, University of Oxford, Oxford, UK

Pernille E. Jensen, Civil Engineering, Technical University of Denmark, Lyngby, Denmark

Ahmet Karagunduz, Gebze Institute of Technology, Department of Environmental Engineering, Kocaeli, Turkey

David A. Kessler, Laboratory for Computational Physics and Fluid Dynamics, US Naval Research Laboratory, Washington, DC, USA

Kyoung-Woong Kim, Department of Environmental Science and Engineering, Gwangju Institute of Science and Technology (GIST), Gwangju, Korea

Soon-Oh Kim, Department of Earth and Environmental Sciences and Research Institute of Natural Science, Gyeongsang National University, Jinju, Korea

Vladimir A. Korolev, Department of Engineering and Ecological Geology, Geological Faculty of MSU named M.V. Lomonosov, Moscow, Russia

Reinout Lageman, Holland Environment BV, Doorn, the Netherlands

Keun-Young Lee, Department of Environmental Science and Engineering, Gwangju Institute of Science and Technology (GIST), Gwangju, Korea

MyungHo Lee, Induk Institute of Technology, Nowon-gu, Seoul, Korea

You-Jin Lee, Department of Chemical and Biomolecular Engineering, Korea Advanced Institute of Science and Technology (KAIST), Daejeon, Korea

M.C. Lobo Bedmar, Instituto Madrileño de Investigación y Desarrollo Rural Agrario y Alimentario (IMIDRA), Alcalá de Henares, Spain

Svenja T. Lohner, Department of Environmental Biotechnology, Water Technology Centre (TZW), Karlsruhe, Germany

Xiaohua Lu, Environmental Science Research Institute, Huazhong University of Science and Technology, Wuhan, China

Rod Lynch, Department of Engineering, University of Cambridge, Cambridge, UK

Charles P. Marsh, US Army Corps of Engineers, Engineering Research and Development Center, University of Illinois, Champaign, IL, USA

M.J. Martínez-Iñigo, Instituto Madrileño de Investigación y Desarrollo Rural Agrario y Alimentario (IMIDRA), Alcalá de Henares, Spain

Eduardo P. Mateus, Departamento de Ciências e Engenharia do Ambiente, Universidade Nova de Lisboa, Caparica, Portugal

Sean Morefield, US Army Corps of Engineers, Engineering Research and Development Center, Champaign, IL, USA

Aldo Muntoni, Department of Geoengineering and Environmental Technologies, University of Cagliari, Cagliari, Italy

Anshy Oonnittan, Laboratory of Applied Environmental Chemistry, Kuopio University, Mikkeli, Finland

Lisbeth M. Ottosen, Civil Engineering, Technical University of Denmark, Lyngby, Denmark

Sibel Pamukcu, Department of Civil and Environmental Engineering, Lehigh University, Bethlehem, PA, USA

Randy A. Parker, US Environmental Protection Agency, Cincinnati, OH, USA

A. Pérez-Sanz, Instituto Madrileño de Investigación y Desarrollo Rural Agrario y Alimentario (IMIDRA), Alcalá de Henares, Spain

A. Plaza-Benito, Instituto Madrileño de Investigación y Desarrollo Rural Agrario y Alimentario (IMIDRA), Alcalá de Henares, Spain

Alessandra Polettini, Department of Hydraulics, Transportation and Roads, University of Rome "La Sapienza," Rome, Italy

Ross Pollock, Churngold Remediation Limited, Bristol, UK

Raffaella Pomi, Department of Hydraulics, Transportation and Roads, University of Rome "La Sapienza," Rome, Italy

Wiebe Pool, Holland Environment BV, Doorn, the Netherlands

Krishna R. Reddy, Department of Civil and Materials Engineering, University of Illinois at Chicago, Chicago, IL, USA

Alexandra B. Ribeiro, Departamento de Ciências e Engenharia do Ambiente, Universidade Nova de Lisboa, Caparica, Portugal

José Miguel Rodríguez-Maroto, Department of Chemical Engineering, University of Málaga, Málaga, Spain

B. Schrammel, Echem, Kompetenzzentrum für Angewandte Elektrochemie, Wiener Neustadt, Austria

Mika Sillanpaa, Laboratory of Applied Environmental Chemistry, Kuopio University, Mikkeli, Finland

Gregory J. Smith, Risk Transfer Services, Woodridge, IL, USA

R. Sri Ranjan, Department of Biosystems Engineering, University of Manitoba, Manitoba, Canada

Andreas Tiehm, Department of Environmental Biotechnology, Water Technology Centre (TZW), Karlsruhe, Germany

M. Unterberger, Echem, Kompetenzzentrum für Angewandte Elektrochemie, Wiener Neustadt, Austria

Carlos Vereda-Alonso, Department of Chemical Engineering, University of Málaga, Málaga, Spain

Chih-Huang Weng, Department of Civil and Ecological Engineering, I-Shou University, Kaohsiung, Taiwán

Wolfgang Wesner, Echem, Kompetenzzentrum für Angewandte Elektrochemie, Wiener Neustadt, Austria

Lukas Y. Wick, Department of Environmental Microbiology, Helmholtz Centre for Environmental Research UFZ, Leipzig, Germany

J. Kenneth Wittle, Electro-Petroleum, Inc., Wayne, PA, USA

Gordon C.C. Yang, Institute of Environmental Engineering, National Sun Yat-Sen University, Kaohsiung, Taiwan

Ji-Won Yang, Department of Chemical and Biomolecular Engineering, Korea Advanced Institute of Science and Technology (KAIST), Daejeon, Korea

Jung-Seok Yang, Korea Institute of Science and Technology, Gangneung Institute, Gangwon-do, Korea

Albert T. Yeung, Department of Civil Engineering, University of Hong Kong, Hong Kong

Songhu Yuan, Environmental Science Research Institute, Huazhong University of Science and Technology, Wuhan, China

Lawrence M. Zanko, Natural Resources Research Institute, University of Minnesota Duluth, Duluth, MN, USA

PART I

INTRODUCTION AND BASIC PRINCIPLES

1

OVERVIEW OF ELECTROCHEMICAL REMEDIATION TECHNOLOGIES

KRISHNA R. REDDY AND CLAUDIO CAMESELLE

1.1 INTRODUCTION

Numerous sites worldwide are found contaminated due to improper past waste disposal practices and accidental spills. Contamination of soils and groundwater is well known, but the ever-growing problem of large quantities of contaminated dredged sediments has only received attention recently. Contaminants found include a wide range of toxic pollutants such as heavy metals, radionuclides, and organic compounds. The public and the environment are being exposed to these pollutants through different exposure pathways to unacceptable dosages, leading to intolerable adverse effects on public health and the environment. The remediation of these sites has become an urgent priority to environmentalists and regulatory bodies.

Several different technologies have been developed to remediate soils, sediments, and groundwater based on physicochemical, thermal, and biological principles (Sharma and Reddy, 2004). However, they are often found to be costly, energy intensive, ineffective, and could themselves create other adverse environmental impacts when dealing with difficult subsurface and contaminant conditions. For instance, inadequate remediation has been demonstrated at numerous polluted sites due to the presence of low permeability and heterogeneities and/or contaminant mixtures (multiple contaminants or combinations of different contaminant types such as coexisting heavy metals and organic pollutants). Electrochemical remediation has been recognized as a promising technology to address such difficult contaminated site conditions, leading to several research programs worldwide for the development of this technology.

The purpose of this book is to provide the state of the art on electrochemical remediation of polluted soils, sediments, and groundwater. Specifically, an

---

Electrochemical Remediation Technologies for Polluted Soils, Sediments and Groundwater,
Edited by Krishna R. Reddy and Claudio Cameselle
Copyright © 2009 John Wiley & Sons, Inc.

4 OVERVIEW OF ELECTROCHEMICAL REMEDIATION TECHNOLOGIES

introduction to the electrophenomena in soil, various fundamental and mathematical modeling studies, laboratory investigations, and field demonstration projects have been detailed. In addition, the regulatory and economic considerations are presented. This chapter introduces the content of this book; specifically, electrochemical remediation processes, versatility in implementation, recent advancements, and future research directions are briefly summarized. The reader is referred to various chapters in the book for the detailed information.

1.2 ELECTROCHEMICAL TECHNOLOGIES FOR SITE REMEDIATION

A typical field electrochemical remediation system is shown in Figure 1.1. Initially, wells/drains are configured and drilled so they surround the contaminated

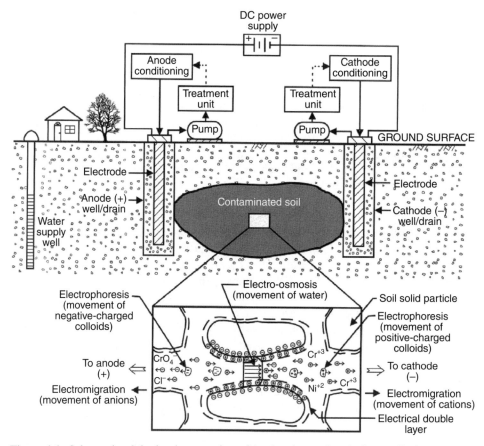

Figure 1.1. Schematic of the implementation of *in situ* electrochemical remediation systems. The electrodes are inserted into the soil and a direct electric field is applied to the contaminated site, which induces the transport of the contaminants toward the electrodes. The electrode solutions are pumped, treated, and circulated for contaminant removal. Selected electrode conditioning solutions may be used to induce favorable chemistry at the electrodes and in the soil.

region. Electrodes are then inserted into each well/drain and a low direct current (DC) or a low potential gradient to electrodes is applied. As a result of the applied electric field, several transport, transfer, and transformation processes are induced, which cause the contaminants to be transported into the electrodes where they can be removed. Alternatively, the contaminants may be stabilized/ immobilized or degraded within the contaminated media. Several patents have been issued that deal with using electrochemical remediation in different creative ways.

Electrochemical remediation is also referred as *electrokinetics, electrokinetic remediation, electroremediation, electroreclamation*, and other such terms in the published literature. It should be noted here that when water alone is used at the electrodes, the process is known as *unenhanced electrochemical remediation*. However, when enhancement strategies (such as using conditioning solutions and ion exchange membranes at the electrodes) are used, then the process is known as *enhanced electrochemical remediation*.

Electrochemical remediation has received tremendous attention from environmental professionals because of its unique advantages over other conventional technologies. These advantages include

- flexibility to use as *ex situ* or *in situ* method;
- applicability to low-permeability and heterogeneous soils (e.g. glacial tills, lacustrine clays and silts, alluvial deposits, saprolitic formations, and loess);
- applicability to saturated and unsaturated soils;
- applicability for heavy metals, radionuclides, and organic contaminants, as well as in any of their combinations (contaminant mixtures); and
- easy integration with conventional technologies, including barrier and treatment systems.

Although implementation of the electrochemical remediation system in the field is relatively simple, its design and operation for successful remediation is cumbersome due to complex dynamic electrochemical transport, transfer, and transformation processes that occur under applied electric potential. In particular, the efficacy of electrochemical remediation depends strongly on contaminated media characteristics such as buffer capacity, mineralogy, and organic matter content, among others. If geochemistry, soil–contaminants interaction, and subsurface heterogeneity are well understood, the electrochemical remediation systems can be engineered to achieve remediation in an effective and economic manner.

Significant advances have been made toward the understanding of the fundamental processes involved in electrochemical remediation through controlled and idealized laboratory experiments. Valuable lessons have been learned from a limited number of documented field applications. Nevertheless, numerous research studies have been undertaken recently or ongoing to further understand the processes under field conditions and to develop innovative field systems so that optimized effective electrochemical remediation systems can be implemented at actual field sites.

1.3 ELECTROCHEMICAL TRANSPORT, TRANSFER, AND TRANSFORMATION PROCESSES

Upon electric field application, decomposition of water (electrolysis reactions) occurs at the electrodes. The electrolysis reactions generate oxygen gas and hydrogen ions (H^+) due to oxidation at the anode and hydrogen gas and hydroxyl (OH^-) ions due to reduction at the cathode as shown by the following reactions:

At anode (oxidation):

$$2H_2O \rightarrow O_{2(gas)} + 4H^+_{(aq)} + 4e^- \quad E^0 = -1.229 \text{ V} \quad (1.1)$$

At cathode (reduction):

$$4H_2O + 4e^- \rightarrow 2H_{2(gas)} + 4OH^-_{(aq)} \quad E^0 = -0.828 \text{ V} \quad (1.2)$$

Essentially, acid is produced at the anode and alkaline solution is produced at the cathode; therefore, pH in the cathode is increased, while pH at the anode is decreased. The migration of H^+ from the anode and OH^- from the cathode into the soil leads to dynamic changes in soil pH during the initial stages of electric potential application (Acar and Alshawabkeh, 1993).

The implications of these electrolysis reactions are enormous in that they impact transport, transformation, and degradation processes that control the contaminant migration, removal, and degradation during electrochemical treatment. The different transport, transfer, and transformation processes induced by the applied electric field and how these processes are impacted by the electrolysis reactions at the electrodes are fundamental to the understanding of the electrochemical remediation technologies and are briefly presented in this section.

1.3.1 Electrochemical Transport Processes

For low-permeability media, advective transport due to hydraulic flow is negligible. The application of an electric field induces the transport of contaminants and water through the contaminated media toward the electrodes due to the following transport processes (Probstein, 2003): *electromigration* (*ionic migration*), *electroosmosis*, *electrophoresis*, and *diffusion*.

1.3.1.1 Electromigration (Ionic Migration) Electromigration, also known as ionic migration, is the movement of the dissolved ionic species present in the pore fluid toward the opposite electrode. Anions move toward the anode and cations move toward the cathode. The degree of electromigration depends on the mobility of ionic species. Electromigration is the major transport process for ionic metals, polar organic molecules, ionic micelles, and colloidal electrolytes. The transport of H^+ and OH^- ions generated by electrolysis reactions is also attributed to electromigration.

The extent of electromigration of a given ion depends on the conductivity of the soil, soil porosity, pH gradient, applied electric potential, initial concentration of the specific ion, and the presence of competitive ions. The electromigrative velocity

(v_{em}) of an ion is proportional to the ion charge and the local electric gradient given by

$$v_{em} = u_i \, z_i \, n \, \tau \, F \, E, \qquad (1.3)$$

where u_i is the ion mobility (m²/V·s), z_i is the ionic valence, n is the porosity, τ is the tortuosity, F is the Faraday's constant (96,487 C/mol electrons), and E is the electric field strength (V/m). Ionic mobility (u_i) is defined as the velocity of the ionic species under the effect of unit electric field and is estimated using the Nernst–Einstein–Townsend relation:

$$u_i = \frac{D_i \, |z_i| \, F}{RT} \qquad (1.4)$$

where D_i is the molecular diffusion coefficient, R is the universal gas constant (8.314 J/K·mol), and T is the absolute temperature (K). The effective mobility u_i^* is defined by Equation 1.5 and is considered the movement of a given ion in a porous matrix with a tortuous path. The effective ionic mobility of a specific ion is a function of its molecular diffusion coefficient, soil porosity, tortuosity factor, and charge.

$$u_i^* = n \, \tau \, u_i. \qquad (1.5)$$

1.3.1.2 Electroosmosis Advective flow occurs due to hydraulic gradient and electrical gradient. The hydraulic flow (q_h) due to hydraulic gradient (i_h) is defined by Darcy's law:

$$q_h = k_h \, i_h, \qquad (1.6)$$

where k_h is the hydraulic conductivity. This flow is significant for permeable soils such as sand, which possess hydraulic conductivity greater than 10^{-3} cm/s; however, this flow in clayey soils is negligible due to very low hydraulic conductivity (10^{-6}–10^{-9} cm/s).

Electroosmosis is induced under electric gradient and it is the movement of the pore fluid, which contains dissolved ionic and nonionic species, relative to the stationary soil mass, toward the electrodes. Generally, soil particle surfaces are charged (generally negatively charged) and counterions (positive ions or cations) concentrate within a diffuse double layer region adjacent to the particle surface. Under an electric potential, the locally excess ions migrate in a plane parallel to the particle surface toward the oppositely charged electrode. As they migrate, they transfer momentum to the surrounding fluid molecules via viscous forces, producing electro-osmotic flow. Electroosmosis is the dominant transport process for both organic and inorganic contaminants that are in dissolved, suspended, emulsified, or such similar forms.

In 1879, Helmholtz introduced one of the first theories concerning electroosmosis, and Smoluchowski modified it in 1914. According to the Helmholtz–Smoluchowski theory (H-S theory), the electro-osmotic flow velocity (v_{eo}) is directly proportional to the applied voltage gradient (E_z), zeta potential (ζ), and dielectric constant (D) of the fluid, and it is inversely proportional to the fluid viscosity (η):

$$v_{eo} = -\frac{D\zeta}{\eta} E_z. \quad (1.7)$$

The H-S equation can also be expressed in terms of the volumetric flow rate (q_{eo}):

$$q_{eo} = nA \frac{D\zeta}{\eta} E_z. \quad (1.8)$$

In this equation, n is the porosity and A is the cross-sectional area of the soil. Furthermore, it should be noted that the use of the effective porosity (n_e) or the porosity (n) divided by the tortuosity squared (n/τ) may be more accurate than using the porosity (n). For civil engineering applications, the electro-osmotic permeability coefficient (k_{eo}) is computed as follows:

$$k_{eo} = n \frac{D\zeta}{\eta}. \quad (1.9)$$

For example, if $k_{eo} = 4.1 \times 10^{-5}$ (V/s) for fine sand, and for this same soil the approximate $k_h = 10^{-4}$ cm/s, which are reasonably close so electroosmosis will not greatly enhance the hydraulic flow through this soil. However, in lower-permeability soils such as kaolin, $k_{eo} = 5.7 \times 10^{-5}$ (V/s) and the approximate $k_h = 10^{-7}$ cm/s, so a very high hydraulic gradient of about 600 would be needed. Thus, electro-osmotic flow though low-permeability regions is significantly greater than the flow achieved by an ordinary hydraulic gradient. Electroosmosis facilitates advective transport of the solubilized contaminants toward the electrodes for removal.

As seen from Equation 1.7, the electro-osmotic flow depends on the dielectric constant and viscosity of pore fluid, as well as the surface charge of the solid matrix represented by the zeta potential (the electric potential at the junction between the fixed and mobile parts in the double layer). The zeta potential is a function of many parameters, including the types of clay minerals and ionic species that are present, as well as the pH, ionic strength, and temperature. If the cations and anions are evenly distributed, an equal and opposite flow occurs, causing the net flow to be zero. However, when the momentum transferred to the fluid in one direction exceeds the momentum of the fluid traveling in the other direction, electro-osmotic flow is produced.

The pH changes induced in the soil by the electrolysis reactions affect the zeta potential of the soil particles, and thereby affect the electro-osmotic flow. The low pH near the anode may be less than the point of zero charge (PZC) of the soil and the soil surfaces are positively charged, while high pH near the cathode may be higher than the PZC of the soil, making the soil more negative. Electro-osmotic flow may be reduced and even ceased as the soil is acidified near the anode. If the majority of the soil is acidified, the electro-osmotic flow direction may even be reversed, from typical anode to cathode to cathode to anode. This phenomenon is known as *electroendosmosis*. Understanding of such electro-osmotic flow variations is critical when remediating organic pollutants.

1.3.1.3 Electrophoresis Electrophoresis (also known as cataphoresis) is the transport of charged particles of colloidal size and bound contaminants due to the application of a low DC or voltage gradient relative to the stationary pore fluid. Compared with ionic migration and electroosmosis, mass transport by electrophoresis is negligible in low-permeability soil systems. However, mass transport by electrophoresis may become significant in soil suspension systems, and it may also be a dominant transport mechanism for biocolloids (i.e. bacteria) and micelles.

1.3.1.4 Diffusion Diffusion refers to the ionic and molecular constituent forms of the contaminants moving from areas of higher concentration to areas of lower concentration because of the concentration gradient or chemical kinetic activity. Estimates of the ionic mobilities from the diffusion coefficients using the Nernst–Einsetin–Townsend relation indicates that ionic mobility of a charged species is much higher than the diffusion coefficient (about 40 times the product of its charge and the electrical potential gradient). Therefore, diffusive transport is often neglected.

1.3.1.5 Relative Contribution of Transport Processes For soluble ionized inorganic contaminants (such as metal cations, metal anions, nitrates, and phosphates), electromigration is the dominant transport mechanism at high concentrations of ionic species, while electroosmosis is dominant at lower concentrations. For readily soluble organic compounds (such as benzene, toluene, xylene, phenolic compounds, and chlorinated solvents), electroosmosis is the dominant transport process in electrochemical remediation. Relative contribution of electroosmosis and ion migration to the total mass transport varies according to soil type, water content, ion species, and their concentration. In silts and low-activity clays, hydraulic flow is negligible in comparison with electro-osmotic flow. For ionic species, the mass transport by ionic migration is from 10 to 300 times greater than the mass transport by electro-osmotic advection. Furthermore, electroosmosis decreases when the pH and zeta potential drop in the later stages of continuous electrochemical process under a constant electric potential.

When micelles (charged aggregate of molecules or particles) are formed with other species in the processing fluid, or when we deal with slurries, electrophoresis may become significant. Diffusion is an important transport mechanism, but it is a very slow process, so it is estimated to only have a minor influence on contaminant transport during electrochemical remediation.

1.3.2 Electrochemical Mass Transfer Processes

The protons (H^+) and hydroxyl (OH^-) ions generated by electrolysis reactions (Eqs. 1.1 and 1.2) migrate toward the oppositely charged electrode. Acar *et al.* (1995) determined that, generally, H^+ is about twice as mobile as OH^-, so the protons dominate the system and an acid front moves across the soil until it meets the hydroxyl front in a zone near the cathode, where the ions may recombine to generate water. Thus, the soil is divided into two zones with a sharp pH jump in between: a high-pH zone close to the cathode and a low-pH zone on the anode side. The actual soil pH values will depend on the extent of transport of H^+ and OH^- ions and the geochemical characteristics of the soil.

During the initial stages of electric potential application, the soil pH changes spatially and temporally, which leads to dynamic geochemistry, leading to mass transfer from one form (phase) to the other (solid/precipitated, sorbed, dissolved, and free phases) and changes in chemical speciation. The most important geochemical reactions that must be considered include

- sorption–desorption reactions;
- precipitation–dissolution reactions; and
- oxidation–reduction reactions.

Sorption refers to the partitioning of the contaminants from the solution or pore fluid to the solid phase or soil surface. Sorption includes adsorption and ion exchange, and it is dependent on (a) the type of contaminant, (b) the type of soil, and (c) the pore fluid characteristics. Desorption is the reverse process and is responsible for the release of contaminants from the soil surface. Both sorption and desorption are affected by soil pH changes caused by the migration of H^+ and OH^- ions, which are produced by the electrolysis reactions. The pH-dependent sorption–desorption behavior is generally determined by performing batch experiments using the soil and contaminant of particular interest.

The precipitation and dissolution of the contaminant species during the electrokinetic process can significantly influence the removal efficiency of the process. The solubilization of precipitates is affected by the hydrogen ions generated at the anode migrating across the contaminated soil, favoring the acidification of soil and the dissolution of metal hydroxides and carbonates, among others. However, in some types of soils, the migration of the hydrogen ions will be hindered due to the relatively high buffering capacity of the soil. The presence of the hydroxyl ions at the cathode will increase the pH value (pH = 10–12) in the electrode solution and in the soil area close to the cathode. In a high-pH environment, heavy metals will precipitate, and the movement of the contaminants will be impeded. During electrokinetic treatment, heavy metals migrates toward the cathode until reach the high-pH zone, where heavy metals accumulate and eventually precipitate, clogging soil pores and hindering the remediation process. For efficient contaminant removal, it is essential to prevent precipitation and to have the contaminants in dissolved form during the electrokinetic process.

The high pH and the low heavy metals concentration condition at the cathode may also lead to the formation of a negatively charged complex species at the cathode compartment. The movement of these negatively charged complex species toward the anode and of the heavy metals toward the cathode relies upon the relative mobility of the hydrogen and hydroxyl ions.

Oxidation and reduction reactions are important when dealing with metallic contaminants such as chromium. Chromium exists most commonly in two valence states: trivalent chromium [Cr(III)] and hexavalent chromium [Cr(VI)]. Cr(III) exists in the form of cationic hydroxides such as $Cr(OH)_2^{-1}$ and it will migrate toward the cathode during electrokinetic remediation. However, Cr(VI) exists in the form of oxyanions such as CrO_4^{2-}, which migrate toward the anode. The valence state depends on the soil composition, especially the presence of reducing agents such as organic matter and Fe(II) and/or oxidizing agents such as Mn(IV), so it is important to know the valence state of metals and their possible redox chemistry to know the chemical speciation of the contaminants and their movement across the soil.

The geochemical processes are affected by the soil composition and any enhancement solutions used at the electrodes. In particular, acid buffering capacity of the soil affects the changes in soil pH. It is found that in high acid buffering soils, pH is not lowered near the anode due to buffering of the acid produced at the anode by the carbonates present in the soil, but pH increases near the cathode as OH^- ions migrate easily. Therefore, the most prominent geochemical processes during electrochemical remediation of contaminated soils include the following: (a) generation of pH gradient and buffering capacity of the soil; (b) change of zeta potential of soil particle surfaces; (c) sorption and desorption of contaminants onto/from soil particle surfaces; (d) complexation; (e) oxidation–reduction (redox) reactions; and (f) interactions of these processes. More details on these processes are presented in Chapter 3.

1.3.3 Electrochemical Transformation Processes

Chapter 2 describes the electrochemical transformation processes that occur at microscale under electric fields. Recently, it has been presented that the application of a low-intensity electric field to a soil can induce electrical transformations on clay surfaces. Those transformations can make the clay particles to act as microelectrodes that provoke redox reaction in the contaminants, especially organic pollutants. Such transformations are mainly attributed to the Faradaic current passage orthogonal to the planes in the electric double layer of clay particles, inducing redox reactions on clay surfaces. Clay particles are conceived as "microelectrodes" possessing a compact Stern layer and a diffuse layer, which mediates Faradaic reactions. As the donated (or accepted) electrons pass across the electrical double layer into and out of the bulk fluid, available species are converted into others via oxidation–reduction reactions. This effect may become significant for polarizable surfaces due to strong adsorption of the cations, anions, and molecules with electrical dipoles within the double layer, resembling the case of contaminated sediments.

Additionally, electrochemical transformations may occur when the contaminants enter into the anode or the cathode, particularly chlorinated organic compounds, which are shown to undergo reductive dechlorination at the cathode and oxidative dechlorination at the anode. Such transformations should also be considered based on the redox conditions in the electrodes and the contaminant characteristics.

1.4 ELECTROCHEMICAL REMOVAL OF INORGANIC POLLUTANTS

Inorganic pollutants include (a) cationic heavy metals such as lead, cadmium, and nickel, (b) anionic metals and inorganics such as arsenic, chromium, selenium, nitrate and fluoride, and (c) radionuclides such as strontium and uranium. The geochemistry of these pollutants can widely vary and it depends on the specific pollutant type and soil/sediment properties. The speciation and transport of these pollutants also depend on the dynamic changes in the pH and redox potential of the soil that occurs under applied electric potential. The dominant transport process is electromigration, and the soil pH changes induced by the electric field complicate the geochemistry and inorganic pollutant removal (refer to Part II).

1.4.1 Cationic Heavy Metals

Numerous studies are reported on the electrokinetic removal of heavy metals from soils (Chapter 4). Many of these studies used ideal soils, often kaolinite, as a representative low-permeability soil, which were spiked with a selected single cationic metal (such as lead and cadmium) in predetermined concentration. The spiked soil is loaded in a small-scale electrokinetic test setup and electric potential is applied. The transport and removal of the metal after specified test duration are determined. It is shown that cationic metals exist in soluble ionic form due to reduced pH near the anode regions and they are transported toward the cathode. However, when they reach near the cathode, they get sorbed or precipitated due to increased pH resulting from OH^- transport from the cathode. The actual removal from the soil is often negligible.

Several studies reported similar results based on testing of field soils. The field soils possess complex mineralogy, organic content, and buffering capacity, which results in relatively low removal of metals. One of the most important considerations is the acid buffering capacity of the soil. If the soil possesses higher acid buffering capacity, soil pH does not reduce near the anode, but it increases near the cathode. Furthermore, testing of aged field contaminated soils showed a very low metal removal due to complex soil composition and strong sorption to the soil constituents.

The major hindering factors for the removal of contaminants are sorption and precipitation of the contaminants resulting from the changes in soil pH. Therefore, in order to enhance metal removal, several enhancement strategies have been employed. These include (a) using organic/mineral acids at the cathode to reduce pH, (b) using ion exchange membranes between soil specimen and electrodes to control H^+ and OH^- ions transport into the soil so that favorable soil pH is maintained, and (c) using chelating agents that can form soluble complexes with the metals at different pH conditions. These enhancement studies have shown high removal efficiencies of cationic metals. In some cases such as elemental mercury, oxidizing agents are first introduced into the soil to oxidize mercury and transform it into ionic form and then to be able to be removed by the electromigration process.

1.4.2 Anionic Metals and Other Anionic Species

Experiments with anionic metals (such as chromium and arsenic) showed behavior opposite to that of cationic metals. The anionic metals are found to exist in soluble ionic form near the cathode and are transported toward the anode. However, when they reach near the anode, they are adsorbed due to low-pH conditions existing due to electrolysis reactions. Nevertheless, the actual removal of anionic metals was found to be greater than that of cationic metals. The use of alkaline solution to increase soil pH near the anode is found to enhance the removal of anionic metals (Chapter 4).

In addition to anionic metals, the problem of groundwater contamination with excess nitrate and fluoride is well recognized (Chapter 6). These anionic species are not as highly toxic as other anionic metals such as chromium and arsenic, but they are shown to have some adverse effects on public health and the environment. Few earlier studies investigated the removal of nitrates from the soils and groundwater using electrochemical methods, but attention on fluoride and other similar contaminants has received attention only recently. The behavior of these contaminants under electric field is similar to that of anionic metals; they electromigrate toward the anode, in opposite direction to the electro-osmotic flow. Low removal is expected near the anode due to low-pH conditions; therefore, anode conditioning with alkaline solution (e.g. NaOH) is used to increase soil pH near the anode and enhance the removal. The effects of anode conditioning on electroosmosis should be considered as it will impact contaminant removal. Alternatively, some researchers used zero-valent iron (ZVI) in the anode to increase soil pH and also to transform nitrate into nitrogen within the anode. Such strategy can also be implemented in an electrokinetic barrier system. It should also be pointed out here that nitrate is delivered into the soil purposely to enhance biostimulation in some studies, and the lessons learned from the studies dealing with the removal of nitrate can be useful for this purpose. The excess amount of nitrate should be avoided as it may be treated as contamination.

1.4.3 Radionuclides

Radioactive contamination at several sites due to improper handling of nuclear wastes and production and operation of nuclear fuel and nuclear reactors is well known (Chapter 5). The principal among radioactive materials are the following isotopes: ^{60}Co, ^{90}Sr, ^{90}Y, ^{106}Ru, ^{137}Cs, ^{144}Ce, ^{147}Pm, $^{238, 239, 240}Pu$, ^{226}Ra, and so on, and they are found to exist in near-surface soils, posing great hazard to public health and the environment. Electrochemical remediation of radionuclides is proposed to either contain within the soil or remove from soils. Electrokinetic containment is applied for preventing radioactive nuclide migration from the polluted region. Similar to heavy metals, electrochemical removal of radionuclides requires enhancement strategies. Specifically, a high-pH environment retards the removal; therefore, enhancement solutions such as acetate buffer solution ($CH_3COONa + CH_3COOH$) is injected into the anode, and acetic acid is injected periodically into the cathode to control any pH increase. Electromigration is the dominant mechanism for the removal of these contaminants. Other enhancement solutions such as NH_4NO_3 or KNO_3 are pumped into the anode and nitric acid or other acidic solution near the cathode to prevent the segregation and precipitation of metal hydroxides near or in the cathode.

1.5 ELECTROCHEMICAL REMOVAL OF ORGANIC POLLUTANTS

Earlier efforts proved that the volatile and/or soluble organic contaminants are relatively easily removed by electroremediation, as well as by other conventional remediation technologies. Recent focus has been on electroremediation of hydrophobic and persistent (hard to degrade) toxic compounds such as *polycyclic aromatic hydrocarbons (PAHs), polychlorinated organic compounds [e.g. polychlorinated biphenyls (PCBs)], pesticides, herbicides*, and *energetic compounds* in soils. Several studies investigated the electrokinetic removal of the organic pollutants as presented in Part III.

1.5.1 PAHs

Electrokinetic removal of PAHs has received greater attention (Chapter 9). In essence, the transport and removal of PAHs under electric field is found to be limited due to limited amounts these pollutants present in dissolved form in pore fluid. PAHs are hydrophobic and are sorbed to soil (especially organic matter). One fundamental requirement to enhance the removal is to solubilize the PAHs using surfactants, biosurfactants, cosolvents, and cyclodextrins. The addition of these solubilizing agents change the pore fluid properties such as dielectric constant, pH, and viscosity, as well as the surface characteristics of soil particles (e.g. use of ionic surfactants and high-pH cosolvent). These changes in fluid and soil surface properties affect the electro-osmotic flow, and thereby, removal efficiency. Electro-osmotic advection is the dominant contaminant transport and removal process; therefore, it is critical to ensure adequate electro-osmotic flow while flushing these solubilizing agents. The pH adjustment and the application of periodic electric potential are found to result in sustained electro-osmotic flow and higher removal efficiency.

1.5.2 Chlorinated Aliphatic Hydrocarbons, Chlorophenols, and Chlorobenzenes

Earlier studies focused on the removal of trichloroethylene (TCE), which is relatively more soluble than other chlorinated compounds. Maintaining adequate electro-osmotic flow was the main consideration in achieving higher removal efficiency. Anode buffering with an alkaline solution was used to maintain electro-osmotic flow. Recently, attention is focused on hydrophobic chlorinated organic compounds (Chapter 10) including chlorinated aliphatic hydrocarbons [e.g. pentachloroethylene (PCE), trichloroacetate (TCA), and TCE], chlorophenols (e.g. pentachlorophenol), and chlorobenzenes (e.g. PCBs). The removal of these pollutants is complicated by their sorption to the soil as well as their potential dissociation characteristics. Therefore, both electroosmosis and electromigration transport processes play a role in the transport and removal of these pollutants. The enhanced removal is accomplished by combinations of using solubilizing agents such as surfactants, cosolvents, and cyclodextrins and buffering anode pH to achieve higher removal by the combined electromigration and electroosmosis processes. The solubilizing agents should be carefully assessed to determine their effect on the surface charge of the soil surfaces and pH of the pore fluid and consequent impact on both electromigration and electroosmosis processes.

1.5.3 Chlorinated Pesticides and Herbicides

Sites contaminated by chlorinated pesticides (e.g. dichlorodiphenyltrichloroethane (DDT), aldrin, dieldrin, and endrin) and herbicides (e.g. atrazine, molinate and bentazone) due to agricultural activities and accidental spills have received little attention (Chapters 11 and 12). These contaminants are nonpolar in characteristics and sorb strongly to the soil. Similar to other hydrophobic organic compounds, desorption using solubilizing agents such as surfactants and cosolvents as well as buffering the anode pH are implemented to achieve high removal. One important characteristic of pesticides is that their solubility is rate limited, which causes lower aqueous-phase concentrations. Periodic electric potential application in such cases may be beneficial, but it has not been tested. Electrokinetic removal of herbicides is similar to that of pesticides and it is possible to remove these pollutants by controlling the pH both at the anode and cathode to result in favorable soil pH for desorption and electro-osmotic advection (Chapter 12).

1.5.4 Nitroaromatic and Other Energetic Compounds

The problem of soils contaminated by nitroaromatic and other energetic compounds (e.g. trinitrotoluene (TNT), dinitrotoluene (DNT), and cyclotrimethylenetrinitramine (RDX)) due to manufacturing and use of munitions has received attention only recently. These compounds are low-polarity organic molecules that exhibit low water solubility and strong affinities for complex formation with clay minerals and organic matter. Therefore, these contaminants must be desorbed, and some attempts have been made to use cosolvents, surfactants, and cyclodextrins to enhance their solubility and transport, but the removal was low. Proper consideration should be given to the charge of the specific solubilizing agents used and their complexes, if formed, as well as the potential electrochemical degradation processes that may occur within the soil or at the electrodes during the electrochemical treatment of these specific compounds (Chapter 13).

1.6 ELECTROCHEMICAL REMOVAL OF CONTAMINANT MIXTURES

Many of the fundamental and laboratory studies often deal with a selected single contaminant, either a heavy metal or an organic compound. Although these studies helped understand the fundamental processes and operational variables, the direct application of these study results to design systems for actual contaminated sites is often questioned. In addition to the soil compositional differences, contamination found at actual contaminated sites consists of multiple contaminants such as multiple heavy metals, multiple organic compounds, or a combination of heavy metals and organic compounds. Although few, there are sites contaminated in combinations of heavy metals, organic compounds, and radionuclides [e.g. Department of Energy (DOE), USA]. Synergistic effects and removal of multiple contaminants (multiple heavy metals, multiple organics, or multiple metals and organics) have been evaluated in a limited number of studies (Chapters 14 and 15).

1.6.1 Heavy Metals Mixtures

The heavy metals are affected by the various geochemical processes due to change in the soil pH as presented in Parts I, II, and IV. The sorption and desorption processes of coexisting multiple metals is quite complicated. Some of the heavy metals are tightly held than the others. If cationic and anionic metals exist, the sorption behavior of them can be quite the opposite. Therefore, the extent of transport of multiple metals depends on their aqueous concentrations and ionic mobilities. In general, the applied electric field is distributed among the multiple metals, thereby resulting in lower transport of a metal in the mixture as compared with the transport

16 OVERVIEW OF ELECTROCHEMICAL REMEDIATION TECHNOLOGIES

observed in the case containing only that specific metal. The dominant transport process for the removal of heavy metals is electromigration.

Several studies have been conducted on soils contaminated with multiple metals (Chapter 14). The removal of cationic metals is hindered by the sorption and precipitation near the cathode, and therefore require the lowering of soil pH near the cathode using weak organic acids, forming soluble complexes at high pH using chelating agents, or preventing high pH generation by the use of ionic exchange membranes. In case of anionic metals, pH near the anode should be increased using alkaline solutions such as NaOH to reduce sorption of these metals to the soil. When the heavy metals exist in soluble form, they are transported and removed predominantly by the electromigration process. Ion exchange membranes can also be used in the electrodes to control the soil pH to the desired level. When cationic and anionic heavy metals are found to coexist, a sequential electrochemical treatment that involves, first, the removal of cationic metals using cathode conditioning with a weak organic acid and then the removal of anionic metals using anode conditioning with an alkaline solution may be needed.

1.6.2 Heavy Metals and Organic Pollutant Mixtures

The problem of soils contaminated with mixed heavy metals and organic compounds is even complex because of the different chemistry of heavy metals and organic compounds (Chapter 15). Some studies have shown that there may be some synergistic effects that retard the contaminant transport and removal, but few other studies show the behavior of heavy metals and organic compounds similar to that observed with either heavy metals or organic compounds.

The heavy metals are removed predominantly by the electromigration process, while the organic contaminants are removed by electroosmosis. The presence of heavy metals causes the zeta potential of the soil to be less negative and even result in a positive value, affecting the electro-osmotic flow and sorption of the contaminants.

The heavy metals are affected by the various geochemical processes due to change in the soil pH as presented in Parts I, II, and IV. The removal of cationic metals is hindered by the sorption and precipitation near the cathode, and therefore require lowering of soil pH near the cathode using weak organic acids, forming soluble complexes at high pH using chelating agents, or preventing high pH generation by the use of electrode membranes. In case of anionic metals, pH near the anode should be increased using alkaline solutions such as NaOH to reduce sorption of these metals to the soil. When the heavy metals exist in soluble form, they are transported and removed predominantly by the electromigration process.

For the simultaneous removal of organic compounds, these compounds are solubilized using different solubilizing agents (surfactants, cosolvents, and cyclodextrins). They are then transported and removed mainly by the electroosmosis process. It is essential to maintain all of the contaminants in soluble form and maintain electro-osmotic flow for the removal of both heavy metals and organic compounds.

Sequential approaches are developed where (a) anionic metals are removed first and then cationic metals when mixed metal contamination is present and (b) organic compounds are removed first followed by the removal of heavy metals when coexisting heavy metal and organic contaminants are found. For example, the simultaneous electrokinetic removal of inorganic and organic pollutants (SEKRIOP) technology was developed by Elektorowicz and Hakimipour (2002), which uses ethylenediaminetetraacetic acid (EDTA) for metals mobility and zwitterionic surfactants for hydrocarbons. Cationic reactive membranes are also incorporated into this technology to capture free metallic ions generated by electrokinetic phenomena before their precipitation in the cathode area. The capture of metal–EDTA complexes was done on anionic reactive membranes.

High removal of multiple metals or coexisting metals and organic compounds (e.g. PAHs) is demonstrated in spiked soil conditions, but the performance was found inadequate in field soils from actual contaminated sites. Proper consideration should be made to reduce the treatment duration, proper handling of secondary liquid waste, and total cost in evaluating the electrochemical removal of contaminant mixtures.

1.7 SPECIAL CONSIDERATIONS IN REMEDIATING POLLUTED SEDIMENTS

Electrochemical remediation of soils, particularly clays, has been studied extensively. However, a very limited number of studies is conducted to evaluate the applicability of electrochemical remediation for the contaminated sediments. Huge amounts of contaminated sediments need to be dredged, dewatered, and treated before reuse or final disposal. Around 500 million m^3 of sediments are dredged each year for navigational purposes, and roughly 1%–4% requires dewatering and treatment prior to disposal, increasing the cost of dredging by a factor of 300–500. Sediments are characterized by a possible simultaneous presence of several types of pollutants that interact with different constituents of solid matrices. Other parameters, including moisture and salt content, may also have an influence on the suitability of different treatment options. While in soils the contaminated fine fraction typically accounts for less than 50% of total solids, in sediments this figure may be as high as 80%–95%, and due to the high surface area may contain the highest contaminant load. Electrokinetic remediation represents a promising option for contaminated sediments because it is capable of treating fine- and low-permeability materials, can also be applied *in situ*, and may represent a possible single-stage option to achieve dewatering, consolidation, and removal of organic and inorganic pollutants as well as salts.

Electrochemical removal of heavy metals in sediments often proved to be ineffective due to complex chemical composition, higher buffering capacity, and different metal speciation as compared to soils. Enhancement strategies using nitric acid, EDTA, and citric acid is found to significantly improve removal efficiency. Enhanced electrochemical removal of organic pollutants, particularly PAHs, is also studied. Generally, solubilization and removal of PAHs proved to be heavily hindered by high organic contents of sediment and the low liquid/solid ratio achievable as compared with chemical washing (Chapter 7).

1.8 ELECTROKINETIC BARRIERS FOR POLLUTION CONTAINMENT

The basic purpose of electrokinetic barriers is to prevent the migration of contaminants from its current location. These barriers are similar to traditional passive containment barriers (such as vertical slurry walls) for soil and groundwater pollution containment and active containment barriers (such as pumping systems and drainage systems) for groundwater pollution containment (Sharma and Reddy, 2004). Such barriers are often used as an interim measure prior to implementing a permanent treatment system.

Electrokinetic barriers consist of a row of electrodes bordering a high-concentration area or polluted groundwater plume. The row of electrodes is set perpendicular to the groundwater flow direction, while the depth of electrodes coincides with the lowest depth where pollutants are found. Electric current is induced into the ground by alternating anodes and cathodes. Anodes and cathodes are connected to separate closed-loop pump systems and are used to circulate electrolytes. Depending on the contaminant, pH can be controlled by conditioning the electrolytes. Periodically, contaminants from the electrolytes are also removed by different techniques such as sorption and ion exchange. More details on electrokinetic barriers can be found in Chapters 16 and 17.

1.9 COUPLED (OR INTEGRATED) ELECTROCHEMICAL REMEDIATION TECHNOLOGIES

Conventional electrochemical remediation essentially refers to the removal of contaminants from the contaminated media (soil, sediment, and groundwater), also known as *electrokinetic extraction* or *electrokinetically enhanced flushing*. As presented in Parts II, III, and IV, removal of contaminants (including contaminant mixtures) from spiked soils is possible using adaptive enhancement strategies. Despite high removal, practical implementation of such removal strategy is limited due to (a) regulatory constraints on injecting the selected enhancement solutions into the subsurface, (b) high cost, and (c) longer treatment time. Recently, electrochemical remediation is combined with other remediation technologies in order to overcome these issues as well as known deficiencies of the conventional remediation methods. Such integrated or coupled technologies investigated include

- electrokinetic biobarriers;
- electrolytic reactive barriers;
- electrokinetic–permeable reactive barriers (PRBs);
- electrokinetic–chemical oxidation/reduction;
- electrokinetic–bioremediation;
- electrokinetic–phytoremediation;
- electrokinetic–stabilization; and
- electrokinetic–thermal treatment.

A brief explanation of these integrated technologies is presented in this section.

1.9.1 Electrokinetic Biobarriers

Electrokinetic biobarriers (also known as electrokinetic biofences) are modified electrokinetic barrier used to contain and biodegrade organic pollutants in groundwater. A row of alternating anode and cathode electrodes is installed perpendicular to the groundwater flow direction. Anodes and cathodes are integrated into separate closed-loop pump systems and are used to circulate electrolytes. Electrolytes collected from anodes and cathodes are mixed above ground and circulated to maintain neutral pH. A row of infiltration filters are used to inject nutrient solutions such as nitrogen, phosphorous, and oxygen donors. These nutrients are generally electrically charged, and they can be dispersed homogeneously through the soil by electromigration. The organic pollutants transported by the groundwater are degraded by enhanced microbial activity within the downstream of the zone. Such electrokinetic biobarrier (electrokinetic biofence) has been successfully used at a site in Europe and detailed information on this field application is provided in Chapter 17.

1.9.2 Electrolytic Reactive Barriers (e⁻ Barriers)

Electrolytic reactive barriers (also known as e⁻ barriers) consist of closely spaced permeable electrodes installed in a trench perpendicular to the direction of groundwater flow intercepts a groundwater contaminant plume, similar to PRBs. Figure 1.2 shows a schematic of implementation of this remedial approach. A low electric potential applied to the electrodes induces oxidizing conditions at the anode electrodes and reducing conditions at the cathode electrodes. Using electrodes to deliver and recover electrons, thermodynamic conditions are shifted to drive transformation of target compounds to nontoxic products. A wide range of redox-sensitive contaminants such as arsenic, chlorinated ethenes (TCE, TCA), and energetic compounds (TNT and RDX), including these mixtures (difficult to treat with other technologies), may be treated using the electrolytic barriers. This remedial approach offers several advantages, including (a) effective degradation of contaminants and reaction intermediates through sequential oxidation and reduction, (b) control

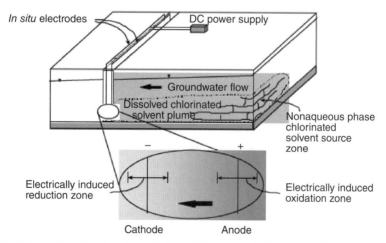

Figure 1.2. Schematic of the implementation of the electrolytic reactive barrier system (Sale, Petersen, and Gilbert, 2005). The barrier is installed perpendicular to the flow of groundwater direction and low electric current is applied to induce oxidation and reduction of contaminants at the electrodes.

accumulation of mineral precipitates via periodic reversal of electrode potentials, (c) no chemicals are introduced and no special permits required, (d) simple operation, and (e) low operation and maintenance costs. Laboratory tests and a field demonstration performed at a DOE site demonstrate that this technology is a viable option to remediate contaminated groundwater. More details on this technology can be found in Sale, Petersen, and Gilbert (2005).

1.9.3 Coupled Electrokinetic–PRBs

PRBs have been extensively used for the remediation of inorganic and organic pollutants in groundwater. Basically, PRBs consist of digging a trench in the path of flowing groundwater and then filling it with a selected permeable reactive material. As the contaminated groundwater passes through the PRB, organic contaminants may be degraded or sequestered and inorganic contaminants are sequestered, and clean groundwater exits the PRB. The reactive materials commonly considered include iron filings, limestone, hydroxyapatite, activated carbon, and zeolite. Monitoring data from several field PRB projects showed that the reactive material is clogged due to mineral precipitation resulting from flow of a high concentration of dissolved inorganic species. In addition, there may be decrease in the reactivity of the material used in the PRB. Coupling electrokinetics with PRBs is conceived to eliminate clogging of the PRB system caused by mineral precipitation and improve the long-term performance of PRBs. More research is needed toward developing a combined electrokinetic–PRB system where adaptive electrokinetic system is used to induce favorable geochemical conditions within the PRB as needed during the course of the remediation process (Chapter 23).

1.9.4 Coupled Electrokinetic–Chemical Oxidation/Reduction

It is possible to remove a wide range of organic contaminants from soils using solubilizing agents such as surfactants, biosurfactants, cosolvents, and cyclodextrins (Part III). However, there may be regulatory objections for injecting these solubilizing agents into the subsurface. In addition, posttreatment of the extracted solutions at the electrodes makes this treatment very costly. An alternative approach is to degrade the organic contaminants within the soil by injecting oxidants (e.g. hydrogen peroxide, permanganate, or persulfate) or reductants (ZVI in the form of nanoscale iron particles). The principles of chemical oxidation and reduction processes have been used for wastewater treatment for decades. However, it is challenging to introduce the oxidants/reductants into low-permeability clay soils. Combining electrokinetics with chemical oxidation/reduction facilitates delivery of the oxidants and reductants as well as increase contaminant availability in low permeability soils. Oxidants such as Fenton's reagents (H_2O_2 and Fe^{2+}) produce hydroxyl radicals, which break C-H bonds of organics into environmentally benign end products. Electrokinetics will also allow control of the soil pH and potential increase in temperature to create optimal conditions to achieve maximum oxidation (Chapters 21 and 22).

It should be noted that the electrokinetic–chemical reduction principles are the same as electrokinetic–PRB using iron filing as reactive media, but one has to wait for the contaminated water to pass through the PRB for the remediation to occur.

However, in electrokinetic–chemical reduction approach, the nanoscale iron particles are introduced into the contaminant source zone itself, thereby reducing the treatment time.

Generally, oxidants are stable only for a short period of time; therefore, electrosynthesis methods have been developed to produce oxidants on-site. Ultrasonic methods involving the application of high-intensity ultrasound are used to cause sonolysis of water to produce hydrogen peroxide and hydroxyl radicals. It should be noted that a simple combination of electrokinetics and ultrasonic waves is also investigated by some researchers, but the purpose was to enhance contaminant removal, not degradation.

Another technology known as electrochemical geo-oxidation (ECGO) is proposed, which involves application of low voltage and amperage to induce reduction–oxidation reactions at the microscale. This technology is based on the premise that soil particles act as microcapacitors that charge and discharge in a cyclic fashion. Even though low voltage and amperage are used, the energy burst on discharge at the microscale is intense, resulting in destruction of organic contaminants, theoretically to carbon dioxide and water. ECGO self-generates the agents for reduction (H as ion or radical) and oxygen (O elemental, OH and its radicals, HO_2 and its radicals) and when combing the generation of H_2O_2 with the corrosion products of steel anodes results in Fenton's reagent (refer to Chapter 32 for details). This technology is particularly attractive because no external oxidants need to be added.

1.9.5 Coupled Electrokinetic–Bioremediation

Bioremediation, which involves degradation of organic compounds using microbes, has received great attention because it is environmentally friendly, inexpensive, and requires low energy. However, it is a slow remediation processes and its effectiveness depends on the availability of nutrients, bioavailability of contaminants, and physical conditions such as temperature and moisture. Coupling electrokinetics with bioremediation (also known as electrobioremediation or electrobioreclamation) can facilitate injection of nutrients, electron acceptors, or microbes (if needed) and increase the bioavailability of contaminants, especially in low-permeability soils where hydraulic delivery techniques are ineffective. Interestingly, electrolysis reactions at the electrodes may be used to provide electron acceptors and donors. Hydrogen produced at the cathode may be used as electron donor for reductive degradation processes, while oxygen produced at the anode may be used for oxidative biodegradation. Although biodegradation is often applied to organic pollutants, few studies reported on biological immobilization heavy metals enhanced by electrokinetics. Advances are being made through additional research to optimize the electrokinetic effects on the microbial activity to achieve efficient contaminant degradation (Chapters 18 and 19).

1.9.6 Coupled Electrokinetic–Phytoremediation

Phytoremediation involves the use of living plants and their associated microorganisms to remove, degrade, or sequester inorganic and organic pollutants from soil, sediment, and groundwater. The main advantages of this method are low cost and ecological friendliness. However, this method is limited to shallow depths (limited by the root depth), slow plant growth, and also solubility and availability of the pollutant. Coupling electrokinetics with phytoremediation is aimed at increasing the availability of the contaminants and also facilitating their transport toward the root zone. The effects of electric field on soil pH, availability of nutrients, and so on may also help plant growth. Electrodes are placed strategically and a low direct current or voltage gradient is applied and the contaminants are transported by electromigration and/or electroosmosis processes toward the plant root zone. Electrode solutions of reduced toxicity toward plants can be used to enhance solubilization of the contaminants. Small-scale experiments showed that the plants are not affected by the exposed electric fields and that overall contaminant removal efficiency is controlled by different geochemical reactions. More research is needed to address organic contaminants and contaminant mixtures and possible effects on soil quality and biology (Chapter 20).

1.9.7 Coupled Electrokinetic–Stabilization

The electrochemical removal of contaminants may not be always possible or practical. For instance, a site may be too polluted to be treated to the acceptable level by any of the technologies, but it is critical to reduce the risk posed by the site contamination. The common approach used for risk reduction is stabilization and solidification (or immobilization) technology. In this approach, contaminants are transformed into a form that does not allow them to be released into the environment. Electrochemical approach may be used to stabilize the contaminated soils at a low cost and it will serve as an interim or pretreatment process to permanent treatment technologies.

Electrochemical stabilization (Chapter 8) focuses more on heavy metals than on organic pollutants, and it typically consists of converting the mobile contaminants into precipitates by injecting conditioning solutions such as alkaline solutions or reducing agents depending on the specific contaminant conditions. For example, reduction of Cr(VI) to Cr(III) by the delivery of iron (Fe^0, Fe^{2+}) is fairly well documented.

Alternatively, iron-rich sacrificial electrodes, which dissolve under acidic conditions generated at the anode by the application of electric field, may be used. The dissolved iron, in cationic form, migrates toward the cathode and then precipitates as iron-rich mineral phases (ferric iron oxyhydroxides, hematite, goethite, magnetite, and ZVI) near the cathode due to high-pH conditions. Contaminants such as Cr(VI) can react with this iron and reduce into Cr(III). Cr(VI) transport may be limited by high sorption under low-pH conditions; therefore, alkaline solution may be injected from the anode to increase the soil pH, and thereby reduce sorption and increase transport of Cr(VI) to react with iron.

1.9.8 Coupled Electrokinetic–Thermal Treatment

Thermal effects during electrochemical remediation are being studied. *In situ* thermal methods such as electrical resistance heating (ERH) have been used for site remediation, but electromotive forces that may occur during ERH are not studied. Heating resulting from electrochemical treatment involves the resistance to the passage of electrical current through soil moisture. It is this resistance to electrical flow that results in increases in temperature. Heat transport from this joule heating takes place mainly through conduction and convection. The increase in temperature decreases the viscosity of pore fluid, decreases the sorption of contaminants (increasing bioavailability), and increases the volatilization of organic contaminants. Heating results in lowering the oxidation–reduction potential in water, thereby affecting redox reactions. Elevated temperatures within the tolerable limits of microorganisms are also conducive to higher metabolic activity and enhanced biodegradation of contaminants. Proper control of increased temperature during electrochemical remediation may be exploited to enhance the overall remedial efficiency (Chapter 24).

1.10 MATHEMATICAL MODELING OF ELECTROCHEMICAL REMEDIATION

Mathematical models are useful to better understand the processes that occur under electric field and predict remedial performance in field application. Compared with laboratory studies, only few studies have been reported on the mathematical modeling of electrochemical processes and remediation. Generally, electrochemical remediation models should incorporate the contaminant transport, transfer, and transformation processes and dynamic changes in electrical conductivity, pH, and geochemical reactions. Recognizing this as a complex task, researchers have developed some simple models based on a set of simplified assumptions (Chapters 25 and 26).

First, the modeling of two dominant transport processes, electromigration (ionic migration) and electroosmosis, is critical in any electrochemical remediation modeling. Several models are reported to predict ion transport under electric field, assuming dilute solutions, rapid dissociation–association chemical reactions, and small double layer thickness. When charged ions transport under the influence of an externally applied electrical field, their concentration distributions change with time, which lead to a change in local electrical conductivity. The change in local electric conductivity directly alters the value of potential gradient at that specific point. Hence, the changing electric conductivity and electrical field describe the transport process of the species implicitly. The migration of the ions in the bulk fluid are modeled taking into consideration the changing electric field due to migration, as well as other effects such as retardation and electropheretic effects that reduce ionic mobility. These models appeared to simulate well the long-term ion distribution in the soil as the conductivity and the electric field vary in time and space.

The classical H-S equation is used to predict the electro-osmotic velocity of the fluid as a function of the electric field and the electrokinetic potential of the clay. Both of these parameters vary during electrokinetic transport, and result in a non-linear process. New models have been developed that uncouple the electro-osmotic velocity from the applied field taking that surface conductivity and the resulting proportion of the current transferred over the solid-liquid interface are used as intrinsic properties of the clay to describe the velocity (Chapter 2). The pH changes affect the zeta potential, and thereby electro-osmotic conductivity. Thus, electro-osmotic conductivity changes as the dynamic changes in soil pH occur.

Modeling of electrolysis reactions and the corresponding changes in soil pH and geochemical reactions require knowledge of electrochemistry and geochemistry. The modeling task is exacerbated spatially and temporally due to dynamic changes in geochemical conditions. The existing chemical speciation models such as MINTEQA2 (Chapter 25) cannot be easily integrated with models developed for electrochemical remediation. Generally, modelers used the same approach followed in the general geochemical models to develop a simplified geochemical model based on the specific problem being investigated and incorporated into their transport model.

Examples of one-dimensional and two-dimensional models to predict the transport of heavy metals under constant DC current are explained in Chapter 25. This model ignores hydraulic advection, electrophoresis, diffusion, and electroosmosis processes and considers only electromigration. Electrical potential distribution is assumed to be a function of the electrical resistance of the soil and depends on the instantaneous local concentration and mobility of all the ions existing in the pore water of the soil. Local chemical equilibrium is assumed to calculate the concentration of chemical species. Validation of these and other developed models based on laboratory and field test results is critical to gain confidence in the accuracy of the model predictions.

Mathematical models are also developed to assess the performance of electrokinetic barriers (Chapter 26). In general, hydraulic advection, electro-osmotic advection, and electromigration processes are incorporated in these models. Future work needs to incorporate biochemical reactions into the modeling of the behavior of electrokinetic reactive barriers. Similarly, a suite of mathematical models is needed for predicting the performance of other integrated electrochemical remediation systems.

1.11 ECONOMIC AND REGULATORY CONSIDERATIONS

1.11.1 Economic Considerations

The general perception among environmental professionals is that electrochemical remediation is very costly. However, one must understand the site conditions where electrochemical remediation is applied and the other potential competitive technologies applicable for the same site conditions. As noted earlier, electrochemical remediation is applicable to difficult sites where low-permeability and heterogeneous soils and/or complex contaminants exist. Many conventional remediation technologies are not even applicable for such site conditions. Therefore, cost comparison is often made with the option of excavation and disposal of hazardous waste.

Unfortunately, only limited data are available on the costs associated with electrochemical remediation technologies. Laboratory studies may not provide an accurate cost estimate as compared with field pilot-scale or full-scale applications. Many field applications have been implemented in Europe, while very few have been implemented in the USA. The reported cost data are based on the specific site conditions encountered at these applications. It is determined that electrochemical remediation costs US$115–US$400 per cubic meter of soil if it is contaminated with inorganic pollutants, and US$90–US$275 per cubic meter of soil if it is contaminated with organic pollutants, and the average cost is about US$200 per cubic meter of soil (equates roughly to US$90 per ton of soil) for inorganic or organic pollutants in saturated clayey soil. These estimated costs are less than the cost of excavation and disposal as hazardous waste. As the technology matures, confidence in designing the field systems will increase, resulting in lower cost. Refer to Chapter 27 for the basis and a detailed breakdown of the costs involved in electrochemical remediation.

1.11.2 Regulatory Considerations

Similar to any other technology, the design, implementation, and monitoring of electrochemical remediation must meet all applicable environmental laws and regulations, which vary from country to country. The environmental laws and regulations may be promulgated at the national, regional, and local levels, and it is required that any field application satisfies all of these laws and regulations for any field application. In the USA, the most relevant regulations addressing the remediation of contamination are (a) the Resource Conservation and Recovery Act, (b) the Comprehensive Environmental Response, Compensation and Liability Act (CERCLA, also known as Superfund), (c) the Clean Air Act, (d) the Clean Water Act, and (e) the Safe Drinking Water Act. These regulations provide the general cleanup process protocol, transport, and storage requirements and cleanup standards. More detailed information is presented in Chapter 28.

1.12 FIELD APPLICATIONS AND LESSONS LEARNED

Documented field applications provide invaluable information to design, assess, and further develop electrochemical remediation technologies. Over 75 field applications have been implemented in Europe, while few field applications have been implemented in the USA. The field demonstration and full-scale field implementation projects in other countries are being initiated.

1.12.1 Field Applications in the USA

The following field applications of electrochemical remediation have been reported addressing metal pollutants (Chapter 29):

- In 1989, the United States Environmental Protection Agency (USEPA) evaluated the remediation of a Superfund site heavily polluted with chromium.
- In 1994, Isotron Corporation implemented the *Electrosorb* process to evaluate the removal of uranium and organic contaminants.
- In 1996, Sandia National Laboratory addressed hexavalent chromium in an unsaturated soil at a chemical waste landfill site in Albuquerque, New Mexico.
- In 1998–2000, the US Army Environmental Center treated chromium and cadmium at NAWS Point, Mugu, California.
- In 2001–2002, Electrochemical Design Associates examined the removal of lead from soil at a Pearl Harbor naval shipyard and intermediate maintenance facility in Honolulu, Hawaii.

Overall, the field applications demonstrate that the electrokinetic treatment is an adequate tool for the remediation of polluted sites with heavy metals. However, the effectiveness of the process largely depends on the geochemical characteristics of the soil and the interaction of the pollutants with the soil. Therefore, a detailed study of each case is necessary to establish the most adequate operating conditions. It includes current intensity or voltage drop, electrode disposition, and chemical conditioning of electrode solutions. Laboratory studies can be carried out prior to the field operation in order to determine the effect of those operating conditions in the solubilization and removal of heavy metals, but it must be considered that the results at field scale can differ from those obtained in the laboratory experiments due to the change in scale.

A well-documented field application of electrochemical remediation is reported to address the problem of chlorinated solvent (TCE) in clay soil at the DOE site in Paducah, Kentucky. This process is known as Lasagna™ (Terran Corporation, Beavercreek, OH) and it combines the electro-osmotic transport of TCE in pore water and degrades it in vertical curtains installed along the flow path within the soil that are filled with iron filings and kaolin clay. Pore water accumulated at the cathode is recycled by gravity back to the anode as makeup water and neutralize the acid formed at the anode. Overall, the treatment is found to be effective. The implementation and performance results are presented in detail in Chapter 30. The same Lasagna™ process is implemented very recently at another site contaminated with TCE in Fonde du Lac, Wisconsin, and performance is being monitored (Chapter 30).

The electrolytic reactive barrier (electrolytic redox barrier or e^- barrier) concept is demonstrated at F.E. Warren Air Force Base in Wyoming in 2002 to remediate groundwater plume contaminated with approximately $300\,\mu g/l$ of TCE. The details of barrier design and installation are provided by Sale, Petersen, and Gilbert (2005). After installation, the barrier is allowed to equilibrate with the contaminant plume for 5 months, then low-voltage DC is applied continuously for about 18 months. Under the maximum potential of 6.5 V, TCE flux is reduced by 90% with no adverse intermediates. Complementary studies are proposed at other sites to treat energetic compounds that are difficult to treat with conventional technologies.

Recently, a field demonstration of the ECGO to remediate contaminated dredged harbor sediments from Lake Superior in Duluth, Minnesota, is completed. The sediments were contaminated by PAHs, PCBs, mercury, and other miscellaneous contaminants. The dredged sediments were planned to be disposed in confined disposal facilities (CDFs). Instead of CDFs serving as perpetual containment systems, a new approach being proposed is to convert CDFs to storage, handling, and treatment facilities. ECGO demonstration was performed within this context to remediate dredged sediments within CDFs. One control cell and one demonstration cell were used. Electrodes (steel sheet pile or steel pipe) were installed at a spacing of 5–15 m in the demonstration cell, and voltages and electric currents that vary from 20 to 100 V and approximately 0.05–50 A were applied. The contaminant concentrations were monitored over a period of 4 months. Field observations showed that reactions are taking place, but concentration data were inconclusive with regard to contaminant reduction. A detailed explanation of the findings can be found in Chapter 32.

1.12.2 Field Applications in Europe

Over 75 full-scale field systems are implemented in Europe, predominantly in the Netherlands. Chapter 33 presents the generalized procedures followed for the design, implementation, and monitoring of the field systems. The major components of field systems include electrodes, electrolyte management system, and electrolyte purification. Electrochemical remediation can be implemented in batch (*ex situ*) or *in situ* modes. Predesign investigations include laboratory soil tests such as cation exchange and anion exchange tests and two special tests known as "turbo test" and "standard test" (see Chapter 33 for details). It is also recommended to conduct field geoelectrical survey to measure the electrical resistivity of the soils at the site. Moreover, remediation equipment and electrode materials are selected based on the site conditions: chemical nature and concentration of contaminants, soil type, and geochemical interactions. Performance data from large sites contaminated with inorganic and/or organic pollutants are presented in Chapter 33, and it shows that electrochemical remediation is an effective remedial method.

1.12.3 Field Application in Korea

The coupled electrokinetic–PRB concept was demonstrated at a landfill site where groundwater was contaminated by the uncontrolled release of landfill leachate (Chapter 31). Two different configuration of electrode wells were tested: The first configuration consisted of anode electrode well in the middle and the cathode wells located radially at a distance of approximately 1 m from the anode well; the second configuration consisted of one row of cathode wells in the middle of two rows of anode electrode wells at a distance of 2 m. ZVI, zeolite, slag, tires, and sand were tested for PRB materials. From the preliminary field investigations, the coupled technology of the electrokinetic–PRB system would be effective in remediating contaminated grounds, and the extraction of pollutants from the subsurface was not necessary due to the reactions between the reactive materials and the contaminants.

1.13 FUTURE DIRECTIONS

As reflected by the content of this book, many significant advances have been made toward establishing electrochemical remediation as a practical remediation technology. The field application of this technology lags far behind the laboratory and other research studies. The following issues should be addressed in order for this technology to be used in field applications and becoming an generalized commercial technology for the remediation of soils, sediments, and groundwater:

- Limit the small-scale laboratory experiments and encourage large-scale pilot tests in laboratory or in the field
- Limit the single contaminant experiments in model soils spiked in laboratory and promote the tests with actual contaminated soils with aged and multiple contaminants
- Extend the electrochemical technology to other contaminated porous matrices such as industrial wastes in order to achieve their sustainable reuse
- Promote the analysis of experimental results based on the physicochemical properties of contaminants, geochemistry of soil, reaction kinetics, equilibrium constants, and transport parameters, rather than a phenomenological analysis of contaminant removal

- Advance the fundamental research and develop predictive modeling tools; in particular, geochemical reactions in various soil and contaminant conditions should be properly characterized
- Establish quality control and quality assurance protocols for laboratory and field studies
- Assess the impacts of electrochemical remediation on soil quality and ecology
- Investigate the short-term and long-term effects on electrochemical processes
- Develop new and innovative approaches of electrochemical remediation that are less expensive and practical
- Develop electrochemical remediation systems that are ecologically safe based on sustainability considerations
- Perform and disseminate results of well-monitored field pilot-scale demonstrations
- Establish communication among the researchers and identify areas that require future research
- Develop guidance documents for the design, installation, and operation of typical electrochemical remediation systems

In summary, the importance of full-scale field demonstration projects cannot be overemphasized. The lessons from these projects are invaluable in identifying the advantages and limitations of this technology and in developing effective and economical adaptive field systems based on site-specific conditions.

REFERENCES

Acar YB, Alshawabkeh AN. (1993). Principles of electrokinetic remediation. *Environmental Science and Technology* **27**(13):2638–2647.

Acar YB, Gale RJ, Alshawabkeh AN, Marks RE, Puppala S, Bricka M, Parker R. (1995). Electrokinetic remediation: Basics and technology status. *Journal of Hazardous Materials* **40**(2):117–137.

Elektorowicz M, Hakimipour M. (2002). Electrical field applied to the simultaneous removal of organic and inorganic contaminants from clayey soil. 18th Eastern Canadian Research Symposium on Water Quality, October 18, Montreal, Quebec, Canada: CAWQ.

Probstein RF. (2003). *Physicochemical Hydrodynamics: An Introduction*, 2nd ed. Hoboken, NJ: John Wiley & Sons.

Sale T, Petersen M, Gilbert D. (2005). Electrically Induced Redox Barriers for Treatment of Groundwater. Environmental Security Technology Certification Program (ESTCP). *Project CU-0112. Final Report.*

Sharma HD, Reddy KR. (2004). *Geoenvironmental Engineering: Site Remediation, Waste Containment, and Emerging Waste Management Technologies*. Hoboken, NJ: John Wiley & Sons.

2

ELECTROCHEMICAL TRANSPORT AND TRANSFORMATIONS

SIBEL PAMUKCU

2.1 INTRODUCTION

The transport of charge from one location to another is a fundamental mechanism underlying many physical and chemical phenomena in natural and synthetic systems. The coupling of multiple physical and chemical processes in electrochemical transport makes it a fascinating and complicated diffusion event to understand (Rubinstein, 1990).

The theoretical development of electrochemical transport has had a long and distinguished tradition that includes the works of Quinke (1861), Helmholtz (1879), Nernst (1888), Planck (1890), and Warburg (1899). Over the years, theoretical understanding has been applied successfully to many electrochemical systems in different areas such as energy storage and conversion, water treatment, microfluidics, metallurgy, and semiconductor devices. Among these, electrokinetically driven mass transport in porous medium has been of special interest to soil scientists and environmental and geotechnical engineers because of its high potential to help strip and extract contaminating chemicals from low-porosity geological materials such as clays.

In this chapter, we will review the fundamental principles of electrochemical transport as it applies to a medium of saturated clay and show some viable results of electroremediation of clays based on the understanding of electrochemical transport and transformation in such media.

2.2 OVERVIEW

As in many electrochemical systems, the flow of electric current through a network of a multiphase system occurs in different phases simultaneously: in the bulk liquid

---

Electrochemical Remediation Technologies for Polluted Soils, Sediments and Groundwater,
Edited by Krishna R. Reddy and Claudio Cameselle
Copyright © 2009 John Wiley & Sons, Inc.

(electrolyte in the pores), on the surface of the solid (clay particles), and in the interface layer(s) between the solid and the liquid. The flow of the current can be achieved by ionic conduction through the liquid phase and electronic conduction through the solid phase and the interface layer(s). The electronic conduction orthogonal to and along the interface layer(s) takes place via charge transfer. In the classical treatise of "electrokinetic phenomenon" in colloidal systems (Lyklema, 1995; Hunter, 2001), it is this interface, known as the *electric double layer*, that plays a critical role in the coupling between the ion motion and the fluid flow. The double layer intrinsically connects the solid and the liquid phases, and mediates the relative motion between the liquid and solid phases through (a) accumulation of charge density, (b) transport of charge and ions along surface, and (c) passage of charge to the surrounding electrolyte (Bard and Faulkner, 1980).

The bulk transport of ions in electrochemical systems without the contribution of advection is described by Poisson–Nernst–Planck (PNP) equations (Rubinstein, 1990). The well-known Nernst–Planck equation describes the processes of *diffusion*, the process that drives the ions from regions of higher concentration to regions of lower concentration, and *electromigration* (also referred to as *migration*), the process that launches the ions in the direction of the electric field (Bard and Faulkner, 1980). Since the ions themselves contribute to the local electric potential, Poisson's equation that relates the electrostatic potential to local ion concentrations is solved simultaneously to describe this effect. The electroneutrality assumption simplifies the mathematical treatise of bulk transport in most electrochemical systems. Nevertheless, this "no charge density accumulation" assumption does not hold true at the interphase regions of the *electric double layer* between the solid and the liquid, hence the cause of most electrokinetic phenomena in clay–electrolyte systems.

2.3 ELECTROCHEMICAL TRANSPORT IN BULK FLUID

The analysis of mass transport by diffusion under chemical ($\partial C/\partial x$) and by migration under electrical ($\partial \Phi/\partial x$) gradients in dilute solutions—for which the interactions between individual species can be neglected—is described by the Nernst–Planck (Eq. 2.1) and the Poisson's (Eq. 2.2), together referred to as the PNP equations:

$$\frac{\partial C_i}{\partial t} = \nabla \left(D_i^* \nabla C_i + u_i z_i F C_i \nabla \Phi \right) \tag{2.1}$$

$$-\varepsilon_s \nabla^2 \Phi = \rho = \sum_i z_i F C_i \tag{2.2}$$

where C, D^*, u, and z are the concentration, diffusion coefficient, mobility, and charge number of a single species i, respectively; F is the Faraday's constant (a mole of charge); Φ is the electrostatic potential; ε_s the permittivity of the solvent; and ρ the charge density.

For many electrochemical systems, the local electroneutrality condition is used, which sets the left-hand side of Equation 2.2 to zero for zero charge density. The mathematical implication is then that the electrical potential satisfies the Laplace equation ($\nabla^2 \Phi = 0$), hence the uniform concentration distribution and uniform conductivity within the electrochemical cell. Yet in real electrochemical systems with concentration gradients, the current density of the system can be affected to the point that it would cause current flow in the opposite direction of the electric field (Newman, 1991). Using the PNP equations and the electroneutrality condition, it can be shown mathematically that concentration gradients give rise to spatial variation of conductivity, where a *diffusion potential* arises to ensure ion movement at the same speed to overcome the charge separation and violation of electrical neutrality (Newman, 1991).

The charge density accumulation cannot be neglected for interphase layers such as the electrical double layer. Using a nondimensionalized form of Equation 2.2, it was shown that the electroneutrality condition is a direct result of the Debye screening length, where the ratio of the Debye length λ_D to the field length L is described as follows (Eq. 2.3) (Chu, 2005):

$$-\varepsilon \nabla^2 \phi = \sum_i z_i c_i \quad (2.3)$$

where $\varepsilon = \dfrac{\lambda_D}{L}$; $\phi = \dfrac{\Phi}{RT/F}$; $c_i = \dfrac{C_i}{C_{eq}}$; $\lambda_D = \sqrt{\left(\dfrac{\varepsilon_s RT}{F^2 C_{eq}}\right)}$

where R = universal gas constant, (8.3144 J/K mol); T = absolute temperature (K); F = Faraday constant (96,485 C/mol electrons); and C_{eq} = equilibrium concentration in bulk.

As shown in Equation 2.3, unless the Laplace compliance holds true—which is not the case for most electrochemical systems owing to the presence of concentration gradients—the electroneutrality of the system can be satisfied when the Debye length λ_D is so small compared with L (such that $\varepsilon \ll$) that the left-hand side of the equation becomes zero. Asymptotic analysis of PNP equations numerically solved at several values of ε showed that for small values of ε—as would be in macroscopic systems—the charge density ρ is zero for majority of an electrochemical cell of length L, except at the boundaries where faradic reactions and Stern-layer capacitance are considered (Chu, 2005). As the Debye length approaches to the width of the electric double layer, nonzero charge density distribution appears in the entire electrochemical cell (Chu, 2005). Hence, the electroneutrality in an electrochemical system will hold when the charge density is small compared with the total ion concentration, C_{eq}, of the bulk fluid; that is, $|C^+ - C^-| \ll C^+ + C^-$. As the charge density approaches to total concentration within the electrochemical cell (e.g. electric double layer for which Debye length is the width of the layer), electroneutrality can be achieved considering Nernstian boundaries and faradic reactions. This will be discussed further in a later section on "Electrochemical Transformations."

2.4 ELECTROCHEMICAL TRANSPORT IN CLAYS IN THE DIRECTION OF APPLIED ELECTRIC FIELD

Electrically induced migration of ions and water is a proven method of externally forced mass transport in clay soils for contaminant remediation purposes (Pamukcu and Wittle, 1992; Acar and Alshawabkeh, 1993, 1996; Lageman, 1993; Probstein and

Hicks, 1993; Acar et al., 1994; Eykholt and Daniel, 1994; Hicks and Tondorf 1994; Shapiro, Probstein, and Hicks, 1995; Alshawabkeh and Acar, 1996; Electorowicz and Boeva, 1996; Yeung, Hsu, and Menon, 1996; Dzenitis, 1997; Reddy and Parupudi, 1997). While electroosmosis is analogous to soil washing, ion migration is probably the primary mechanism of mass transport when the contaminants are ionic or surface charged.

Relative contribution of electroosmosis and ion migration to total mass transport varies according to soil type, water content, ion species, and their concentrations. In silts and low-activity clays, electroosmotic flow reaches maximum in comparison with hydraulic flow. But the mass transport by ionic migration is always much higher than the mass transport by electroosmotic advection (Acar and Alshawabkeh, 1993). The effect of electroosmosis decreases significantly when pH and zeta (ζ) potential drops in the later stages of a sustained electrokinetic process under a constant electric potential (Hamed, Acar, and Gale, 1991; Pamukcu, Weeks, and Wittle, 1997). When micelles (charged aggregate of molecules or particles) are formed with other species in the processing fluid, or when we deal with slurries, electrophoresis may become significant (Pamukcu, Filipova, and Wittle, 1995).

2.4.1 Electroosmotic Transport

In 1809, Reuss observed the electrokinetic phenomena when a direct current (DC) was applied to a clay–water mixture. Water moved through the capillary toward the cathode under the electric field. When the electric potential was removed, the flow of water immediately stopped. In 1861, Quincke found that the electric potential difference across a membrane resulted from streaming potential. Helmholtz first treated electroosmotic phenomena analytically in 1879, and provided a mathematical basis. Smoluchowski (1914) later modified it to also apply to electrophoretic velocity, also known as the Helmholtz–Smoluchowski (H-S) theory. The H-S theory describes under an applied electric potential the migration velocity of one phase of material dispersed in another phase. The electroosmotic velocity of a fluid of certain viscosity and dielectric constant through a surface-charged porous medium of zeta or electrokinetic potential (ζ), under an electric gradient, E, is given by the H-S equation as follows:

$$v_{eo} = \frac{\varepsilon_s \zeta}{\eta} \cdot \frac{\partial \Phi}{\partial x} = k_{eo} E, \qquad (2.4)$$

where

v_{eo} = electroosmotic (electrophoretic) velocity;
ε_s = permittivity of the solvent or the pore fluid;
η = viscosity of the solvent or the pore fluid;
k_{eo} = coefficient of electroosmotic conductivity; and
$\partial \Phi / \partial x = E$ = electric field.

The ζ potential in Equation 2.4 has been shown to vary with the pH and ionic concentration of the pore fluid, as well as the electric field, and therefore is not constant during electroosmotic transport in clay medium (Probstein and Hicks, 1993). This makes it difficult to assess a velocity term for temporal and spatial distribution predictions of electroosmotic transport. The nonlinearity and nonuniformity associated with the electroosmotic velocity become apparent when experiments showed that v_e/E increased with E, hence rendering the other parameters of Equation 2.4 to be functions of the electrical field (Ravina and Zaslavsky, 1967). In a given clay deposit in the field, pore cross-sections, the electrical double layer properties, fluid viscosity, dielectric constant, solute distribution, and electrical field can vary significantly from one point to another, along both the flow path and the direction normal to it. Hence, averaging microscopic parameters (e.g. ζ potential) to use along with macroscopic parameters (e.g. E) in the H-S equation can lead to the nonlinear effects.

The ζ potential is the electrical potential at the junction between the fixed and mobile parts of the electrical double layer. The ζ potential is influenced by the type and concentration of the electrolytes added to the particle suspension (Kruyt, 1952; Smith and Narimatsu, 1993). For clay soils, ζ potential is usually negative because of the net negative charge on clay particle surfaces. The magnitude and sign of this potential highly depends on pore fluid chemistry. As the hydrogen and hydroxyl ions are the potential determining ions, the lower pH will reduce the ζ potential in magnitude for most clay. At low enough pH, ζ potential may become positive. Hunter and James (1992) observed that adsorption of partially hydrolyzed metal cations such as Co^{2+}, Cd^{2+}, and Cu^{2+} cause ζ potential reversals for kaolinite. As the concentration of hydrolyzable metal ions increases, ζ potential becomes more positive at low pH levels due to the accumulation of the cations in a compressed electrical double layer bearing a larger charge than is present on the solid surface (Kruyt, 1952). The effect is largest at an intermediate pH, slightly above the value at which precipitation of the metal hydroxide would be expected in the bulk solution. Due to the influence of sorption of hydrolyzable metal ions, the sign reversal of the ζ potential can make the net electroosmotic flow insignificant in clay soils with high pore fluid electrolyte concentrations. In this case, and also for saline soils, electromigration becomes the dominant mechanism of electrokinetic transport.

Recognizing the inability to uncouple v_{eo} from E in the classical H-S equation for macroscopic systems, Khan (1991) proposed a modified theory of electroosmotic flow through clay soil. In Khan's work, the ionic conductance through the bulk fluid and the electronic conduction through the electric double layer at the solid–liquid interface were lumped in series, representing resistances in parallel or conductance in series. When modeling mass flow in the direction of the electric field, it is intuitive to line up the conducting elements in the direction of the electric field. The electrokinetic transport theory clearly shows that it is the electronic conductance at the electric double layer that induce electroosmotic transport and that it is the ionic conductance in the bulk fluid that result in the ionic transport. Hence, the following relations are expected to hold:

$$i = i_s + i_b \tag{2.5}$$

$$E = \frac{R_s i_s}{L} = \frac{R_b i_b}{L} \tag{2.6}$$

$$v_{eo} = k_{eo} \frac{R_s i_s}{L} = \left[\frac{k_{eo}\tau}{\sigma_s n}\right] \cdot \frac{i_s}{A} \tag{2.7}$$

$$v_{eo} A = Q_{eo} = K i_s \tag{2.8}$$

where k_{eo} is the H-S coefficient of electroosmotic conductivity; i, i_s, and i_b are the total, surface, and bulk currents (ampere or C/s), respectively; R_s and R_b (ohm) are the surface and the bulk resistances, respectively; and σ_s is the surface conductivity (siemens/l or l^{-1} ohm^{-1}, where l represents the unit of length); τ is the tortuosity; n is the porosity; and A is the total area ($A_{flow} = nA$). The relation between the resistance and conductivity is given as $R_s = \dfrac{L'}{\sigma_s A_{flow}}$, and the tortuosity term τ is included to augment L ($L' = L\tau$) since the electric currents may follow a tortuous path through the length, L, of the electrochemical cell for which the pore walls are assumed lined up with connected electric double layers. In this model, K emerges as a factor that relates volume rate of electroosmotic flow of fluid movement per electrical charge.

Shang, Lo, and Inculet (1995) looked at the relation of conductivity and apparent polarization in several wet clay systems, where they found that the surface conductivity and the apparent permittivity decreased significantly when the volume fraction of the particles was increased beyond 0.3. They also found that the conductivity of the bulk fluid was at least an order of magnitude higher than the surface conductivity ($R_b/R_s \cong 0.1$), and that the surface conductivity increased when the clay was remolded ($R_b/R_{sr} \cong 0.2$–0.5). These findings support Khan's (1991) "parallel resistances" model when the flow direction is implicitly assumed to be the direction of the electric current flow. Because the electroosmotic velocity representation using the surface conductivity, σ_s, would require fully developed electric double layers with connectivity, increasing the dry clay content would inhibit the connectivity of the electric double layers, and hence cause the reduction in the measurement of the surface conductivity for the overall system. Remolding of wet clay pastes has been shown to reduce the repulsion between particles and cause them align in a face-to-face orientation (Quigley, 1980; Mitchell, 1993), therefore possibly increasing the connectivity of the electric layers in edge-to-edge correspondence. It follows that the preferential alignment of the particles in the direction of the electrical field may help increase surface conductivity as was shown in the Shang, Lo, and Inculet study (1995).

Equation 2.8 can further be simplified to determine a characteristic K value. Recognizing resistances in parallel or conductance in series once again, the expression for i_s [$i_s = (R_b/(R_s + R_b)) \cdot i$] can be simplified as $i_s = (R_b/R_s) \cdot i$ for concentrated bulk liquid systems where $R_b \ll R_s$ and that their ratio remain approximately constant. Alternatively, $i_s \cong 0.5i$ in dilute systems when $R_b \cong R_s$. Hence, based on the data reported by Shang, Lo, and Inculet (1995) and the treatise above, it can be said that $0.1i \leq i_s \leq 0.5i$ in most natural clay–electrolyte systems, with the assumption that surface conductivity can never be larger than the bulk conductivity.

Through laboratory experiments, K has been shown to be a constant, independent of the electric field and soil mineralogy for similar major ion concentrations of the bulk fluid (Pamukcu and Wittle, 1992; Wittle and Pamukcu, 1993). In these studies, the volume of electroosmotic flow (V, cc) per electric charge (either in

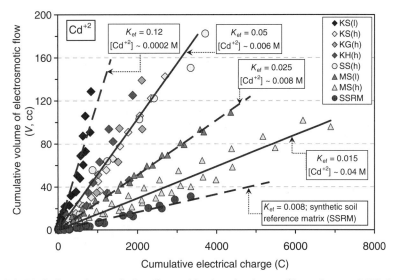

Figure 2.1. Variation of cumulative flow with electric charge (Pamukcu and Wittle, 1992). l: low concentration; h: high concentration.

coulombs or moles of electrons) referred to as the *electroosmotic efficiency coefficient*, K_{ef}, were determined from numerous electroosmotic tests of different clay and solute combinations. The relation between the cumulative electrical charge $\left(\sum_i i_i \cdot t_i\right)$ versus the cumulative volume of electroosmotic flow $\left(V = \sum_i Q_{eoi} \cdot t_i\right)$ was found to be linear for a given clay type (e.g. kaolinite, montmorillonite, and sand–clay mixtures), with a given initial concentration of different ion species (As, Cd, Co, Cr, Cs, Hg, Ni, Pb, Sr, U, and Zn) all applied separately in the pore fluid.

Figure 2.1 illustrates the results of cumulative electrical charge versus cumulative volume of flow for various different clay and clay mixtures when the major contributing ion in the soil matrix was cadmium. In these experiments, the initial concentration of Cd^{+2} in each medium was determined after mixing clay slurry with a concentrated solution of cadmium salt, which was left to consolidate to a constant effective stress over a few weeks, and measuring the total cadmium retained in the compressed clay matrix. This procedure allowed uniform distribution of the ion in the clay, the time necessary for potential ion exchanges to take place, and more importantly, the full development of the electric double layer at the ion concentration retained within the clay bulk fluid.

In these series of experiments (Wittle and Pamukcu, 1993), the clay matrixes used were pure kaolinite (KS), montmorillonite (MS), kaolinite with simulated groundwater electrolytes (KG), kaolinite with humic substances (KH), and a mixture of fine sand and 10% by weight montmorillonite (SS). In all cases, the transport experiments were conducted over 48 h with a constant applied field of 4 V/cm across the electrodes. The clay specimens were sandwiched between two electrode chambers housing a set of graphite electrodes in tap water. As observed in Figure 2.1, the $K_{ef}\left(=\left(\sum_i Q_{eoi} \cdot t_i\right) \Big/ \left(\sum_i i_i \cdot t_i\right)\right)$ is demonstrated to be a constant factor for the soil type, but changes with the initial concentration of the major ion, in this

36 ELECTROCHEMICAL TRANSPORT AND TRANSFORMATIONS

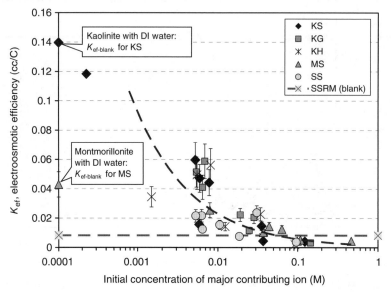

Figure 2.2. Variation of electroosmotic efficiency K_{ef} with ionic concentration.

case Cd^{+2}. Also shown on the graph is data from a reference soil matrix [synthetic soil reference soil matrix (SSRM)], which represented a typical well-graded reference soil matrix (obtained from United States Environmental Protection Agency (USEPA)). The behavior of K_{ef} shown in Figure 2.1 is typical of all the other major contributing ions tested in these series of experiments (Pamukcu and Wittle, 1992, 1993).

The constancy of K_{ef} with soil type or clay mineral demonstrates that the assumptions made in developing Equation 2.8 are valid. The following treatise strengthens further the validity of these assumptions. Figure 2.2 presents the variation of the measured K_{ef} for all soils and selected six major contributing ions (Cd, Cs, Cr, Pb, Sr, and Zn) in the bulk fluid, all applied separately. There is a clear trend of decreasing K_{ef} factor with increasing ion concentration, where at lower concentrations the data displays larger scatter than at higher concentrations. Highly dependent on concentration and type of ion at higher values, K_{ef} becomes independent of concentration at lower values.

To further analyze K_{ef}'s dependency on the ion type, Figure 2.3 is plotted where the normalized K_{ef} of different soil types with similar molar concentrations of the different ions are shown. The values are normalized by the measured K_{ef} values of respective "blank" clays, those mixed and consolidated only with distilled water. At low soil concentrations of Cs, Cd, and Sr, ($\leq 0.01\,M$), the concentration and the type of ion appear to control the K_{ef} factor as all soil types (except montmorillonite for Sr) show very close to constant normalized values of K_{ef} for the three ions involved. Subsequently, as the concentrations of the contributing ions increase beyond 0.01 M, fewer soil types comply with the categorization, as the mineralogy of the soil, hence its unique chemical interaction with the dominant ion comes into effect. Because of the uniqueness of this interaction, the normalized values show greater variation for each ion involved at high concentration (Cr, Pb, Zn), then the low concentration (Cs, Cd, Sr), group.

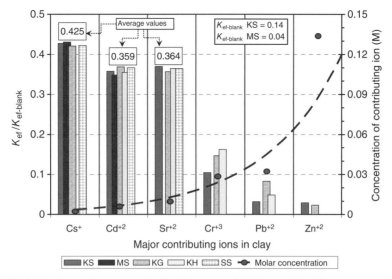

Figure 2.3. Variation of measured K_{ef} factors with ion and soil type at the same concentration.

The following conclusions are drawn from these observations presented in Figures 2.1 through 2.3.

1. K_{ef} factor is a phenomenological parameter, controlled by the type and concentration of the major contributing ion in the bulk fluid and the composition and mineralogy of the soil matrix.
2. K_{ef} is dependent on the *type* and *concentration* of the major contributing ion in the bulk fluid (that which contributes to the formation of the electric double layer) in dilute systems where $R_b \cong R_s$.
3. K_{ef} factor decreases with increasing ion concentration in the bulk fluid and as the $R_b \ll R_s$ condition remains valid, but it becomes independent of the concentration above a threshold value when the *chemistry* of the bulk fluid and the *composition* and *mineralogy* of the soil appear to play a dominant role in the variation of K_{ef}.

K_{ef} is computed easily by measuring in the laboratory the total volume of flow through the clay matrix and the corresponding total current. In Equation 2.5, the total current, i, is given as the summation of the bulk current, i_b, and the surface current, i_s. Unless the contribution of the surface current is uncoupled, the measured K_{ef} based on total current, i, will always display high dependency on soil composition and mineralogy and the type and concentration of the major contributing ion. During electrokinetic transport, as the bulk fluid chemistry changes so does the surface and electric double layer properties of the clay medium. Equation 2.7, derived earlier, shows a ratio of the H-S k_{eo} and the surface conductivity, σ_s, which should remain constant as they are uniquely related.

In an attempt to demonstrate the constancy of the K factor in Equation 2.8 ($K = Q_{eo}/i_s$), the data presented in Figure 2.1 is utilized once again. As discussed

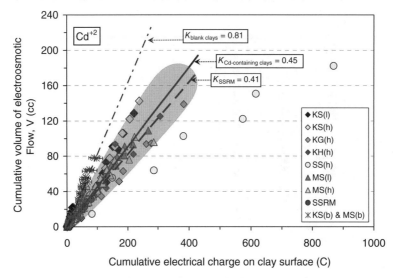

Figure 2.4. Variation of cumulative flow with electrical charge on clay surface. l: low concentration; h: high concentration; b: blank.

previously, assigning $i_s = 0.1i$ for the most dilute system data (0.0002 M) and $i_s = 0.5i$ for the most concentrated system (0.04 M) in the category of cadmium-spiked clay specimens, the volume of flow is replotted against the surface current i_s, as shown in Figure 2.4. The distribution of the surface current corresponding to the in-between concentration values of 0.006 and 0.008 M followed a power curve, which was a plausible distribution based on the concentration versus K_{ef} in Figure 2.2. As observed in Figure 2.4, the average slope of the volume of flow versus the electrical charge transferred over the clay surface is 0.45 ± 0.1 cc/C for all the Cd-containing clay soils except for sand with 10% montmorillonite composition. When the SSRM data was plotted, it fell right into the cluster of clay data with a slope value of 0.41 cc/C.

The blank kaolinite and montmorillonite clays computed a close average value of $K = 0.81$ cc/C with 20% error bars displayed on the graph. The behavior presented in Figure 2.4 shows that the proportionality between flow and surface current for electroosmotic transport in clay systems can be expressed as a constant, independent of soil mineralogy and major ion concentration in the bulk fluid. Previous findings for pure clay systems reported a constant K value of 0.75 cc/C for blank muscovite and kaolinite clays measured at electric fields varying from 10 to 80 V/m by Khan (1991). This value is in close agreement with 0.081 cc/C found as the average K for montmorillonite and kaolinite clays as shown in Figure 2.4.

It is important to note here that the use of Equation 2.8 necessitates the presence of distributed surface conductivity that enables the surface currents to travel within the electrochemical system for electroosmosis to occur. Yin, Finno, and Feldkamp (1995) backed up Khan's theory, which relate electroosmotic mobility and surface conductivity. They found that there is no apparent relationship between electroosmotic mobility and the applied electric field. The term electroosmotic mobility, defined as the average velocity achieved by the pore water relative to the solid skeleton due to an externally applied electrical field of unit strength (cm^2/s-V), appeared to be directly proportional to the specific conductance (mho/cm) of the specimens at different water saturations in independent tests (Yin, Finno, and Feldkamp, 1995).

In summary, the modeling of the electroosmotic component of the electrochemical transport is dependent on the electroosmotic velocity of the fluid flow. The classical H-S equation expresses this parameter as a function of the field gradient. Due to the tight coupling between the ion concentrations and electric potential—as the ions contribute to the local electric potential themselves—the use of H-S electroosmotic velocity in transport determination in clay soils may result in nonlinear predictions (Ravina and Zaslavsky, 1967; Chu, 2005). Hence, uncoupling this parameter from the electric potential using the surface conductivity σ_s, and the resulting proportion of the current transferred over the solid–liquid interface i_s, should provide an intrinsic electroosmotic velocity dependent on clay surface properties only, as first introduced by Khan (1991) in Equation 2.8.

2.4.2 Ion Transport

In the past two decades, many researchers studied and proposed theoretical models of ion transport under electric field as it applied to contaminated clays. In most cases, dilute solutions, rapid dissociation–association chemical reactions, and small double layer thickness were considered. These models were all based on the Nernst–Planck equation, and each was calibrated to the experimental findings (Mitchell and Yeung, 1991; Shapiro and Probstein, 1993; Alshawabkeh and Acar, 1996; Denisov, Hicks, and Probstein, 1996; Cao, 1997). In 1997, Cao modeled ion transport considering the effect of changing electrical field due to redistribution of charge concentration. The effect was later shown in a rigorous mathematical analysis by Chu (2005), where concentration gradients give rise to spatial variation of conductivity to overcome the violation of electroneutrality of the bulk fluid.

In the basic governing equation of advection–diffusion, dispersion refers to the movement of species under the influence of gradient of chemical potential, while advection is the stirring or hydrodynamic transport caused by density gradient or forced convection. A general one-dimensional mass transfer to an electrode is governed by the Nernst–Planck equation:

$$J_i(x) = -D_i^* \frac{\partial C_i(x)}{\partial x} - u_i z_i F C_i \frac{\partial \Phi(x)}{\partial x} - C_i v(x), \qquad (2.9)$$

where

J_i = total flux of species i ($Mt^{-1}l$);
$u_i = D_i^*/RT$ = mobility of species i (Nernst–Einstein relation); and
$v(x)$ = advective velocity [= $v_{eo}(x)$, the electroosmotic velocity].

The mobility of an ion, u_i, is defined as the limiting velocity of an ion moving in an electric field of unit strength. The minus sign arises because the direction of flux opposes the direction of increasing C_i.

Applying Fick's second law, we arrive at another form of the Nernst–Planck equation (also given earlier in Equation 2.1) with the added advection term

$$\frac{\partial C_i}{\partial t} = -\nabla J_i(x) \qquad (2.10)$$

$$\frac{\partial C_i}{\partial t} = \nabla \left(D_i^* \nabla C_i + u_i z_i F C_i \nabla \Phi \right) + (u_i z_i F \nabla \Phi + v) \nabla C_i. \qquad (2.11)$$

Equation 2.11 is the basic mass transfer equation for an electrochemical system under an electric field. In Equation 2.11, the first term is the diffusion, the second term is the migration, and the third term is the advection contribution to the total mass transport of the species i in an electrochemical system. In clay soils where the hydraulic advection is negligible compared with the electroosmotic advection, the velocity term v is simplified as the electroosmotic velocity v_{eo}.

The charge flux J_i, according to its definition, is equivalent to a current density, which is defined as the flow of molar concentration of charge through a unit cross sectional area ($Mt^{-1}l^{-2}$). Relative contributions of diffusion, electromigration, and advection to the flux of a species in bulk fluid, and hence to the bulk current, differ at a given time for different locations in an electrochemical cell. Considering j_b as the bulk current density ($= i_b/A$), the following can be written:

$$j_b = j_d + j_{em}, \qquad (2.12)$$

where j_d is the current density due to diffusional flux of ions and j_{em} is the current density due to the electromigrational flux of ions. They are vector quantities, which may be in the same or opposite direction, depending on the charge on the electroactive species and the direction of the advective flow. The two important system parameters that contribute to the ion flux, and hence to the distribution of current density in the system, are the electromigration, v_{em}, and electroosmotic, v_{eo}, velocities given in summation as the advective velocity, v_{adv}:

$$v_{adv} = v_{em} + v_{eo}, \qquad (2.13)$$

where $v_{em} = u_i z_i F \nabla \Phi$ and, from Equations 2.4 and 2.7, $v_{eo} = \left[\dfrac{\varepsilon_s \zeta \tau}{\eta \sigma_s n} \right] \cdot \dfrac{i_s}{A}$.

Equation 2.13 does not consider the effect of electroosmosis explicitly since in electrokinetic processing, the ion flux would be affected by the electroosmotic transport, as well as the excess water transport in the direction of the positive concentration gradient. Yet, since in high ionic concentration cases the electromigration flux can be several orders of magnitude higher than that contributed by electroosmosis, hence electroosmotic flux may safely be neglected (Acar and Alshawabkeh, 1993).

The electromigration velocity in Equation 2.13 is the speed of ion movement in the pore water caused by an electric field in infinitely dilute solutions. In the pore fluids with finite ion concentrations, which more closely resemble the electrochemical systems of contaminated clays, influence of the interionic attraction should be considered. Generalized ion mobility that account for the possibility of interactions between the ionic species can be used to modify the electrochemical potentials of individual ions in the flux term (Equation 2.9) when strong chemical reactions between ionic species are considered (Newman, 1991). The resulting relative electric field due to the retardation effects of a dissymmetrical ionic atmosphere during ion transport was considered in Cao's model (1997) as described below.

The increasing concentration of the ions leads to a decrease in equivalent conductance in their migration direction (Kortüm and Bockris, 1951). According to Kohlrausch's law of the independent migration of ions, the equivalent conductance of an electrolyte is additively composed of the ionic conductivities of the constituent ions. As ions move under an applied electric field at their terminal ionic velocity (u_i) of infinite dilution, two other phenomena occur that retard ion velocity and effectively reduce the bulk conductivity. These are called the "retardation" and "electrophoretic" effects. The retardation effect comes about when the ionic atmosphere becomes unsymmetrical about an ion in motion as the charge density decrease in front of the ion and increase behind it. For a moving positive ion, a net negative charge trail behind it, exerting an electrostatic force in the opposite direction. The electrophoretic effect occurs as the negative ions in the ionic atmosphere around the positive ions migrate in the opposite direction, taking their solvent sheaths with them. The positive ion therefore travels against a medium moving in the opposite direction, hence a viscous drag. Debye, Hückel, and Onsager gave a quantitative formulation of ion mobility, relating the magnitudes of the electrostatic retardation force and the viscous drag force directly to the radius of the ionic atmosphere (Kortüm and Bockris, 1951). Their results provided a theoretical basis for the well-known empirical relation known as the "Kohlrausch square root law," given below:

$$\Lambda_v = \Lambda_\infty - (A'\Lambda_\infty + A'')\sqrt{\alpha z n_e C}, \qquad (2.14)$$

where

- Λ_v = equivalent conductance (l^2 ohm^{-1});
- Λ_∞ = equivalent conductance at infinite dilution (l^2 ohm^{-1});
- α = electrolyte dissociation constant ($\cong 1$ for strong electrolytes or high dilutions);
- D = dielectric constant of the bulk fluid;
- n_e = electrochemical valency (i.e. for CaCl$_2$, z^+ = 2, z^- = 1, while $n_e = z^+ \cdot 1 = z^- \cdot 2 = 2$);

and

$$A' = \frac{(8.205 \times 10^5) z^2}{(DT)^{3/2}} \text{ and } A'' = \frac{(82.48) z}{\eta (DT)^{1/2}} \qquad (2.15)$$

The equivalent conductance, Λ_v, is also the sum of individual ionic conductivities ($F.u_+$ for positive ions and $F.u_-$ for negative ions) as below:

$$\Lambda_v = F(u_+ + u_-). \qquad (2.16)$$

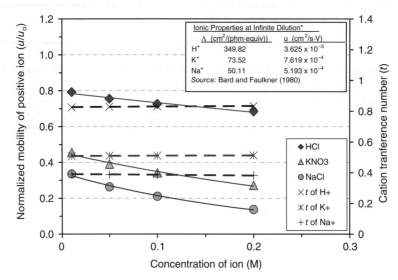

Figure 2.5. Variation of mobility and transference with concentration of selected ions.

Hence, the ionic mobility of a given ion can be expressed in terms of the equivalent conductance (Bard and Faulkner, 1980):

$$u_i = \frac{\Lambda_v t_i'}{F}, \tag{2.17}$$

where $t_i' = u_i/(u_+ + u_-)$ is the transference number of the species i. Substituting Equation 2.14 into Equation 2.17, we get:

$$u_i = \left(\Lambda_\infty - (A'\Lambda_\infty + A'')\sqrt{\alpha z n_e C}\right) \cdot \left(t_i'/F\right). \tag{2.18}$$

Equation 2.18 effectively incorporates the retardation effects into the mobility determination for high concentration solutions. As an example, for aqueous solution at room temperature (T = 298K), using D = 78.56 and η = 0.008948, the variation of the mobility of the positive ion with concentration in 1,1 valency electrolytes of HCl, KNO$_3$, and NaCl are plotted in Figure 2.5 according to Equation 2.18. The variation of the transference numbers of the cations with the concentration are also plotted to discern its effect on the mobility of each ion. As observed, the square root model represents the reduction of the mobility of each ion with increasing concentration, where the reduction appear to be mostly dependent on Λ_v.

As discussed earlier, the charge flux, J_i, is equivalent to a current density, which is defined as the flow of molar concentration of electrons through a unit cross-sectional area. The bulk current density in a clay electrochemical system can be expressed as in Equation 2.12. The electromigrational component j_d is closely related to the conductivity of the bulk solution. The *conductivity*, σ_b, which is an intrinsic property of a solution, is the sum of all the contributing ionic movement of different species in the solution, and hence proportional to the concentration of the ions as given below:

ELECTROCHEMICAL TRANSPORT IN CLAYS

$$\sigma_b = F \sum_i |z_i| u_i C_i \tag{2.19}$$

Following, the current density due to ionic motion under an electric field of $E(x)$ ($=\partial \Phi / \partial x$) can be expressed as

$$j_{em} = \left(F \sum_i |z_i| u_i C_i \right) \cdot \left(\frac{\partial \Phi}{\partial x} \right) = \sigma_b E. \tag{2.20}$$

Similarly, the current density due to diffusion motion of the ions under a concentration gradient of $\partial C / \partial x$ can be expressed as

$$j_d = F \sum_i z_i D_i^* \frac{\partial C_i}{\partial x}. \tag{2.21}$$

Combining Equations 2.20 and 2.21 into the expression for bulk fluid current density, we obtain the following expression for $E(x,t)$, which can be solved numerically as the concentration $[C_i(x,t)]$ and the mobility $[u_i(x,t)]$ of the ions, and the current densities ($j_b = j_d + j_{em}$) of the system are updated using the appropriate equations, from 2.11 through 2.21:

$$E(x,t) = \left(\frac{j_b(t) + F \sum_i z_i D_i^* \left(\frac{\partial C_i(x,t)}{\partial x} \right)}{\sigma_b(x,t)} \right) \tag{2.22}$$

2.4.3 Model Predictions of Conductivity and Electric Field

The model predicted distribution and evolution of bulk conductivity, σ_b, and field strength, E, in kaolinite clay, containing $Pb(NO_3)_2$ at an initial concentration of 0.05 M in its pore fluid are presented in Figures 2.6 and 2.7, respectively (Cao, 1997). In Figure 2.6, the normalized distribution of the conductivity (normalized by initial conductivity of 0.28 siemens/m) shows a consistent decrease in the conductivity at the cathodic region and a steady spread of the lower conductivity toward the anode area over time. As will be shown later in the mass transport prediction, part of this result is attributed to the decrease of dissolved lead concentration, which prevail over the increase in H^+ concentration. The change in the conductivity is influenced by a combination of three factors: (a) the initial concentrations of the species; (b) the spatial and temporal distribution of ionic mobility of each species; and (c) the production rate of H^+ at the anode.

As conductivity varies, the lower conductive area requires higher voltage difference over it to keep a consistent current with the other parts of the soil column, resulting in a nonlinear electric potential profile across the cell. This tendency is observed in Figure 2.7, where the time evolution of the electric potential distribution is presented. As the electrochemical transport continues, the potential curve becomes less steep except in a narrow region adjacent to the cathode, where the electrical potential is likely to have the largest gradient. This tendency agrees well

44 ELECTROCHEMICAL TRANSPORT AND TRANSFORMATIONS

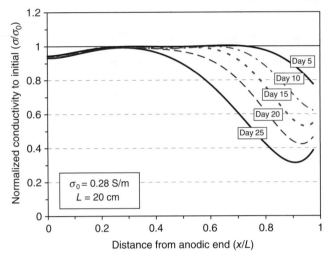

Figure 2.6. Spatial and temporal variations of conductivity in Pb-contaminated kaolinite under electrochemical transport as predicted by a model (Cao, 1997).

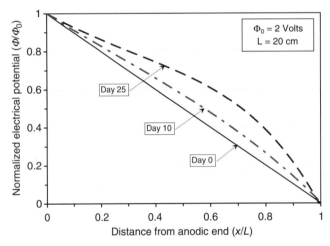

Figure 2.7. Spatial and temporal variation of electric potential in Pb-contaminated kaolinite as predicted by a model (Cao, 1997).

with the experimental data reported by Wittle and Pamukcu (1993) and Acar and Alshawabkeh (1996), among many others.

2.4.4 Model Predictions of Ion Transport

When charged ions transport under the influence of an externally applied electrical field, their concentration distributions change with time, which lead to a change in local electrical conductivity. The change in local electric conductivity directly alters the value of the potential gradient at that specific point. Hence, the changing electric conductivity and electrical field describe the transport process of the species implicitly.

The Cao model was used to predict the spatial and temporal distributions of four ionic species—Pb^{+2}, NO_3^-, H^+, and OH^-—in kaolinite clay. Four one-dimensional partial differential equations (Equation 2.11) are developed, one for each ionic species. Chemical reactions, including production of lead hydroxide, water autoionization, and lead sorption, were described in a set of algebraic equations. Electrical potential, conductivity, and field strength were evaluated using previously presented equations. The boundary conditions applied at the inlet and outlet of the soil column maintained the equality between the flux of solute at the inside and immediately outside of the column.

When solving flow problems numerically, a Neumann boundary is described as an insulated boundary (or impermeable boundary), which means that there is no flux at the boundary, while a Dirichlet boundary indicates that the value of head (potential, concentration, etc.) is constant at the boundary. Constant boundary conditions are not able to describe the nature of the electrokinetic transport realistically due to the existence of flux boundaries caused by the electrode reactions and advection of fluid. In Cao's model, the boundary conditions applied at the inlet and outlet of the soil column maintained the equality between the flux of solute at the inside of the column and the flux of solute immediately outside of the column. The following boundary condition was used at the inlet (Lafolie and Hayot, 1993):

$$-D_i^* \frac{\partial C_i}{\partial x} + v C_i \big|_{x=0} = v C_{i0}(t), \quad (2.23)$$

where $C_{i0}(t)$ is the concentration of the incoming solution, a Dirac or step function. The boundary condition at the outlet of the column was similar to the inlet boundary. If the diffusion and dispersion inside the exit reservoir are assumed negligible and the concentration across the boundary is continuous, then the outlet boundary becomes a Neumann boundary. Using only the electroosmotic advection immediately outside the soil, the following boundary equations were utilized in the model:

$$-D_i^* \frac{\partial C_i}{\partial x} + v_{adv} C_i \big|_{x=0} = v_{eo} C_{i0}(t) + \frac{j_{em}}{z_i F} = J\big|_{x=0} \quad (2.24a)$$

$$-D_i^* \frac{\partial C_i}{\partial x} + v_{adv} C_i \big|_{x=L} = v_{eo} C_{i0}(t) + \frac{j_{em}}{z_i F} = J\big|_{x=L}, \quad (2.24b)$$

where v_{adv} and j_{em} are given in Equations 2.13 and 2.20, respectively.

The above boundary conditions can also be obtained by analyzing the equivalence of the charge flux. The right-hand side of Equations 2.24a and 2.24b is actually the charge flux of species j at any position inside the soil column. It should be equal to the charge flux at the location immediately outside of the soil column. The validity of the boundary condition at the outlet side ($x = L$) is questionable due to the assumption of continuity of the concentration across the boundary. However, this assumption does not significantly affect the predictions when the column Peclet number P_e is high ($= vL/D^* > 20$) (Parker and van Genuchten, 1984), which is always maintained in the electrokinetic processing of soils due to relatively high ion migration velocity.

Application of DC to a saturated soil column always involves the electrode reactions, which produces H^+ at the anode due to oxidation and OH^- at the cathode due to reduction of water.

$$2H_2O - 4e^- \Rightarrow O_2 \uparrow + 4H^+ \text{ (anode)} \\ 4H_2O + 4e^- \Rightarrow 2H_2 \uparrow + 4OH^- \text{ (cathode)} \quad (2.25)$$

The generated concentrations of H^+ and OH^- move toward oppositely charged electrodes under the driving force of the applied electric potential. This movement results in the redistribution of pH in the soil between the electrodes. The hydrogen/hydroxyl transport and the evolution of the acid/base environment are important features in electrokinetic transport. The pH change influences the chemical reactions, especially for heavy metal transport. The cation exchange capacity (CEC) of clay, sorption and desorption reactions, and chemical productions are all directly influenced by the pH environment in the soil. The basic mass transport model used to describe the movement of H^+ and OH^- include the aqueous-phase reactions involving these two ions such as water autoionization, and the precipitation reactions of OH^- with Pb^{+2} in the form of algebraic equations. Similarly, the sorption and precipitation of Pb^{+2} with OH^- were incorporated in the mass transport of Pb^{+2} as algebraic reactions:

$$R^a_{H^+} = R^a_{OH^-} = \Delta C \\ C_{H^+} \cdot C_{OH^-} = K_w = 10^{-14}. \quad (2.26)$$

The existence of hydrogen and hydroxyl ions in the solution must satisfy the equilibrium state as given in Equation 2.26. In the model, H^+ and OH^- transport independently, which raises the importance of the incorporation of water autoionization reactions into the model to guarantee the equilibrium state of the two ions in the aqueous phase. This reaction will generate or consume an equal amount of hydrogen and hydroxyl ions. Knowing the concentrations of H^+ and OH^- at a given time, the fraction of hydrogen or hydroxyl ions that participate in these reactions can be computed for the time step.

Assuming the electrode reactions to only include the oxidation or reduction of water, the boundary conditions at the anode, where $C^+_{H_{x=0^-}}$ is the concentration of H^+, which migrate into the soil from the anolyte, dictate that all current is expended in the generation of hydrogen at the anode. Hence, for $z = 1$ and $j_{H^+} = i/A_e$ (where i is the total current and A_e is the electrode surface area), the flux at the anode boundary can be expressed as

$$J_{H^+}\big|_{x=0} = v_{eo} C_{H^+}\big|_{x=0^-} + \frac{i}{FA_e}. \quad (2.27a)$$

Since continuous flux is assumed at the cathode side, the advective flux of H^+ into the cathode provides the following expression, where C_{H^+} is the concentration of H^+ in the soil at $x = L^-$:

$$J_{H^+}|_{x=L} = v_{eo}C_{H^+}|_{x=L^-} \tag{2.27b}$$

Similarly, at the cathode, if all current is spent in the generation of OH⁻, then the flux of OH⁻ at the cathode boundary is

$$J_{OH^-}|_{x=L} = v_{eo}C_{OH^-}|_{x=0^-} - \frac{i}{FA_e}, \tag{2.28a}$$

and at the anode, it is

$$J_{OH^-}|_{x=0} = v_{eo}C_{OH^-}|_{x=0^-}. \tag{2.28b}$$

The one-dimensional mass transport equation of OH⁻ includes two additional reaction terms, where $R^a_{OH^-}$ denotes the concentration of OH⁻ participating in water autoionization reaction, and $R^p_{OH^-}$ represents the part that is consumed in reaction with lead. Equation 2.29 presents lead hydroxide precipitation/dissolution reaction in aqueous phase. pH is the major factor that influences the reaction direction and production rate. The equilibrium requires 2 mol of OH– per 1 mol of Pb^{+2} to generate 1 mol of $Pb(OH)_2$. This means that the change in hydroxyl molar concentration due to the precipitation reaction is twice as the change in Pb^{+2}'s molar concentration.

$$Pb^{+2} + 2OH^- \Leftrightarrow Pb(OH)_2$$
$$R^p_{Pb^+} = 2R^p_{OH^-} \tag{2.29}$$

The one-dimensional mass transport equation for Pb^{+2} includes R^p_{Pb}, which reflects the fraction of lead precipitated, while R^s_{Pb} is due to the fraction of lead consumed in the sorption/desorption onto/from soil particle surfaces. Note that no electrode reaction was considered for Pb(II), therefore boundary conditions for lead transport are set as

$$J_{Pb^{+2}}|_{x=0} = 0 \text{ and } J_{Pb^{+2}}|_{x=L} = v_{eo}C_{Pb^{+2}}|_{x=L^-}, \tag{2.30}$$

where the concentration of Pb^{+2} in the anode compartment remains zero.

The mass transport equation for NO_3^- and the boundary conditions are given similarly as that of the other three species. The NO_3^- is the least reactive species compared with the other three. Although the fate of NO_3^- was not sought, its contribution to the charge concentration was used to achieve the electrical neutrality and in the distribution of the electrical field.

Finally, the effects of retardation and sorption/desorption were included in the model. Kaolinite has a low CEC (about 0.01 mg/g) and low buffering capacity (Mitchell, 1993), hence retardation for H⁺ transport may not be significant. However, due to the change in pH environment, the tortuosity and the presence of other species, there may be a significant influence of retardation to the transport of H⁺ (Acar and Alshawabkeh, 1993). An experimentally obtained retardation factor of R_d was applied to the rate of concentration change term for the mass transport equation of H⁺.

Lead ion, Pb(II) present in pore fluids, is highly adsorbed by clay minerals. Selectivity of kaolinite for Pb is higher than Ca, Cu, Zn, and Cd (Alloway, 1990). For a given type of clay, lead adsorption is mainly controlled by its concentration and the pH value of the solution. Past results showed that the ratio of adsorbed lead to total lead in kaolinite increases linearly with pH up to 5 (Yong et al., 1990). The following empirical relation developed by Alshawabkeh and Acar (1996) was used to represent the lead adsorption of kaolinite at different pH values:

$$R_{Pb}^s = 0.27 C_{Pb}(pH-1) \quad 1.0 \leq pH \leq 4.7$$
$$R_{Pb}^s = 0 \quad\quad\quad\quad\quad\quad\quad pH < 1.0 \quad , \quad (2.31)$$

where C_{Pb} is the total lead concentration including both the adsorbed and solution concentration in M. When pH is larger than 4.7, the adsorbed lead no longer fits the experimental data obtained in a dilute concentration (Yong et al., 1990). The largest possible adsorbed lead is limited by the CEC of the clay. Hence, according to an empirical relation developed by Yong et al. (1990) for kaolinite,

$$R_{Pb}^s = 0.5 \text{CEC}(207.2 \times 10^6) \quad pH > 4.7, \quad (2.32)$$

where R_{Pb}^s is given in mg/kg of soil.

Figures 2.8 and 2.9 present the spatial distribution of the total lead and the pH for 1-day and 35-day runs of the Cao model. Superimposed on the numerical solutions are results from laboratory tests of $Pb(NO_3)$-contaminated kaolinite clay subjected to long-term electrokinetic treatments. Although the initial concentrations of the lead used in the numerical simulator and the experiments were different (0.05 M in the numerical and 0.15 M in the experiment), as well as the length of the

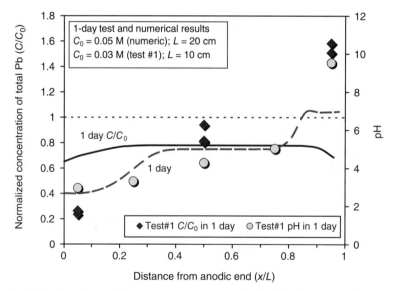

Figure 2.8. Variation of total Pb concentration and pH distribution as predicted by Cao's (1997) model and measured by experiment for a 1-day duration.

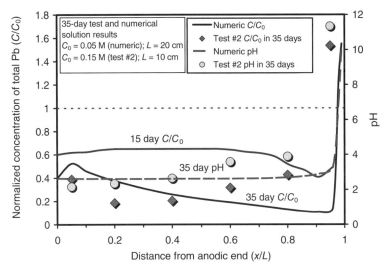

Figure 2.9. Variation of total Pb concentration and pH distribution as predicted by Cao's (1997) model and measured by experiment for a 35-day duration.

physical and numerical specimens employed, the normalized distributions are comparable. As shown in Figure 2.8, the pH distribution of the model agrees well with the experimental pH after 1 day of treatment, but the distribution of lead does not agree for the same duration. The numerical model estimates a conservative uniform concentration distribution of the metal ion at around 80% of the original, while the available data at three points along the soil column indicate a substantial reduction at the anode and a substantial accumulation at the cathode. When we observe the pH prediction at the cathode end, we again find that the model makes a conservative prediction of pH at the cathode end, hence a lower concentration of lead, not adequately capturing its accumulation by precipitation.

Figure 2.9 shows the long-term model predictions and the matching experimental data from the same series of tests for 35-day treatment. The long-term behavior of the both the lead transport and the pH distribution is predicted well by the model, including the accumulation of the metal in a *narrow zone adjacent to the cathode*. The model prediction for the 15-day and the 35-day distribution of lead concentration indicates that the migration velocity of the ions become affected by the changing potential, and hence the electrical field distribution over time. The reduction in electrical conductivity at the cathode zone results in a higher electrical field in that region. The high electrical gradient hurls the ions to the narrow high-pH zone to be accumulated by precipitation, hence evacuating the region immediately behind it. The progression of this effect is clearly seen by the evolution of lead distribution from 15 to 35 days as the distribution toward the cathode regions dips. The distribution of lead at the anode region does not dip at the same rate because the conductivity remains high and the electric potential distribution tends to flatten. Consequently, the ions slow down, responding to a lower electric field. Figures 2.6 and 2.7, which display the time variation of the model-predicted conductivity and electric potential provide evidence for these conclusions. Other factors such as sorption, retardation, and precipitation obviously influence the model predictions, but their relative contributions may be small, other than the narrow high-pH zone near the cathode.

In summary, an electrochemical model (Cao, 1997), which considers the evolving change of spatial distribution of conductivity and electric field, in addition to a limited number of chemical reactions, appear to simulate closely the long-term behavior of Pb(II) transport in kaolinite clay. The model can be calibrated using long-term field data to study the relative influence of all the parameters used.

2.5 ELECTROCHEMICAL TRANSFORMATIONS

2.5.1 Treatise of Electrically Induced Transformations on Clay Surface

Electrochemical transport in clay–electrolyte media is complicated because of the coupling between multiple physical and chemical processes, where bulk liquid and surface processes occur simultaneously. In the bulk, the driving force for ion motion comes from two fundamentally different sources: electroosmosis and ion migration. At electrochemical surfaces, such as that of a colloid or clay particle, double layer charging, electrochemical reactions, and surface conduction can take place simultaneously (Kortüm and Bockris, 1951).

Clay particles can be thought of as "microelectrodes" possessing a compact Stern layer and a diffuse layer, which mediates faradic reactions, as depicted in Figure 2.10. As the donated (or accepted) electrons pass across the electrical double layer into and out of the bulk fluid, available species are converted into others via oxidation–reduction reactions (Grahame, 1951, 1952). This effect may become significant

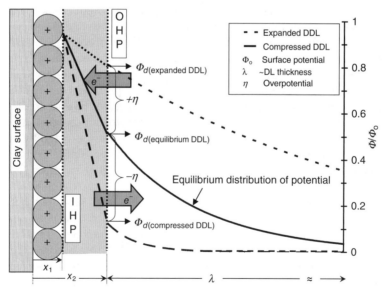

Figure 2.10. The concept of faradic current development in the GCSG model of clay DDL (Pamukcu, Hannum, and Wittle, 2008).

for polarizable surfaces due to strong adsorption of the cations, anions, and molecules with electrical dipoles within the double layer, resembling the case of contaminated sediments.

Three interface layers occur within the electrical or the diffuse double layer (DDL) of a clay particle: the inner Helmholtz plane (IHP); the outer Helmholtz plane (OHP) with constant thicknesses of x_1 and x_2, respectively; and third is the plane of shear where the *electrokinetic potential* is measured (Fig. 2.10). This plane of shear is sometimes assumed to coincide with the OHP plane. The IHP is the outer limit of the specifically adsorbed water, molecules with dipoles, and other species (anions or cations) on the clay solid surface. The OHP is the plane that defines the outer limit of the Stern layer, the layer of positively charged ions that are condensed on the clay particle surface. In this model, known as the Gouy–Chapman–Stern–Grahame (GCSG) model, the diffuse part of the double layer starts at the location of the shear plane or the OHP plane (Hunter, 1981). The electric potential drop is linear across the Stern layer that encompasses the three planes (IHP, OHP, and shear planes); and it is exponential from the shear plane to the bulk solution, designated as the reference zero potential.

When an external electric field is applied to water-saturated clay of high ionic concentration in the pore fluid, given the incompatibility between the conductivity of two conducting layers in the mixture, (a) the DDL of clay particles with low conductivity (σ_s) and (b) the surrounding electrolyte solution (bulk solution or pore fluid) with high conductivity (σ_b), a large electrical potential is induced across the DDL. This results in the compression of the DDL, hence the DDL becomes a capacitor (Bard and Faulkner, 1980). The compression of the DDL occurs in the diffused section, in front of the shear plane toward the pore fluid, hence resulting in a higher charge density in this region. The charge density of the compact Stern layer (IHP $\Rightarrow$ OHP) remains the same. But because of the compression of the diffuse layer, the potential distribution shifts back and downward, lowering the electrokinetic potential at the shear plane (assumed to coincide with the OHP plane here). This, in turn, causes an increase in the intensity of the electric field between the IHP and the OHP, as shown by the steeper slope of the potential drop in Figure 2.10.

In a dielectric interface, like the DDL, the material tends to undergo densification in regions of high-intensity electrical field, as sketched in Figure 2.11. This is an irreversible process known as *electrostriction*. Here, the phenomenon of electrostriction is thought to play a role for DDL to react as a capacitor when an external potential is applied to clay–electrolyte systems. The analogy of electrostriction in dielectric materials is analogous to volumetric compression of mechanical systems,

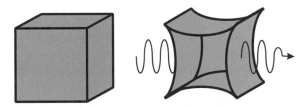

Figure 2.11. Electrostriction in a dielectric material (Pamukcu, Hannum, and Wittle, 2008).

where the stiffness or the elastic spring constant of the material increases as the material becomes denser. In this analogy, the spring constant, k (= force/displacement) of the mechanical system is analogous to the elastance, $1/C$ (= potential/charge) of a dielectric, where C is the capacitance.

It follows then that as the electrical field intensity increases across a dielectric, the "stiffness" or the elastance of the dielectric increases. This comes about by physical compaction of the particles in the diffuse layer and by strong orientation of the dipoles or the polarization of charges in the compact layer. Hence, as the electrical field intensity is increased by compression of DDL, the capacitance of the system decreases due to polarization. An example of a similar behavior has been shown in thin electrolyte films between silica surfaces, where the dielectric permitivity decreases with increasing electrolyte concentration, resulting in increasing field strength across the thin film (Israelachvili and Adams, 1978; Pashley, 1981; Basu, 1993).

The change in capacitance of the Stern layer causes electron transfer across the OHP toward the solution, giving rise to a faradic "cathodic" current, and hence a cathodic reaction in the diffuse layer and its vicinity. Assuming that the anions in the system do not participate in the electrode reactions, the faradic current is driven by the electrodeposition of the cation at the cathode and the electrodissolution of the cation at the anode as follows:

$$M^+ + e^- \Leftrightarrow M \quad \text{(reduction of } M^+ \text{ in the diffuse layer)}$$
$$M \Leftrightarrow M^+ + e^- \quad \text{(oxidation of M in the diffuse layer)} \quad (2.33)$$

Accordingly, as the DDL compresses or expands, faradic currents (cathodic or anodic) can generate across the OHP, whereby the concentration gradients are overcome and the electrical equilibrium is restored.

The excess potential that causes the faradic current flow is defined as the "overpotential," η, which is a measure of the degree of polarization or the departure of the electrode potential from the Nernstian (equilibrium) value (Bard and Faulkner, 1980). The Nernstian equilibrium applies to reversible (nonpolarizable) electrode surfaces, where dynamic equilibrium is established rapidly. When a large electrical field is applied, a large current flows, and the equilibrium at the electrodes cannot adjust quickly to overcome the electrode reactions, thus giving rise to irreversible electrodes. To maintain the current flow, the electrodes adapt a new potential, differing from the equilibrium by the overpotential, η. As electrons migrate and reduction–oxidation (redox) reactions take place, the equilibrium potential is restored.

2.5.2 Electrochemical Model

It is hypothesized that the above treatise of DDL interactions in the presence of an electrical field is a viable model for the explanation of enhanced oxidation–reduction in clay–electrolyte systems and may also help explain the beneficial use of electricity in the release of strongly held compounds on clay surfaces (Shi, Hauke, and Wick, 2008).

Assuming the clay surfaces act as microelectrodes with fully developed electric double layers, as shown in Figure 2.10, the Nerstian equilibrium potential, $\Phi_{d(eq)}$

can be interpreted as the equilibrium potential at the OHP plane in the absence of an applied current. Φ_o is the potential on clay surface or the IHP plane and η is the overpotential that drives the current for redox reactions. This model of the electrical double layer can be explained well with GCSG assumptions, where Nernst relation gives the surface potential and the diffuse part of the double layer starts at the location of the OHP plane (Hunter, 1981). The basic equations for GSCG model are

$$\frac{1}{C_{ddl}} = \frac{1}{C_i} + \frac{1}{C_d}, \qquad (2.34)$$

where, C_i and C_d are the capacitance of the compact (IHP $\Rightarrow$ OHP), and the diffuse layers, respectively. Except when Φ_d and/or the ionic concentration are very small, C_d is normally so very large that the integral capacitance, C_{ddl} is essentially equivalent to C_i (Hunter, 1981). The potential difference, $\Delta\Phi$, across the entire DDL can then be written as

$$\Delta\Phi = \frac{q_i}{C_{ddl}} + \Phi_{d(eq)} \qquad (2.35)$$

$$q_m + q_i + q_d = 0, \qquad (2.36)$$

where q_i is the charge under IHP $\Rightarrow$ OHP, q_d is the charge under the diffuse layer, and q_m is the surface charge of the clay. The q_i is the majority of the charge that counters the surface charge of the clay particle, q_m. Under the assumptions of constant surface charge and electrical neutrality, $q_i + q_d$ remains constant. Hence, the total potential difference $\Delta\Phi$ across the DDL is constant. Furthermore, the OHP plane is considered to coincide with the shear plane, where the electrokinetic potential develops. Then, according to Equation 2.35, any change in shear plane potential brings about a change in the capacitance term, C_{ddl}, or vice versa in order to preserve the constancy of $\Delta\Phi$.

Considering the Nerstian assumptions of the GCSG model of the clay surface, when the electrokinetic potential at the OHP, $\Phi_{d(eq)}$, drops to lower negative values or increases to higher negative ones, the capacitance of the compact region will change to compensate the equilibrium given in Equation 2.35. Reduction in capacitance indicates polarization of dipoles in the dielectric to reduce the induced electrical field (Kortüm and Bockris, 1951). Then, the change in capacitance can be expressed by the measure of the degree of polarization, or the overpotential η, as follows:

$$\Delta\Phi = \frac{q_i}{C_{ddl}} + (\Phi_d \pm \eta), \qquad (2.37)$$

where Φ_d is the new adjusted OHP potential, determined by the degree of polarization under the dielectric. As discussed before, η can be interpreted as the potential that drives the faradic currents across the OHP to restore the OHP potential back to its equilibrium value, $\Phi_{d(eq)}$. Hence, as the overpotential drives the charges across the OHP, the Nerstian equilibrium or the equilibrium potential distribution of the DDL is restored, as shown in the conceptual sketch in Figure 2.10.

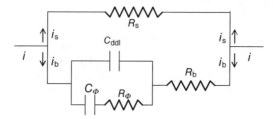

Figure 2.12. Conceptual electrical circuit model for clay–electrolyte response when subjected to external potential.

The driving force for all electrochemical processes at conventional electrodes (i.e. mass transport to electrodes and electrochemical reactions) is the difference between the electrode potential at operation and at equilibrium, which is the overpotential η. It is theoretically a difficult term to work with because it combines the effects of many physical processes (e.g. Stern layer potential gradient change and concentration polarization) into a single quantity. Yet it is accessible experimentally, and hence a practical quantity for the better understanding of complicated electrochemical systems such as clays.

2.5.3 Equivalent Electrical Circuit Model

Considering the overall conversion, the externally applied potential, Φ_{app}, necessary to drive the redox reactions at the polarizable surfaces of clay particles can be written simply as

$$\Phi_{app} = \Phi_r + (\Phi_d \pm \eta), \tag{2.38}$$

where Φ_r ($= i_b R_b$) is the ohmic potential in the bulk liquid or the pore fluid. An equivalent electrical circuit, as shown in Figure 2.12, can be used to model the soil electrolyte system described by Equation 2.38.

When representing a saturated clay–electrolyte system with an equivalent circuit, certain elements can be modeled in series, while others in parallel. In this case, we consider a saturated clay matrix whose pores are filled with an electrolyte, and the pore walls are lined with electric double layer. In Figure 2.12, the compact layer is modeled with capacitance C_i ($=C_{ddl}$) in series with the pore fluid resistance R_b. The capacitive behavior of the interphase due to charge transfer is described by two elements, C_ϕ and R_ϕ, in series, where C_ϕ is a "pseudocapacitance" (Taylor and Gileadi, 1995) describing the faradic charge transfer under overpotential ($=\Delta q/\eta$) and R_ϕ is a faradic or polarization resistance that is an integral part of the phenomenon that gives rise to C_ϕ. Since the capacitance of the DDL is essentially equal to C_i (Equation 2.34), this circuit element is renamed as C_{ddl}. The resistance of the diffuse layer $R_s(=\Phi_0/i_s)$, which was shown to be responsible for electroosmotic flow earlier (Equations 2.5 through 2.8), is modeled in series with the pore fluid resistance, R_b. The current at the source is the sum of the currents traversing the individual branches (Kirchoff's current law) of the circuit: $i = i_s + i_b$. This equation is the same as Equation 2.5, hence the model complies with the earlier assumption of additive current of parallel elements discussed in the treatise of electroosmotic velocity. This model neglects electron transfer along the solid surface, while although small, electron transfer and exchange of specifically adsorbed ions may allow current flow along the solid surfaces and over the particle contact points.

The series depiction of the pore fluid resistance with that of DDL supports the behavior for mixed systems of different conductivities as discussed above. The C_ϕ can be evaluated for a direct current (DC) or a low-level alternating current (AC). In the former one, a transient (Faradic) current is passed through until enough charge is transferred to recover the equilibrium potential, after which the capacitance presents infinite resistance to DC. Considering the capacitive behavior of C_ϕ in response to AC, it behaves as diffusional impedance with a phase retardation of $-\pi/2$ between potential and current. As the frequency of the sinusoidal potential increases, the C_ϕ approaches to zero. If mass transport limitations are considered, then Warburg impedance, $[Z_w(\omega); \omega =$ angular frequency] may replace the C_ϕ (Grahame, 1952). Warburg impedance, which represents resistance to diffusion-controlled mass transfer, becomes negligible when the ohmic resistance of bulk fluid (R_b) is small. In this discussion, we limit the application of the equivalent electric circuit model given in Figure 2.12 to only DC cases.

2.5.4 Evidence of Electrochemical Transformations on Clay Surfaces

We hypothesized that the above treatise of DDL interactions in the presence of an electrical field is a viable model for the explanation of enhanced oxidation–reduction in clay–electrolyte systems. Electrolytic transformations of selected chlorinated hydrocarbons (CHCs) and polyaromatic hydrocarbons (PAHs) have been demonstrated successfully in water and wastewater (Franz, Rucker, and Flora, 2002; Pulgarin et al., 1994). There has been field and laboratory evidence that these transformations can also take place in porous media (Banarjee et al., 1987; Pamukcu, Weeks, and Wittle, 2004; Alshawabkeh and Sarahney, 2005; Pamucku, Hannum, and Wittle, 2008). As discussed previously, faradic reactions do take place on clay particle surfaces when current pass in the pathways of the DDLs (Grahame, 1951, 1952). Hence, external supply of electrical energy can help drive favorable oxidation–reduction reactions in contaminated clays not only in the bulk fluid but also on clay surfaces, as well as on where most of the contaminants tend to reside because of adsorption or exchange.

In field applications, evidence of enhanced reduction of Cr(VI) by a DC electric field was first reported in 1987 by Banarjee et al. (1987) at a Superfund site in Corvallis, Oregon. Later, controlled laboratory experiments of kaolinite clay injected with Fe(II) showed that an externally applied electric field caused an additional "cathodic current" that drive forth the reduction of Cr(VI) in clay (Pamukcu, Weeks, and Wittle, 2004). These transformations were characterized as to have benefited the capacitive changes on the clay surfaces. The results in these experiments showed that the system oxidation–reduction potential (ORP) increased by a positive shift from the standard solution ORP in the presence of the clay medium and the induced electrical field. Figure 2.13 shows the ORP measurements plotted against the reaction quotient of the Nernst relation, where the data is categorized by pH. Under anoxic conditions and acidic environments, Fe(II) can be the dominant reductant of Cr(VI), as given by the following redox reaction:

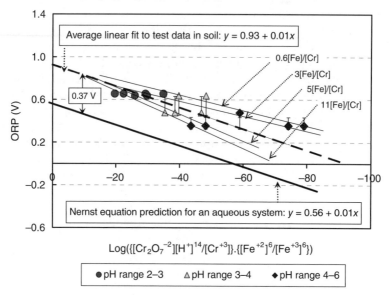

Figure 2.13. Measured and predicted redox potential variation with reaction quotient of measured species concentration in the Nernst relation (average potential shift 0.37 V) (Pamukcu, Weeks, and Wittle, 2004).

$$Cr_2O_7^{-2} + 14H^+ + Fe^{+2} \Leftrightarrow 2Cr^{+3} + Fe^{+3} + 7H_2O. \tag{2.39}$$

The potential for the reaction given in Equation 2.39 can be written in the form of a Nernst expression as follows:

$$\Phi = \Phi_0 + \left(\frac{2.3RT}{6F}\right) \cdot \text{Log}\left(\frac{[Cr_2O_7^{-2}][H^+]^{14}[Fe^{+2}]^6}{[Cr^{+3}]^2[Fe^{+3}]^6}\right), \tag{2.40}$$

where Φ_0 = standard potential = 0.56 V.

As observed in Figure 2.13, the low pH range (pH range 2 to 3) data showed the best agreement with the linear fit. The scatter of data at higher pH values was attributed to the influence of pH on the degree of electric layer polarization. At low pH, the DDL is compressed with a higher ion concentration, higher electric field (Israelachvili and Adams, 1978), and hence, higher degree of polarization. At this state, the overpotential is closer to a maximum as the measured values tend to cluster, with little deviation from the mean value. At higher pH, the DDL is expanded with a reduced degree of polarization, and consequently, smaller overpotential. The ions in the expanded DDL are now less restricted to move; therefore, discernable fluctuations are likely to occur in the potential development across the OHP interface, where the cathodic current is diffusion controlled.

More evidence for electrokinetically assisted surface transformations in kaolinite clay was found when a clean specimen of the clay was permeated with polymer-coated dispersed nanoiron particles of positive surface charge (Sun et al., 2006) under an electrical gradient of 0.1 V/cm (Pamukcu, Hannum, and Wittle, 2008).

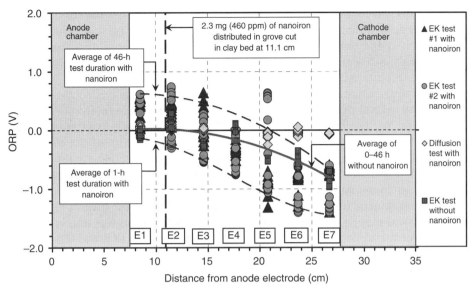

Figure 2.14. Distribution of the ORP across a saturated clay specimen measured with and without nanoiron during electrokinetic transport (Pamukcu, Hannum, and Wittle, 2008).

When there are no large concentrations of dominant ions in the clay matrix (i.e. contaminating heavy metals), the primary electron receptors are water and residual dissolved oxygen. Hence, the chemical reactions, which oxidize the nanoiron are

$$Fe^0(s) + 2H_2O(aq) \rightarrow Fe^{+2}(aq) + H_2(g) + 2OH^-(aq) \quad (2.41)$$

$$2Fe^0(s) + 4H^+(aq) + O_2(aq) \rightarrow 2Fe^{+2}(aq) + 2H_2O(l). \quad (2.42)$$

Equation 2.41 is the dominant redox process in presence of overwhelming concentration of water as the solvent. As hydrogen ions are utilized in Equation 2.42, the pH would increase at corrosion locations, resulting in a simultaneous decline in the solution ORP. The Nernst relation can be applied to the oxidation–reduction reaction of iron as follows:

$$\Phi = -0.440 + 0.030 \log(Fe^{+2}), \quad (2.43)$$

where the standard electrode potential for pure iron at 1 M concentration and at 25 °C is −0.44 V, and the concentration for a pure solid phase is taken as 1.

The transient ORP measured at discrete locations across a thin clay bed (2 mm) are shown in Figure 2.14, along with those in the control tests of diffusion and electrokinetics without the nanoiron. As observed, the ORP was reduced across the soil bed from anode to cathode in all but the diffusion tests. The initial soil pH varied from 6.1 to 6.3; which went up to 6.8 with the introduction of nanoiron. Based on the water chemistry of nanoiron (Sun *et al.*, 2006), if the only contributor to the ORP was nanoiron, the ORP of the system would be expected to vary between +50 to −100 mV between the pH units of 6.0 and 6.5. The ORP measurements for the diffusion tests did remain between +50 and −50 mV over a 46-h duration without the electric field application. At the end of this same duration, both the control (diffusion only) and the electrokinetic test specimens showed uniform distribution of iron along the 20-cm-long thin clay bed, assuring the presence of the iron at all measurement locations.

The pH distribution in specimens with the nanoiron displayed higher values, shifted up by several pH units (1 to 4), than those without the nanoiron (Pamukcu, Hannum, and Wittle, 2008). The increased pH in the nanoiron experiments were expected to result in lower ORP, as was the case for the first 6 h of the treatment. However, soon after the first 6 h, the ORP of the system increased above the average distribution measured in electrokinetic tests without the nanoiron, while the pH did not change as appreciably to explain this increase. The rate of positive shift of the ORP in electrokinetic tests appeared to decrease with time, while little or no change in the ORP was observed across the specimens when nanoiron was simply allowed to diffuse for the same duration of 46 h. A clear shift in the ORP was observed in regions of low pH, which is above the ORP level expected from the water chemistry of nanoiron alone, and also for the clay under the electric field alone. Hence, clay surface polarization, as discussed above, is suggested to offer an explanation for the enhanced activity, especially in the regions of low pH range (2 to 4).

The voltage distribution across the clay bed and its temporal variation are plotted in Figure 2.15. As observed, the electrical gradient fluctuated between adjacent electrode locations, which would have influenced the velocity of the migrating particles from one station to the next, causing accumulation in some locations and evacuation in others. The voltage distribution appeared to follow spatial and temporal oscillations. The magnitude of these oscillations ceased in time, attaining a more or less constant gradient in soil beyond 17.5 h. The largest change in magnitude occurred consistently at around electrode station E5. The isoelectric pH of the

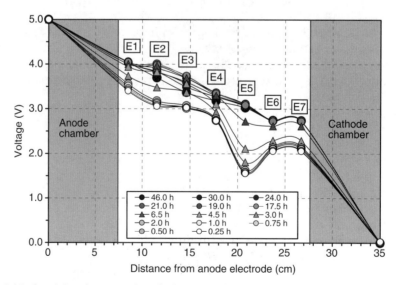

Figure 2.15. Spatial and temporal variations of voltage distribution across a clay specimen (Pamukcu, Hannum, and Wittle, 2008).

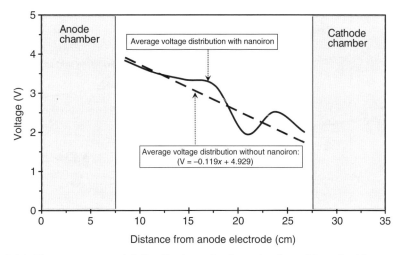

Figure 2.16. The average spatial distribution of voltage in clay with and without nanoiron (Pamukcu, Hannum, and Wittle, 2008).

nanoparticles, determined as 8.3, was very close to the pH of approximately 8.0 measured between stations E5 and E6. Part of the reason for the larger voltage fluctuations at station E5 may be the occurrence of soil pH at the vicinity of the nanoiron isoelectric pH, where the ζ potential of the migrating nanoiron would change sign.

The spatial and temporal oscillations of the potential, and hence ORP, were also discussed in terms of possible oscillatory collapse and restoration of DDL due to passing fronts of high charge density in the pore fluid as the nanoiron migrated through the clay matrix. Figure 2.16 depicts the difference between the average voltage distribution of the electrokinetic tests in kaolinite clay with and without nanoiron. While the clay-only tests show a constant potential gradient of 0.1 V/cm (equation of the linearly fit line), the oscillatory distribution of the potential for the nanoiron case may indicate a charged particle front in motion.

Theoretically, the particle velocity would decrease at regions of low electrical fields, causing accumulation and higher charge density, but pick up at regions of higher electric fields, hence separation and low charge density. This process is likely to have a progressive nature that would allow for self-correction by diffusion potential and Faradic currents to maintain the electrical neutrality. Such a process could also result in the oscillation of the voltage to remain stationary, with only the magnitudes of the peaks changing (as seen in Fig. 2.15). For example, if initially, an accumulation was triggered by a physical obstruction (i.e. agglomeration of the particles in small clay pore throats), the voltage gradient at that locale would increase because of reduced charge conduction. The increased charge density would then cause surface capacitance and restore faradic reactions in the same vicinity. In the meantime, the increased voltage gradient would tend to impart force on the particles to move and vacate the location, subsequently causing a drop in charge density locally and reversing the surface reactions. It is obviously difficult to capture such continuous variations, if they occur, through sporadic voltage and ORP

60 ELECTROCHEMICAL TRANSPORT AND TRANSFORMATIONS

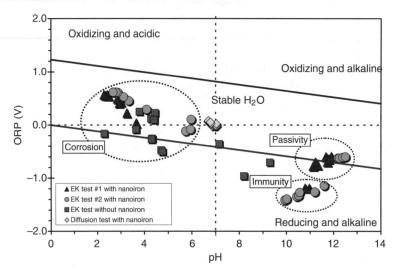

Figure 2.17. pH–ORP (Eh) variation of clay with and without nanoiron tests in electrokinetic transport (Pamukcu, Hannum, and Wittle, 2008).

measurements in a typical electrokinetic transport testing of clay. Nevertheless, based on the data collected and evaluated, the empirical evidence appears to support that the presence of clay and the ensuing surface interactions enhance the ORP under a DC electrical field.

In the diffusion test specimens, despite the measured similarity of the total iron distributions as those in the electrokinetic tests, it was difficult to discern the activation of the iron without the evidence of color change. The color change to brown tones was most evident in the electrokinetic tests, indicating corrosion products of iron. In order to clarify the visual observations, the ORP versus pH were plotted, as shown in Figure 2.17. The chart depicts the approximate regions of corrosion, passivity, and immunity for iron (Davis, 2000). The cluster of data that fell within the *immunity zone* in Figure 2.17 corresponded to the first 6 h of electrokinetic transport, where the ORP at the cathode end was below −1.0 V. This indicated that the nanoiron particles had already been transported to the cathode side of the clay bed by the end of 6 h or earlier. However, better evidence that corroborate with the visual observation of color change was the cluster of data that fell within the *passivity zone*. This data corresponded to the measurements made later, from 6 to 46 h. By the end of 46 h, the corrosion products with *higher* ORP (the passivity zone) appear at the cathode end of the clay bed, as was evidenced by the brown tone color change in that region of the test bed.

2.6 SUMMARY

The fundamental principles of electrochemical transport and transformations applied to electrokinetic processing of saturated clays were reviewed, and some new findings were also presented.

The classical H-S equation expresses the electroosmotic velocity of the fluid as a function of the electric field and the electrokinetic potential of the clay. Both of these parameters vary during electrokinetic transport and result in a nonlinear process. It was shown analytically and experimentally that the electroosmotic velocity could be uncoupled from the applied electric field when surface conductivity, σ_s, and the resulting portion of the current transferred over the solid–liquid interface, i_s, are used as intrinsic properties of the clay to describe the velocity, as first introduced by Khan (1991).

The migration of the ions in the bulk fluid were modeled, taking into consideration the changing electric field due to migration and other effects such as retardation and electrophoretic effects that reduce ion mobility. The model appeared to simulate well the long-term ion distribution in the soil as the conductivity and the electric field varied in time and space.

Finally, the experimental results of electrically enhanced transformations on clay surfaces were presented and discussed in the framework of an electrochemical model. The data appeared to support the hypothesis that faradic current passage orthogonal to the planes in the electric double layer of clay particles may drive forth redox reactions on clay surfaces.

REFERENCES

Acar YB, Alshawabkeh AN. (1993). Principles of electrokinetic remediation. *Environmental Science and Technology* **27**(13):2638–2647.

Acar YB, Alshawabkeh AN. (1996). Electrokinetic remediation I: Pilot scale tests with lead spiked kaolinite. *Journal of Geotechnical and Geoenvironmental Engineering* **122**(3): 173–185.

Acar YB, Hamed J, Alshawabkeh A, Gale R. (1994). Cd(II) Removal from saturated kaolinite by application of electrical current. *Geotechnique* **44**(3):239–254.

Alloway BJ. (1990). *Heavy Metals in Soils*. New York: John Wiley & Sons.

Alshawabkeh AN, Acar YB. (1996). Electrokinetic remediation II: Theoretical Model. *Journal of Geotechnical and Geoenvironmental Engineering* **122**(3):186–196.

Alshawabkeh AN, Sarahney H. (2005). Effect of current density on enhanced transformation of naphthalene. *Environmental Science and Technology* **39**:5837–5843.

Banerjee S, Horng J, Ferguson JF, Nelson PO. (1987). *Field Scale Feasibility Study of Electrokinetic Remediation*, CR811762-01. Cincinnati, OH: Risk Reduction Engineering Laboratory, Office of Research and Development, US EPA.

Bard AJ, Faulkner LR. (1980). *Electrochemical Methods: Fundamentals and Applications*. New York: John Wiley & Sons.

Basu S. (1993). Theory for Hydration Forces in Thin Films of Aqueous Electrolytes. MS Thesis, University of Texas at Austin, Austin, TX.

Cao X. (1997). Numerical Modeling of Electrokinetically Enhanced Transport Mechanism in Soils. MS Thesis, Department of Civil and Environmental Engineering, Lehigh University, Bethlehem, PA.

Chu KT. (2005). Asymptotic Analysis of Extreme Electrochemical Transport. PhD Dissertation, Department of Mathematics, MIT, Boston, MA.

Davis JR. (2000). *Corrosion: Understanding the Basics*. Materials Park, OH: ASM International.

Denisov G, Hicks RE, Probstein RF. (1996). On the kinetics of charged contaminant removal from soils using electric fields. *Journal of Colloid and Interface Science* **178**:309–323.

Dzenitis JM. (1997). Soil chemistry effects and flow prediction in electroremediation of soil. *Environmental Science and Technology* **31**(4):1191–1197.

Electorowicz M, Boeva V. (1996). Electrokinetic supply of nutrients in soil bioremediation. *Environmental Technology* **17**:1339–1349.

Eykholt GR, Daniel DE. (1994). Impact of system chemistry on electroosmosis in contaminated soil. *Journal of Geotechnical and Geoenvironmental Engineering* **120**(5):797–815.

Franz AJ, Rucker JW, Flora JRV. (2002). Electrolytic oxygen generation for subsurface delivery: effects of precipitation at the cathode and an assessment of side reactions. *Water Research* **36**:2243–2254.

Grahame DC. (1951). The role of the cation in the electrical double layer. *Journal of The Electrochemical Society* **98**:343–350.

Grahame DC. (1952). Mathematical theory of the Faradaic admittance. I. Pseudocapacity and polarization resistance. *Journal of The Electrochemical Society* **99**:370C–385C.

Hamed J, Acar YB, Gale RJ. (1991). Pb(II) Removal from kaolinite by electrokinetics. *Journal of Geotechnical Engineering* **117**(2):241–271.

Helmholtz H. (1879). Studien uber electrische grenzschichten. *Annual Review of Physical Chemistry* **7**(Ser. 3):337–382.

Hicks EE, Tondorf S. (1994). Electrorestoration of metal contaminated soils. *Environmental Science and Technology* **28**(12):2203–2210.

Hunter RJ. (1981). *Zeta Potential in Colloidal Science Principles and Applications*. New York: Academic Press.

Hunter RJ. (2001). *Foundations of Colloid Science*. Oxford: Oxford University Press.

Hunter RJ, James M. (1992). Charge reversal of kaolinite by hydrolysable metal ions: An electroacoustic study. *Clays and Clay Minerals* **40**(6):644–649.

Israelachvili JN, Adams GE. (1978). Measurements of forces between two silica surfaces in aqueous electrolyte solutions in the range of 0–100 nm. *Journal of the Chemical Society, Faraday Transactions* **74**(4):975–1001.

Khan, LI. (1991). Study of Electroosmosis in Soil: A Modified Theory and Its Applications in Soil Decontamination. PhD Dissertation, Department of Civil and Environmental Engineering, Lehigh University, Bethlehem, PA.

Kortüm G, Bockris JOM. (1951). *Textbook of Electrochemistry*, Vol. 2. Amsterdam, the Netherlands: Elsevier.

Kruyt HR. (ed.) (1952). *Colloidal Science*. Amsterdam, the Netherlands: Elsevier.

Lafolie F, Hayot Ch. (1993). One-dimensional solute transport modeling in aggregated porous media. Part 1: Model description and numerical solution. *Journal of Hydrology* **143**:63–83.

Lageman R. (1993). Electro-reclamation: Applications in the Netherlands. *Environmental Science and Technology* **27**(13):2648–2650.

Lyklema J. (1995). *Fundamentals of Interface and Colloid Science. Volume II: Solid-Liquid Interfaces*. San Diego, CA: Academic Press.

Mitchell JK. (1993). *Fundamentals of Soil Behavior*, 2nd ed. New York: John Wiley & Sons.

Mitchell JK, Yeung TC. (1991). Electro-kinetic flow barriers in compacted clay. *Transportation Research Record* **1288**:1–9.

Nernst W. (1888). Zur kintetik der in losung befindlichen korper. *Zeitschrift fur Physikalische Chemie* **2**:613–637.

Newman J. (1991). *Electrochemical Systems*, 2nd ed. Englewood Cliffs, NJ: Prentice-Hall.

Pamukcu S, Filipova I, Wittle JK. (1995) The role of electroosmosis in transporting PAH compounds in contaminated soils. In *Proceedings of the Symposium on Electrochemical Technology Applied to Environmental Problems* (eds. EW Brooman, JM Fenton, C Hamel). Pennington, NJ: The Electrochemical Society, PV 95–12, pp 252–266.

Pamukcu S, Hannum L, Wittle KJ. (2008). Delivery and activation of nano-iron by DC electric field. *Journal of Environmental Science and Health* **A43**:934–944.

Pamukcu S, Weeks A, Wittle JK. (1997). Electrochemical separation and stabilization of selected inorganic species in porous media. *Journal of Hazardous Materials* **55**: 305–318.

Pamukcu S, Weeks A, Wittle KJ. (2004). Enhanced reduction of Cr (VI) by direct electric field in a contaminated clay. *Environmental Science and Technology* **38**(4): 1236–1241.

Pamukcu S, Wittle JK. (1992). Electrokinetic removal of selected heavy metals from soil. *Environmental Progress* **11**(3):241–250.

Pamukcu S, Wittle JK. (1993). Electrokinetically enhanced in situ soil decontamination, Chapter 13. In *Remediation of Hazardous Waste Contaminated Soils* (eds. DL Wise, DJ Trantolo). New York: Marcel Dekker, pp. 245–298.

Parker JC, van Genuchten MT. (1984). Flux-averaged and volume-averaged concentrations in continuum approaches to solute transport. *Water Resources Research* **22**:399–407.

Pashley RM. (1981). Hydration forces between mica surfaces in aqueous electrolyte solutions. *Journal of Colloid and Interface Science* **80**(1):153–162.

Planck M. (1890). Ueber die erregung von elektricitat und warme in electrolyten. *Annual Review of Physical Chemistry* **39**(Ser. 3):161–186.

Probstein RF, Hicks RE. (1993). Removal of contaminants from soils by electric fields. *Science* **260**:498–503.

Pulgarin C, Adler N, Peringer P, Comninellis C. (1994). Electrochemical detoxification of a 1,4-benzoquinone solution in wastewater treatment. *Water Research* **28**:887–893.

Quigley RM. (1980). Geology, mineralogy and geochemistry of Canadian soft soils: A geotechnical perspective. *Canadian Geotechnical Journal* **20**(2):288–298.

Quinke G. (1861). Ueber die Fortfuhrung materieller Teilchen durch stromende Electrizitat. *Annalen der Physik und Chemie*. **113**(8):33.

Ravina I, Zaslavsky D. (1967). Non-linear electrokinetic phenomena I: Review of literature. *Soil Science* **106**(1):60–66.

Reddy KR, Parupudi US. (1997). Removal of chromium, nickel, and cadmium from clays by in-situ electrokinetic remediation. *Journal of Soil Contamination* **6**(4):391–407.

Reuss FF. (1809). Sur un novel effet de l'electricite galvanique. *Memories de la Societe Imperial des Naturalistes de Moscow* **2**:327.

Rubinstein I. (1990). *Electro-Diffusion of Ions. SIAM Studies in Applied Mathematics*. Philadelphia: SIAM.

Shi L, Hauke H, Wick LY. (2008). Electroosmotic flow stimulates the release of alginate-bound phenanthrene. *Environmental Science and Technology* **42**(6):2105–2110.

Shang JQ, Lo KY, Inculet II. (1995). Polarization and conduction of clay-water-electrolyte systems. *Journal of Geotechnical Engineering* **121**(3):243–248.

Shapiro AP, Probstein RF. (1993). Removal of contaminants from saturated clay by electro-osmosis. *Environmental Science and Technology* **27**:283–291.

Shapiro AP, Probstein RF, Hicks RE. (1995). Removal of cadmium (II) from saturated kaolinite by application of electric current. *Geotechnique* **45**:355–359.

Smith RW, Narimatsu Y. (1993). Electrokinetic behavior of kaolinite in surfactant solutions as measured by both the microelectrophoresis and streaming potential methods. *Minerals Engineering* **6**(7):753–763.

Smoluchowski M. (1914). *Handbuch der Electrizitat und des Magnetismus*, II. (ed. S Graetz). **2**:336. Leipzig, Germany: J.A. Barth.

Sun YP, Li XQ, Cao J, Zhang WX, Wang HP. (2006). Characterization of zero-valent iron nanoparticles. *Advances in Colloid and Interface Science* **120**:47–56.

Taylor SR, Gileadi E. (1995). Physical interpretation of the Warburg impedance. *Corrosion Science* **51**(9):664–671.

Warburg E. (1899). Ueber das verhalten sogenannter unpolarisirbarer elektroden gegen wechselstrom. *Annalen der Physik und Chemie* **67**(Ser. 3):493–499.

Wittle JK, Pamukcu S. (1993). *Electrokinetic Treatment of Contaminated Soils, Sludges and Lagoons*. DOE/CH-9206, No. 02112406. Chicago, IL: Argonne National Laboratories.

Yeung AT, Hsu CN, Menon R. (1996). EDTA-enhanced electrokinetic extraction of lead. *Journal of Geotechnical Engineering* **122**(8):666–673.

Yin J, Finno RJ, Feldkamp JR. (1995). Electro-osmotic mobility measurement for kaolinite clay. *Proceedings of the Specialty Conference on Geotechnical Practice in Waste Disposal, ASCE, Part 2 (of 2)*, New Orleans, LA. New York: ASCE. February 24–26, pp. 1550–1563.

Yong RN, Warkentin BP, Phadungchewit Y, Galves R. (1990). Buffer capacity and lead retention in some clay materials. *Water, Air, and Soil Pollution* **53**:53–67.

3

GEOCHEMICAL PROCESSES AFFECTING ELECTROCHEMICAL REMEDIATION

ALBERT T. YEUNG

3.1 INTRODUCTION

Electrochemical remediation of polluted soils may include electrochemical mobilization of contaminants, electrokinetic extraction of contaminants from fine-grained soils, electrochemical injection of decontamination agents, electrochemical transformation of contaminants in soils, and their various combinations and modifications. There are always geochemical processes associated with these processes. These geotechnical processes may enhance or retard the electrochemical remediation processes. The pertinent geochemical processes, including (a) generation of pH gradient, (b) change of zeta potential of soil particle surfaces, (d) change of direction of electroosmotic flow, (d) sorption and desorption of contaminants onto/from soil particle surfaces, (e) buffer capacity of soil, (f) complexation, (g) oxidation–reduction (redox) reactions, and (h) interactions of these processes, that affect the electrochemical remediation of contaminated soils are thus discussed in this chapter to facilitate further elaboration of the electrochemical remediation processes and applications in other chapters of this book. Techniques to model these reactions are beyond the scope of this chapter. However, interested readers should refer to Bethke (2008).

Remediation of contaminated fine-grained soils is still an unresolved challenge to geoenvironmental professionals because of the extremely low hydraulic conductivity and large specific area of fine-grained soils, and the existence of many contaminant-specific, environment-dependent, dynamic, and reversible geochemical processes and soil–contaminant interactions (Yeung, 2008). Electrokinetics

---

Electrochemical Remediation Technologies for Polluted Soils, Sediments and Groundwater,
Edited by Krishna R. Reddy and Claudio Cameselle
Copyright © 2009 John Wiley & Sons, Inc.

provides an effective driving force for the migration of pore fluid and chemicals in fine-grained soils to facilitate remediation of the contaminated soil and many other applications in geotechnical and geoenvironmental engineering (Yeung, 1994; Azzam and Oey, 2001). More importantly, electrokinetics can drive more uniform flows of pore fluid and chemicals in hydraulically heterogeneous soils, thus providing a more uniform remediation of soils polluted by multiple contaminants (Szpyrkowicz et al., 2007). Electrochemical reactions and geochemical processes associated with the process, if they are properly controlled and utilized, may enhance the remediation process (Yeung, 2006a). Although the potential of electrochemical remediation of soil is evident, the geochemical processes and the electrochemical reactions associated with the process have to be fully understood before they can be properly controlled and fully utilized.

Soil is fundamentally a three-phase material composed of air, water, and soil solid particles. When a soil is contaminated, it becomes a multiphase material, as the organic and/or inorganic contaminants in soil may exist in different phases and chemical states. They may exist as a gaseous phase in air voids, an immiscible liquid phase in pore fluid, a dissolved phase in pore fluid, a sorbed phase on surfaces of colloidal-size particles suspending in pore fluid, and/or a sorbed phase on soil particle surfaces. The existing phases and chemical states of contaminants may be contaminant specific, dynamic, reversible, and heavily dependent on environmental conditions, in particular the pH of the environment. Therefore, they are critically interrelated with the geochemical processes associated with the electrochemical remediation processes. More importantly, the existing phases of contaminants in soil dominate the mobility of the contaminants in soil, and thus the remediation efficiency of most *in situ* subsurface remediation technologies, and electrochemical remediation is no exception. For example, the mechanisms of metal retention in soils can be classified into two major categories: (a) sorption of ions onto the surface of soil components, for example, clay and organic matter; and (b) precipitation of discrete metal compounds, for example, oxides, carbonates, hydroxides, sulfides, and so on (Evans, 1989). It is extremely important to solubilize metals in sorbed and precipitated states for successful removal of metals from soils. The chemical states of the contaminants also determine their toxicity and hazards to living organisms.

Electrochemical remediation of contaminated soils involves, basically, the application of a direct current (DC) electric field across the soil and the utilization of the resulting electrokinetic flow processes, geochemical processes, and electrochemical reactions to remediate the contaminated soils. These electrokinetic flow processes, geochemical processes, and electrochemical reactions include flow of electric current through the soil; electroosmotic migration of pore fluid; electromigration of ions, charged particles, colloids, and bacteria; electrolysis of pore water in soil at the electrodes and subsequent migration of hydrogen and hydroxide ions into the soil, resulting in a spatial and temporal change of soil pH; gas generation at electrodes; development of nonuniform electric field; occurrence of reverse electroosmotic flow; changes in electrokinetic properties of soil; hydrolysis; phase change of contaminants; soil–contaminant interactions such as sorption and desorption of contaminants onto and from soil particle surfaces; formation of complexes of contaminants; precipitation of contaminants; and so on, and the interactions of these processes.

The migration of pore fluid, ions, charged particles, colloids, and bacteria can be utilized to remove contaminants from polluted soil and/or to inject enhancement agents, nutrients, and so on to facilitate various remediation processes. The geochemical processes can be used to provide the necessary environmental conditions to control the direction of electroosmotic flow and to solubilize contaminants in the soil, so as to enhance the efficiency of the electrochemical remediation processes.

3.2 SOIL–FLUID–CHEMICAL SYSTEM AS ACTIVE ELECTROCHEMICAL SYSTEM

A soil polluted by organic and/or inorganic contaminants is basically a multiphase soil–fluid–chemical system. Therefore, it is important to understand the nature of the system prior to the detailed discussion on the occurrence of geochemical processes within the system during electrochemical remediation.

Soil has long been considered as a chemical system due to its semipermeability to chemicals, bioactivity, interactions with chemicals, and so on. As a result, soil has been idealized as a leaky semipermeable membrane in chemical osmosis to explain various abnormal transport phenomena of water and chemicals in soil (Hanshaw, 1972; Marine and Fritz, 1981; Fritz and Marine, 1983; Yeung, 1990; Keijzer, Kleingeld, and Loch, 1999); as a Donnan membrane (Donnan, 1924) to examine the influences of soil type, water content, electrolyte concentration, and the cation and anion distribution in pore fluid on electroosmotic flow of fluid in soil (Gray and Mitchell, 1967); as a bioreactor to evaluate the impact of oxygen transfer on efficiency of bioremediation (Woo and Park, 1997); and so on.

When a DC electric field is imposed on a soil–fluid–chemical system, simultaneous electrokinetic flows occur through the system. Yeung (1990) idealized the system as a homogeneous charged membrane system out of thermodynamic equilibrium and formulated the general simultaneous coupled transport equations of water, electricity, and ions induced by simultaneously imposed hydraulic, electrical, and chemical gradients using the formalism of nonequilibrium thermodynamics. The soil–fluid–chemical system or the membrane was considered as a transition zone of variable ion concentrations, hydraulic potential, and electrical potential, with boundaries merging with the solutions of the two compartments adjoining the soil–fluid–chemical system. The two compartments are of different ion concentrations, hydraulic potentials, and electrical potentials separated by the membrane. The general coupled flow equations describing the simultaneous flows of water, electricity, and ions induced by hydraulic, electrical, and chemical gradients were derived. The validity of the equations is supported theoretically by the agreement of the results of special cases of the general coupled flow equations with the existing solutions of well-known phenomena such as diffusion potential, advection–dispersion equation, van't Hoff equation, and osmosis and ultrafiltration (Yeung, 1990). Validity of the equations is also substantiated by the experimental results presented by Yeung and Datla (1996).

As a chemical system allowing the conduction of electricity, a soil–fluid–chemical system may behave as a passive electrical conductor of finite impedance that can be simply idealized as a combination of resistors, capacitors, and/or inductors, and electrical energy is dissipated or stored by the system when an electric current flows through it. On the contrary, the system may behave as an active electrochemical system.

Does a soil–fluid–chemical system behave as an active electrochemical system or a passive electrical conductor under the influence of a DC electric field? This is a fundamental question of significant implications. The evaluation criterion that can be used to differentiate the two systems of completely different nature is vested in Faraday's laws of electrolysis, as the transfer of electrons from the electrodes to the system and vice versa in an ideal electrochemical system is invariably associated with chemical reactions obeying Faraday's laws of electrolysis (Antropov, 1972). The two important fundamental laws of electrolysis can be simply expressed as follows: (a) the amount of chemical deposition is proportional to the quantity of electric charges flowing through the system in an electrolytic process, and (b) the masses of different species deposited at or dissolved from electrodes by the same quantity of electric charges are directly proportional to their equivalent weights (Crow, 1979).

When a direct electric current flows through an active electrochemical system, the products of electrolysis depend on the redox or electrode potentials of the chemicals in the system and the chemical properties of the electrodes (Antropov, 1972). The products of electrolysis in soil saturated by dilute chemical solutions using graphite or most commonly used metal electrodes are generally oxygen and hydrogen, generated at the anode and the cathode, respectively, as given in these reactions of electrolytic decomposition of water:

$$\text{oxidation at the anode} \quad 2H_2O - 4e^- = O_2 + 4H^+ \quad (3.1)$$

$$\text{reduction at the cathode} \quad 4H_2O + 4e^- = 2H_2 + 4OH^- \quad (3.2)$$

It can be observed from the electron transfer indicated in Equations 3.1 and 3.2 that oxygen is oxidized at the anode and that hydrogen is reduced at the cathode. Theoretically, when 4 mol of electrons flow through the system, 1 mol of oxygen and 2 mol of hydrogen are generated at the anode and the cathode, respectively. Therefore, comparison of the gas generation rates at the electrodes and the electric current flowing through the soil–fluid–chemical system provides a valid evaluation of the electrochemical behavior of the system.

The experimental apparatus specifically designed, fabricated, and assembled by Yeung et al. (1997) as shown in Figure 3.1 facilitates the measurement of volumes of gas generated at the electrodes during the electrochemical remediation processes. The gas generated at the electrodes bubbles through the reservoir fluid to enter the gas volume measurement devices, and displace the reservoir fluid downward. The volume of gas generated can be determined after making the necessary adjustments for the volume change due to varying gas pressure and the volume of gas dissolved in the reservoir fluid during the bubbling process. The hydrogen generation rate (mol/s) versus $I/2F$, where I = electric current (A or C/s) and F = Faraday constant (96,493 C/mol), at different stages of all these experiments are depicted in Figure 3.2. The straight line represents the theoretical generation rate of hydrogen obtained using Faraday's laws of electrolysis. It can be observed that the experimental data points are in close proximity of the theoretical line, indicating that Faraday's laws of electrolysis are obeyed and the electrochemical behavior of the soil–fluid–

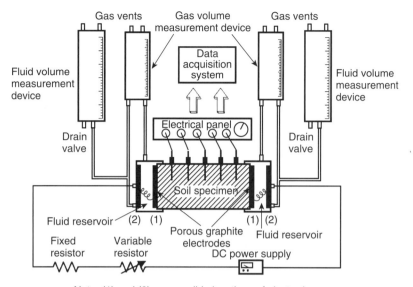

Figure 3.1. Schematic of the electrokinetic extraction apparatus.

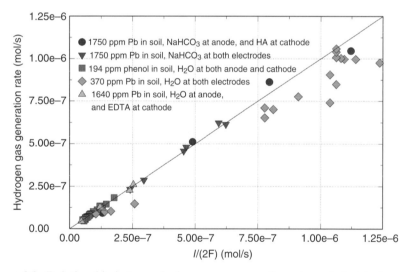

Figure 3.2. Relationship between hydrogen gas generation rate and electrical current.

chemical system thus resembles that of an active electrochemical system. Therefore, it can be concluded evidently by the experimental results of Yeung (2006b) and many others obtained using different contaminants, different soils, and different operating parameters in experiments on electrochemical remediation of soils that soil–fluid–chemical systems are active electrochemical systems. The behavior is independent of soil type, chemicals in the system, pore fluid chemistry, magnitude of the DC electric field applied across the system, electric current density flowing through the system, and chemistry of reservoir fluid, provided the products of electrolysis are not changed by any variation of these parameters.

The recognition of soil–fluid–chemical systems as active electrochemical systems is very important in the development of a better understanding of the physics and chemistry involved in electrochemical remediation of soils. It laid the foundation to support the use of established principles in electrochemistry to explain the phenomena observed during electrochemical remediation of soils, and to explore the use of different enhancement techniques to improve the efficiency of the technology.

3.3 GENERATION OF pH GRADIENT

Among all the geochemical processes, the generation of pH gradient in contaminated soil has the most profound impact on the electrochemical remediation process, as the change of pore fluid pH has a significant influence on the direction of electroosmotic flow of pore fluid, and on the mechanism and degree of sorption and desorption of contaminants onto and from soil particle surfaces, and complexes formation and precipitation of chemical species, dissociation of organic acids, and so on, thus affecting the feasibility and efficiency of electrochemical remediation of contaminated soils.

As depicted in Equations 3.1 and 3.2, hydrogen and hydroxide ions are generated at the anode and the cathode, respectively. As a result, an acidic environment is developed in the vicinity of the anode and an alkaline environment in the vicinity of the cathode. Although the mechanism of the migration of the acid front is not fully understood, most researchers observe the propagation of an acid front from the anode toward the cathode. However, the acid front has never been observed to flush through the specimen to generate a uniform acidic pore fluid environment throughout the specimen. Some researchers describe the acid front propagation process by applying the advection–dispersion equation to the transport of hydrogen and hydroxide ions using the measured ionic mobilities and diffusivities of hydrogen and hydroxide ions (Eykholt and Daniel, 1994; Alshawabkeh and Acar, 1996). However, it is well known in physical chemistry that the measured ionic mobilities of hydrogen and hydroxide ions are exceptionally high in comparison with those of other ions. In fact, these ions are transferred along a series of hydrogen-bonded water molecules by the rearrangement of hydrogen bonds, as shown in Figure 3.3 (Silbey, Alberty, and Bawendi, 2005). Water molecules are oriented under the influence of the imposed DC electric field. The initial bonding between a hydrated hydrogen ion and a group of oriented water molecules is shown in Figure 3.3a, and the final rearranged bonding is shown in Figure 3.3b, resulting in the apparent migration of the hydrogen ion. Similarly, the transfer of a hydrated hydroxide ion in the opposite direction is shown in Figure 3.3c,d. Therefore, the measured ionic mobilities of these ions are not related to their actual ionic migration velocities. The phenomena cast serious doubt on the validity of applying the advection–dispersion equation to describe the migration of hydrogen and hydroxide ions in contaminated soil and to compute the resulting pH distribution in the soil.

As soil–fluid–chemical systems must be electrically neutral, charges cannot be added to, formed in, or removed from a system without the simultaneous addition,

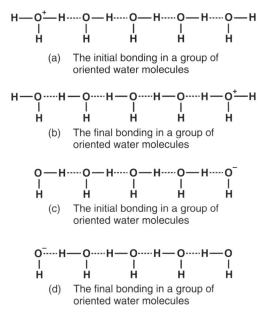

Figure 3.3. Transfer mechanisms of hydrogen and hydroxide ions in water.

formation, or removal of an equal number of the opposite charge (Snoeyink and Jenkins, 1980). Using the general coupled flow equation and maintenance of charge balance throughout the system at all times, Yeung and Datla (1996) proposed an electrokinetic transport model that the electrical neutrality of the system is restored by the immediate availability of hydrogen and hydroxide ions. Their model is able to explain some of the phenomena observed in experiments such as that the acid front has never been observed to flush through the specimen to develop a uniform acidic pore fluid environment throughout the specimen.

Nonetheless, the spatial and temporal changes in soil pH induced by the electrochemical remediation process are evident. As a result, physical and chemical changes of the soil and the contaminants associated with the change in soil pH have to be carefully taken into account to maximize the efficiency of electrochemical remediation of contaminated soil.

Moreover, it should be noted that soil pH can be changed by the introduction of enhancement agents during electrochemical remediation. The change is enhancement agent specific. The experimental results of Reddy et al. (2006) on a sandy soil with nonplastic fines contaminated with polycyclic aromatic hydrocarbon (PAH) compounds and heavy metals demonstrate that the addition of 20% *n*-butylamine into the system increased the soil pH, while 3% Tween 80, 5% Igepal CA-720, and 10% hydroxypropyl-β-cyclodextrin (HPCD) did not cause substantial change in soil pH. The results of Reddy and Ala (2005, 2006) on a high-plasticity clay from Chicago, Illinois, contaminated with lead, cadmium, mercury, and arsenic also indicate the distributions of soil pH along the length of specimens after electrochemical remediation are dependent on the type of enhancement agent introduced into the system during the electrochemical remediation process.

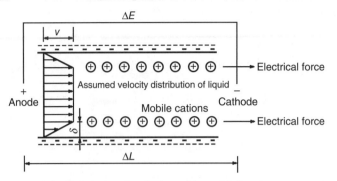

Figure 3.4. Helmholtz–Smoluchowski model for electroosmosis.

3.4 CHANGE OF ZETA POTENTIAL OF SOIL PARTICLE SURFACES

The Helmholtz–Smoluchowski model, as shown in Figure 3.4, is one of the earliest but still widely used theoretical formulations of electrokinetic phenomena, in particular for electrokinetic phenomena in soil (Yeung, 1994). The soil–fluid interface is treated as an electrical condenser with ions of one sign (co-ions) near to or on the soil particle surface and mobile ions of opposite sign (counterions) in the fluid concentrated in a layer at a very small distance from the soil particle surface. The surface is usually negatively charged in soil due to isomorphic substitution and presence of broken bonds. Therefore, there will be a net excess of cations in the fluid forming the mobile cation shell, as shown in Figure 3.4. Concentration distribution of the cations in the fluid can be mathematically described by the Stern–Gouy–Chapman diffuse double layer theory. The derivation of related equations in the International System of Units is given by Yeung (1992).

It should be noted that there are two principal origins of surface charges on soil particles: (a) isomorphic substitutions among ions of different valences in soil minerals, such as silicon positions occupied by aluminum ions or aluminum positions occupied by magnesium or iron ions, and presence of broken bonds; and (b) reactions of surface functional groups with ions in soil pore fluid (Sposito, 1989). The net total particle charge is composed of four different types of surface charges: (a) permanent structural charge σ_0, resulting from isomorphic substitutions or broken bonds in soil; (b) net proton charge σ_H, that is, the difference between the number of moles of hydrogen and that of hydroxide ions complexed by surface functional groups; (c) inner-sphere complex charge σ_{IS}, that is, the net total charge of ions other than H^+ and OH^- bound into inner-sphere surface complexes; and (d) outer-sphere complex charge σ_{OS}, that is, the net total charge of ions other than H^+ and OH^- bound into outer-sphere surface complexes. The algebraic sum of σ_0 and σ_H is denoted as the intrinsic surface charge of a soil as they arise principally from the ionic constituents of soil minerals and surface OH groups on inorganic and organic solids. The relative contribution of σ_0 and σ_H to the intrinsic surface charge of a soil depends on the extent of weathering of its minerals and organic matter content. Since the dominant source of permanent structural surface charge is the siloxane cavity, it is evident that σ_0 is the dominant contribution in soils at the early and intermediate stages of weathering such as those containing characteristic minerals of feldspars, illite, vermiculite, chlorite, smectites, and so on. Therefore, these soils are termed permanent-charge soils. Soils at the advanced stage of weathering are enriched in minerals bearing reactive OH groups such as kaolinite, gibbsite, goethite, hematite, and so on. In these soils, σ_H dominates σ_0, and changes in soil pH change the development of surface charges in these soils. Therefore, they are termed variable-charge soils. It should be noted that σ_0 is virtually always negative in soil; however, σ_H, σ_{IS}, and σ_{OS} can be positive, zero, or negative, depending on soil pH. As a result, the net total particle charge, that is, the sum of σ_0, σ_H, σ_{IS}, and σ_{OS}, can be positive, zero or negative, depending on soil pH and the relative dominance of σ_0, which depends on the extent of weathering.

When a DC electric field is imposed on the soil, it will move the mobile cation shell toward the cathode and drag the fluid between the shearing surfaces by a plug flow mechanism, resulting in the electroosmotic flow of fluid from the anode toward the cathode. The shearing surface in the fluid may be taken as a plane parallel to the surface at a distance δ (m) from it. If ε (F/m) is the permittivity of the fluid between these hypothetical plates, then it is known from electrostatics that

$$\varphi_o - \varphi_s = -\zeta = \frac{-\sigma\delta}{\varepsilon}, \qquad (3.3)$$

where φ_o and φ_s are the electrical potentials in the fluid and at the shearing surface, respectively, (V) and σ = surface charge density (C/m^2). If the potential in the fluid is taken as the reference or zero potential, the potential at the shearing surface is called electrokinetic or zeta potential, and is usually denoted by the Greek symbol ζ. In general, if the sign of the surface charge of soil particles is negative and that of the mobile ions is positive, as in most soils, then ζ (V) is negative and vice versa.

The coefficient of electroosmotic conductivity k_e of soil is defined by

$$Q = k_e i_e A, \qquad (3.4)$$

where Q = volume flow rate (m^3/s) and A = gross total cross-sectional area perpendicular to the flow direction through which the flow occurs (m^2). It is related to the zeta potential of the soil particle surface by the Helmholtz–Smoluchowski model by Yeung (1994),

$$k_e = -\frac{\varepsilon \zeta n}{\eta}, \qquad (3.5)$$

where n = porosity of soil and η = viscosity of the fluid (Ns/m^2). It can be observed in Equation 3.5 that when ζ is negative, as in most soils under normal conditions, then k_e is positive, that is, electroosmotic flow is from the anode toward the cathode.

However, the surface charge of soil particles is dependent on soil pH, so are the zeta potential of soil particles and the coefficient of electroosmotic coefficient. In general, the zeta potential of soil particles increases with acidity, that is, becoming less negative with the decrease in pH value. At the point of zero charge, the zeta potential vanishes. The point of zero charge is defined to be the pH value at which there is no net charge contributed by the ions in the diffuse double layer, that is, the net total particle charge of soil is zero (Sposito, 1989). It can be determined experimentally by measuring the pH at which soil particles do not move in an applied DC electric field, that is, when the electrophoretic velocity of soil particles u is zero. As

$$u = u_E \times \left(-\frac{\Delta E}{\Delta L}\right), \qquad (3.6)$$

where u_E = electrophoretic mobility of soil particles and $(-\Delta E/\Delta L)$ = the applied electric field; the electrophoretic mobility of soil particles is zero at the point of zero charge. There are different theoretical models relating the electrophoretic mobility of soil particles to their zeta potential (Yeung, 1994). However, all the models are in the form of

$$u_E = C\frac{\varepsilon\zeta}{\eta}, \qquad (3.7)$$

where C = a constant. If the soil particle surface is negatively charged and the counterions are positively charged, then both ζ and u_E are negative. It can then be observed from Equations 3.6 and 3.7 that soil particles are moving in the direction of increasing electrical potential, that is, from the cathode towards the anode. Otherwise, a positive electrophoretic mobility indicates that the soil particles are positively charged and will move in the direction of decreasing electrical potential. More importantly, a negative electrophoretic mobility indicates a positive electro-osmotic flow, that is, from the anode toward the cathode, if the movement of soil particles is restricted and vice versa. In fact, most zeta potential analyzers measure the electrophoretic mobility of fine particles to determine their zeta potentials using Equation 3.7. Therefore, it is more useful to report electrophoretic mobility than zeta potential.

The variations of the zeta potential of three different soils as a function of pH are presented in Figure 3.5 to illustrate the effect of soil type on zeta potential.

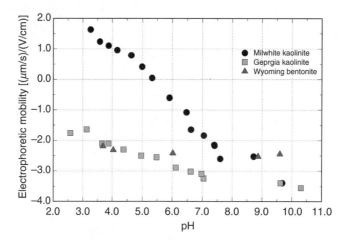

Figure 3.5. Electrophoretic mobility of different soils (background electrolyte = 0.001 M NaCl).

Detailed experimental procedure is given in Hsu (1997). It can be observed that the electrophoretic mobility of Wyoming bentonite, a permanent-charge soil, does not vary much with pH. Moreover, the electrophoretic mobility of Wyoming bentonite is always negative, indicating that electroosmotic flow in Wyoming bentonite is always from the anode toward the cathode. The variation of electrophoretic mobility with pH is larger in Georgia kaolinite, a high-purity and well-characterized kaolinitic soil from Georgia, USA. Its properties are readily available in the literature (van Olphen and Fripiat, 1979). Although Georgia kaolinite is in the advanced stage of weathering, σ_0 still dominates σ_H because of its high purity. As a result, the electrophoretic mobility of Georgia kaolinite is always negative, indicating that electroosmotic flow is always from the anode toward the cathode, as observed by many researchers in electrochemical remediation of fine-grained soil. Milwhite kaolinite is a commercially refined product originating from Bryant, Arkansas. It is reddish brown in color and contains approximately 4.3% iron oxides as Fe_2O_3 as well as other impurities such as SiO_2 and TiO_2. X-ray diffraction results indicate it is predominantly composed of the mineral kaolinite, with a trace of chlorite. The cation exchange capacity of the soil is measured to be 29 mmol/kg. The variation of electrophoretic mobility with pH of Milwhite kaolinite is much larger, with a point of zero charge at approximately 5.3. When the soil pH is lower than 5.3, the electrophoretic mobility becomes positive and the coefficient of electroosmotic conductivity becomes negative, indicating a reverse electroosmotic flow, that is, from the cathode toward the anode. It should also be noted that the variation between electrophoretic mobility and pH for a particular soil can be modified by the presence of other chemicals in the system, as shown in Figure 3.6. The impact of the chelating agent ethylenediaminetetraacetic acid (EDTA) on the electrokinetic properties of Milwhite kaolinite is evident. When EDTA is introduced into the system, the electrophoretic mobility of Milwhite kaolinite is always negative and will not reverse in a low-pH environment.

The particular geochemical process of change of zeta potential of soil particle surfaces as a function of soil pH and pore fluid chemistry poses significant complexity to

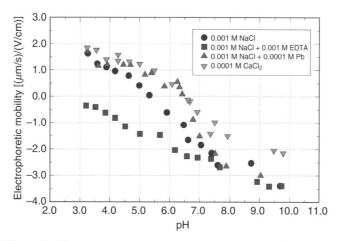

Figure 3.6. Effect of pH and electrolyte species on electrophoretic mobility of Milwhite kaolinite.

the electrochemical remediation process. When the soil pH is high, most metals exist in the form of precipitates and become immobile. It needs a low-pH environment to solubilize metallic contaminants. However, a low-pH environment may reverse the direction of electroosmotic flow, thus diminishing the removal efficiency of ionic migration of cations. Nonetheless, the migration of ions is dictated by electromigration, and the diminishing effect may be minimal. A low-pH environment prohibits the dissociation of organics, but the reverse electroosmotic flow promotes the migration of the dissociated organic ions toward the anion. A high-pH environment promotes the dissociation of organics, but the forward electroosmotic flow retards the migration of the dissociated organic ions toward the cathode.

Different techniques have been attempted to control the pH in the vicinity of electrodes so as to control the pH of the contaminated soil being remediated. These techniques include (a) the use of different agents to depolarize the cathode reaction given in Equation 3.2 (Rødsand, Acar, and Breedweld, 1995; Nogueira et al., 2007); (b) the use of sodium hydroxide to neutralize the anode reaction given in Equation 3.1 (Saichek and Reddy, 2003); (c) the use of an ion-selective membrane to prevent hydroxide ion migration into the soil from the cathode (Rødsand, Acar, and Breedweld, 1995; Li, Yu, and Neretnieks, 1998); (d) rinsing away the hydroxide ions generated at the cathode (Hicks and Tondorf, 1994); (e) the use of buffer solutions in the reservoirs and an appropriate electrode configuration (Ko, Schlautman, and Carraway, 2000; Yeung and Hsu, 2005; Khodadoust, Reddy, and Narla, 2006); (f) circulation of electrolyte solution from the cathode reservoir to the anode reservoir (Lee and Yang, 2000; Chang and Cheng, 2007); and (g) the use of special electrodes to control the electroosmotic fluid flow direction (Leinz, Hoover, and Meier, 1998; Mattson, Bowman, and Lindgren, 2000). These pH-control techniques are discussed in detail by Yeung (2006a) and other authors of this book. Therefore, they are not repeated here.

3.5 CHANGE IN DIRECTION OF ELECTROOSMOTIC FLOW

It can be observed in Equation 3.5 that the direction of electroosmotic flow in soil is dictated by the sign of zeta potential on soil particle surfaces, which is controlled by the extent of weathering and pH of the soil. When the electroosmotic flow is in the positive direction, that is, from the anode toward the cathode, the ionic migration of cationic contaminants is enhanced by the electroosmotic flow, while the ionic migration of anionic contaminants and negatively charged colloids or bacteria are retarded by the electroosmotic flow and vice versa. It has been well established, experimentally, that the ionic migration velocities of chemicals and the electrophoretic velocities of particles in soil are always greater than the advective velocity of chemicals or particles generated by the electroosmotic flow of pore fluid. Nonetheless, the most effective operation of electrochemical remediation is to generate a positive electroosmotic flow to remove cationic contaminants at the cathode, or to generate a negative electroosmotic flow to remove anionic contaminants at the anode. It should be noted that the polarity and chemical state of contaminant ions can be modified by the addition of enhancement agents (Manna, Sanjay, and Shekhar, 2003; Ottosen et al., 2005; Yeung and Hsu, 2005; Mukhopadhyay, Sundquist, and Schmitz, 2007; Pazos et al., 2007).

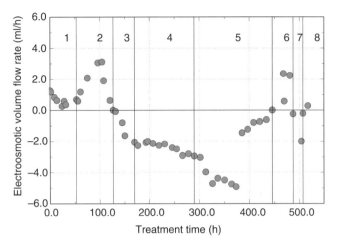

Figure 3.7. Variation of electroosmotic volume rate with time.

The direction of electroosmotic flow can be controlled by changing the reservoir chemistry (Dzenitis, 1997; Yeung and Hsu, 2005). The results of an interesting experiment are presented in Figure 3.7 as an example (Yeung, 2006b). The experiment performed on a Milwhite kaolinite specimen can be divided into eight distinct stages. Each stage was identified by either the introduction of a new chemical into a reservoir or a change in the electroosmotic flow direction. Details of the experiment are given in Table 3.1. During stage 1, 0.25 M $NaHCO_3$ solution of pH 9 was introduced into the anodic reservoir to initiate a forward electroosmotic flow, that is, from the anode toward the cathode, and to maintain the flow for 52.93 h. The anodic reservoir solution was then replaced by deionized water in stage 2. The forward electroosmotic flow lasted for another 73.87 h in stage 2 before the direction of electroosmotic flow reversed. After reverse electroosmotic flow had continued for 42.95 h in stage 3, $NaHCO_3$ solution was introduced into the anodic reservoir in stage 4 in an attempt to change the direction of electroosmotic flow to forward. However, the reverse electroosmotic flow continued for another 120.06 h in stage 4, indicating that addition of a high-pH solution to the anodic reservoir during reverse electroosmotic flow is not an effective technique for changing the electroosmotic flow direction to forward. $NaHCO_3$ solution was then introduced into the cathodic reservoir in stage 5. After another 155.84 h of reverse electroosmotic flow in stage 5, the flow direction changed to forward, which was sustained for 41.63 h in stage 6. An acidic environment was then generated by replacing the $NaHCO_3$ solution in the anodic reservoir with 0.01 M acetic acid in stage 7 to confirm the impact of soil pH on electroosmotic flow direction. The electroosmotic flow direction was reversed almost immediately. After the reverse electroosmotic flow had been sustained for 20.43 h in stage 7, the anodic reservoir was again filled with $NaHCO_3$ solution again in stage 8. In contrast with the observation made in stage 4, the flow direction was changed to forward in this stage. However, it should be noted that the pH in most parts of the specimen was already high, as $NaHCO_3$ solution had been kept in the cathodic reservoir for a long time. The experiment was terminated after 517.46 h of electrokinetic treatment. The results from this

TABLE 3.1. Details of the Experiment on Controlling of Electroosmotic Flow Direction

Stage	Treatment Time (h)	Treatment Duration (h)	Reservoir Solution[a] Anode	Reservoir Solution[a] Cathode	DC Electric Field Applied (V/m)	Electroosmotic Flow Volume (ml)[b]	Average Coefficient of Electroosmotic Conductivity (k_e; m^2/Vs)
1	0–52.93	52.93	NaHCO$_3$	H$_2$O	157.41	33.49	2.45×10^{-10}
2	52.93–126.80	73.87	H$_2$O	H$_2$O	157.35	124.50	6.52×10^{-10}
3	126.80–169.75	42.95	H$_2$O	H$_2$O	157.21	−48.19	-4.3×10^{-10}
4	169.75–289.81	120.06	NaHCO$_3$	H$_2$O	157.28	−290.87	-9.4×10^{-10}
5	289.81–445.65	155.84	NaHCO$_3$	NaHCO$_3$	129.92	−387.21	-1.2×10^{-9}
6	445.65–487.18	41.63	NaHCO$_3$	NaHCO$_3$	65.35	20.93	4.69×10^{-10}
7	487.18–507.61	20.43	CH$_3$COOH	NaHCO$_3$	105.38	−8.37	-2.14×10^{-10}
8	507.61–517.46	9.85	NaHCO$_3$	NaHCO$_3$	262.47	4.19	9.87×10^{-11}

[a]Concentration of NaHCO$_3$ is 0.25 M adjusted to pH 9 by 0.1 M NaOH; concentration of CH$_3$COOH is 0.01 M.
[b]Electroosmotic flow from the anode toward the cathode is defined as forward and vice versa.

experiment indicate that the soil pH is the dominant factor controlling the direction of electroosmotic flow. Once the flow direction is reversed, it will take a high-pH solution entering the soil from the cathode for a much longer period to restore the forward flow direction, probably because it takes time for the high-pH solution to travel to the vicinity of the anode. However, the impact of the anodic solution on electroosmotic flow direction can be observed almost instantaneously. It can be observed that the direction of electroosmotic flow is primarily controlled by the soil pH in the immediate vicinity of the anode.

3.6 SORPTION AND DESORPTION OF CONTAMINANTS ONTO/FROM SOIL PARTICLE SURFACES

Depending on environmental conditions, contaminants can be sorbed on soil particle surfaces. In general, sorption refers to the transfer of mobile contaminant ions from the pore fluid to soil particle surfaces, rendering them immobile. It can occur via cation exchange on clays and humus, and can also take place by specific adsorption, including (McBride, 1994) (a) cation complexation with organic functional groups and bonding on variable-charge minerals and (b) anion selective bonding (chemisorption) at variable-charge mineral surfaces and layer silicate particle edges. The immobility of contaminants becomes an obstacle to most contaminant removal technologies.

If the contaminants can be solidified or stabilized permanently, chemical hazards to the public are minimized. As a result, long-term solidification or stabilization of contaminated soil may be an effective contaminant containment technology by isolating the contaminants from human contacts. However, natural attenuation and/or sorption of contaminants in soil cannot be considered as an effective contaminant containment technology, as soil has a limited capacity on natural attenuation of contaminants and sorption and desorption are chemical-specific, soil-specific, dynamic, reversible, and pH-dependent geochemical processes. On the other hand, if the sorbed contaminants can be desorbed from soil particle surfaces and solubilized in the pore fluid, they can be removed effectively. Electrokinetics introduces an additional complexity or advantage, if controlled properly, as the generation of pH gradient is a geochemical process implicitly associated with the electrochemical remediation process.

Moreover, the sorption and desorption characteristics of contaminants on soil particle surfaces can be altered by the addition of other chemicals, such as surfactants or cleanup enhancement agents, to the system (Yeung, Hsu, and Menon, 1996; Ottosen et al., 2005; Yeung and Hsu, 2005; Mukhopadhyay, Sundquist, and Schmitz, 2007; Pazos et al., 2007). Addition of these chemicals adds another variable to the complex situation.

The results obtained from sorption/desorption edge experiments on Milwhite kaolinite and cadmium are presented in Figure 3.8 as an illustration. The testing parameters of these experiments are tabulated in Table 3.2, and the experimental procedure is detailed in Yeung and Hsu (2005). The low concentration of cadmium in the dissolved phase used in the experiments precludes the formation of cadmium precipitates. Therefore, the reduction in cadmium concentration in the dissolved phase was caused by the sorption of cadmium onto Milwhite kaolinite particle

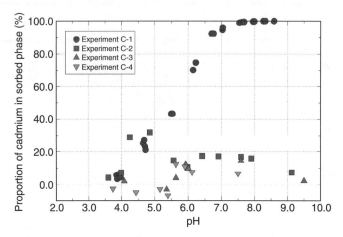

Figure 3.8. Effect of EDTA concentration on desorption of cadmium from Milwhite kaolinite surfaces.

TABLE 3.2. Testing Parameters of Sorption/Desorption Experiments on Cadmium

Experiment Number	Concentration of Milwhite Kaolinite (g/L)	Initial Concentration of Cadmium (mg/l)	Initial Concentration of Ethylenediaminetetraacetic Acid (M)
C-1	100	26.26	0
C-2	100	24.74	0.01
C-3	100	24.74	0.05
C-4	100	24.74	0.1

surfaces. The results of experiment C-1 give the sorption characteristics of cadmium onto Milwhite kaolinite particle surfaces. When the pH is lower than 4, all the cadmium in the system is in the dissolved phase, and the proportion of cadmium in the sorbed phase increases with pH. When the pH is higher than approximately 7, all the cadmium in the system is practically sorbed on soil particle surfaces and becomes immobile. The practical consequence is that the metallic contaminant cannot be removed unless it can be desorbed from the soil particle surfaces and solubilized in the pore fluid. Effects of the enhancement agent EDTA on the desorption of sorbed cadmium from Milwhite kaolinite particle surfaces are illustrated by the results of experiments C-2–C-4. Cadmium had been allowed to sorb onto Milwhite kaolinite particle surfaces for 24 h before EDTA and pH adjustment solutions were added to simulate a cleanup process. The introduction of EDTA changed the distribution of cadmium between the dissolved and sorbed phases drastically, as EDTA desorbed the cadmium sorbed on soil particle surfaces and kept it in the dissolved phase. The proportion of cadmium desorbed increases with the concentration of EDTA solution, as presented in Figure 3.8. Moreover, the EDTA makes the desorption process less pH dependent.

Hydrophobic organic contaminants, such as PAHs, are a large group of persistent and toxic contaminants in the subsurface. They repel water, and thus have low water solubility. Moreover, they are very stable and resistant to chemical attack (Gillette *et al.*, 1999). As they are not appreciably diluted or transported by precipitation, surface water, or groundwater, they remain concentrated in the environment. They are also extremely difficult to clean up because of their hydrophobicity and persistence. As a result, they become widespread contaminants in the subsurface.

There are various mechanisms for hydrophobic organic contaminants to sorb onto soils, including (a) absorption into amorphous or "soft" natural organic matter or nonaqueous-phase liquid (NAPL), (b) absorption into condensed or "hard" organic polymeric matter or combustion residue, (c) adsorption onto water-wet organic surfaces, (d) adsorption to exposed water-wet mineral surfaces, and (e) adsorption into microvoids or microporous minerals with unsaturated porous surfaces. The sorption processes of mechanisms (a), (c), and (d) may occur in less than a minute, while those of mechanisms (b) and (e) may take days (Luthy *et al.*, 1997). Moreover, the resistance of chemicals to desorption increases with soil–chemical contact time, that is, aging (Loehr and Webster, 1996; Li and Liu, 2005). Micropore blockage by precipitated minerals also contributes to slow desorption (Farrell, Grassian, and Jones, 1999).

A systematic evaluation program of surfactants/cosolvents for desorption/solubilization of phenanthrene in clayey soils was conducted by Saichek and Reddy (2004). Phenanthrene, an uncharged organic compound that consists of three aromatic rings, was chosen as a representative PAH compound for the research program. Numerous sites throughout the USA have been contaminated by PAHs. As these compounds may be toxic, mutagenic, and/or carcinogenic, they are threatening the environment and public health. Therefore, the need for prompt remediation of sites contaminated by PAHs is evident. As the success of electrokinetically enhanced *in situ* flushing depends heavily on the effective desorption and solubilization of PAHs from soil particle surfaces through micellization and reduction of surface tension through the proper use of surfactants/cosolvents, and there are more than 700 different types of commercially available surfactants (Myers, 1988), the evaluation program was conducted to select appropriate surfactants/cosolvents for the electrochemical remediation process.

Six surfactants/cosolvents were selected for the evaluation program on the basis of (a) solution chemistry, (b) proven ability to desorb/solubilize PAHs from soil particle surfaces in previous studies, (c) human health and environmental protection, and (d) compatibility with *in situ* electrochemical remediation technique. The chosen surfactants/cosolvents were (a) 3% Igepal CA-720, (b) 5% Igepal CA-720, (c) 5% Triton X-100, (d) 3% Tween 80, (e) 40% ethanol, and (f) a mixture of 40% ethanol and 5% Igepal CA-720. Two clayey soils, kaolin and glacial till, were selected for the study. Kaolin consists mainly of kaolinite clay mineral, while glacial till consists of a combination of different soil minerals including quartz, feldspar, carbonates, illite, chlorite, vermiculite, and trace amounts of smectite.

The two soils were artificially contaminated (spiked) with phenanthrene at initial target concentrations of 100, 300, 500 and 1000 μg/g. Batch desorption experiments were conducted using a soil concentration of 5 g per 25 ml of solution mixed in vials sealed with Teflon screw-type tops. Each vial was shaken by hand for approximately a minute to ensure that the soil was fully saturated with the solution, and the vials were then placed on a rotary shaker table at 250 rpm for 24 h. Afterwards, the concentration of phenanthrene in the supernatant was determined by liquid–liquid

TABLE 3.3. Results of Desorption Experiments on Phenanthrene (After Saichek and Reddy, 2004)

Surfactant/Cosolvent	$K_d = C_s/C_w$ (ml/g)	
	Kaolin	Glacial Till
Deionized water	—	210
5% Igepal CA-720	1.35	9.86
3% Igepal CA-720	7.14	18.50
5% Triton X-100	1.47	21.92
3% Tween 80	2.85	10.15
40% Ethanol	11.44	20.42
40% Ethanol and 5% Igepal CA-720	2.15	13.31

C_s, concentration of phenanthrene sorbed on soil particle surfaces (μg/g); C_w, total concentration of phenanthrene dissolved in aqueous solution and solubilized into surfactant micelles, if they are present (μg/ml).

extraction followed by gas chromatography/mass spectrometry (GC-MS) chemical analysis. Details of the testing procedure are given in Saichek (2001), and the results are summarized in Table 3.3.

It can be observed from the results that soil composition has a profound influence on the desorption and solubilization behaviors of phenanthrene. It is much more difficult to desorb and solubilize phenanthrene from glacial till than from kaolin, probably due to the strong binding characteristics of phenanthrene to organic matter. The hydrophilicity and concentration of the surfactant/cosolvent affects phenanthrene desorption and solubilization significantly. As hydrophilic surfactants are less prone to sorption onto soil particle surfaces, more surfactant molecules are available for micellar solubilization, resulting in better performance. In addition, more micelles are available for phenanthrene solubilization at higher surfactant concentrations.

It can be concluded that surfactant/cosolvent chemistry and soil properties and composition affect the desorption and solubilization behaviors of PAHs significantly. The realistic performance of the surfactant/cosolvent selected for a particular remediation project should be verified by the results of systematic laboratory experiments using the specific soil and contaminant of the problem in hand.

3.7 BUFFER CAPACITY OF SOIL

The buffer capacity or buffer intensity of a system is defined to be the amount of strong base (strong acid) that when added to the system causes a unit increase (decrease) in pH. It is an indicator of the capacity of a system in resisting the pH change resulting from the addition of strong base or strong acid. Important chemical processes that affect soil pH are carbonic acid dissociation, acid–base reactions of soil humus and aluminum hydroxyl polymers, and mineral weathering reactions (Sposito, 1989). Therefore, the composition of a soil has a significant influence on its buffer capacity. For example, the presence of calcium carbonate or other compounds such as magnesium carbonate or sodium carbonate causes high acid/base

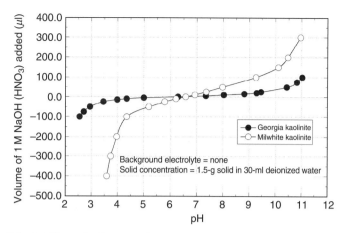

Figure 3.9. Acid/base buffer capacity curves of Georgia and Milwhite kaolinite.

buffer capacity of a natural soil obtained from a contaminated industrial site near Chicago, Illinois (Reddy and Ala, 2005). Most natural soils also contain impurities, such as iron oxides, quartz, titanium oxides, and so on, that increase their acid/base buffer capacity and affinity for heavy metals (Ulrich and Sumner, 1991). Moreover, the buffer capacity is also pH dependent (Sposito, 1989; Yeung, Hsu, and Menon, 1996; Yeung and Hsu, 2005).

The acid/base buffer capacities of Georgia and Milwhite kaolinite as a function of pH are presented in Figure 3.9 as an illustration. Experimental details are given in Yeung, Hsu, and Menon (1996) and Hsu (1997). The volume of 1 M HNO_3(aq) added to soil mixtures of concentration of 1.5 g soil in 30 ml of deionized water is taken to be negative, consistent with the definition of buffer capacity. The slopes of the curves give the buffer capacities of the two kaolinites at different pH levels. It is obvious that Milwhite kaolinite has a much higher acid/base buffer capacity than Georgia kaolinite at all pH levels. It should be noted that the buffer capacity of soil is pH dependent. The buffer capacity of Milwhite kaolinite is very high at low pH, probably due to the presence of Al oxides and Fe oxides in the soil (Ulrich and Sumner, 1991). Therefore, it is extremely difficult to adjust the pH of Milwhite kaolinite to any value lower than 3.5, rendering the removal of some metals very difficult without the use of an appropriate enhancement agent.

3.8 COMPLEXATION

Complexation is the binding of monodentate or multidentate ligands to one or more central ions to form complexes. The central ions are usually metal ions. The ligands, which can be ions or molecules, are called chelants, chelators, chelating agents, or sequestering agent. The complex can be nonionic, cationic, or anionic, depending on the charges of the central ions and the ligands. The central ions and ligands can exist individually and be combined in complexes. Ligands that bind to the central metal ion at only one point, such as H_2O, OH^-, Cl^-, and CN^-, are called monodentate ligands. Ligands that bind at two or more sites are called multdentate ligands or

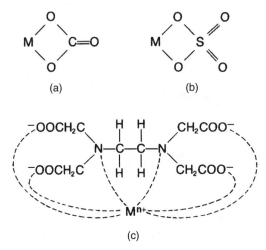

Figure 3.10. Configuration of metal complexes. (a) Carbonate; (b) Sulfate; and (c) EDTA.

chelating agents. For example, chloride forms a monodentate complex with mercuric ion, and carbonate and sulfate can form bidentate complexes with a central metal ion M as shown in Figure 3.10a,b, respectively. The chelating agent ethylenediaminetetraacetate ion EDTA^{4-} can bind a metal ion at six sites, as each of the acetate groups and the two nitrogen atoms have free electron pairs necessary for the formation of coordinate bonds, as shown in Figure 3.10c. The configuration contributes to the high stability of metal–EDTA complexes (Snoeyink and Jenkins, 1980).

The formation of complexes modifies the metal species dissolved in pore fluid and those sorbed on soil particle surfaces interactively. The formation of complexes reduces the concentration of free metal ions in the dissolved phase. As a result, concentration-dependent properties of metals are altered. The effects include (a) modification of solubility, toxicity, and possibility biostimulatory properties of metals; (b) modification of surface properties of solids; and (c) sorption of metals from solution (Snoeyink and Jenkins, 1980). As a result, chelating agents can enhance the electrochemical remediation of subsurface contamination.

The chelating agents used by researchers to enhance electrochemical remediation include EDTA, citric acid, oxalic acid, ammonia, iodide/iodine, potassium iodide solution, sodium chloride solution, 1-hydroxyethane-1,1-diphosphonic acid (HEDPA), HPCD, and so on. EDTA is the most frequently used chelating agent in electrochemical remediation. The complexation chemistry of EDTA is briefly outlined as an illustration.

Let Y denote the ethylenediaminetetraacetate ion EDTA^{4-}. With a metal ion M, it can form a complex MY, a protonated complex MHY, a hydroxo complex MY(OH)$_n$, and a mixed complex MYX, where X is a monodentate ligand. The complexation reactions involved are

$$M^{n+} + Y^{4-} = MY^{n-4} \tag{3.8}$$

$$M^{n+} + H^+ + Y^{4-} = MHY^{n-3} \tag{3.9}$$

$$M^{n+} + OH^- + Y^{4-} = MY(OH)^{n-5} \tag{3.10}$$

and the stability constants of these reactions are defined by

$$K_{MY} = \frac{[MY^{n-4}]}{[M^{n+}][Y^{4-}]} \quad (3.11)$$

$$K_{MHY} = \frac{[MHY^{n-3}]}{[M^{n+}][H^+][Y^{4-}]} \quad (3.12)$$

$$K_{MY(OH)} = \frac{[MY(OH)^{n-5}]}{[M^{n+}][OH^-][Y^{4-}]}, \quad (3.13)$$

where the unit of concentration as denoted by [ ] is mol/l.

In the laboratory-scale experimental study performed by Wong, Hicks, and Probstein (1997), EDTA was added to the catholyte for injection into sandy soil specimens to solubilize precipitated lead and zinc. The resulting complexes were then transported to the anode with metal removal efficiencies approaching 100%. The results of Reddy and Ala (2005) also demonstrate that EDTA is more effective than diethylene triamine pentaacetic acid (DTPA), potassium iodide solution, and HPCD for the simultaneous removal of a variety of metals from an industrial site contaminated with lead, cadmium, mercury, and arsenic on mass basis. During the electrochemical remediation of a manganese-contaminated kaolinite, manganese ions were migrated to the cathode. However, the alkaline front generated by the reduction of water at the cathode promoted the precipitation of manganese hydroxide, thus preventing the removal of manganese ions in the cathode chamber. The introduction of EDTA to form complexes or chelates with Mn^{2+} ions enabled 42% of manganese to reach the cathode. The balance of manganese concentrated in the soil near the cathode (Nogueira *et al.*, 2007).

The complicating factors of using EDTA to enhance electrochemical remediation of clayey soils contaminated by multiple heavy metals are identified by Reddy, Danda, and Saichek (2004). They conducted batch and electrokinetic experiments to investigate the removal of three different heavy metals, chromium (VI), nickel (II), and cadmium (II), from a clayey soil by using EDTA as the complexing agent. The results of batch experiments reveal that 62%–100% removal of these heavy metals is possible by using either a 0.1 or 0.2 M EDTA over the pH range of 2–10. However, the results of electrokinetic experiments using EDTA at the cathode indicate a low heavy metal removal efficiency. Using EDTA at the cathode along with a pH control at the anode using NaOH increases the pH throughout the soil and achieves 95% Cr(VI) removal, but the removal of Ni(II) and Cd(II) was limited due to the precipitation of these metals near the cathode. The low mobility of EDTA and its migration direction opposite to that of the electroosmotic flow may have prohibited the occurrence of EDTA complexation. Moreover, many complicating factors that affect EDTA-enhanced electrochemical remediation have been identified, and further research is necessary to optimize the process to achieve better contaminant removal efficiency.

The results of Suer and Lifvergren (2003) indicate that mercury was removed from a field-contaminated soil by a combination of redox and complexation processes with iodide/iodine and electrokinetic mobilization. Iodide added to the cathode compartment was transported into the soil and oxidized to iodine near the anode. Mercury–iodide complex was mobilized and transported to the anode. After 5 days of treatment, some 50% of the initial mercury contaminant was migrated to the anode compartment, and another 25% was recovered from the soil water near the anode. No volatile mercury was formed. It can be observed that electromigration is the dominant transport mechanism for the charged mercury–iodide complex, as electroosmotic flow of pore fluid would have migrated the mercury toward the cathode. The results of Reddy, Chaparro, and Saichek (2003) indicate that potassium iodide solution is a more effective complexing agent than Na-EDTA for mercury in both kaolin and glacial till undergoing electrochemical remediation. For kaolin, electrochemical remediation enhanced by potassium iodide solution removed approximately 97% of the initial mercury contaminant, that is, 500 mg/kg, leaving a residual concentration of 16 mg/kg of Hg in the soil. For glacial till, only 56% of the initial mercury contaminant was removed, leaving a residual concentration of 220 mg/kg of Hg in the soil. The lower mercury removal from glacial till is attributed to the presence of organic matter, which promotes mercury adsorption or formation of insoluble mercury surface complexes. The effectiveness of iodide/iodine complexing in the remediation of a mercury-contaminated soil from a chlor-alkali factory is also demonstrated experimentally by Hakansson et al. (2008).

The experimental results of Popov et al. (2001) reveal the influence of the chelating agent HEDPA on the electroosmotic flow rate in a natural sod-podzolic soil contaminated with 0.003 g phenol per gram of dry soil. The introduction of a small amount of HEDPA into the system increases the coefficient of electroosmotic conductivity of the soil from 4×10^{-9} to 11.2×10^{-9} m^2/V-s. Up to 80%–95% of the phenol was removed after 30–50 h of treatment, and the average electroosmotic flow rate remained constant throughout the tests.

The effect of ammonia as a complexing agent to enhance the efficiency of electrochemical remediation was evaluated by Vereda-Alonso et al. (2007). Laboratory-scale experiments on copper-spiked kaolin were carried out to explore if metal precipitation could be prohibited at neutral and alkaline pH. Ammonia was added at the anode compartment and migrated into the soil as ammonium by electromigration. Two set of experiments were performed: (a) using ammonia solution as anolyte and sodium nitrate as catholyte, and (b) combining the ammonia enhancement with the addition of acetic acid into the cathode (weak-acid enhancement). The results indicate that metal recovery with only ammonia was less than 40%, while the combination of ammonia and acid enhancements achieved a copper recovery of more than 90%.

The effects of solubility and transport competition among several metals, that is, manganese, iron, copper, and zinc, during their transport through polluted clay were evaluated by Pazos et al. (2008). The results of unenhanced electrochemical remediation indicate a limited removal of the metals because they were retained in the kaolinite specimen by the migration of the alkaline front. Metals exhibited a degree of removal in accordance with the solubility of the corresponding hydroxide and its formation pH. After 7 days of treatment, the removal results were as follows: 75.6% of Mn, 68.5% of Zn, 40.6% of Cu, and 14.8% of Fe. Two different techniques to diminish the negative effects of the basic front generated at the cathode were evaluated: (a) addition of citric acid as complexing agent to the contaminated kaolinite specimen, and (b) use of citric acid to control the pH of the cathode chamber. Both techniques are based on the capability of citric acid to act as a complexing and neutralizing agent. Almost complete removal of Mn, Cu, and Zn was achieved when citric acid was used as the neutralizing or complexing agent. However, only 33% of Fe was removed as it formed a negatively charged complex with citrate that retarded its migration to the cathode.

On the removal of carcinogenic and mutagenic PAHs and heavy metals from the aged contaminated soil obtained from a former manufactured gas plant in Chicago, Illinois, and contaminated sediment of high organic content at Indiana Harbor, Indiana, four enhancement agents, including 3% Tween 80 (surfactant), 5% Igepal CA-720 (surfactant), 20% n-butylamine (cosolvent), and 10% HPCD (cyclodextrin), were used to enhance the solubilization of PAHs and heavy metals in the soil (Reddy and Ala, 2005, 2006; Reddy et al., 2006). Their experimental results demonstrate that Igepal CA-720 and 20% n-butylamine yielded the highest removal efficiency due to partial solubilization of PAHs, causing some PAHs to migrate toward the cathode. However, the removal efficiency of PAHs from the sediment is very low as PAHs may be strongly sorbed to the organic content in the sediment. Removal of heavy metals is negligible due to strong sorption and precipitation of metals.

3.9 OXIDATION–REDUCTION (REDOX) REACTIONS

Oxidation–reduction reactions or redox reactions are defined as a family of reactions in which electrons are transferred between species. The species that receives electrons is reduced and that donates electrons is oxidized. Similar to acid–base reactions, redox reactions are always a matched pair of half-reactions. An oxidation reaction cannot occur without a reduction reaction occurring simultaneously.

Similar to the concept of pH on the overall acidity/alkalinity of a system, reduction potential (also known as redox potential, oxidation/reduction potential or ORP) is an intensity parameter of the overall redox reaction potential of a system. It measures the tendency of a system to either gain or lose electrons when it is subject to change by the introduction of a new species. A system with a higher (more positive) reduction potential than the new species will have a tendency to gain electrons from the new species, that is, to be reduced by oxidizing the new species. A system with a lower (more negative) reduction potential will have a tendency to lose electrons to the new species, that is, to be oxidized by reducing the new species. Similar to the fact that transfer of hydrogen ions between chemical species determines the pH of a system, the transfer of electrons between chemical species determines the reduction potential of a system. However, it does not characterize the capacity of the system for oxidation or reduction, similar to the fact that pH does not characterize the buffer capacity of a system.

Oxidation–reduction reactions play a very important role in geochemical processes, particularly in the remediation processes of subsurface contamination. The reactions determine the mobility of many inorganic contaminants, as well as of many biologically important elements such as nitrogen and sulfur in the subsurface, and the toxicity of some metallic contaminants (Vance, 1996). For example, chromium can exist in the subsurface in valence states ranging from –2 to +6, with the trivalent Cr(III) and hexavalent Cr(VI) states being the most stable in the surface and subsurface environments. Cr(III) can exist in the form of cation Cr^{3+} or hydroxo complexes such as $CrOH^{2+}$, $Cr(OH)_3$, and $Cr(OH)_4^-$. The cationic Cr(III) species such as Cr^{3+} and $CrOH^{2+}$ are less mobile in the subsurface as they are sorbed onto negatively charged soil particle surfaces and have limited solubility in pore fluid. Cr(VI) can exist as oxyacids or anionic species such as hydrochromate $HCrO_4^-$, dichromate $Cr_2O_7^{2-}$, and chromate CrO_4^{2-}. These anionic Cr(VI) species are quite soluble in pore fluid, and are thus mobile in the subsurface. The Cr(VI) species are much more toxic than the Cr(III) species because they are carcinogenic, irritative, and corrosive. Therefore, there exists a greater health hazard if existing Cr(III) is oxidized into Cr(VI) under certain environmental conditions in the subsurface.

During electrochemical remediation of chromium-contaminated soils, anionic Cr(VI) species are migrated toward the anode, while cationic Cr(III) species are migrated toward the cathode. Therefore, the efficiency of electrochemical remediation of chromium-contaminated soils depends on the presence of naturally occurring oxidizing or reducing agents. Naturally occurring reducing agents in the subsurface may include organic matter, ferrous iron, and sulfides, and naturally occurring reducing agents may include manganese (Chinthamreddy and Reddy, 1999; Reddy and Chinthamreddy, 1999). The experimental results of Pamukcu, Weeks, and Wittle (2004) demonstrate the viability of injecting Fe(II) into a Cr(VI)-contaminated kaolinite clay by electrokinetics to reduce Cr(VI) to the less toxic and less mobile Cr(III).

In the literature, reduction potential is often reported as Eh, that is, the potential generated between a platinum electrode and a standard hydrogen electrode when placed in a system. Eh–pH or Pourbaix diagrams depict the thermodynamically form of an element as a function of reduction potential and pH. They are commonly used in mining and geochemistry for the assessment of the stability conditions of minerals and dissolved species. For example, the Eh–pH diagram for chromium species assuming the activity of dissolved Cr to be 10^{-6} is shown in Figure 3.11. It can be observed that much of the diagram is occupied by insoluble Cr_2O_3. However, the species dissolves to form $CrOH^{2+}$ when the pH is slightly lower than 5, and to form CrO_2^- when the pH is higher than 13.5. Cr(III) oxidizes to form Cr(VI) as $HCrO_4^-$ and CrO_4^{2-} ions at high Eh. From an environmental risk assessment standpoint, it should be noted that Cr(VI) species occupy fairly large Eh–pH fields. Various chromates, such as crocoite, and so on, are known to exist in nature. However, Cr_2O_3 is rare in nature as most Cr(III) exist in chromites or other chromium spinels. The thermodynamic data for chromium used to develop the Eh–pH diagram shown in Figure 3.11 are given in Brookins (1988).

Eh–pH diagrams can be very useful in geochemistry. However, it should also be noted that the measurement of redox potential in natural soil–water systems can be difficult, as the platinum electrode is not responsive to many reactions involving solid phases, or to some of the soluble-phase redox couples common in groundwater systems such as O_2–H_2O, SO_4^{2-}–H_2S, CO_2–CH_4, NO_3^-–N_2, and N_2–NH_4^+. When there is more than one redox couple in the system, the redox potential cannot be accurately measured unless the reactions are at thermodynamic equilibrium. However, most environmental processes are in thermodynamic disequilibrium due to (a) flowing groundwater condition, (b) biological activity, (c) redox reactions of light bioactive elements (C, H, O, and S) involving the breaking of covalent bonds—a very slow chemical process, and (d) electrochemical reactions of exposed active mineral surfaces (Vance, 1996). Therefore, extreme caution has to be exercised

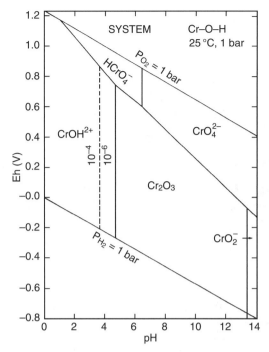

Figure 3.11. Eh–pH diagram for part of the system Cr–O–H, assuming activity of dissolved Cr = 10^{-6} (after Brookins, 1988).

when the measured redox potential is used to design an *in situ* electrochemical remediation process and to predict the possible outcome of the treatment process.

3.10 INTERACTIONS OF GEOCHEMICAL PROCESSES

The geochemical processes discussed in previous sections are interrelated and interdependent. If they are properly understood and utilized, they can enhance the efficiency of electrochemical remediation. Examples are given to demonstrate the importance of these interactions of geochemical processes in electrochemical remediation of contaminated fine-grained soils.

As illustrated in Figures 3.6 and 3.8, the addition of EDTA, a chelating agent, can change the electrokinetic properties of soil and the desorption characteristics of the contaminant from soil particle surfaces. The results of Yeung and Hsu (2005) demonstrate that by using an appropriate enhancement agent to modify and control the electrokinetic properties of the contaminated soil so as to change the direction of electroosmotic flow, the desorption characteristics of the contaminant from the soil particle surfaces to solubilize the contaminant, and the chemical state of the contaminant and the charge of the solubilized contaminant, the contaminant can be removed effectively from the soil.

Another innovative application of the interactions of geochemical processes during electrochemical remediation of soils is demonstrated by Faulkner, Hopkinson,

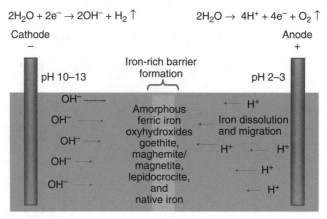

Figure 3.12. Principle of iron-rich barrier generation by electrokinetics (after Faulkner, Hopkinson, and Cundy, 2005).

and Cundy (2005). They have successfully generated subsurface barriers of continuous iron-rich precipitates *in situ* by electrokinetics in their laboratory-scale experiments. The barrier is generated using sacrificial iron electrodes installed on either side of a soil/sediment mass, as shown in Figure 3.12. The applied DC electric field dissolves the sacrificial anode and injects the iron ions into the system, and the reprecipitation of the iron ions in an alkaline environment forms the barrier. Continuous vertical and horizontal iron-rich bands up to 2-cm thick have been generated by applying an electrical voltage of less than 5 V over a period of 300–500 h, with an electrode separation of between 15 and 30 cm. The iron-rich barrier is composed of amorphous iron, goethite, lepidocrocite, maghemite, and native iron. The barrier may function as an impervious barrier to contaminant transport or as a reactive barrier to degrade contaminants by a Fenton process (Kim, Kim, and Han, 2005; Kim *et al.*, 2006). By monitoring the DC electric current intensity passing through the barrier, the integrity of the iron-rich band may be assessed. Moreover, any defects in the barrier during operation may be repaired by the continuing application of a DC electric current.

3.11 SUMMARY

During electrochemical remediation of contaminated soils, there are many geochemical processes associated with the remediation process. If they are properly understood and utilized, they can be used to enhance the remediation process. Otherwise, they may become obstacles. It is impossible to discuss all the complex geochemical processes in detail in a book chapter. Therefore, the most prominent geochemical processes during electrochemical remediation of contaminated soils, including (a) generation of pH gradient, (b) change of zeta potential of soil particle surfaces, (c) change in direction of electroosmotic flow, (d) sorption and desorption of contaminants onto/from soil particle surfaces, (e) buffer capacity of soil, (f) complexation, (g) oxidation–reduction (redox) reactions, and (h) interactions of these processes, are presented in this chapter to facilitate further elaboration of the processes and applications in other chapters of this book.

The importance of these geochemical processes cannot be overemphasized. More importantly, they are contaminant specific, soil specific, dynamic, reversible, and pH dependent. Therefore, it may be dangerous to generalize the implications of these processes and to extrapolate previous experimental results to other cases without a detailed study. However, the complex nature of these processes may lead to new avenues for innovative applications of electrokinetics in geotechnical and geoenvironmental engineering, in particular electrochemical remediation of contaminated soils.

REFERENCES

Alshawabkeh AN, Acar YB. (1996). Electrokinetic remediation. 2. Theoretical model. *ASCE Journal of Geotechnical Engineering* **122**(3):186–196.

Antropov L. (1972). *Theoretical Electrochemistry*. Moscow, Russia: Mir.

Azzam R, Oey W. (2001). The utilization of electrokinetics in geotechnical and environmental engineering. *Transport in Porous Media* **42**(3):293–314.

Bethke CM. (2008). *Geochemical and Biogeochemical Reaction Modeling*, 2nd ed. Cambridge, UK: Cambridge University Press.

Brookins DG. (1988). *Eh-pH Diagrams for Geochemistry*. Berlin, Germany: Springer-Verlag.

Chang JH, Cheng SF. (2007). The operation characteristics and electrochemical reactions of a specific circulation-enhanced electrokinetics. *Journal of Hazardous Materials* **141**(1): 168–175.

Chinthamreddy S, Reddy KR. (1999). Oxidation and mobility of trivalent chromium in manganese-enriched clays during electrokinetic remediation. *Journal of Soil Contamination* **8**(3):197–216.

Crow DR. (1979). *Principles and Applications of Electrochemistry*, 2nd ed. London: Chapman & Hall.

Donnan FG. (1924). The theory of membrane equilibria. *ACS Chemical Reviews* **1**(1): 73–90.

Dzenitis JM. (1997). Soil chemistry effects and flow prediction in electroremediation of soil. *Environmental Science & Technology* **31**(4):1191–1197.

Evans LJ. (1989). Chemistry of metal retention by soils. *Environmental Science & Technology* **23**(9):1046–1056.

Eykholt GR, Daniel DE. (1994). Impact of system chemistry on electro-osmosis in contaminated soil. *ASCE Journal of Geotechnical Engineering* **120**(5):797–815.

Farrell J, Grassian D, Jones, M. (1999). Investigation of mechanisms contributing to slow desorption of hydrophobic organic compounds from mineral solids. *Environmental Science & Technology* **33**(8):1237–1243.

Faulkner DWS, Hopkinson L, Cundy AB. (2005). Electrokinetic generation of reactive iron-rich barriers in wet sediments: Implications for contaminated land management. *Mineralogical Magazine* **69**(5):749–757.

Fritz SJ, Marine IW. (1983). Experimental support for a predictive osmotic model of clay membranes. *Geochimica et Cosmochimica Acta* **47**:1515–1522.

Gillette JS, Luthy RG, Clemett SJ, Zare RN. (1999). Direct observation of polycyclic aromatic hydrocarbons on geosorbents at the subparticle scale. *Environmental Science & Technology* **33**(8):1185–1192.

Gray DH, Mitchell JK. (1967). Fundamental aspects of electro-osmosis in soils. *ASCE Journal of the Soil Mechanics and Foundations Division* **93**(SM6):209–236.

Hakansson T, Suer P, Mattiasson B, Allard B. (2008). Sulphate reducing bacteria to precipitate mercury after electrokinetic soil remediation. *International Journal of Environmental Science & Technology* **5**(2):267–274.

Hanshaw BB. (1972). Natural-membrane phenomena and subsurface waste emplacement. In *Underground Waste Management and Environmental Implications. Memoir 18* (ed. TD Cook). Tulsa, OK: American Association of Petroleum Geologists, pp. 308–317.

Hicks RE, Tondorf S. (1994). Electrorestoration of metal contaminated soils. *Environmental Science & Technology* **28**(12):2203–2210.

Hsu C. (1997). Electrokinetic Remediation of Heavy Metal Contaminated Soils. PhD Dissertation, Texas A&M University, College Station, TX.

Keijzer ThJS, Kleingeld PJ, Loch JPG. (1999). Chemical osmosis in compacted clayey material and the prediction of water transport. *Engineering Geology* **53**(2):151–159.

Khodadoust AP, Reddy KR, Narla O. (2006). Cyclodextrin-enhanced electrokinetic remediation of soils contaminated with 2,4-dinitrotoluene. *ASCE Journal of Environmental Engineering* **132**(9):1043–1050.

Kim JH, Han SJ, Kim SS, Yang JW. (2006). Effect of soil chemical properties on the remediation of phenanthrene-contaminated soil by electrokinetic-Fenton process. *Chemosphere* **63**(10):1667–1676.

Kim SS, Kim JH, Han SJ. (2005). Application of the electrokinetic-Fenton process for the remediation of kaolinite contaminated with phenanthrene. *Journal of Hazardous Materials* **118**(1–3):121–131.

Ko SO, Schlautman MA, Carraway ER. (2000). Cyclodextrin-enhanced electrokinetic removal of phenanthrene from a model clay soil. *Environmental Science & Technology* **34**(8):1535–1541.

Lee HH, Yang JW. (2000). A new method to control electrolytes pH by circulation system in electrokinetic soil remediation. *Journal of Hazardous Materials* **77**(1–3):227–240.

Leinz RW, Hoover DB, Meier AL. (1998). NEOCHIM: An electrochemical method for environmental application. *Journal of Geochemical Exploration* **64**(1–3):421–434.

Li A, Liu XY. (2005). Combined effects of aging and cosolvents on sequestration of phenanthrene in soils. *ASCE Journal of Environmental Engineering* **131**(7):1068–1072.

Li Z, Yu J-W, Neretnieks I. (1998). Electroremediation: Removal of heavy metals from soils by using cation selective membrane. *Environmental Science & Technology* **32**(3):394–397.

Loehr RC, Webster MT. (1996). Behavior of fresh vs. aged chemicals in soil. *Journal of Soil Contamination* **5**(4):361–383.

Luthy RG, Aiken GR, Brusseau ML, Cunningham SD, Gschwend PM, Pignatello PM, Reinhard M, Traina SJ, Weber WJ Jr., Westall JC. (1997). Sequestration of hydrophobic organic contaminants by geosorbents. *Environmental Science & Technology* **31**(12):3341–3347.

Manna M, Sanjay K, Shekhar R. (2003). Electrochemical cleaning of soil contaminated with a dichromate lixiviant. *International Journal of Mineral Processing* **72**(1–4):401–406.

Marine IW, Fritz SJ. (1981). Osmotic model to explain anomalous hydraulic heads. *Water Resources Research* **17**:73–82.

Mattson ED, Bowman RS, Lindgren ER. (2000). Electrokinetic remediation using surfactant-coated ceramic casings. *ASCE Journal of Environmental Engineering* **126**(6):534–540.

McBride MB. (1994). *Environmental Chemistry of Soils*. New York: Oxford University Press.

Mukhopadhyay B, Sundquist J, Schmitz RJ. (2007). Removal of Cr(VI) from Cr-contaminated groundwater through electrochemical addition of Fe(II). *Journal of Environmental Management* **82**(1):66–76.

Myers D. (1988). *Surfactant Science and Technology*. New York: VCH Publishers.

Nogueira MG, Pazos M, Sanromán MA, Cameselle C. (2007). Improving on electrokinetic remediation in spiked Mn kaolinite by addition of complexing agents. *Electrochimica Acta* **52**(10):3349–3354.

van Olphen H, Fripiat JJ. (eds.) (1979). Data summaries—Clay Minerals Society. In *Data Handbook for Clay Materials and Other Nonmetallic Minerals*. New York: Pergamon Press, pp. 13–37.

Ottosen LM, Pedersen AJ, Ribeiro AB, Hansen HK. (2005). Case study on the strategy and application of enhancement solutions to improve remediation of soils contaminated with Cu, Pb and Zn by means of electrodialysis. *Engineering Geology* **77**(3–4):317–329.

Pamukcu S, Weeks A, Wittle JK. (2004). Enhanced reduction of Cr(VI) by direct electric current in a contaminated clay. *Environmental Science & Technology* **38**(4):1236–1241.

Pazos M, Gouveia S, Sanroman MA, Cameselle C. (2008). Electromigration of Mn, Fe, Cu and Zn with citric acid in contaminated clay. *Journal of Environmental Science and Health Part A. - Toxic/Hazardous Substances & Environmental Engineering* **3**(8):823–831.

Pazos M, Ricart MT, Sanroman MA, Cameselle C. (2007). Enhanced electrokinetic remediation of polluted kaolinite with an azo dye. *Electrochimica Acta* **52**(10):3393–3398.

Popov K, Kolosov A, Yachmenev VG, Shabanova N, Artemyeva A, Frid A, Kogut B, Vesnovskii S, Sukharenko V. (2001). A laboratory-scale study of applied voltage and chelating agent on the electrokinetic separation of phenol from soil. *Separation Science and Technology* **36**(13):2971–2982.

Reddy KR, Ala PR. (2005). Electrokinetic remediation of metal-contaminated field soil. *Separation Science and Technology* **40**(8):1701–1720.

Reddy KR, Ala PR. (2006). Electrokinetic remediation of contaminated dredged sediment. In *Contaminated Sediments: Evaluation and Remediation Techniques*, ASTM Special Technical Publication 1482 (ed. M Fukue). West Conshohocken, PA: ASTM, pp. 254–267.

Reddy KR, Ala PR, Sharma S, Kumar SN. (2006). Enhanced electrokinetic remediation of contaminated manufactured gas plant soil. *Engineering Geology* **85**(1–2):132–146.

Reddy KR, Chaparro C, Saichek RE. (2003). Removal of mercury from clayey soils using electrokinetics. *Journal of Environmental Science and Health Part A. Toxic/Hazardous Substances & Environmental Engineering* **38**(2):307–338.

Reddy KR, Chinthamreddy S. (1999). Electrokinetic remediation of heavy metal-contaminated soils under reducing environments. *Waste Management* **19**(4):269–282.

Reddy KR, Danda S, Saichek RE. (2004). Complicating factors of using ethylenediamine tetraacetic acid to enhance electrokinetic remediation of multiple heavy metals in clayey soils. *ASCE Journal of Environmental Engineering* **130**(11):1357–1366.

Rødsand T, Acar YB, Breedveld G. (1995). Electrokinetic extraction of lead from spiked Norwegian marine clay. In *Characterization, Containment, Remediation, and Performance in Environmental Geotechnics*, Geotechnical Special Publication No. 46, Vol. **2** (eds. YB Acar, DE Daniel). New York: ASCE, pp. 1518–1534.

Saichek RE. (2001). Electrokinetically Enhanced In-Situ Flushing for HOC-Contaminated Soils. PhD Dissertation, University of Illinois at Chicago.

Saichek RE, Reddy KR. (2003). Effect of pH control at the anode for the electrokinetic removal of phenanthrene from kaolin soil. *Chemosphere* **51**(4):273–287.

Saichek RE, Reddy KR. (2004). Evaluation of surfactants/cosolvents for desorption/solubilization of phenanthrene in clayey soils. *International Journal of Environmental Studies* **61**(5):587–604.

Silbey RJ, Alberty RA, Bawendi MG. (2005). *Physical Chemistry*, 4th ed. Hoboken, NJ: John Wiley & Sons.

Snoeyink VL, Jenkins D. (1980). *Water Chemistry*. New York: John Wiley & Sons.

Sposito G. (1989). *The Chemistry of Soils*. New York: Oxford University Press.

Suer P, Lifvergren T. (2003). Mercury-contaminated soil remediation by iodide and electroreclamation. *ASCE Journal of Environmental Engineering* **129**(5):441–446.

Szpyrkowicz L, Radaelli M, Bertini S, Daniele S, Casarin F. (2007). Simultaneous removal of metals and organic compounds from a heavily polluted soil. *Electrochimica Acta* **52**(10):3386–3392.

Ulrich B, Sumner ME. (eds.) (1991). *Soil Acidity*. Berlin, Germany: Springer-Verlag.

Vance DB. (1996). Redox reactions in remediation. *Environmental Technology* **6**(4):24–25.

Vereda-Alonso C, Heras-Lois C, Gomez-Lahoz C, Garcia-Herruzo F, Rodriguez-Maroto JM. (2007). Ammonia enhanced two-dimensional electrokinetic remediation of copper spiked kaolin. *Electrochimica Acta* **52**(10):3366–3371.

Wong JSH, Hicks RE, Probstein RF. (1997). EDTA-enhanced electroremediation of metal-contaminated soils. *Journal of Hazardous Materials* **55**(1–3):61–79.

Woo SH, Park JM. (1997). Estimation of oxygen transfer in soil slurry bioreactor. *Biotechnology Techniques* **11**(10):713–716.

Yeung AT. (1990). Coupled flow equations for water, electricity, and ionic contaminants through clayey soils under hydraulic, electrical and chemical gradients. *Journal of Non Equilibrium Thermodynamics* **15**(3):247–267.

Yeung AT. (1992). Diffuse double-layer equations in SI units. *ASCE Journal of Geotechnical Engineering* **118**(12):2000–2005.

Yeung AT. (1994). Electrokinetic flow processes in porous media and their applications. In *Advances in Porous Media*, Vol. 2 (ed. MY Corapcioglu). Amsterdam, the Netherlands: Elsevier, pp. 309–395.

Yeung AT. (2006a). Contaminant extractability by electrokinetics. *Environmental Engineering Science* **23**(1):202–224.

Yeung AT. (2006b). Fundamental aspects of prolonged electrokinetic flows in kaolinites. *Geomechanics and Geoengineering: An International Journal* **1**(1):13–25.

Yeung AT. (2008). Electrokinetics for soil remediation. In *Environmental Geotechnology and Global Sustainable Development 2008* (eds. AT Yeung, IMC Lo). Hong Kong: Advanced Technovation, pp. 16–25.

Yeung AT, Datla S. (1996). Fundamental formulation of electrokinetic extraction of contaminants from soil: Reply. *Canadian Geotechnical Journal* **33**(4):682–684.

Yeung AT, Hsu C. (2005). Electrokinetic remediation of cadmium-contaminated clay. *ASCE Journal of Environmental Engineering* **131**(2):298–304.

Yeung AT, Hsu C, Menon RM. (1996). EDTA-enhanced electrokinetic extraction of lead. *ASCE Journal of Geotechnical Engineering* **122**(8):666–673.

Yeung AT, Scott TB, Gopinath S, Menon RM, Hsu C. (1997). Design, fabrication, and assembly of an apparatus for electrokinetic remediation studies. *ASTM Geotechnical Testing Journal* **20**(2):199–210.

PART II

REMEDIATION OF HEAVY METALS AND OTHER INORGANIC POLLUTANTS

4

ELECTROKINETIC REMOVAL OF HEAVY METALS

LISBETH M. OTTOSEN, HENRIK K. HANSEN, AND PERNILLE E. JENSEN

4.1 INTRODUCTION

The movement of heavy metals in soil in an applied electric field was first reported in 1980 (Segall *et al.*, 1980), by researchers who attempted to dewater a dredged material disposal site and its embankment foundations in order to increase the capacity of the disposal site. In the liquid samples from the process, they found various heavy metals, and even though this was only noted briefly in the actual work, the finding may have inspired other researchers to use an applied electric field for soil remediation because in the end of the 1980s different teams started to develop electrochemical remediation methods independently of each other. The majority of the published research conducted in the late 1980s and early 1990s originated from research teams at Geokinetics (Lageman, 1989), Louisiana State University, (Acar *et al.*, 1990), Lehigh University (Pamukcu, Khan, and Fang, 1990), and Massachusetts Institute of Technology (Probstein and Renaud, 1987).

Many studies were made of electrokinetic (EK) removal of heavy metals from spiked model soils—mainly kaolinite (e.g. Hamed, Acar, and Gale, 1991; Acar *et al.*, 1994; Kim, Moon, and Kim, 2001; Reddy, Chinthamreddy, and Al-Hamdan, 2001). In general, very high removal efficiency was obtained with the spiked soils, and the duration of the treatment is short. These works provide the basic insight into the electrochemical processes responsible for remediation; however, the remediation results are generally not transferable to industrially and aged polluted soils (Cox, Shoesmith, and Ghosh, 1996; Ottosen, Lepkova, and Kubal, 2006; Jensen, Ottosen, and Harmon, 2007c). To successfully remove heavy metals from soil by an electrochemical method, it is necessary to desorb/dissolve the heavy metals during the process, and this way make them available for transport by either electromigration or in an electroosmotic flow (Ribeiro and Mexia, 1997). In spiked/model soils,

Electrochemical Remediation Technologies for Polluted Soils, Sediments and Groundwater,
Edited by Krishna R. Reddy and Claudio Cameselle
Copyright © 2009 John Wiley & Sons, Inc.

the heavy metals are adsorbed less strongly than in industrially polluted soils, and thus the remediation results of such may be misleading.

In this chapter, we provide a qualitative overview of the EK soil remediation experiences reported from laboratory and field with industrially polluted soils regarding the most common heavy metals and soil types. Based on the results reported in the literature, a general guideline of suitable remediation conditions and possible enhancements is given. Lastly, a short overview of published works on EK remediation (EKR) removal of heavy metals from waste materials other than soil is given, highlighting the similarities and differences among material characteristics that influence remediation.

4.2 PRINCIPLE OF EK REMOVAL OF HEAVY METALS FROM SOILS

When an electric field is applied to a moist porous material such as soil, the electric current is carried by ions in the pore solution. Electromigration is the term for such movement of individual ions in a solution induced by an applied electric direct current (DC) field. The direction of the ionic electromigration is toward the corresponding electrode. Anions will move toward the anode and cations will move toward the cathode. Good control of the flow direction for the electromigrating ions can thus be achieved because the ions move along the field lines that are defined by the electrode placement. If the polluting heavy metals in the soil are present as ions or ionic species in the pore solution, these will electromigrate toward the electrodes, where they concentrate. However, most often, the heavy metals from industrial pollution do not enter the soil in the form of heir soluble salts but rather in the form of some more or less soluble chemical substance. Furthermore, aging of the pollution imply that the heavy metals adsorbed to the soil particles or precipitated in the form of various mineral, thus they must be desorbed/dissolved before they can be removed in the applied electric field. The desorption/dissolution process is often the rate-limiting step for remediation when dealing with industrially polluted soils, and this is the major reason why work with spiked model soils cannot be compared directly with soil remediation.

Electromigration is not dependent on the pore size so the process can be used in both coarse- and fine-grained soils. If the heavy metal species present in the pore solution are uncharged, they cannot be removed by electromigration. In such case, electroosmotic flushing can be utilized in fine-grained soils. Electromigration is, however, demonstrated to be the major transport mechanism for charged species under electrical fields (Acar and Alshawabkeh, 1993).

Several investigators have tried different experimental setups in order to perform the best controlled heavy metal removal. Figure 4.1 shows the generalized EK soil remediation setup in principle. Typically, in the laboratory, the soil is placed in the middle of the cell either in cylindrical (e.g. Hansen et al., 1997; Ottosen et al., 1997; Kim, Kim, and Kim, 2005b; Ricart et al., 2005) or in rectangular (e.g. Maini et al., 2000; Kim, Kim, and Kim, 2005b) cells. These cells are either closed or open (Eid et al., 2000; Suer and Lifvergren, 2003) systems. The configuration of the electrode chambers could include physical (and passive) separators toward the soil compartment, allowing the transport of liquid and ions in both directions; or the separators could have direct influence on the transport processes such as ion exchange membranes (Hansen et al., 1997; Ottosen et al., 1997; Lara, Rodriguez-Postigo, and

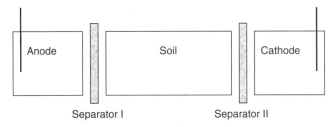

Figure 4.1. Typical laboratory test setup.

Garcia-Herruzo, 2005; Wieczorek *et al.*, 2005; Amrate, Akretche, and Innocent, 2006; Castillo, Soriano, and Delgado, 2008). When using ion exchange membranes, separator I is an anion exchange membrane and separator II is a cation exchange membrane. This setup is often known as electrodialytic remediation (Hansen *et al.*, 1997; Ottosen *et al.*, 1997).

Most investigators agree on the necessity of avoiding the development of an alkaline front from the cathode compartment into the soil since most heavy metals precipitate/adsorb to the soil in the alkaline environment, and the result is that the remediation process ceases, as shown in many of the early works on EK soil remediation (Kim and Kim, 2001). Most commonly, pH control maintains acidic conditions in the cathode compartment, and thus efficiently prevents the alkaline front from developing into the soil (Hicks and Tondorf, 1994; Wieczorek *et al.*, 2005). Implementation of ion exchange membranes as mentioned above is another way to prevent the alkaline front (Hansen *et al.*, 1999). In laboratory experiments, the soil is commonly homogenized and the water is saturated. In full-scale experiments, the situation is different: Here the electrode compartments are placed directly into the inhomogeneous soil, whose humidity is decided by the natural precipitation, as these systems are open and most often *in situ*. This could be one major reason why the scale up from laboratory cell experiments to full-scale remediation is so difficult (Hansen *et al.*, 1997; Ottosen *et al.*, 1997).

The acidic front developing from the anode into the soil most commonly aid the mobilization of heavy metals, as most heavy metal cations are dissolved in acidic conditions (Alloway, 1995). Acid front development, as well as heavy metal desorption/dissolution, depends on many factors and the extent to which the factors have influence on the remediation action are dependent on the soil type and the heavy metal itself. The pH and redox conditions in the soil are both important factors affecting heavy metal retention, and a change of these parameters may be beneficial to the remediation. Soil pH affects both the adsorption of heavy metals in exchange sites, the specific adsorption of heavy metals, as well as many dissolution processes. Some heavy metals are removed at higher pH (i.e. slightly acidic) than others. The order of removal of different heavy metals in the acidic front has been reported as follows: Ni ≈ Zn > Cu > Cr in a soil polluted from a chlor-alkali factory (Suer, Gitye, and Allard, 2003) and Zn > Cu > Pb or Cd > Zn > Cu > Pb > Ni Cr in different industrial polluted soils (Ottosen *et al.*, 2001a; Jensen, 2005).

4.3 HEAVY METAL AND SOIL TYPE

Table 4.1 outlines the experimental results obtained with the typical setup (Fig. 4.1), where the soil is acidified from the anode side and the heavy metals are mobilized

TABLE 4.1. Heavy Metal Removal by Electric Fields. Experiences and Results of Selected Studies for Conventional Nonenhanced Treatment

Heavy Metal	Soil Type	Origin of Pollution	Full (F)/Pilot (P)/Laboratory (L)	Current density (mA/cm^2)/Voltage (V/cm)/Duration (days)	pH Initial/Final	Concentration Initial/Final (mg/kg)	Removed (%)	Reference
As	Soil (56 g Ca/kg)	Gasworks	P (46.7 kg dry weight (dw))	0.128–0.32/—/112	8/2.6	138/158	0	Maini et al. (2000)
	Heavy clay soil	Timber plant	F (191 m^3)	unknown	unknown	250/30	88	Clarke et al. (1996)
	Tailing soil	Gold mine	L	1.4/<20/6	3.5/4.8	3210/2700	16	Ricart et al. (2005)
	Weathered top soil	Wood preservation	L	0.2/<1.7/42	5.8/4.0	900/580	38	Ottosen et al. (2000)
	Loamy sand	Wood preservation	L	0.2/<1.5/35	7.2/3.5	770/380	51	Ribeiro (1994)
	Loamy sand	Wood preservation	L	0.2/—/125	7.0/3.5	770/290	62	Ribeiro et al. (1998)
	Loamy sand	Wood preservation	L	0.4/—/16	6.0/5.5	1900/590	69	Ribeiro (1994)
Cd	Unknown	Electroplating and metal finishing	L	0.5/—/48	unknown/unknown	140/140	0	Gent et al. (2004)
	Tailing soil (<180 μm)	Mining area	L	1.2/—/5	7.5/3.5	179/157	12	Kim and Kim (2001)
	Silt	Military area	L	3–4/3/25	—	55/40	27	Giannis and Gidarakos (2005)
	Unknown	Disposal for electroplating and metal finishing	L	0.5/—/~50	unknown/unknown	145/75	48	Gent et al. (2004)
	Sludge	Municipal wastewater treatment	P (0.06 m^3)	1/—/5	8/1.5	6.8/1.7	75	Kim et al. (2005)
	Agrillaceous sand	Temporary landfill	F (5505 m^3)	unknown	unknown	>180/<40	>78	Clarke et al. (1995)
	Clay	Military airbase	F (1925 m^3)	unknown	unknown	660/47	93	Clarke et al. (1995)

Cr	<1 mm	Filling material from chromate production	L	0.025–0.015/1.2 V/46	—	25.3/24.5	3	Weng, Takiyama, and Huang (1994)
	Clay ~19%, silt ~64%	Electroplating	L	—/2.3/59	7.1/unknown	890/744	16	Wieczorek et al. (2005)
	Clay	Military airbase	F (1925 m^3)	unknown	unknown	7300/755	90	Clarke et al. (1996)
	Loamy sand	Wood preservation	L	0.2/—/85	7.0/3.0	880/700	20	Ribeiro (1994)
	Loamy sand	Wood preservation	L	0.4/—/16	6.0/3.8	550/380	31	Ribeiro (1994)
	Unknown	Electroplating and metal finishing	L	0.5/—/~50	unknown	900/600	33	Gent et al. (2004)
	Sludge	Municipal wastewater treatment	P (0.06 m^3)	1/—/5	8/1.5	116/75	35	Kim et al. (2005)
	Clay ~67%, silt ~13%	Electroplating	L	—/2.3/59	5.5/unknown	1464/589	60	Wieczorek et al. (2005)
	Sand (0.85–0.42 mm)	Spill of chemicals	L	0.015/0.9–1.2/9	unknown	14.2/5.1	64	Bibler, Osteen, and Meaker (1993)

TABLE 4.1. (Continued)

Heavy Metal	Soil Type	Origin of Pollution	Full (F)/Pilot (P)/Laboratory (L)	Current density (mA/cm²)/Voltage (V/cm)/Duration (days)	pH Initial/Final	Concentration Initial/Final (mg/kg)	Removed (%)	Reference
Cu	Clay ~19%, silt ~64%	Electroplating	L	—/2.3/59	7.1/unknown	860/916	0	Wieczorek et al. (2005)
	Peat/clay soil	Paint factory	F (229 m³)	unknown	unknown	1220/200	84	Clarke et al. (1996)
	Clay	Military airbase	F (1925 m³)	unknown	unknown	770/98	87	Clarke et al. (1996)
	Soil (56 g Ca/kg)	Gasworks	P (47 kg dw)	0.128–0.32/—/112	8/2.6	434/443	0	Maini et al. (2000)
	Tailing soil (<180 µm)	Mining area	L	1.2/—/5	7.5/3.5	210/180	15	Kim and Kim (2001)
	Clay ~67%, silt ~13%	Electroplating	L	—/2.3/59	5.5/unknown	171/124	27	Wieczorek et al. (2005)
	Sludge	Municipal wastewater treatment	P (0.06 m³)	1/—/5	8/1.5	340/240	28	Kim, Kim, and Kim (2005a)
	Soil	Accumulator renewal	L	0.1/—/11	unknown	480/270	44	Jensen, Kubes, and Kubal (1994)
	Fine sand (carbonate 4.9%, organic matter 4.4%)	Unknown	L	0.2/<3.8/40	7.3/3.6	490/200	49	Ottosen et al. (2001b)
	Sand (carbonate content 11%)	Cable production	L	0.5/—/26	7.1/3.4	20,000/4000	80	Ottosen et al. (1998)
	Loamy sand	Wood preservation	L	0.2/—/85	7.0/3.0	310/60	81	Ribeiro (1994)
	Loamy sand	Wood preservation	L	0.4/—/16	6.0/3.8	2500/420	83	Ribeiro (1994)
	Weathered top soil	Wood preservation	L	0.2/<1.7/42	5.8/3.0	830/40	95	Ottosen et al. (2000)
	Loamy sand	Wood preservation	L	0.2/<2.7/70	5.8/3.0	1360/20	99	Ottosen et al. (1997)

Metal	Soil type	Source	Setup	Current/Voltage	pH i/f	Conc. i/f	%	Reference
Hg	Sand	Accidental spill with elemental Hg	L	0.07/<10V/54	8.7/3.5	1920/1920	0	Thoming, Kliem, and Ottosen (2000)
Ni	Silty sand (+organic matter)	Unknown	L	0.11/—/22	7.5/unknown	97/78	20	Mohamed (1996)
	Clay	Military air base	F (1925 m³)	unknown	unknown	860/80	91	Clarke et al. (1996)
	Sandy silt	Unknown	L	0.11/—/64	7.4/unknown	570/460	20	Mohamed (1996)
	Clay ~19%, silt ~64%	Electroplating	L	—/2.3/59	7.1/unknown	1251/969	23	Wieczorek et al. (2005)
	Clay ~67%, silt ~13%	Electroplating	L	—/2.3/59	5.5/unknown	560/189	66	Wieczorek et al. (2005)
	Clay	Air base		unknown	unknown	860/80	91	Clarke et al. (1995)
Pb	Soil (56g Ca/kg)	Gasworks	P (47 kg dw)	0.128–0.32/—/112	8/2.6	1040/1230	0	Maini et al. (2000)
	Soil (56g Ca/kg)	Gasworks	L	0.37/—/23	7.8/4.5	2300/2150	7	Maini et al. (2000)
	Tailing soil (<180 μm)	Mining area	L	1.2/—/5	7.5/3.5	5180/4650	10	Kim and Kim (2001)
	Sludge	Municipal wastewater treatment	P (0.06 m³)	1/—/5	8/1.5	63/37	41	Kim et al. (2005)
	Peat/clay soil	Paint factory	F (229 m³)	unknown	unknown	3780/280	92	Clarke et al. (1996)
	Clay	Military air base	F (1925 m³)	unknown	unknown	730/108	85	Clarke et al. (1996)
	Sludge + sand (mixed 11:89) 3.5% organic	Millpond + foundry	L	—/20/7	8/2.6	178/95	47	Khan and Alam (1994)
	Silty soil (carbonates 0.4%)	Production of electronic devices	L	0.2/—/76	6.1/2.6	1090/20	98	Ottosen et al. (2005)

TABLE 4.1. (Continued)

Heavy Metal	Soil Type	Origin of Pollution	Full (F)/Pilot (P)/Laboratory (L)	Current density (mA/cm^2)/Voltage (V/cm)/Duration (days)	pH Initial/Final	Concentration Initial/Final (mg/kg)	Removed (%)	Reference
Zn	Tailing soil (<180 μm)	Mining area	L	1.2/—/5	7.5/3.5	7600/6280	17	Kim and Kim (2001)
	Clay soil	Galvanizing plant	F (38 m^3)	unknown	unknown	1400/600	57	Clarke et al. (1996)
	Clay	Military air base	F (1925 m^3)	unknown	unknown	2600/289	89	Clarke et al. (1996)
	Clay ~67%, silt ~13%, limestone	Electroplating	L	—/2.3/59	5.5/unknown	180/146	19	Wieczorek et al. (2005)
	Clay ~19%, silt ~64%, limestone	Electroplating	L	—/2.3/59	7.1/unknown	308/240	22	Wieczorek et al. (2005)
	Soil (56 g Ca/kg)	Gasworks	P (47 kg dw)	0.128–0.32/—/112	8/2.6	830/600	27	Maini et al. (2000)
	Loamy sand	Wood preservation	L	0.2/<1.5/35	7.2/3.5	200/140	30	Ribeiro (1994)
	Soil (56 g Ca/kg)	Gasworks	L	0.37/—/23	7.8/4.5	1600/960	40	Maini et al. (2000)
	Sandy soil (carbonates 9.7%)	Cable production	L	0.2/<6.6/28	7.1/6.2	260/130	50	Ottosen et al. (2005)
	Fine grained (63% less than 63 μm)	Production of electronic devices	L	0.4/<22.4/76	6.1/2.8	340/120	65	Ottosen et al. (2005)
	Sludge + sand (mixed 11:89)	Millpond + foundry	L	—/16/7	8/2.6	1880/530	72	Khan and Alam (1994)
	Sediment	From disposal site, 5.2% carbonates	L	<2.5/—/21	6.4/3.1	780/200	74	Samson, Tack, and Verloo (2003)
	Fine sand (carbonates 4.9%, organic matter 4.4%)	Unknown	L	0.2/<3.8/40	7.3/3.6	810/200	75	Ottosen et al. (2001b)
	Loamy sand (carbonates 3.7%, organic matter 3.7%)	Car painting facility	L	0.2/<6.4/42	7.1/3.0	500/80	84	Ottosen et al. (2001b)
	Loamy sand. (carbonates 9.7%, organic matter 3.3%)	Cable production	L	0.2/<5.9/102	7.3/3.9	400/20	95	Ottosen et al. (2001b)
	Soil (0.25–1 mm?)	Accumulator renewal	L	—/0.1/7–11	unknown	1430/16	99	Jensen, Kubes, and Kubal (1994)

due to soil acidification. Table 4.1 gives the best result obtained in the cited literature, meaning that it is the lowest final concentration that is given, and the concentration is not necessarily obtained throughout the whole soil volume. In the table, the soil type and origin of pollution are given, as are the initial and final pH of the soil for evaluation of the result. Furthermore, the current density and/or voltage and duration of the actual experiment are listed. The results given in the table are probably not the best results that can be obtained in every case (e.g. longer duration or higher current density might give an improvement of the remediation result for some of the examples). In Table 4.2, the results obtained with systems that are enhanced additionally to control pH by, for example, the addition of enhancement solutions for formation of charged complexes that can increase mobilization. Both tables include laboratory-scale studies as well as pilot- and full-scale studies, but as may be found from the tables, numerous laboratory studies exist, while only very limited pilot- or field-scale studies have been reported.

In general, the electroremediation of heavy metals and metalloids in Table 4.1 are dependent mainly on the development of an acidic front through the soil since the acidification aids mobilization. In most of the experiments, the acidification had not reached through the whole soil specimen during the remediation period, and the remediation percentages given in the table were obtained only in a short distance from the anode.

From Table 4.1, the following observations can be made for different metal contaminants:

- Arsenic (As): The success with removal of As was limited. The highest removal percentage was 62%, and this was reached after a long period of 125 days with applied current. The chemistry of As differs significantly from the other trace elements of this investigation. Desorption of As is highly dependent on both redox potential and pH. The primary forms of As in soils are arsenate As(V) and arsenite As(III), and under moderately reducing conditions, As(III) is the predominant form, whereas at higher redox levels the predominant form is As(V). The experiments reported in Table 4.1 were conducted in closed laboratory cells, and As(III) is expected to be the primary form and the main stable species of As is uncharged (H_3AsO_3), and thus not mobile for electromigration (Ottosen et al., 2000).
- Cadmium (Cd): For Cd, both high and low remediation efficiencies have been reported for unenhanced treatment. It seems that the removal success is highly dependent on site and speciation. Low pH in the soil clearly favors the removal.
- Chromium (Cr): The Cr removal was within the range of 3%–64% and the majority of the experiments resulted in removal less than 35%, making Cr one of the more difficult elements to remove. Cr exists mainly as Cr(III) or Cr(VI) in soils dependent on the prevailing redox potential. Cr(VI) adsorption increases with decreasing pH, whereas on the contrary, Cr(III) desorption is most significant at low pH values. It is not possible to extract from Table 4.1 whether the main reason for the poor removal of Cr is due to the adsorption of Cr(VI) at low pH or because the pH reached was not low enough to desorb Cr(III)

TABLE 4.2. Heavy Metal Removal by Electric Fields. Experiences and Results of Selected Studies for Enhanced Treatment

Heavy Metal	Soil Type	Origin of Pollution	Full (F)/ Pilot (P)/ Laboratory (L)	Enhancement	Current density (mA/cm^2)/Voltage (V/cm)/Duration (days)	pH Initial/Final	Concentration Initial/Final (mg/kg)	Removed (%)	Reference
As	Digested sludge	Industrial wastewater sewage sludge	L	pH adjusted to 2.4 in soil	—/1.25/7	2.4/2	190/160	14	Wang et al. (2004)
	Tailing soil	Gold mine	L	Catholyte KH_2PO_4	1.4/<2/6	3.5/2.0	3210/1930	40	Kim, Kim, and Kim (2005a)
	Mine tailing (>50% calcium oxide)	Abandoned mine (iron ore)	L	Anolyte kept alkaline with 0.5 M NaOH	—/0.25/28	8.5/10.9	83/55	44	Baek et al. (2009)
	Mine tailing (>50% calcium oxide)	Abandoned mine (iron ore)	L	Catholyte kept acidic with 0.1 M HNO_3	—/1.6/28	8.5/8.5	83/32	62	Baek et al. (2009)
	Loamy sand	Wood preservation	L	Addition of 0.1 M $K_4P_2O_7$ to soil	0.2/<6.7/35	8.6/2.6	770/270	65	Ribeiro et al. (1998)
	Tailing soil	Gold mine	L	Catholyte NaOH; Anolyte Na_2CO_3	1.4/6.7–13/6	3.5/12.2	3210/610	81	Kim, Kim, and Kim (2005a)
	Weathered top soil	Wood preservation	L	Addition of 2.5% ammonia to soil	0.6–1.0/<2.6/69	9.7/9.6	900/90	90	Ottosen et al. (2000)
		Production of green color agent	L	Tartaric acid at anode	0.3/—/50	unknown	5000/400	92	Goldmann, Schlösinger, and Rauner (1996)
		Wood preservation (fraction <63 μm)	L	Stirred suspension (L:S 10)	0.4/—/10	unknown	9260/1950	79	Jensen (2005)
		Wood preservation (fraction <63 μm)	L	Stirred suspension (L:S 10)	0.4/—/10	unknown	3030/1730	43	Jensen (2005)

Cd	Soil with brackish pore water	Electroplating and metal finishing	F (64 m³)	Addition of citric acid	1–1.7/—/180	4–8/4–6.5	5–20/<3	40–85	Gent et al. (2004)
	Unknown	Disposal for electroplating and metal finishing	L	Citric acid	0.5/—/~50	unknown	80/35	56	Gent et al. (2004)
	Soil (fraction <63 μm)	Unknown	L	Stirred suspension (L:S 10)	0.4/—/10	unknown	43/3.4	92	Jensen (2005)
Cr	Digested sludge	Industrial wastewater sewage sludge	L	pH adjusted to 2.4	—/1.25/7	2.4/2	1900/770	40	Wang et al. (2004)
	Unknown	Disposal for electroplating and metal finishing	L	Citric acid	0.5/—/~50	unknown	860/500	42	Gent et al. (2004)
	Unknown	Electroplating and metal finishing	F	Addition of citric acid	1–1.7/—/180	4–8/4–6.5	180–1100/<20	89–98	Gent et al. (2004)
	Soil (fraction <63 μm)	Wood preservation	L	Stirred suspension (L:S 10)	0.4/—/10	unknown	2310/1039	55	Jensen (2005)

TABLE 4.2. (Continued)

Heavy Metal	Soil Type	Origin of Pollution	Full (F)/ Pilot (P)/ Laboratory (L)	Enhancement	Current density (mA/cm²)/Voltage (V/cm)/ Duration (days)	pH Initial/ Final	Concentration Initial/Final (mg/kg)	Removed (%)	Reference
Cu	Sandy soil (carbonates 9.7%)	Cable production	L	5% ammonia to soil	0.9/<12.4/22	9.0/unknown	5870/3520	60	Ottosen et al. (2005)
	Sandy soil (carbonates 12%)	Cable production	L	1 M ammonium citrate (pH = 10) to soil	0.8/<1.1/21	9.0/unknown	2730/630	77	Ottosen et al. (2005)
	Unknown	Production of green color agent	L	Tartaric acid at anode	0.3/—/~50	unknown	4360/800	82	Goldmann Schlösinger, and Rauner (1996)
	Sand (carbonates 11%)	Cable production	L	Addition of 2.5% ammonia	0.65/—/21	Alkaline	17,000/1200	93	Ottosen et al. (1998)
	Digested sludge	Industrial wastewater sewage sludge	L	pH adjusted to 2.4	—/1.25/7	2.4/2	7760/360	95	Wang et al. (2004)
	Soil (fraction <63 μm)	Wood preservation	L	Stirred suspension (L:S 10)	0.4/—/10	unknown	6820/141	95	Jensen (2005)
Hg	Sandy loam, (organic 11%)		L	I_2/I^- addition	—/0.39/14	unknown	250/unknown	84	Cox, Shoesmith, and Ghosh (1996)
	Soil (illitic clay, organic 2.5%)	Chlor-alkali plant	L	Cl^- addition	—/1.11/182	7/1.5	89/89	0	Suer and Allard (2003)
	Soil (illitic clay, org. 2.5%)	Chlor-alkali plant	L	I^- addition	—/1.11/182	7/1.5	89/3	97	Suer and Allard (2003)

	Matrix	Source	L/P	Treatment				Reference
Ni	Sludge (90.6%, <50 μm)	Incinerator sludge from a gas plant	L	pH adjusted with HNO3 to 4.5	—	4.5/2	460/460	0	Zagury, Dartiguenave, and Setier (1999)
	Digested sludge	Industrial wastewater sewage sludge	L	pH adjusted to 2.4	—/1.25/7	2.4/2	2050/260	87	Wang et al. (2004)
	Soil (fraction <63 μm)	Metal foundry	L	Stirred suspension (L:S 10)	0.4/—/10	unknown	75/36	52	Jensen (2005)
Pb	Calcareous soil	Leaded gasoline	L	Acetic acid	0.5/—/133	8.5/—	110,000/—	~88	Alshawabkeh et al. (1998)
	Digested sludge	Industrial wastewater sewage sludge	L	pH in catholyte adjusted to 2	—/1.25/7	2.4/2	580/480	19	Wang et al. (2004)
	Sandy soil (carbonates 6.8%)	Unknown	L	1M ammonium citrate (pH = 10) to soil	0.8/<0.7/21	9.0/unknown	1060/760	28	Ottosen et al. (2005)
	Calcareous soil	Battery manufactory	L	0.1M EDTA	0.02–1.8/1.0/10	7.3/unknown	4432/—	36	Amrate et al. (2005)
	Calcareous soil	Battery manufactory	L	0.1M EDTA	—/1.0/40	7.3/unknown	4432/—	80	Amrate and Akretche (2005)
	Calcareous soil	Battery manufactory	L	0.1M EDTA; cation exchange membrane	—/1.0/20	7.3/unknown	4432/—	56	Amrate, Akretche, and Innocent (2006)
	Clay	Shooting range	P	Citric acid or EDTA supplied at the anode	—/2.5/148	unknown	Mean 4350/—	~90	Kim et al. (2006)
	Calcareous soil (fraction <63 μm)	Unknown	L	Stirred suspension (L:S 10)	0.8/<14/39	7.5/1.4	673/27	96	Jensen, Ottosen, and Ferreira (2006)

TABLE 4.2. (Continued)

Heavy Metal	Soil Type	Origin of Pollution	Full (F)/ Pilot (P)/ Laboratory (L)	Enhancement	Current density (mA/cm²)/Voltage (V/cm)/Duration (days)	pH Initial/Final	Concentration Initial/Final (mg/kg)	Removed (%)	Reference
Zn	Sandy soil (carbonates 9.7%)	Cable production	L	5% ammonia to soil	0.9/<12.4/21	9.9/unknown	2760/2050	25	Ottosen et al. (2005)
	Sandy soil (carbonates 12%)	Cable production	L	1 M ammonium citrate (pH = 10) to soil	0.8/<1.1/21	9.0/unknown	260/170	35	Ottosen et al. (2005)
	Sediment	From disposal site, 3.0% carbonates	L	Addition of HNO_3 to soil	<2.5/—/21	6.4/2.8	780/200	74	Samson, Tack, and Verloo (2003)
	Digested sludge	Industrial wastewater sewage sludge	L	pH adjusted to 2.4	—/1.25/7	2.4/2	18,060/400	98	Wang et al. (2004)
	Soil (fraction <63 μm)	Unknown	L	Stirred suspension (L:S 10)	0.4/—/10	unknown	7210/793	89	Jensen (2005)

since the speciation of Cr in the initial soil is not known. A conclusion though is that Cr is difficult to remove without enhancement.

- Copper (Cu): Good removal percentages of up to between 80% to 99% have been obtained in about half of the soils, whereas poor remediation (between 0% and 50% removed) was the result in the other half of the experiments. In general, the best results are obtained after a long period of applied current of more than 1 month of treatment, and all the best results have been obtained using the electrodialytic method with ion exchange membranes (see above).
- Mercury (Hg): It was shown by Thoming, Kliem, and Ottosen (2000) that the acidic front resulted in an increased oxidation rate of elemental Hg, but the oxidation process was slow and no Hg was removed from the soil in total.
- Nickel (Ni): General low removal efficiency without any enhancement even for 2 months' processing. Only Clarke, Lageman, and Smedley (1997) showed that a high removal efficiency could be achieved but remediation time and conditions were not mentioned, so it could not be evaluated if the remediation was enhanced or not.
- Lead (Pb): Some really low remediation percentages (0%–10%) were obtained in studies of calcareous soil (Maini *et al.*, 2000) and tailing soil (Kim and Kim, 2001), whereas around 50% removal was obtained in studies of different sludges (Khan and Alam, 1994; Kim *et al.*, 2005). Highly successful removals (92%–98%) were obtained in full scale, as well as in a study of noncalcareous soil (Clarke *et al.*, 1996; Ottosen *et al.*, 2005). Apart from confirming the fact that acidification, and thus buffer capacity, is a determinant of remediation success, it is difficult to deduct any conclusions from the results since Clarke *et al.* (1996) give no detailed information about remediation conditions. However, it seems that long remediation times are necessary for successful removal (Ottosen *et al.*, 2005).
- Zinc (Zn): Remediation results between 17% and 99% have been obtained. Most results are good, with >70% removal. The low removals were obtained in experiments of either short duration (Kim and Kim, 2001) or with calcareous soils (Maini *et al.*, 2000; Ottosen *et al.*, 2005; Wieczorek *et al.*, 2005). This corresponds well to Zn being among the easiest heavy metals to mobilize by EKR (Ottosen *et al.*, 2001a; Suer, Gitye, and Allard, 2003).

4.4 ENHANCEMENT METHODS

From Table 4.1, it is generally seen that in order to obtain a successful removal, it is necessary to apply the electric field for quite a long time (sometimes several months), even in laboratory experiments. It is also seen that a very long duration is not always enough to reach low concentrations. Based on these findings, researchers have suggested different enhancement methods, with the aim of obtaining lower concentrations in the soil after remediation and/or in a shorter duration. Enhancement is typically based on reagent addition (see Section 4.4.1) or on a combination of different remediation methods (see Section 4.4.3).

4.4.1 Enhancement Techniques Based on Reagent Addition

Enhancement techniques are mainly directed toward simultaneously controlling pH and maintaining or bringing contaminants into solution by the addition of enhancement solutions with the ability to form complexes with the actual heavy metals in the soil (Page and Page, 2002). The use of enhancement solutions is beneficial, or even necessary, for the removal of heavy metals from, for example, high-buffering-capacity soils or for some combinations of pollutants within a reasonable remediation time. The buffering capacity against the developing acidic front from the anode end of the soil is high in calcareous soils, and thus the remediation will progress very slowly if the acidic front is used solely for the desorption of heavy metals. Also, manipulation of the redox state of the soil for heavy metal mobilization is used as an enhancement. The most important chemical elements that undergo oxidation–reduction reactions in soils are C, N, O, S, Mn, and Fe. For contaminated soils, Cr, Cu, As, Ag, Hg, and Pb could be added to the list. Thus, there are two ways in which the redox reactions can influence the chemical forms of the heavy metals in the soil: directly through a change in oxidation state of the heavy metal itself or indirectly through a change in oxidation state of a ligand that can form chemical bonds with the heavy metal (Sposito, 1984). The oxidation state can thus, in some cases, be expected to play a major role in the mobility of heavy metals in an electric field.

Another group of enhancement methods aims at shortening the transport distance for the heavy metals, and thereby shortening the treatment time. The transport rate for heavy metals in soil in an applied electric field is most often significantly less than 1 cm/day (Ottosen *et al.*, 2008), which means that in order to avoid a long remediation time, the electrodes should be placed relatively close or concentration compartments could be placed between the electrodes where the heavy metals are trapped, as suggested by Ottosen *et al.* (1999). Another tested solution to overcome long treatment time is to treat the soil as a suspension (Jensen, Ottosen, and Ferreira, 2007).

Table 4.2 outlines the results obtained with enhanced EK soil remediation systems. For comparison, the same parameters are given in Tables 4.1 and 4.2— Table 4.2 is extended to include information of the type of enhancement.

For each of the pollutants in Table 4.2, the following observations can be made:

- Arsenic (As): The removal percentages were generally improved in the enhanced experiments as compared with the unenhanced systems (Table 4.1). One way to enhance As removal is to add an alkaline enhancement solution (e.g. NaOH or ammonia) to ensure that As(III) is present as a charged species, or to oxidize the As(III) species to As(V) by the addition of an oxidizing agent $K_4P_2O_7$, or treatment of the soil as a stirred suspension for oxidation through contact with air.
- Cadmium (Cd): Citric acid seems to be a suitable leaching/complexing agent when treating soil with cadmium. Up to 85% of cadmium was removed when adding citric acid (Gent *et al.*, 2004). On a site used for disposal of electroplating and metal finishing waste, some improvement was found on the removal efficiency when adding citric acid either before or during remediation.

- Chromium (Cr): Even though Cr is one of the most difficult of the heavy metals to remove with the EK system, the research on enhancement is very limited. Addition of citric acid gave good results in a field-scale action.
- Copper (Cu): Very good removal percentages of up to 99%. Cu was obtained in unenhanced systems (Table 4.1); however, the duration of the successful experiments was very long. Enhancement in the case of copper is mainly focused on a faster acidification of the soil, and thus remediation. Citric acid showed good results. The acid demand for soils with high buffer capacity is high, and in such soils the enhancement may be the addition of a complex binder for Cu so the remediation can occur at neutral to alkaline pH. An example of this is ammonia.
- Mercury (Hg): For the EK removal of Hg in ionic form, iodine crystals (Cox, Shoesmith, and Ghosh, 1996), sodium ethylenediaminetetraacetic acid (Na-EDTA), and potassium iodide (KI) (Reddy, Chaparro, and Saichek, 2003) and iodide (Suer and Lifvergren, 2003) have been tested as enhancement chemicals. With the use of iodine crystals, 99% of Hg was removed from spiked loam, but only 6% was removed from an industrially polluted sandy loam; the main reason for the decreased success with the industrially polluted soil was thought to be the higher organic content, and thus higher demand for iodine crystals (Cox, Shoesmith, and Ghosh, 1996). Suer and Allard showed, by sequential leaching, that Hg was more mobile in a soil polluted with Hg from a chlor-alkali factory after iodide addition and application of an electric field than before treatment. Reddy, Chaparro, and Saichek (2003) found that iodide was a better enhancement solution than EDTA for Hg removal from spiked kaolinite and spiked glacial till, even though the removal performed much better in the kaolinite than in the till.
- Nickel (Ni): No special process enhancement has been reported yet when focusing on nickel removal—only general acid addition. With a lowering of pH to around 2, the electric field could remove around 87% of Ni from digested wastewater sludge. This could indicate that a simple pH lowering could be sufficient for proper Ni removal.
- Lead (Pb): Most effort has been given to enhance remediation of Pb from recalcitrant calcareous soils. The results listed in Table 4.2 suggest that EDTA, acetic acid, and treatment of soil fines in suspension successfully reduce remediation times, while treatment with ammonium citrate at pH 10 was less successful.
- Zinc (Zn): Enhancement of Zn remediation was mainly attempted using two strategies: complexation with ammonia and removal at alkaline pH in order to avoid disintegration of the soil, and lowering of pH with the addition of acid. Of these two strategies, the latter was clearly the most successful. In addition, the treatment of soil fines in suspension was efficient.

4.4.2 Situations Where Enhancement is Necessary

The use of enhancement solutions is beneficial, or even necessary, for the removal of heavy metals from, for example, high-buffering-capacity soils since power consumption for remediation of such soils is high. The buffering capacity against the developing acidic front from the anode end of the soil is high in calcareous soils, and thus the remediation will progress very slowly if the acidic front is used for desorbing the heavy metals. An example is given in Lageman (1993) with a pilot test at a galvanizing plant where the soil contained ammonia and ammonium chloride, resulting in high buffering capacity and slow acidification and, consequently, slow Zn removal.

Also, some specific single pollutants require enhancement in order to be mobilized in the electric field, in particular Cr(III) and Hg^0.

For some combinations of heavy metals, it is also necessary to use enhancement solutions to ensure the simultaneous removal of all pollutants (Ottosen et al., 2003). Especially, the presence of As in the soil necessitate alternative solutions to the acidic front since As generally has low mobility under acidic conditions, whereas As is more mobile under alkaline conditions, where most heavy metals are not mobile (Le Hecho, Tellier, and Astruc, 1998; Ottosen et al., 2000). Le Hecho, Tellier, and Astruc (1998) conducted laboratory experiments with spiked soils, where the pollutants were As and Cr. Successful remediation was obtained in the developing alkaline front combined with the injection of sodium hypochlorite. As was mobile in the alkaline environment, and Cr(III) was oxidized to Cr(VI) by hypochlorite and mobilized in the alkaline environment. In loamy sand polluted with Cu and As from wood preservation, As and Cu were mobile simultaneously after the addition of NH_3 to the soil (Ottosen et al., 2000). As was mobile due to the alkaline environment and Cu formed charged tetra-ammine complexes. For the simultaneous mobilization and electrochemical removal of Cu, Cr, and As, ammonium citrate has shown to be successful (Ottosen et al., 2003).

Enhancement may also be necessary when the soil is polluted with a mixture of heavy metals and organics. Reddy et al. (2006) showed that Igepal CA-720 improved the removal efficiency of polycyclic aromatic hydrocarbons (PAHs), but the removal of heavy metals was negligible using this flushing agent.

4.4.3 Combination of EK with Other Soil Remediation Methods and Changes in Process Design

Another way to improve the remediation efficiency is to combine the EK principle with other methods, or to manipulate the electric field in such a way that the efficiency is increased. The main reasons for this combination/manipulation are (a) to reduce the concentration gradients generated during the process with a constant electric field, (b) to speed up the desorption/dissolution of the heavy metals, or (c) to decrease the heavy metal migration distance.

The overall idea of treatment of the soil in suspension is to combine electroremediation with the soil washing process, as these processes complement each other for the different fractions of the soil. Electroremediation is suitable, in particular, for the treatment of fine-grained materials (Page and Page, 2002), whereas the soil washing method, which is a mechanical and thus less complex and expensive method, manages the decontamination of the coarse fractions of the soil while leaving a highly contaminated fine-grained sludge (VanBenschoten, Matsumoto, and Young, 1997; Mann, 1999). In combination, these methods provide a possible continuous treatment of contaminated soils, with remediation times of hours as compared with months and easy application of various necessary enhancement solutions (Jensen, Ottosen, and Ferreira, 2006; Jensen, Ottosen, and Ahring, 2007; Jensen, Ottosen, and Ferreira, 2007). The benefits of *in situ* technology are however lost.

Some experiments have been made to evaluate if the application of electric fields could enhance phytoremediation, or plant uptake of heavy metals, of the contaminated sites (O'Connor *et al.*, 2003; Zhou *et al.*, 2007). Cu uptake was significantly higher with ryegrass root when stimulated by an electric field. On the other hand, the Cd uptake seemed to be unaffected by the electric field. Anyway, it was concluded that the combination of the two techniques represented a very promising approach to the decontamination of metal-polluted soils, which would require validation in field conditions. It was also possible to incorporate sulfur-oxidizing bacteria and EK to enhance copper removal from contaminated sulfur-containing soil (Maini *et al.*, 2000).

Recently, the use of ultrasound in EK soil remediation has been tested (Chung and Kamon, 2005; Sandoval-Gonzales, Silva-Martinez, and Blass-Amador, 2007). The idea is to increase the desorption and dissolution of contaminants during the EKR process. Leaching when applying ultrasound was increased remarkably (Sandoval-Gonzales, Silva-Martinez, and Blass-Amador, 2007). In soils that are contaminated with heavy metals and organics, this could be an option since the EK technique was applied to remove mainly heavy metals, and the ultrasonic technique was applied to remove mainly organic compounds (Chung and Kamon, 2005).

Some experiences have been obtained in changing the applied electric field from a DC field to either a cyclic or a pulsed field (Reddy and Saichek, 2004; Kornilovich, Mishchuk, and Abbruzzese, 2005; Hansen and Rojo, 2007). This was tested to minimize or eliminate concentration gradients that could form during the application of DC electric fields. For copper removal from mine tailings, pulsed electric fields seemed to increase the removal efficiency. When treating spiked clays, a cyclic electric field also improved the removal efficiency.

In order to decrease heavy metal migration distance and minimize power consumption, moving anodes and bipolar electrodes have been tested. In the case of moving anodes, the anode is moved together with the migration of the heavy metal cations. In that way, the distance of the soil to be treated is reduced continuously, as with power consumption. During the moving anode test, the removal efficiency of Cd in soils was enhanced by 54.9% compared with that of the fixed anode tests (Chen *et al.*, 2006). The idea to implement bipolar electrodes—or iron plates—in the soil was developed to reduce and collect Cu at these plates, and Cu^{2+} would not have to migrate all the distance to the cathode compartment (Hansen, Rojo, and Ottosen, 2007c). The results from copper mine tailings remediation showed that the copper removal was increased from 8% (applying 20V for 8 days in sulfuric acidified tailings) without bipolar electrodes to 42% when bipolar electrodes were implemented.

Related to this last process enhancement is the use of an EK barrier, or fence. The EK barrier has two different objectives depending on its use: (a) to precipitate or transform heavy metals transported by electric fields; or (b) to "stop" a hydraulic generated flow of liquid with contaminants by the application of an electric field in the opposite direction. In the first approach, zero-valent iron has mainly been used to fix either arsenic (Yuan and Chiang, 2007) or chromium (Weng et al., 2007) transported by electric fields. Implementing slag in the soil during EK processing showed an increase in the remediation efficiency of groundwater (Chung and Lee, 2007). Lynch et al. (2007) showed that it was possible to control a hydraulic flow of contaminants. In their work, it was found that an electric field of 125 V/m was sufficient to prevent significant copper incursion from a contaminant flow generated at an old mining district under a hydraulic gradient of 1.3. In addition, see Chapter 19 for more details about this type of enhancement.

4.5 REMEDIATION OF MINE TAILINGS, ASHES, SEDIMENTS, AND SLUDGE

Following the success of remediation of heavy metal-contaminated soil, the EK method has been tested for the removal of heavy metals from various heavy metal-containing waste products (Ottosen et al., 2003) such as impregnated waste wood (Ribeiro et al., 2000; Christensen et al., 2006), fly ash from municipal solid waste incineration (MSWI) (Pedersen, Ottosen, and Villumsen, 2003; Ferreira, Ribeiro, and Ottosen, 2005; Ferreira et al., 2005), fly ash from biomass combustion [straw (Ottosen et al., 2007), wood (Pedersen, Ottosen, and Villumsen, 2003)], bottom ash (Traina, Morselli, and Adorno, 2007), ash from combustion of chromated copper arsenate (CCA)-treated wood (Pedersen and Ottosen, 2006), harbor sediment (Nystroem et al., 2005b), wastewater sludge (Kim et al., 2002; Jakobsen et al., 2004), anaerobic granular sludge (Jakobsen et al., 2004; Virkutyte, Sillanpaa, and Lens, 2006), manganese-contaminated industrial sludge (Ricart et al., 2005), and mine waste and tailing/soils (Kim and Kim, 2001; Hansen, Rojo, and Ottosen, 2005; Rojo, Hansen, and Ottosen, 2006). The experiences with these waste products are obtained in laboratory scale, and in general good results have been obtained meaning that the mobilization of heavy metals was successful.

4.5.1 Mine Tailings

Metal sulfide-based mining produces huge amounts of solid waste, where the most concerning are the mine tailings. For example, in Chile, copper mining is the main industry. Here approximately 50% of mining is based on exploiting copper sulfide ores with a copper content of around 1%. Copper is concentrated to 30%–35% by flotation processes. The tailings from the flotation process end up in deposits in amounts corresponding to more than 96% of the original mineral exploited. These tailings contain considerable amounts of copper and other heavy metals (Dold and Fondbote, 2001; Hansen et al., 2007). Particularly in the north of Chile, the arsenic content is of great concern (Dold, 2006).

Recently, copper mine tailings have been treated by the EKR and electrodialytic remediation (EDR) methods in laboratory scale by different investigators (Kim and Kim, 2001; Hansen, Rojo, and Ottosen, 2005).

Hansen et al. (2007) found that recently produced tailings were much more difficult to remediate than tailings deposited more than 30 years ago. The important difference between the two tailing samples is the pH: the fresh tailings are approximately neutral, while the old deposited tailings are acidic. This is due to the oxidation of pyrite—the main residual mineral in sulfide tailings—which releases protons due to the overall reaction (Kontopoulus *et al.*, 1995):

$$4\,FeS_2(s) + 15\,O_2 + 14\,H_2O \rightarrow Fe(OH)_3(s) + 8\,SO_4^{2-} + 16\,H^+. \tag{4.1}$$

In the old tailings, copper was removed easily from the tailings due to the dissolution of the copper sulfides. This corresponds well with the findings from the sequential analysis of these tailings (Hansen, Yianatos, and Ottosen, 2005), where the mobility of copper in old tailings was found to be highest.

Comparing the EKR and the EDR of fresh and deposited copper mine tailings, the main findings were that copper could be removed easily by both EKR and EDR without pretreatment or enhancement from the deposited—or aged—tailings (Hansen *et al.*, 2007). On the other hand, fresh tailings seemed to be difficult to treat without lowering the pH. Both sulfuric and citric acids were tested, and the complexing effect of citric acid seemed to enhance the process a little better.

Hansen *et al.* (2008) evaluated an enhancement system, including an airlift stirring of suspended fresh tailings in dilute sulfuric acid. The tailings could be remediated in suspended EDR more efficiently than in static EDR. Eighty per cent of the copper was removed when suspending the tailings by airlift during EDR. In contrast, only 15% was removed in static EDR with similar operation conditions. The liquid-to-solid ratio (L:S ratio) was analyzed, and in the case of copper mine tailings, a suitable L:S ratio seems to be around 6–9 ml/g. Furthermore, if no stirring was applied—maintaining the same L:S ratios—no copper removal was observed, indicating that the electric current passes in the stagnant liquid above the settled particles. Initial experiments showed that pH did not seem to be the most important parameter for copper removal in suspended tailings.

4.5.2 Freshwater and Saline Sediments

Sediments and soils have many characteristics in common. However, commonly, sediments are fine grained, rich in organic matter, and more homogeneous than surface soils. Furthermore, sediments of saline waters are rich in salts (chloride), while freshwater sediments of eutrophic waters are rich in phosphate.

Electrodialysis was tested for remediation of harbor sediments in a number of works (Nystrom, Ottosen, and Villumsen, 2003, 2005; Nystroem, Ottosen, and Villumsen, 2005a,b; Nystroem *et al.*, 2006; Ottosen *et al.*, 2007a), in which the potential was documented (Nystrom, Ottosen, and Villumsen, 2003). Like for soils, remediation was shown to be faster for noncalcareous sediments compared with calcareous ones (Nystrom, Ottosen, and Villumsen, 2005). Furthermore, remediation of sediment in suspension was more efficient than remediation of sediment in a solid column (Nystrom, Ottosen, and Villumsen, 2005). It was shown that the addition of HCl, lactic acid, citric acid, NaCl, and ammonium citrate as enhancing agents reduced remediation efficiency. The highest removals obtained were 67%–87% Cu, 79%–98% Cd, 90%–97% Zn, and 91%–96% Pb regardless of the initial heavy metal concentration (Nystroem, Ottosen, and Villumsen, 2005b). One possible drawback that should be paid attention to concerning the treatment of sediments is the leaching behavior of toxic elements from contaminated sediments after treatment, which was documented to be equal to or larger than that before treatment (Gardner, Nystroem, and Aulisio, 2007). EK was also tested for sediment contaminated by organics as well as by heavy metals. Various surfactants were tested for enhancement, however, found to be inefficient for the enhancement of heavy metal removal. Rather low removal efficiencies were found for heavy metals, and this was attributed to the high organic content and buffering capacity of the sediment (Acar *et al.*, 1994; Reddy and Ala, 2006).

Recently, the potential of using electrochemical methods for treatment of freshwater sediments was documented in an electrochemical cell, where the metals were transported from the acidified sediment in which carbon rod anodes were placed directly, and into the catholyte separated from the sediment by a cellulose filter (Matsumoto, Uemoto, and Saiki, 2007). Removal percentages of 18, 21, 53, 81, 86, and 98 for Pb, Cu, Ni, Cr, and Zn, respectively, were obtained after 10 days of treatment at $2.9\,mA/cm^2$. Another work proved that Cu can be removed (up to 85% after 14 days with $0.15\,mA/cm^2$) from artificially contaminated lake sediments, and that the use of nylon membranes and cation exchange membranes as barriers between sediment and cathode improves the treatment (Virkutyte and Sillanpaa, 2007). By means of the electrodialytic method also used for treatment of harbor sediments, it was shown that removal of Pb, Zn, Cu, Cr, and Ni could be obtained from industrially contaminated millpond sediment, with removals of approximately 95%, 85%, 75%, 65%, and 55%, respectively, after 14 days of treatment at $0.8\,mA/cm^2$ (Jensen, Ottosen, and Villumsen, 2007).

4.5.3 Different Fly Ashes

A serious drawback of MSWI is the production of chemically unstable flue gas purification products that are rich in heavy metals. Biofuels such as wood and straw are becoming increasingly important worldwide as alternatives to fossil fuels. After combustion of biofuels, the remaining fly ashes have a relatively high Cd concentration, and a high percentage of this Cd is highly mobile (Hansen *et al.*, 2001). At present, both MSWI and bio ashes are most often disposed at special landfills as hazardous waste, but there is an increasing demand for methods to treat the ashes as opposed to disposal.

Fly ashes are fine-grained material containing toxic elements, which can, to some extent, be compared with heavy metal-polluted soils. Accordingly, it has been evaluated whether EKR can be used for treatment of different fly ashes (Pedersen, Ottosen, and Villumsen, 2003; Ferreira, Ribeiro, and Ottosen, 2005; Ferreira *et al.*, 2005; Christensen *et al.*, 2006). However, even though the two materials seem comparable overall, there are many differences of importance when it comes to EKR. The major difference is that the ashes contain a high water soluble fraction (mainly salts), which makes it difficult to treat fly ash in the traditional cell as shown in Figure 4.1. Hansen, Ottosen, and Villumsen (2004) found that about 2/3 wt% straw ash was dissolved during electrodialytic treatment, and thus the transference number of the pollutants was very low and the process to control was difficult due to the significantly decreasing volume of the ash.

It proved beneficial to prewash the ash in water to remove the soluble parts before treatment (Pedersen, 2003). The advantage is that less current is wasted on the removal of harmless ions and the volume loss during treatment is less. Furthermore, by prewashing the residues, the production of chlorine gas at the anode was reduced. For the optimization of the process, it was also found to be highly beneficial to treat the ash in a stirred suspension as compared with its treatment as water-saturated matrix (Pedersen, 2003).

Some remediation results obtained with the treatment of different ashes in the stirred electrodialytic cell are as follows:

- 86% Cd, 20% Pb, 65% Zn, 81% Cu, and 44% Cr was removed from a MSWI fly ash during 70 days of treatment where 0.25 M ammonium citrate was used as the assisting agent (Pedersen, Ottosen, and Villumsen, 2005).
- 66% Zn and 88% Cd were removed from MSWI fly ash (mixture of samples from boiler, scrubber, and baghouses) (Ferreira *et al.*, 2008).
- 70% Cd removed from wood ash in 21 days. A mixture of 0.25 M ammonium citrate and 1.25% NH_3 was used as the assisting agent (Pedersen, 2003).
- 75% Cd removed from straw ash in 14 days. The ash was suspended in distilled water (Ottosen *et al.*, 2007b).

4.6 SUMMARY

The major challenge to overcome when remediating a heavy metal-polluted soil by an EK method is to bring the pollutants to ionic form in the soil solution. In general, heavy metals adsorb quite strongly to the soils, and desorption is necessary in order to obtain a successful remediation.

The simplest way to apply EK soil remediation to a heavy metal-polluted soil is to utilize the acidic form that develops from the anode side to obtain desorption (and prevent the alkaline front from the cathode). This system works well in general for Cd, Cu, and Zn in most soils.

Enhancement of the remediation is necessary in various situations where the acidic front is not sufficient to obtain the crucial desorption. These situations are

- soils with a high buffering capacity toward acidification; in such soils, the unenhanced remediation is unrealistically time consuming;
- soils polluted with Cr(VI), As(III), and Hg^0;
- heavy metal present in compounds, which are insoluble or little soluble in acid; and
- mixed contaminations.

Enhancement can involve the addition of additives to the soil for chemical manipulation of the redox condition or to aid desorption by formation of complexes with the heavy metals. Enhancement can also be focused at shortening the duration of the action either by placement of the electrodes or by combining the EK soil remediation method with other methods.

Remediation of heavy metal-polluted soil is complicated since a huge variety of adsorption types and pollution origins exist. However, research has overcome some major difficulties and it is possible to remediate many soil and pollution types, as well as various waste products by means of EKR.

REFERENCES

Acar, YB, Alshawabkeh, A. (1993). Principles of electrokinetic remediation. *Environmental Science & Technology* **27**(13):2638–2647.

Acar YB, Gale RJ, Hamed J, Putnam GA. (1990). Acid/base distributions in electrokinetic soil processing. *Transportation Research Record* **1228**:23–34.

Acar YB, Hamed JT, Alshawabkeh AN, Gale RJ. (1994). Removal of cadmium (II) from saturated kaolinite by the application of electrical-current. *Geotechnique* **44**(2):239–254.

Alloway BJ. (1995). *Heavy Metals in Soils*. Glasgow, Scotland: Blackie Academic & Professional.

Alshawabkeh AN, Ozsu-Acar E, Gale RJ, Puppala SK. (1998). Remediation of soils contaminated with tetraethyl lead by electric fields. *Transportation Research Record* **1615**:79–85.

Amrate S, Akretche DE. (2005). Modeling EDTA enhanced electrokinetic remediation of lead contaminated soils. *Chemosphere* **60**:1376–1383.

Amrate S, Akretche DE, Innocent C. (2006). Use of cation-exchange membranes for simultaneous recovery of lead and EDTA during electrokinetic extraction. *Desalination* **193**(1–3):405–410.

Amrate S, Akretche DE, Innocent C, Seta P. (2005). Removal of Pb from a calcareous soil during EDTA-enhanced electrokinetic extraction. *Science of the Total Environment* **349**:56–66.

Baek K, Kim DH, Park SW, Ryu BG, Bajargal T, Yang JS. (2009). Electrolyte conditioning-enhanced electrokinetic remediation of arsenic-contaminated mine tailing. *Journal of Hazardous Materials* **161**:457–462.

Bibler JP, Osteen AB, and Meaker TF. (1993). Application of electrokinetic migration technology for removal of chromium and uranium from unsaturated soil at SRS (U). *Proceedings of Waste Management 93*, 28 February–4 March, Tucson, AZ: WM Symposia, pp. 855–860.

Castillo AMN, Soriano JJ, Delgado RAG. (2008). Changes in chromium distribution during the electrodialytic remediation of a Cr (VI)-contaminated soil. *Environmental Geochemistry and Health* **30**(2):153–157.

Chen XJ, Shen ZM, Yuan T, Zheng SS, Ju BX, Wang WH. (2006). Enhancing electrokinetic remediation of cadmium-contaminated soils with stepwise moving anode method. *Journal of Environmental Science and Health Part A* **41**:2517–2530.

Christensen IV, Pedersen AJ, Ottosen LM, Ribeiro A. (2006). Electrodialytic remediation of CCA-treated waste wood in a $2 m^2$ pilot plant. *Science of the Total Environment* **364**:45–54.

Chung HI, Kamon M. (2005). Ultrasonically enhanced electrokinetic remediation for removal of Pb and phenanthrene in contaminated soils. *Engineering Geology* **77**(3–4):233–242.

Chung HI, Lee M. (2007). A new method for remedial treatment of contaminated clayey soils by electrokinetics coupled with permeable reactive barriers. *Electrochimica Acta* **52**(10):3427–3431.

Clarke RL, Kimmel S, Lageman R, Smedley SI. (1996). Electrokinetic remediation of soils, sludges and groundwater. *Proceedings of the American Power Conference 1996*, April 9–11, Chicago, pp. 347–352.

Clarke RL, Lageman R, Pool W, Clarke SR. (1995). Electrochemically aided biodigestion of organic materials in contaminated soil. Using power supply to drive current between electrodes between which electrolyte is circulated. US Patent 5,846,393, issued December 8, 1998.

Clarke RL, Lageman R, Smedley SI. (1997). Some practical applications of integrated electrochemical techniques used on remediation, recycling and resource recovery. 1. Electrokinetic treatment of soils and sediments. Fourth European Electrochemical Processing Conference: Electrochemical Processing—The Versatile Solution, April 14–18, Barcelona, Spain.

Cox CD, Shoesmith MA, Ghosh MM. (1996). Electrokinetic remediation of mercury-contaminated soils using iodine/iodide lixiviant. *Environmental Science & Technology* **30**(6):1933–1938.

Dold B. (2006). Element flows associated with marine shore mine tailings deposits. *Environmental Science & Technology* **40**(3):752–758.

Dold B, Fondbote L. (2001). Element cycling and secondary mineralogy in porphyry copper tailings as a function of climate. primary mineralogy, and mineral processing. *Journal of Geochemical Exploration* **77**:3–55.

Eid N, Elshorbagy W, Larson D, Slack D. (2000). Electro-migration of nitrate in sandy soil. *Journal of Hazardous Materials* **79**(1–2):133–149.

Ferreira C, Ribeiro A, Ottosen L. (2005). Effect of major constituents of MSW fly ash during electrodialytic remediation of heavy metals. *Separation Science and Technology* **40**(10): 2007–2019.

Ferreira C, Jensen P, Ottosen L, Ribeiro A. (2005). Removal of selected heavy metals from MSW fly ash by the electrodialytic process. *Engineering Geology* **77**(3–4):339–347.

Ferreira C, Jensen PE, Ottosen LM, Ribeiro AB. (2008). Preliminary treatment of MSW fly ash as a way of improving electrodialytic remediation. *Journal of Environmental Science and Health Part A* **43**(8):837–843.

Gardner KH, Nystroem GM, Aulisio D. (2007). Leaching properties of estuarine harbor sediment before and after electrodialytic remediation. *Environmental Engineering Science* **24**(4):424–433.

Gent DB, Bricka RM, Alshawabkeh AN, Larson SL, Fabian G, Granade S. (2004). Bench- and field-scale evaluation of chromium and cadmium extraction by electrokinetics. *Journal of Hazardous Materials* **110**(1–3):53–62.

Giannis A, Gidarakos E. (2005). Washing enhanced electrochemical remediation for removal cadmium from real contaminated soil. *Journal of Hazardous Materials* **123**:165–175.

Goldmann T, Schlösinger F, Rauner T. (1996). Versuche zur elektrokinetischen Sanierung eines mit Kupfer und Arsen belasteten Bodens. *TerraTech* **2**:55–60.

Hamed J, Acar YB, Gale RJ. (1991). Pb(II) removal from kaolinite by electrokinetics. *Journal of Geotechnical Engineering* **117**(2):241–271.

Hansen HK, Ottosen LM, Hansen L, Kliem BK, Villumsen A, Bech-Nielsen G. (1999). Electrodialytic remediation of soil polluted with heavy metals—Key parameters for optimization of the process. *Chemical Engineering Research & Design* **77**(A3):218–222.

Hansen HK, Ottosen LM, Kliem BK, Villumsen A. (1997). Electrodialytic remediation of soils polluted with Cu, Cr, Hg, Pb and Zn. *Journal of Chemical Technology and Biotechnology* **70**(1):67–73.

Hansen HK, Ottosen LM, Villumsen A. (2004). Electrodialytic removal of cadmium from straw combustion fly ash. *Journal of Chemical Technology and Biotechnology* **79**(7):789–794.

Hansen HK, Pedersen AJ, Ottosen LM, Villumsem A. (2001). Speciation and mobility of cadmium in straw and wood combustion fly ash. *Chemosphere* **45**:123–128.

Hansen HK, Ribeiro AB, Mateus EP, Ottosen LM. (2007). Diagnostic analysis of electrodialysis in mine tailing materials. *Electrochimica Acta* **52**(10):3406–3411.

Hansen HK, Rojo A. (2007). Testing pulsed electric fields in electroremediation of copper mine tailings. *Electrochimica Acta* **52**(10):3399–3405.

Hansen HK, Rojo A, Ottosen LM. (2005). Electrodialytic remediation of copper mine tailings. *Journal of Hazardous Materials* **117**(2–3):179–183.

Hansen HK, Rojo A, Ottosen LM. (2007). Electrokinetic remediation of copper mine tailings—Implementation of bipolar electrodes. *Electrochimica Acta* **52**(10):3355–3359.

Hansen HK, Rojo A, Pino D, Ottosen LM, Ribeiro AB. (2008). Electrodialytic remediation of suspended mine tailings. *Journal of Environmental Science and Health Part A* **43**(8): 832–836.

Hansen HK, Yianatos JB, Ottosen LM. (2005). Speciation and leachability of copper in mine tailings from porphyry copper mining: Influence of particle size. *Chemosphere* **60**:1497–1503.

Hicks RE, Tondorf S. (1994). Electrorestoration of metal-contaminated soils. *Environmental Science & Technology* **28**(12):2203–2210.

Jakobsen MR, Fritt-Rasmussen J, Nielsen S, Ottosen LM. (2004). Electrodialytic removal of cadmium from wastewater sludge. *Journal of Hazardous Materials* **106**(2–3):127–132.

Jensen JB, Kubes V, Kubal M. (1994). Electrokinetic remediation of soils polluted with heavy metals. Removal of zinc and copper using a new concept. *Environmental Technology* **15**:1077–1082.

Jensen PE. (2005). Application of microbial products to promote electrodialytic remediation of heavy metal contaminated soil. PhD Thesis, Technical University of Denmark, Lyngby, Denmark.

Jensen PE, Ottosen LM, Ahring BK. (2007). Application of organic acids to promote electrodialytic remediation of Pb-contaminated soil fines (<63 my) in suspension. *Journal of Chemical Technology and Biotechnology* **82**:920–928.

Jensen PE, Ottosen LM, Ferreira C. (2006). Kinetics of electrodialytic extraction of Pb and soil cations from contaminated soil fines in suspension. *Journal of Hazardous Materials* **B138**:493–499.

Jensen PE, Ottosen LM, Ferreira C. (2007). Electrodialytic remediation of Pb-polluted soil fines (<63 my) in suspension. *Electrochimica Acta* **52**:3412–3419.

Jensen PE, Ottosen LM, Harmon TC. (2007). The effect of soil type on the electrodialytic remediation of lead-contaminated soil. *Environmental Engineering Science* **24**(2): 234–244.

Jensen PE, Ottosen LM, Villumsen A. (2007). Electrodialytic removal of toxic elements from sediments of eutrophic fresh waters. *Abstracts of the 6th Symposium on Electrokinetic Remediation* (ed. C. Cameselle). June 12–15, Vigo, Spain: University of Vigo, pp. 25–26.

Khan LI, Alam MS. (1994). Heavy metal removal from soil by coupled electro-hydraulic gradient. *Journal of Environmental Engineering* **120**(6):1524–1543.

Kim SO, Kim KW. (2001). Monitoring of electrokinetic removal of heavy metals in tailing-soils using sequential extraction analysis. *Journal of Hazardous Materials* **85**(3):195–211.

Kim SO, Kim WS, Kim KW. (2005a). Evaluation of electrokinetic remediation of arsenic-contaminated soils. *Environmental Geochemistry and Health* **27**:443–453.

Kim SO, Moon SH, Kim KW. (2001). Removal of heavy metals from soils using enhanced electrokinetic soil processing. *Water Air and Soil Pollution* **125**(1–4):259–272.

Kim SO, Moon SH, Kim KW, Yun ST. (2002). Pilot scale study on the ex situ electrokinetic removal of heavy metals from municipal wastewater sludges. *Water Research* **36**: 4765–4774.

Kim SS, Han SJ, Kang MS, Kim NY. (2006). Applicability of in-situ hybrid electrokinetic remediation technologies on a lead-contaminated "CLAY" shooting range. *5th*

ICEG Environmental Geotechnics: Opportunities, Challenges and Responsibilities for Environmental Geotechnics—Proceedings of the ISSMGE 5th International Congress, June 26–30, Cardiff, UK: Thomas Telford, pp. 302–309.

Kim SY, Tanaka N, Matsuto T, Tojo Y. (2005). Leaching behavior of elements and evaluation of pre-treatment methods for municipal solid waste incinerator residues in column leaching tests. *Waste Management & Research* **23**(3):220–229.

Kim WS, Kim SO, Kim KW. (2005b). Enhanced electrokinetic extraction of heavy metals from soils assisted by ion exchange membranes. *Journal of Hazardous Materials* **118**(1–3):93–102.

Kontopoulus A, Komnitsas K, Xenidis A, Papassiopi N. (1995). Environmental characterization of the sulphidic tailings in Lavrion. *Minerals Engineering* **8**(10):1209–1219.

Kornilovich B, Mishchuk N, Abbruzzese K. (2005). Enhanced electrokinetic remediation of metals-contaminated clay. *Colloids and Surfaces A* **265**(1–3):114–123.

Lageman R. (1989). Theory and Praxis of Electro-reclamation. NATO/CCMS Study. Demonstration of Remedial Action Technologies for Contaminated Land and Groundwater. May 8–9, Copenhagen, Denmark: ATV Denmark.

Lageman R. (1993). Electroreclamation. *Environmental Science & Technology* **27**(13):2648–2650.

Lara R, Rodriguez-Postigo J, Garcia-Herruzo F. (2005). Decontamination of soils by membrane processes: Characterization of membranes under working conditions. *Industrial & Engineering Chemistry Research* **44**(2):400–407.

Le Hecho I, Tellier S, Astruc M. (1998). Industrial site soils contaminated with arsenic or chromium: Evaluation of the electrokinetic method. *Environmental Technology* **19**(11):1095–1102.

Lynch RJ, Muntoni A, Ruggeri R, Winfield KC. (2007). Preliminary tests of an electrokinetic barrier to prevent heavy metal pollution of soils. *Electrochimica Acta* **52**(10):3432–3440.

Maini G, Sharman AK, Knowles CJ, Sunderland G, Jackman SA. (2000). Electrokinetic remediation of metals and organics from historically contaminated soil. *Journal of Chemical Technology and Biotechnology* **75**(8):657–664.

Mann MJ. (1999). Full-scale and pilot-scale soil washing. *Journal of Hazardous Materials* **66**(1–2):119–136.

Matsumoto N, Uemoto H, Saiki H. (2007). Case study of electrochemical metal removal from actual sediment, sludge, sewage and scallop organs and subsequent pH adjustment of sediment for agricultural use. *Water Research* **4**:2541–2550.

Mohamed AMO. (1996). Remediation of heavy metal contaminated soils via integrated electrochemical processes. *Waste Management* **16**(8):741–747.

Nystroem GM, Ottosen LM, Villumsen A. (2005a). Acidification of harbour sediment and removal of heavy metals induced by water splitting in electrodialytic remediation. *Separation Science and Technology* **40**(11):2245–2264.

Nystroem GM, Ottosen LM, Villumsen A. (2005b). Electrodialytic removal of Cu, Zn, Pb, and Cd from harbor sediment: Influence of changing experimental conditions. *Environmental Science & Technology* **39**(8):2906–2911.

Nystroem GM, Pedersen AJ, Ottosen LM, Villumsen A. (2006). The use of desorbing agents in electrodialytic remediation of harbour sediment. *Science of the Total Environment* **357**(1–3):25–37.

Nystrom GM, Ottosen LM, Villumsen A. (2003). The use of sequential extraction to evaluate the remediation potential of heavy metals from contaminated harbor sediment. *Journal de Physique* **107**:975–978.

Nystrom GM, Ottosen LM, Villumsen A. (2005). Test of experimental set-ups for electrodialytic removal of Cu, Zn, Pb and Cd from different contaminated harbor sediments. *Engineering Geology* **77**(3–4):349–357.

O'Connor CS, Lepp NW, Edwards R, Sunderland G. (2003). The combined use of electrokinetic remediation and phytoremediation to decontaminate metal-polluted soils: A laboratory-scale feasibility study. *Environmental Monitoring and Assessment* **84**(1–2): 141–158.

Ottosen LM, Christensen IV, Rörig-Dalgaard I, Jensen PE, Hansen HK. (2008). Utilization of electromigration in civil and environmental engineering. Processes, transport rates and matrix changes. *Journal of Environmental Science and Health Part A* **43**(8):795–809.

Ottosen LM, Hansen HK, Bech-Nielsen G, Villumsen A. (2000). Electrodialytic remediation of an arsenic and copper polluted soil—Continuous addition of ammonia during the process. *Environmental Technology* **21**(12):1421–1428.

Ottosen LM, Hansen HK, Hansen L, Kliem BK, Bech-Nielsen G, Pettersen B, Villumsen A. (1998). Electrodialytic soil remediation—Improved conditions and acceleration of the process by addition of desorbing agents to the soil. *Proceedings of Contaminated Soil '98*, the Sixth International FZK/TNO Conference on Contaminated Soil. May 17–21, London: Thomas Telford, pp. 471–478.

Ottosen LM, Hansen HK, Laursen S, Villumsen A. (1997). Electrodialytic remediation of soil polluted with copper from wood preservation industry. *Environmental Science & Technology* **31**(6):1711–1715.

Ottosen LM, Hansen HK, Ribeiro AB, Villumsen A. (2001a). Removal of Cu, Pb and Zn in an applied electric field in calcareous and non-calcareous soils. *Journal of Hazardous Materials* **85**(3):291–299.

Ottosen LM, Kristensen IV, Pedersen AJ, Hansen HK, Villumsen A, Ribeiro AB. (2003). Electrodialytic removal of heavy metals from different solid waste products. *Separation Science and Technology* **38**(6):1269–1289.

Ottosen LM, Lepkova K, Kubal M. (2006). Comparison of electrodialytic removal of Cu from spiked kaolinite, spiked soil and industrially polluted soil. *Journal of Hazardous Materials* **137**:113–120.

Ottosen LM, Nystroem GM, Jensen PE, Villumsem A. (2007a). Electrodialytic extraction of Cd and Cu from sediment from Sisimiut Harbour, Greenland. *Journal of Hazardous Materials* **140**:271–279.

Ottosen LM, Pedersen AJ, Hansen HK, Ribeiro AB. (2007b). Screening the possibility for removing cadmium and other heavy metals from wastewater sludge and bio-ashes by an electrodialytic method. *Electrochimica Acta* **52**(10):3420–3426.

Ottosen LM, Pedersen AJ, Ribeiro AB, Hansen HK. (2005). Case study on the strategy and application of enhancement solutions to improve remediation of soils contaminated with Cu, Pb and Zn by means of electrodialysis. *Engineering Geology* **77**(3–4):317–329.

Ottosen LM, Ribeiro A, Hansen HK, Kliem BK, Hansen L, Bech-Nielsen G. (1999). Important parameters for electrodialytic removal of heavy metals from a polluted soil matrix. *Electrochemical Society Proceedings '99, 196th Meeting of the Electrochemical Society*. October 17–22. Pennington, NJ: The Electrochemical Society, pp. 151–161.

Ottosen LM, Villumsen A, Hansen HK, Ribeiro, AB, Jensen PE, Pedersen AJ. (2001b). Electrochemical soil remediation—Accelerated soil weathering? In *EREM 2001 3rd Symposium and Status Report on Electrokinetic Remediation* (eds. C Czurda, R Haus, C Kappeler, R Zorn). Angewandte Geologie Karlsruhe, pp. 5.1–5.14.

Page MM, Page CL. (2002). Electroremediation of contaminated soils. *Journal of Environmental Engineering* **128**(3):208–219.

Pamukcu S, Khan LI, Fang H-Y. (1990). Zinc detoxification of soils by electro-osmosis. *Transportation Research Record* **1288**:41–46.

Pedersen AJ. (2003). Characterization and electrodialytic treatment of wood combustion fly ash for the removal of cadmium. *Biomass Bioenergy* **25**:447–458.

Pedersen AJ, Ottosen LM. (2006). Elemental analysis of ash residue from combustion of CCA treated waste wood before and after electrodialytic extraction. *Chemosphere* **65**:110–116.

Pedersen AJ, Ottosen LM, Villumsen A. (2003). Electrodialytic removal of heavy metals from different fly ashes—Influence of heavy metal speciation in the ashes. *Journal of Hazardous Materials* **100**(1–3):65–78.

Pedersen AJ, Ottosen LM, Villumsen A. (2005). Electrodialytic removal of heavy metals from municipal solid waste incineration fly ash using ammonium citrate as assisting agent. *Journal of Hazardous Materials* **122**(1–2):103–109.

Probstein RF, Renaud PC. (1987). Electroosmotic control of hazardous wastes. *Journal of Physicochemical Hydrology* **9**:345–360.

Reddy KR, Ala PR. (2006). Electrokinetic remediation of contaminated dredged sediment. *Journal of ASTM International* **3**(6):254–267.

Reddy KR, Ala PR, Sharma S, Kumar SN. (2006). Enhanced electrokinetic remediation of contaminated manufactured gas plant soil. *Engineering Geology* **85**:132–146.

Reddy KR, Chaparro C, Saichek RE. (2003). Removal of mercury from clayey soils using electrokinetics. *Journal of Environmental Science and Health Part A* **38**(2):307–338.

Reddy KR, Chinthamreddy S, Al-Hamdan A. (2001). Synergistic effects of multiple metal contaminants on electrokinetic remediation of soils. *Remediation Journal* **11**(3):85–109.

Reddy KR, Saichek RE. (2004). Enhanced electrokinetic removal of phenanthrene from clay soil by periodic electric potential application. *Journal of Environmental Science and Health Part A* **39**(5):1189–1212.

Ribeiro AB. (1994). Electrokinetic removal of heavy metals from a polluted soil. In *Comett Course—Pollution Control and Removal of Pollutants (Nitrates, Nitrites, Heavy Metals) from surface and groundwaters*. March 21–23, Lisbon, Portugal: Instituto Superior Tecnico.

Ribeiro AB, Mateus EP, Ottosen LM, Bech-Nielsen G. (2000). Electrodialytic removal of Cu, Cr, and As from chromated copper arsenate-treated timber waste. *Environmental Science & Technology* **34**(5):784–788.

Ribeiro AB, Mexia JT. (1997). A dynamic model for the electrokinetic removal of copper from a polluted soil. *Journal of Hazardous Materials* **56**(3):257–271.

Ribeiro AB, Villumsem A, Bech-Nielsen G, Réfega A, Vieira e Silva J. (1998). Electrodialytic remediation of a soil from a wood preservation industry polluted by CCA. *Proceedings of the 4th International Symposium on Wood Preservation*. paper no 50101:14, February 2–3, Cannes-Mandelieu, France. Stockholm: IRG Secretariat.

Ricart MT, Hansen HK, Cameselle C, Lema JM. (2005). Electrochemical treatment of a polluted sludge: Different methods and conditions for manganese removal. *Separation Science and Technology* **39**(15):3679–3689.

Rojo A, Hansen HK, Ottosen LM. (2006). Electrodialytic remediation of copper mine tailings: Comparing different operational conditions. *Minerals Engineering* **19**:500–504.

Samson D, Tack F, Verloo M. (2003). Electroremediation study on heavy metal contaminated sediment by using ion selective membranes. (eds. GR Gobran and N Lepp) *Proc. Ext. Abst. 7th. Int. Conf. Biogeochemistry of Trace Elements*, June 15–19, Uppsala, Sweden. SLU Service/Repro, pp. 280–281.

Sandoval-Gonzales A, Silva-Martinez S, Blass-Amador G. (2007). Ultrasound leaching and electrochemical treatment combined for lead removal soil. *Journal of New Materials for Electrochemical Systems* **10**(3):195–199.

Segall BA, O'Bannon CE, Matthias JA. (1980). Electro-osmosis chemistry and water quality. *Journal of Geotechnical Engineering Division* **106**(GT10):1148–1152.

Sposito G. (1984). *The Surface Chemistry of Soils*. Oxford: Oxford University Press.

Suer P, Allard B. (2003). Mercury transport and speciation during electrokinetic soil remediation. *Water Air and Soil Pollution* **143**(1–4):99–109.

Suer P, Gitye K, Allard B. (2003). Speciation and transport of heavy metals and macroelements during electroremediation. *Environmental Science & Technology* **37**(1):177–181.

Suer P, Lifvergren T. (2003). Mercury-contaminated soil remediation by iodide and electroreclamation. *Journal of Environmental Engineering* **129**(5):441–446.

Thoming J, Kliem BK, Ottosen LM. (2000). Electrochemically enhanced oxidation reactions in sandy soil polluted with mercury. *Science of the Total Environment* **261**(1–3):137–147.

Traina G, Morselli L, Adorno GP. (2007). Electrokinetic remediation of bottom ash from municipal solid waste incinerator. *Electrochimica Acta* **52**(10):3380–3385.

VanBenschoten JE, Matsumoto MR, Young WH. (1997). Evaluation and analysis of soil washing for seven lead-contaminated soils. *Journal of Environmental Engineering* **123**(3):217–224.

Virkutyte J, Sillanpaa M. (2007). The hindering of experimental strategies on advancement of alkaline front and electroosmotic flow during electrokinetic lake sediment treatment. *Journal of Hazardous Materials* **143**:673–681.

Virkutyte J, Sillanpaa M, Lens P. (2006). Electrokinetic copper and iron migration in anaerobic granular sludge. *Water Air and Soil Pollution* **117**(1–4):147–168.

Wang JY, Zhang DS, Stabnikova O, Tay JH. (2004). Processing dewatered sewage sludge using electrokinetic technology. *Water Science and Technology* **50**(9):205–211.

Weng C, Lin Y, Lin T, Kao CM. (2007). Enhancement of electrokinetic remediation of hyper-Cr(VI) contaminated clay by zero-valent iron. *Journal of Hazardous Materials* **149**(2):292–302.

Weng CH, Takiyama LR, Huang CP. (1994). Electro-osmosis for the in-situ treatment of chromium-contaminated soil. *Hazardous and Industrial Wastes* **26**:496–505.

Wieczorek S, Weigand H, Schmid M, Marb C. (2005). Electrokinetic remediation of an electroplating site: Design and scale-up for an in-situ application in the unsaturated zone. *Engineering Geology* **77**:203–215.

Yuan C, Chiang TS. (2007). The mechanisms of arsenic removal from soil by electrokinetic process coupled with iron permeable reaction barrier. *Chemosphere* **67**(8):1533–1542.

Zagury GJ, Dartiguenave Y, Setier J-C. (1999). Ex situ electroreclamation of heavy metals contaminated sludge: Pilot scale study. *Journal of Environmental Engineering* **125**(10): 972–978.

Zhou DM, Chen HF, Chang L, Wang Y. (2007). Ryegrass uptake of soil Cu/Zn induced by EDTA/EDDS together with a vertical direct-current electrical field. *Chemosphere* **67**(8):1671–1676.

5

ELECTROKINETIC REMOVAL OF RADIONUCLIDES

VLADIMIR A. KOROLEV

5.1 INTRODUCTION

A number of serious problems occur in the process of atomic energy enterprises operations that are related to the pollution of certain sites with radioactive materials. The sources of the radioactive contamination of soils are the sites where different atomic objects are located, for example, plants for processing of nuclear wastes, atomic power plants, plants for the production of nuclear fuel and nuclear reactors, and others. Radioactive contamination enters into the soils as a result of different emergencies concerning the storage, transport, and use of nuclear fuel or nuclear waste. Nuclear tests were important and dangerous sources of the entrance of radioactive contamination into the environment prior to their prohibition across the world.

The principal among radioactive materials are the following isotopes: ^{60}Co, ^{90}Sr, ^{90}Y, ^{106}Ru, ^{137}Cs, ^{144}Ce, ^{147}Pm, ^{238,239,240}Pu, ^{226}Ra, and so on. The groups of toxic radioactive elements are presented in Table 5.1. Approximately 95% of radioactive pollutants are concentrated in the upper soil layer. Individual radioactive nuclides are present in the soil in different forms. Depending on the soil composition, 8%–30% of radioactive nuclides are present in exchange forms, 2%–10% of them are in water-soluble forms, and 60%–85% in tightly bounded forms (Table 5.2).

Bednar *et al.* (2004) describe the processes affecting thorium mobility in semiarid soil, which has implications for future remedial action. Aqueous extraction and filtration experiments have demonstrated the colloidal nature of thorium in the soil, due in part to the low solubility of thorium oxide. Colloidal material was defined as that removed by a 0.22-μm or smaller filter after being filtered to a nominally dissolved size (0.45 μm). Additionally, association of thorium with natural organic

---

Electrochemical Remediation Technologies for Polluted Soils, Sediments and Groundwater,
Edited by Krishna R. Reddy and Claudio Cameselle
Copyright © 2009 John Wiley & Sons, Inc.

TABLE 5.1. Groups of Toxic Radioactive Elements

Index of Groups	Group of Toxicity	Radioactive Elements
A	Particularly high	^{210}Pb, ^{210}Po, ^{226}Ra, ^{228}Th, ^{232}Th, ^{232}U, ^{237}Np, ^{238}Pu, ^{239}Pu, ^{241}Am, ^{242}Cm
B	High	^{90}Sr, ^{106}Ru, ^{124}Sb, ^{126}I, ^{129}I, ^{144}Ce, ^{170}Tm, ^{210}Bi, ^{223}Ra, ^{224}Ra, ^{227}Th, ^{234}Th, ^{230}U, ^{233}U, ^{234}U, ^{235}U, ^{241}Ru
C	Middle	^{22}Na, ^{24}Na, ^{32}P, ^{35}S, ^{36}Cl, ^{54}Mn, ^{56}Mn, ^{59}Fe, ^{60}Co, ^{82}Br, ^{89}Sr, ^{90}Y, ^{91}Y, ^{95}Nb, ^{95}Zr, ^{105}Ru, ^{125}Sb, ^{132}I, ^{133}I, ^{134}I, ^{134}Cs, ^{137}Cs, ^{141}Ce, ^{171}Tm, ^{203}Pb, ^{206}Bi, ^{231}Th, ^{239}Np
D	Small	^{14}C, ^{38}Cl, ^{55}Fe, ^{64}Cu, ^{69}Zn, ^{71}Ge, ^{97}Zr, ^{131}Cs, ^{136}Cs
E	Low	^{3}H

TABLE 5.2. Contents of the Different Forms of ^{137}Cs in Soils

Soil	Forms of ^{133}Cs (%)		
	Exchange Forms	Water-Soluble Forms	Tightly Bounded Forms
Clay sand	28.3	10.3	61.4
Loam	21.3	5.2	73.5
Gray forest	9.3	5.8	84.9
Meadow	8.6	6.3	85.1
Chernozem	14.7	3.5	81.8
Sod-podzol	19.6	6.3	85.1

matter is suggested by microfiltration and ultrafiltration methods and electrokinetic data, which indicate thorium migration as a negatively charged particle or an anionic complex with organic matter. Soil fractionation and digestion experiments show a bimodal distribution of thorium in the largest and smallest size fractions, most likely associated with detrital plant material and inorganic oxide particles, respectively. Plant uptake studies suggest this could also be a mode of thorium migration as plants grown in thorium-containing soil had higher thorium concentration than those grown in control soils. Soil erosion laboratory experiments with wind and surface water overflow were performed to determine bulk soil material movement as a possible mechanism of mobility. Information from these experiments is being used to determine viable soil stabilization techniques at the site to maintain a usable training facility with minimal environmental impact (Bednar et al., 2004).

Therefore, the problem of effective remediation of soils of radioactive nuclides is very relevant. Most technologies of soil decontamination that are currently developed are based on flushing soils with various chemicals and include processes of chemical leaching and selective extraction of radioactive nuclides (Prozorov et al., 2000; Shevtsova, 2003). The electrokinetic technique is a new and perspective method of soil remediation, whose main advantage consists in its applicability to decontaminating soils with low filtration ability directly at the site of its local pollution (e.g. in a rock mass) (Pamukcu and Wittle, 1992; Acar et al., 1993, 2001; Janosy and Piot, 1998; Korolev, 2001, 2006; Korolev, Barkhatova, and Shevtsova, 2007).

Electrokinetic technologies are applicable in various options for radioactive nuclide removal. The first option includes methods of electrokinetic localization of radioactive pollution foci with the purpose of preventing farther diffusion of the pollution into the environment. This option incorporates methods of protective electrokinetic barriers (protective screens) and technologies of electrokinetic restoration of protective functions of these sorption barriers (screens). The second option includes methods of soil cleaning from radioactive nuclides by means of electrokinetic technologies. These technologies may vary depending on the applied electrokinetic process (with or without radioactive nuclide leaching). This chapter discusses these options in more detail.

5.2 ELECTROKINETIC LOCALIZATION OF RADIOACTIVE NUCLIDE POLLUTION

5.2.1 Electrokinetic Antiradioactive Nuclide Screens

Electrokinetic screens are applied for preventing radioactive nuclide diffusion from the pollution foci into the environment (groundwater, soil, and other subsurface materials), and they play the role of protective screens against possible migration of radioactive nuclides. Figures 5.1 and 5.2 show schemes of such electrokinetic

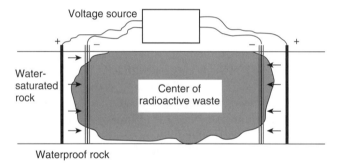

Figure 5.1. Protective electrokinetic screen around the radioactive waste.

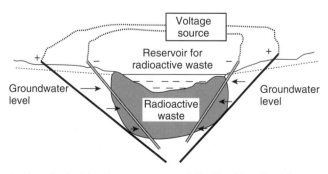

Figure 5.2. Protective electrokinetic screen around the liquid radioactive waste depositary.

protective screens. Figure 5.1 shows a section of an electrokinetic protective screen that is created around a focus of radioactive nuclide pollution in a case when the focus lies at a shallow depth and is underlain by an aquiclude. The electric field generated around the focus by means of rows of anode and cathode wells is oriented in the direction opposite to possible radioactive nuclide migration and forces the radioactive nuclides to migrate toward the center of the focus and not away from it.

Figure 5.2 shows another scheme of an electrokinetic protective screen. This is applicable in a case when a focus of radioac5tive nuclide pollution is located on the ground surface and is an accumulating pond of liquid radioactive waste. In a case when no protective screen exists at the base of this pond, the radioactive nuclides will then migrate downward and away from the pond with a flow of percolating water. An inclined protective screen consisting of rows of anode and cathode wells should be constructed along the perimeter of this accumulating pond with the purpose of preventing this migration. The electric field generated by these wells will localize the radioactive nuclides within the accumulating pond and will prevent pollution of adjacent soils and groundwater.

The presented schemes of screens are applicable at emergency sites and in other foci of radioactive pollution.

5.2.2 Electrokinetic Restoration of Protective Screens

Restoration of the protective properties of artificial sorption screens created on the ways of radioactive nuclide migration from accumulating ponds is no less an important problem. Usually, clay soils with high sorption ability and containing such minerals as smectite and illite are applied for constructing such screens. In addition, artificial protective screens made of other disperse soils (from sand to loam) are applied; the sorption ability of such soils is artificially enhanced beforehand with strengthening compoundssuch as oxalic silica-alumina gel. These compounds decrease the filtration ability of the screen and raise its sorption characteristics.

Colloid formation of uranium, thorium, radium, lead, polonium, strontium, rubidium, and cesium in briny (high ionic strength) groundwater is studied to predict their capability as vectors for transporting radionuclides (Maiti, Smith, and Laul, 1989). This knowledge is essential in developing models to infer the transport of radionuclides from the source region to the surrounding environment. Except for polonium, based on the experimental results, colloid formation of uranium, thorium, radium, lead, strontium, rubidium, and cesium is unlikely in brines with compositions similar to the synthetic Palo Duro Basin brine. This observation of no colloid formation is explained by electrokinetic theory and inorganic solution chemistry (Maiti, Smith, and Laul, 1989).

However, the protective properties of such screens deteriorate with time, and the screens themselves accumulate considerable amount of radioactive nuclides. Application of electrokinetic technology is very efficient for restoring the protective properties of such screens. Its application can help not only raise the sorption properties of a protective screen but also clean it from accumulated radioactive nuclides without extracting the polluted ground.

We performed special laboratory investigations for examining the possibility of restoring protective properties of a sand screen strengthened with oxalic silica-

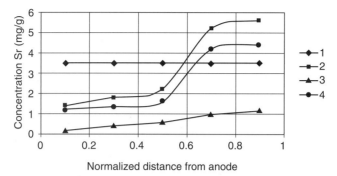

Figure 5.3. Redistribution of ^{88}Sr in different forms in the sand–gel soil: 1—initial concentration; 2—end total concentration; 3—end concentration in free form; 4—end concentration in sorbed form.

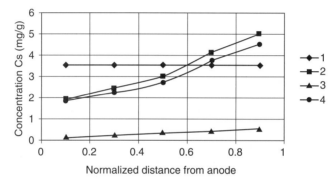

Figure 5.4. Redistribution of ^{133}Cs in different forms in the sand–gel soil: 1—initial concentration; 2—end total concentration; 3—end concentration in free form; 4—end concentration in sorbed form.

alumina gel. Sand specimens strengthened with oxalic silicate-alumina gel contained 3.5 mg/g ^{88}Sr stable isotopes and the same proportion of ^{133}Cs as radioactive nuclide. They underwent electrokinetic cleaning without application of leaching agents.

We found out that the ^{88}Sr concentration decreased in a cleaned specimen from 3.5 to 1.2 mg/g in the anode zone and increased to 4.4 mg/g in the cathode zone (Fig. 5.3). In much the same way, the ^{133}Cs concentration changed from 1.8 to 4.4 mg/g in the same zones (Fig. 5.4).

These results show that the total concentration of strontium and cesium decreased in the absorbing complex of the ground owing to electrokinetic cleaning and that radioactive nuclides migrated from the anode to the cathode. Electrokinetic processes in ground strengthened with oxalic silica-alumina gel are similar to those that were recorded in clay ground. Both the pore fluid with dissolved radioactive nuclides and the pollutant sorbed on the surface of gel and clay particles became involved in electrokinetic migration. In addition, it is known that strontium is most mobile in an acid medium (pH < 5.5) and becomes least mobile in weakly acid (5.5 < pH < 7.5) and alkali (pH > 7.5) media resultant from precipitation processes. This geochemical

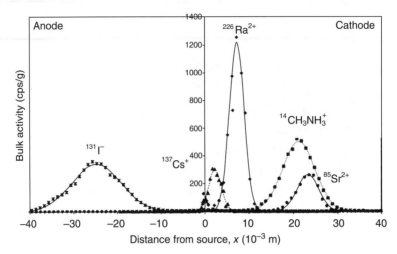

Figure 5.5. Typical radionuclide distribution profiles in the clay cores obtained after electromigration (Maes *et al.*, 2001).

factor does not prevent electrokinetic cleaning since electric migration of Sr^{2+} and Cs^+ is most active in the acid anodic region.

Figures 5.3 and 5.4 show the results of determining electrokinetic migration of Sr^{2+} and Cs^+ in different forms in the sand–gel ground: in free, dissolved, and sorbed forms. These graphs show that both free (dissolved) and sorbed (bound) radioactive nuclides in the sand ground strengthened with oxalic silica-alumina gel become involved in electrokinetic migration. Hence, electrokinetic treatment helps clean the screens and restore their protective ability.

To develop an effective assessment of protective clay screens, the great importance of the diffusion of radionuclides should be considered. Maes *et al.* (1999, 2001) give an overview of the quantitative and qualitative use of electromigration as a powerful technique to study radionuclides migration (^{137}Cs and ^{226}Ra) in clays. Typical diffusion patterns in the clay cores for some studied radionuclides are presented in Figure 5.5. The method is fully assessed for the determination of migration parameters (diffusion coefficient and over) of radionuclides with a simple chemistry. Electromigration is a fast technique, making it possible to obtain migration parameters for strongly retarded species such as Cs^+, Ra^{2+} and over radionuclides in a relatively short period of time.

5.3 ELECTROKINETIC CLEANING OF GROUND FROM RADIOACTIVE NUCLIDES

The cleaning of radionuclides from the soil is a very important practical problem. This problem has become particularly relevant after the accident at the Chernobyl nuclear power plant in 1986 (Fig. 5.6). As a result of the accident, land in the vast territory of Ukraine, Belarus, Russia, and some neighboring states were contaminated with radionuclides (^{137}Cs, ^{90}Sr, $^{239,240}Pu$; Tables 5.3–5.5). Cesium isotopes were the main polluter after the accident. Some data on contaminated land by ^{137}Cs in

Figure 5.6. The accident at the Chernobyl nuclear power plant in 1986.

TABLE 5.3. The Area of Land Contaminated by ^{137}Cs in Some European Countries after the Accident at the Chernobyl Nuclear Power Plant in 1986

Countries	Square of 10^3 km^2		Chernobyl Nuclear Fallout	
	Country	Territory with Pollutions More Than 1 Ci/km^2	kCi	% of Summary Nuclear Fallout in Europe
Austria	84	11.08	42.0	2.4
Belorussia	210	43.5	400.0	23.4
Great Britain	240	0.16	14.0	0.83
Germany	350	0.32	32.0	1.8
Greece	130	1.24	19.0	1.1
Italy	280	1.35	15.0	0.9
Norway	320	7.18	53.0	3.1
Poland	310	0.52	11.0	0.63
Russia (European part)	3800	59.30	520.0	29.7
Romania	240	1.20	41.0	2.4
Slovakia	49	0.02	4.7	0.28
Slovenia	20	0.61	8.9	0.52
Ukraine	600	37.63	310.0	18.8
Finland	340	19.0	83.0	4.8
Czechia	79	0.21	9.3	0.54
Switzerland	41	0.73	7.3	0.43
Sweden	450	23.44	79.0	4.6
Europe (total)	9700	207.5	1700.0	100.0
Whole world			2100.0	

TABLE 5.4. The Area of Land Contaminated by ^{137}Cs (in Thousands of Hectares) in Some Regions of Russia after the Accident at the Chernobyl Nuclear Power Plant in 1986 (by 1998)

Region	Total (Thousands of Hectares)	Density of Soil Pollution (Ci/km^2)		
		1–5	5–15	15–40
Bryansk	168.8	103.1	39.7	26.0
Kaluga	177.8	132.6	43.8	1.4
Orel	97.4	95.6	1.5	—
Ryazan	70.3	70.2	0.1	—
Kursk	21.3	21.2	0.1	—
Penza	148.4	148.4	—	—

TABLE 5.5. The Area of Land Contaminated by Radionuclides in the Ural Regions of Russia

Region	Total (Thousands of Hectares)	Including in Area Extent Contaminated Soils by ^{137}Cs (Ci/km^2)			
		1–5	5–15	15–40	>40
Chelyabinsk	18.65	18.57	—	—	0.08
Kurgan	0.76	0.68	0.08	—	—
Total	19.41	19.25	0.08	—	0.08

Source: Federal Forestry Agency of the Russian Federation, 2007.

different areas of Russia are listed in Table 5.5. In Russia, the largest areas of radioactive pollution are located in the Bryansk and Tula areas. In these areas after the accident, the raised meanings of γ-radiation are registered, which vary a little from year to year.

On the Russian territory, some lands in the Urals and in Siberia are also contaminated by radionuclides. In the Ural region, heavily contaminated soils can be found along the valley of the river Techa (Fig. 5.7). Significant territory is contaminated by radionuclides around the manufacturing association "Mayak" (Table 5.6). Electrokinetic removal of the radionuclides on these territories is very urgent.

5.3.1 Electrokinetic Cleaning of Ground

Cleaning of ground, including soils, from radioactive nuclide pollutants is another relevant problem. The electrokinetic method is the most efficient of the available methods of removing radioactive nuclides from the ground; its application has two options—with and without flushing.

We examined in our experiments the factors that influence electrokinetic cleaning of different clay grounds (from sandy loam to clay) from radioactive nuclides containing isotopes of ^{90}Sr, ^{88}Sr, ^{137}Cs, ^{133}Cs, and so on. Of them, isotopes of ^{90}Sr and ^{137}Cs are the most biologically harmful, having half-value periods of 28 and 30 years, respectively. These and other radioactive isotopes are present in both free (dissolved) and sorbed forms on particle surfaces (Cornell, 1993).

Figure 5.7. Floodlands of the river Techa (Ural region) contaminated by the radionuclides as a result of emergency on the manufacturing association "Mayak." The zone of the radioactive contamination of the soil on ^{137}Cs composes 1–5 Ci/km^2, and on ^{90}Sr composes 0.15–3.0 Ci/km^2 (Federal Forestry Agency of the Russian Federation, 2007).

TABLE 5.6. The Area of Land Contaminated by Radionuclides in the Territory of Manufacturing Firms in Some Regions of Russia (by 1998)

Manufacturing Firm	Region	Area of Contaminated Land (ha)
Manufacturing association "Mayak"	Chelyabinsk, Ural	168,000
Siberian chemical industrial complex	Tomsk, West Siberia	1039
Mining industrial complex	Chita, Buryatia	842
Mining industrial complex	Krasnoyarsk, Central Siberia	442
National enterprise "Diamond"	Stavropol, North Caucasus	134

The results of electrokinetic cleaning of sandy loam from ^{88}Sr isotope present in the ground in different forms are shown in Figure 5.8. The experiments were performed without flushing (closed system). It follows from the obtained data that the strontium concentration decreased in the anode zone nearly twice because of cleaning. At that, strontium migrated from the anode to the cathode, where its concentration conversely gradually grew. We also found that the ground cleaning proceeded largely at the expense of distributing sorbed strontium, whereas its redistribution is insignificant in the free form. Thus, the electrokinetic technique helps significantly clean the anode zone of the ground from radioactive nuclides.

5.3.2 Electrochemical Leaching of Radioactive Nuclides

The most efficient electrokinetic removal of radioactive nuclides from polluted ground can be achieved in combination with electrochemical leaching, that is, with

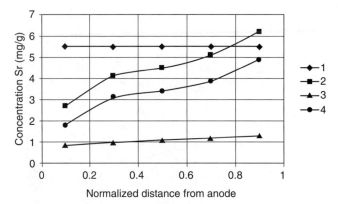

Figure 5.8. Redistribution of ^{88}Sr in different forms in the loamy soil: 1—initial total concentration; 2—end total concentration; 3—end concentration in free form; 4—end concentration in sorbed form.

ground flushing. In this case, we applied a stripping anolyte filtered through the polluted ground and facilitating a more complete electrokinetic removal of radioactive nuclides with a catholyte (Shevtsova, 2003; Korolev, Barkhatova, and Shevtsova, 2007).

The experiment for removing cesium and cobalt from the contaminated soil around the TRIGA (Training, Research, Isotopes, General Atomics) reactor in Korea using the electrokinetic method was conducted (Kim et al., 2003). To increase the removal efficiency by the electrokinetic method, a device to restrain the pH increase in the soil column was suggested. An acetate buffer solution (CH_3COONa + CH_3COOH) was injected into the soil column, and acetic acid was periodically injected into the cathode reservoir to restrain any pH increase. Many ^{137}Cs and ^{60}Co were transferred by electromigration rather than by electroosmosis during the initial remediation period, and no precipitate was formed in the soil column. Twenty-five per cent of cesium radioactivity in the soil column was removed after 10 days, while 94% of cobalt radioactivity was removed. Furthermore, the remaining radioactive concentrations predicted by the developed numerical model were similar to those obtained by the experiment (Kim et al., 2003).

The NPO Radon organization (Moscow, Russia) has developed various methods to manage waste from various nuclear facilities (Prozorov et al., 2000; Perera, 2002). The liquid waste treatment includes effluent processing, cementation, bituminization, and vitrification. The solid waste treatment includes compaction and plasma incineration. The Radon organization has developed an electrokinetic method of decontamination based on ion transport in a direct current field that can be used at the sites. The purpose of the investigations was to determine the best conditions of electrokinetic soil decontamination. Impacts of the different parameters of electric field, such as current, voltage between the electrodes, and power, on the dynamics of ^{137}Cs extraction were studied. The investigations have demonstrated the main possibilities of the electrokinetic remediation of loamy soil from radionuclide contamination (Prozorov et al., 2000; Shevtsova, 2003).

We examined jointly with NPO Radon the specimens of moraine sandy loam polluted with technogenic ^{137}Cs waste resultant from a spill of liquid radioactive

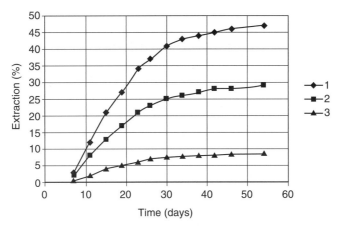

Figure 5.9. Kinetics of solvent exhaustion ^{137}Cs from loam by anolytes: 1—NH$_4$NO$_3$; 2—KNO$_3$; 3—H$_2$O (Shevtsova, 2003).

waste with a specific ^{137}Cs activity of 80–200 kBq/kg (Korolev, Barkhatova, and Shevtsova, 2007). For the experiments, we applied a three-celled electrokinetic installation, in which the anode and cathode cells were separated by neutral baffles and by a drainage filter filled with polymer from the central cell filled with polluted ground. Stainless steel (cathodes) and platinum-coated titanium (anode) plates were applied as electrodes. The prepared paste of the test ground was placed in the central cell of the electrokinetic installation. Stripping agents (NH$_4$NO$_3$ or KNO$_3$ and others) were pumped into the anode cell in a concentration of 1 mol/l. We supported pH ≤ 1 for neutralizing the alkali that occurred in the catholyte at water electrolysis and for preventing segregation and precipitation of metal hydroxides in the cathode cells; this was achieved by adding concentrated nitric acid. We removed the ^{137}Cs nuclide supplied into the cathode cell by means of changing the catholyte. The kinetics of ^{137}Cs removal from a sandy loam with different anolyte solutions is shown in Figure 5.9.

In the next series of experiments, we set a goal of raising the efficiency of electrochemical removal of ^{137}Cs by selecting an agent for transforming radioactive nuclides into a more mobile state. Simultaneously, research studies were aimed at lowering the ground acidity and selecting more sparing, ecologically safe, deactivation conditions that could cut down the expenses for the subsequent neutralization of acidity, that is, for a more complete rehabilitation of polluted territories.

Upon investigating the ^{137}Cs leaching from the ground, we performed experiments aimed at comparing the efficiency of application of nitric and phosphoric anolytes for the electrokinetic purification technique. Shevtsova (2003) found that application of mixed monomolar solutions of acids and their ammonia salts as anolyte is the most advisable: a monomolar solution of H$_3$PO$_4$ + NH$_4$H$_2$PO$_4$ and a monomolar solution of HNO$_3$ + NH$_4$NO$_3$ (Fig. 5.10).

The results of these laboratory experiments demonstrated that cation NH$_4^+$ is more efficient than cation K$^+$ as a desorbent in the electrochemical cleaning of ground. Extraction of ^{137}Cs is 1.6 times more efficient when using NH$_4$NO$_3$ solutions than when using KNO$_3$ solutions. In this case, the total extraction of ammonium

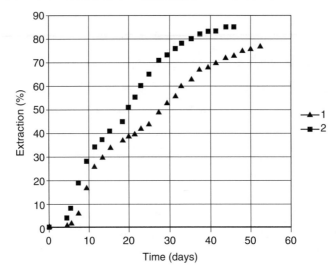

Figure 5.10. Kinetics of solvent exhaustion ^{137}Cs from loam by anolytes: 1—HNO_3 + NH_4NO_3; 2—H_3PO_4 + $NH_4H_2PO_4$.

nitrate and potassium nitrate was approximately 50% and 30%, respectively, which is 5.4 and 3.3 times higher than when without salt application. The highest cleaning indices were achieved with a complex agent of a monomolar solution of H_3PO_4 + $NH_4H_2PO_4$. These leaching agents can be applied as the most efficient ones for electrokinetic cleaning of ground from radioactive nuclides.

5.4 SUMMARY

Electrokinetic technologies can be successfully applied for localization of radioactive pollution foci, cleaning polluted ground from radioactive nuclides, and restoring protective functions of sorption screens on the way of radioactive nuclide migration. Electromigration is fully assessed for the determination of the migration parameters of different radionuclides in soils.

REFERENCES

Acar YB, Gale NJ, Ugaz A, Marks R. (1993). Feasibility of removing uranyl, thorium and radium from kaolinite by electrokinetics. *Proceedings of the 19th Annual RREL Hazardous Waste Research Symposium.* US Environmental Protection Agency, EPA/600/R-93/040, pp.161–165.

Acar YB, Gale R, Marks RE, Ugaz A. (2001). *Feasibility of Removing Uranium, Thorium, and Radium from Kaolinite by Electrochemical Soil Processing.* US Environmental Protection Agency Report No. 009-292. Baton Rouge, LA: Electrokinetics.

Bednar AJ, Gent DB, Gilmore JR, Sturgis TC, Larson SL. (2004). Mechanisms of thorium migration in a semiarid soil. *Journal of Environmental Quality* **33**(6):2070–2077.

Cornell RM. (1993). Adsorption of cesium on minerals: A review. *Journal of Radioanalytical and Nuclear Chemistry* **2**(171):483–500.

Federal Forestry Agency of the Russian Federation. (2007). Overcoming the consequences of emergency on CHAES in Russia. Safeguard measure in forestry. http://www.rosleshoz.gov.ru.

Janosy LA, Piot P. (1998). PRBs on the way of migration of radioactive wastes. *Oak Ridge National Laboratory Inquirer* **8**:92.

Kim G-N, Oh W-Z, Won H-J, Kim M-G. (2003). TRIGA soil decontamination by electrokinetic method. *Proceedings of the International Conference on Radioactive Waste Management and Environmental Remediation, ICEM '03*, Oxford, UK, September 21–25, pp. 1491–1497.

Korolev VA. (2001). *Cleaning Soils from Pollution*. Moscow: MAIK/Interperiodika (in Russian).

Korolev VA. (2006). Electrokinetic remediation of a contaminated land in cities. *Proceedings of the 10th IAEG Congress, "Engineering Geology for Tomorrow's Cities"*, September 6–10, 2006, Nottingham, UK. (E-version, Paper 134). http://www.iaeg.info/iaeg2006/PAPERS/IAEG_134.PDF.

Korolev VA, Barkhatova YE, Shevtsova EV. (2007). Elektrokinetic remediation of the radionuclide-contaminated soils. *Proceedings of the 6th Symposium on Electrokinetic Remediation (EREM-2007)*, June 12–15, 2007, Vigo, Spain: University of Vigo, pp. 169–170.

Maes N, Moors H, Dierckx A, Aertsens M, Wang L, De Canniere P, Put M. (2001). Studying the migration behaviour of the radionuclides in boom clay by electromigration. 3rd Symposium and Status Report on Electrokinetic Remediation (EREM-2001). *Schriftenreihe Angewandte Geologie Karlsruhe* **63**:35-1–35-21.

Maes N, Moors H, Dierckx A, De Canniere P, Put M. (1999). The assessment of electromigration as a new technique to study diffusion of radionuclides in clayey soils. *Journal of Contaminant Hydrology* **36**:231–247.

Maiti TC, Smith MR, Laul JC. (1989). Colloid formation study of U, Th, Ra, Pb, Po, Sr, Rb, and Cs in briny (high ionic strength) groundwaters: Analog study for waste disposal. *Nuclear Technology* **84**(1):82–87.

Pamukcu S, Wittle JK. (1992). Electrokinetics for removal of low-level radioactivity from soil. *Fourteenth Annual Department of Energy Low-Level Radioactive Waste Management Conference*, November 18–20, 1992, Phoenix, AZ. Conf-921137-Proc, pp. 256–278.

Perera J. (2002). Waste control and management in Moscow. *Nuclear Engineering International* **47**(581):39–41.

Prozorov LB, Shcheglov MY, Nikolaevsky VB, Shevtsova EV, Korneva SA. (2000). The influence of electric parameters on the dynamics of the electrokinetic decontamination of soils. *Journal of Radioanalytical and Nuclear Chemistry* **246**(3):571–574.

Shevtsova EV. (2003). *Research and Development of Physical-Chemical Bases of Electrokinetic Technology of Soils Clearing from Radionuclides*. Moscow: NPO "Radon" (in Russian).

6

ELECTROKINETIC REMOVAL OF NITRATE AND FLUORIDE

KITAE BAEK AND JUNG-SEOK YANG

6.1 INTRODUCTION

The major components of most soils are silicate and aluminosilicate minerals. Under natural conditions, these components have negative charges. Anionic pollutants are less attracted to negatively charged surfaces. The high solubility of anionic pollutants and low adsorptive capacity of soils can result in the persistence of high concentration of anionic pollutants, including Cl^-, $Cr(VI)$, NO_3^-, $Se(VI)$, $Se(IV)$, $As(V)$, $As(III)$, $Tc(VII)$, $Mo(VI)$, PO_4^-, F^-, ClO_4^-, and SO_4^{2-} (Blowes et al., 2000).

Most pollution by inorganic compounds is caused by both natural and industrial sources. For example, arsenic is a naturally occurring element present in rocks and soil. Arsenic and fluoride contamination of groundwater may come from the geological characteristics of the areas through which it passes. Nitrate is also a common groundwater pollutant even though it does not come from naturally occurring rocks or minerals. Nitrate contamination is usually the result of human activities such as land cultivation, with chemical fertilizer representing a major source of nitrate pollution. Table 6.1 shows the representative anionic pollutants, their regulation, and their health effects.

In civil engineering, electrokinetic techniques have been applied to the desalination of concrete, the realkalization of carbonated concrete, the removal of salts from brick, crack closure in concrete, and so on (Ottosen et al., 2008). Chloride ions, which exhibit interesting anion-binding activities when binding with cement, have been shown to threaten the stability of concrete buildings by causing corrosion as the bound chloride anions are released over time. Electromigration, which is an electrokinetic transport mechanism, has been successfully applied to mitigate concrete corrosion by removing the anions and transporting chloride, carbonate,

---

Electrochemical Remediation Technologies for Polluted Soils, Sediments and Groundwater,
Edited by Krishna R. Reddy and Claudio Cameselle
Copyright © 2009 John Wiley & Sons, Inc.

TABLE 6.1. Permissible Limits and Health Effects of Representative Anionic Pollutants

Pollutant	Permissible Limit[a]			Health Effects[e]
	Korea[b]	USEPA[c]	WHO[d]	
Arsenic	0.05 (6)	0.01	0.01	Skin damage or problems with circulatory systems; may have increased risk of getting cancer
Chromium	0.05 (4[f])	0.1	0.05	Allergic dermatitis
Fluoride	1.5 (400)	4	1.5	Bone disease (pain and tenderness of the bones); children may get mottled teeth
Nitrate	10[g]	10[g]	50	Infants below the age of 6 months who drink water containing nitrate in excess of the MCL could become seriously ill and, if untreated, may die; symptoms include shortness of breath and blue baby syndrome
Selenium	0.01	0.05	0.01	Hair or fingernail loss; numbness in fingers or toes; circulatory problems

[a]Drinking water standards (mg/l).
[b]Number in parenthesis indicates the Korean standard for soil contamination (warning limit, mg/kg).
[c]United States Environmental Protection Agency (minimum contamination level, MCL, mg/l).
[d]World Health Organization.
[e]USEPA (http://www.epa.gov/safewater/contaminants/index.html#primary).
[f]As Cr(VI).
[g]As NO_3–N.

hydroxyl, and borate ions to effect crack closure. The use of electromigration in civil engineering was well summarized by Ottosen et al. (2008).

Nitrate and fluoride are common contaminants (Baek et al., 2009; Kim et al., 2009). Although their toxicity is not as severe as that of anionic metals such as arsenic and chromate, contamination with nitrate and fluoride is related to agricultural land use, and thus, to human health. Unlike the case of anionic metals, there are only few reports on the use of electrokinetics to remove nitrate and fluoride, and so we focus on them in this chapter.

6.2 POLLUTION AND HEALTH EFFECTS OF ANIONIC POLLUTANTS

6.2.1 Nitrate

Saline and/or sodic soils are major agricultural problems throughout the world. High salt content in the soil is the dominant problem related to agricultural land use. Excessive amounts of salts have adverse effects on the physiological and chemical properties of soil, microbiological processes, and plant growth.

Recently, cultivation in hothouses or greenhouses has become a general trend, and various vegetables such as squash, red peppers, lettuce, and cucumber are cultivated in greenhouses. Excessive use of chemical fertilizer has become prevalent, and residual fertilizer not used by crops tends to remain in the soil. This residual fertilizer is further concentrated by the continuous use of chemical fertilizer.

Fertilizer concentration is also a problem in greenhouses. A high concentration of salts in the soil inhibits water movement from soil to plant due to high osmotic pressure between pore water and plant root.

Nitrate and its movement to surface and groundwater following extensive nitrogen fertilizer application to agricultural fields is a critical problem worldwide. Nitrates are an important nutrient and, as such, play a vital role in plant growth. However, when used excessively, nitrates can easily move within the soil and reach the groundwater (Manokararajah and Ranjan, 2005a). An overly high nitrate concentration in soil inhibits plant growth, and a high level of nitrate in drinking water represents a significant risk to human health. Nitrates in drinking water are directly responsible for methemoglobinemia in infants (blue baby syndrome) and may play a role in the development of some forms of cancer (Manokararajah and Ranjan, 2005c).

6.2.2 Fluoride

Fluoride is a fairly common element that does not occur in its elemental state in nature because of its high reactivity. Naturally occurring fluorides in groundwater are a result of the dissolution of fluoride-containing rock minerals. The most common fluoride-bearing minerals are fluorite (CaF_2), fluorapatite [$Ca_5(PO_4)_3F$], cryolite (Na_3AlF_6), villiaumite (NaF), topaz [$Al_2(SiO_4)F_3$], and clays (Ozsvath, 2009). Soil fluoride is derived from the weathering of parent rock, volcanic activity, emissions from industry, and additives in phosphorus fertilizers (Loganathan et al., 2006). Whereas the fluoride content of most rocks ranges from 100 to 1300 mg/kg, soil concentrations vary between 20 and 500 mg/kg (Ozsvath, 2009).

At low-level concentrations, fluoride can reduce the risk of dental cavities. Exposure to somewhat higher amounts of fluoride can cause dental fluorosis, which, in its mildest form, can result in the discoloration of teeth. Severe dental fluorosis produces pitting and alteration of tooth enamel. Higher intake of fluoride, taken over a long period of time, can result in changes to bone and in a condition known as skeletal fluorosis. The effects of skeletal fluorosis include joint pain, restriction of mobility, and a possible increase in the risk of some bone fractures.

6.3 REMOVAL OF ANIONIC POLLUTANTS BY ELECTROKINETICS

6.3.1 Removal Mechanisms of Nitrate and Fluoride in Soils

In electrokinetic processes, there are two major transport mechanisms: electromigration and electro-osmosis. Generally, in an electrical field, electromigration causes cationic metals such as cadmium, zinc, lead, nickel, and copper to move from the anode toward the cathode; in electro-osmosis, the direction of movement of the pore water is toward the cathode when the zeta potential of the soil surface is negative. This can result in an enhanced removal of metals because the direction of transport of the ions in both mechanisms is the same. However, the direction of electromigration for anionic pollutants is toward the anode and that for electro-osmosis is from anode to cathode, as stated previously. The opposite direction of movement means that the removal rate of anionic pollutants could be reduced.

Soil pH plays an important role in the adsorption and desorption of pollutants and, ultimately, influences transport. Generally, the desorption of cationic metals would be enhanced by acidic conditions because of ion exchange reactions between hydrogen ions and cationics. However, anionic pollutants would be desorbed more easily in an alkaline condition than in an acidic environment, indicating that the process fluid should be changed for the electrokinetic removal of anionic pollutants.

6.3.2 Removal of Nitrate by Electrokinetics

Electrokinetics has been applied to remove nitrate from soil to protect groundwater and to enhance bioremediation of soil contaminated by other pollutants. In this section, nitrate is treated as a pollutant; in Section 6.3.3, it is treated as a nutrient.

Electrokinetic removal of nitrate from saline soil has been reported by several research groups. Application of an electric potential over a saline soil field resulted in increased salt content in drainage water and salt accumulation near the cathode (Gibbs, 1966). In addition, it was found that solute movement increased with greater electrical voltage. Cairo, Larson, and Slack (1996) reported that nitrate concentration increased from anode to cathode in saturated soil as solute flowed toward a horizontal drainage tube located adjacent and parallel to the cathode after application of a direct current (DC) electrical potential in field lysimeter experiments. However, after the soil became unsaturated, continued electrical input resulted in nitrate movement toward the anode.

Eid et al. (1999, 2000) and Eid, Slack, and Larson (2000) found that an electrokinetic technique was effective for concentrating and retaining nitrate close to the anode at all hydraulic flow velocities. Nitrate movement toward the anode was directly related to the applied electricity (current and duration). The migration rate of nitrate was higher under conditions of greater electric intensity and lower initial nitrate concentration. Eid et al. also indicated that the movement of nitrate was strongly influenced by the pH gradient (the pH distribution was related to flow velocity), and a mathematical relationship was established to relate these parameters. Jia et al. (2005b) reported results from a study on nitrate movement and pH changes in a vertical, partially saturated sandy soil column subjected to an electrical current. As electrical input increased, nitrate concentration adjacent to the anode in sandy soil became extremely high, and consequently, the nitrate content near the cathode reduced to a very low level. Also, as is usually found in electrokinetic experiments, increased electrical input reduced the pH of the soil near the anode significantly and increased the pH near the cathode. In other research on clay and sandy soils, Jia et al. (2005a) showed that (a) in sandy soil, nitrates were easily moved due to lack of nitrate adsorption by the soil particles, (b) in clayed soil, ions were moved slowly and less effectively than in the sandy soil, and (c) the electrokinetic effect on nitrate movement in the clayed soil was minimal (9h of electrical input). Manokararajah and Ranjan (2005a,b,c) investigated an electrokinetic barrier for nitrate ions in a silty loam soil. They assessed the feasibility of electrokinetic methods to denitrify a nitrate-contaminated, one-dimensional horizontal soil column subjected to a 1.25 hydraulic gradient to generate water movement. The results showed that nitrates in soil could be denitrified to other forms of nitrogen at the near cathode using electrokinetic techniques.

Some researchers have used a catalyst to remove nitrate from soil. Chew and Zhang (1998, 1999) reported nitrate–nitrogen transformation using electrokinetics coupled with zero-valent iron (ZVI). They said that the placement of ZVI near the anode increased the pH in that region, resulting in the conversion of nitrate–nitrogen to ammonia–nitrogen and nitrogen gas. Nitrate transformation was about 60% when an iron wall was placed near the cathode and was about 95% when the iron wall was placed near the anode. Yang, Hung, and Tu (2008) used nanosized bimetallic iron (Pd/Fe) slurry to evaluate the efficiency of nitrate removal and degradation in a bench-scale horizontal soil column in an electrokinetic processing system. In the system, 0.05 wt% of negatively charged, polyallylamin (PAA)-modified nanosized Pd/Fe bimetal was transported by electrophoresis, and over 99% of nitrate was removed by chemical reduction. Nitrate has also been removed from bricks effectively in this manner.

6.3.3 Transport of Nitrate by Electrokinetics

Some researchers have reported the feasibility of supplying nitrate or ammonium as a nitrogen source to stimulate bioremediation in low-permeable soils by electrokinetic processes. In bioremediation of polluted soils, the availability of nutrients and electron accepters is a limiting factor of microbial degradation. Elektorowicz and Boeva (1996) introduced nitrate to soil by electrokinetics for the purpose of soil bioremediation. They reported that nitrate concentration was highest in the anode area 27 days after operation, and injection of lower concentrations of nutrients yielded a better relative distribution within the electrical field. Furthermore, Elektorowicz and Boeva found that the required quantity of nitrate could be supplied to soil by carefully regulating electrical parameters and operation time. Thevanayagam and Rishindran (1998) injected nutrients and terminal electron acceptors into the mineral kaolin using electrokinetic methods. Utilizing theoretical analysis and experiment results, they asserted that electromigration was a significant factor compared with advective transport rate by electro-osmosis. Budhu et al. (1997) reported that nitrates or ammonium were able to be transported through low-permeability soils (kaolinite) using electrokinetics. Their results showed that nitrate concentration decreased in the first one-third of the column nearest the cathode, reached a minimum concentration between one-third and one-half of the column, and then increased slightly toward the anode. However, a steady-state condition of nitrate concentration was not reached during 3 weeks of operation. Rawahy et al. (2003) tried to attract and retain nitrates in the root zone of drip-irrigated barley in small lysimeter trials to prevent groundwater contamination. Although nitrate content near the anode was higher during a portion of the test, electrical input had little effect on nitrate distribution at other times due to careful management of water and nitrate inputs. Schmidt, Barbosa, and de Almeida (2007) studied the electrokinetic injection of nitrate and phosphate into an organic clayey soil having high electrical conductivity, low hydraulic conductivity, low density, large buffering capacity, and high cation exchange capacity in bioremediation of field soil contaminated with crude oil. While nitrates showed high mobility, phosphates did not, due to reactions with calcium carbonate and subsequent precipitation. Lohner, Katzoreck, and Tiehm (2008a,b) applied electrokinetic migration in synthetic media (silica sand) to enhance microbial pollution degradation by effecting the transport of nitrates and sulfates as electron acceptors and ammonium and phosphate as nutrients. It was determined that a linear relationship existed between voltage gradients and ion transport rate, and that ion distribution in the soil was significantly affected by soil pH and voltage gradient. In a comparison between nitrate transport in groundwater and in salt solutions using demineralized water, the transfer rate and mass transfer of nitrate in groundwater was higher than that in demineralized water because of the buffer capacity of groundwater.

6.3.4 Removal of Fluoride by Electrokinetics

Fluoride mobility in soil is highly dependent on the soil's sorption capacity, which varies with pH, the types of sorbents present, and soil salinity (Cronin et al., 2000). The overall water quality (e.g. pH, hardness, and ionic strength) also plays an important role in fluoride solubility through its influence on mineral solubility, complexation, and sorption/exchange reactions (Ozsvath, 2009). Fluoride mobility reaches a minimum at pH 6.0–6.5. As pH rises above 6.5, colloid surfaces release adsorbed fluoride as it is displaced by increasing OH^- concentrations, resulting in high fluoride concentrations in water (up to 30 mg/l). At pH values less than 6, the formation of a cationic AlF^{2+} complex inhibits adsorption of fluoride ions. High salinity (ionic strength) affects fluoride mobility by enhancing the potential for fluoride complexes to form and by increasing the number of ions that compete for soil sorption sites (Ozsvath, 2009). Maximum dissolved fluoride levels are usually controlled by the solubility of fluorite (CaF_2).

Fluoride has a negative charge; desorption of fluoride from soil is enhanced by alkaline conditioning (Pomes et al., 1999; Costarramone et al., 2000; Kim et al., 2009). In alkaline conditions, the formation of primarily iron and aluminum complexes was avoided, and fluoride complexes were the predominant species evident in pore water. In electrokinetic remediation, fluoride ions could be transported by electromigration and electro-osmosis (Costarramone et al., 2000). Pomes et al. (1999) reported that anolyte conditioning with 0.1 N NaOH increased fluoride recovery 1.8 times over recovery by washing. Additionally, because alkaline earth fluorides (MgF_2 and CaF_2) exhibit positive surface charge in their saturated brines (Hu et al., 1997), there is a possibility that alkaline earth fluorides can be moved to the area near the cathode by electromigration. Kim et al. (2009) reported that anolyte conditioning with a strong base (1.0 M NaOH) was very effective in maintaining alkaline conditions of soils in electrokinetic cells. Anolyte conditioning sharply increased the electro-osmotic flow, which resulted in a dramatically increased electro-osmotic transport of fluoride.

6.4 SUMMARY

Electrokinetics is a very effective technique to transport nitrate and fluoride, although the transport characteristics of anionic contaminants such as nitrate and fluoride are quite different from the transport of cationic metals. In electrokinetic restoration of saline soil, the electrokinetic transport of cationic salts, as well as nitrates, should be considered. Changes in pH near electrodes after electrokinetic treatment of saline soil are the disadvantages of electrokinetic restoration; however, the problem can be fixed by neutralization. Even though alkaline conditioning by anolyte enhanced the desorption of fluoride, it increased electro-osmotic flow from anode to cathode; however, the overall system removal efficiency could be significantly influenced by electromigration and by counteracting electro-osmotic flow in the opposite direction. Generally, agricultural land area is very large compared with areas with metal-contaminated soil. Therefore, treatment goals for agricultural areas should not be the removal of nitrate or salts but the restoration of saline soil since the soil will be used as agricultural land after treatment. To achieve this goal, further field research is necessary.

REFERENCES

Baek K, Kim D-H, Park S-W, Ryu B-G, Bajargal T, Yang J-S. (2009). Electrolyte conditioning-enhanced electrokinetic remediation of arsenic-contaminated mine tailing. *Journal of Hazardous Materials* **161**:457–462.

Blowes DW, Ptacek CJ, Benner SG, McRae CWT, Bennett TA, Puls RW. (2000). Treatment of inorganic contaminants using permeable reactive barriers. *Journal of Contaminant Hydrology* **45**:123–137.

Budhu M, Rutherford M, Sills G, Rasmussen W. (1997). Transport of nitrates through clay using electrokinetics. *Journal of Environmental Engineering* **123**:1251–1253.

Cairo G, Larson D, Slack D. (1996). Electromigration of nitrates in soil. *Journal of Irrigation and Drainage Engineering* **122**:286–290.

Chew CF, Zhang TC. (1998). In-situ remediation of nitrate-contaminated ground water by electrokinetics iron wall processes. *Water Science and Technology* **38**:135–142.

Chew CF, Zhang TC. (1999). Abiotic degradation of nitrates using zero-valent iron and electrokinetic processes. *Environmental Engineering Science* **16**:389–401.

Costarramone N, Tellier S, Grano B, Lecomte D, Astruc M. (2000). Effect of selected conditions on fluorine recovery from a soil, using electrokinetics. *Environmental Technology* **21**:789–798.

Cronin SJ, Manoharan V, Hedley MJ, Loganathan P. (2000). Fluoride: A review of its fate, bioavailability, and risks of fluorosis in grazed-pasture systems in New Zealand. *New Zealand Journal of Agricultural Research* **43**:295–321.

Eid N, Elshorbagy W, Larson D, Slack D. (2000). Electro-migration of nitrate in sandy soil. *Journal of Hazardous Materials* **79**:133–149.

Eid N, Larson D, Slack D, Kiousis P. (1999). Nitrate electromigration in sandy soil in the presence of hydraulic flow. *Journal of Irrigation and Drainage Engineering* **125**:7–11.

Eid N, Slack D, Larson D. (2000). Nitrate electromigration in sandy soil: Closed system response. *Journal of Irrigation and Drainage Engineering* **126**:389–397.

Elektorowicz M, Boeva V. (1996). Electrokinetic supply of nutrients in soil bioremediation. *Environmental Technology* **17**:1339–1349.

Gibbs HJ. (1966). Research on electroreclamation of saline-alkali soils. *Transactions of the American Society of Agricultural Engineers* **9**:164–169.

Hu YH, Lu YJ, Veeramasuneni S, Miller JD. (1997). Electrokinetic behavior of fluoride salts as explained from water structure considerations. *Journal of Colloid and Interface Science* **190**:224–231.

Jia X, Larson D, Slack D, Walworth J. (2005a). Electrokinetic control of nitrate movement in soil. *Engineering Geology* **77**:273–283.

Jia X, Larson DL, Zimmt WS, Walworth JL. (2005b). Nitrate pollution control in different soils by electrokinetic technology. *Transactions of the ASAE* **48**:1343–1352.

Kim D-H, Jeon C-S, Baek K, Ko S-H, Yang J-S. (2009). Electrokinetic remediation of fluorine-contaminated soil: Conditioning of anolyte. *Journal of Hazardous Materials* **161**:565–569.

Loganathan P, Gray CW, Hedley MJ, Roberts AHC. (2006). Total and soluble fluorine concentrations in relation to properties of soils in New Zealand. *European Journal of Soil Science* **57**:411–421.

Lohner ST, Katzoreck D, Tiehm A. (2008a). Electromigration of microbial electron acceptors and nutrients: (I) Transport in synthetic media. *Journal of Environmental Science and Health Part A* **43**:913–921.

Lohner ST, Katzoreck D, Tiehm A. (2008b). Electromigration of microbial electron acceptors and nutrients: (II) Transport in groundwater. *Journal of Environmental Science and Health Part A* **43**:922–925.

Manokararajah K, Ranjan R. (2005a). Creation of a barrier for nitrate ions using electrokinetic methods in a silty loam soil. *Canadian Biosystems Engineering* **47**:15–21.

Manokararajah K, Ranjan RS. (2005b). Electrokinetic denitrification of nitrates in a nitrate contaminated silty loam soil. *Applied Engineering in Agriculture* **21**:541–549.

Manokararajah K, Ranjan RS. (2005c): Electrokinetic retention, migration and remediation of nitrates in silty loam soil under hydraulic gradients. *Engineering Geology* **77**:263–272.

Ottosen LM, Christensen IV, Rorig-Dalgard I, Jensen PE, Hansen HK. (2008). Utilization of electromigration in civil and environmental engineering—Processes, transport rates and matrix changes. *Journal of Environmental Science and Health Part A* **43**:795–809.

Ozsvath DL. (2009). Fluoride and environmental health: A review. *Reviews in Environmental Science and Biotechnology* **8**:59–79.

Pomes V, Fernandez A, Costarramone N, Grano B, Houi D. (1999). Fluorine migration in a soil bed submitted to an electric field: Influence of electric potential on fluorine removal. *Colloids and Surfaces A: Physicochemical and Engineering Aspects* **159**:481–490.

Rawahy S, Larson DL, Walworth J, Slack DC. (2003). Effect of an electrical input with drip irrigation on nitrate distribution in soil. *Applied Engineering in Agriculture* **19**:55–58.

Schmidt CAB, Barbosa MC, de Almeida MDSS. (2007). A laboratory feasibility study on electrokinetic injection of nutrients on an organic, tropical, clayey soil. *Journal of Hazardous Materials* **143**:655–661.

Thevanayagam S, Rishindran T. (1998). Injection of nutrients and TEAs in clayey soils using electrokinetics. *Journal of Geotechnical and Geoenvironmental Engineering* **124**:330–338.

Yang GCC, Hung CH, Tu HC. (2008). Electrokinetically enhanced removal and degradation of nitrate in the subsurface using nanosized Pd/Fe slurry. *Journal of Environmental Science and Health Part A* **43**:945–951.

7

ELECTROKINETIC TREATMENT OF CONTAMINATED MARINE SEDIMENTS

Giorgia De Gioannis, Aldo Muntoni, Alessandra Polettini, and Raffaella Pomi

7.1 INTRODUCTION

It has been estimated that around 5% of watersheds in industrialized countries have health- and environment-threatening sediments and that 10% of marine and estuarine sediments are potentially hazardous for the aquatic environment. Indeed, industrial activities are often located near the seashore; therefore, it is not unusual to find marine sediments heavily contaminated by either heavy metals or organic contaminants. Dredging and management operations on contaminated sediments are also related to marine trade routes maintenance; although contaminated sediments may be left in place covered by a low permeability and erosion-resistant capping, dredging may still be required for harbor/routes maintenance purposes. In such cases, huge amounts of (possibly) contaminated sediments need to be dredged, dewatered, and treated before reuse or final disposal. Around 500 million cubic meters of sediments are dredged each year for navigational purposes and roughly 1%–4% requires dewatering and treatment prior to disposal, increasing the cost of dredging by a factor of 300–500.

Despite decades of research, surprisingly little is known about successful treatment of contaminated sediments. Presently, the most common approach to management of dredged sediment is generally limited to land (or aquatic) disposal. Confined aquatic disposal (CAD) involves underwater contaminated sediments capping with clean sediments or sand, geotextiles, liners, or even reactive barriers, either with or without lateral confinement. Sediments are deposited in natural or artificial depression, usually by means of a clamshell or a hydraulic pipeline. Capping may also shift

---

Electrochemical Remediation Technologies for Polluted Soils, Sediments and Groundwater, Edited by Krishna R. Reddy and Claudio Cameselle
Copyright © 2009 John Wiley & Sons, Inc.

the conditions of the upper contaminated sediment layer from oxidizing to anoxic, which may change the solubility of metals and the possibility of microbial degradation of the organic contaminants (United States Environmental Protection Agency [USEPA], 2005). Although cost-effective, CAD cells have a number of relevant shortcomings: (a) currents can affect the accuracy of deposition as well as the thickness and integrity of the capping layer; (b) sandy caps may have a high permeability which, in turn, may result in significant contaminant fluxes: the release of several contaminants from CAD cells has been measured to reach tons per year (Shine, 2000). Confined disposal facilities (CDFs) are fully engineered sites where contaminated sediments can be disposed of; they can be land disposal sites, or be constructed partially in water near the shore or completely surrounded by water (USEPA, 1993). Coastal CDFs may represent a chance for construction of structures or even coastal landscaping, although the influence of wind and waves should be carefully evaluated. Concerns arose in many cases with regard to the risks of contaminant seepage and/or uptake by plants and animals growing and feeding nearby, and the need for defining construction standards and contaminant concentration limits for sediments allowed for coastal CDF disposal is well acknowledged.

With regard to the subject of this chapter, the configuration of CAD and CDF cells, under some circumstances, allows for the application of several treatment techniques (Gardner, 2005), among which electrochemical remediation is believed to be particularly suited for (Acar, 1992; Acar and Alshawabkeh, 1993).

In terms of contamination characteristics, sediments are characterized by a possible simultaneous presence of several types of pollutants, which interact with different constituents of solid matrices. The mobility of target contaminants from the solid material is strongly related to such properties as grain size distribution, permeability, organic matter content, which will affect the efficiency of contamination treatment. Other parameters, including moisture and salts content, may also have an influence on the suitability of different treatment options. While in soils the contaminated fine fraction typically accounts for less than 50% of total solids, in sediments this figure may be as high as 80%–95%. For finely grained matrices, a number of treatment technologies have proved to be ineffective for the purpose of remediation. Up to now, sediment treatment has not been frequent due to the huge volumes involved, the high costs, and the risk of modest effectiveness. The USEPA (2005) emphasizes that too few widely tested and accepted sediment cleanup techniques are presently available, no defined performance standards exist to assist in remedy selection, clear process performance criteria are still lacking, and there is only a little experimental data inventory and difficulty of finding appropriate treatment methods for extremely large volumes of, sometimes, low-level contaminated sediment.

For the reasons provided above, sediment decontamination is a complex technical issue, which requires the study of different treatment alternatives and likely the application of a multiple-stage treatment sequence. In this framework, it is a general opinion that technically and economically feasible treatment strategies should be identified; therefore, treatment processes allowing for further utilization of decontaminated sediment are receiving renewed attention by authorities, technicians, and researchers and, among them, electrokinetic remediation deserves particular attention.

7.2 CONTAMINATED SEDIMENT TREATMENT OPTIONS

Techniques for *in situ* sediment treatment are still in an early stage of development, and the very few methods currently commercially available are usually restricted to situations where the low water levels allow for water flow diversion. Potential *in situ* treatment methods include biological, chemical, and stabilization processes. The main advantage relies upon avoiding sediment removal and, therefore, the related risk of spreading the contamination. However, technical limitations to the application of *in situ* treatments still exist, mainly because the effective delivery of such process reagents as substrates, nutrients, and chemicals, as well as mixing with the contaminated material, may be problematic.

Ex situ treatment usually requires a preliminary dewatering stage (Mulligan, Yong, and Gibbs, 2001) to be performed in temporary storage facilities, which also allows for a kind of equalization; the resulting water flow typically needs a proper treatment prior to discharge. Any other pretreatment steps may involve coarser fractions removal as well as coagulants addition so as to enhance settling and/or adjusting pH in view of the subsequent treatment stages.

The most widely studied and applied *ex situ* treatments are solidification/stabilization, thermal oxidation and desorption, soil washing (mechanical and chemical), and bioremediation. Solidification/stabilization treatments are mainly applied to sediments contaminated by heavy metals; they entail the increase of the volume to deal with and can also be affected by the sulfate and chloride content, as well as by relevant concentrations of heavy metals. Thermal oxidation allows for high destruction efficiencies of organic pollutants, but is affected by high costs; it is sometimes applied together with thermal desorption. An effective preliminary desaturation is mandatory in order to reduce the costs of thermal oxidation and desorption, and the efficiency of the latter can be hindered by the high porosity of sediment. Mechanical washing is probably the most applied *ex situ* technique. The different units of a mechanical washing process are combined to deal with sediments eventually characterized by a multiple contamination (heavy metals, halogenated, and aromatic compounds, including total petroleum hydrocarbons [TPHs], polycyclic aromatic hydrocarbons [PAHs], polychlorinated biphenyls [PCBs], etc.). Although sometimes used as a mere pretreatment, mechanical washing is capable of achieving high heavy metal removal efficiencies, especially in the presence of primary metal ores. In general, better results can be attained for sediments with moderate silt contents.

Chemical washing is one of the most promising techniques. A number of extracting agents have been tested in order to remove inorganic and organic pollutants from sediments: ethylenediaminetetraacetic acid (EDTA), nitrilotriacetic acid (NTA), citric and acetic acid, ethylenediaminedisuccinic acid (EDDS), HNO_3, H_2SO_4, HCl, hexane, methanol, propane, butane, and so on. Multiple washing stages are often required to achieve the treatment targets; the separation of silt particles from the extracting agent may also be required, which may represent a critical issue. Development of processes based on the use of nontoxic and biodegradable extracting agents is advisable to avoid the increase in ecotoxicity of treated sediment. Recovery of the extracting agent from the exhaust solution is also an objective that should be pursued to limit reagent consumption and the associated treatment costs.

Biological treatments have been applied on either slurry or solid phase. Since the remediation performance can be reduced by the presence of chloride and high heavy metal contents, a pretreatment (e.g. mechanical or chemical washing) may be necessary.

The degradation of organic compounds can also be achieved through electrochemical oxidation, which seems to be a promising tool for TPH, PAH, and PCB removal as demonstrated by different experimental studies carried out using hydrogen peroxide, permanganate, ozone, and Fenton's reagents (Huang et al., 1999; Andreottola et al., 2008).

7.3 ELECTROKINETIC TREATMENT OF SEDIMENTS

Among the available *in situ* and *ex situ* treatment options, electrokinetic remediation is recognized as a promising technique, successfully applied to remove a number of inorganic and organic pollutants from soil and sludge (Lageman, Pool, and Seffinga, 1989; Pamucku and Wittle, 1992; Acar and Alshawabkeh, 1993; Probstein and Hicks, 1993; Shapiro and Probstein, 1993; Hicks and Tondorf, 1994; Muniram et al., 1997; Pamucku and Weeks, 1997; Reddy, 1999; Reddy and Chinthamreddy, 2004; Yeung and Hsu, 2005). A number of patents for electrokinetic treatment have been granted for the treatment of unsatured and satured soils, using one or more electrodes (Probstein, Renaud, and Shapiro, 1991; Lindgren and Mattson, 1995; Pool, 1995, 1996; Doring and Doring, 1997; Wittle and Pamukcu, 1997).

The efficacy of an electrokinetic remediation process strongly relies upon sediment properties such as buffer capacity, mineralogy, and organic matter content. Surface complexation, sorption, and ion exchange mechanisms strongly counteract mobilization of metals from the negatively charged particles. In order to control system chemistry and promote contaminant solubilization and transport, chemical agents often need to be added to the process solution (Yeung, Hsu, and Menon, 1996; Wong, Hicks, and Probstein, 1997; Chung and Kang, 1999; Yeung and Hsu, 2005) or proper pretreatments need to be performed (e.g. ultrasonic treatment; Chung and Kamon, 2005).

The ever-growing attention deserved by electrokinetic sediment decontamination can be explained considering the advantages of the process. First of all, electrokinetic remediation can be applied *in situ*, and may consequently also be adopted where decontamination is required but dredging is not. Furthermore, the technique is generally able to treat fine and low-permeability materials, and treatment costs can be limited by a proper selection of a low current/voltage application. Finally, it may represent a possible single-stage option to achieve dewatering, consolidation, and removal of inorganic and organic pollutants as well as (for marine sediments) salts. On the other hand, the theoretical remediation performance of the electrokinetic process can be altered by side reactions, competition between species/elements, detrimental effects on electrodes, precipitation reactions, and parasite current generation. Thus, deeper investigation of the complex reactions taking place in real contaminated sediment and the identification of the most appropriate enhancing chemicals are issues of critical concern for ongoing research on electrokinetic sediment decontamination.

Until now, relatively few applications of electrokinetic remediation to sediments, at either laboratory or full scale, have been documented. Many of the available data relates to the application of electrokinetics to achieve dewatering and speed up sediment consolidation (Mohamedelhassan and Shang, 2001; Reddy, Urbanek, and Khodadoust, 2006). Promoting fast consolidation may be important when CDFs are used for dredged sediment management in order to achieve adequate mechanical properties (i.e. shear strength) and to provide enough room for sediment disposal. Usually, disposal in CDFs may not be as rapid as desired due to the slow consolidation caused by the high water content and the low hydraulic conductivity of sediment. Reddy, Urbanek, and Khodadoust (2006) report that, considering an annual CDF disposal of a 0.9-m thick layer of sediment, for a real harbor, a period of 10 years would be required to manage the whole amount of dredged sediment. A feasibility study was performed on sediment samples in order to investigate the possibility of accelerating dewatering and consolidation. The results showed that, due to the induced electroosmotic flow, sediment consolidation was increased by an order of magnitude as compared to the compaction degree achieved by gravity, applying a voltage gradient of 1.0V/cm for less than 10 days; the addition of polymers at low concentrations (0.5%–1% dry wt.) enhanced this effect.

In the framework of a land reclamation project, electrokinetics was applied in combination with preloading and prefabricated vertical drains to improve the mechanical strength of marine sediments; intermittent current and noninsulated electrodes were adopted (Lo *et al.*, 2000). The results showed that the mechanical properties of sediment were significantly improved in the vicinity of the electrodes after 22–34 days, by applying voltage gradients ranging between 9.5 and 28.2V/m.

Micic *et al.* (2001) performed cell tests on marine sediments, which indicated that a combination of electrokinetic treatment and preloading is effective in promoting dewatering and improving the mechanical properties of sediment: The increase in the average undrained shear strength of sediment was up to 145% greater than that achieved by preloading alone.

Electrokinetics has also been tested as a means to increase the free settling velocity and the solids concentration of river sediments (Buckland, Shang, and Mohamedelhassan, 2000). The best results were obtained through a combination of electrokinetics and conditioning with $FeCl_3$. It was concluded that electrokinetic sedimentation is more effective than chemical coagulation.

As far as electrokinetic sediment remediation is concerned, some authors underscore that unenhanced electrokinetics can hardly result in significant mobilization of heavy metals, mainly due to the more complex chemical composition of sediments, higher buffering capacity, and different metal speciation as compared to soils.

Reddy and Chinthamreddy (2004) performed unenhanced electrokinetic remediation tests on high acid buffering spiked glacial till soils. Negligible removal of cationic contaminants due to precipitation phenomena under the high pH conditions was observed; similar results were also obtained by De Gioannis *et al.* (2008) in unenhanced tests on marine sediments, as also shown in the following. Grundl and Reese (1997) observed that the presence of calcite buffers the system, preventing pH from shifting to the acidic range, which would be more favorable to metal detachment from the solid matrix.

Nevertheless, promising results in terms of heavy metal removal can be attained using an appropriate enhancing solution, to be selected on the basis of contaminant characteristics and sediment composition. Removal efficiencies up to 80% for the above-mentioned glacial till soils spiked with Cr(VI), Ni(II), and Cd(II) were achieved by Reddy and Chinthamreddy (2004) when purging solutions were adopted; the highest removal of Ni(II) and Cd(II) was obtained using 1.0 M acetic acid (1.0 V/cm voltage gradient); adopting 0.1 M NaCl as the anolyte and 0.1 M EDTA as the catholyte resulted in significant contaminant migration toward the electrodes. Chung and Kang (1999) applied electrokinetic treatment to a marine clay spiked with lead. Three different chemicals were used: nitric acid, EDTA, and acetic acid; results showed that such chemicals expedited the remediation process and lead to a Pb removal higher than 88% (up to 94% with acetic acid enhancement).

Enhanced electrokinetic remediation has been studied in order to remove organic pollutants as well. Bench-scale electrokinetic experiments were conducted by Reddy and Ala (2006) on lake sediments characterized by an organic content of 19%, a hydraulic conductivity of 3.3×10^{-7} cm/s, a pH of 7, and a high acid buffering capacity, as well as a wide range of PAHs and heavy metals at high concentrations. Tests were conducted at a voltage gradient of 2 V/cm using two surfactants (5% Igepal CA-720, produced by Sciencelab.com Inc., Houston, TX, and 3% Tween 80, produced and supplied by several chemical companies, i.e. Sigma-Aldrich Corp., St. Louis, MO), a cosolvent (20% n-butylamine), and cyclodextrin (10% HP-β-CD) to produce solubilization/desorption of PAHs. The results showed that 20% n-butylamine and 5% Igepal 720 were capable of yielding partial solubilization of PAHs, while both 3% Tween 80 and 10% HP-β-CD systems were ineffective. None of the selected flushing agents was capable of removing heavy metals. The low removal efficiencies attained for both PAHs and heavy metals were related both to the high organic content and to the acid buffering capacity of the sediment.

As far as *in situ* application of electrokinetic sediment remediation is concerned, Shresta, Fischer, and Rahner (2003) and Shresta, Fischer, and Sillanpa (2007) proposed an innovative electrode arrangement in which one electrode was placed in the sediment and the other one was immersed in water, in order to promote heavy metals mobilization/accumulation through electrochemically initiated processes and reactions at—or nearby—the electrodes. Tests were carried out with a natural heavy metal contaminated river sediment using a vertical polyvinylchloride (PVC) column and by varying position and distance between the electrodes. The results showed that, for cationic contaminants, the best configuration involves placing the anode in the sediment and the cathode in water; under this condition, heavy metals were mobilized by the effect of the anodic polarization and transported by migration from sediment toward the sediment-water interface. By creating a pH barrier at the sediment–water interface, those metals were precipitated at the steep pH gradient, resulting in accumulation at the sediment–water interface (Shresta, Fischer, and Rahner, 2003). For anionic contamination, the best immobilization was obtained positioning cathode into the sediment (Shresta, Fischer, and Sillanpa, 2007).

In general, a dramatic reduction of the full-scale performance of electrokinetic sediment treatment may be observed for *in situ* applications. An evaluation project was developed at a marine embayment used as a former storage and handling area receiving facility effluents and stormwater runoff (USEPA, 2007). The goal of the demonstration was to assess the possibility of reducing the concentrations of Hg,

PAHs, and phenolic compounds through mineralization of the organic contaminants and complexation, mobilization and removal of the metal. Nine anodic (steel plates) and nine cathodic (graphite plates) electrodes were placed in the sediments according to two parallel, 9 m distant rows over an area of approximately 15 × 15 m. A monitoring campaign showed no significant decrease in Hg and PAHs concentration over time, while measurements of the electrical parameters showed a steady reduction in performance during the test. The reason for the poor results was mainly found in the complete corrosion of connections between the electrical supply and anodes. The cost for treatment of the 225 m^2 area down to a 1.5 m depth over a 6-month period was estimated to be approximately 1150 USD/m^3.

The high treatment costs and the inadequate removal efficiencies attained claimed for the study of alternative, low-tech, and low-energy applications. The Ferric Iron Remediation and Stabilisation (FIRS) process was presented by Cundy and Hopkinson (2005) and Faulkner, Hopkinson, and Cundy (2005); the technique was based on the application of a low-voltage gradient (typically less than 0.2 V/cm) between two or more sacrificial, Fe-rich electrodes, and was capable of producing a strong pH and Eh gradient as well as precipitation at an Fe-rich barrier. Therefore, contamination control was achieved through mobilization of metals and concentration on and around the Fe-rich barrier. Arsenic-contaminated sediments are particularly suited for, due to the high pH at the cathode and the affinity of As for the precipitated Fe oxides and oxyhydroxides. Tests performed on estuarine sediments showed promising results (Cundy and Hopkinson, 2005).

7.4 CASE STUDY: TESTS ON ELECTROKINETIC REMEDIATION OF SEA HARBOR SEDIMENTS

In the following, a case study of lab-scale application of an assisted electrokinetic process for heavy metal removal from real contaminated sediments is presented. The process made use of both chemical and physical enhancing methods, including addition of chelating and acidic agents, as well as the application of a hydraulic gradient. Lab-scale electrokinetic runs were applied on two different dredged sediments varying the applied voltage gradient and the treatment duration.

In the framework of a research program of national interest, electrokinetic remediation was applied to two different sea sediments dredged from industrial harbors located in Northeastern Italy (namely, sediment V) and Southwestern Sardinia (namely, sediment S). Both areas were subjected for several years to industrial pressure of chemical and mining type, and the analysis performed on the two sediment samples indicated significant heavy metal content; high contents of As and Zn (181 and 3720 mg/kg, respectively) were detected in sediment V, as were Pb and Zn (4970 and 5210 mg/kg, respectively) in sediment S (see Table 7.2).

A number of experiments were performed on both sediments by varying the specimen length (12, 20, and 30 cm), the applied voltage (12, 18, 30, and 60 V), the treatment duration (7, 14, 20, 22, and 30 days), and the type of conditioning agents used at the electrodes (EDTA, nitric acid, and citric acid). A summary of the different conditions applied is reported in Table 7.1.

At the end of each experiment, the material was demoulded from the apparatus and cut into a number of slices to evaluate the removal/accumulation of species as

TABLE 7.1. Summary of the Experimental Runs

Run	Sediment	Test Duration (day)	Applied Voltage (V)	Applied Voltage Gradient (V/cm)	Specimen Length (cm)	Catholyte	Anolyte	Additional Conditions
W1	V	20	12	0.6	20	Deionized water	Deionized water	—
W2	V	7	30	2.5	12	Deionized water	Deionized water	—
W1-S	S	14	30	1.0	30	Deionized water	Deionized water	—
W2-S	S	14	30	1.0	30	Deionized water	Deionized water	Washed sediment
CIT1	V	20	12	0.6	20	0.2M Citric acid	0.2M Citric acid	—
CIT2	V	22	12	1.0	12	0.2M Citric acid	0.2M Citric acid	—
CIT3	V	22	18	1.5	12	0.2M Citric acid	0.2M Citric acid	—
CIT4	V	7	30	2.5	12	0.2M Citric acid	0.2M Citric acid	—
EDTA1	V	7	30	2.5	12	0.2M EDTA	0.2M EDTA	—
EDTA2	V	7	30	2.5	12	0.2M EDTA	0.2M EDTA	Washed sediment
EDTA3	V	7	30	2.5	12	0.2M EDTA	Deionized water	—
EDTA4	V	7	30	2.5	12	0.2M EDTA + 0.1M HNO$_3$[a]	0.2M EDTA	—
EDTA5	V	7	30	2.5	12	0.2M EDTA	0.2M EDTA	Hydraulic gradient[b]
EDTA-S	S	14	30	1.0	30	0.2M EDTA	0.2M EDTA	Washed sediment
AC-S	S	14	30	1.0	30	Deionized water + 0.1M HNO$_3$[c]	Deionized water	Washed sediment
NITAC-S	S	30	60	2.0	30	Deionized water + 1M HNO$_3$[c]	Deionized water + 0.1M HNO$_3$[c]	Washed sediment[d]

[a]HNO$_3$ was added periodically to maintain the pH at around 3 units.
[b]A hydraulic gradient of 8.3 m/m was applied in the direction cathode → anode.
[c]HNO$_3$ was added periodically to maintain the pH at around 3 units.
[d]Sieving was performed before the electrokinetic treatment in order to remove particles >500μm.

TABLE 7.2. Elemental Composition of Sediments

Element	Sediment V	Sediment S
	Concentration (mg/kg dry wt)	
As	180.9	30.0
Ca	94080	62000
Cd	30.7	54.3
Cr	108.1	60.8
Cu	323.0	169.2
Fe	41110	15000
Hg	17.3	—
K	13420	1230
Mg	28790	13400
Mn	411.3	300
Mo	209.0	—
Ni	83.1	124.1
Pb	886.2	4969.8
Sb	7.7	—
Si	141190	17000
Zn	3723.0	5211.6

a function of the distance from the electrodes; each slice was analyzed for pH and total metal content. In a number of cases where formation of a precipitate on the electrode surface was visually detected, this was recovered and analyzed for metal content. Heavy metal concentrations were also measured in the anodic and cathodic chambers.

The electrokinetic treatments were performed using an experimental setup, manufactured by the researchers, consisting of a Plexiglas cell, a power source capable of applying a constant voltage and a multimeter used to monitor the voltage and measure the current flow through the specimen during testing. Tests on sediment S were performed adopting a cylindrical cell (Fig. 7.1), 30 cm in length and 6.2 cm in diameter, fitted with platinized titanium electrodes at both ends; the anodic and cathodic chambers were separated from the soil by means of porous stones and provided with vents for evacuation of the electrolysis gases. In order to better investigate the effect of different operating conditions on the electrokinetic treatment performance, tests on sediment V were performed using an improved 6-cm diameter cell capable to allow for the variation of specimen length between 12 and 20 cm (Fig. 7.2).

Preliminary operations involved compacting the material in the cell, according to a standardized procedure. In order to assess the effect of the natural salt content of sediment on the process, a preliminary washing stage was adopted for some experimental runs using deionized water repeatedly as long as electrical conductivity was equal to that of the washing liquid.

The sediments, homogenized to obtain representative subsamples, were subjected to measurement of physical and chemical properties, including particle size distribution, water content, pH, buffer and cation exchange capacity (CEC), electrical conductivity, elemental and mineralogical composition (assessed by X-ray

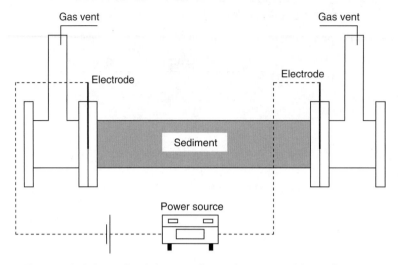

Figure 7.1. Schematic of the experimental setup used for sediment S.

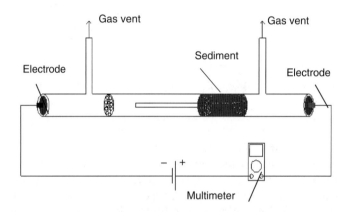

Figure 7.2. Schematic of the improved experimental setup used for sediment V.

diffraction (XRD)), heavy metal distribution and speciation, total organic carbon (TOC), and hydraulic conductivity.

According to grain size analysis, both sediments are potentially suitable for electrokinetic remediation: sediment V is composed of 2% gravel, 28% sand, 63% silt, and 7% clay, thus it can be classified as a silty loam (Ceremigna et al., 2005, 2006), while sediment S is a silty sand having a grain size distribution characterized by 70% sand, 26% silt, and 4% clay.

The analysis of metal partitioning indicated that for sediment V metals, including As, Cd, Cu, Pb, Mn, and Zn are associated to exchangeable + carbonate-bound fractions and Fe/Mn oxides-bound fraction in proportions ranging from 25% to 85%; Ni and Cr are mainly associated to the residual fraction of the material. Sediment S showed a different metal partitioning; Cd, Zn, and Pb are mainly (~60%) associated to exchangeable + carbonate-bound fractions and Fe/Mn oxides-

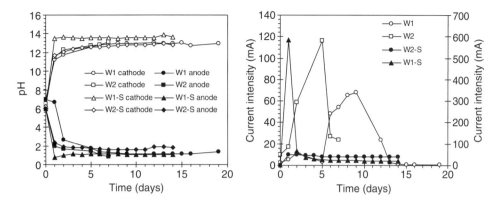

Figure 7.3. Evolution of (a) pH at the electrodes and (b) current intensity as a function of time, current intensity values for run W1-S on right axis (sediments *V* and *S*; process fluid: deionized water).

bound fraction, while the remaining portion is associated to sulfides; most of the Cu content (90%) is associated to sulfides and the remaining portion to the residual fraction; 50%–60% of the Cr and Ni content is found in the residual fraction and 25%–35% is present as sulfides.

7.4.1 Unenhanced Electrokinetic Experiments: Deionized Water as Process Fluid

The first runs were performed on both sediments using deionized water as the process fluid (Fig. 7.3). It was found that water hydrolysis reactions were quite fast, producing sharp pH variations at the electrodes over time. For the lower applied voltage gradient (run W1, 0.6 V/cm), the maximum current intensity was displayed after 10 days of treatment, decreasing dramatically afterward down to very low values after 13 days of treatment. An increase in the applied voltage, together with a decrease in specimen length (run W2, 2.5 V/cm), resulted in faster stabilization of pH at the electrodes, as expected on the basis of the faster generation rate of H^+ and OH^- ions. Furthermore, such experimental conditions produced both an increase in maximum current intensity and a reduction in the time required for the current peak to appear. The observed decrease in current intensity observed after certain treatment durations can be explained by various mechanisms, including gradual depletion of solutes (Yu and Neretnieks, 1997), precipitation of nonconductive solids (Acar and Alshawabkeh, 1993; Acar *et al.*, 1995), and polarization of electrodes (Kornilovich *et al.*, 2005). The induced current intensity was characterized by a similar trend in run W1-S (1 V/cm) on sediment *S* (Fig. 7.3). When the same treatment conditions were applied to the prewashed sediment *S* (run W2-S, 1 V/cm), the effect of the salt content on electric current intensity was evident: Run W2-S was characterized by lower and roughly constant intensity values (10 mA as compared, e.g. to the peak of 600 mA for run W1-S) over the 14 days of treatment (Fig. 7.3). The salt content also influenced the cumulative electroosmotic flow, with a value in run W2-S six times higher than for run W1-S.

160 ELECTROKINETIC TREATMENT OF CONTAMINATED MARINE SEDIMENTS

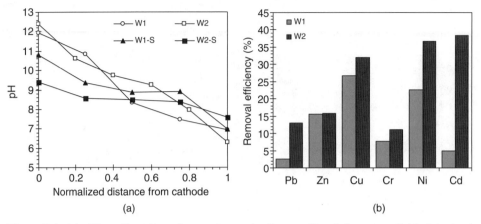

Figure 7.4. (a) pH profiles along the specimens (sediments V and S; process fluid: deionized water) and (b) total metal removal efficiency (sediment V).

Figure 7.4 reports the trend of sediment pH as a function of distance from cathode and the total metal removal efficiency at the end of the test. As far as sediment V was concerned, an increase in the applied voltage gradient from 0.6 (run W1) to 2.5 V/cm (run W2) promoted, as expected, the migration of the acidic front along the specimen, lowering sediment pH in the sections closer to the anode (Fig. 7.4a). As for sediment S, the application of a 1 V/cm voltage gradient yielded significant pH changes only in the specimen sections closer to the electrodes. As indicated by the data depicted in Figure 7.4b, the increased voltage applied per unit length of material improved total metal removal efficiency from sediment V, even more if the lower treatment duration of run W2 (7 days) as compared to run W1 (20 days) is considered.

The mechanisms of metal removal during electrokinetic treatment of sediment V can be explained considering the final distribution of residual contaminants observed along the specimen, reported in Figure 7.5 (run W2) for Pb, Zn, Cu, and Ni. A first, general feature was that appreciable metal portions were detected in the anodic chamber under both the experimental conditions tested on sediment V with deionized water as the process fluid. Although several literature studies (Acar and Alshawabkeh, 1993; Acar et al., 1995; Suèr, Gitye, and Allard, 2003; Kornilovich et al., 2005) indicate electromigration of metals in the form of cationic aqueous complexes toward the cathode, accumulation in either the anodic solution or the cell sections close to the anode has also been reported by some investigators (Acar et al., 1995; Reddy and Chinthamreddy, 2004). It is well established that the migration of the acidic front across the material from the anode to the cathode leads to desorption/detachment of contaminants from the active surface sites of the solid (Acar et al., 1995; Li, Yu, and Neretnieks, 1996; Reddy and Shirani, 1997; Wong, Hicks, and Probstein, 1997). The species mobilized from the material, which are presumably in the cationic form under the relatively low pH conditions at which desorption occurs, tend to migrate toward the cathode. However, depending on the location at which the H^+ and OH^- ions fronts meet and the extent of pH change at that location, which, in turn, are affected by the buffer capacity of sediment as

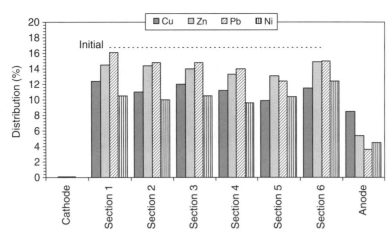

Figure 7.5. Final distribution of Pb, Zn, Cu, and Ni along the electrokinetic cell (sediment V; process fluid: deionized water, run W2).

reported also by Li, Yu, and Neretnieks (1996), metal complexes may become negatively charged due to the excess of OH⁻ ions in the alkaline region, and thus start migrating backward to the anode. As the newly formed anionic metal complexes move toward regions of lower pH, they will tend to speciate back to cationic forms. According to the proposed mechanism, the existence of zones with strong differences in pH will cause an alternating movement of metal ions across the cell, reducing the efficiency of the electroremediation process. The occurrence of metals in the anolyte can be explained by the fact that even under acidic conditions, although at low concentrations, negatively charged aqueous metal complexes including $Me(OH)_3^-$ and $Me(OH)_4^{2-}$ can still be present. The fact that in this study no metal migration toward the cathode was observed may tentatively be ascribed to a different ionic mobility of negatively and positively charged metal species. In the case of typically strong amphoteric metals such as Pb, Zn, and partly Cu, such hypotheses are supported by the experimental data. However, the proposed mechanisms alone are not able to explain the observed behavior of Cd and Ni, since in this case, the mass balance analysis yielded an error higher than 20%. It may be hypothesized that the lower solubility of such metals in the alkaline regions may have lead to precipitation of insoluble forms (not visually detectable) that were separated from the liquid solution during the filtration step prior to the analytical determination. The relatively high removal efficiencies (38% for Cd and 37% for Ni; Fig. 7.4b) attained in run W2 should also be interpreted considering the effect of precipitation at the cathode.

In the case of sediment S, the unenhanced electrokinetic treatment, as expected, was not effective toward heavy metal mobilization, since the process was unable to induce significant pH changes along the whole specimen. The high electric current intensities observed in run W1-S were caused by ion migration related to the soluble salt content, as confirmed by the drastic reduction in electric conductivity with time as soon as such species were depleted from the material, with a final value very close to that of the prewashed material (~1 mS/cm). Small fractions of Cu and Ni

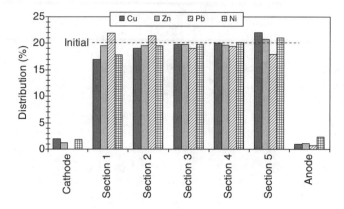

Figure 7.6. Final distribution of Pb, Zn, Cu, and Ni along the electrokinetic cell (sediment S; process fluid: deionized water, run W1-S).

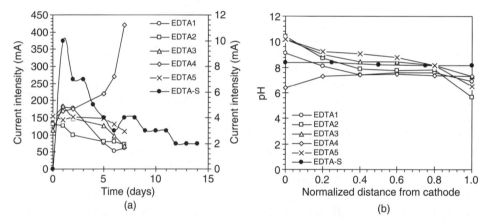

Figure 7.7. (a) Evolution of current intensity as a function of time, current intensity values for run EDTA-S on right axis, and (b) profile of sediment pH along the specimen (sediments V and S; process fluid: EDTA).

were removed at both the cathodic and anodic chambers, and Pb showed a trend toward slight accumulation into the sections close to the cathode (Fig. 7.6).

7.4.2 EDTA-Enhanced Electrokinetic Experiments

On the basis of the results of the first series of experiments, further runs were conducted on both sediments using EDTA as conditioning agent, with the purpose of forming soluble metal-chelant complexes and promoting heavy metal removal.

For sediment V, the addition of EDTA produced increased current intensities due to the increased ionic strength of the circulating solution as compared to the experiments using deionized water only (Fig. 7.7). A different result was obtained for run EDTA-S performed on the prewashed material, where current intensity attained the same value as recorded in the unenhanced test (run W2-S), but then decreased sharply. A similar feature was observed by Reddy, Danda, and Saicheck (2004) during EDTA-enhanced treatment of contaminated kaolin, and may have been caused by the formation of a precipitate onto the anodic electrode which was not observed in run W2-S.

A number of considerations can be derived comparing the results obtained under different experimental conditions. Considering the two runs EDTA1 (on unwashed sediment) and EDTA2 (on prewashed sediment), it is confirmed that soluble salts in sediment provide a large contribution to the amount of charged species circulating in the system, as already observed for runs W1-S and W2-S. The initial current intensities measured during run EDTA2 were lower than the corresponding values for run EDTA1. It was also found that the use of EDTA promoted migration of the acidic front across the specimen, with an associated decrease in sediment pH in the sections closer to the cathode. As expected, such a decrease was particularly significant when cathode depolarization was performed by means of periodical HNO_3 addition during the experiment. It was found that the use of 0.2 M EDTA as the catholyte (run EDTA1) affected sediment pH near the cathode, so that the pH of the section adjacent to the cathode decreased from 12.4 for run W2 to 9.1 for run EDTA1.

It was also noticed that the application of a washing pretreatment (run EDTA2) and the use of deionized water as the anolyte (run EDTA3) resulted in increased sediment pH in the regions close to the cathode. Considering the much lower sediment pH attained in run EDTA4, it is clear that for runs EDTA1–EDTA3 the use of EDTA as the catholyte was not adequate to produce cathode depolarization for the entire duration of the experiments. Periodic addition of HNO_3 at the cathode was more effective in preventing the OH^- ions generated therein from migrating toward the anode, thus keeping sediment pH in the acidic range along the whole test cell.

Concerning sediment S, the pH profile in the test on prewashed sediment (run EDTA-S) confirmed the difficulties in reducing pH experienced in runs W1-S and W2-S; the use of 0.2 M EDTA decreased pH in the anodic and cathodic chambers only, but did not lead to any substantial pH change throughout the specimen (Fig. 7.7).

Figure 7.8 depicts the results from the five experimental runs on sediment V using EDTA in terms of final metal distribution along the cell. The overall metal removal efficiencies for such runs are reported in Figure 7.9. The electromigration behavior observed for runs W1 and W2 was enhanced in the presence of EDTA. Trace metals including Pb, Zn, and Cu were found to accumulate in either the anodic solution or in the section close to the anode, as also reported by other investigators (Wong, Hicks, and Probstein, 1997). This can easily be explained by the fact that metal complexes with EDTA are typically negatively charged. In the case of Cr, not reported in Figure 7.8, a fraction varying from about 2% to 6% depending on the experimental conditions was found to be present in the anodic solution. Although EDTA is not known to form aqueous complexes with Cr(VI), the observed behavior is in agreement with the chemical characteristics of this metal, since aqueous hexavalent Cr complexes, which are expected to be present in the vicinity of the anode due to the oxidizing conditions existing in this zone, are typically oxyanionic. Although a number of negatively charged complexes of EDTA with trivalent Cr are reported (including $Cr\text{-}EDTA^-$, $CrOH\text{-}EDTA^{2-}$), their presence in the anodic

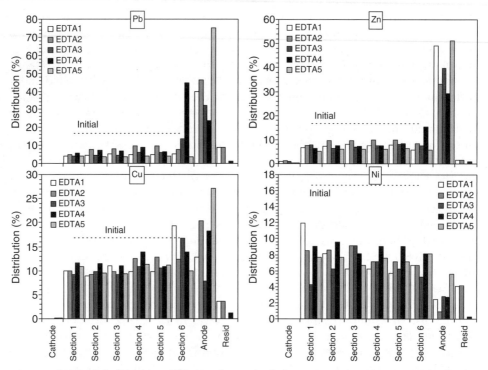

Figure 7.8. Final distribution of Pb, Zn, Cu, and Ni along the electrokinetic cell (sediment V; process fluid: EDTA). Note: "Resid" refers to the precipitate electrodeposited at the anode.

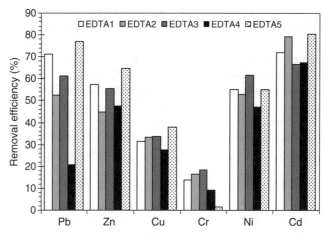

Figure 7.9. Total metal removal efficiency (sediment V; process fluid: EDTA).

region is highly improbable due to the oxidizing conditions. The reduction of Cr(VI) to Cr(III) can explain the presence of the small amounts of this metal detected in the cathodic solution, as also noted by other authors (Reddy *et al.*, 1997; Reddy, Xu, and Chinthamreddy, 2001; Reddy and Chinthamreddy, 2004). However, looking at Figure 7.9, the overall removal efficiency was quite low for Cr (max. 18%), indicating that at the end of the electrokinetic treatment, Cr was still almost uniformly distributed along the specimen. The low Cr removal efficiency can be explained considering that, according to the analysis of the initial distribution among the different sediment fractions, Cr was found to be mainly associated to the detrital (i.e. the less labile) portion of the material. In the case of Ni and Cd (the latter being not shown in Fig. 7.8), the distribution in the test cell again indicated some migration toward the anode. However, as outlined before for runs W1 and W2, the mass balance analysis suggested that a portion of Ni and Cd may also have precipitated at the cathode.

As for the overall metal removal efficiency, the best results were obtained for Pb and Zn as well as Cd (if precipitation at the cathode is considered), with more than 60% of each metal being mobilized as a result of the electrokinetic process. This result is likely related to multiple mechanisms, including speciation of trace metals in the original sediment, thermodynamic stability, and formation kinetics of metal-EDTA complexes, as well as ionic mobility of individual trace metal ions. However, due to the presumably interconnected effects of the mentioned mechanisms, a precise quantification of their individual contribution to trace metal removal from sediment is hard to be attained.

Among the different runs on sediment V using EDTA, the higher removal was displayed when a hydraulic gradient was applied, while the depolarization of the cathode with HNO_3 did not result in improved efficiency. This may be explained considering that the main effect of EDTA on metal removal from sediment was the formation of soluble metal-chelant complexes (Wong, Hicks, and Probstein, 1997); in the presence of high H^+ concentrations as in run EDTA4, as discussed in a previous study on the speciation of metal complexes during assisted sediment washing (Polettini, Pomi, and Rolle, 2007), it is likely that most of the chelating agent was in the form of H-EDTA complexes including H-EDTA^{3-}, H_2EDTA^{2-}, and H_3EDTA$^-$, with a reduced amount of chelant available for heavy metal complexation. In the case of run EDTA5, it is interesting to emphasize that promoting fluid flow across sediment by means of a hydraulic gradient significantly contributed to the migration of contaminants.

In the case of EDTA-enhanced test on prewashed sediment S (run EDTA-S), the results in terms of heavy metal mobilization were modest, with a slight accumulation of Cd and Pb in the central sections of the specimen, probably as a result of electromigration of negatively charged EDTA metal complexes toward the anode and an opposite mass transport caused by the electroosmotic flow toward the cathode (Fig. 7.10); however, the low metal mobilization may also derive from the lower voltage gradient applied (1 V/cm) if compared to tests on sediment V (2.5 V/cm).

7.4.3 Citric Acid-Enhanced Electrokinetic Experiments

The use of citric acid to enhance the electrokinetic treatment of sediment V was not capable of significantly affecting current intensity if compared to the other

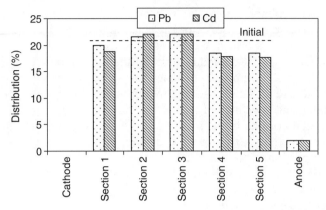

Figure 7.10. Final distribution of Pb and Cd along the electrokinetic cell (prewashed sediment S; process fluid: EDTA).

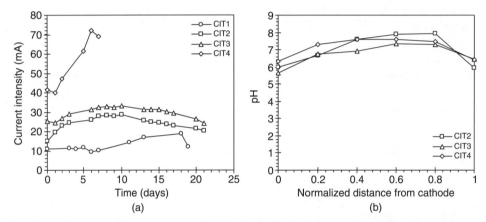

Figure 7.11. (a) Evolution of current intensity as a function of time and (b) profile of sediment pH along the specimen (sediment V; process fluid: citric acid).

process fluids analyzed. Current intensity was related to specimen length, as evident from Figure 7.11 when comparing the results obtained for runs CIT1 and CIT2. The pH profile across sediment was similar for all runs, with pH values below 8 units along the whole specimen. The low pH values attained, particularly near the cathode, can be explained as a result of depolarization of the electrode. This allowed farther migration of the acidic front toward the cathode. The fact that current intensity was not affected by the higher concentrations of H^+ ions along the specimen was unexpected and cannot be explained with the results obtained.

The distribution of Pb, Zn, Cu, and Ni along the test cell for runs CIT2, CIT3, and CIT4 is reported in Figure 7.12 (the results for run CIT1 were not plotted since the final specimen was cut into five sections only instead of six). Pb and Zn were found to be slightly mobilized toward the anode; however, as shown in Figure 7.13, the overall metal removal was lower than 20% and 30%, respectively, showing a

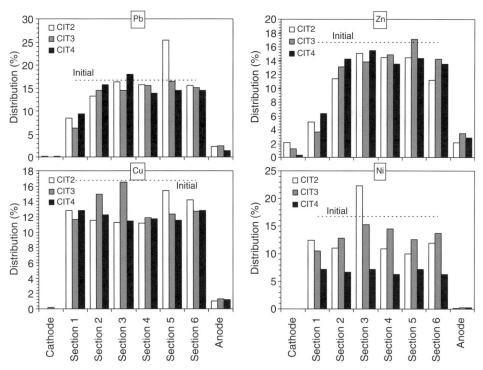

Figure 7.12. Final distribution of Pb, Zn, Cu, and Ni along the electrokinetic cell (sediment V; process fluid: CIT).

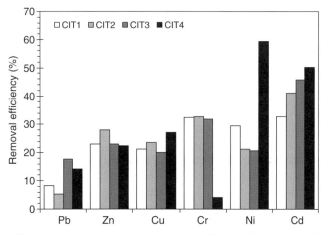

Figure 7.13. Total metal removal efficiency (sediment V; process fluid: CIT).

notably lower remediation performance if compared to EDTA. Higher removal yields were measured for Ni and Cd; however, as noted previously for EDTA, since the mass balance analysis for these metals at the end of the tests yielded on average recoveries of ~75% for Ni and ~60% for Cd, it is not possible to assess whether the observed removal was associated to precipitation phenomena in the electrode chambers, or was rather due to the inhomogeneity of sediment.

The different behavior of Pb and Zn in electrokinetic tests using EDTA and CIT (namely runs EDTA1 and CIT4, which were conducted under the same operating conditions) can hardly be explained on the basis of the final pH profile established along the specimen, which was very similar for the two runs under concern. In addition, the results from a previous work (Polettini, Pomi, and Rolle, 2007) focusing on the performance of sediment washing using different chelating agents indicated very similar extraction efficiencies of Pb and Zn by 0.2 M CIT and 0.2 M EDTA, with only a slight effect of solution pH. In the same work, deeper investigation of metal speciation in the exhaust solutions by means of geochemical modeling showed that with EDTA as the extracting agent, both Pb and Zn were almost entirely present as Me-EDTA^{2-} complexes; when the chelating agent used was CIT, the two metals were extracted in the form of Me-CIT$_2^{4-}$ (75%–95%), with smaller amounts of Me-CIT$^-$. It is tempting to hypothesize that the complexes Pb-CIT$_2^{4-}$ and Zn-CIT$_2^{4-}$ have a lower ionic mobility than Pb-EDTA^{2-} and Zn-EDTA^{2-}, which may result into an enhanced electrochemical mobilization of Pb and Zn in the presence of EDTA as opposed to CIT.

7.4.4 Nitric Acid-Enhanced Electrokinetic Experiments

The results obtained for sediment S using either deionized water or EDTA as process fluids suggested the opportunity of enhancing acidification of sediment to improve mobilization of cationic contaminants, making them available to either electromigration or electroosmotic dragging. Additional issues for a successful treatment were assumed to include periodic depolarization of the electrodes and prevention of cathode coating caused by the formation of a precipitate layer, thus pH was controlled through periodical HNO$_3$ additions into both the electrode chambers in order to maintain the pH at around 3 units (run AC-S). This required continuous conditioning of the catholyte since pH tended to readily increase after each acid addition and the formation of a white precipitate on the cathode was often observed. As shown in Figure 7.14, acidic conditioning strongly increased current intensity if compared to the other experiments on sediment S; nevertheless, this was not effective in lowering the pH along the specimen and this again resulted in poor heavy metal mobilization. The high current intensities were due to dissolution of Ca and Mg compounds and subsequent transport of the corresponding ionic species toward the cathode, where precipitates formed. This also affected the performance of the electrode itself, requiring re-dissolution through HNO$_3$ additions or even mechanical removal. The analysis performed on the precipitate confirmed that it mainly contained Ca and Mg, while the heavy metal content was not detectable.

A second attempt (run NITAC-S) was based on more aggressive process conditions. In order to reduce the buffer capacity of sediment, the 500 μm oversize fraction, which had high calcium carbonate content due to the presence of numerous shells fragments, was preliminarily removed through sieving. This fraction accounted

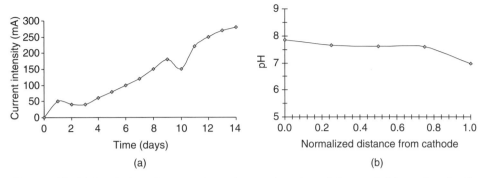

Figure 7.14. (a) Evolution of current intensity as a function of time and (b) profile of sediment pH along the specimen (sediment S; process fluid: deionized water with periodic HNO$_3$ additions into the electrode chambers).

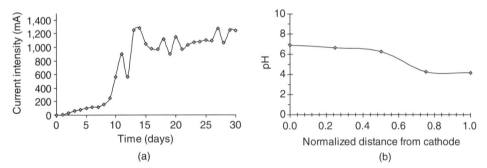

Figure 7.15. Evolution of (a) current intensity as a function of time and (b) profile of sediment pH along the specimen (sediment S; catholyte: 1 M HNO$_3$; anolyte: 0.1 M HNO$_3$).

for 1.7% of the sediment mass and contained 30% and 2% of calcium and total heavy metals, respectively. Figure 7.15 shows that the preliminary size separation, the higher voltage gradient applied (2 V/cm), the more aggressive acidic conditioning (1 and 0.1 M HNO$_3$ at the cathode and anode, respectively), as well as the prolonged treatment duration, were capable of modifying the pH of sediment S. Acidification of the material was also evident during the test, since the electroosmotic flow was reversed after only 4 days of treatment. The system achieved high electric current intensities after about 15 days of treatment that were maintained until the end of the test run.

The treatment was capable of mobilizing Ca and Mg to a considerable extent, promoting migration mainly toward the cathode, where periodic renewal of the process fluid and mechanical cleaning of the electrode from precipitates were required. The precipitates removed from the cathode were analyzed for Ca, Mg, and heavy metal contents. Appreciable removal of heavy metals was achieved and, in this case, appreciable portions of these were also found in the precipitate.

Figure 7.16 shows the heavy metal distribution along the specimen at the end of the test and the corresponding removal efficiency. The final distribution indicates

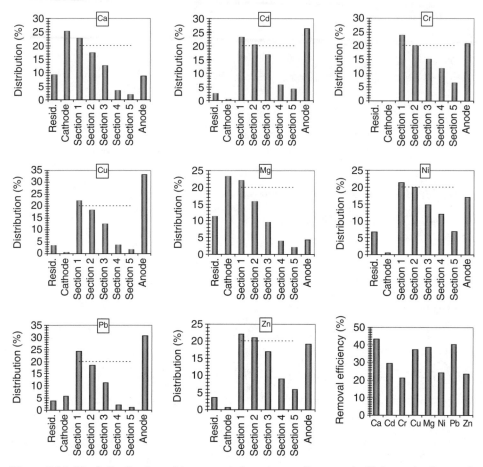

Figure 7.16. Final distribution of heavy metals and overall removal efficiency (sediment S; catholyte: 1 M HNO_3; anolyte: 0.1 M HNO_3). Note: "Resid" refers to the precipitate electrodeposited at the cathode.

that migration occurred mostly toward the anode for all the trace metals. This is probably due to the synergistic effect of the transport of dissolved species caused by the reversed electroosmotic flow and the electromigration of negatively charged aqueous metal complexes, including hydroxide [$Me(OH)_x^{(2-x)}$], chloride [$MeCl_x^{(2-x)}$], or carbonate [$Me(CO_3)_2^{2-}$] forms; the formation of chloride and carbonate complexes was hypothesized by Nystrom, Ottosen, and Villumsen (2005) to explain the observed migration of metals toward the anode. Small portions of trace metals (2%–7%) were also found in the cathodic chamber, with the exception of Cr, probably due to the predominance of oxyanionic hexavalent aqueous complexes of this metal formed under the oxidizing conditions at the anode. Migration of Ca and Mg occurred mostly toward the cathode, where in total 35% (~10% in the precipitate) of their initial mass was collected; this indicates, as obvious on the basis of chemical speciation considerations, the cationic nature of complexes of such elements in the system.

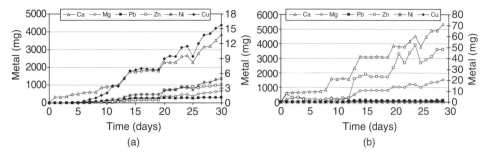

Figure 7.17. Cumulative mass of Ca, Mg, Pb, Zn, Ni, Cu accumulated over time in the (a) anodic (Ni and Cu on right axis) and (b) cathodic (Ni, Cu, and Zn on right axis) chambers (sediment S; catholyte: 1 M HNO_3; anolyte: 0.1 M HNO_3).

The overall heavy metal removal ranged from 21% (Cr) to 40% (Pb). The results from sequential extraction explain the low removal efficiencies attained for Cr and Ni (21% and 24%), which were mainly associated to the detrital portion in the untreated sediment, as well as the highest removal yield displayed by Pb, which was initially mainly associated to exchangeable and carbonate-bound fractions.

Figure 7.17 shows the cumulative mass of Ca, Mg, Pb, Zn, Ni, and Cd over time measured in the cathodic and anodic chambers. As mentioned, Ca and Mg mostly accumulated at the cathode, and their migration toward the electrodes clearly started readily at the beginning of the test; conversely, the presence of trace metals in the chambers could only be detected after 5–10 days of treatment. The onset of heavy metal mobilization was found to correspond to ~10% electro-removal of Ca from sediment. It should be emphasized that this adds to ~30% of Ca removal accomplished through the preliminary mechanical separation of the 500-μm oversize fraction.

7.4.5 Comparison of Removal Efficiencies and Estimation of Energy Consumption

Figure 7.18 compares the metal removal efficiencies attained for sediments V and S using different process fluids. It was found that the best overall metal removal yield was displayed by EDTA for sediment V, while the same agent was not effective for sediment S, although in this case, the lower voltage applied may have negatively affected the results. The use of nitric acid as the conditioning agent on prewashed sediment S was particularly effective in terms of Pb removal and also had a positive effect on Cu mobilization. The use of citric acid for sediment V did not allow for significant improvement in treatment efficiency if compared to deionized water, with the only exception of Ni. As outlined before in the case of EDTA, it is likely that the use of 0.2 M citric acid as the catholyte may have shifted the speciation of the chelating agent toward H-Cit complexes, reducing the amount available for complexation of trace metals.

Another critical issue that must be taken into account for the purpose of process scale-up is the estimation of energy consumption during the electrokinetic treatment. The overall energy consumption per unit volume of sediment (E_u) was calculated as follows:

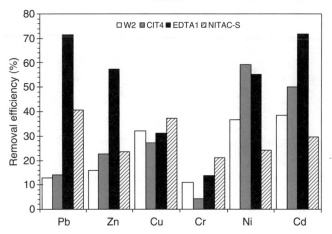

Figure 7.18. Metal removal efficiency using different process fluids; sediments V (W2, CIT4, and EDTA1) and S (NITAC-S).

$$E_u = \frac{1}{V_s}\int VI\,dt \qquad (7.1)$$

where V_s = volume of treated sediment; V = applied voltage; I = current intensity.

The effect of the salt content on energy consumption can be judged comparing the results obtained for tests W1-S (E_u = 741 kWh/m^3) and W2-S (prewashed sediment; E_u = 94 kWh/m^3), as well as EDTA1 (E_u = 1805 kWh/m^3) and EDTA2 (prewashed sediment, E_u = 1415 kWh/m^3). It is clearly evident that, as expected, a high soluble salt content results in high energy consumption, thus affecting treatment cost; for this reason, prewashing appears to be advisable. The energy consumption values for EDTA-tests on sediment V indicate that some enhancement strategies can sometimes result in higher treatment costs without any appreciable gain in removal efficiency; this can be inferred considering that run EDTA4 (EDTA + nitric acid) was characterized by a large energy consumption (E_u = 3200 kWh/m^3), although no improvement in removal efficiency was attained as compared to runs EDTA1 (E_u = 1805 kWh/m^3) and EDTA3 (E_u = 1905 kWh/m^3). Coupling the application of a hydraulic gradient with EDTA conditioning (run EDTA5) proved to be more economically convenient, since energy consumption was lower than for run EDTA4 (E_u = 2130 vs. 3200 kWh/m^3), while treatment efficiency was the highest among all the EDTA-runs. In general, the electrokinetic treatment seems not to be suitable for sediment S, since an aggressive and energy-intensive (E_u = 35,000 kWh/m^3) approach was necessary in order to achieve fair removal efficiencies.

7.5 SUMMARY

Little is known about successful treatment of contaminated sediments, although huge amounts of contaminated sediments need to be dredged, dewatered, and treated before reuse or final disposal. Contaminated sediments are often characterized by the simultaneous presence of several types of pollutants that interact with the different constituents of the solid matrices, as well as by high fine fractions, moisture and, likely, salt content. A number of treatment technologies usually applied for contaminated soils have proved to be ineffective for the purpose of sediment remediation which, therefore, represents a complex technical issue requiring the study of different treatment alternatives and, presumably, the application of a multiple-stage treatment sequence.

In this framework, electrokinetic treatment/remediation can be considered as a promising technique, since it is able to treat fine and low-permeability materials, can also be applied *in situ* (although detrimental influence of specific site conditions on system performance should be carefully considered), and may represent a possible single-stage option to achieve dewatering, consolidation, and removal of organic and inorganic pollutants as well as (for marine sediments) salts.

In situ treatment is a perspective of particular interest since it may be applied where decontamination is required but dredging is not and, most of all, would avoid sediment removal and, therefore, the related risk of spreading the contamination. To this regard, innovative approaches involve placing one electrode in the sediment and the other one in water in order to promote heavy metal mobilization/accumulation, or degradation of organic compounds through electrochemical oxidation.

The application of electrokinetic treatment to sediments permanently or temporarily located in CDF cells can also be considered a sort of *in situ* treatment. The goals to be achieved may involve dewatering, speeding up consolidation (the compaction degree can be increased by an order of magnitude if compared to that achieved by gravity), increasing the shear strength (in particular when CDF have to be used as construction or coastal landscaping elements), and, of course, removing contaminants and salts. Consolidation and dewatering appear to benefit from combined application of electrokinetic treatment and chemical conditioning. Contaminant removal and dewatering, in the case of temporary storage, could be aimed at pretreating the sediment in view of *ex situ* processes that could be affected by relevant water, sulfate, chloride, and heavy metal content (solidification/stabilization, bioremediation, thermal oxidation, or desorption).

Unenhanced electrokinetics often proved to be ineffective for mobilization of heavy metals due to the more complex chemical composition of sediments, higher buffering capacity, and different metal speciation as compared to soils. However, positive results (heavy metal removal higher than 90%) can be achieved using an appropriate enhancing solution. The results of the presented case study on application of assisted electrokinetics using nitric acid, EDTA, and citric acid for the removal of heavy metals from two real contaminated sea sediments, indicated that the remediation efficiency can be significantly affected by speciation and partitioning of metals in the less mobile fractions as well as buffer capacity of the material.

Enhanced electrokinetic remediation has been studied in order to remove organic pollutants as well, PAHs in particular. Solubilization and removal of PAHs proved to be heavily hindered by high organic contents of sediment, as well as by the low liquid/solid ratio achievable as compared to chemical washing, as reported in Andreottola *et al.* (2008) who also underlined the negative influence exerted by the high viscosity of the chemicals adopted on the process yield.

In general, full-scale performance of electrokinetic sediment treatment could be drastically reduced, in case of *in situ* applications, by heavy corrosion phenomena of the connections between the electrical supply and anodes.

In order to overcome the critical issues limiting the technical and economical feasibility of electrokinetic sediment treatment, a deeper investigation of the effect exerted by different enhancing conditions (i.e. nature, dosages, and mode of application of the agents promoting contaminants migration and/or degradation, preliminary separation of hindering/interfering components or species, pH control, and regulation) should be carried out. At the same time, the most appropriate operating conditions (hydraulic gradient application, electrodes geometry, and configuration, current intensity, etc.) need to be investigated as well.

REFERENCES

Acar YB. (1992). Electrokinetic cleanups. *Civil Engineering* **62**:58–60.

Acar YB, Alshawabkeh AN. (1993). Principles of electrokinetic remediation. *Environmental Science Technology* **27**(13):2638–2647.

Acar YB, Gale RJ, Alshawabkeh AN, Marks RE, Puppala S, Bricka M, Parker R. (1995). Electrokinetic remediation: Basics and technology status. *Journal of Hazardous Material* **40**:117–137.

Andreottola G, Bonomo L, De Gioannis G, Ferrarese E, Lavagnolo MC, Muntoni A, Polettini A, Pomi R, Saponaro S. (2008). Marine, lagoon, and river sediment remediation. In *Proceedings of I2SM, International Symposium on Sediment Management* (eds. N-E Abriak, D Damidot, R Zentar), Lille, France, July 9–11, pp. 29–37.

Buckland DG, Shang JQ, Mohamedelhassan E. (2000). Electrokinetic sedimentation of contaminated Welland river sediment. *Canadian Geotechnical Journal* **37**(4):735–747.

Ceremigna D, Polettini A, Pomi R, Rolle E, De Propris L, Gabellini M, Tornato A. (2005). Comparing sediment washing yields using traditional and innovative biodegradable chelating agents. In *Proceedings of Third International Conference on Remediation of Contaminated Sediments* (eds. RF Olfenbuttel, PJ White), January 24–27, New Orleans, LA, on CD-ROM.

Ceremigna D, Polettini A, Pomi R, Rolle E, De Propris L, Gabellini M, Tornato A. (2006). A kinetic study of chelant-assisted remediation of contaminated dredged sediment. *Journal of Hazardous Material* **137**:1458–1465.

Chung HI, Kamon M. (2005). The coupled effect of electrokinetic and ultrasonic remediation for the treatment of contaminated sediments. In *Proceedings of the International Offshore and Polar Engineering Conference*, June 19–24 (eds. JS Chun, SW Hong, J Koo, T Komai, W Koterayama). Seoul, Korea: International Society of Offshore and Polar Engineers, pp. 652–657.

Chung HI, Kang BH. (1999). Lead removal from contaminated marine clay by electrokinetic soil decontamination. *Engineering Geology* **53**(2):139–150.

Cundy AB, Hopkinson L. (2005). Electrokinetic iron pan generation in unconsolidated sediments: Implications for contaminated land remediation and soil engineering. *Applied Geochemistry* **20**(5):841–848.

De Gioannis G, Muntoni A, Polettini A, Pomi R. (2008). Enhanced electrokinetic treatment of different marine sediments contaminated by heavy metals. *Journal of Environmental Science and Health* **43**(8):852–865.

Doring FR, Doring N (inventors). (1997). Method and device for the elimination of toxic materials from, in particular, top soil. US Patent 5,595,644. P + P Geotechnik GmbH, assignee.

Faulkner DWS, Hopkinson L, Cundy AB. (2005). Electrokinetic generation of reactive iron-rich barriers in wet sediments: Implications for contaminated land management. *Mineralogical Magazine* **69**(5):749–757.

Gardner K. (2005). Electrochemical Remediation and Stabilization of Contaminated Sediments. Report submitted to the NOAA/UNH Cooperative Institute for Coastal and Estuarine Environmental Technology (CICEET), NOAA grant number NA17OZ2507.

Grundl T, Reese C. (1997). Laboratory study of electrokinetic effects in complex natural sediments. *Journal of Hazardous Materials* **55**(1):187–201.

Hicks RE, Tondorf S. (1994). Electrorestoration of metal contaminated soils. *Environmental Science and Technology* **28**(12):2203–2210.

Huang CP, Cha D, Chang J, Qiang Z, Sung Me, Chiang YC. (1999). Electrochemical Processes for In-Situ Treatment of Contaminated Soils. Final Progress Report, September 1998 to May 1999, Project ID: 54661, Department of Energy.

Kornilovich B, Mishchuk N, Abbruzzese K, Pshinko G, Klishchenko R. (2005). Enhanced electrokinetic remediation of metals-contaminated clay. *Colloids and Surfaces A* **265**:114–123.

Lageman R, Pool W, Seffinga G. (1989). Contaminated soil: Electro-reclamation: Theory and practice. *Chemistry and Industry* **18**:585–590.

Li Z, Yu JW, Neretnieks I. (1996). A new approach to electrokinetic remediation of soils polluted by heavy metals. *Journal of Contaminant Hydrology* **22**:241–253.

Lindgren R, Mattson ED (inventors). (1995). Electrokinetic electrode system for extraction of soil contaminants from unsaturated soils. US Patent 5,435,895. United States, assignee.

Lo KY, Micic S, Shang JQ, Lee YN, Lee SW. (2000). Electrokinetic strengthening of a soft marine sediment. *International Journal of Offshore and Polar Engineering* **10**(2):137–144.

Micic S, Shang JQ, Lo KY, Lee YN, Lee SW. (2001). Electrokinetic strengthening of a marine sediment using intermittent current. *Canadian Geotechnical Journal* **38**(2):287–302.

Mohamedelhassan E, Shang JQ. (2001). Analysis of electrokinetic sedimentation of dredged Welland river sediment. *Journal of Hazardous Material* **85**(1–2):91–109.

Mulligan CN, Yong RN, Gibbs BF. (2001). Remediation technologies for metal-contaminated soils and groundwater: An evaluation. *Engineering Geology* **60**(1–4):193–207.

Muniram B, Rutherford M, Sills G, Rasmussen W. (1997). Transport of nitrates through clay using electrokinetics. *Journal of Environmental Engineering* **123**(12):1251–1253.

Nystrom GM, Ottosen LM, Villumsen A. (2005). Test of experimental set-ups for electrodialytic removal of Cu, Zn, Pb and Cd from different contaminated harbour sediments. *Engineering Geology* **77**:349–357.

Pamukcu S, Weeks A. (1997). Electrochemical extraction and stabilization of selected inorganic species in porous media. *Journal of Hazardous Material* **55**:305–318.

Pamukcu S, Wittle JK. (1992). Electrokinetic removal of selected heavy metals from soil. *Environmental Progress* **11**(3):241–250.

Polettini A, Pomi R, Rolle E. (2007). The effect of operating variables on chelant-assisted remediation of contaminated dredged sediment. *Chemosphere* **66**:866–877.

Pool W (inventor). (1995). Process for the electroreclamation of soil material. US Patent 5,433,829.

Pool W (inventor). (1996). Process for the electroreclamation of soil material. US Patent 5,589,056 (continuation of 5,433,829).

Probstein RF, Hicks RE. (1993). Removal of contaminants from soils by electric fields. *Science* **260**:498–503.

Probstein RF, Renaud PC, Shapiro AP (inventors). (1991). Electroosmosis techniques for removing materials from soil. US Patent 5,074,986. Massachusetts Institute of Technology, assignee.

Reddy K, Urbanek A, Khodadoust AP. (2006). Electroosmotic dewatering of dredged sediments: Bench-scale investigation. *Journal of Environmental Management* **78**:200–208.

Reddy KR. (1999). Preliminary assessment of electrokinetic remediation of soil and sludge contaminated with mixed waste. *Air & Waste Management Association* **49**:823–830.

Reddy KR, Ala PR. (2006). Electrokinetic remediation of contaminated dredged sediment. *Journal of ASTM International* **3**(6):254–267.

Reddy KR, Chinthamreddy S. (2004). Enhanced electrokinetic remediation of heavy metals in glacial till soils using different electrolyte solutions. *Journal of Environmental Engineering-ASCE* **130**:442–455.

Reddy KR, Danda S, Saicheck RE. (2004). Complicating factors of using ethylenediamine tetraacetic acid to enhance electrokinetic remediation of multiple heavy metals in clayey soils. *Journal of Environmental Engineering-ASCE* **130**(11):1357–1366.

Reddy KR, Parupudi US, Devulapalli SN, Xu CY. (1997). Effects of soil composition on the removal of chromium by electrokinetics. *Journal of Hazardous Material* **55**:135–158.

Reddy KR, Shirani AB. (1997). Electrokinetic remediation of metal contaminated glacial tills. *Geotechnical and Geological Engineering* **15**:3–29.

Reddy KR, Xu CY, Chinthamreddy S. (2001). Assessment of electrokinetic removal of heavy metals from soils by sequential extraction analysis. *Journal of Hazardous Material* **84**:279–296.

Shapiro AP, Probstein RF. (1993). Removal of contaminants from saturated clay by electroosmosis. *Environmental Science Technology* **27**(2):283–291.

Shine J. (2000). The use of confined aquatic disposal (CAD) cells to manage contaminated sediments in ports and harbors: Chemical migration. *Conference on Dredged Material Management: Options and Environmental Considerations*, December 3–6, MIT, Cambridge, MA, USA.

Shresta R, Fischer R, Rahner D. (2003). Behaviour of cadmium, lead and zinc at the water-sediment interface by electrochemically initiated process. *Colloids and Surfaces A: Physicochemical and Engineering Aspects* **222**(1–3):261–271.

Shresta R, Fischer R, Sillanpa M. (2007). Investigation on different position of electrodes and their effects on the distribution of Cr at the water sediment interface. *International Journal of Environmental Science and Technology* **4**(4):413–420.

Suèr P, Gitye K, Allard B. (2003). Speciation and transport of heavy metals and macroelements during electroremediation. *Environmental Science Technology* **37**:177–181.

United States Environmental Protection Agency (USEPA). (1993). Selecting Remediation Techniques for Contaminated Sediments. Report EPA-823-B93-001.

United States Environmental Protection Agency (USEPA). (2005). Contaminated Sediment Remediation—Guidance for Hazardous Waste Sites. Report EPA-540-R-05-012.

United States Environmental Protection Agency (USEPA). (2007). Electrochemical Remediation Technologies (ECRTs)—In Situ Remediation of Contaminated Marine Sediments. Report EPA/540/R-04/507.

Wittle JK, Pamukcu S (inventors). (1997). Electrochemical system and method for the removal of charged species from contaminated liquid and solid wastes. US Patent 5,614,077. Electro-Petroleum, Inc. and Lehigh University, assignees.

Wong JSH, Hicks RE, Probstein RF. (1997). EDTA-enhanced electroremediation of metal-contaminated soils. *Journal of Hazardous Material* **55**:61–79.

Yeung AT, Hsu CN. (2005). Electrokinetic remediation of cadmium-contaminated clay. *Journal of Environmental Engineering* **131**(2):298–304.

Yeung AT, Hsu CN, Menon RM. (1996). EDTA-enhanced electrokinetic extraction of lead. *Journal of Geotechnical Engineering* **122**(8):666–673.

Yu JW, Neretnieks I. (1997). Theoretical evaluation of a technique for electrokinetic decontamination of soils. *Journal of Contaminant Hydrology* **26**:291–299.

8

ELECTROKINETIC STABILIZATION OF CHROMIUM (VI)-CONTAMINATED SOILS

LAURENCE HOPKINSON, ANDREW CUNDY, DAVID FAULKNER, ANNE HANSEN, AND ROSS POLLOCK

8.1 INTRODUCTION

Land contaminated with heavy metals constitutes a multifaceted scientific challenge. While the selection of remediation method(s) is dependent on site characteristics, concentration, pollutant types, eventual land use, and associated financial considerations (e.g. Mulligan, Yong, and Gibbs, 2001), the physical separation and extraction of contaminants is not always possible or practicable. For instance, an individual site may be so grossly impacted with hazardous waste so as to render any meaningful remediation to a benchmark target value unobtainable. Alternately, site-specific conditions may prohibit effective cleanup, or else individual sites may be placed in a larger prioritization program. Hence, risk abatement approaches may include solidification, stabilization, site isolation, and/or toxicity reduction. For the purposes of this chapter, stabilization is defined as fixing the contaminant in place, thereby rendering it less likely to move elsewhere under ambient hydrogeological conditions (Haran et al., 1995; Pamukcu, Weeks, and Wittle, 1997).

The utilization of electrokinetics in civil and environmental engineering is wide ranging (e.g. Ottosen et al., 2007). Large-scale field applications of electrokinetics have been employed in civil engineering projects for a significant length of time. For instance, geotechnical stabilization of railway cuttings for some 850 m employing sacrificial aluminum electrodes has been conducted (e.g. Casagrande, 1947; Casagrande et al., 1983). Furthermore, electro-osmotic dewatering has been utilized to improve the mechanical strength of soils and slurries (e.g. Lamont-Black, 2001). In addition, strategic subsurface mobilization of grouts and cements has been

Electrochemical Remediation Technologies for Polluted Soils, Sediments and Groundwater,
Edited by Krishna R. Reddy and Claudio Cameselle
Copyright © 2009 John Wiley & Sons, Inc.

documented (e.g. Ingles and Metcalf, 1972). In recent years, electrokinetic soil remediation has undergone rapid development, with some full-scale trials having been undertaken in the USA and, to a lesser extent, Europe (Acar and Alshawabkeh, 1993; Probstein and Hicks, 1993; Reddy *et al.*, 1997; Virkutyte, Sillanpaa, and Lastostenmaa, 2002). Less attention has been focused on the use of electrokinetics to stabilize contaminated soils, although electrokinetic techniques potentially offer a low-cost tool to abate immediate environmental risk for a range of contaminants. It is important to note that electrochemical stabilization, in many cases, represents an interim or pretreatment process to permanent treatment technologies (e.g. Pamukcu, Weeks, and Wittle, 1997).

Chromium has found extensive application in a variety of industries (e.g. Batchelor *et al.*, 1998; Graham *et al.*, 2006; Mukhopadhyay, Sundquist, and Schmitz, 2007; Bini, Maleci, and Romanin, 2008; Moghaddam and Mulligan, 2008). The carcinogenic nature of hexavalent chromium [Cr(VI)] means that it represents a first-order environmental issue where soils and groundwater have become contaminated, as illustrated by the World Health Organization and by the European Union's 0.05 mg/l maximum contaminant level ceiling on Cr(VI) levels in drinking water (e.g. Mukhopadhyay, Sundquist, and Schmitz, 2007).

A number of techniques have been proposed for the stabilization and remediation of Cr(VI)-contaminated soils. In particular, the reduction of Cr(VI) to trivalent chromium [Cr(III)], which is considered essentially immobile in the environment and of low toxicity, is a critical step in the stabilization/remediation of Cr(VI)-contaminated sites (e.g. Reddy *et al.*, 2003; Bewley, 2007; Bini, Maleci, and Romanin, 2008). The chemical reduction of Cr(VI) to Cr(III) occurs through different groups of compounds: sulfur compounds, iron-based compounds, and various organic compounds. (e.g. Eary and Rai, 1988; Su and Ludwig, 2005; Bewley, 2007). In addition, granular activated carbon has been employed to remove Cr(VI) from aqueous solutions (Frank and McMullen, 1996). Many of these techniques employ a two-step process in which Cr(VI) is transformed to Cr(III) by a reducing agent, at an acidic pH, and then precipitated as an insoluble hydroxide at an alkaline pH (Mukhopadhyay, Sundquist, and Schmitz, 2007). Furthermore, many of the techniques necessitate physical mixing of the reducing agent with the contaminated soil.

Electrokinetics has proven effectiveness for *in situ* treatment of Cr(VI)-contaminated groundwater (e.g. Mukhopadhyay, Sundquist, and Schmitz, 2007) and potential effectiveness for soil (Bewley, 2007). Conversely, many commercial electrokinetic systems are technically complex and energy intensive operating under very specific field or laboratory-based conditions (e.g. Acar and Alshawabkeh, 1993; Virkutyte, Sillanpaa, and Latostenmaa, 2002). Employing electrokinetics to stabilize Cr(VI)-contaminated soils by inducing the reduction of Cr(VI) to Cr(III) and simultaneously modifying hydrological conditions to promote isolation of contaminated soils from the immediate environment may well have site-specific applications.

It is well established that the application of a low-magnitude direct electric potential between opposite-polarity iron-rich electrodes emplaced in soil or sediment results in the generation of a strong pH (and Eh) gradient, dissolution of the anodic electrode(s), and precipitation of iron-rich mineral phases, including ferric iron oxyhydroxides, hematite, goethite, magnetite, and zero-valent iron, at the interface of the anodic and cathodic domains (e.g. Röhrs, Ludwig, and Rahner,

2002, Faulkner, Hopkinson, and Cundy, 2005). Because of the adverse effect of OH⁻ on soil remediation, due to the immobilization of many metal ions by precipitation in alkalinized soils, and the reduced efficiency of electrokinetic remediation when sacrificial iron-rich electrodes are employed (e.g. Leinz, Hoover, and Meier, 1998), noncorrosive electrodes and techniques to minimize soil alkalinization are generally employed for electrokinetic remediation (e.g. Röhrs, Ludwig, and Rahner, 2002; Virkutyte, Sillanpaa, and Latostenmaa, 2002). However, low adsorption of Cr(VI) in soils occurs in alkaline conditions, whereas high adsorption of Cr(VI) is favored in acidic conditions (Reddy *et al.*, 1997). Furthermore, the reduction of Cr(VI) to Cr(III) by the delivery of iron (Fe⁰, Fe²⁺) is fairly well documented (Rai, Sass, and Moore, 1987; Eary and Rai, 1991; Haran *et al.*, 1995; Powell *et al.*, 1995; Pamukcu, Weeks, and Wittle, 1997; Batchelor *et al.*, 1998; Reddy *et al.*, 2003). Accordingly, under an applied direct current (DC) electric field, stabilization of Cr(VI)-contaminated soils may potentially be achieved where oxidative dissolution of iron-rich anodic electrodes provides $Fe^{2+}_{(aq)}$ to react with the anode-bound migration of Cr(VI). Hence, the use of iron-rich sacrificial electrodes and soil alkalinization may find application in the electrokinetic stabilization of Cr(VI)-contaminated soils. This concept is explained in this chapter based on the results of laboratory stabilization experiments on three Cr(VI)-impacted soils taken from three sites within the UK.

8.2 MATERIALS AND METHODS

The experiments were undertaken on 50-kg samples of the contaminated soils. In an effort to emulate as closely as possible how stabilization may be actualized in the real environment the soils were not sieved. The soils were placed in Perspex cells (700 × 300 × 300 mm) along with 1500 ml of tap water. The Perspex cells were fabricated at the University of Brighton, UK. Five cast-iron electrodes [fabricated at the University of Brighton, UK, from Meehanite gray iron (flake) grade 250 + 260; West Yorkshire Steel Company Limited, UK] measuring 25 mm in diameter and 250 mm in length were inserted in the soil at opposite ends of the experimental vessel, giving an opposite-polarity electrode spacing of 500 mm. Two-centimeter-wide holes, filled with quartz-rich medium-grained sand, were inserted in the soil behind each electrode at the two opposite extremities of the experimental vessel to facilitate the addition of tap water to the anodic compartment and the collection of effluent from the cathodic compartment. The electrodes were wired to a Thurlby Thandar Instruments Limited (Huntingdon, Cambridgeshire, UK) (EX752M) power pack.

A potential difference of 75 V (1.5 V/cm) was applied for 1008 h (42 days) to the Warwick soil (experiment A). The Lanarkshire soil (experiment B) and Glasgow soil (experiment C) were subjected to 150 V (3 V/cm) for 1008 h. Fifty to seventy-five milliliters of tap water was added every 24 h to the sand behind the anodic electrodes. Fifty to one hundred milliliters of pore water was collected daily from the sand lining the wall of the vessel behind the cathode electrodes. No measurable soil heating occurred over the duration of the experiments. The pH was measured intermittently using a Hanna Instruments Limited (Bedfordshire, UK) 9024 pH meter.

Chemical analyses were conducted at a certified laboratory. Total chromium [Cr(T)] was measured by microwave-assisted acid digestion of a sample aliquot (2 g) followed by determination of metals in the extract (50 ml) by inductivity coupled

TABLE 8.1. Total Chromium, Cr(T), and Hexavalent Chromium, Cr(VI), Data for the Three Experiments

Distance from Anode (cm)	Sample Depth (cm)	Experiment A		Experiment B		Experiment C	
		Cr(VI) (mg/kg)	Cr(T) (mg/kg)	Cr(VI) (mg/kg)	Cr(T) (mg/kg)	Cr(VI) (mg/kg)	Cr(T) (mg/kg)
0	0–5	290	24,000	130	2,600	230	11,000
	5–10	200	13,000	28	1,800	200	13,000
	10–15	100	16,000	57	2,600	270	9,400
25	0–5	130	8,600	25	1,700	210	9,200
	5–10	110	8,500	26	1,600	230	9,300
	10–15	110	8,500	30	1,800	210	9,200
50	0–5	39	6,300	12	1,400	93	12,000
	5–10	49	6,500	18	1,200	76	9,300
	10–15	62	6,700	15	2,000	77	11,000
Pretreatment concentration.	—	740	10,900	460	1,855	506	10,378

Experiment A was conducted on contaminated soil from Warwick (England). Experiment B was conducted on contaminated soil from Lanarkshire (Scotland). Experiment C was conducted on contaminated soil from Glasgow (Scotland).

plasma mass spectrometry using a Perkin Elmer Elan 6100 (Waltham, MA). For soil Cr(VI) content, 10 g of dried soil sample were ground into 0.1 M HCl (50 ml) at 37 °C. Analyses were performed by manual colorimetry using 1,5-diphenylcarbazide (Hach colorimeter, model DR/890, Loveland, CO). At the cessation of the experiment, analyses were performed on soil samples taken from three depth intervals (0–5, 5–10, and 10–15 cm) at 0-, 25-, and 50-cm distance from the anode (Table 8.1). It should also be noted that due to the differences in methods used to determine Cr(T) and Cr(VI), disparities can emerge and that, in general, compared with inductively coupled plasma spectrometry, the colorimetric method is less accurate (e.g. Mukhopadhyay, Sundquist, and Schmitz, 2007). Electron microscopic analyses were conducted with a Jeol, (Tokyo, Japan) scanning electron microscope (SEM) with Oxford Instruments (Abingdon, Oxfordshire, UK) energy dispersive x-ray spectrometry. SEM analyses were conducted at the University of Brighton, UK.

8.3 EXPERIMENTAL RESULTS

8.3.1 Experiment A (Warwick Soil Sample)

The Warwick soil was collected from 40- to 80-cm depth and consisted of sandy red clay with small stones, pieces of brick, and concrete. Prior to the experiment, a 5-g subsample of soil mixed with 10 ml of deionized water yielded a pH of 7.7 (measured at 25 °C). Cr(T) was measured at 10,900 mg/kg, and Cr(VI) at 740 mg/kg. SEM analysis of the untreated soil showed a fairly homogeneous clay matrix, with disseminated silicate sand grains. Figure 8.1a shows the emergence of the imposed pH gradient over 21 days, beyond which no measurable reduction in pH in the anodic compartment occurred. Iron staining of the soil within the anodic compartment, evident after 7 days, projected 25 cm from the anodes at the termination of the experiment. The current fell from 0.67 to 0.19 A during the first 7 days, remaining

EXPERIMENTAL RESULTS 183

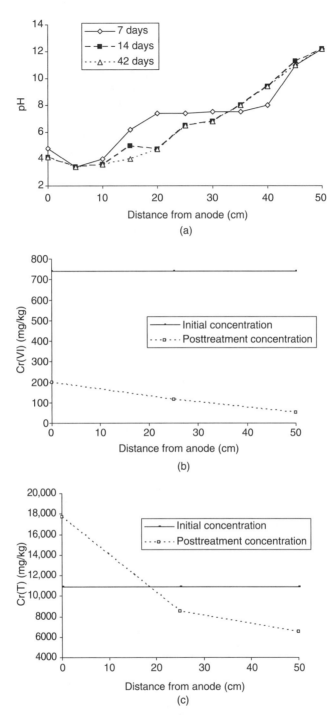

Figure 8.1. Experiment A (Warwick soil). (a) The evolution of the pH gradient between anode and cathode arrays as a function of experimental duration. (b) Cr(VI) concentration measured with 0-, 25-, and 50-cm distance from the anodes. Note that each measurement is the mean value of readings taken at the three depth intervals (0–5, 5–10, and 10–15 cm; see Table 8.1 for individual depth interval values). (c) Cr(T) concentration measured 0-, 25-, and 50-cm distance from the anodes. Note that each measurement is the mean value of readings taken at the three depth intervals (0–5, 5–10, and 10–15 cm), see Table 8.1.

at broadly similar values for the remainder of the experiment, and was measured at 0.17 A at the end of the experiment.

Analyses of samples taken from nine set depths within the experimental cell indicate a pronounced reduction in Cr(VI) concentration relative to the 740 mg/kg pretreatment level, with mean values at anode and cathode of 197 and 50 mg/kg, respectively (Fig. 8.1b), giving an average 93.2% reduction of Cr(VI) at the cathode and a 73.4% reduction at the anode. Cr(T) at the anode is 17,666 mg/kg, and 6500 mg/kg at the cathode. These values, compared with the preexperimental Cr(T) value, indicate a 40.4% depletion in Cr(T) at the cathode, and concomitantly, a 62.1% enrichment around the anodic electrodes (Fig. 8.1c). SEM analysis of soil samples taken from iron-stained sections of the anodic compartment soils reveal abundant quantities of Cr-bearing iron oxide particles. The iron particles are platy in character, with longest dimensions of 1–8 μm, and locally cement clay minerals and larger detrital silicate grains (Fig. 8.2a).

8.3.2 Experiment B (Lanarkshire Soil Sample)

The Lanarkshire soil sample consisted of gray clay with small stones and some bright yellow nodules. SEM analysis indicated the presence of a Cr-bearing silicate mineral phase, chromite, and traces of calcium carbonate within the clay fraction. Prior to the experiment, a 5-g subsample of soil mixed with 10 ml of deionized water yielded a pH of 8.9 (measured at 25 °C). Cr(T) and Cr(VI) were measured at 1855 and 460 mg/kg, respectively (Table 8.1). After 14 days, a discernable pH gradient developed, with pH levels of 4 and 11 in the anodic and cathodic compartments, respectively, emerging by 42 days (Fig. 8.3a). During the first week of the experiment, the current dropped from 0.33 to 0.23 A; there was a progressive decrease in the current during the remainder of the experiment. The current was measured at 0.14 A at the termination of the experiment. Iron migration was noted after 1 week, with iron migration to approximately 10 cm from the anode by week 6. At the cessation of the experiment, the concentration of Cr(VI) was, on average, 71.66 and 15 mg/kg near the anode and cathode, respectively, indicating an average 96.7% reduction of Cr(VI) at the cathode and an 84.4% reduction at the anode. Cr(T) was 2333 and 1533 mg/kg near the anode and cathode, respectively. This represents a 25.8% increase in Cr(T) around the anodic array, and a complementary 17.4% reduction in Cr(T) around the cathode array (Fig. 8.3b,c). In common with the Warwick soil, SEM analysis of the iron-stained soils from the anodic domain revealed the presence of abundant fine-grained iron oxide/oxyhydroxides, which contain accessory quantities of chromium (Fig. 8.2b).

8.3.3 Experiment C (Glasgow Soil Sample)

The Glasgow soil was collected from between 1- and 3-m depth and consists of brown clay, contains building debris, and shows centimeter-sized nodules of chromium ore processing residue (COPR). SEM analysis of the untreated material revealed an array of Cr-bearing silicate mineral phases and traces of chromite (Fig. 8.2c). Traces of manganese and iron oxide were also identified. COPR-contaminated soils can show a varied and complex mineralogy (Graham et al., 2006), including Cr(III) hosted in chromite and brownmillerite, with hydrogarnets and hydrocalu-

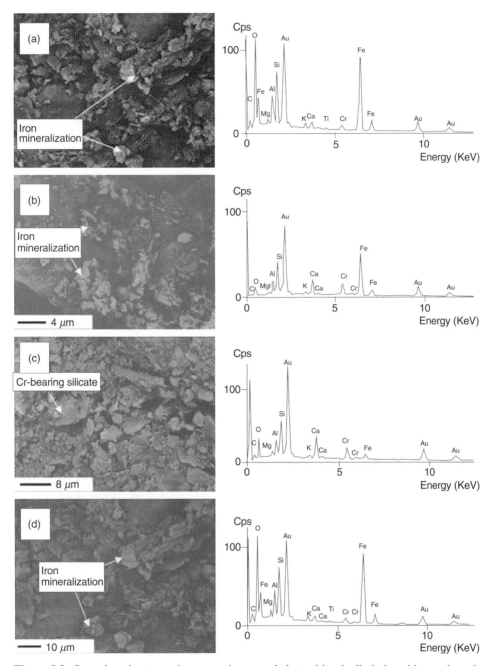

Figure 8.2. Scanning electron microscope images of electrokinetically induced iron mineralization in (a) anode-zone soil from Warwick, (b) anode-zone soils from Lanarkshire, (c) untreated COPR-contaminated Glasgow soil, and (d) postexperiment anode-zone iron mineralized COPR-contaminated Glasgow soil.

186 ELECTROKINETIC STABILIZATION

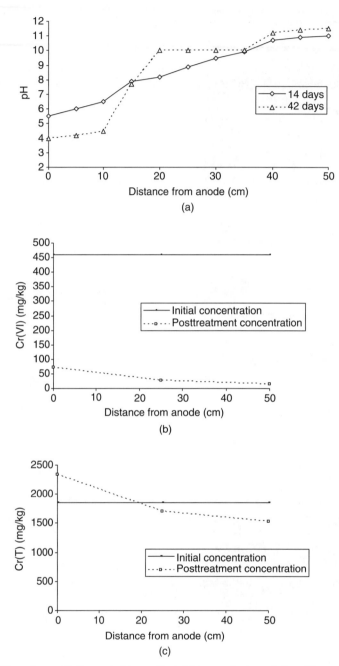

Figure 8.3. Experiment B (Lanarkshire soil). (a) The evolution of the pH gradient between anode and cathode arrays as a function of experimental duration. (b) Cr(VI) concentration measured with 0-, 25-, and 50-cm distance from the anodes. Note that each measurement is the mean value of readings taken at the three depth intervals (Table 8.1). (c) Cr(T) concentration measured with 0-, 25-, and 50-cm distance from the anodes. Each measurement is the mean value of readings taken at the three depth intervals (0–5, 5–10, and 10–15 cm) reported in Table 8.1.

mite acting as potential sources of Cr(VI). Moreover, Cr(III) organic complexes have also been identified in groundwater at one location (Graham *et al.*, 2006).

At the outset of the experiment, Cr(T) and Cr(VI) were measured at 10,378 and 506 mg/kg, respectively (Table 8.1). A 5-g subsample of soil mixed with 10 ml of deionized water yielded a pH of 9.9 (measured at 25 °C). COPR-contaminated soils are known to be intrinsically alkaline, with a strong buffering capacity (e.g. Bewley, 2007). Consequently, the soil proved resistant to developing an acid and alkali front, in comparison with the Warwick soil, with a pH gradient emerging after 21 days but with no sharp pH jump developing (Fig. 8.4a). During the first 7 days of the experiment, the current dropped from 0.13 to 0.11 A; after 42 days the current was measured at 0.06 A. Iron staining was evident within the anodic portion of the experimental cell, extending ~10 cm from the electrodes by the end of the experiment. At which point, the concentration of Cr(VI) was, on average, 233 and 82 mg/kg near the anodes and cathodes, respectively, yielding an average 83.8% reduction in Cr(VI) concentration at the cathode and a 54.0% reduction at the anode (Fig. 8.4b). Cr(T) was measured at 11,333 and 10,766 mg/kg at the anode and cathode, respectively, representing an increase in Cr(T) of 7.3% and 3.7% around the anodic and cathodic electrodes, respectively. An 11% decrease in Cr(T) relative to pretreatment measured value is present, equidistant between the opposite-polarity electrode arrays (Fig. 8.4c). In common with the posttreatment anodic-zone soils from Warwick and Lanarkshire, SEM analysis of the iron-mineralized COPR-contaminated soil indicate the widespread precipitation of fine-grained (<10 μm) Cr-bearing iron oxide/oxyhydroxides (Fig. 8.2d).

8.4 DISCUSSION

In all three experiments, Cr(VI) underwent a substantive decrease in concentration, and the distribution of Cr(T) was modified, indicating the widespread mobilization of chromium and the extensive coeval reduction of Cr(VI) to Cr(III). Accumulation of chromium in the anode chamber shows that electromigration was the predominant driving force for the transport of ions (e.g. Mukhopadhyay, Sundquist, and Schmitz, 2007). Generally, under neutral or high pH conditions, Cr(VI) exists as the soluble and mobile CrO_4^{2-} ion (Reddy *et al.*, 2003) and, to a lesser extent, the dichromate ion $Cr_2O_7^{2-}$. Reaction of the dichromate ion and the chromate ion to chromic acid occurs only under strongly acidic conditions (e.g. Mukhopadhyay, Sundquist, and Schmitz, 2007) and were most likely not attained in this study.

The high pH and reducing conditions resulting from the electrolysis reaction at the cathode favors Cr(VI) solubility and electromigration (Reddy *et al.*, 2003). This is evidenced in the experiments by the greatest decline in Cr(VI) concentration occurring in the cathode zones of the three cells. The low pH and oxidizing conditions resulting from the electrolysis reaction at the anode tend to cause Cr(VI) adsorption, precipitation, or reduction (Reddy *et al.*, 2003). In an iron-rich environment, afforded by the sacrificial electrodes, the anode-bound electromigration of chromate and dichromate ions in the solution in the presence of $Fe^{2+}_{(aq)}$ facilitates Cr reduction (e.g. Rai, Sass, and Moore, 1987; Batchelor *et al.*, 1998; Reddy *et al.*, 2003; Mukhopadhyay, Sundquist, and Schmitz, 2007):

188 ELECTROKINETIC STABILIZATION

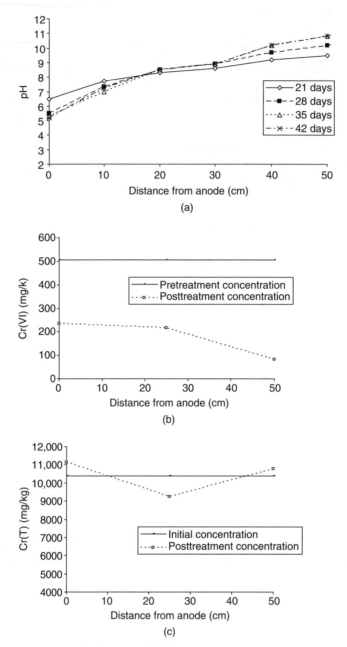

Figure 8.4. Experiment C (Glasgow soil). (a) The evolution of the pH gradient between anode and cathode arrays as a function of experimental duration. (b) Cr(VI) concentration measured with 0-, 25-, and 50-cm distance from the anodes. (c) Cr(T) concentration measured with 0-, 25-, and 50-cm distance from the anodes. Note that plotted Cr(VI) and Cr(T) values are the mean values of readings taken at the three depth intervals reported in Table 8.1.

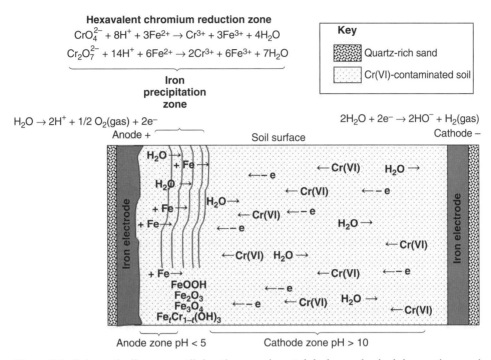

Figure 8.5. Schematic diagram outlining the experimental design and principle reactions and mineral products associated with the electrokinetic stabilization of Cr(VI) with sacrificial iron electrodes.

$$CrO_4^{2-} + 8H^+ \, 3Fe^{2+} \rightarrow Cr^{3+} + 3Fe^{3+} + 4H_2O \tag{8.1}$$

$$Cr_2O_7^{2-} + 14H^+ + 6Fe^{2+} \rightarrow 2Cr^{3+} + 6Fe^{3+} + 7H_2O \tag{8.2}$$

Electrokinetically driven iron mineralization originates when Fe(III) combines with OH⁻ ions produced at the cathode to form insoluble ferric hydroxides [Fe(OH)$_{3(s)}$], hematite (α-Fe$_2$O$_3$) and goethite (FeOOH) (e.g., Faulkner, Hopkinson, and Cundy, 2005; Mukhopadhyay, Sundquist, and Schmitz, 2007). SEM observations of soil samples taken from the anodic zones of the experimental cells reveals a ubiquitous association between the iron minerals and subsidiary quantities of chromium. Cr(III) can substitute for Fe(III) in the FeOOH structure (Eary and Rai, 1988), and Cr(VI) reduction by Fe^{2+} leads to the development of solids at near-neutral pH showing mixed iron/chromium solid solution of the form Fe$_x$Cr$_{1-x}$(OH)$_3$ (Eary and Rai, 1988; Fendorf and Li, 1996). Such conditions would have been met at the interface between the acidic and alkali portions of the experimental cells (Fig. 8.5). When Fe(III) is produced solely from the stochiometric reaction with chromate, the value of x is 0.75 (Batchelor et al., 1998):

$$4H_2O + CrO_4^{2-} + 3Fe^{2+} + 4OH^- \rightarrow 4Fe_{0.75}Cr_{0.25}(OH)_3 \, (s) \tag{8.3}$$

However, in natural environments, soil-specific characteristics means that fractions of Fe$^{2+}_{(aq)}$, liberated by anode dissolution, will be rendered unavailable to react with Cr(VI) due to other interactions with other solid phases (e.g. Batchelor et al., 1998). This means that site-specific models require development for prediction of the iron requirement to stabilize chromium under *in situ* conditions (e.g. Batchelor et al., 1998).

The highest residual traces of Cr(VI) occur in the anodic sections of the experimental cells. Cr(VI) removal from aqueous solutions is enhanced by the presence of ferric iron oxyhydroxide phases, as Cr(VI) adsorbs onto FeOOH (e.g. Aoki and Munemori, 1982; Mesuere and Fish, 1992a,b; Mukhopadhyay, Sundquist, and Schmitz, 2007). The amount of $Fe^{2+}_{(aq)}$ released by anodic electrode dissolution primarily depends on the applied current and the duration of the passage of the current through the electrodes (e.g. Mukhopadhyay, Sundquist, and Schmitz, 2007). Differences in the lateral extent of iron mineralization in the three experiments illustrate that the buffering capacity of the soils influenced the spatial extents of the zone of Cr(VI) reduction and complementary alkaline zone. The Warwick soil (experiment A) operated at half the applied voltage to experiments B and C, experienced the furthest advance of iron mineralization from the anode array, quickly developed a sharp pH jump, and attained the most acidic conditions. Collectively, these attributes indicate that the Warwick soil had a comparatively low buffering capacity relative to the other two soils examined.

The high buffering capacity of the Lanarkshire soil (experiment B) compared with experiment A, is demonstrated by the limitation of acidification and associated iron staining to within ~10cm of the anode. However, a pH jump did emerge after 3 weeks, with alkaline conditions prevailing in ~60% of the experimental cell. The 96.8% reduction in the Cr(VI) content of the cathode zone, as well as the 84.4% decrease in the anodic zone, suggests that stabilization was advanced. The complementary 25.8% increase in Cr(T) around the anodic array, as well as the associated 17.4% reduction in Cr(T) around the cathode array (Fig. 8.3), is consistent with the anodic domain operating as an efficient localized zone of Cr(VI) reduction to Cr(III).

COPR is a complex material, and previous studies have highlighted the limitations of the more conventional approaches for Cr(VI) remediation (Bewley, 2007). The strong buffering capacity of the COPR-contaminated soil (experiment C) is evidenced by the duration of time required for an acid front to develop, the limitation of iron mineralization to within ~10cm of the anode electrodes, and the absence of the emergence of a sharp pH jump in the experimental duration (Fig. 8.3a). Nevertheless, a 53.4% and 83.8% decrease in Cr(VI) was attained in the acidic and alkaline portions of the cell, respectively (Fig. 8.3b). The experimental results show depletion of Cr(T) at near-neutral pH at the center of the cell, with a 7.3% and 3.7% enrichment at the anodic and cathodic electrodes, respectively. Therefore, results suggest coeval electromigration of chromium to opposite-polarity electrodes (Fig. 8.3c). Because Cr(III) exists in cationic form, it may undergo migration to the cathode at comparatively slow rates with respect to Cr(VI) (Reddy and Parupudi, 1997). Accordingly, the apparent Cr(T) depletion within the center of the cell may reflect the cumulative effect of anodic bound electromigration of Cr(VI) and coeval cathode-bound mobilization of Cr(III). The later may be arrested by high adsorption and precipitation in high-pH regions (Griffin, Frost, and Shimp, 1976; Reddy and Parupudi, 1997).

Overall, experimental results suggest that electrokinetically mediated iron mineralization, combined with purposeful neglect to control soil alkalization, provides a potentially viable mechanism to stabilize Cr(VI)-contaminated soils. If this stabilization mechanism represents an interim or pretreatment process to permanent treatment technologies, the question to be addressed concerns the longevity of the stabilization process. Cr(III) shows only low solubility, especially in the presence of Fe(III) at neutral pH (Batchelor et al., 1998). In addition, consideration of iron and chromium speciation in the Eh–pH space indicates that the $Fe_xCr_{1-x}(OH)_3$ solid phases are stable under neutral and alkaline, as well as moderately to strongly oxidizing, conditions. Furthermore, any scenario involving reductive dissolution of the mineral phases under acidic conditions means that $Fe^{2+}_{(aq)}$ is available to react with Cr(VI), or else chromium is already present as Cr(III). It is also important to note that chromium–iron hydroxide solid solutions have lower equilibrium solution activities than pure solid phases (Powell et al., 1995).

Several experiments have shown that electrokinetically mediated iron mineralization is, in essence, a self-limiting process in which the current falls as an essentially impermeable (1×10^{-7} m/s or less) iron-rich lithified soil mass develops (e.g. Hopkinson and Cundy, 2003; Cundy and Hopkinson, 2005). In the context of Cr(VI) stabilization, the generation of low-permeability zones, in which chemically stabilized Cr(III) is locked in iron-rich mineral phase(s), means that stabilized zones are likely to be resilient to reactive flow, with the low permeability also serving to limit potential loss of adsorbed Cr(VI) residue on the iron mineral phase(s).

8.5 SUMMARY

The experiments documented here appear to demonstrate a comparatively simple, low-energy, and low-cost approach to abate immediate environmental risk from Cr(VI). Although it should be noted that the experimental data reported here only relates to the final chromium content in the soil materials, the data does suggest that electrokinetically mediated iron mineralization may potentially serve dual purposes, namely, *in situ* construction of physical barriers and coeval stabilization of Cr(VI)-contaminated soils, by its reduction to Cr(III) and coeval incorporation in newly formed ferric iron mineral phase(s). Furthermore, given that reactions involving iron play a major role in the environmental cycling of a wide range of important organic, inorganic, and radioactive contaminants (e.g. Cundy, Hopkinson, and Whitby, 2008), it is suggested that electrokinetically driven iron mineralization may find application in a range of areas.

ACKNOWLEDGMENTS

The authors are grateful to Knowledge Transfer Partnership (award KTP000871). We also thank Engineering and Physical Sciences Research Council UK (EPSRC) (grant number GR/S27924/01) for funding the exploration of heavy metal stabilization by ferric iron mineralization. The technique and ideas described here are the subject of European Patent EP03750979.1. We express our thanks for the constructive comments made by the reviewers.

REFERENCES

Acar YB, Alshawabkeh AN. (1993). Principles of electrokinetic remediation. *Environmental Science and Technology* **27**:2638–2642.

Aoki T, Munemori M. (1982). Recovery of chromium (VI) from waste waters with iron (III) hydroxide. *Water Research* **16**:793–796.

Batchelor B, Schlautman M, Hwang I, Wang R. (1998). Kinetics of Chromium (VI) Reduction by Ferrous Iron. Amarillo National Resource Centre for Plutonium. *Report ANRCP-1998-13*.

Bewley R. (2007). Treatment of Chromium Contamination and Chromium Ore Processing Residue. CL:AIRE Technical Bulletin. *TB14:1-4*.

Bini C, Maleci L, Romanin A. (2008). The chromium issue in soils of the leather tannery district in Italy. *Journal of Geochemical Exploration* **96**:194–202.

Casagrande L. (1947). The Application of Electro-Osmosis to Practical Problems in Foundations and Earthworks. Department of Scientific and Industrial Research. Building Research, London, UK. *Technical. Paper No. 30*.

Casagrande L, Wade N, Wakely M, Loughney R. (1983). Electro-osmosis projects, British Columbia, Canada. *Proceedings of the International Conference on Soil Mechanics and Foundation Engineering. Stockholm* **3**:607–610.

Cundy AB, Hopkinson L. (2005). Electrokinetic iron pan generation in unconsolidated sediments: implications for contaminated land remediation and soil engineering. *Applied Geochemistry* **20**:841–848.

Cundy AB, Hopkinson L, Whitby RLD. (2008). Use of iron-based technologies in contaminated land and groundwater remediation: A critical review. *Science of the Total Environment* **400**:42–51.

Eary LE, Rai D. (1988). Chromate removal from aqueous waste by reduction with ferrous iron. *Environmental Science and Technology* **22**:972–977.

Eary LE, Rai D. (1991). Chromate reduction by surface soils under acidic conditions. *Soil Science Society of America Journal* **55**:676–683.

Faulkner D, Hopkinson L, Cundy AB. (2005). In situ electrokinetic generation of reactive iron barriers in sediment: implications for contaminated land. *Mineralogical Magazine* **69**(5):749–757.

Fendorf SE, Li G. (1996). Kinetics of chromate reduction by ferrous iron. *Environmental Science and Technology* **30**:1614–1617.

Frank WL, McMullen MD. (1996). Removing chromium from groundwater and process wastewater. *The National Environmental Journal* **March–April**:36–39.

Graham MC, Farmer JG, Anderson P, Paterson E, Hillier S, Lumsdon DG, Bewley RJF. (2006). Calcium polysulfide remediation of hexavalent chromium contamination from chromite ore processing residue. *Science of the Total Environment* **364**:32–44.

Griffin RA, Frost RR, Shimp NF. (1976). Effect of pH on removal of heavy metals from leachates by clay minerals. Residual management by land disposal. *Proceedings of the Hazardous Waste Research Symposium* (ed. WH Fuller), EPA-600/9-76-015. February 2–4, Cincinnati, OH: U.S. Environmental Protection Agency, pp. 259–268.

Haran BS, Zheng G, Popov BN, White RE. (1995). Electrochemical decontamination of soils: development of a new electrochemical method for decontamination of hexavalent chromium from sand. *Proceedings of the Electrochemical Society* **12**:227–251.

Hopkinson L, Cundy AB. (2003). FIRS (ferric iron remediation and stabilisation): A Novel Electrokinetic Technique for Soil Remediation and Engineering. CL:AIRE Research Bulletin. *RB2:1-4*.

Ingles OG, Metcalf JB. (1972). *Soil Stabilization: Principles and Practice*. London: Butterworth.

Lamont-Black J. (2001). E.K.G: The next generation of geosynthetics. *Ground Engineering* **34**(10):22–23.

Leinz RW, Hoover DB, Meier AL. (1988). NEOCHIM: An electrochemical method for environmental application. *Journal of Geochemical Exploration* **64**:421–434.

Mesuere K, Fish W. (1992a). Chromate and oxalate adsorption on goethite. 1. Calibration of surface complexation models. *Environmental Science and Technology* **26**:2357–2364.

Mesuere K, Fish W. (1992b). Chromate and oxalate adsorption on goethite. 2. Surface complexation modelling of competitive adsorption. *Environmental Science and Technology* **26**:2365–2408.

Moghaddam AH, Mulligan CN. (2008). Leaching of heavy metals from chromated copper arsenate (CCA) treated wood after disposal. *Waste Management* **28**:628–637.

Mulligan CN, Yong RN, Gibbs BF. (2001). Remediation technologies for metal-contaminated soils and ground water: An evaluation. *Engineering Geology* **60**:193–207.

Mukhopadhyay B, Sundquist J, Schmitz RJ (2007). Removal of Cr(VI) from Cr-contaminated groundwater through electrochemical addition of Fe(II). *Journal of Environmental Management* **82**:66–76.

Ottosen LM, Christensen IB, Rörig-Dalgård I., Jensen PE. (2007). Utilization of electromigration in civil and environmental engineering. *6th Symposium on Electokinetic Remediation* (ed. C Cameselle), June 12–15, Vigo, Spain: University of Vigo, pp. 5–6.

Pamukcu S, Weeks A, Wittle JK. (1997). Electrochemical extraction and stabilization of selected inorganic species in porous media. *Journal of Hazardous Materials* **55**:305–318.

Powell RM, Puls RW, Hightower SK, Sabatini DA. (1995). Coupled iron corrosion and chromate reduction: Mechanisms for subsurface remediation. *Environmental Science and Technology* **29**:1913–1922.

Probstein RF, Hicks RE. (1993). Removal of contaminants from soils by electric fields. *Science* **260**:498–503.

Rai D, Sass BM, Moore DA. (1987). Chromium (III) hydrolysis constants and solubility of chromium(III) hydroxide. *Inorganic Chemistry* **26**:345–349.

Reddy KR, Chinthamreddy S, Saichek RE, Cutright TJ. (2003). Nutrient amendment for the bioremediation of a chromium contaminated soil by electrokinetics. *Energy Sources* **25**:931–943.

Reddy KR, Parupudi US. (1997). Removal of chromium, nickel and cadmium from clays by in-situ electrokinetic remediation. *Journal of Soil Contamination* **6**:391–407.

Reddy KR, Parupudi US, Devulapalli SN, Xu CY. (1997). Effects of soil composition on the removal of chromium by electrokinetics. *Journal of Hazardous Materials* **55**:135–158.

Röhrs J, Ludwig G, Rahner D. (2002). Electrochemically induced reactions in soils-a new approach to the in-situ remediation of contaminated soils? Part 2: Remediation experiments with natural soil containing highly chlorinated hydrocarbons. *Electrochimica Acta* **47**:1405–1414.

Su C, Ludwig RD. (2005). Treatment of hexavalent chromium in chromite ore processing residue solid waste using a mixed reductant solution of ferrous sulphate and sodium dithionite. *Environmental Science and Technology* **39**:6208–6216.

Virkutyte J, Sillanpaa M, Latostenmaa P. (2002). Electrokinetic soil remediation—A critical overview. *Science of the Total Environment* **289**:97–121.

PART III

REMEDIATION OF ORGANIC POLLUTANTS

9

ELECTROKINETIC REMOVAL OF PAHs

JI-WON YANG AND YOU-JIN LEE

9.1 INTRODUCTION

Since the mid-1980s, electrokinetics (EK) has been widely used as a soil remediation method, especially for low-permeability soils. Numerous studies have demonstrated the effectiveness of EK in the removal of soil contaminants. At first, most investigations focused on metals and only a few targeted relatively soluble organic pollutants, such as gasoline hydrocarbons, phenols, and trichloroethylene. In the case of hydrophobic organic compounds (HOCs) with low solubility in water, but with a high tendency to be adsorbed onto soil, electrokinetic remediation was previously considered as "not applicable" because transport by electroosmosis and/or electrophoresis was not to be expected (Acar et al., 1995; Virkutyte, Sillanpaa, and Latostenmaa, 2002). Therefore, methods to increase the solubility of HOCs had to be coupled with electrokinetic remediation.

Polycyclic aromatic hydrocarbons (PAHs), representative HOCs, are very hydrophobic and have quite low aqueous solubility. The solubilization/desorption and partitioning of PAHs in soil-water systems have been extensively studied using solubility-enhancing solutions such as surfactants and cosolvents to achieve effective removal of PAHs from contaminated sites. Recently, a number of laboratory studies on the electrokinetic removal of PAHs have appeared, evaluating the effect of enhancing solution and electrokinetic variables. The field remains under-researched in comparison with metal removal studies.

The electrokinetic process can assist in elimination of PAHs from soil when combined with other technologies such as biodegradation (Kim et al., 2005; Niqui-Arroyo and Ortega-Calvo, 2007) and chemical oxidation (Kim et al., 2006; Park et al., 2005). However, the present chapter focuses on the electrokinetic removal of PAHs using solubility-enhancing agents. Other integrated electrokinetic

Electrochemical Remediation Technologies for Polluted Soils, Sediments and Groundwater,
Edited by Krishna R. Reddy and Claudio Cameselle
Copyright © 2009 John Wiley & Sons, Inc.

technologies for PAH removal are discussed in other chapters. They include EK-bioremediation (Chapters 18 and 19) and EK-chemical oxidation (Chapter 21).

In this chapter, background information on PAHs and EK-facilitating agents is briefly provided, and earlier research papers are discussed, with reference to types of enhancing solutions and process variables. The chapter continues with a description of recent research trends and conclusions.

9.2 BACKGROUNDS

9.2.1 PAHs

PAHs, a group of stable chemical compounds including two or more fused benzene rings, are ubiquitous organic contaminants found naturally in the environment. PAHs are of concern because of their cytotoxic, mutagenic, and carcinogenic effects. They belong to a class of persistent organic pollutants (POPs) and 16 PAH compounds have been identified as priority pollutants by the US Environmental Protection Agency and the European Union (Fig. 9.1) (Ravindra, Sokhi, and Van Grieken, 2008).

PAHs arise from natural and/or anthropogenic processes, but usually from the incomplete combustion of fossil fuel and other organic matter. At high temperature, organic compounds are pyrolyzed into unstable fragments, such as radicals, and then recombine to form relatively stable PAHs. Oil spill accidents and wastewater discharge can also be sources of PAHs (Wilcke, 2000; Rivas, 2006).

Because of their nonpolar and stable chemical structure, PAHs are characterized by high hydrophobicity. PAHs show low aqueous solubility, low vapor pressure, and tend to be rapidly adsorbed onto soil particles. They are thus preserved as solids in the environment. Most PAH properties are related to the molecular weights and structures of the compounds. Relevant properties include solubility, melting and boiling points, vapor pressure, and octanol-water partitioning coefficients (K_{ow}). For example, PAH solubility ranges from 31.5 mg/l for naphthalene, the lowest molecular weight PAH, down to 1.4×10^{-4} mg/l for coronene, which has the highest molecular weight among PAHs of environmental interest (Saichek and Reddy, 2005a). As PAH molecular weight increases, carcinogenicity also rises, and acute toxicity decreases (Wilcke, 2000; Rivas, 2006; Ravindra, Sokhi, and Van Grieken, 2008).

Although PAHs are barely soluble in water and strongly adsorbed onto soil particles, they can be transmitted to subsoils indicating that PAH leaching occurs. Jones *et al.* (1989) found that leaching rates for PAHs in archived soil samples ranged from 0.009 to 0.14 mg/m and did not decrease with increase in PAH molecular weight. This indicated that high molecular-weight PAHs (and not only low molecular-weight compounds) can be transported in association with dissolved organic matter or organic particles (Wilcke, 2000). Such leaching means that PAHs pose a long-term threat to human health.

As soil contamination with PAHs has accelerated in many countries, particularly in urban regions (because of increased consumption of fuel), quick and effective removal of PAHs from the environment, thus minimizing PAH adverse effects, has emerged as a critical issue. PAHs are typically associated with soil matrices. Because

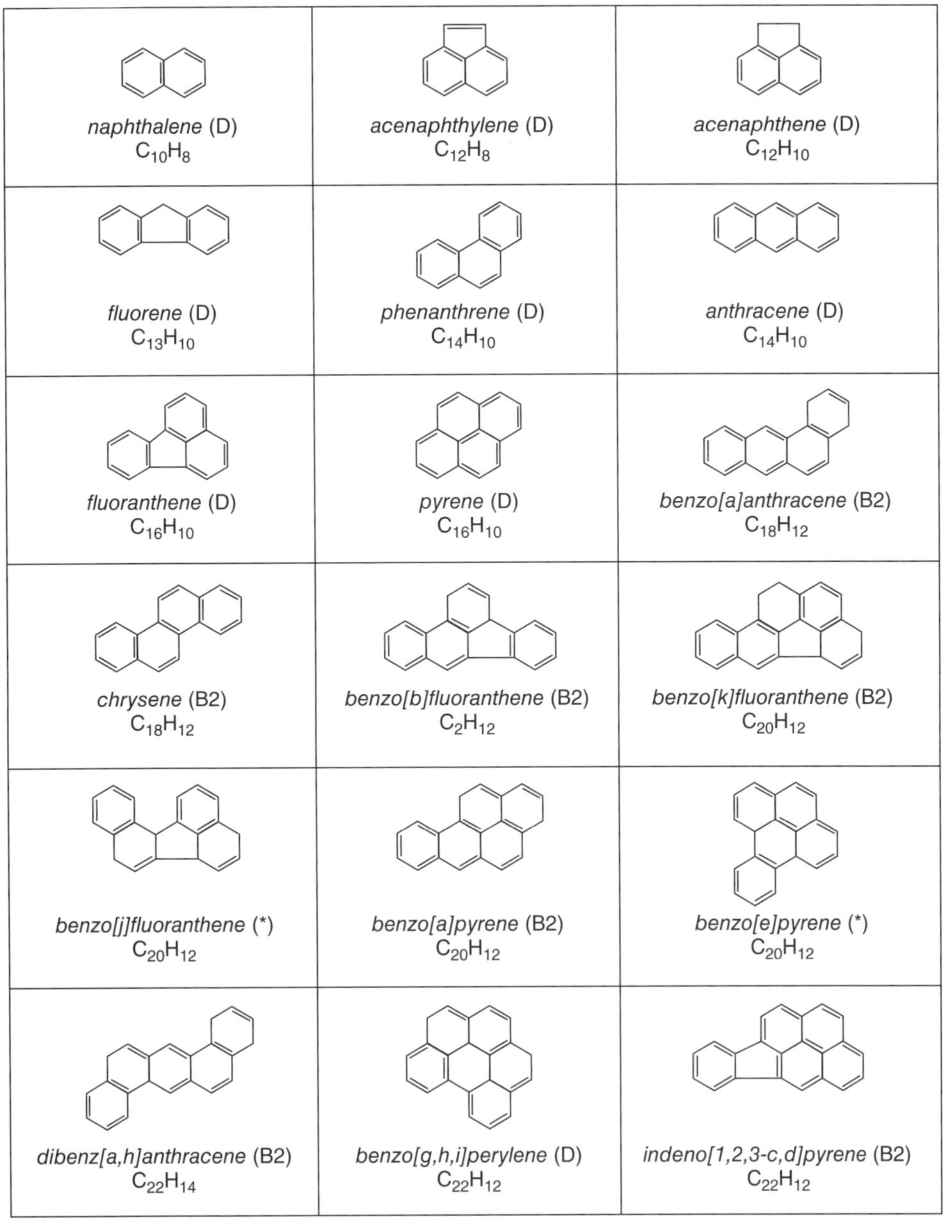

naphthalene (D) $C_{10}H_8$	acenaphthylene (D) $C_{12}H_8$	acenaphthene (D) $C_{12}H_{10}$
fluorene (D) $C_{13}H_{10}$	phenanthrene (D) $C_{14}H_{10}$	anthracene (D) $C_{14}H_{10}$
fluoranthene (D) $C_{16}H_{10}$	pyrene (D) $C_{16}H_{10}$	benzo[a]anthracene (B2) $C_{18}H_{12}$
chrysene (B2) $C_{18}H_{12}$	benzo[b]fluoranthene (B2) C_2H_{12}	benzo[k]fluoranthene (B2) $C_{20}H_{12}$
benzo[j]fluoranthene (*) $C_{20}H_{12}$	benzo[a]pyrene (B2) $C_{20}H_{12}$	benzo[e]pyrene (*) $C_{20}H_{12}$
dibenz[a,h]anthracene (B2) $C_{22}H_{14}$	benzo[g,h,i]perylene (D) $C_{22}H_{12}$	indeno[1,2,3-c,d]pyrene (B2) $C_{22}H_{12}$

Figure 9.1. Priority-listed PAHs (*, Not included in priority list; D, no information on human carcinogenicity; B2, probable human carcinogen) (Ravindra, Sokhi, and Van Grieken, 2008).

of their low aqueous solubility and high adsorption characteristics, it is difficult to remove them from subsurface environments using traditional technologies such as pump-and-treat, soil-vapor extraction, or bioremediation. One approach is to enhance the apparent solubility of PAHs by employing facilitating agents such as solvent mixtures (including ethanol/water or methanol/water) and surfactants

(Badr, Hanna, and de Brauer, 2004), thereby overcoming the mass-transfer limitation of the biodegradation process and the desorption stage of *in situ* soil flushing.

9.2.2 Facilitating Agents

As explained above, PAHs are barely soluble in water and it is thus common to use facilitating agents which can improve the solubilization/desorption of PAHs from soil. There are four groups of relevant additives—surfactants, cosolvents, cyclodextrins, and biosurfactants. Here, the characteristics of and general information on each type of facilitating agent are introduced.

9.2.2.1 Surfactants Surfactants are amphiphilic compounds with both hydrophilic and hydrophobic moieties. Because of their amphiphilic nature, surfactants accumulate at interfaces and thus minimize system-free energies. Surfactants increase PAH solubility by lowering interfacial tension as well as by accumulating the hydrophobic materials in micelles (the micellar solubilization) (Rosen, 1978; West and Harwell, 1992).

Depending on the hydrophilic head groups, surfactants are typically classified as cationic, anionic, nonionic, or zwitterionic (both cationic and anionic head groups present). Cationic surfactants are considered inappropriate additives for soil remediation because of low biodegradability and a tendency toward high adsorption to negatively charged soil surfaces. Rather, anionic surfactants are usually selected for surfactant-based remediation because of their lower degree of adsorption compared to cationic materials. However, anionic surfactants may precipitate and their critical micelle concentrations (CMCs) are higher than those of cationic and nonionic surfactants, so relatively large amounts of anionic surfactants are needed to achieve desired remediation levels. In general, nonionic surfactants are relatively nontoxic and economical because they show low CMCs and high solubilization capacities (West and Harwell, 1992).

Overall, desirable characteristics of surfactants for soil remediation purposes include effectiveness, biodegradability, low toxicity, good solubility at groundwater temperature, low adsorption to soil, low surface tension, and a low CMC (West and Harwell, 1992; Mulligan, Yong, and Gibbs, 2001; Lee, 2004).

9.2.2.2 Cosolvents The addition of polar organic solvents (cosolvents) that are completely miscible with, or highly soluble in, water, can significantly change solution-phase activities, such as solubility, sorption kinetics, and transport velocity of organic compounds; this is termed the "cosolvency effect" (Wood *et al.*, 1990; Brusseau, Wood, and Rao, 1991; Li, Cheung, and Reddy, 2000; Bouchard, 2002).

When a miscible solvent is added to water, the solvent increases the solubility and decreases the sorption of nonionic organic compounds by reducing the net polarity of the solution. The effect of a cosolvent depends on the nature of the solute and the solvent–cosolvent system, but there is usually a log-linear relationship between chemical aqueous solubility and the volumetric fraction of water-miscible solvent (Wood *et al.*, 1990; Bouchard, 2002).

Lower alcohols (e.g. methanol, ethanol, propanol), ketones (e.g. acetone), and alkyl amines are often used in solvent-flushing processes (Li, Cheung, and Reddy,

2000). However, it has been reported that alcohol concentrations greater than 3% might be toxic to microorganisms (Kim *et al.*, 2005). Therefore, the environmental properties of cosolvents, especially their effects on microbes, must be assessed prior to use for *in situ* soil remediation.

9.2.2.3 Cyclodextrins Cyclodextrins are cyclic oligosaccharides formed from enzymatic degradation of starch by bacteria. Recently, cyclodextrins have been proposed as alternative solubility-enhancing agents for HOCs because of the ability of cyclodextrins to form inclusion complexes with a wide range of substrates in aqueous solution. Unlike surfactant micelles, cyclodextrins are nontoxic and biodegradable, are less adsorbed onto solid phases over a wide range of pH values, and do not show critical concentrations for HOC solubilization, whereas surfactants enhance HOC solubility via micellar solubilization only at concentrations above their CMCs (Ko, Schlautman, and Carraway, 2000; Maturi and Reddy, 2006). However, cyclodextrins show lower solubilization power than do conventional surfactants because the cyclodextrin cavity is not as hydrophobic as the interior of a surfactant micelle, and another component such as an alcohol or surfactant may be required for good HOC solubilization. In addition, some cyclodextrins are too expensive for practical use (Saichek and Reddy, 2005a). Several papers have discussed how cyclodextrin use can be enhanced by overcoming cost barriers, but, to date, such examples are limited to laboratory-scale experiments (Blanford *et al.*, 2001).

9.2.2.4 Biosurfactants Many biological molecules are amphiphilic and partition preferentially at interphases. Microbial compounds that exhibit particularly high surface activity and emulsifying activities are termed biosurfactants (Cameotra and Bollag, 2003). These compounds are produced mainly by hydrocarbon-utilizing microorganisms that exhibit surface activity, such as some bacteria, yeasts, and fungi (Mulligan, Yong, and Gibbs, 2001; Saichek and Reddy, 2005a). Biosurfactants are typically glycolipids, lipopeptides, phospholipids, fatty acids, or neutral lipids. The lipophilic portion is usually a hydrocarbon (alkyl) tail containing one or more fatty acids, which may be saturated, unsaturated, hydroxylated, or branched. Each fatty acid is linked to the hydrophilic group by a glycosidic ester or amide bond. Most biosurfactants are either anionic or neutral, and only a few are cationic, such as those containing amine groups (Mulligan, Yong, and Gibbs, 2001; Cameotra and Bollag, 2003).

Biosurfactants have certain advantages over synthetic surfactants, including high specificity, biodegradability, and biocompatibility (Mulligan, Yong, and Gibbs, 2001; Park *et al.*, 2002; Saichek and Reddy, 2005a). A certain rhamnolipid was four times more effective in the column test of the hexadecane removal (Saichek and Reddy, 2005a) and glycolipids from *Rhodococcus* species 413A were 50% less toxic in naphthalene solubilization tests than was Tween 80 (Kanga *et al.*, 1997).

Many potential applications of biosurfactants have been considered and some of biosurfactants can be produced economically in commercial quantities. However, much work is still needed for process optimization, at both the biological and engineering levels, to improve the economics of biosurfactants' use (Bognolo, 1999).

9.2.3 Solubilization and Desorption of PAHs

A number of batch and column studies have sought to enhance the solubilization and desorption of PAHs from soil. Surfactants are commonly used to remediate PAH-contaminated soil. Two methods are employed: micellar solubilization and PAH mobilization by reduction of interfacial tension (West and Harwell, 1992). As surfactant toxicity became a significant issue, biodegradable and biocompatible surfactants have been more widely examined. For example, food-grade surfactants such as Tergitol 15-S-X (X = 7, 9, and 12) (Li and Chen, 2002) and other surfactants with indirect food additive status, such as alkyl diphenyl disulfonate (DOWFAX) (Deshpande *et al.*, 2000), have been investigated for use in solubilization/desorption of single PAHs or PAHs mixtures from contaminated soils.

In *in situ* soil-flushing processes, nonionic polyoxyethylenes (POEs) are widely used because of high solubilization capacity and biodegradability (Joshi and Lee, 1996; Mulligan, Yong, and Gibbs, 2001), but POEs can be adsorbed onto the soil matrix, resulting in the partitioning of PAHs into immobile adsorbed surfactants (Haigh, 1996; Ko, Schlautman, and Carraway, 1998). Edwards, Adeel, and Lirthy (1994) reported partitioning of a nonionic surfactant, Triton X-100, and phenanthrene, in a soil-water system. The results demonstrated that adsorbed surfactants acted as a phenanthrene sorbent and were much more effective than was organic matter. Both surfactant sorption and PAH distribution depend on a number of factors, mainly surfactant concentration and the nature of the solid sorbent. Moreover, for POE surfactants, sorption generally increases with decrease in the number of oxyethylene groups (Rosen, 1978). These data indicate that larger amounts of surfactants are required for effective micellization because of surfactant loss through adsorption (Edwards, Adeel, and Lirthy, 1994; Ko, Schlautman, and Carraway, 1998).

The use of anionic-nonionic mixed surfactants has been suggested to not only solve the nonionic surfactant sorption problem but also the precipitation of anionic surfactant. Several batch and column studies have shown a synergistic effect on PAH solubilization when SDS (sodium dodecyl sulfate)-Triton X-100, SDS-Triton X-305, SDS-Brij 35, and SDS-Tween 80 mixtures were used. This was attributed to the formation of mixed micelles, lower CMCs of the mixed surfactant solutions, and increases in solute molar solubilization ratios (Zhu and Feng, 2003; Mohamed and Mahfoodh, 2006; Zhou and Zhu, 2008). In addition, Zhou and Zhu (2008) found that sorption of Triton X-100 onto soil was very much reduced with increasing mass fraction of SDS in mixed surfactant solutions. Therefore, SDS–Triton X-100 mixed surfactants removed more phenanthrene than did the individual surfactants in column-flushing experiments. Mixed surfactants are consequently expected to improve surfactant-enhanced soil remediation by reducing required surfactant levels, thus lowering remediation cost.

Miscible solvents can significantly enhance PAH desorption and solubilization from the environment. According to Li, Cheung, and Reddy (2000), each of three cosolvents, n-butylamine, acetone, and tetrahydrofuran, increased phenanthrene solubility by more than five orders of magnitude, compared to a cosolvent-free control. Moreover, Peters and Luthy (1993) found that n-butylamine more readily solubilized coal tar into the cosolvent–water phase than did isopropyl alcohol or acetone. This indicates that cosolvents of higher molecular weight, and with few polar groups (such as amines), may be optimal in facilitating solubilization of organic compounds. In the solvent-flushing experiments performed by Augustijn et al. (1994), naphthalene and anthracene were removed from fine sand using different methanol-water mixtures. As the cosolvent fraction rose, contaminant removal improved. However, as the toxicity of a cosolvent solution increases with cosolvent concentration, the cosolvent level should be cautiously determined for *in situ* applications.

The solubility of PAHs can be enhanced by β-, γ-cyclodextrin, and modified cyclodextrins, termed hydroxypropyl-β-cyclodextrin (HPCD) and carboxymethyl-β-cyclodextrin (CMCD). Shixiang et al. (1998) reported that β-CD and CMCD impressively increased the aqueous solubility of six PAHs by formation of inclusion complexes with the materials. Hence, solubilization power was enhanced when PAH structures were well fitted within the CD cavities. This explains why the solubilization power of β-CD and CMCD for phenanthrene (0.58 nm in width, 0.78 nm in length) and 2-methylphenanthrene (0.58 nm in width, 0.98 nm in length) are much greater than for fluorene and 1-methylfluorene. Other larger or smaller compounds, or those inappropriate structure, could achieve only partial inclusion, resulting in decreased solubilization effects (Shixiang et al., 1998).

At low concentrations, β-CD showed a higher solubilization power than did CMCD. This might be caused by substitution of the carboxymethyl group outside the cavity of CMCD, which may impede diffusion of a PAH molecule into the cavity. Unfortunately, β-CD has only limited water solubility (18.7 g/l at 23 °C); this means that CMCD is the more attractive solubilization reagent for PAHs (Shixiang et al., 1998).

In batch desorption tests using two PAHs, naphthalene and phenanthrene, HPCD showed a higher removal capacity than did β-CD. Cyclodextrin sorption onto soil was relatively low and depended on the content of soil organic matter. This suggests that PAH desorption may be inhibited by competitive hydrophobic interactions of pollutant with organic matter and CD molecules, and by their co-sorption (Badr, Hanna, and de Brauer, 2004). In soil-flushing column tests, the addition of cyclodextrins such as β-CD, HPCD, and methyl-β-cyclodextrin (MCD) greatly improved the removal of PAHs from aged industrial contaminated soil and removal efficiency was enhanced with increases in cyclodextrin concentration (Viglianti et al., 2006). It may thus be concluded that cyclodextrin-enhanced soil flushing is feasible for practical PAH removal.

Some biosurfactants show better efficiency in PAH desorption than do synthetic surfactants, which are not biodegradable. Rhamnolipids enhanced the solubilization of four-ring PAHs more effectively than that of three-ring PAHs. The rhamnolipids were fivefold more effective than SDS in this regard (Mulligan, Yong, and Gibbs, 2001). Biosurfactants produced from *Pseudomonas aeruginosa* strain ATCC 9027 and another *Pseudomonas* strain isolated in the laboratory, *P. aeruginosa* P-CG3, also effectively enhanced the solubilization and desorption of phenanthrene and pyrene in a soil-aqueous system under thermophilic conditions (55 °C). Notably, the P-CG3 biosurfactant exhibited better performance than did Tween 80 or the other biosurfactant with a 28-fold increase in the apparent solubility of phenanthrene at a concentration of 10 times of its CMC (Cheng, Zhao, and Wong, 2004). Even the biosurfactant produced by *Rhodococcus* strain H13-A significantly increased the solubility of three- and four-ring PAHs. This biosurfactant was 2.2–35-fold more effective than Tween 80 in this regard (Page et al., 1999).

The effectiveness of biosurfactants was demonstrated in soil-washing tests using two soil samples contaminated with PAHs over several decades. Biosurfactants from urban waste compost (cHAL) enhanced PAH desorption from one soil 2–4-fold, relative to SDS. In addition, cHAL showed much less absorption than did SDS (12%–54% vs. 68%–95%). The results indicate that biosurfactants isolated from compost, instead of synthetic surfactants, can be used for soil remediation (Montoneri et al., 2009).

9.3 ELECTROKINETIC REMOVAL OF PAHs USING FACILITATING AGENTS

Many studies on the electrokinetic removal of PAHs have used various solubilizing agents to achieve good solubilizing capacity or high removal efficiency in batch or column desorption tests (Table 9.1). In most cases, batch results are not equivalent to the electrokinetic data because the mass transport and contaminant removal mechanisms differ considerably between batch and electrokinetic tests. For effective removal of PAHs by electrokinetic flushing, it is necessary to sustain not only sufficient soil-solution-contaminant interaction but also the mobilization of solubilized contaminants. In EK, the phenomenon of electroosmosis makes it possible to increase soil-solution-contaminant interaction in low-permeability regions thereby reducing tailing, remediation time, and cost (Saichek and Reddy, 2005a).

Below, laboratory studies on the electrokinetic removal of PAHs are discussed according to the types of solubility-enhancing agents used and the effects of electrokinetic variables on contaminant removal efficiency.

9.3.1 Effects of Facilitating Agents

9.3.1.1 Surfactant-Enhanced EK Among surfactants, nonionic materials have been widely used to transport nonpolar organics by electroosmotic flow. Tested material include Tween 80 [polyoxyethylene (20) sorbitan monooleate] (Reddy and Saichek, 2003; Saichek and Reddy, 2003a); Brij 30 (polyoxyethylene-4-lauryl ether) (Park et al., 2007); APG (alkyl polyglucosides) (Yang et al., 2005; Park et al., 2007); Igepal CA-720 (Saichek, 2002; Saichek and Reddy, 2003b; Reddy and Saichek, 2004); and the Tergitol series (Lee, 2004). Also, some researchers have investigated anionic surfactants with low soil surface adsorption tendencies. Examples of surfactants used in PAH-removal electrokinetic studies are listed in Table 9.2.

According to the Helmholtz–Smoluchowski (H–S) theory, electroosmotic flow is directly proportional to the dielectric constant of the fluid, the zeta potential, and the voltage gradient, and inversely proportional to fluid viscosity. In general, the surfactant solution and the cosolvent have lower dielectric constants than water, and the surfactant solution may even have a much higher viscosity than that of water. Therefore, the addition of facilitating agents results in a reduction of electroosmotic flow as well as an increase in PAH solubility.

When 3% of Tween 80, a nonionic surfactant commonly used in soil-flushing processes, was applied to an electrokinetic system used to remove phenanthrene from kaolin and glacial till, significant desorption, solubilization, and migration of the pollutant occurred near the anode. The phenanthrene concentrations in the

TABLE 9.1. Examples of Electrokinetic Tests Using Facilitating Agents

Facilitating agent	Soil	PAH	Remarks	Ref.
APG, Brij 30, SDS	Kaolin	Phenanthrene	APG 5 g/l—the highest removal (>75%) Brij30 5 g/l—removal was enhanced by acetate buffer (>71%) SDS 5 g/l—44-57% removal	(Park et al., 2007)
APG, Calfax 16L-35	Kaolin	Phenanthrene	5-30g/L APG enhanced PAH removal (63%-98%) but anionic Calfax 16L-35 did not (17%-24%). >68% of Calfax 16L-35 remained in soil.	(Yang et al., 2005)
Igepal CA-720	Kaolin	Phenanthrene	Moderate removal (significant near anode). A high surfactant concentration (5%) exhibited greater desorption. No advantages of a larger voltage gradient (2V/m), or a high NaOH concentration (0.1M) were observed.	(Saichek and Reddy, 2003b)
Tween 80, ethanol	Kaolin, glacial till	Phenanthrene	Moderate removal near anodic region. Low removal from glacial till because of high organic content.	(Reddy and Saichek, 2003)
Tergitol 15-S-7, Tergitol 15-S-12	Kaolin	Phenanthrene	Tergitol 15-S-7, a more hydrophobic surfactant, achieved lower removal than did Tergitol 15-S-12 because of adsorption.	(Lee, 2004)
n-butylamine, acetone, tetrahydrofuran	Glacial till	Phenanthrene	n-butylamine: 43% removal because of high pH Acetone, tetrahydrofuran: negligible removal	(Li et al., 2000)
HPCD	Kaolin	Phenanthrene	The higher the HPCD concentration, the greater the removal—56% at 6.85mM.	(Ko et al., 2000)
mannosylerythritol lipid (MEL), BS-UC	Kaolin	Phenanthrene	Na_2CO_3 buffer enhanced removal efficiency (75%). Higher removal efficiency (16%-17%) than shown by synthetic surfactants (6%-11%).	(Park et al., 2002)

TABLE 9.2. Examples of Surfactants Used in Electrokinetic Studies

Surfactant	Structure	Trade name
Polyoxyethylene alcohol	(structure)	Brij series
Polyoxyethylene p-t-octyl phenol	(structure)	Igepal CA series
Polyoxyethylene sorbitan esters	(structure)	Tween series
Polyoxyethylene isoalcohol	(structure)	Tergitol x-S-n series
Alkyl polyglucosides (APG)	(structure)	—
Sodium lauryl sulfate (sodium dodecyl sulfate, SDS)	(structure)	—

central and cathodic region were, however, much higher than the initial concentration because of insufficient electroosmotic flow and surfactant adsorption, resulting in changes in the soil/solution chemistry (Reddy and Saichek, 2003; Saichek and Reddy, 2003a). Surfactant has also been employed for the electrokinetic remediation of dredged sediment (Reddy and Ala, 2006) and gas plant soil (Reddy et al., 2006), both of which were contaminated with a wide range of PAHs and heavy metals. In these tests, PAH removal occurred only around anodic regions and it appeared that 3% of Tween 80 proved ineffective for desorption/solubilization of PAHs from soils of high organic content (Reddy and Ala, 2006; Reddy et al., 2006). Thus, although the results were promising near the anode, it was concluded that 3% of Tween 80 was not adequate to enhance phenanthrene solubility. The use of higher Tween 80 concentrations, or pH control, is necessary to achieve complete electrokinetic treatment of PAH-contaminated soil.

Saichek (2002) screened various surfactant/cosolvent mixtures in 96 batch tests and showed that a nonionic surfactant, Igepal CA-720, was the most promising flushing solution for the electrokinetic removal of PAHs. When the surfactant was applied in electrokinetic tests at concentrations of 3% and 5%, smaller amounts of phenanthrene removal were seen than were obtained in batch tests (Saichek and Reddy, 2003b). This is because *in situ* conditions under an electric field are quite different to those of batch tests. Although overall removal efficiency was not high, the solubilization and migration of phenanthrene near the anode were observed; the migration accelerated with increase in surfactant concentration. The results indicated that Igepal CA-720 could be a suitable facilitating agent for electrokinetic flushing. In another study of Reddy and Saichek (2004), the surfactant solution permitted good PAH removal efficiency (over 90%) when a high voltage gradient was applied periodically.

Park et al. (2007) introduced two nonionic surfactants, APG and Brij30, and one anionic surfactant, SDS, to remediate phenanthrene-spiked kaolinite. These authors found that APG produced the highest electroosmotic flow and showed the best removal efficiency among tested surfactant solutions. An electrokinetic test using Brij 30 showed relatively low phenanthrene removal because of low electroosmotic flow, even though the surfactant showed better solubilization than did APG. However, electroosmotic flow using Brij 30 could be enhanced by addition of acetate buffer, thus enhancing removal efficiency.

With an eye to environmental concerns, Lee (2004) conducted electrokinetic experiments using two Tergitol surfactants, Tergitol 15-S-7 and 15-S-12 (both are mixtures of secondary alcohol ethoxylates), which are readily biodegradable and have been approved for general-purpose cleaning or as an ingredient of general-purpose cleaner for use in federally inspected meat and poultry-processing plants by the Food Safety and Inspection Service of the US Department of Agriculture (Li and Chen, 2002). In the electrokinetic tests with high concentrations, 7.5 g/l, of these surfactants, over 95% phenanthrene removal was achieved. When Tergitol 15-S-7 was used at low concentrations (1.5–2.5 g/l), however, removal efficiency (22%–24%) fell. But Tergitol 15-S-12 still permitted high phenanthrene removal (80%–98%) at the same low concentration. Tergitol 15-S-7 showed surfactant loss during the electrokinetic test; the material adsorbed onto soil particles. This was confirmed by examining the distribution of surfactant concentrations in the electrokinetic test cell. These results are strong evidence that surfactants with hydrophobic structures may be ineffective for removal of organic compounds from soil because only a small proportion of surfactant molecules actually contributes to the mobilization of organics. Consequently, it appears that Tergitol 15-S-12 is more useful in electrokinetic remediation than is Tergitol 15-S-7.

One trial coupled EK with a cationic surfactant, cetyltrimethyl ammonium bromide (CTAB), to remediate a model diesel-contaminated soil artificially spiked with BTEX (benzene, toluene, ethyl benzene, and xylene) and three PAHs (Ranjan, Qian, and Krishnapillai, 2006). Although cationic surfactants are generally not used for soil flushing because of their toxicity and high adsorption onto negatively charged soil surfaces, it was anticipated that a cationic surfactant would be better than a nonionic or anionic surfactant in this instance because electrophoretic transport of micelles toward the cathode would encourage increased electroosmotic flow production. However, the results showed that CTAB retarded the electrokinetic transport of both PAHs and BTEX despite higher electroosmotic flow. It may be concluded that cationic surfactants are unsuitable in electrokinetic processes even if micellar charge is such that electroosmotic flow is facilitated.

Many studies have employed anionic surfactants to dissolve organic pollutants such as PAHs, but the use of such surfactants remains controversial in EK. Park *et al.* (2007) reported that SDS solutions injected into the anodic compartment could be transported toward the cathode and phenanthrene removal was as effective as that seen with use of a nonionic surfactant, APG. These researchers hypothesized that migration of SDS micelles in the direction opposite to that of electric charge was caused by a decrease in apparent SDS charge density at low pH. However, when Calfax 16L-35, a form of disulfonated surfactant, was used as a flushing agent, phenanthrene removal was barely enhanced even at the highest surfactant concentration tested (30 g/l). After the electrokinetic operation, most surfactant supplied from the anodic reservoir remained in the soil cell (64%) and the anodic compartment (21%); there was little surfactant in the effluent (0.2%). The authors concluded that anionic surfactants migrating in the direction opposite to electroosmotic flow were not helpful in electrokinetic removal of PAHs (Yang *et al.*, 2005).

On the other hand, tests with sodium dodecylbenzene sulfonate showed that PAH desorption could be accomplished by electromigration of anionic micelles in the direction of the anode; the surfactant was injected at the cathodic side of the electrokinetic cell. PAH removal of 90% was seen in the cathodic region (Pamukcu, 1994). Consequently, it is apparent that anionic surfactants migrate against the electroosmotic flow, and hence, are less useful in EK than are nonionic surfactants, even though the anionic surfactants are less adsorbed onto soil than are nonionic materials.

A recent study on electrokinetic removal of ethyl benzene from contaminated clay showed a promising use of anionic-nonionic mixed surfactants. Surfactant addition resulted in 1.6–2.4-fold more removal than afforded by EK alone, and a mixed surfactant system, including 0.5% SDS and 2.0% PANNOX 110 (nonyl phenol polyethylene glycol ether) permitted optimal ethyl benzene removal (98%). This indicated that, in the presence of mixed surfactant micelles, the zeta potential of the soil particles significantly increased compared with that seen when anionic surfactant micelles were formed, and electrolytic mobility was thus enhanced (Yuan and Weng, 2004). The use of anionic-nonionic mixed surfactants in EK will also improve the desorption and migration of PAHs.

9.3.1.2 Cosolvent-Enhanced EK The effect of cosolvents on electrokinetic removal of phenanthrene from glacial till was investigated by Li, Cheung, and Reddy (2000). These authors conducted batch and electrokinetic experiments using three cosolvents: n-butylamine, acetone, and tetrahydrofuran, all of which have higher solubilizing capacities for PAHs than do alcohols such as ethanol, methanol, or iso-propanol. Although batch desorption tests indicated that addition of 20% cosolvent greatly enhanced phenanthrene desorption from soil, only n-butylamine had a significant effect on the electrokinetic transport of phenanthrene (43% removal) because high solution pH maintained high electroosmotic flow. The effects of acetone and tetrahydrofuran were negligible because of low electroosmotic flow. Hence, the addition of cosolvents in EK could be evaluated as "moderately successful" only in the case of n-butylamine; the results emphasize the importance of electroosmotic flow for effective cosolvent flushing.

After these feasibility test data appeared (Li, Cheung, and Reddy, 2000), n-butylamine was employed in several electrokinetic tests as a flushing solution (Reddy and Ala, 2006; Reddy *et al.*, 2006; Maturi and Reddy, 2008). In a study by Maturi and Reddy (2008), the concentration of n-butylamine greatly influenced the amount of electroosmotic flow as well as phenanthrene mobility. When n-butylamine was present at 20%, greater phenanthrene removal was seen than when the cosolvent was present at 10% because of increased phenanthrene solubilization. Electroosmotic flow decreased, however, with increases in cosolvent concentration, resulting in an overall rather limited removal of phenanthrene. Therefore, optimization of cosolvent concentration is required to increase contaminant solubilization while permitting sustained electroosmotic flow.

The chemical n-butylamine has been also used in electrokinetic tests employing contaminated sediment and gas plant soil. The cosolvent contributed to the solubilization/desorption of PAHs. In the sediment test, 20% n-butylamine was more effective for partial solubilization of PAHs than was 3% Tween 80, and PAH concentrations in the anodic region were reduced to levels equivalent to those seen when 5% Igepal CA-720 was employed (Reddy and Ala, 2006). Also, 20% n-butylamine did not effectively transport PAHs in gas plant soil, even though PAHs were solubilized in effluent samples. This low removal efficiency might be attributable to the high organic content of the soil (Reddy *et al.*, 2006).

Ethanol has also been investigated as a solubility-enhancing agent in several electrokinetic studies. With 40% ethanol, slight phenanthrene solubilization, desorption, and migration occurred in soil nearest to the anode, but not as much as seen when a surfactant solution, 3% Tween 80, was employed (Reddy and Saichek, 2003; Saichek and Reddy, 2003a).

Overall, it appears that PAH desorption in an electrokinetic cell could be enhanced by use of n-butylamine or ethanol as a cosolvent, but actual removal efficiencies are very low. Therefore, further study is required to achieve high removal efficiencies in cosolvent-enhanced systems. Moreover, the application of cosolvent requires care because cosolvents are flammable and explosive, which could pose particular dangers in electrokinetic applications (Saichek and Reddy, 2005a).

9.3.1.3 Cyclodextrin-Enhanced EK A modified cyclodextrin termed hydropropyl β-cyclodextrin (HPCD) may be an appropriate facilitating agent for electrokinetic removal of PAHs because of low soil adsorption affinity, nontoxicity, biodegradability, and good PAH solubilization capacity. Electrokinetic experiments were conducted using HPCD concentrations of 1.37 to 6.85 mM. The removal of phenanthrene from model clay was affected principally by HPCD concentration, and less so by longer operation or cumulative volume of electroosmotic flow. When HPCD was used at 6.85 mM, greater phenanthrene removal (56%) over 6 days was seen than when HPCD at 1.37 mM was employed over 14 days, despite the lower electroosmotic flow achieved over the 6-day period. The highest overall removal (about 75%) was obtained using the highest concentration of HPCD, with a buffer solution, to increase electroosmotic flushing (Ko, Schlautman, and Carraway, 2000).

In addition, the use of cyclodextrin in simultaneous electrokinetic removal of heavy metals and PAHs has been investigated. When 1% and 10% HPCD were used in removal of phenanthrene and nickel from kaolinite, phenanthrene removal was poor. The low concentration of HPCD (1%) was insufficient to increase phenanthrene solubilization despite high electroosmotic flow, whereas, with 10% HPCD, phenanthrene mobilization occurred only near the anode because of poor electroosmotic flow (Maturi and Reddy, 2006). In other tests, 10% HPCD solution contributed to partial solubilization of PAHs in soils contaminated with various PAHs and heavy metals but overall removal efficiencies were insignificant compared to those achieved using nonionic surfactants or cosolvents (Reddy and Ala, 2006; Reddy et al., 2006). Complete pollutant removal was thus not seen but removal efficiency would be improved by sustaining a high electroosmotic flow in the presence of high cyclodextrin concentration.

9.3.1.4 Biosurfactant-Enhanced EK Only a few studies using biosurfactants as solubility-enhancing agents in electrokinetic processes have appeared. Park et al. (2002) used two biosurfactants, mannosylerythritol lipid (MEL) and BS-UC, produced by *Candida antarctica* JWSH-112, to remediate phenanthrene-spiked kaolinite. The feasibility of biosurfactant use was evaluated in comparison with other additives such as synthetic surfactants and hydroperoxide. MEL and BS-UC yielded higher removal efficiencies, 16.2% and 17.4%, respectively, than seen when three commercial surfactants (APG, Brij 30, and SDS) were employed (removal proportions of 6.3%–11.1%). It is expected that biosurfactants of low toxicity and good biodegradability can assist in effective electrokinetic flushing.

Biosurfactants are useful in contaminant biodegradation because biosurfactants enhance HOC bioavailability by increasing HOC aqueous solubility. Ju and Elektorowicz (2000) used rhamnolipids produced by *P. aeruginosa* to enhance electrokinetic remediation of phenanthrene-contaminated soil. This study showed the potential for on-site production of biosurfactant that was then directly introduced to the electrokinetic cell. Biosurfactants thus show promise as enhancers of biodegradation and for the solubilization and desorption of PAHs in electrokinetic processes.

9.3.2 Effects of Electrokinetic Variables

As described above, electrokinetic remediation has many interrelated variables determining overall remediation efficiency. These include pH, ionic strength, soil characteristics, voltage gradient, and zeta potential. Electroosmosis is very important especially in remediation of nonpolar organic compounds, such as PAHs, and solubility-enhancing solutions are often required. Researchers have therefore examined the effects of variables on electrokinetic phenomena.

9.3.2.1 Effect of pH Control According to the H–S theory, the velocity of electroosmotic flow is proportional to the zeta potential (a function of mineral surface charge that may be influenced by changes in solution pH). As sustained high electroosmotic flow is very important for removal of nonpolar organics, a neutral or basic pH is preferable in an electrokinetic process, to maintain a negative zeta potential. As electroosmotic flow depends on many variables (H–S theory), however, the velocity of electroosmotic flow is generally not uniform throughout an electrokinetic system, resulting in a low hydraulic conductivity zone in the middle of the reactor. It is therefore crucial to uniformly control soil pH to support constant high electroosmotic flow.

pH control is typically accomplished by using bases or pH buffering solutions in anolytes but such materials may change the interaction of solubility-enhancing solutions with PAHs and soil. In work using 3% Tween 80 and pH control, it was observed that 0.01M NaOH solution caused a reduction in the CMC of the nonionic surfactant, and/or an increase in aggregation number, which commonly occurs following addition of electrolytes. This resulted in a higher electroosmotic flow, by lowering the degree of surfactant adsorption, than was seen in the control test without NaOH. The results indicated that control of pH was helpful in increasing contaminant solubilization and migration from the soil region adjacent to the anode (Saichek and Reddy, 2003a). Moreover, a high NaOH concentration in the anodic compartment may assist in counteracting the electrolysis reaction (Saichek and Reddy, 2003b).

Acetate buffer (Park *et al.*, 2007) and carbonate buffer solutions (Ko, Schlautman, and Carraway, 2000) have often been used to enhance electroosmotic flow in surfactant or cyclodextrin-enhanced electrokinetic systems. The effect of pH buffer is more clearly seen in soils of low carbonate content; pH control significantly improves electroosmotic flow under such conditions. However, it was reported that a combination of buffer solution with a surfactant/cosolvent solution only marginally improved electroosmotic flow (Saichek, 2002). Consequently, pH control should be considered together with other variables influencing electroosmotic flow.

9.3.2.2 Effect of Electrolyte In electrokinetic processes, electrolyte is frequently employed to induce the electrical current to pass through the pore fluid. Electrolyte concentration is associated with electrical potential and power consumption, and affects the zeta potential of soil, thus also influences electroosmotic flow. In general, the presence of an electrolyte reduces the CMC of a surfactant because of a solution-depolarizing effect, so greater micellization and less surfactant adsorption can be expected (Saichek and Reddy, 2003a).

When 0.001–0.1 M NaCl solutions were employed as electrolytes, most tests showed maintenance of low electrical potential gradients under constant current conditions with consequent decreases in power consumption. However, because increments in ionic strength induce decreases in soil surface zeta potential, the electroosmotic flow decreased, and phenanthrene removal was barely enhanced

(Lee, 2004). Therefore, an optimum electrolyte concentration is required for simultaneous maintenance of enhanced micellization and high electroosmotic flow.

9.3.3.3 Effect of Soil Properties Kaolin, as a clay representative, has been widely used in laboratory electrokinetic experiments. Kaolin is composed of kaolinite and has a much lower cation exchange capacity compared to other clay minerals such as montmorillonite and illite (Pierzynski, Sims, and Vance, 1994). Kaolin serves as a model soil in electrokinetic tests because experimental variations arising from soil heterogeneity can be minimized as kaolin has a low organic content, uniform mineralogy, nonreactivity, and a low cation exchange capacity (Reddy and Saichek, 2003). However, kaolin is not an appropriate representative of field soils, which are often quite heterogeneous, with high organic matter concentrations and complex mineralogy.

Although most early experiments were conducted using kaolinite (Ko, Schlautman, and Carraway, 2000; Saichek and Reddy, 2003b; Lee, 2004; Yang et al., 2005), recent studies have used other clays or real soils contaminated with various PAHs, in attempts to assess the potential of electrokinetic technology for field application (Maini et al., 2000; Reddy and Saichek, 2003; Reddy and Ala, 2006; Reddy et al., 2006).

Reddy and Saichek (2003) compared two clayey soils, kaolin and glacial till, in electrokinetic tests of phenanthrene removal. Compared to kaolin, glacial till has more carbonates and a more diverse mineralogy. The carbonate ions contributed to the higher electrical current and conductivity of glacial till compared to kaolin, resulting in significantly higher cumulative electroosmotic flow in tests using water, surfactant, and cosolvent. However, phenanthrene removal from glacial till was more difficult than from kaolin because PAHs often bind to and associate with organic matter. Besides, the relatively high organic content (3%) of glacial till induced adsorption of surfactant monomers and, consequently, a decrease in surfactant micelles in solution. This led to a reduction in ion solubilization.

Similarly, in tests with dredged sediment and gas plant soil, substantial electroosmotic flow was produced because of high acid buffering capacity, but phenanthrene removal efficiency was very low because PAHs were strongly adsorbed to organic matter. The organic matter proportions were 19% and 4% in the sediment and gas plant soil, respectively (Reddy and Ala, 2006; Reddy et al., 2006). Particularly, in tests with cyclodextrin solution, it was found that the low PAH removal efficiency was caused by adsorption of cyclodextrin-PAH complexes onto minerals and organic matter in the sediments (Reddy and Ala, 2006).

Maini et al. (2000) attempted to remediate historically contaminated soil from a former gasworks site in East London, UK. The organic content of the soil was not reported but the soil contained a range of heavy metals, PAHs, and BTEX. Although no significant metal removal was observed, PAHs were effectively (over 90%) removed by electroosmosis, with or without surfactant, in both small- and large-scale reactor studies. This shows that electrokinetic soil flushing shows promise for field application.

Some studies investigated the effects of heterogeneous soil conditions on surfactant-enhanced EK (Pamukcu, 1994; Saichek and Reddy, 2005b). A series of bench-scale experiments was conducted using two soils (sand and kaolin) displaying various heterogeneities such as layered, lens, and mixed soil configurations. In heterogeneous samples, lower remediation rates were observed when greater proportions of kaolin were used but, ultimately, most phenanthrene was removed from all heterogeneous soils using the nonionic surfactant Igepal CA-720 (Saichek and Reddy, 2005b). This shows that *in situ* electrokinetic processes may be very useful to remove PAHs from heterogeneous soil showing layering or lensing, or from soil mixtures with areas of high and low hydraulic conductivity.

9.3.3.4 Effect of Voltage Gradient It is obvious that the velocity of electroosmotic flow is directly related to the potential gradient; a higher voltage causes generation of higher electroosmotic flow. On the other hand, the periodic application of electrical potential was suggested to increase micellar solubilization and to enhance remedial efficiency because mass transfer is limited by diffusion of PAHs from the soil matrix (Yeom, Ghosh, and Cox, 1996). A periodic voltage mode was applied with a 5-day period of continuous application followed by 2 days of "down time." In tests employing a high (2.0VDC/cm) voltage gradient, and a periodic application mode, more contaminant removal was achieved compared to that seen with the continuous mode. The enhanced removal was caused by an increase in phenanthrene solubilization and mass transfer because of reduced flow of the bulk solution during the down time, and improved flushing action resulting from pulsed electroosmotic flow (Reddy and Saichek, 2004).

9.4 SUMMARY

In this chapter, previous studies on electrokinetic PAH removal have been thoroughly reviewed, with particular focus on the use of facilitating solutions, including surfactants, cosolvents, cyclodextrins, and biosurfactants. Among these enhancing agents, surfactants have been most extensively examined and it has been found that some nonionic surfactants, such as Tergitol 15-S-12, APG, and Igepal CA-720, can significantly enhance electrokinetic PAH removal. With cosolvent use, PAH desorption near anodic regions is enhanced by n-butylamine or ethanol. A modified cyclodextrin, HPCD, and some other biosurfactants, have also proved useful in desorption and mobilization of PAHs in EK. In some cases, electrokinetic performance enhancements were as great as shown by synthetic nonionic surfactants.

Facilitating agents change solution properties, such as dielectric constant, pH, and viscosity, and influence electroosmotic flow velocity, thereby impacting on PAH removal efficiency. The effect of electrokinetic variables on PAH removal has therefore been investigated to optimize operating conditions. pH adjustment and periodic voltage applications positively affect PAH removal efficiency.

Previous reports have generally described laboratory-scale experiments and there were few field demonstrations or full-scale applications. Kaolin was used as a model soil in most studies. However, recent electrokinetic tests of PAH removal have been conducted with real heterogeneous contaminated soils and have demonstrated the potential of EK.

The above review suggests that EK, enhanced by facilitating solutions, has great potential for removal of PAHs from low-permeability soils. Further research on effective field applications is required. Moreover, any facilitating agent should be selected with consideration of both environmental properties and capability to effect contaminant mobilization.

REFERENCES

Acar YB, Gale RJ, Alshawabkeh AN, Marks RE, Puppala S, Bricka M, Parker R (1995). Electrokinetic remediation: Basics and technology status. *Journal of Hazardous Materials* **40**(2):117–137.

Augustijn DCM, Jessup RE, Rao PSC, Wood AL (1994). Remediation of contaminated soils by solvent flushing. *Journal of Environmental Engineering-ASCE* **120**(1):42–57.

Badr T, Hanna K, de Brauer C (2004). Enhanced solubilization and removal of naphthalene and phenanthrene by cyclodextrins from two contaminated soils. *Journal of Hazardous Materials* **112**(3):215–223.

Blanford WJ, Barackrnan ML, Boing TB, Klingel EJ, Johnson GR, Brusseau ML (2001). Cyclodextrin-enhanced vertical flushing of a trichloroethene contaminated aquifer. *Ground Water Monitoring and Remediation* **21**(1):58–66.

Bognolo G (1999). Biosurfactants as emulsifying agents for hydrocarbons. *Colloids and Surfaces A—Physicochemical and Engineering Aspects* **152**(1–2):41–52.

Bouchard DC (2002). Cosolvent effects on sorption isotherm linearity. *Journal of Contaminant Hydrology* **56**(3–4):159–174.

Brusseau ML, Wood AL, Rao PSC (1991). Influence of organic cosolvents on the sorption kinetics of hydrophobic organic chemicals. *Environmental Science & Technology* **25**(5):903–910.

Cameotra SS, Bollag J-M (2003). Biosurfactant-enhanced bioremediation of polycyclic aromatic hydrocarbons. *Critical Reviews in Environmental Science and Technology* **30**(2):111–126.

Cheng KY, Zhao ZY, Wong JWC (2004). Solubilization and desorption of PAHs in soil-aqueous system by biosurfactants produced from *Pseudomonas aeruginosa* P-CG3 under thermophilic condition. *Environmental Technology* **25**(10):1159–1165.

Deshpande S, Wesson L, Wade D, Sabatini DA, Harwell JH (2000). DOWFAX surfactant components for enhancing contaminant solubilization. *Water Research* **34**(3):1030–1036.

Edwards DA, Adeel Z, Lirthy RG (1994). Distribution of nonionic surfactant and phenanthrene in a sediment/aqueous system. *Environmental Science & Technology* **28**(8):1550–1560.

Haigh SD (1996). A review of the interaction of surfactants with organic contaminants in soil. *Science of the Total Environment* **185**(1–3):161–170.

Jones KC, Stratford JA, Tidridge P, Waterhouse KS, Johnston AE (1989). Polynuclear aromatic hydrocarbons in an agricultural soil: Long-term changes in profile distribution. *Environmental Pollution* **56**(4):337–351.

Joshi MM, Lee SG (1996). Optimization of surfactant-aided remediation of industrially contaminated soils. *Energy Sources* **18**(3):291–301.

Ju L, Elektorowicz M (2000). In-situ phenanthrene removal provoked by electrokinetic transport of on-site produced biosurfactants, Annual conference abstracts, Canadian Society for Civil Engineering, May, London, ON, p. 135.

Kanga SA, Bonner JS, Page CA, Mills MA, Autenrieth RL (1997). Solubilization of naphthalene and methyl-substituted naphthalenes from crude oil using biosurfactants. *Environmental Science & Technology* **31**(2):556-561.

Kim J-H, Han S-J, Kim S-S, Yang J-W (2006). Effect of soil chemical properties on the remediation of phenanthrene-contaminated soil by electrokinetic-Fenton process. *Chemosphere* **63**(10):1667–1676.

Kim S-J, Park J-Y, Lee Y-J, Lee J-Y, Yang J-W (2005). Application of a new electrolyte circulation method for the ex situ electrokinetic bioremediation of a laboratory-prepared pentadecane contaminated kaolinite. *Journal of Hazardous Materials* **118**(1–3):171–176.

Ko SO, Schlautman MA, Carraway ER (1998). Partitioning of hydrophobic organic compounds to sorbed surfactants. 1. Experimental studies. *Environmental Science & Technology* **32**(18):2769–2775.

Ko SO, Schlautman MA, Carraway ER (2000). Cyclodextrin-enhanced electrokinetic removal of phenanthrene from a model clay soil. *Environmental Science & Technology* **34**(8):1535–1541.

Lee YJ (2004). Effect of System Variables on Surfactant-Enhanced Electrokinetic Remediation of Clayey Soil Contaminated with PAHs. MS Thesis, Korea Advanced Institute of Science and Technology, Daejeon, South Korea.

Li A, Cheung KA, Reddy KR (2000). Cosolvent-enhanced electrokinetic remediation of soils contaminated with phenanthrene. *Journal of Environmental Engineering-ASCE* **126**(6):527–533.

Li JL, Chen BH (2002). Solubilization of model polycyclic aromatic hydrocarbons by nonionic surfactants. *Chemical Engineering Science* **57**(14):2825–2835.

Maini G, Sharman AK, Knowles CJ, Sunderland G, Jackman SA (2000). Electrokinetic remediation of metals and organics from historically contaminated soil. *Journal of Chemical Technology and Biotechnology* **75**(8):657–664.

Maturi K, Reddy KR (2006). Simultaneous removal of organic compounds and heavy metals from soils by electrokinetic remediation with a modified cyclodextrin. *Chemosphere* **63**(6):1022–1031.

Maturi K, Reddy KR (2008). Cosolvent-enhanced desorption and transport of heavy metals and organic contaminants in soils during electrokinetic remediation. *Water Air and Soil Pollution* **189**(1–4):199–211.

Mohamed A, Mahfoodh A-SM (2006). Solubilization of naphthalene and pyrene by sodium dodecyl sulfate (SDS) and polyoxyethylenesorbitan monooleate (Tween 80) mixed micelles. *Colloids and Surfaces A—Physicochemical and Engineering Aspects* **287**(1–3):44–50.

Montoneri E, Boffa V, Savarino P, Tambone F, Adani F, Micheletti L, Gianotti C, Chiono R (2009). Use of biosurfactants from urban wastes compost in textile dyeing and soil remediation. *Waste Management* **29**(1):383–389.

Mulligan CN, Yong RN, Gibbs BF (2001). Surfactant-enhanced remediation of contaminated soil: A review. *Engineering Geology* **60**(1–4):371–380.

Niqui-Arroyo JL, Ortega-Calvo JJ (2007). Integrating biodegradation and electroosmosis for the enhanced removal of polycyclic aromatic hydrocarbons from creosote-polluted soils. *Journal of Environmental Quality* **36**(5):1444–1451.

Page CA, Bonner JS, Kanga SA, Mills MA, Autenrieth RL (1999). Biosurfactant solubilization of PAHs. *Environmental Engineering Science* **16**(6):465–474.

Pamukcu S (1994) Electrokinetic Removal of Coal Tar Constituents from Contaminated Soils. Final Report. Palo Alto, CA: Electric Power Research Institute; Report No: EPRI TR-103320, Project 2879-21.

Park JY, Chen Y, Chen J, Yang JW (2002). Removal of phenanthrene from soil by additive-enhanced electrokinetics. *Geosciences Journal* **6**(1):1–78.

Park JY, Kim SJ, Lee YJ, Baek K, Yang JW (2005). EK-Fenton process for removal of phenanthrene in a two-dimensional soil system. *Engineering Geology* **77**(3–4):217–224.

Park JY, Lee HH, Kim SJ, Lee YJ, Yang JW (2007). Surfactant-enhanced electrokinetic removal of phenanthrene from kaolinite. *Journal of Hazardous Materials* **140**(1–2): 230–236.

Peters CA, Luthy RG (1993). Coal tar dissolution in water-miscible solvents: Experimental evaluation. *Environmental Science & Technology* **27**(13):2831–2843.

Pierzynski GM, Sims JT, Vance GF (1994). *Soil and Environmental Quality*. Boca Raton, FL: Lewis Publishers/CRC Press, Inc.

Ranjan RS, Qian Y, Krishnapillai M (2006). Effects of electrokinetics and cationic surfactant cetyltrimethylammonium bromide [CTAB] on the hydrocarbon removal and retention from contaminated soils. *Environmental Technology* **27**(7):767–776.

Ravindra K, Sokhi R, Van Grieken R (2008). Atmospheric polycyclic aromatic hydrocarbons: Source attribution, emission factors and regulation. *Atmospheric Environment* **42**(13):2895–2921.

Reddy KR, Ala PR (2006). Electrokinetic remediation of contaminated dredged sediment. *Journal of ASTM International* **3**(6):254–267.

Reddy KR, Ala PR, Sharma S, Kumar SN (2006). Enhanced electrokinetic remediation of contaminated manufactured gas plant soil. *Engineering Geology* **85**(1–2):132–146.

Reddy KR, Saichek RE (2003). Effect of soil type on electrokinetic removal of phenanthrene using surfactants and cosolvents. *Journal of Environmental Engineering-ASCE* **129**(4): 336–346.

Reddy KR, Saichek RE (2004). Enhanced electrokinetic removal of phenanthrene from clay soil by periodic electric potential application. *Journal of Environmental Science and Health part A—Toxic/Hazardous Substances & Environmental Engineering* **39**(5):1189–1212.

Rivas FJ (2006). Polycyclic aromatic hydrocarbons sorbed on soils: A short review of chemical oxidation based treatments. *Journal of Hazardous Materials* **138**(2):234–251.

Rosen MJ (1978). *Surfactant and Interfacial Phenomena*, 2nd ed. New York: John Wiley & Sons, Inc.

Saichek RE (2002). Electrokinetically enhanced in-situ flushing for HOC-contaminated soils. PhD Thesis, University of Illinois, Chicago.

Saichek RE, Reddy KR (2003a). Effect of pH control at the anode for the electrokinetic removal of phenanthrene from kaolin soil. *Chemosphere* **51**(4):273–287.

Saichek RE, Reddy KR (2003b). Effects of system variables on surfactant enhanced electrokinetic removal of polycyclic aromatic hydrocarbons from clayey soils. *Environmental Technology* **24**(4):503–515.

Saichek RE, Reddy KR (2005a). Electrokinetically enhanced remediation of hydrophobic organic compounds in soils: A review. *Critical Reviews in Environmental Science and Technology* **35**(2):115–192.

Saichek RE, Reddy KR (2005b). Surfactant-enhanced electrokinetic remediation of polycyclic aromatic hydrocarbons in heterogeneous subsurface environments. *Journal of Environmental Engineering Science* **4**(5):327–339.

Shixiang G, Liansheng W, Qingguo H, Sukui H (1998). Solubilization of polycyclic aromatic hydrocarbons by [beta]-cyclodextrin and carboxymethyl-[beta]-cyclodextrin. *Chemosphere* **37**(7):1299–1305.

Viglianti C, Hanna K, de Brauer C, Germain P (2006). Removal of polycyclic aromatic hydrocarbons from aged-contaminated soil using cyclodextrins: Experimental study. *Environmental Pollution* **140**(3):427–435.

Virkutyte J, Sillanpaa M, Latostenmaa P (2002). Electrokinetic soil remediation—Critical overview. *Science of the Total Environment* **289**(1–3):97–121.

West CC, Harwell JH (1992). Surfactants and subsurface remediation. *Environmental Science & Technology* **26**(12):2324–2330.

Wilcke W (2000). Polycyclic aromatic hydrocarbons (PAHs) in soil—A review. *Journal of Plant Nutrition and Soil Science* **163**(3):229–248.

Wood AL, Bouchard DC, Brusseau ML, Rao PSC (1990). Cosolvent effects on sorption and mobility of organic contaminants in soils. *Chemosphere* **21**(4–5):575–587.

Yang JW, Lee YJ, Park JY, Kim SJ, Lee JY (2005). Application of APG and Calfax 16L-35 on surfactant-enhanced electrokinetic removal of phenanthrene from kaolinite. *Engineering Geology* **77**(3–4):243-251.

Yeom IT, Ghosh MM, Cox CD (1996). Kinetic aspects of surfactant solubilization of soil-bound polycyclic aromatic hydrocarbons. *Environmental Science & Technology* **30**(5):1589–1595.

Yuan C, Weng C-H (2004). Remediating ethylbenzene-contaminated clayey soil by a surfactant-aided electrokinetic (SAEK) process. *Chemosphere* **57**(3):225–232.

Zhou W, Zhu L (2008). Enhanced soil flushing of phenanthrene by anionic-nonionic mixed surfactant. *Water Research* **42**(1–2):101–108.

Zhu L, Feng S (2003). Synergistic solubilization of polycyclic aromatic hydrocarbons by mixed anionic-nonionic surfactants. *Chemosphere* **53**(5):459–467.

10

ELECTROKINETIC REMOVAL OF CHLORINATED ORGANIC COMPOUNDS

XIAOHUA LU AND SONGHU YUAN

10.1 INTRODUCTION

Chlorinated organic compounds (COCs) refer to the substitution of one or more hydrogen in aliphatic and aromatic hydrocarbons and their derivatives by chlorine. COCs are widely used in the fields of chemistry, medicine, electronics, pesticides, etc. Many COCs are endocrine disturbance substances, show carcinogenic effects, and have been listed as priority pollutants by the US Environmental Protection Agency (USEPA). When released into the environment, COCs are transported in both air and water. However, COCs are chemically stable and difficult to destroy, and they are eventually deposited in soils and sediments due to their hydrophobicity. Soils and sediments contaminated with COCs are long-term sources of pollutants and pose great threats to human health and ecosystems. Therefore, remediation of these contaminated soils and sediments is of great importance.

In this chapter, we will describe the electrokinetic (EK) removal of three kinds of representative COCs from soils and sediments: chlorinated aliphatic hydrocarbons (CAHs), chlorophenols (CPs), and chlorobenzenes (CBs).

10.2 ELECTROKINETIC REMOVAL OF CHLORINATED ALIPHATIC HYDROCARBONS

10.2.1 Characteristics of CAHs

CAHs, a family of compounds that are commonly used as organic solvents, have been increasingly detected in soil and in groundwater. The most prevalent CAHs

Electrochemical Remediation Technologies for Polluted Soils, Sediments and Groundwater,
Edited by Krishna R. Reddy and Claudio Cameselle
Copyright © 2009 John Wiley & Sons, Inc.

are perchlorethylene (PCE), trichloroethylene (TCE), and trichloroethane (TCA). Since the 1960s, these solvents have been used for degreasing in industries such as dry cleaning, electronics, industrial manufacturing, and machine maintenance. Due to leakage of these CAHs from storage tanks and machinery, dissolved PCE, TCE, and TCA now appear in soil and groundwater at concentrations that have been proved to be unhealthy and even carcinogenic (Regenesis.com, 2003). The low interfacial tension between liquid CAHs and water allows the CAHs to easily enter small fractures and pore spaces. However, the high relative solubilities of CAHs and the low partitioning of CAHs to soil materials imply that soils will not significantly retard the movement of CAHs (Pankow and Cherry, 1996). The low sorption of CAHs onto soil matrix suggests that CAHs are mainly in the dissolved state in soil pore solution, the movement of which is relatively easy. As CAHs are not prone to dissociate, their movement in soils under EK is via electro-osmosis. As a consequence, maintaining a high electro-osmotic flow (EOF) is essential in the remediation process.

10.2.2 Electrokinetic Removal of CAHs

Among all the CAHs, TCE is universally found in contaminated sites. In the US, about 50% of the superfund sites on the National Priority List contain TCE (Kovalick, 2000). For the EK removal of TCE from soils, Bruell, Segall, and Walsh (1992) reported that the EK process had the potential to remove TCE from kaolin clay. Twenty-five percent removal was achieved after a 3-day treatment under 0.4 V/cm, and 99% removal was predicted to be achieved after a 405-day treatment. The researchers concluded that the EK process is capable of removing chemicals with relatively high aqueous solubility and low distribution coefficients (Bruell, Segall, and Walsh, 1992), and the removal efficiency of TCE can be enhanced by increasing the electric gradient (Weng, Lin, and Hsieh, 2000) (Fig. 10.1a). Adjusting the processing fluid pH to 12.0 led to higher electro-osmotic (EO) permeability and consequently a higher TCE removal than that without controlling the pH (8.3) (Fig. 10.1b). Up to 91.2%, 75.1%, and 53.8% of TCE were removed from kaolin with initial concentrations of 5, 84, and 137 mg/kg, respectively, under a voltage of 1 V/cm for 3 days (Fig. 10.1c).

The series-EK system has been developed to enhance the EK removal of TCE from soils (Weng, Yuan, and Tu, 2003) (Fig. 10.2). The TCE removal increased with increased electric gradients and processing time. A higher TCE removal was observed in series-EK systems than in the nonseries systems. A buffer solution of sodium acetate combined with a working-solution circulation system (WSCS) was also designed to enhance the EK removal of PCE, TCE, CCl_4, and $CHCl_3$ from unsaturated soils (Chang, Qiang, and Huang, 2006). All target CAHs could be effectively removed from the soil with removal efficiencies ranging from 85 to 98% after 2 weeks of treatment. The mobility of CAHs in soils increased with the increase of their water solubilities, i.e., following the sequence of $CHCl_3 > CCl_4 > TCE > PCE$ (Fig. 10.3). The removal kinetics can be approximately described by a mathematical diffusion–advection–sorption model with a linear sorption isotherm.

To enhance the remediation of CAHs, the EK process coupled to other technologies has been developed. For instance, the EK process was used to transport Fenton agents for the remediation of TCE-contaminated soils (Yang and Liu, 2001). EK–

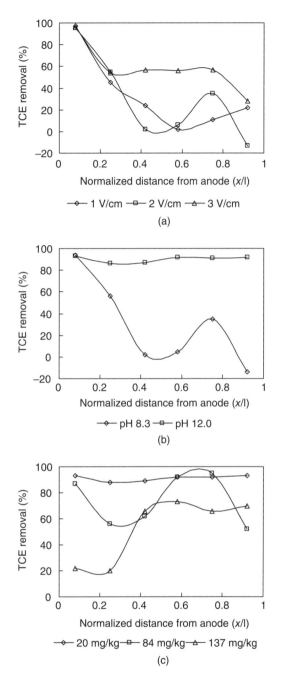

Figure 10.1. TCE removal versus normalized distance from anode for the effect of (a) electric gradients (processing fluid pH 12.0), (b) processing fluid pH, and (c) initial TCE concentration in the clay (Weng, Lin, and Hsieh, 2000).

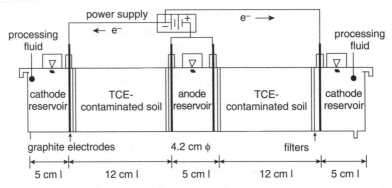

Figure 10.2. Schematic of series-EK decontamination process (Weng, Yuan, and Tu, 2003).

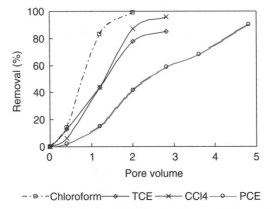

Figure 10.3. Removal efficiency of CAHs as a function of pore volume. Experimental conditions: electrolyte concentration = 10^{-2} mol/l CH_3COONa, water content = 20% (w/w), potential gradient = 1.2 V/cm (Chang, Qiang, and Huang, 2006).

Fenton showed reliable performance and was proved to be cost effective in TCE removal. Ho *et al.* (1999) developed a novel *in situ* remediation technology called Lasagna for cleaning up TCE-contaminated low-permeability soils. TCE in the soils was transported into carbon-containing treatment zones where it was trapped. Over 98% of TCE was removed from the contaminated soils, with most treatment samples showing a removal of over 99%. Chang and Cheng (2006) designed an integrated technique, which combined the EK process with zero-valent metal (ZVM), to remediate PCE-contaminated soils. The removal efficiency reaches 99 and 90% for decontamination of PCE in the pore-water and soil, respectively, after 10 days of treatment. More details of the EK process coupled with other technologies can be seen in other chapters in this book.

10.2.3 Summary of Electrokinetic Removal of CAHs

Generally, significant removal of CAHs from soils can be achieved by the EK process. The EK removal of CAHs was affected by both the sorption behavior of CAHs on soils and the EO flow (EOF). On one hand, the sorption of CAHs on soils is heavily dependent on their aqueous solubilities, so it is easier to flush out CAHs with higher solubilities. On the other hand, it is critical to maintain a high EOF in order to achieve a high removal efficiency in a short time. In general, the pH of the anodic purging solution should be controlled at a relatively high level as a decrease in pH led to a decrease in EOF. The removal of CAHs from soils can be promoted by the EK process coupled to other processes.

10.3 ELECTROKINETIC REMOVAL OF CHLOROPHENOLS

10.3.1 Characteristics of CPs

CPs are a group of phenols with one to five chlorine atoms added on the benzene rings, comprising 19 congeners ranging from monochlorophenol to pentachlorophenol (PCP). Because of their broad-spectrum antimicrobial properties, CPs have been used as preservative agents for wood, paints, vegetable fibers and leather, and as disinfectants. In addition, they are used as herbicides, fungicides, and insecticides, and as intermediates in the production of pharmaceuticals and dyes. The toxic effect of CPs depends upon the degree of chlorination and the position of the chlorine atoms. Bioaccumulation of CPs appears to be moderate, and most bioconcentration factors (BCFs) fall roughly between 100 and 1000 (WHO, 1989).

CPs may be released into water either during their manufacture and use or through degradation of other chemicals (e.g., phenoxyalkanoic acids). CPs may also be formed as a result of the chlorination of humic matter or natural carboxylic acids during the chlorination of municipal drinking water. Once released into the environment, most CPs will enter the water, with a minority entering the air. CPs that do not degrade in water will be incorporated into soils and sediments. The persistence of CPs in soils and sediments depends on their adsorption/desorption characteristics. An increase in chlorination leads to a decrease in solubility and an increase in K_{ow} (Westall, Leuenberger, and Schwarzenbach, 1985). As a result, CPs with lower chlorination have higher solubilities and adsorbed less on soils. Additionally, it is notable that CPs are weak organic acids, and their acidity increases with the increase in chlorination. Thus, CPs with more chlorine atoms are more likely to be ionized at environmental pH values (around 7). The pK_a values of CPs range from 4.75 to 9.20, which indicates the state proportion is sensitive to pH. For example, the proportion of dissociated PCP increases with the increase of pH, and nearly all of the PCP dissociates at pH above 7 (Fig. 10.4). Generally, CPs in the molecular state show great adsorption on soil materials, while those with a dissociated state show minimal sorption. Therefore, CPs can be largely adsorbed on acidic soils and be minutely adsorbed on basic soils (Choi and Aomin, 1972). In addition, soil organic matters are important adsorbents for the sorption of CPs (Choi and Aomin, 1974).

10.3.2 Electrokinetic Removal of CPs

In the EK removal of CPs from soils, the transport of CPs is relatively complicated. Molecular CPs are neutrally charged, so they can be moved only by the EOF, normally toward the cathode. The dissociated CPs are negatively charged, and thus are

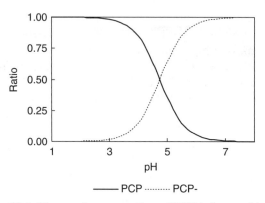

Figure 10.4. The species proportion of PCP influenced by pH.

mainly moved via electromigration toward the anode. The proportion of molecular and dissociated CPs depends on soil pH, which varies largely from anode to cathode in the EK cell. In sections near the anode with low pH values, CPs are neutral and move with EOF toward the cathode. Conversely, in sections near the cathode with high pH values, CPs are dissociated and move mainly by electromigration toward the anode. Thus, when CPs are moved from sections near the anode to sections near the cathode, the molecular CPs will be transformed into dissociated CPs, and the direction of movement will be reversed. This phenomenon is particularly notable for the CPs with a pK_a value around the original pH of the soil.

It has been addressed that the adsorbed phenol at a level of 500 mg/kg on dry kaolin could be efficiently removed by the EK process (Acar, Li, and Gale, 1992). The removed phenol was found mostly in the effluent, and no retardation of the phenol removal in effluent profiles was exhibited. Jackman *et al.* (2001) reported that the application of a current density of $3.72 A/m^2$ led to migration of 2,4-dichlorophenoxyacetic acid (2,4-D) toward the anode at a rate of 4 cm/day. 2,4-D has a pK_a of 2.64 and is negatively charged at neutral pH values, so it was moved toward the anode by electromigration during the EK process (Fig. 10.5). A definite movement of PCP in soil, mainly in areas closest to the electrodes, was revealed (Harbottle *et al.*, 2001). The PCP concentration experienced a significant depletion in the area near the anode (Fig. 10.6) as a result of transporting molecular PCP via EOF toward the cathode. PCP concentration near the cathode also experienced a depletion via electromigration of the dissociated PCP toward the anode (at a greater speed than that via electro-osmosis). The dissociated PCP would associate with H^+ produced at the anode to form molecular PCP, which conversely migrated toward the cathode via electro-osmosis only. This made the movement of PCP complicated. The combination of PCP and EK exposure could greatly reduce microbial counts, respiration, and carbon substrate utilization potential as a consequence of the increased toxicity of PCP at lower soil pH (Lear *et al.*, 2007). The EK removal of phenol and PCP from soils was significantly influenced by applied voltage and current, the type of purging solutions, soil pH, permeability, and zeta potentials of soil (Kim, Moon, and Kim, 2000). When CPs were transported through an electrode in the EK process, the electrochemical degradation at the electrode would happen (Cong, Ye, and Wu, 2005).

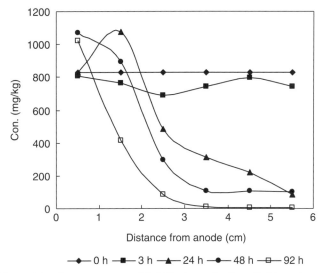

Figure 10.5. Electrokinetic movement of 2,4-D through silt soil. Experimental conditions: water content = 30% (w/w), current density = 3.72 A/m^2 (Jackman et al. 2001).

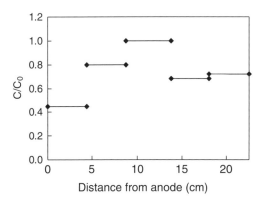

Figure 10.6. PCP profile after EK treatment (Harbottle et al., 2001).

An accelerated desorption and movement of phenol and 2,4-dichlorophenol (2,4-DCP) in unsaturated soils was achieved by using nonuniform electrokinetics (Luo et al., 2005). Electromigration and EOF were the main driving forces, and their roles in the mobilization of phenol and 2,4-DCP varied with soil pH. In sandy loam, 2,4-DCP moved between 1.0 and 1.5 cm/(day V) slower toward the anode than in the kaolin soil, and about 0.5 cm/(day V) greater than phenol in the sandy loam (Fig. 10.7). When the sandy loam was adjusted to pH 9.3, the movements of phenol and 2,4-DCP toward the anode were about two and five times faster than those at pH 7.7, respectively. The movement of phenol and 2,4-DCP in soils can be easily controlled by regulating the operational mode of electric field. The nonuniform electric field could enhance the *in situ* bioremediation process by promoting the mass

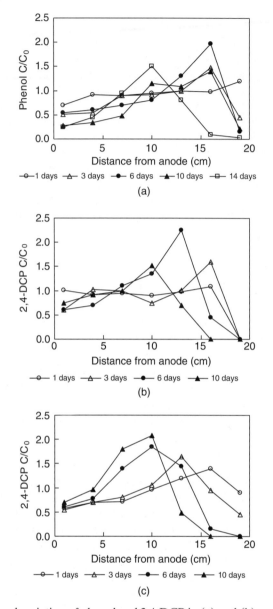

Figure 10.7. Temporal variation of phenol and 2,4-DCP in (a) and (b) sandy loam, (c) clayed soil. Potential gradient = 1 V/cm (Luo et al., 2005).

transfer of organics to degrading bacteria under *in situ* conditions (Fan et al., 2007). The combination of EK removal and a permeable reactive barrier (PRB) could not only remove organochlorine compounds in soil but also convert these pollutants into other compounds through dechlorination (Tomoyuki and Shunitsu, 2006). Residue 4-chlorophenol (4-CP) levels in an EK system coupled with an iron reactive wall was lower than that in an EK system alone. The combination of an iron reactive wall with EK remediation faciliated the removal of organochlorine compounds such as TCE and 4-CP from soils.

10.3.3 Summary of Electrokinetic Removal of CPs

CPs, particularly those with a pK_a around the original pH of the soil, are generally difficult to remove from soils. The EK movement of CPs in soils is heavily dependent on the states (molecular or dissociated), which is related to pH. Due to the temporary and spatial variation of soil pH in EK processes, molecular and dissociated CPs transformed mutually between the anode and the cathode. To achieve a high removal efficiency, it is important to ensure that the majority of the CPs are in a single state, either molecular or dissociated, so that only one directional driving force (electro-osmosis or electromigration) contributes to the movement. Concretely, controlling the soil pH above the pK_a of CPs leads to the removal of dissociated CPs by electromigration; controlling the soil pH below the pK_a leads to the removal of molecular CPs by electro-osmosis. However, pH should be controlled above the point of zero charge (PZC) of soils to maintain the EO flow in one direction.

10.4 ELECTROKINETIC REMOVAL OF CHLOROBENZENES

10.4.1 Characteristics of CBs

CBs are a family of cyclic aromatic compounds with one or more hydrogen atoms of the benzene ring substituted by chlorine atoms. Some of their congeners are designated as priority or "Red List" pollutants by environmental protection agencies in many countries including the UK, the USA, and Canada due to their toxicity and carcinogenicity (Lee and Fang, 1997). CBs are much less reactive than the corresponding chlorinated derivatives of alkyl compounds and are similar in reactivity to the vinyl halides. They are inert to nucleophilic attack while relatively reactive to electrophilic attack. The higher chlorinated benzenes tend to resist electrophilic substitution. The ability of CBs to undergo wide varieties of chemical reactions makes CBs useful as reactants in numerous commercial processes to produce varied products (USEPA, 1994). Mono-CB is used as a solvent for some pesticide formulations, as a degreaser, and a source of other chemicals. CBs are used mainly as intermediates in the synthesis of pesticides and other chemicals. Hexachlorobenzene (HCB) is an important member of the CB family and is extremely insoluble in water (0.005 mg/l at 25°C) with a high log K_{ow} value of 5.5 (CEPA, 1993). In industry, HCB has been used directly in the manufacture of pyrotechnics, tracer bullets, and as a fluxing agent in the manufacture of aluminum (CEPA, 1993).

CBs are a group of hydrophobic organic compounds (HOCs) with a wide spectrum of K_{ow} (log K_{ow} between 2.90 and 5.73). Most CBs, particularly highly chlorinated benzenes, tend to adhere to suspended particles through hydrophobic interactions and accumulate in biota. Among all the CBs, HCB is the most stable and nonreactive. It is one of the most persistent environmental pollutants. Oliver and Charlton (1984) estimated that 80% of the HCB loaded into Lake Ontario in 1982 was lost via volatilization, with the remainder removed by sedimentation (15%) and outflow to the St. Lawrence River (5%). However, HCB in resuspended bottom sediments would release and serve as an important and continuous source of HCB into water, even though inputs to the system cease (Oliver, Charlton, Durham, 1989). In the troposphere, HCB is transported long distances by virtue of its persistence, but undergoes slow photolytic degradation, or is removed from the air phase via atmospheric deposition to water and soil (Lane, Johnson, Hanley,

1992). High HCB levels have been found near point sources. For example, HCB levels ranging from 19 to 273 mg/kg (dry weight) were reported in 1979 for sediment samples of the East River (Nova Scotia) near a chlor-alkali plant discharge (MacLaren Marex Inc., 1979). In the St. Clair River, HCB levels in the 5-km stretch downstream of the Dow Chemical sewer discharges were found as high as 24,000 mg/kg in 1984 (mean 5200 µg/kg) and 280,000 mg/kg in 1985 (dry weight) (Oliver and Pugsley, 1986).

Sorption and desorption have been found to be governing processes for the availability and mobility of HOCs in the environment. In many remediation processes, it is a preliminary requirement that contaminants should be desorbed from soils. Contaminants that show significant sorption on soils are generally difficult to remove. The organic carbon-normalized linear equilibrium partitioning model has long been proposed and widely used in predicting the sorption and desorption of HOCs in soils and sediments. However, the sorption was suggested by later researchers to consist of multiple compartments: a linear sorption compartment and one or more nonlinear sorption compartments (Cornelissen et al., 2000). The nonlinear compartment probably shows slower sorption than the linear compartment, and the nonlinearity of the sorption isotherm increases as a function of time (Xing and Pignatello, 1996). The linearly sorbed part has been hypothesized to reside in amorphous, more flexible organic matter (Weber and Huang, 1996). Multiphasic desorption kinetics for soil- or sediment-bound HOCs has been also observed, including an initial rapid and reversible desorption fraction followed by slow desorbing, or desorption-resistant, fraction (Cornelissen et al., 1997). Desorption is rapid from linearly sorbing organic matter, and slow and extremely slow from nonlinearly sorbing sites (Cornelissen et al., 2000).

The sorption of mono-CB on marine sediment required 3 hours to reach a sorption/desorption equilibration (Zhao et al., 2001). Compared with sediment organic carbon content, surface and microporosity of the sediment might have a more important effect on the sorption of mono-CB (Zhao et al., 2001). Temperature does not have great influence on the adsorption behavior of mono-CB in seawater, but the saturate sorption capacity decreases as the temperature increases. The Tenax-mediated desorption of HCB from four freshly spiked, artificial sediments exhibited biphasic kinetics; the first-order rate constants of the fast desorbing phase and slow desorbing phase were 9.6×10^{-3} and $7.2 \times 10^{-3} h^{-1}$, respectively (Chai et al., 2007). The fast desorption fractions for the four sediments varied from 41.2 to 68.8% (Chai et al., 2007).

As considerable desorption of HOCs from soils is critical in order to achieve large EK removal, it is necessary to ensure the high desorption efficiency. Facilitating agents, such as surfactants, cosolvents, and cyclodextrins have shown great potential to increase the desorption of HOCs. When surfactants were used to enhance the desorption of HOCs such as HCB, the desorption efficiency was dependent on the aqueous micelle concentration of surfactants (Yuan et al., 2007). Although reliable desorption of HCB was obtained by the addition of a high concentration of surfactants, the surfactants showed great sorption loss in the process. In contrast, although cyclodextrins contribute to less desorption of HOCs from soils compared with surfactants, the sorption loss of cyclodextrins is minimal (Ko, Schlautman, Carraway, 1999), thus making cyclodextrins more suitable for the enhanced desorption process.

10.4.2 Electrokinetic Removal of CBs

As known, the aqueous solubilities of CBs decrease with the increase in chlorination thereby resulting in an increase in sorption of CBs on soils. It can be inferred that the transport of CBs becomes difficult with the increase of chlorination, and the EK removal of low chlorinated CBs is easier than that of high chlorinated ones. Information on the EK removal of CBs is sparse. The representative results were demonstrated by Yuan and his group.

Yuan, Wan, and Lu (2007) compared the EK behavior of multiple low chlorinated CBs, including 1,2,3,4-tetrachlorobenzene (TeCB); 1,2,4,5-tetrachlorobenzene (i-TeCB); and 1,2,3-trichlorobenzene (TCB), in spiked kaolin. The addition of β-cyclodextrin (β-CD) in quantities of 2 or 5 mmol/l in anodic purging solution inhibited the EK removal of CBs (Fig. 10.8). The largest removal was obtained when an Na_2CO_3/$NaHCO_3$ buffer (0.025 mol/l) was used as the anodic purging solution. The removal efficiencies were related to the aqueous solubilities of CBs. With the same cumulative EO flow, greater solubility of CBs led to higher removal efficiency. As a consequence, the removal of the three CBs conforms to the sequence of TCB > TeCB > iso-TeCB. It was found that the removal of CBs was inhibited by the addition of 5 mmol/l β-CD into the anodic flushing solution. The formation of the less soluble inclusion compounds between β-CD and CBs was responsible for the lower removal efficiency. It was proved feasible for the EK technology to remove CBs from kaolin.

Yuan, Tian, and Lu (2006) further investigated the EK movement of HCB in kaolin (Fig. 10.9). Negligible HCB movement was achieved when the Na_2CO_3/$NaHCO_3$ buffer solution was used as an anodic flushing solution. Although β-CD led to less desorption of HCB from kaolin than Tween 80, the removal of HCB in the presence of β-CD was much higher than that by Tween 80. The sorption loss of Tween 80 is much larger than that of β-CD, resulting in a much less efficient fraction to enhance the desorption of HCB in the movement. The mechanism of HCB movement was proposed as the enhanced desorption of HCB from soil, the dissolution of HCB in the soil pore fluid, and the movement of HCB with the EOF.

For the EK movement of HCB in real contaminated sediments, a comparative study was conducted to quantify the effect of organic cosolvents and cyclodextrins on both the EK parameters and enhancement in HCB movement (Wan et al., 2009). Results revealed that ethanol had a more negative effect on EOF than methyl-β-cyclodextrin (MCD). The movement of HCB in sediments was observed to increase with increasing concentrations of ethanol or MCD. Ethanol at the fraction of 50% exhibited the highest performance, followed by MCD at 50 g l^{-1} (Fig. 10.10). The different performances of HCB removal for tests with varied solubilizing agents was found to be a combined effect of the distribution of solubilizing agents in sediments, the dissolution of HCB by pore liquid, and the quantity of cumulative EOF.

10.4.3 Summary of Electrokinetic Removal of CBs

CBs, especially highly chlorinated ones, are typically hydrophobic and prone to adsorb and bind tightly to the soils. The removal of CBs from soils by nonenhanced

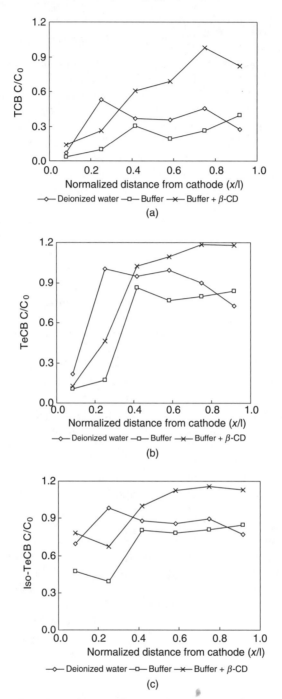

Figure 10.8. Distribution of CBs in the soils after EK treatment. (a) TCB, (b) TeCB, (c) Iso-TeCB. The buffer refers to 0.025 mol/l $Na_2CO_3/NaHCO_3$ solution; the concentration of β-CD is 5 mmol/l. The test was conducted at a potential gradient of 1 V/cm for 11 days (Yuan, Wan, and Lu, 2007).

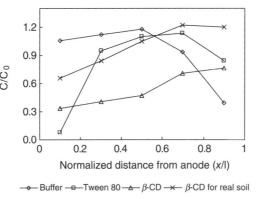

Figure 10.9. Distribution of HCB in the soils. The buffer refers to 0.005 mol/l Na_2CO_3/$NaHCO_3$ solution; the concentration of Tween 80 and β-CD is 1%. The test was conducted at a potential gradient of 1 V/cm for about 11 days (Yuan, Tian, and Lu, 2006).

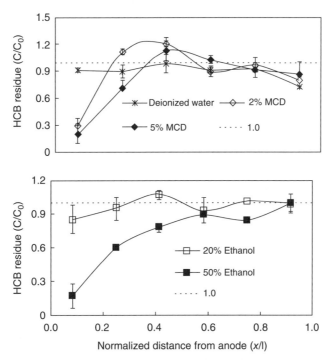

Figure 10.10. Distribution of HCB in sediment matrix. The pH of purging solution controlled at above 9 by the addition of NaOH. The test was conducted at a potential gradient of 2 V/cm for 14 days (Wan et al., 2009).

EK removal has proved to be inefficient. However, EK processes enhanced by solubilizing agents, including organic cosolvents and cyclodextrins, show reliable enhancement in CB movement across the soils. The low chlorinated CBs are easier to remove from soils when subjected to an enhanced EK. A higher solubilizing ability of the agents and a higher EOF in the EK process were both found to be critical for CBs removal by EK.

10.5 SUMMARY

In summary, the EK removal of COCs was influenced by the sorption/desorption of COCs on the interface of soil and pore solution, and the EK movement of COCs in soil pore solution. Among COCs, the sorption/desorption of CPs on the soil varied due to the change of soil characteristics (pH) that resulted from the electrode reactions. The EK movement of CPs was affected by electromigration and electroosmosis. The removal efficiency stemmed from the net effect of the two mechanisms. For the COCs that cannot be dissociated, the sorption/desorption was slightly affected by the change of soil characteristics. As a result, it is necessary to maintain a considerable desorption efficiency of COCs for EK removal. Sometimes facilitating agents such as surfactants, cosolvents, and cyclodextins are added to enhance the removal of hydrophobic COCs. It is also important to achieve a high EOF in the process. It is possible to use the EK process to remove COCs, particularly soluble COCs, from contaminated soils. Soils with a high contaminant concentration and in a small area are more suitable for EK removal.

REFERENCES

Acar YB, Li H, Gale RJ. (1992). Phenol removal from kaolinite by electrokinetics. *Journal of Geotechnical Engineering* **118**(11):1837–1852.

Bruell CJ, Segall BA, Walsh MT. (1992). Electroosmotic removal of gasoline hydrocarbons and TCE from clay. *Journal of Environmental Engineering-ASCE* **118**(1):68–83.

Canadian Environmental Protection Act (CEPA). (1993). *Priority Substances List Assessment Report:* Hexachlorobenzene. Ottawa, Canada: Canada Communication Group Publishing.

Chai YZ, Qiu XJ, Davis JW, Budinske RA, Bartels MJ, Sahhir SA. (2007). Effects of black carbon and montmorillonite clay on multiphasic hexachlorobenzne desorption from sediments. *Chemosphere* **69**:1204–1212.

Chang JH, Cheng SF. (2006). The remediation performance of a specific electrokinetics integrated with zero-valent metals for perchloroethylene contaminated soils. *Journal of Hazardous Materials B* **131**:153–162.

Chang JH, Qiang ZM, Huang CP. (2006). Remediation and stimulation of selected chlorinated organic solvents in unsaturated soil by a specific enhanced electrokinetics. *Colloids and Surfaces A* **287**:86–93.

Choi J, Aomin S. (1972). Effects of the soil on the activity of pentachlorophenol. *Soil Science and Plant Nutrient* **8**(6):255–260.

Choi J, Aomin S. (1974). Adsorption of pentachlorophenol by soils. *Soil Science and Plant Nutrient* **20**(2):135–144.

Cong YQ, Ye Q, Wu ZC. (2005). Electrokinetic behaviour of chlorinated phenols in soil and their electrochemical degradation. *Process Safety and Environmental Protection* **83**(B2): 178–183.

Cornelissen G, Rigterink H, van Noort PCM, Govers HAJ. (2000). Slowly and very slowly desorbing organic compounds in sediments exhibit Langmuir-type sorption. *Environmental Toxicology and Chemistry* **19**:1532–1539.

Cornelissen G, Rigterink H, Vrind BA, ten Hulscher DTEM, Ferdinandy MMA, van Noort PCM. (1997). Two-stage desorption kinetics and *in situ* partitioning of hexachlorobenzene and dichlorobenzenes in a contaminated sediment. *Chemosphere* **35**: 2405–2416.

Fan XY, Wang H, Luo QS, Ma JW, Zhang XH. (2007). The use of 2D non-uniform electric field to enhance *in situ* bioremediation of 2,4-dichlorophenol-contaminated soil. *Journal of Hazardous Materials B* **148**:29–37.

Harbottle MJ, Sills GC, Thompson IP, Jackman SA. (2001). Movement of pentachlorophenol in unsaturated soil by electrokinetics. *Proceedings of the Third Symposium and Status Report on Electrokinetic Remediation*, Karlsruhe, Germany, April 17. pp. 1–13.

Ho SV, Athmer C, Sheridan PW, Hughes BM, Orth R, Mckenzie D, Brodsky PH, Shapiro A, Thornton R, Salvo J, Schultz D, Landis R, Griffith R, Shoemaker S. (1999). The lasagna technology for *in situ* soil remediation. 1. Small field test. *Environmental Science & Technology* **33**(7):1086–1091.

Jackman SA, Maini G, Sharman AK, Sunderland G, Knowles DJ. (2001). Electrokinetic movement and biodegradation of 2,4-dichlorophenoxyacetic acid in silt soil. *Biotechnology and Bioengineering* **74**(1):40–48.

Kim SO, Moon SH, Kim KW. (2000). Enhanced electrokinetic soil remediation for removal of organic contaminants. *Environmental Technology* **21**(4):417–426.

Ko SL, Schlautman MA, Carraway ER. (1999). Partitioning of hydrophobic organic compounds to hydroxypropyl-β-cyclodextrin: Experimental studies and model predictions for surfactant-enhanced remediation applications. *Environmental Science and Technology* **33**(16):2765–2770.

Kovalick, Jr. WW. (2000). *Perspectives on innovative characterization and remediation technologies for contamination sites*. http://www.clu-in.com/tiopersp. Accessed Sept. 27, 2001.

Lane DA, Johnson MJJ, Hanley WH. (1992). Gas- and particle-phase concentrations of alpha-hexachlorocyclohexane, gamma-hexachlorocyclohexane, and hexachloro-benzene in Ontario air. *Environmental Science & Technology* **26**:126–133.

Lear G, Harbottle M, Sills G, Knowles CJ, Semple KT, and Thompson IP. (2007). Impact of electrokinetic remediation on microbial communities with PCP contaminated soil. *Environmental Pollution* **146**:139–146.

Lee CL, Fang MD. (1997). Sources and distribution of chlorobenzenes and hexachlorobutadiene in surficial sediments along the coast of southwestern Taiwan. *Chemosphere* **35**: 2039–2050.

Luo QS, Zhang XH, Wang H, Qian Y. (2005). Mobilization of phenol and dichlorophenol in unsaturated soils by non-uniform electrokinetics. *Chemosphere* **59**:1289–1298.

MacLaren Marex Inc. (1979). Report on an Environmental Survey for Chlorobenzenes at Four Coastal Sites in Nova Scotia, prepared for Environmental Protection Service, Environment Canada, Dartmouth, Nova Scotia.

Oliver BG, Charlton MN. (1984). Chlorinated organic contaminants on settling particulates in the Niagara River vicinity of Lake Ontario. *Environmental Science & Technology* **18**:903–908.

Oliver BG, Charlton MN, Durham RW. (1989). Distribution, redistribution, and geochronology of polychlorinated biphenyl congeners and other chlorinated hydrocarbons in Lake Ontario sediments. *Environmental Science & Technology* **23**:200–208.

Pankow JF, Cherry JA. (1996). *Dense Chlorinated Solvents and Other DNAPLS in Groundwater*. Hove, UK: Waterloo Press.

Oliver BG, Pugsley CW. (1986). Chlorinated contaminants in St. Clair River sediments. *Water Pollution Research Journal of Canada* **21**:368–379.

Regenesis.com. (2003). *The nature of the chlorinated aliphatic hydrocarbons*. http://www.regenesis.com/HRCtech/hrctb111.htm. Accessed 2003.

Tomoyuki K, Shunitsu T. (2006). Removal of organochlorine compounds in contaminated soil by electrokinetic remediation combined with iron reaction wall. *Journal of Japan Society on Water Environment* **29**(2):101–105.

United States Environmental Protection Agency (USEPA). (1994). Location and Estimating Air Emission from Source of Chlorobenzenes (Revised). Office of Air Quality Planning and Standards.

Wan JZ, Yuan SH, Chen J, Li TP, Lin L, Lu XH. (2009). Solubility-enhanced electrokinetic movement of hexachlorobenzene in sediments: A comparison of cosolvent and cyclodextrin. *Journal of Hazardous Materials* **166**:221–226.

Weber, Jr. WJ, Huang W. (1996). A distributed reactivity model for sorption by soils and sediments. 4. Intraparticle heterogeneity and phase-distribution relationships under non-equilibrium conditions. *Environmental Science & Technology* **30**:881–888.

Weng CH, Lin YH, Hsieh YH. (2000). Electrokinetic remediation of trichloroethylene contaminated kaolinite. *Journal of the Chinese Institute of Environmental Engineering* **10**(4):279–289.

Weng CH, Yuan C, Tu HH. (2003). Removal of trichoroethylene from clay soil by series-electrokinetic process. *Practice Periodical of Hazardous, Toxic, and Radioactive Waste Management* **7**(1):25–30.

Westall JC, Leuenberger C, Schwarzenbach RP. (1985). Influence of pH and ionic strength on the aqueous–nonaqueous distribution of chlorinated phenols. *Environmental Science & Technology* **19**(2):193–198.

World Health Organization (WHO). (1989). *Chlorophenols Other Than Pentachlorophenol*. Environmental Health Criteria 093. Geneva: WHO.

Xing B, Pignatello JJ. (1996). Time-dependent isotherm shape of organic compounds in soils organic matter: implications for sorption mechanism. *Environmental Toxicology and Chemistry* **15**:1282–1288.

Yang GCC, Liu CY. (2001). Remediation of TCE contaminated soils by in situ EK-Fenton process. *Journal of Hazardous Materials* **85**:317–331.

Yuan SH, Shu Z, Wan JZ, Lu XH. (2007). Enhanced desorption of hexachlorobenzene from soils by single and mixed surfactants. *Journal of Colloid and Interface Science* **314**(1):167–175.

Yuan SH, Tian M, Lu XH. (2006). Electrokinetic movement of hexachlorobenzene in contaminated soils enhanced by Tween 80 and β-cyclodextrin. *Journal of Hazardous Materials* **137**(2):1218–1225.

Yuan SH, Wan JZ, Lu XH. (2007). Electrokinetic movement of multiple chlorobenzenes in contaminated soils in the presence of β-cyclodextrin. *Journal of Environmental Science* **19**(8):968–976.

Zhao XK, Yang GP, Wu P, Li NH. (2001). Study on adsorption of chlorobenzene on marine sediment. *Journal of Colloid and Interface Science* **243**:273–279.

11

ELECTROKINETIC TRANSPORT OF CHLORINATED ORGANIC PESTICIDES

AHMET KARAGUNDUZ

11.1 INTRODUCTION

Contamination due to agricultural activities has become a significant environmental problem and has been given more attention since the adverse health affects of pesticides were discovered. Eight of the chlorinated pesticides (i.e., DDT, aldrin, dieldrin, endrin, heptachlor, mirex, and taxophene) were listed in 12 priority persistent organic lists in the 2001 Stockholm Convention on Persistent Organic Pollutants. In addition, HCB is also listed as an industrial chemical in the list, because it had been commonly used in agricultural activities as a fungicide. Several important properties of these pesticides as well as some of their by-products are presented in Table 11.1. The most frequently reported chlorinated pesticides are p,p′-DDE, p,p′-DDT, dieldrin, and mirex. DDT was used excessively since its discovery toward the end of World War II and remains persistent in the environment today. Restrictions were imposed on its use once the adverse health affects of DDT were discovered. In 1960s, DDT use reached peak and concentrations at µg/g and even tens of µg/g were found in soils. The use of DDT was forbidden in 1972 in the USA. The half-life of DDT ranges from 2 to 30 years depending on environmental conditions. The chlorine atom in DDT was thought to be why DDT was an effective insecticide. The production of similar chlorinated pesticides, mainly chlordane, aldrin, dieldrin, and mirex followed (Zitko, 2003). Samples taken from 38 agricultural soil samples and two garden soil samples from the Corn Belt of the USA have shown total DDT concentrations (p,p'-DDT+ p,p'-DDD+ p,p'-DDE+ o,p'-DDT) varying from a few ng/g to 11846 ng/g. Elevated concentration of DDT was observed in a soil containing 33.7% of organic carbon. The geometric mean of the total DDT concentration was 9.63 ng/g for soils with organic contents varying from 0.74

Electrochemical Remediation Technologies for Polluted Soils, Sediments and Groundwater,
Edited by Krishna R. Reddy and Claudio Cameselle
Copyright © 2009 John Wiley & Sons, Inc.

TABLE 11.1. Properties of Selected Chlorinated Pesticides

Pesticide	Molecular Weight	Solubility in Water (mg/l)	Henry's Law Constant (Pa/m3/mole)	Log Kow	Log Koc
p,p-DDT[†]	355	5	1.80	6.19	5.39
o,p-DDT[†]	355	85	0.75	6.79	6.62
p,p-DDE[†]	318	11	25.00	7.00	6.64
o,p-DDE[†]	318	140	20.20	6.00	5.78
DDD[†]	320	45	1.28	6.02	5.89
Aldrin[†]	365	30	10.70	5.30	4.98
Dieldrin[†]	381	135	1.13	3.50	3.23
Endrin[†]	381	250	1.10	5.20	4.93
Mirex[†]	546	6	3.64	6.89	7.38
Chlordane[‡]	409.8	56	4.91	5.54	3.49–4.65
Heptachlor[‡]	373	50	29.79	6.10	4.34
Taxophene[‡]	414	3000	6383.48	3.30	2.47
Hexachlorobenzene[‡]	284.8	6	58.77	5.73	5.22

[†]Zitko (2003).
[‡]ATSDR (2008).

to 3.2 (Aigner, Leone, and Falconer, 1998). In the same study, mean concentrations of total chlordane and dieldrin were found to be 1.4 and 1.0 ng/g, respectively. Concentrations up to 750 ng/g of total chlordane and 4250 ng/g of dieldrin were observed in agricultural soils with high organic carbon content. Greater concentrations of DDT were also reported in agricultural soils around the world. DDT concentrations in 17 agricultural soils in Beijing, China, were reported to range from 0.0072 to 2.91 mg/kg, with a geometric mean of 0.3813 mg/kg (Shi et al., 2005).

The removal of chlorinated compounds from soils has long been a challenging task. Conventional remediation technologies such as pump-and-treat and natural attenuation have failed to clean up most contaminated sites to desired levels. Therefore, innovative technologies such as electrokinetic remediation have been developed; the latter has been considered for the removal of contaminants from the soils with low hydraulic permeability.

11.2 ELECTROKINETIC REMOVAL OF CHLORINATED PESTICIDES

The transport of contaminants under an electrical field in soil is governed by several processes, including electromigration, electro-osmosis, diffusion, sorption, solubilization, and chemical and biological reactions.

Electromigration is the movement of charged species under the electrical field. Anions move toward the anode and cations move toward the cathode. The degree of the electromigration depends on the mobility of the ionic species. Electromigration is negligible for most chlorinated pesticides since most of them are not ionized.

Advective flow occurs either by *electro-osmosis* or *hydraulic gradient*. The flow due to hydraulic gradient is represented by Darcy's Law.

$$q_d = K_h i_h, \quad (11.1)$$

where, q_h is the Darcy's velocity (L/T), K_h is the hydraulic conductivity of soil (L/T), and i_h is the hydraulic gradient (L/L). The hydraulic conductivity of soils decreases substantially as the pore size decreases. Typical hydraulic conductivity values for fine sand varies from 2×10^{-7} to 2×10^{-4} m/s; however, these values are in between 1×10^{-11} and 4.7×10^{-9} m/s for clay (Domeico and Schwartz, 1990). Therefore, significant hydraulic flow cannot occur in soils with low permeability if a high hydraulic gradient is not provided. In most cases, that is neither practical nor feasible. Electro-osmosis, however, is relatively independent of pore size. The water of hydration is transported with ions while they move to the respective electrodes. Usually, more cations are present in soil solutions since most soils are negatively charged. In addition, electrolysis yields more hydrogen ions (in the anode region) than the hydroxyl ions (in the cathode region) due to the stoichiometry. Therefore, a net water flow occurs from the anode to the cathode, which is called electro-osmotic flow (EOF) (Mitchell, 1993). EOF can be represented as follows:

$$q_e = k_e i_e, \quad (11.2)$$

where, q_e is the electro-osmotic velocity (L/T), k_e is electro-osmotic permeability (L^2/T/V), and i_e is the applied voltage gradient (V/L). Based on Helmholtz-Smoluehowski theory, electro-osmotic permeability is linearly proportional to the dielectric constant of the pore fluid, zeta potential and the porosity of the soil, and inversely proportional to the viscosity of the pore fluid. Typical k_e values for water in soils are between 1×10^{-8} and 1×10^{-9} m^2/s/V, and they vary little for different soils (Mitchell, 1993). Applied voltage in typical electrochemical applications varies from 1 to 2 V/cm. Based on 1 V/cm, EOF velocity becomes in at between 1×10^{-6} and 1×10^{-7} m/sec. Hydraulic flow due to hydraulic gradient at that degree is only possible if the hydraulic gradient is more than 1000 m/m. Therefore, electro-osmosis can be considered the main transport mechanism for uncharged species in low-permeability soils. The presence of chlorinated pesticides in soil can be assumed not to alter the EOF rate. As discussed below, chlorinated pestices partition into soil organic matter and therefore do not change the surface characteristics substantially. Furthermore, their solubilities in water are relatively low. Fluid properties (e.g., viscosity and dielectric constant) are not changed significantly.

Although some degree of advective flow is generated by electro-osmosis, the low solubilization of chlorinated pesticides in the aqueous phase and low desorption rates from the soils limit the movement of contaminants in soils. As represented in Table 11.1, solubilities of the chlorinated pesticides are very low. Furthermore, the solubility of organic pesticides in most cases is rate limited, which causes even lower aqueous phase concentrations.

The *sorption* and *desorption* of chlorinated pesticides limits the potential success of electro-osmotic transport. The sorption of hydrophobic organic compounds (HOCs) is a partitioning type of process and can be represented by linear isotherms until aqueous phase solubility is approached (Chiou, Peters, and Freed, 1979). Karickhoff (1981) reported that the sorption isotherm was linear for HOCs up to half of the aqueous phase solubility level of the solute. Although HOCs adsorb to mineral surfaces mainly due to van der Waals forces (Schwarzenbach, Gschwend, and Imboden, 1993), the overall amount is quite insignificant if the soil contains no organic matter. HOC sorption is considered to be a partitioning process involving soil organic matter. A linear sorption isotherm can be represented as follows:

$$S_{HOC} = K_D C_{HOC}, \qquad (11.3)$$

where, S_{HOC} is the sorbed solute mass (M/M), K_D is the solute distribution coefficient between water and sorbate, and C_{HOC} is the solute aqueous phase concentration (M/L^3). The linear sorption coefficient, K_D, is related to organic carbon–water partition coefficient (K_{oc}) and soil organic matter (Schwarzenbach, Gschwend, and Imboden, 1993):

$$K_D = f_{oc} K_{oc}, \qquad (11.4)$$

where, f_{oc} is the fraction of the organic carbon of a soil of interest. Given that approximately half the soil organic matter is made up of carbon, f_{oc} can be assumed to be half of f_{om}. If K_{oc} is known for a specific compound, K_D can be approximately calculated for any soil with the relationship given in Eq. 11.4. Slight variations may be observed in the value of K_{oc} because of the nature of the soil organic matter.

Mobilization of HOCs is minimal by EOF, since the degree of partitioning of these compounds into the soil organic matter is very high. For example, Maturi and Reddy (2008) found that 10.6 pore volume of deionized water flow occurred in a kaolin soil when 2 V/cm gradient was applied in 190 days (in cycles of 5 days on, 2 days off). Approximately 5 m^3 of water is yielded for 1 m^3 of contaminated soil with a porosity of 0.47. The sorption coefficient of DDT to a soil with 0.2% organic carbon can be determined as 490 ml/g using the K_{oc} data presented in Table 11.1. Then, the maximum aqueous concentration of DDT for a soil containing 10 ng/g of DDT can be determined as 0.02 µg/l. Therefore, 5 m^3 of water can solubilize 102 µg of DDT, which is only 0.6% of the total mass of DDT in the soil. If soils contain greater fractions of organic material, the solubilized fraction of DDT becomes lower. For the same soil, the retardation coefficient, R, can be determined approximately as 1700 pore volume. This means that the contaminant can be moved if more than 1700 pore volume of water is flushed through the soil. Therefore, extended periods of times are necessary to be able to mobilize bound HOCs. In conclusion, conventional electrokinetic methods are limited in transporting uncharged highly bounded HOCs in soils and sediments. Limitations in the electrokinetic removal of chlorinated compounds may be reduced by using surfactants and other cosolvents.

The use of surfactants in remediation technologies to enhance the process due to their properties to increase solubility has long been investigated. The removal of nonaqueous phase liquids (NAPLs) using surfactants (pump-and-treat) has been extensively studied, and dramatic increases in removal efficiencies were observed (Pennell, Abriola, and Weber, 1993). Surfactant-enhanced bioremediation studies yielded mixed results, showing enhancement, reduction, or no change in biodegradation of HOCs (Rouse et al., 1994; Yeh, 2000). Such conflicting results may be due to the distribution of contaminants among water, monomers, micelles, soil, and microbial populations. Surfactant-enhanced electrokinetic remediation of organic contaminants has been given more attention in recent years (Ranjan, Qian, and

Krishnapillaj, 2006; Saichek and Reddy, 2005; Karagunduz, Gezer, and Karasuloglu, 2007; Saichek and Reddy, 2003; Yuan and Weng, 2004).

11.3 SURFACTANT-ENHANCED ELECTROKINETIC REMEDIATION OF CHLORINATED PESTICIDES

The term "surfactants" derives from the phrase "surface active agents." Surfactants are amphiphilic compounds that have two moieties: a hydrophilic head portion (polar) and a hydrophobic (nonpolar) tail portion. The hydrophobic portion of the surfactants is usually a long chain of hydrocarbons, and the hydrophilic portion of surfactants is usually ionic or highly polar groups (Rosen, 1989). Depending on the type of ionic groups, surfactants can be classified as cationic, ionic, nonionic, and zwitterionic. Surfactants may attract or repel each other in water since they posses both hydrophobic and hydrophilic forces (Porter, 1994). If the surfactants are charged, their hydrophilic groups repel each other. When the charge density of the polar group is higher, the repulsion force becomes greater. However, in contrast to hydrophilic groups, hydrophobic groups of surfactants attract each other. At low surfactant concentration, the interaction among monomers is minimal in aqueous solutions. As surfactant concentration increases, surfactant molecules form aggregates. Surfactant molecules orient their hydrophobic groups away from water and form micelles. The surfactant concentration at which micelle formation begins is called the critical micelle concentration (CMC).

Surfactants influence electrokinetic remediation in various ways. They adsorb to soils and alter their surface properties, and as a result, EOF and sorption of hydrophobic organics to the soils are affected. Surfactants also increase the aqueous phase concentration of organics via micellar solubilization. Depending on the type of surfactants, micelles, and therefore organic contaminants within the micelles, may be transported toward the anode or cathode. A simplified conceptual model of a surfactant-enhanced electrokinetic process is presented in Figure 11.1. The effects of surfactants on these processes are discussed below in more detail.

11.3.1 Effects of Surfactants in Sorption and Solubilization of Chlorinated Pesticides

Surfactants exist as monomers in solution and have a minimal effect on the solubility of most organic compounds at low concentrations (Pennell, Abriola, and Weber, 1993). However, as the concentration of the surfactant is increased above the CMC, a linear enhancement can be observed in the solubility of the organics. As the surfactant concentration is increased, the hydrophobic moieties of the surfactant monomers tend to associate with one another to form micelles. A linear enhancement in the solubility of organic compounds is typically observed above the CMC, which is attributed to the incorporation of the organic species within the hydrophobic core of the surfactant micelle (Rosen, 1989).

The solubility potential of a surfactant can be represented using the weight solubilization ratio (WSR).

$$\text{WSR} = \frac{C_{\text{HOC}} - C_{\text{HOC(CMC)}}}{C_s - \text{CMC}}, \qquad (11.5)$$

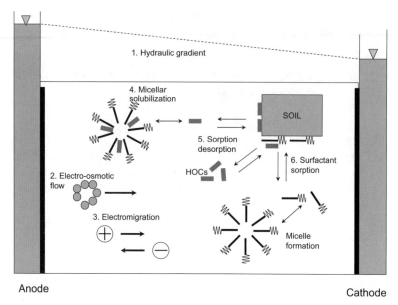

Figure 11.1. Conceptual model of surfactant-enhanced electrokinetic transport.

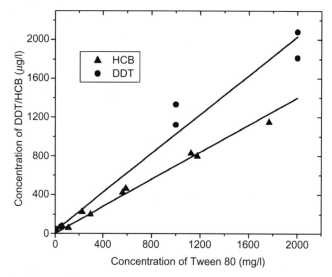

Figure 11.2. Micellar solubilization of HCB and DDT by Tween 80 (HCB data adapted from Karagunduz 2002; DDT data adapted from Karagunduz, Gezer, and Karasuloglu, 2007).

where, C_{HOC} is the concentration of solubilized HOC (M/L^3), $C_{HOC(CMC)}$ is the concentration of HOC at CMC (M/L^3), C_s is the surfactant concentration (M/L^3), and CMC is the critical micelle concentration (M/L^3). Higher WSR yields greater solubilization of HOCs. Solubilization of hexachlorobenzene (HCB) and DDT by a nonionic surfactant (Tween 80) is presented in Figure 11.2. The WSR for DDT and HCB is 1.0×10^{-3} g DDT/g Tween 80 and 0.7×10^{-3} g HCB/g Tween 80, respectively. Since nonionic surfactants have relatively low CMC values, micellar solubilization

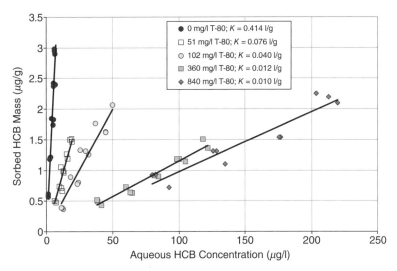

Figure 11.3. HCB sorption isotherms as a function of Tween 80 concentration (Appling soil, organic matter content = 0.78%) (adapted from Karagunduz, 2002).

occurs even at lower surfactant concentrations. For example, the solubility of HCB in water was only 5 µg/l; however, it increased to 300 µg HCB/l and 1150 µg HCB/l in the presence of 300 and 1700 mg/l of Tween 80, respectively.

The solid–liquid distribution coefficient of HOCs in the presence of surfactants can be represented as follows (Sun et al., 1995):

$$K^* = \frac{\text{Mass of sorbed HOC per mass solid–phase}}{\text{Mass of DDT insolution–phase per bulk liquidvolume}}$$

$$= \frac{S^*_{HOC}}{C^*_{HOC}} = \frac{K_D + S_{ss}K_{ss}}{1 + C_{mn}K_{mn} + C_{mc}K_{mc}}, \quad (11.6)$$

where, K^* is the apparent solute soil–water distribution coefficient (L^3/M), C_{mn} is the concentration of surfactant monomers (M/L^3), C_{mc} is the concentration of surfactant micelles (M/L^3), $C_{aq,HOC}$ is the HOC concentration in the aqueous (water) phase (M/L^3), K_{mn} is the solute distribution coefficient between surfactant monomers and water (L^3/M), and K_{mc} is the solute distribution coefficient between the surfactant micelle and water (L^3/M). S_{ss} is the mass of sorbed surfactant (M/M), and K_{ss} is the HOC distribution coefficient between the sorbed surfactant and water (L^3/M).

K^* tends to decrease at high surfactant concentrations. As the surfactant concentration increased, the partition coefficient of HOCs decreases. The distribution coefficient of HCB as a function of a nonionic surfactant, Tween 80, is presented in Figure 11.3 as an example. Although K_D was 0.414 g/l in the absence of surfactant, it decreased substantially to 0.010 l/g as the aqueous Tween 80 concentration was increased.

The partitioning coefficient decreases as the surfactant concentration increases above the CMC. However, the opposite may occur below the CMC. The sorbed

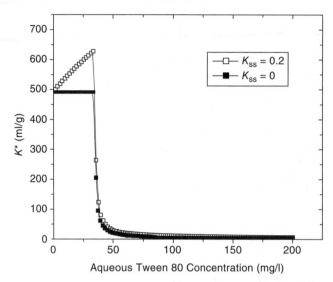

Figure 11.4. Change of linear partitioning coefficient with Tween 80 concentration. (K_D = 490 ml/g, CMC = 35 mg/l, OM = 0.17%, S_m = 12.5 mg/g, b = 0.007 l/mg, K_{mn} = 1.4 × 10^3 l/g).

surfactant molecules to the soil surface may act as soil organic matter. Therefore, additional organics partition into the sorbed surfactant phase, causing even greater sorption capacity. The effect of sorption of HOCs to the surfactant phase is presented in Figure 11.4. Below the CMC, the partition coefficient increased to 650 ml/g when K_{ss} was assumed to be 0.2 ml/mg. It followed a sudden decrease due to the micellar solubilization. In the absence of an additional sorption term (K_{ss} = 0 ml/mg), K^* remains the same as K_D, followed by a greater decrease. Above the CMC, both curves are almost identical.

The phase distribution of an HOC at varying surfactant concentrations can be estimated using the parameters discussed above. The total concentration of HOC, $M_{T,HOC}$, in a batch reactor in the presence of surfactants and soil can be represented as follows:

$$M_{T,HOC} = M_{s,HOC} + M_{aq,HOC} + M_{mn,HOC} + M_{mc,HOC}, \quad (11.7)$$

where, $M_{s,HOC}$, $M_{aq,HOC}$, $M_{mn,HOC}$, and $M_{mc,HOC}$ represent the mass of HOC in sorbed phase, water, surfactant monomer, and surfactant micelles, respectively. The total HOC mass ($C_{T,HOC}$) can be represented as a function of the aqueous (water) phase HOC concentration by substituting the corresponding relationships for HOC associated with soil, micelles, and monomers as follows:

$$\begin{aligned} C_{T,HOC} &= \rho C_{s,HOC} + \theta C_{aq,HOC} + \theta C_{mn,HOC} + \theta C_{mc,HOC} \\ &= \rho(K_D + S_{ss}K_{ss})C_{aq,HOC} + \theta(1 + K_{mn}C_{mn} + K_{mc}C_{mc})C_{aq,HOC} \end{aligned} \quad (11.8)$$

where, ρ and θ represent bulk density (M/L^3) and volumetric water content (L^3/L^3) of the porous media, respectively.

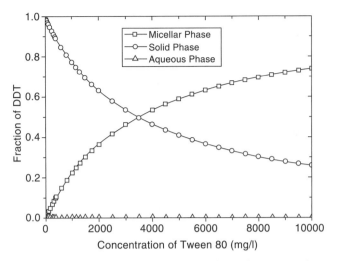

Figure 11.5. Phase distribution of DDT in micellar, solid, and aqueous phases as a function of Tween 80 (Model parameters: solid/liquid = 1590 kg/5m^3, K_{mn} = 10$^{4.77}$, K_{mc} = 10$^{4.77}$, CMC = 35 mg/l, K_{ss} = 0, b = 0.009 l/mg, S_m = 8.0 mg/g, K_D = 0.49 l/g).

The predicted phase distribution of DDT mass among the soil, micelles, and water are presented in Figure 11.5 as a function of aqueous Tween 80 concentration. The initial mass of the soil was assumed to be 10 ng/g. In the absence of Tween 80, 99.4% of the total mass of DDT was associated with the solid phase, and only 5% of the mass was in the aqueous phase. As the surfactant concentration increased, the micellar solubilization became more dominant. While DDT mass in the micellar phase was increased, DDT associated with the soil decreased drastically. For example, in the presence of 10,000 mg/l of Tween 80, 74.1% of DDT was associated with surfactant micelles, and only 25.2% was associated with the solid phase. The substantial increase of HOCs in the aqueous phase (within micelles) allows combining surfactants with various remediation technologies including electrokinetics.

11.3.2 Effects of Surfactants in Electro-osmosis and Electromigration

As discussed before, EOF is proportional to the zeta potential and porosity of the soil and the dielectric constant of the pore fluid. It has been reported that the zeta potential of the soil became more positive in the presence of cationic surfactants. However, anionic surfactants decreased the zeta potential, and nonionic surfactants caused slight increase (Kaya and Yukselen, 2005). The influence of surfactants on zeta potential is closely related to the sorption of surfactants in soils. The sorption of surfactants onto the solid phase is affected by various factors, including solid and surfactant types and environmental conditions such as pH, temperature, and electrolytes (Rosen, 1989).

Cationic surfactants strongly bind to the soil particles. As the surfactant concentration increases, hemimicelle formation on the sorbed surface cause charge reversal, which may eventually result in a reverse EOF (Li and Gale, 1996; Kaneta, Tanaka, and Taga, 1993). The problem of using cationic surfactants in electrokinetic remediation is the strong association of surfactant monomers with soils. This substantially hinders the desorption of HOCs from soils. The effect of cetyltrimethylammonium bromide (CTAB), a cationic surfactant, on hydrocarbon removal by electrokinetics was investigated by Ranjan, Qian, and Krishnapillaj (2006). It was reported that CTAB coupled with the electrokinetic method retarded the removal of hydrocarbons. This can be attributed to the strong interaction of CTAB with soil particles.

Anionic surfactants do not adsorb to soils nearly as much as cationic surfactants. However, the degree of sorption is sufficient to reduce the zeta potential even further. Because of the reduction in zeta potentials, higher EOFs are expected. However, contradictory results have been reported in the literature (Park et al., 2007). This can be attributed to several factors. First, because of the electrolysis, H^+ is produced in the anode chamber, and it moves toward the cathode. This may cause protonation of the surfaces; as a result, lesser negative charged surfaces are formed. Second, the EOF also depends on the dielectric constants of the pore fluid and is inversely proportional to viscosity. Surfactant solutions have lower dielectric constants than water, and they have slightly higher viscosities. These two factors contribute to the reduction in EOF.

Nonionic surfactants yield varying results depending on the soil and other environmental conditions (e.g., pH). It was reported that EOF rates were almost the same for water and 3% Tween 80 surfactant for glacial till; however, significantly reduced rates were obtained for Tween 80 in kaolinite clay (Reddy and Saichek, 2003). This was attributed to the differences in soil properties. Significant reductions in EOF were also observed when Igepal CA-720 was used (Saichek and Reddy, 2003). Reduction in EOF was attributed to the low dielectric constant of the surfactant and to the change of surface properties (e.g., charge density) of the soil due to sorption. The type of the surfactant also affects the reduction of EOFs in different degrees. For example, a greater EOF was obtained in the presence of alkyl polyglucoside than with the Brij 30 surfactant (Park et al., 2007). Both surfactants yielded lower EOFs than that of water. Uncharged chlorinated pesticides are not mobile under an electrical field. However, ionic surfactants have electrophoretic mobility. The use of surfactants in capillary electrophoresis dates back to 1984. Micellar electrokinetic chromatography was initially developed for the separation of uncharged species. Usually, anionic surfactants are used to retain compounds against the strong EOF so that the separation occurs (Terabe, 2008). Electrophoretic mobility of surfactants can be used to effectively transport chlorinated compounds in electrokinetic remediation. The electrophoretic mobility of micelles is linearly proportional to its charge and inversely proportional to its radius based on Debye–Huckel–Henry Theory.

11.3.3 Surfactant-Coupled Electrokinetic Transport

Surfactant-coupled electrokinetic methods involve various complex and related processes. However, the limiting step for the electrokinetic transport of chlorinated pesticides and other organics with low aqueous solubility is the desorption process. Surfactants may improve the desorption of HOCs substantially if sufficient flow can be generated. The advective flow is mainly due to EOF in low-permeability soils. However, reductions may occur in EOF in the presence of surfactants. For nonionic

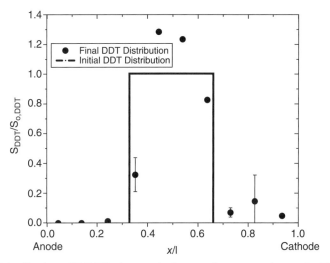

Figure 11.6. Distribution of DDT before and after a column experiment for Tween 80 (20V, 20 days) (adapted from Karagunduz, Gezer, and Karasuloglu, 2007).

surfactants, EOF is usually from anode to cathode. Neither nonionic surfactants nor chlorinated pesticides move under the electrical field (e.g., electromigration and electrophoresis). The EOF is the only driving force in transport. However, the affinity of soils to nonionic surfactants is significant especially for soils with high organic matter content. Therefore, transporting the necessary amount of surfactants may be a problem. For example, Karagunduz, Gezer, and Karasuloglu (2007) conducted a column experiment, in which DDT-contaminated soil was placed in the middle one-third of the column. A solution of 7500 mg/l of Tween 80 was introduced from the anode compartment. The result of this experiment is presented in Figure 11.6. It was observed that only a slight transport toward cathode was observed. This was attributed to the fact that the amount of surfactant transported through EOF was not sufficient to mobilize the DDT. Only 225 ml of EOF was generated in 20 days. The surfactant mass carried with the EOF could only fill up the half of the available sites of the soil for sorption. Therefore, the aqueous phase of Tween 80 remained in low concentrations, which had little effect on desorption of DDT.

Anionic surfactants may be preferable for use in electrokinetic remediation since the degree of association between soil and surfactant is smaller. Although anionic surfactants usually have greater CMC values, their WSRs are comparable to those of nonionic surfactants. In addition to the solubilization potential of anionic surfactants, they carry a net surface charge. Micelles can be considered small colloids in the aqueous phase and have electrical mobility under an electrical field. However, the direction of EOF and the electromobility of anionic surfactants are opposite. Anionic micelles move toward the anode, while EOF is usually toward the cathode. Karagunduz Gezer, and Karasuloglu (2007) found that anionic micelle migration overcame the EOF causing the transport of DDT toward the anode. This is illustrated in Figure 11.7. The performance of electrokinetic transport with anionic surfactants depends on minimizing the EOF. Controlling pH both in anode and cathode can enhance the process.

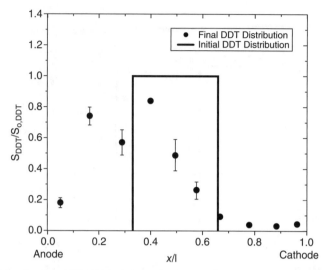

Figure 11.7. Distribution of DDT before and after a column experiment for an anionic surfactant, SDBS (20 V, 30 days) (adapted from Karagunduz, Gezer, and Karasuloglu, 2007).

The use of cationic surfactants in electrokinetics is limited due to the strong interactions of cationic surfactants and the soil. The sorbed cationic surfactants behave as natural soil organic matter. Therefore, the degree of sorption of HOCs in the soil matrix increases.

11.4 COSOLVENT-ENHANCED ELECTROKINETIC REMEDIATION OF CHLORINATED PESTICIDES

Cosolvent-enhanced electrokinetic remediation of chlorinated pesticides has yet to be explored. Few studies have shown enhancements in the transport of polycyclic aromatic hydrocarbons (PAHs) when cosolvents were used (Maturi and Reddy, 2008; Reddy and Saichek, 2003). The potential success of cosolvents depends on their ability to mobilize the HOCs and to increase the EOF. It was shown that 10% *n*-butylamine generated the greatest EOF, followed by 20% *n*-butylamine and water (Maturi and Reddy, 2008). The sorption of solvents to the soil matrix may not be as high as the surfactants, which cause a better solubilization with the same amount of electro-osmotic transport.

11.5 SUMMARY

Remediation of chlorinated organic compounds by using conventional electrokinetic method is limited because of the limitations of caused by desorption and solubilization. These compounds bind strongly to the soil organic matter. Therefore, very low aqueous phase concentrations are reached at equilibrium. Furthermore, low solubilities of HOCs limit the concentrations in water. Therefore, either surfactant or cosolvent enhancements are required to increase the solubilization as well as the desorption of HOCs. The potential success of surfactants depends on the type of surfactants, soil properties, and other factors (i.e., pH, presence of other cations, etc.). Micellar solubilization substantially increases the rate of desorption of HOCs from the soil. Due to the low hydraulic permeabilities of soils, EOF is the main cause for water movement. In the presence of nonionic surfactants, EOF rates may decrease, resulting in lower surfactant mass transport within the soil matrix. When anionic surfactants are used, the EOF may increase or decrease. However, since anionic surfactant micelles have electrophoretic mobility, the electromigration of micelles and, therefore, solubilized HOCs, toward the anode also becomes an important driving force for the transport. For this case, the EOF from anode to cathode should be minimized. Cationic surfactants have limited applications in electrokinetics due to their strong interactions with soil matrix. Cosolvents have potential use in electrokinetics. However, their use in chlorinated organic pesticides should be explored further.

REFERENCES

Aigner EJ, Leone AD, Falconer RL. (1998). Concentrations and enantiomeric ratios of organochlorine pesticides in soils from the U.S. Corn Belt. *Environmental Science and Technololy* **32**:1162–1168.

Agency for Toxic Substances and Disease Registry (ATSDR). (2008). http://www.atsdr.cdc.gov/. Accessed April 10, 2008.

Chiou CT, Peters LJ, Freed VH. (1979). A physical concept of soil–water equilibria for nonionic organic compounds. *Science* **206**:831–832.

Domeico PA, Schwartz FW. (1990). *Physical and Chemical Hydogeology*. New York: John Wiley and Sons, Inc.

Kaneta T, Tanaka S, Taga M. (1993). Effect of cetyltrimethylammonium chloride on electro-osmotic and electrophoretic mobilities in capillary zone electrophoresis. *Journal of Chromatography A* **653**(2):313–319.

Karagunduz A. (2002). Influence of surfactants on the sorption and transport of contaminants in saturated and unsaturated soils. PhD Dissertation. Georgia Institute of Technology, Atlanta, GA.

Karagunduz A, Gezer A, Karasuloglu G. (2007). Surfactant enhanced electrokinetic remediation of DDT from soils. *The Science of the Total Environment* **385**(1–3):1–11.

Karickhoff SW. (1981). Semi-empirical estimation of sorption of hydrophobic pollutants on natural sediments and soils. *Chemosphere* **10**:833–846.

Kaya A, Yukselen Y. (2005). Zeta potential of soils with surfactants and its relevance to electrokinetic remediation. *Journal of Hazardous Materials* **120**:119–126.

Li H, Gale RJ. (1996). The role of surfactants in capillary electro-osmosis and electrophoresis. *Journal of Environmental Science and Health Part A* **31**(9):2363–2379.

Maturi K, Reddy KR. (2008). Cosolvent-enhanced desorption and transport of heavy metals and organic contaminants in soils during electrokinetic remediation. *Water Air and Soil Pollution* **189**:199–211.

Mitchell JK. (1993). *Fundamentals of Soil Behavior*. New York: John Wiley and Sons, Inc.

Pennell KD, Abriola AM, Weber JR. (1993). Surfactant enhanced solubilization of residual dodecane in soil columns 1. Experimental investigations. *Environmental Science and Technology* **27**:2332–2340.

Park JY, Lee HH, Kim SJ, Lee YJ, Yang JW. (2007). Surfactant enhanced electrokinetic removal of phananthrene from kaolinite. *Journal of Hazardous Materials* **140**:230–236.

Porter MR. (1994). *Handbook of Surfactants*. New York: Chapman Hall.

Ranjan RS, Qian Y, Krishnapillaj M. (2006). Effects of electrokinetics and cationic surfactant cetyltrimethylammonium bromide [CTAB] on the hydrocarbon removal and retention from contaminated soils. *Environmental Technology* **27**:767–778.

Reddy KR, Saichek R. (2003). Effects of soil type on electrokinetic removal of phanenthrene using surfactants and cosolvents. *Journal of Environmental Engineering* **129**(4):336–346.

Rosen MJ. (1989). *Surfactant and Interfacial Phenomena*, 2nd ed. New York: John Wiley and Sons, Inc.

Rouse JD, Sabatini DA, Suflita JM, Harwell JH. (1994). Influence of surfactants microbial degredation of organic compounds. *Critical Reviews in Environmental Science and Technology* **24**(4):325–370.

Saichek RE, Reddy KR. (2003). Effects of system variables on surfactant enhanced electrokinetic removal of polycyclic aromatic hydrocarbons from clayey soils. *Environmental Technology* **24**:503–515.

Saichek RE, Reddy KR. (2005). Electrokinetically enhanced remediation of hydrophobic organic compounds in soils: a review. *Critical Reviews in Environmental Science and Technology* **192**:35–115.

Schwarzenbach RP, Gschwend PM, Imboden DM. (1993). *Environmental Organic Chemistry*. New York: John Wiley and Sons, Inc.

Shi Y, Meng F, Guo F, Lu Y, Wang T, Zhang H. (2005). Residues of organic chlorinated pesticides in agricultural soils of Beijing, China. *Archives of Environmental Contamination and Toxicology* **49**:37–44.

Sun S, Inskeep WP, Boyd SA. (1995). Sorption of nonionic organic compounds in soil–water systems containing a micelle-forming surfactant. *Environmental Science and Technology* **29**:903–913.

Terabe S. (2008). Micellar electrokinetic chromatography. In *Handbook of Capillary and Microchip Electrophoresis and Associated Microtechniques*, 3rd ed. (ed. JP Landers). Boca Raton: CRC Press, pp. 3–74.

Yeh DH. (2000). Influence of Nonionic Surfactants on the Bioavailability and Microbial Reductive Dechlorination of Hexachlorobenzene. PhD Dissertation. Georgia Institute of Technology, Atlanta, GA.

Yuan C, Weng CH (2004). Remediating ethylbene-contaminated clayey soil by a surfactant aided electrokinetic (SAEK) process. *Chemosphere* **57**:225–232.

Zitko V. 2003. Chlorinated pesticides: aldrin, DDT, endrin, dieldrin, mirex. In *The Handbook of Environmental Chemistry, Vol. 3, Part O, Persistant Organic Pollutants* (ed. H Fiedler). Berlin: Springer-Verlag, pp. 47–89.

12

ELECTROKINETIC REMOVAL OF HERBICIDES FROM SOILS

ALEXANDRA B. RIBEIRO AND EDUARDO P. MATEUS

12.1 INTRODUCTION

Agriculture depends on soil for water and nutrients supply, as well as for root fixation. Soil is a vital natural resource, nonrenewable at a human scale, that performs a number of key environmental, social, and economic functions. Soil is an interface reactor, performing storage, filtering, buffering, and transformation functions, thus playing a central role in water protection and in the exchange of gases with the atmosphere. It is also a habitat and gene pool, an element of the landscape and cultural heritage, as well as a provider of raw materials. In order to perform its main functions, it is essential to maintain soil quality, which is crucial for sustainability and for sustaining agricultural productivity (Blum, 2005; Silva et al., 2006). However, soil is under increasing threat from a wide range of human activities, which degrade it and are undermine its long-term availability and viability [Blum, 2005; Santos, 2008; Commission of the European Communities (2002); Commission of the European Communities (2006)].

Environmental contamination with pesticides due mainly to agricultural practices and accidental spills leads to soil, surface, and groundwater quality degradation, which may have negative impacts on public health and biodiversity (Correia et al., 2006; Santos, 2008). This issue of increasing concern has lead to many studies describing and quantifying contamination with pesticides (Thorstensen et al., 2001; Boivin, Cherrier, and Schiavon, 2005; Silva et al., 2006; Abreu, 2008). Pesticides are toxic compounds deliberately released into the environment to fight plant pests and diseases. Their introduction may result in the damage or loss of some or several functions of soil and in the possible cross-contamination of water. They can accumulate in the soil, leach to the groundwater, and evaporate into the air, from which

---

Electrochemical Remediation Technologies for Polluted Soils, Sediments and Groundwater,
Edited by Krishna R. Reddy and Claudio Cameselle
Copyright © 2009 John Wiley & Sons, Inc.

further deposition onto soil can take place. They may also affect biodiversity and enter the food chain, entailing multiple negative consequences for human health, ecosystems, and other natural resources [Commission of the European Communities (2002); Commission of the European Communities (2006)].

Rice is one of the major crops cultivated worldwide (Chang, 2000). Among the major herbicides used to control plagues in rice fields are atrazine, molinate, and bentazone (Castro et al., 2005; Correia et al., 2006).

The food demand of an increasing world population challenges the scientific community to pay attention to how food production leads to environmental pressures and to accelerate efforts to develop methodologies to minimize their impacts on human health and the environment (Santos, 2008).

The electrokinetic process is a remediation technique for the removal of contaminants from polluted sites. It uses a low-level direct current (DC), as the cleaning agent, to transport contaminants out of the soil toward one of the electrode compartments, from where they can be removed. Several authors have critically reviewed its state of knowledge (Pamukcu and Wittle, 1992; Acar et al., 1995; Ottosen, 1995; Yeung and Datla, 1995; Page and Page, 2002; Virkutyte, Sillanpaa, and Latostenmaa, 2002; Ribeiro and Rodríguez-Maroto, 2006).

This chapter reports on the application of the electrokinetic process to remove herbicides such as atrazine, molinate, and bentazone from soils. The discussion is focused on (a) the assessment of the behavior of atrazine, molinate, and bentazone in soils when submitted to an electric field, and (b) the evaluation of the applicability of the technique to remove atrazine, molinate, and bentazone from soils.

12.2 HERBICIDES

12.2.1 Atrazine

Atrazine (2-chloro-4-ethyl-amino-6-isopropylamino-s-triazine) is a selective triazine herbicide used pre- and postemergence with restricted permitted uses to control broad-leaved and grassy weeds in agriculture and in conifer reforestation plantings (FOOTPRINT Pesticide Properties Database). Atrazine is highly persistent in soil (its main physicochemical properties are presented in Table 12.1). Chemical hydrolysis, followed by degradation by soil microorganisms, accounts for most of the breakdown of atrazine. Hydrolysis is rapid in acidic or basic environments but is slower at neutral pH levels. Addition of organic material increases the rate of hydrolysis. Atrazine can persist for longer than 1 year under dry or cold conditions. Atrazine is moderately to highly mobile in soils with low clay or organic matter content. Because it does not adsorb strongly to soil particles and has a lengthy half-life (60 to >100 days), it has a high potential for groundwater contamination despite its moderate solubility in water (EXTOXNET, 1996). As a consequence, atrazine is one of the most frequently detected contaminants in the waterbodies in Europe (Gascón, Dubina, and Barceló, 1997; Carabias-Martínez et al., 2002).

Additionally, atrazine has been reported as a potential endocrine disruptor (Renner, 2000, 2002).

TABLE 12.1. Physicochemical Properties of the Herbicides Atrazine, Molinate, and Bentazone

Parameter	Atrazine	Molinate	Bentazone
Structure diagram	(triazine structure)	(thiocarbamate structure)	(benzothiazinone structure)
CAS	1912-24-9	2212-67-1	25057-89-0
EC number	217-617-8	218-661-0	246-585-8
Chemical group	Triazine	Thiocarbamate	Benzothiazinone
Chemical formula	$C_8H_{14}ClN_5$	$C_9H_{17}NOS$	$C_{10}H_{12}N_2O_3S$
Molecular mass (g/mol)	215.68	187.3	240.3
Physical state	White crystals	Clear liquid	Dirty white crystals
Bulk density (g/ml)/ specific gravity	1.23	1.06	1.41
Solubility in water at 20 °C (mg/l)	35	1100	570
Octanol–water partition coefficient (LogP)	2.7	2.86	−0.46
Dissociation constant (pKa at 25 °C)	1.7	Not applicable (neutral)	3.28
Vapor pressure at 25 °C (mPa)	0.04	500	0.17
Henry's constant at 25 °C (Pa·m³/mol)	$1.50E^{-04}$	$6.87E^{-01}$	$7.20E^{-05}$
Soil degradation DT50 (days)	75	28	13
GUS leaching potential index	3.75	2.49	2.55
Organic carbon sorption constant K_{oc} (ml/g)	100	190	51

Source: FOOTPRINT Pesticide Properties Database.

12.2.2 Molinate

Molinate is a selective herbicide that belongs to the thiocarbamate chemical class (EXTOXNET, 1996). It is used worldwide to control germinating broad-leaved and grassy weeds in rice paddies (Barreiros et al. 2003; WHO, 2003). Molinate is rapidly taken up by plant roots and transported to the leaves. In the leaves, molinate inhibits leaf growth and development. It is rapidly metabolized in nonsusceptable plants (EXTOXNET, 1996). Several studies reported molinate to be found in groundwater (Castro et al., 2005), and toxicological studies showed that molinate could lead to adverse effects such as reproductive toxicity in mammals (Cochran et al., 1997; APVMA, 2003). Molinate is stable under normal temperatures and pressures, but thermal decomposition may release toxic oxides of nitrogen and sulfur (EXTOXNET,

1996). The main physicochemical properties are presented in Table 12.1. Molinate presents high solubility in water (1100 mg/l), is highly volatile and moderately mobile, bioaccumulates moderately (LogP = 2.86) and presents low persistence in the soil.

12.2.3 Bentazone

Bentazone [3-(1-methyltethyl)-1H-2, 1, 3-benzothiadiazin-4(3H)-one 2, 2-dioxide] is a postemergence herbicide used for selective control of broad-leaved weeds and sedges (a weed) in beans, rice, corn, peanuts, mint, and others. Bentazone is a contact herbicide, which means that it causes injury only to the parts of the plant to which it is applied. It interferes with the ability of susceptible plants to use sunlight for photosynthesis. Visible injury to the treated leaf surface usually occurs within 4–8 h, followed by plant death (EXTOXNET, 1996). The sorption and desorption on organic matter is strongly pH controlled for bentazone due to the fact that it shows an acid behavior (Boivin, Cherrier, and Schiavon, 2005). The sorption is stronger for pH 2–3 (Poll and Vink, 1996; Aguilar et al., 1999; Santos, Rocha and Barceló, 2000; Boivin et al., 2004; Fontanals, Marcé, and Borrull, 2004; Boivin, Cherrier, and Schiavon, 2005). The dissociation constant of bentazone is pKa = 3.28, the solubility in neutral water is 570 mg/l, and its log K_{OW} under neutral conditions is −0.318 (FOOTPRINT Pesticide Properties Database). These properties (several others are presented in Table 12.1) show the polar character of bentazone (Aguilar et al., 1999; Fontanals, Marcé, and Borrull, 2004).

12.3 CASE STUDY

The laboratory study procedure and results are presented in this section to demonstrate the electrokinetic removal of herbicides from soils.

12.3.1 Soils

Four types of soil were used. The first, sampled at Valadares (Vale de Milhaço, Portugal), corresponds to a Eutric Regossol (Food and Agriculture Organization/ United Nations Educational, Scientific and Cultural Organization [FAO/UNESCO] soil classification). An average sample was collected at 0- to 15-cm depth; it has a sandy texture, and its characteristics are shown in Table 12.2 (Soil 0). More details related to its physical, chemical, and mineralogical properties can be found in Ribeiro (1992).

The second soil, collected at a depth of 0–20 cm from the rice crop fields of the Ebro Delta area (Tarragona, Spain), where the pesticides are currently applied, was "naturally" contaminated with aged residues corresponding to the last atrazine application carried out 40 days before its sampling. Characteristics of this soil are shown in Table 12.2 (Soil 1), and more details are available in Durand, Forteza, and Barceló (1989).

Soil 2 and soil 3 were sampled at Bico da Barca, an experimental rice site, located at the Rio Mondego valley, Portugal. Several physical and chemical characteristics of these soils (soil 2 and soil 3) are presented in Table 12.2. Soil 2 and soil 3 have different histories of pesticide application: Soil 2 comes from a rice field with no

TABLE 12.2. Physical and Chemical Characteristics of the Soils Used

Parameter	Soil 0	Soil 1	Soil 2	Soil 3
Soil texture (%)				
Coarse sand ($200 < \emptyset < 2000\,\mu m$)	69.8	1.8	1.7	2.1
Fine sand ($20 < \emptyset < 200\,\mu m$)	24.2	42.3	26.5	26.1
Silt ($2 < \emptyset < 20\,\mu m$)	3.5	24.1	51.1	49.7
Clay ($\emptyset < 2\,\mu m$)	2.5	31.8	20.7	22.1
Textural classification	Sandy	Clay loam	Silt loam	Silt loam
pH (H_2O)	6.0	6.5	6.46	5.83
Organic carbon (g/kg)	—	—	21.1	26.8
Organic matter (%)	0.41	1.64	3.64	4.61
Total nitrogen (g/kg)	—	—	1.55	1.72
Exchangeable cations ($cmol_{(c)}/kg$)				
Ca^{2+}	0.35	—	6.44	4.46
Mg^{2+}	0.07	—	1.34	1.32
K^+	0.07	—	0.59	0.61
Na^+	0.07	—	0.89	0.23
Sum of exchangeable cations ($cmol_{(c)}/kg$)	0.54	—	9.26	6.62
Cation exchange capacity ($cmol_{(c)}/kg$)	1.47	12.36	9.26	10.0
Base saturation (%)	36.7	—	100.0	66.2

tradition of specific type of pesticide application, which has been used for biological agriculture, and soil 3 comes from a rice field with a history of traditional herbicide application (bentazone, propanil, and 2-methyl-4-chlorophenoxyacetic acid [MCPA]). However, both soils (soil 2 and soil 3), together with soil 0, were initially free of herbicide residues.

The soils were then spiked with herbicides without residue aging: soil 0 with atrazine, soil 2 with molinate, and soil 3 with molinate, or with bentazone (according to Table 12.3).

12.3.2 Reagents and Chemicals

The enzyme-linked immunosorbent assay (ELISA) kits, Atrazine RaPID Assay® (A00071) and Magnetic Separation Rack (A00004), used were obtained from Strategic Diagnostics Inc. (Newark, DE). All organic solvents used were high-performance liquid chromatography (HPLC) grade and purchased from Merck (Darmstadt, Germany), Sigma-Aldrich (Steinheim, Germany), or Panreac (Barcelona, Spain). The acetonitrite, methanol, and acetone were gradient grade, the methylene chloride was Pestanal grade, and the diethyl ether and pentane were Pro Analysis (PA) grade. The Supelclean ENVI™ 18 Disks, 47 mm in diameter, for solid-phase extraction (SPE), were purchased from Supelco (Bellefonte, PA). All triazine standards, atrazine, simazine, and propazine, as well as molinate and bentazone standards, were Pestanal grade, purchased from Riedel-de-Haën (Seelze, Germany).

All the materials were washed to prevent contamination, being immersed overnight in water with Sodosil from Riedel-de-Haën. After cleaning, the materials were rinsed with water in order to remove detergent residues, followed by rinsing with a 50/50 acetone/distilled water (v/v), and then with distilled water. At the end, the materials were dried in a furnace at 40 °C.

TABLE 12.3. Experimental Conditions for the Electrokinetic Remediation Experiments

Experiment	Soil (Spiked With)	Time (days)	Current Intensity (mA)	Central Slice (g)	Total Soil Mass (g)	Mass of Herbicide in Soil (mg)	pH Adjustment	Flow Rate (ml/min)
A	Soil 0 (atrazine)	9	10	No	342	6.5	Catholyte pH ≈ 3 with HNO_3	1.4
B	Soil 0 (atrazine)	9	10	98	342	4.4	Catholyte pH ≈ 3 with HNO_3	1.4
C	Soil 0 (atrazine)	9	10	103	342	4.6	Catholyte pH ≈ 3 with HNO_3	1.4
D	Soil 1 with atrazine aged residues	9	10	No	214	0.033	Catholyte pH ≈ 3 with HNO_3	1.4
E	Soil 2 (molinate)	13	10	102	379	20.1	No	3.0
F	Soil 2 (molinate)	7	10	121	379	50.1	No	2.5
G	Soil 3 (molinate)	10	10	82	299	52.4	Anolyte pH ≈ 8 with 6 M NaOH	5.0
H	Soil 3 (bentazone)	7	10	92	301	25.1	No	5.0
I	Soil 3 (bentazone)	22	20	110	296	14.6	No	5.0
J	Soil 3 (bentazone)	6	20	111	267	15.2	Anolyte pH ≈ 8 with 6 M NaOH	5.0
K	Soil 3 (bentazone)	10	10	78	337	24.1	Anolyte pH ≈ 8 with 6 M NaOH	5.0

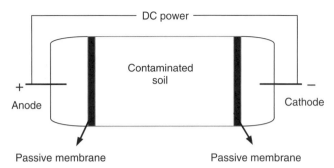

Figure 12.1. Schematic representation of the laboratory cell.

12.3.3 Electrokinetic Laboratory Cell

Eleven electrokinetic experiments were carried out in a laboratory cell schematically presented in Figure 12.1. The cell is divided into three compartments, with an internal diameter of 8 cm, consisting of two electrode compartments ($L = 7.46$ cm) and a central one ($L = 3$ cm), in which the soil, saturated with deionized water according to Rhoades (1982), is placed (Fig. 12.1). A set of five cellulose filters, previously tested and known to work as passive membranes (Ribeiro, 1998), assured the separation between the central compartment and the electrode ones. A power supply (Hewlett Packard E3612A, Palo Alto, CA) maintained at a constant DC and a multimeter (Kiotto KT 1000H, Kiotto) monitored the voltage drop. The electrodes (diameter = 3 mm, $L = 5$ cm) were platinized titanium bars (Bergsøe Anti Corrosion A/S, Herfoelge, Denmark). In all experiments, the electrolyte was a 10^{-2} M $NaNO_3$ solution (with a pH value according to Table 12.3), circulating due to a peristaltic pump (Watson-Marlow 503 U/R, Watson-Marlow Pumps Group, Falmouth, Cornwall, UK) with one head and three extensions.

12.3.4 Electrokinetic Experimental Conditions

The detection and quantification of the initial atrazine in soil 1 and the proof of the nonexistence of atrazine, molinate, and bentazone in soil 0, soil 2, and soil 3 were carried out before all experiments. The polluted soils (spiked and natural) were submitted to the electrokinetic process under different experimental conditions presented in Table 12.3.

The spiking solutions were prepared by dissolving the masses of the herbicide standards, referred to in Table 12.3, in 25 ml of diethyl ether to facilitate the homogeneousness and dispersion of the active principle throughout the soil. The spiking solutions were added to the saturated soil.

Electrolyte samples (catholyte and anolyte) were collected during the experiments for further quantification of the herbicides (further on explained), and the pH and respective volume were registered.

At the end of each experiment, the total soil in the central cell compartment was sectioned in slices, one nearer the anode (AN soil slice), another one nearer the cathode (CAT soil slice), and a central soil slice (central soil slice). Their respective masses were determined, and subsamples were collected to humidity and pH in H_2O (1:2.5) measurements. The rest of the known mass of each "slice" was submitted to extraction three times by sonication with a Bandelin Sonarex Super RK102 H (G H Zeal Ltd, London, UK) for 10 min using 50 ml of fresh methanol in experiments A–F, and using 50 ml of acetone in experiments G and H–K. The soil and passive membrane extract samples were concentrated on a rotavapor Buchi R-205 Rotary Evaporator (Postfach, Switzerland) until 5–10 ml and refrigerated until analysis.

For the determination of the herbicide adsorption to the passive membranes, those were put in 100 ml of methanol (in experiments A–F) or acetone (in experiments G and H–K) and submitted to sonication with a Bandelin Sonarex Super RK102 H for 10 min. The obtained extracts were concentrated to 10 ml and filtered by Acrodisc Gelman (Gelman Sciences, Inc., Arbor, MI) filters (0.45 μm in diameter) before being submitted to analysis by ELISA, in the case of atrazine (experiments A–D), or to gas chromatography with flame ionization detector (GC-FID) in the case of molinate (experiments E and F), or to HPLC with ultraviolet detection (HPLC-UV) in the case of molinate and bentazone (experiments G and H–K).

12.3.5 Herbicide Analysis

Atrazine analysis were performed by ELISA, in soils and electrolyte solutions, according to the procedures stated on the Technical Notes 0003 and A00071 from Strategic Diagnostics, Inc. The absorbance measurements were performed on a spectrophotometer UNICAM Helios α v2.03 (Thermo Scientific Inc., Waltham, MA) at 450 nm. The detection limit achieved for atrazine was 0.05 μg/l (Ribeiro et al., 2005). Confirmation of atrazine detection (positive ELISA) was conducted using a Merck-Hitachi HPLC system (Tokyo, Japan) with a UV detector in 10% of the samples. The analytical separation was performed on a Lichrospher RP-18 (125 mm × 4.6 mm, 5 μm) from Merck. The UV wavelength was setup to 220 nm. The analysis was performed in isocratic mode using water–methanol (65:35), with a flow rate of 1.0 ml/min (Ribeiro et al., 2005). When the atrazine content in an electrolyte sample was below the detection limit of the ELISA method, a SPE step was used to concentrate the samples based on Technical Note T00020 from Strategic Diagnostics, Inc and Supelco application note 59 (Ribeiro et al., 2005). Quantitative analysis of atrazine in soils for each experiment was carried out by extracting twice 5 g of soil with 20 ml of methanol by sonication using a Bandelin Sonarex Super RK102 H for 10 min. Both extracts were collected in conjunction and concentrated to 10 ml under a nitrogen flux. When the atrazine content was smaller than the detection limit of the ELISA method, the extracts were concentrated to 1 ml.

Molinate and bentazone analysis were performed by chromatography. GC was used for molinate determination in experiments E and F electrolyte solutions. HPLC was used for bentazone determination in soils and electrolyte solutions, and for molinate determination in soils and experiment K electrolyte solution.

The GC analysis was performed on a Carbon Erba HRGC 5300 Mega series (Milano, Italy) gas chromatograph equipped with an FID. The capillary column used was a ZB-5 (5% phenyl, 95% dimethylpolysiloxane), 30 m × 0.25 mm in i.d. and 0.25 μm in film thickness, from Phenomenex (Torrance, CA). The sample injection was made on a split-splitless injector, with the slit ratio set to 1.4. The oven temperature was programmed to start at 100 °C for 5 min, after being increased to 120 °C at a rate of 6 °C/min and then increased at 10 °C/min to 290 °C, where it held for 15 min. The temperatures of the injector and detector were 250 and 300 °C, respectively. The carrier gas was hydrogen set at 1.0 ml/min. Data were registered using the Merck-Hitachi integrator D-200 (Tokyo, Japan).

The HPLC analysis was performed on an Agilent 1100 series LC (Santa Clara, CA), equipped with a quaternary pump (G1311A) and a UV multiwave length detector (G1365B), set at 220 and 280 nm. The injector was a Rheodyne manual injector valve, model 7725i (Santa Clara, CA), equipped with a 20-l sample loop. The column temperature was controlled using a ThermaSphere TS-130 oven from Phenomenex set at 40 °C.

Herbicide separations were carried out using a Chromolith Performance RP-18e column with 100 mm × 4.6 mm from Merck, protected using Onyx SecurityGuard cartridges (4 × 3.0 mm) from Phenomenex. All sample analysis were performed with a flow rate of 1 ml/min at gradient mode. The eluents were acetonitrile/water solutions (solution A: 35/65; solution B: 90/10), adjusted to pH = 2.8 using a formic acid solution (Panreac, 98%).

The gradient run was set at 100% A from 0 to 8 min, then to 100% B until 14.5 min, where it held until minute 17. The system reequilibration was performed by changing to 100% A from minute 17 to 18 and remaining with 100% A until minute 20. All operations and data analysis were processed by the Chemstation software v.8.03 (Agilent Technologies, Inc., Santa Clara, CA).

Before analysis, the herbicides were extracted from the electrolyte solutions by means of SPE and by solvent extraction, using sonication, from the soil and passive membranes. The pH values of the collected electrolyte solution was between 4 and 7 for experiments E and F, and below 4 for experiment G. For experiments H–K, the collected electrolyte solutions were adjusted to pH 2 by adding NaOH or HNO_3.

For experiments E, F, and G, the SPE has been performed using Strata C18-T cartridges (Phenomenex). The SPE sorbent amount has been selected according to the sample volume of 500 mg for volumes lower than 300 ml, and 1000 mg when the sample volume was higher than 300 ml. For experiments H–K, the SPE has been performed using 200-mg Strata-X cartridges (Phenomenex). All analytes were eluted after extraction with methylene chloride.

After SPE extraction, the extracts were concentrated under a gentle stream of nitrogen, until 1.0 ml in experiments E and F. In experiments G and H–K, the extracts were evaporated to dryness and resuspended to 2.0 ml with methanol/distilled water (1:1).

After concentration, the samples for GC analysis were transferred to a 2.0-ml vial and kept at −20 °C until analysis. The samples for HPLC analysis, in methanol/water, were kept at 5 °C and analyzed in 24 h.

Molinate and bentazone analysis were performed in soil, passive membranes, and electrolyte sample solutions according to the method developed and validated by Buchholz (2007).

12.3.6 Results and Discussion

The experiments A–H and K, all carried out at 10 mA, presented similar values of voltage and current (results not shown). In the experiments with no pH control in the electrolytes, the catholytes presented predominantly an alkaline pH, and the anolytes an presented acid pH that varied between 3 and 4. The accumulated volumes of the catholyte and anolyte of the experiments presented close values because the electroosmotic flow was low, but the anolyte cumulative volumes were higher than the catholyte ones, particularly in experiments A–D, in which the catholyte pH was adjusted to approximately 3. Experiment B was the one that registered a higher difference between the anolyte and catholyte volumes, and clearly indicates an electroosmotic flow toward the anode (results not shown) at an acidic soil pH.

Figure 12.2 illustrates the cumulative mass of herbicide removed in some experiments: (a) atrazine in experiment C; (b) molinate in experiment E; and (c) bentazone in experiment I. Figure 12.3 presents the quantities of herbicide remaining at the end of all the experiments: in the passive membranes, nearer the cathode (CAT membrane) and nearer the anode (AN membrane); in the soil, in the soil slice near to the cathode (CAT soil slice), in the central soil slide and in the soil slice nearer to the anode (AN soil slice); as well as an estimate of the removed herbicide percentages obtained.

The atrazine quantities removed by the electrokinetic process for the cathode and the anode compartments of the cell showed that in experiments A–D, the highest amounts are in the anolyte, and that 30%–50% of its initial amount is removed from the soils within the first 24 h. In experiment A, the quantity of atrazine in the catholyte is quite significant in relation to the ones in experiments B and C (results not shown). This is due to the fact that in experiment A, all the soil in contact with the membranes was spiked, allowing atrazine to pass into the cathode compartment by diffusion. At the end of the experiments, higher quantities of atrazine were obtained in the central soil slice and in the soil slice nearer to the anode, in experiments B and C (Fig. 12.3), corroborating the results previously obtained in the electrolytes, confirming that under the action of an electric field, atrazine is mainly mobilized toward the anode compartment due to reverse electroosmosis. In experiment A, the quantities of atrazine in the soil slice nearer the cathode and nearer the anode are similar (Fig. 12.3), probably due to the previously pointed out passage of atrazine toward the cathode due to diffusion. In the electrokinetic experiments carried out with atrazine, the removal efficiencies in the soil solution where high, ≥89% (Fig. 12.3). No atrazine was found in the passive membranes.

Experiment G was the one that registered the highest difference between the catholyte and anolyte volumes (results not shown), giving a strong indication of an electroosmotic flow toward the cathode. In fact, there seems to be a relation between the electroosmotic flows observed in the three experiments carried out with molinate (experiments E–G) and the range of pH values obtained at their anolytes. It seems that the more acid pH the anolyte presented, the bigger the tendency to obtain a reversed electroosmosis. This is in accordance with the conditions that determine the reversion of the electroosmotic flow, namely the acidification of the soil provoked by the propagation of an acid front due to the generation of H^+ at the anode reaction. It seems that, independent of the initial amount of molinate and of the flow rate of the electrolyte solution, most of the molinate (>80%) found in the electrolytes is removed in the first 4 days. Molinate seems to be mobilized preferentially to the cathode compartment (e.g. experiment E in Fig. 12.2b). The pH control of the incoming anolyte solution of experiment G (according to Table 12.3) may have contributed to the mobilization of molinate toward the cathode compartment as it counteracted the reversion of the electroosmosis, and therefore impeding

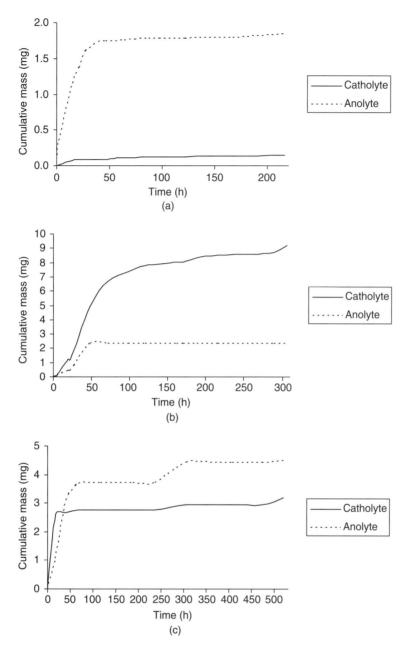

Figure 12.2. Cumulative mass of herbicide removed for the electrolytes: (a) atrazine in experiment C; (b) molinate in experiment E; and (c) bentazone in experiment I.

the acidification of soil, according to Acar *et al.* (1996), Yeung, Hsu, and Menon (1996), and Saichek and Reddy (2003). After the electrokinetic experiments were completed, the mass balance conducted for the molinate in the system, compared with the molinate spiked in the soil, presented very high removal efficiencies, greater than 90% (Fig. 12.3). Most of the molinate was found in the soil (central

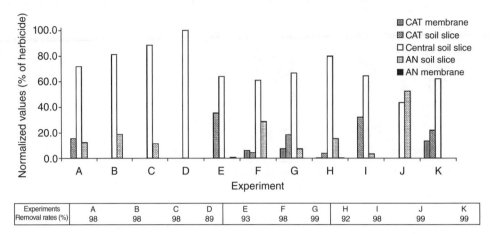

Figure 12.3. Quantities of herbicide remaining at the end of all the experiments: in the passive membranes, nearer the cathode (CAT membrane) and nearer the anode (AN membrane); in the soil, in the soil slice near to the cathode (CAT soil slice), in the central soil slide and in the soil slice nearer to the anode (AN soil slice); as well as an estimate of the removed herbicide percentages obtained.

soil slide) and in the soil slice nearer to cathode (CAT soil slice, in Figure 12.3, for experiment E). These results corroborate the direction of the movement of molinate obtained toward the electrolyte solutions.

Bentazone was mobilized from the soil when submitted to the action of an electric field, toward both electrode compartments, but mostly toward the anolyte (e.g. Fig. 12.2c). The accumulated volumes of the catholyte and anolyte of experiments H–K presented slighter higher values in the catholytes because the electroosmotic flow was toward that compartment (results not shown). The electrokinetic process was able to remove bentazone in the soil solution, reaching a balance after 70 h of the process. At the end of the experiments, more than 90% of the bentazone initial quantities were removed (Fig. 12.3). For the bentazone mobilized toward the cathode compartment, it takes longer to reach the balance (approximately 150 h). It seems that an application of a lower current intensity (10 mA) is related to a higher mobilization of this herbicide toward the cathode compartment, whereas higher current intensities (20 mA) promoted a greater rate of bentazone mobilization also toward the anolyte (Fig. 12.3, for instance, for experiment J, where the soil slice nearer to anode—AN soil slice—presented the highest bentazone concentration; or Fig. 12.2c, for experiment I).

12.3.7 Conclusions from the Case Study

Rice culture contributes to environmental contamination with pesticides, leading to soil, surface, and groundwater quality degradation, which may have negative impacts on public health and biodiversity. Atrazine, molinate, and bentazone are examples of herbicides of concern. The electrokinetic process is able to remove these herbicides from soils, since they are all mobilized in the soil solution, under the action of an electric field. The case study presented showed the following:

- Atrazine was mainly mobilized toward the anode compartment due to reverse electroosmosis, and about 30%–50% of its initial quantity was removed during the first 24 h. This behavior occurred in all experiments (spiked and natural contaminated soils).
- Molinate removal efficiencies were high, above 90%. It seems that, independent of the initial amount of molinate and of the flow rate of the electrolyte solution, most of the molinate (>80%) found in the electrolytes is removed in the first 4 days. Molinate seems to mobilize preferentially to the cathode compartment. The pH control basification at the anolyte may have contributed to the mobilization of molinate toward the cathode compartment since it should counteract the reverse electroosmosis.
- Bentazone was mobilized from the soil when submitted to the action of an electric field, toward both electrode compartments, but mostly toward the anolyte. At the end of the experiments, more than 90% of the bentazone initial quantities were removed. It seems that an application of a lower current intensity (10 mA) is related to a higher mobilization of this herbicide toward the cathode compartment, whereas higher current intensities (20 mA) promoted greater rate of bentazone mobilization also toward the anolyte.

12.4 SUMMARY

The atrazine, molinate, and bentazone behavior in soils when submitted to an electric field is presented as a case study. The applicability of the electrokinetic process in these herbicides' soil remediation is evaluated. Four polluted soils were used, respectively with and without herbicides residues, with the last ones being spiked. Eleven electrokinetic experiments were carried out at a laboratory scale. Determination of the herbicide residues were performed by different methods: by ELISA for atrazine, by GC-FID for molinate, and by HPLC-UV for molinate and bentazone residues. GC hyphenated with mass spectrometry was used to confirm and identify molinate on the samples.

The results show that the electrokinetic process is able to mobilize atrazine, molinate, and bentazone from soils, and efficiently remove them from the soil solution. Atrazine is mainly carried out toward the anode compartment, and estimations show that 30%–50% of its initial amount is removed from the soil within the first 24 h. Molinate seems to be mobilized preferentially to the cathode compartment, with removal efficiencies higher than 90%. Bentazone also shows a high removal from soil, reaching efficiencies above 92%. It is mainly carried out toward the anode compartment.

ACKNOWLEDGMENTS

The authors thank the Environment Quality Group of the Departamento de Ciências e Engenharia do Ambiente (DCEA)/Faculdade de Ciências e Tecnologia (FCT)/Universidade Nova de Lisboa (UNL) for allowing us to use the UV spectrophotometer for the immunoassays readings and Professor Marco Gomes da Silva from REQUIMTE of the Departamento de Química (DQ)/FCT/UNL for sharing the gas chromatograph for the molinate determination; and Professor Damià Barceló of the Environmental Chemistry Department, CID-CSIC, Barcelona, Spain, for supplying the contaminated soil with atrazine, and Professor Olga Nunes of the Chemistry Department, FEUP, Portugal, for supplying soils S2 and S3. Drs. Helena Gomes, Joana Serpa Santos, Carla Abreu, and Christoph Buchholz are thanked for all their experimental and analytical work carried out with atrazine, molinate, and bentazone. This work was partly funded by the Portuguese Project PPCDT/AMB/59836/2004, approved by the FCT.

REFERENCES

Abreu CSV. (2008). Electro-remediação de solos contaminados com pesticidas: caso da bentazona. MSc Dissertation, Faculdade de Ciências e Tecnologia, Universidade Nova de Lisboa (in Portuguese). Lisboa, Portugal.

Acar YB, Gale R, Alshawabkeh A, Marks R, Puppala S, Bricka M, Parker R. (1995). Electrokinetic remediation: Basics and technology status. *Journal of Hazardous Materials* **40**:117–137.

Acar YB, Ozsu E, Alshawabkeh AN, Rabbi FM, Gale R. (1996). Enhance soil bioremediation with electric fields. *Chemtech* **26**(4):40–44.

Aguilar C, Ferrer I, Borrull F, Marcé RM, Barceló D. (1999). Monitoring of pesticides in river water based on samples previously stored in polymeric cartridges followed by on-line solid phase extraction-liquid chromatography-diode array detection and confirmation by atmospheric pressure chemical ionization mass spectrometry. *Analytica Chimica Acta* **386**:237–248.

APVMA. (2003). The reconsideration of approvals and registrations relating to molinate. Review Scope Document, Australian Pesticides and Veterinary Medicines Authority. http://www.apvma.gov.au/chemrev/downloads/molinate_scope.pdf (accessed January 8, 2008).

Barreiros L, Nogales B, Manaia CM, Ferrerira ACS, Pieper DH, Reis MA, Nunes OC. (2003). A novel pathway for mineralization of the thiocarbamate herbicide molinate by defined bacterial mixed culture. *Environmental Microbiology* **5**(10):944–953.

Blum WEH. (2005). Function of soil for society and environment. *Reviews in Environmental and Science Biotechnology* **4**:75–79.

Boivin A, Cherrier R, Perrin-Ganier C, Schiavon M. (2004). Time effect of bentazone sorption and degradation in soil. *Pesticide Management Science* **60**:809–814.

Boivin A, Cherrier R, Schiavon M. (2005). A comparison of five pesticides adsorption and desorption processes in thirteen contrasting field soils. *Chemosphere* **61**:668–676.

Buchholz C. (2007). Chromatographic Method and Sample Preparation for the Analysis of Pesticides. Diploma Thesis, Hochschule Magdeburg-Stendal Institut Chemie/Pharmatechnik Vertiefungsrichtung Analytische Chemie, Faculdade de Ciências e Tecnologia da Universidade Nova de Lisboa. Lisboa, Portugal.

Carabias-Martínez R, Gonzalo ER, Hernandez EH, Roman FJS, Flores MGP. (2002). Determination of herbicides and metabolites by solid-phase extraction and liquid chromatography: Evaluation of pollution due to herbicides in surface and groundwaters. *Journal of Chromatography A* **950**:157–166.

Castro M, Silva-Ferreira AC, Manaia CM, Nunes O. (2005). A case study of molinate application in a Portuguese rice field: Herbicide dissipation and proposal of a clean-up methodology. *Chemosphere* **59**:1059–1065.

Chang TT. (2000). Rice. In *The Cambridge World History of Food*, Chap. II.A.7 (eds. KF Kiple, KC Ornelas). Cambridge University Press. http://www.cambridge.org/us/books/kiple/rice.htm (accessed January 9, 2008).

Cochran RC, Formoli TA, Pfeifer KF, Aldous CN. (1997). Characterization of risks associated with the use of molinate. *Regulatory Toxicology and Pharmacology* **25**:146–157.

Commission of the European Communities. (2002) COM(2002)179. Communication from the Commission to the Council, the European Parliament, the Economic and Social Committee and the Committee of the Regions: Towards a Thematic Strategy for Soil Protection.

Commission of the European Communities. (2006) COM(2006)232. Proposal for a Directive of the European Parliament and of the Council Establishing a Framework for the Protection of Soil and Amending Directive 2004/35/EC.

Correia P, Boaventura RAR, Reis MAM, Nunes OC. (2006). Effect of operating parameters on molinate biodegradation. *Water Research* **40**:331–340.

Durand G, Forteza R, Barceló D. (1989). Determination of chlorotriazine herbicides, their dealkylated degradation products and organophosphorus pesticides in soil samples by means of two different clean up procedures. *Chromatographia* **28**:597–604.

EXTOXNET. (1996). Pesticide information profile: Molinate. Extension Toxicology Network. http://extoxnet.orst.edu/pips/molinate.htm. Accessed 2007 Oct 10.

Fontanals N, Marcé RM, Borrull F. (2004). Solid-phase extraction of polar compounds with a hydrophilic copolymeric sorbent. *Journal of Chromatography A* **1030**:63–68.

FOOTPRINT Pesticide Properties Database. (2007) Molinate properties. http://sitem.herts.ac.uk/aeru/footprint/ (accessed March 7, 2007).

Gascón J, Oubina A, Barceló D. (1997). Detection of endocrine-disrupting pesticides by enzyme-linked immuno-sorbent assay (ELISA): Application to atrazine. *Trends in Analytical Chemistry* **16**:554–562.

Ottosen LM. (1995). Electrokinetic Remediation. Application to Soils Polluted from Wood Preservation. PhD Dissertation, Technical University of Denmark, Lyngby, Denmark.

Page MM, Page CL. (2002). Electroremediation in contaminated soils. *Journal of Environmental Engineering* **128**:208–219.

Pamukcu S, Wittle JK. (1992). Electrokinetic removal of selected heavy metals from soil. *Environmental Progress* **11**:241–249.

Poll JM, Vink M. (1996). Gas chromatographic determination of acid herbicides in surface water samples with electron-capture detection and mass spectrometric confirmation. *Journal of Chromatography A* **733**:361–366.

Renner R. (2000). A new route for endocrine disrupters. *Environmental Science and Technology* **34**:415A–416A.

Renner R. (2002). Atrazine linked to endocrine disruption in frogs. *Environmental Science and Technology* **36**:55A–56A.

Rhoades JD. (1982). Soluble salts. In *Methods of Soil Analysis, Part 2, Chemical and Microbiological Properties* (ed. AL Page) Agronomy Monograph No. 9, 2nd ed. Madison, WI: American Society of Agronomy Publisher, pp. 167–179.

Ribeiro AB. (1992). Contribuição para o estudo da contaminação de solos por metais pesados—Caso do Cobre num Regossolo Psamítico. MSc Dissertation, Faculdade de Ciências e Tecnologia da Universidade Nova de Lisboa (in Portuguese). Lisboa, Portugal.

Ribeiro AB. (1998). Use of Electrodialytic Remediation Technique for Removal of Selected Heavy Metals and Metalloids from Soils. PhD Dissertation, Technical University of Denmark, Denmark. Lisboa, Portugal.

Ribeiro AB, Rodríguez-Maroto JM. (2006). Electroremediation of heavy metal-contaminated soils. Processes and applications. Cap. 18 In *Trace Elements in the Environment: Biogeochemistry, Biotechnology and Bioremediation* (eds. MNV Prasad, KS Sajwan, R Naidu). Boca Raton, FL: Taylor & Francis, CRC Press, pp. 341–368.

Ribeiro AB, Rodríguez-Maroto JM, Mateus EP, Gomes H. (2005). Removal of organic contaminants from soils by an electrokinetic process: The case of atrazine. Experimental and modelling. *Chemosphere* **59**:1229–1239.

Saichek RE, Reddy KR. (2003). Effect of pH control at the anode for the electrokinetic removal of phenanthrene from kaolin soil. *Chemosphere* **51**(4):273–287.

Santos JS. (2008). Electrokinetic Remediation of Rice Field Soils Contaminated by Molinate. MSc Dissertation, Faculdade de Ciências e Tecnologia, Universidade Nova de Lisboa. Lisboa, Portugal.

Santos TCR, Rocha JC, Barceló D. (2000). Determination of rice herbicides, their transformation products and clofibric acid using on-line solid-phase extraction followed by liquid chromatography with diode array and atmospheric pressure chemical ionization mass spectrometric detection. *Journal of Chromatography A* **879**:3–12.

Silva E, Batista S, Viana P, Antunes P, Serôdio L, Cardoso AT, Cerejeira MJ. (2006). Pesticides and nitrates in groundwater from oriziculture areas of the 'Baixo Sado' region (Portugal). *International Journal Environmental Analytical Chemistry* **86**:955–972.

Technical Note A00071. *User Guide—RaPID Assay Atrazine.* Strategic Diagnostics.

Technical Note T00003. *Detection of Atrazine, Alachlor, Cyanazine and Metalochlor in Soil.* Strategic Diagnostics.

Technical Note T00020. *C18 Extraction of Atrazine for Increased Sensitivity in the RaPID Assay.* Strategic Diagnostics.

Thorstensen C, Lode O, Eklo OM, Christiansen A. (2001). Sorption of bentazone, dichlorprop, MCPA, and propiconazole in reference soils from Norway. *Journal of Environmental Quality* **30**:2046–2052.

Virkutyte J, Sillanpaa M, Latostenmaa P. (2002). Electrokinetic soil remediation critical overview. *Science of the Total Environment* **289**:97–121.

WHO. (2003). Molinate in drinking-water—Background document for development of guidelines for drinking-water quality. World Health Organization, Geneva. http://www.who.int/water_sanitation_health/dwq/chemicals/molinate.pdf (accessed October 16, 2007).

Yeung AT, Datla S. (1995). Fundamental formulation of electrokinetic extraction of contaminants from soil. *Canadian Geotechnical Journal* **32**:569–583.

Yeung AT, Hsu C, Menon RM. (1996). EDTA-enhanced electrokinetic extraction of lead. *Journal of Geotechnical Engineering* **122**(8):666–673.

13

ELECTROKINETIC REMOVAL OF ENERGETIC COMPOUNDS

DAVID A. KESSLER, CHARLES P. MARSH AND SEAN MOREFIELD

13.1 INTRODUCTION

In keeping with the theme of the previous four chapters, we discuss the state of the art and ongoing research in the area of electrokinetically enhanced remediation of soils contaminated with a particular class of dangerous organic contaminants: nitroaromatic and other energetic compounds. An increasing amount of attention and regulation has been placed in recent years on the environmental impact of the US military (and those of other nations) both in theaters of operation and long-term permanent installations. In both cases, the potential exists for significant contamination of soil and groundwater with various types of energetic compounds found in typical munitions and formed as by-products during the manufacturing of munitions, such as 2,4,6-trinitrotoluene (2,4,6-TNT), 2,4-dinitrotoluene (2,4-DNT), octahydro-1,3,5,7-tetranitro-1,3,5,7-tetrazocane (HMX), cyclotrimethylenetrinitramine (RDX), and others. The health risks associated with these types of compounds are severe (Rickert, 1985). TNT, for instance, has been shown to be toxic and mutagenic to humans (Kaplan and Kaplan, 1982). It is of great importance then to examine methods of removing these chemicals from our environment.

The presence of contaminants in predominantly clay soils poses a difficult and potentially costly remediation problem. The low hydraulic conductivity of these soils prohibits the use of traditional pump-and-treat remediation techniques. Other treatment methods involving excavation and employing various thermal treatments are costly, energy intensive, and could themselves create other adverse environmental impacts. Incomplete remediation of contaminated clay could lead to posttreatment seepage and recontamination of the remainder of the site. For these reasons, much research has been done to advance the field of electrokinetic (EK)

Electrochemical Remediation Technologies for Polluted Soils, Sediments and Groundwater,
Edited by Krishna R. Reddy and Claudio Cameselle
Copyright © 2009 John Wiley & Sons, Inc.

remediation as a viable cleanup technique, the physics and historical development of which is discussed extensively in the first three chapters of this book. It is not our intent to revisit that information in the present chapter, and accordingly, we only briefly describe the EK process as necessary to provide context for the central topics of this discussion: namely, the chemistry of the clay–energetic compound complex, the use of chemical additives to increase the solubility and bioavailability of the contaminants, and the removal of these solubilized compounds from or the introduction of biological degradation agents to the reaction zone using electrokinetics.

13.2 CHEMISTRY OF CLAY–ENERGETIC COMPOUND COMPLEXES

Soil is a thoroughly complex and varied mixture of organic compounds, inorganic components, and numerous interrelated physical, chemical, and biological interactions that in turn affect the soil's structure. At least three generally accepted soil classification systems exist, but they all share a common feature of characterization via particle size. Neglecting organic content for the moment, all soils can be described by the percentage of dry soil mass that falls into the three particle size categories of sand, silt, and clay. Clays have effective particle diameters of 2.0 μm or less. Sands are the largest particles, with size up to 2000 μm, and particles above this size are considered gravel. The intermediate silt particle size, located between clays and sands, is defined as being between 2.0 and 20–60 μm (Marshall, 1947; Hodgson, 1974). The smallest particles, clays (sometimes called the "active fraction"), have a profound effect on phenomena such as cation exchange, plasticity, and swelling that in turn affects the physical behavior and hydrology of soil.

Clays typically consist of layered crystalline structures in which sheets of octahedrally coordinated oxygen atoms centered on aluminum atoms are sequentially stacked, or are fused with tetrahedrally coordinated layers based on silicon atoms. The varied and differing sequences of layering give rise to specific types of clay. These materials are typically classified as 1:1 (a layer of tetrahedral coordination around Si^{4+} bonded to an octahedrally coordinated layer around Al^{3+} or Mg^{2+}) and 2:1 (a basic layer unit of two tetrahedrally coordinated sheets forming a sandwich with an octahedrally coordinated sheet). The 1:1 layer silicates (e.g. kaolinite, halloysite, dickite, nacrite) are bonded to the next layer at a fixed basal spacing distance of 0.72 nm by hydrogen bonding and therefore display no swelling behavior. The 2:1 layer silicates (e.g. montmorillonite, illite, pyrophyllite, talc, mica, vermiculite) can have fixed basal spacing (e.g. 1 nm for illite via nonexchangeable K^+) or can expand, often to a great degree. Montmorillonite, for example, begins with a basal spacing of about 1.4 nm but can swell to many tens of nanometers into an expanded tactoid structure.

The surface chemistry of these clay minerals has a large and often predominant effect on the chemical and binding processes that occurs in soils. Most of the clay particles have a net negatively charged surface. This arises from isomorphous substitutions of less highly charged positive ions within the clay lattice (i.e. Al^{3+} for Si^{4+}, or Mg^{2+} for Al^{3+}). This negative surface charge is balanced by an excess of cations in the soil solution immediately adjacent to the clay particle surface. These readily exchangeable cations are the source of the term "cation exchange capacity" (CEC).

This characteristic varies by clay type and depends only on surface area and surface charge density.

The active surfaces of the clay layers allow the adsorption of various contaminants, including metals and organic molecules. Nitroaromatic compounds (NACs) display a great deal of variability in the ability to form a complex with the clay surface and the resulting strength of the interaction. Haderlein and Schwarzenbach (1993) studied the adsorption of a wide range of substituted nitrobenzenes and nitrophenols to homoionic kaolinites and found that the strength of interaction depended on the type of cation that is adsorbed to the siloxane surface of the clay mineral. When cations with strong hydration shells such as Li^+, Na^+, Mg^{2+}, Ca^{2+}, and Al^{3+} were present, no specific adsorption of the NACs was observed. For more weakly hydrated cations, the adsorption coefficient was found to increase with decreasing free energy of hydration of the particular cation (Haderlein and Schwarzenbach, 1993). The type of substituted molecular group also affected the adsorption behavior. In particular, the substituents with strong electron-withdrawing and electron-delocalizing properties strongly enhanced sorption of the nitrobenzene molecule. Haderlein and Schwarzenbach (1993) attributed this process to the formation of electron donor–acceptor (EDA) complexes between the nitroaromatic rings of the contaminant molecules and the siloxane surface. They argued that these EDA complexes would be able to form near the ditrigonal cavities where Al^{3+} ions have displaced the Si^{4+} ions, leaving a net negative charge in the cavity. For weakly hydrated exchangeable cations, there should be sufficient space for a complex to form between the electron-donating oxygen atoms at the edge of the cavity and the electron-accepting π-system of the NACs.

Later work showed similar adsorption–desorption behavior for NACs in clay types other than kaolinite. The adsorption constant was found to increase fairly uniformly for all tested NACs in the ratio of 1:6:12 (kaolinite : illite : montmorillonite). This ratio is quite close to the relative surface areas of the three minerals; illite and montmorillonite have 6 and 16 times more available surface area, respectively, than kaolinite (Haderlein, Weissmahr, and Schwarzenbach, 1996). This observation suggests that a similar EDA complexation mechanism controls the adsorption of NACs to these minerals as well. The authors also showed that TNT, 2,4-DNT, and various other NACs adsorbed strongly to all clay surfaces tested, and that the adsorption was completely reversible. Daun *et al.* (1998) performed similar adsorption experiments on TNT and its metabolites to a montmorillonitic clay. They confirmed that the adsorption of TNT and two of its metabolites, 2-amino-4,6-dinitrotoluene (2-ADNT) and 4-amino-2,6-dinitrotoluene (4-ADNT), was completely reversible, but the adsorption of a third degradation product, 2,4-diamino-6-nitrotoluene (2,4-DANT), was only partially reversible, and that of 2,4,6-triaminotoluene (TAT) was completely irreversible. It was hypothesized that hydrophobic interactions were the dominating sorption mechanism for these species. Sheremata *et al.* (1999) also found a significant hysteresis in the adsorption and desorption isotherms for 2,4-DANT in illite and natural topsoil. A slight hysteresis was observed for TNT and 4-ADNT as well, but to a much lesser degree. The adsorption coefficient of an NAC increases with amino group substitution (Daun *et al.*, 1998; Sheremata *et al.*, 1999), and it has been proposed that the amino group in the para position of 2,4-DANT undergoes irreversible reactions with soil organic matter in topsoil (Sheremata *et al.*, 1999). In an illite clay (devoid of organic matter), it is likely that EDA complexation is the controlling mechanism for adsorption, and the observed irreversibility of the process could be due to free radical formation at the para position with subsequent dimer formation (Sheremata et al., 1999).

Organic matter also plays a large role in ion exchange/binding and soil particle aggregation; especially in surface soils. This organic matter consists of plant and animal debris that is in the process of being altered via soil animals, fungi, or bacteria. At any one instant, the entire spectrum of decay products is present, including the end product of humus. These humic materials help bind soil and improve aggregation properties. This effect in turn will impact soil pore structure and stability. Surface soils typically have something less than 10 wt% of organic matter, while subsoils contain 1 wt% or less. The soil organic matter is also critical in interactions with molecular organic contaminants. Organic pollutants covalently bind to the soil organic matter, or they are dissolved in the humic organic phase. Pollutants in these environments have limited accessibility and are exceedingly difficult to remove or degrade without destroying the soil (e.g. through combustion) (Kopinke and Remmler, 1995).

Irreversible sorption of TNT and 2,4-DNT to a natural sandy soil rich in organic matter has also been reported (Yamamoto et al., 2004). Eriksson and Skyllberg (2001) showed that TNT and its degradation products were 6.4–22 times more likely to bind with particulate organic matter than dissolved organic matter but that the binding strength was greater for dissolved organic matter. They concluded that the main source of binding in TNT and soil organic matter systems was its degradation products, and that the interaction with particulate matter was mediated through site-specific (ionic) interactions and that the interaction with dissolved organic matter was nonspecific and possibly hydrophobic in nature (Eriksson and Skyllberg, 2001). Nuclear magnetic resonance spectroscopy studies showed that the binding of TNT to a natural soil closely resembled that of a K^+-montmorillonite clay despite low levels of TNT in the soil sample (Emery et al., 2001).

13.3 REMEDIATION STRATEGIES

13.3.1 Direct Extraction via Solubilization

Explosive compounds are low-polarity organic molecules that exhibit low water solubility and strong affinities for complex formation with clay minerals and organic matter, as discussed in the previous section. Even in "young," laboratory-contaminated soils, it is difficult to extract these molecules. Aged soils can retain these energetic compounds even more readily. Accordingly, soil washing with water alone is an ineffective treatment. The contaminants must first be desorbed from the clay surfaces and then mobilized in solution for any significant transport and subsequent removal to occur. We have already seen that the type of cation present at the surface of the clay particles plays a significant role in the adsorption–desorption behavior of organic molecules. Hence, the introduction of more strongly hydrated ions into the contaminated region could serve to increase the level of purged contaminants up to the solubility limit in the background solution. Cosolvents and surfactants are other potential solubility-enhancing agents. A vast body of literature is available concerning the solubilization of various organic compounds using surfactants, and reviewing all of these is beyond the scope of this chapter. Instead, we point the reader to the review of Saichek and Reddy (2005, pp. 170–175) and the references therein for a thorough discussion of the topic, and choose to focus on a particular study that elucidates the differences between using anionic and cationic surfactants. Cicek and Govind (1997) used electrokinetics in conjunction with a cationic surfactant [cetyltrimethylammonium chloride (CTAC)], an anionic surfactant [sodium dodecyl sulfate (SDS)], and a nonionic surfactant (Triton X-100). They found the cationic surfactant to remove the greatest amount of phenol from kaolin and silica samples, followed by the nonionic and the anionic surfactant, respectively. Other researchers have found cationic surfactants to be unfavorable, however, since the positively charged micelles are also attracted to the negatively charged clay surfaces and could lead to lower contaminant removal (Taha, Gale, and Zappi, 1994; Taha, 1996; Qian, 1998). Taha *et al.* (1997) showed that the use of anionic surfactants (SDS and DOWFAX 8390) enhanced desorption of 2,4,6-TNT from a natural surface soil taken from a contaminated field site. Desorption using cationic surfactants was measured to be less than using water alone (Taha *et al.*, 1997). Surfactants and cosolvents are not environmentally benign, and care must be taken to ensure that all of the chemicals are removed from the treated site and that the effluent is properly collected. In response to these concerns, the use of cyclodextrin (CD) molecules as solubility-enhancing agents has become popular in the literature. These nonhazardous molecules, which will be described in more detail below, have a hydrophobic inner cavity in which complexes with various organic molecules can form. They are water soluble and have shown no propensity to adsorb to clay particles, making them excellent candidates for the remediation of contaminated soils.

13.3.2 CD Chemistry

CD was first discovered in 1891 by Villiers by digesting starch with *Bacillus amylobacter* and *Bacillus macerans* spores (Villiers, 1891; Szejtli, 1998). The structures of CDs can be described as crystalline, homogeneous, nonhygroscopic substances forming toruslike rings built up from glucopyranose units. The three major CDs (α-CD, β-CD, and γ-CD) consist of six, seven, and eight glucopyranose units, respectively. The CD ring is a conical cylinder with secondary hydroxyl groups situated on one edge of the ring and primary hydroxyls situated on the other edge. The cavity is lined by hydrogen atoms and glycosidic oxygen bridges. Thus, CDs exhibit hydrophilic nature on the outside of the rings and hydrophobic characteristics inside the cavity. The cavities are approximately 8 Å in depth and range from 5 to 10 Å in diameter depending on the specific CD and can therefore accommodate different-sized organic "guest" molecules. The hydrophilic exterior provides water solubility for the CD, while the hydrophobic interior provides an energetically favorable site for the adsorption of contaminants to create a water-soluble inclusion complex. Hydrogen bonding, electrostatic forces, and van der Waals forces all serve to attract and hold organic molecules in the CD cavity (Murakami *et al.*, 1996).

Another attractive feature of CD molecules is the availability of numerous hydroxyl groups for chemical modification. The total number of OH groups on each CD molecule is considerable and varies with the size of the specific CD (α-CD = 18, β-CD = 21, γ-CD = 24); thus, the number of possible derivatives is nearly limitless (Szejtli, 1998). Examples of conversion of the OH group to the following functional groups have been reported: iodide, azide, thioacetate, hydroxylamine, alkyl, polyalkylamines (Murakami et al., 1996), monoiodo (Ueno et al., 1987), azido (Melton and Slessor, 1971; Tsujihara, Kurita, and Kawazu, 1977), thio (Griffiths and Bender, 1973; Siegel, 1979), hydroxylamino (Fikes et al., 1992), and alkylamino (Tabushi and Shimizu, 1979; Petter et al., 1990). The chemistry of modification is well characterized, including selective modification at specific sites on the molecule, and the desired CD is limited only by cost and imagination. Both anionic CD and cationic CD have been synthesized (Croft and Bartsch, 1983). In addition, modified CDs exhibit enhanced complexing properties and solubility compared with the unmodified forms.

Current literature has demonstrated the increased interest in the use of CD for remediation of contaminated soils. CD can remove organic contaminants from association with the solid phase by enhancing the solubility of the contaminant in water (Wang and Brusseau, 1993; Brusseau, Wang, and Hu, 1994; Wang and Brusseau, 1995a,b; Hawari et al., 1996; McCray and Brusseau, 1998; McCray et al., 1999; Sheremata and Hawari, 2000). In addition, there has been no observation of sorption, retardation, or pore exclusion of CD during interactions with soils or clays (Brusseau, Wang, and Hu, 1994). Enhanced solubility, and thus effective removal, has been shown for such diverse organic compounds as anthracene, pyrene, chlorobiphenyls, benzene, toluene, ethylbenzene, and xylenes (BTEX) compounds, decane, and dichlorodiphenyltrichloroethene (DDT). Other work discusses the utility of CDs for removal of nonaqueous-phase organic liquids (NAPL) (Crini et al., 1998, 1999; McCray and Brusseau, 1998, 1999; Brusseau et al., 1999; McCray, Boving, and Brusseau, 2000), chlorinated solvents (Bizzigotti, Reynolds, and Kueper, 1997; Fava, De Gioia, and Marchetti, 1998; Boving and Brusseau, 2000), hydrocarbons (Wang and Brusseau, 1995a; Brusseau, McCray, and Wang, 1996; Boving, Wang, and Brusseau, 1998; Gao et al., 1998; Ko, Schlautman, and Carraway, 1999; Bardi et al., 2000; Jiradecha and Kasetsart, 2000; Reid et al., 2000), and pesticides (Dailey, Dowler, and Glaze, 1990; Fenyvesi, Szeman, and Szejtli, 1996) from soil and groundwater. More interesting, CD can increase solubilization and removal of explosive compounds such as TNT (Sheremata and Hawari, 2000) and RDX (Hawari et al., 1996) from soils in batch studies. The implication is that for an aged TNT-contaminated soil, where an appreciable fraction of the TNT is strongly bound to the organic matter, extraction of this fraction is enhanced by the use of CD. Due to the relative sizes of the explosive molecule and the interior volume of the cavity, TNT is best adsorbed by β-CD, while RDX fits better into the γ-CD (Cahill and Bulusu, 1993). Adsorption of toxic heavy metals (e.g. cadmium) has been noted as an additional benefit of CDs (Wang and Brusseau, 1995b). Complexation of cadmium takes place on the exterior of the CD and does not interfere with the formation of the inclusion complex. Therefore, interaction of CD with the soil matrix may provide a uniquely beneficial substrate for the removal of both organic compounds and heavy metal species. The energetic TNT molecule removed from the soil will concentrate in the effluent solution as TNT–CD inclusion complexes. An additional benefit of CDs is their ability to stabilize the energetic molecule so dangers of detonation are eliminated, as demonstrated for nitroglycerin (Stadler-Szoke and Szejtli, 1979).

The electrochemical reduction and oxidation of CDs has not yet been fully investigated. A preliminary study by Kessler et al. (2004) has shown the potential for various types of CD molecules to degrade in the presence of an electric field.

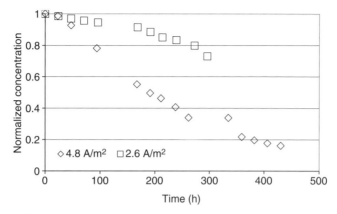

Figure 13.1. Fraction of CD remaining over time in batch degradation experiments for β-CD under two different current densities.

In those experiments, solutions of unmodified α- and β-CDs and a functionally modified amino-β-CD with initial concentrations of 10,000 ppm were subjected to a variable electric field that maintained effective current densities of either 2.6 or 4.8 A/m². The ratios of the CD concentrations (determined using high-performance liquid chromatography) to the initial concentrations are plotted as a function of time for degradation tests of β-CDs in Figure 13.1. The rate of degradation seems to be almost linear in time, and an increase in current density by a factor of 1.85, from 2.6 to 4.8 A/m², resulted in an increase in the degradation rate by a factor of approximately 3 for the β-CD. Similar experiments performed on α-CD solutions yielded almost identical results. The results of initial testing of a negatively charged amino-β-CD solution suggested that the degradation rate was nonlinear, but more research is necessary to substantiate this finding.

The lack of structural stability of the CD molecules is a point of concern when considering an EK remediation process using CDs since it is likely that significant degradation of the solubility-enhancing agent could occur. We note, however, that the electrochemical degradation effect may in fact be useful in a posttreatment process for the purpose of eliminating any remaining CD and contaminants from the effluent stream collected in containment wells. Further research work that results in controlled and variable CD stability could ultimately be a key factor in a well-optimized remediation process. Regardless, CD molecules have shown great promise in enhancing the removal efficiencies of many explosive compounds from contaminated soils and should be considered for field remediation efforts.

13.3.3 Bioremediation

Another viable remediation strategy is the biotic degradation of contaminants by agents such as bacteria and fungi. A major advantage of such a technique is that the process can be performed *in situ*, which limits the physical handling of contaminated effluents. Considerations specific to the bioremediation of energetic compounds must be recognized to ensure successful degradation and mineralization of the by-products. Explosives are recalcitrant to biological degradation under aerobic conditions due to the presence of highly oxidized nitro groups on the aromatic ring. With TNT, instead of being biodegraded, a variety of partially reduced TNT daughter products is produced (Pasti-Grigsby et al., 1996; Fiorella and Spain, 1997; Vorbeck et al., 1998), resulting in the formation of aminodinitrotoluenes, diaminonitrotoluenes, and/or tetranitroazoxytoluenes (Roberts, Ahmad, and Pendharkar, 1996). These metabolites are resistant to further degradation and are therefore considered dead-end products (Vorbeck et al., 1998). Clearly, in the case of multiple degradation pathways, those leading to complete remediation and/or those avoiding more dangerous intermediate compounds are preferable.

Explosives are much more readily biodegraded under anaerobic conditions (McCormick, Cornell, and Kaplan, 1981; Funk et al., 1993; Kitts, Cunningham, and Unkefer, 1994; Roberts, Ahmad, and Pendharkar, 1996; Hawari, 2000). Bacteria from the genus *Clostridium* are considered prime facilitators of TNT and RDX biodegradation in anaerobic environments (Lewis et al., 1996; Huang et al., 2000). The degradation of 2,4-DNT by *Burkholderia* sp. strain DNT was shown first by Spanggord et al. (1991), and subsequent experiments by Ortega-Calvo, Fesch, and Harms (1999) showed that a similar breakdown process occurred when 2,4-DNT was sorbed to a montmorillonite clay surface. The degradation was found to take place in solution away from the clay particles, but fast desorption of the 2,4-DNT was observed. In fact, desorption occurred more readily for solutions containing *Burkholderia* sp. strain DNT than for the background solution alone, possibly due to the production of strongly hydrated ions or local acidification due to CO_2 formation during bacterial excretions (Ortega-Calvo, Fesch, and Harms 1999). Numerous studies have demonstrated that bacteria belonging to the family Enterbacteriaceae are capable of degrading RDX. This includes genera belonging to *Morganella*, *Providencia*, *Citrobacter freundii*, and *Serratia* species (Kitts, Cunningham, and Unkefer, 1994; Young et al., 1997). Several groups have speculated about the involvement of homoacetogens in the anaerobic biodegradation of RDX (Adrian and Lowder, 1999; Hawari et al., 2000; Wildman and Alvarez, 2001; Beller, 2002; Adrian and Arnett, 2004).

Most contaminated soils are inhabited by microbial communities with the genetic capability to degrade the contaminants present. Anaerobic microbial growth requires electron donors, electron acceptors, and various micro- and macronutrients. However, natural attenuation by these communities is usually slow and well below their potential genetic capability due to inadequate growth conditions. Most explosive compounds do not serve as substrates for microbial growth, thus an external cosubstrate is required (Nishino, Spain, and He, 2000). Poor bioavailability, whereby contaminants become so tightly or intimately associated to soil particles that they are unavailable to microbial attack, is another important constraint that limits rates of microbial degradation. Another common limitation of microbial degradation of contaminants can be the requirement for additional nutrients (e.g. N, P, or K) or other substrates, such as inducers (e.g. salicylic acid), that are required for growth or expression of genes responsible for degradation. Typically, hydraulic processes are used to deliver such substances into subsurface environments (Barns and Heaston, 2001), but this method is unsuitable for low-permeability clay soils. The use of electrical fields to deliver the required ingredients for effective biodegradation to subsurface environments would overcome these limitations. Such a process has been employed in the bioremediation of various soils contaminated with copper (Maini et al., 2000), 2,4-dichlorephenoxyacetic acid (Jackman et al., 2001), phenanthrene (Niqui-Arroyo et al., 2006), phenol (Luo et al., 2006), pentachlorophenol (Lear et al., 2007), and polycyclic aromatic hydrocarbons (Niqui-Arroyo and Ortega-Calvo, 2007; Shi et al., 2008).

13.4 ELECTROKINETICS TO ENHANCE REMEDIATION STRATEGIES

The introduction of solubility-enhancing agents as well as targeted bacteria and their proper food supply into predominantly clay subsurface environments can be extraordinarily difficult using conventional techniques due to the low hydraulic permeability of the soil matrix (typically on the order of 10^{-9}–10^{-10} m/s). The process of electroosmosis has been shown to significantly increase the permeability of clay-dominated soil matrices by several orders of magnitude when a voltage gradient is placed across the test regions. Many factors affect the effectiveness of this process, including acid production due to hydrolysis at the anode. These topics are discussed in greater detail in Chapter 3 and in the review article by Yeung (2006). When proper care is taken in the design of an EK system, significant removal efficiencies of energetic compounds is achievable when used in conjunction with a desorbing and solubility-enhancing agent.

A growing body of literature is available on the topic of electrokinetically enhanced remediation of various organic contaminants from surface and subsurface soils as discussed in Chapters 11–14. Considerably less has been written specifically about the remediation of soils contaminated with nitroaromatic and other energetic compounds. Khodadoust, Reddy, and Narla (2006) investigated the use of two purging solutions, deionized water and low-concentration hydroxypropyl-β-CD (HPCD) solutions, combined with an EK flushing process to remove 2,4-DNT from a kaolin clay (0% organic matter) and a glacial till silty soil (2.8% organic matter). For the EK experiments with kaolin soil, a 1% HPCD solution was used, and a 2% solution was used in conjunction with the glacial till. These concentrations were chosen because a further increase in concentration showed negligible improvement in the extraction of 2,4-DNT during batch washing experiments. In the EK experiments, the authors found it much easier, in general, to decontaminate the kaolin soil than the organic matter-rich, natural glacial till. Removal efficiencies of 76.7% and 94.2% of 2,4-DNT from kaolin were obtained using deionized water and 1% HPCD solutions, respectively. Only approximately 21% of the 2,4-DNT was removed from the glacial till using both the deionized water and 2% HPCD purging solutions. No enhancement in removal was observed when using the HPCD solution in this case, and it was inferred that this was due to the strong affinity of 2,4-DNT for soil organic matter. In contrast, the use of the CD purging solution increased the removal of 2,4-DNT by nearly 20% over that obtained using deionized water, an amount roughly equal to the enhancement observed in batch extraction tests. In all cases, no 2,4-DNT was detected in the effluent, suggesting that it was transformed into other compounds after being desorbed from the clay fraction of the soils. The authors suggested that electrochemical degradation of the 2,4-DNT occurred due to the presence of variable redox conditions throughout the test cell. An elemental analysis showed that 80% of the original organic carbon present in the spiked kaolin samples due to the presence of 2,4-DNT remained in the soil after the completion of the EK experiments. The soil samples were also tested for diaminotoluene compounds, degradation products of 2,4-DNT under reductive potentials (Rogers and Bunce, 2001), but none were detected. This suggests that the 2,4-DNT was transformed into products that are complicated and difficult to analyze, and no information is available concerning the toxicity or structural stability of these compounds.

Jiradecha, Urgun-Demirtas, and Pagilla (2006) investigated the ability of several purging solutions containing $NaNO_3$, carboxymethyl-β-CD (CMCD), and a combination of the two to remove naphthalene and 2,4-DNT from a moderate permeability (6.3×10^{-6} m/s) sandy loam soil using a pumped flushing method and electrokinetics. The soil chosen was natural urban Illinois topsoil composed primarily of sand (75.2%) with average clay content of only 7.4%. The soil organic matter content was measured to be 4.9%, a relatively high value. In this case, the soil was spiked in the laboratory by slowly pumping a prepared $NaNO_3$-contaminant solution through the test samples until the contaminant concentrations in the influent and effluent streams were equal (approximately 14 days). A suite of experiments was performed comparing the efficacy of $NaNO_3$ and CMCD as purging solutions. In all experiments, the purging solution to be tested was introduced at the cathode and pumped toward the anode. Hence, in cases where an electric field was applied, the electroosmotic flow (from anode to cathode) was in the opposite direction of the pumped solution. Effluent samples were drawn from the anode chamber. The authors found that a purging solution of 0.01 N $NaNO_3$ removed nearly 40% of the 2,4-DNT from samples using the pumped flow method alone. The combined use of electrokinetics (7.5 and 10 V/cm field strength) and pumped flushing with $NaNO_3$ showed only very slight improvement in removal efficiency. By contrast, a pumped CMCD purging solution (from cathode to anode) increased the removal efficiency to 75%. Increasing the concentration of CMCD from 2 to 5 g/l did not noticeably affect the total amount of 2,4-DNT removed from the samples, but it did increase the rate of removal. Only 100–110 pore volumes of 5 g/l CMCD flowing through the sample were required to reach the maximum level of remediation, compared with approximately 150 pore volumes for the lower concentration. The combined use of electrokinetics (10 V/cm) and pumped 2 g/l CMCD (with 0.01 N $NaNO_3$) solution showed a further improvement in 2,4-DNT removal to 89%, compared with 75% using pumping alone. The authors suggested that this increase could be due to longer transit times of CMCD molecules since the electroosmotic and advected flows from the pumping source were in opposite directions. This longer time could encourage increased CMCD-2,4-DNT complex formation.

Finally, we describe the status of ongoing research at US Army Engineer Research and Development Center-Construction Engineering Research Laboratory (ERDC-CERL) involving the remediation of montmorillonite clay samples contaminated with nitrobenzene and 2,4-DNT. In this work, we consider the ability of CDs and Na^+ ions (dissociated from NaCl and NaOH compounds) to function as solubility-enhancing agents when introduced into the contaminated zone using an EK procedure similar to that of Khodadoust, Reddy, and Narla (2006). Details of the experimental procedure can be found in Kessler et al. (2004).

Figure 13.2 shows the amount of nitrobenzene solubilized by three types of uncharged CD molecules, α-CD, β-CD, and γ-CD in batch experiments. These molecules are functionally identical in shape except for their interior cavity size, which is known to be a critical factor in whether the CD will form a complex with

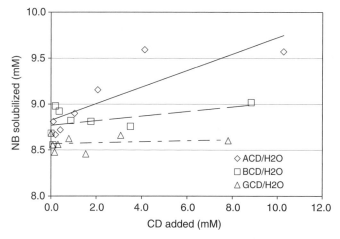

Figure 13.2. Molar fraction of nitrobenzene (NB) solubilized by three different uncharged CDs.

other molecules. The results indicate that the most effective CD for complexation with nitrobenzene is the α-CD, and accordingly, this molecule was chosen as the solubility-enhancing agent for nitrobenzene in the ensuing EK experiments. It has been well documented that β-CD possesses the most suitable cavity size for complexation with 2,4-DNT (Jiradecha, Urgun-Demirtas, and Pagilla, 2006; Khodadoust, Reddy, Narla, 2006;), so we used a modified β-CD with negatively charged functional groups (amino-β-CD) in the 2,4-DNT remediation experiments. This is in contrast to the work described above in which only charge-neutral β-CDs were investigated.

In all experiments, the laboratory-contaminated montmorillonite samples were mixed with 1000 ppm NaCl solution to yield a soil moisture content of 100%. The electrolyte solution was used to increase the availability of free ions in the pore liquid to improve the capacity for electroosmotic flow. Note that in batch experiments, washing of nitrobenzene-contaminated montmorillonite clays with the same NaCl solution resulted in only 2%–4% removal of the nitrobenzene (Fig. 13.3). A constant voltage gradient of 3.25 V/cm was applied across the wetted soil samples. In the first set of experiments, two purging solutions, NaCl and NaCl + α-CD, were used to remove nitrobenzene from laboratory-spiked montmorillonite clay. The concentrations of nitrobenzene, aniline, and α-CD in solution near the anode and cathode are shown as a function of time in Figure 13.4. Significantly higher levels of aniline—one of the common by-products in the reduction pathway of nitrobenzene along with phenylhydroxylamine, azoxybenzene, azobenzene, and nitrosobenzene (Oliveira, 2003)—than nitrobenzene are found in the cathode chamber, suggesting that rapid electrochemical degradation of the nitrobenzene occurs either at the cathode or within the soil matrix. The electrochemical reduction of nitrobenzene has been investigated extensively in the past and has been shown to be sensitive both to the electrode material and the background solution (Baizer 1973; Kokkinidis, Papoutsis, Papanastasiou, 1993, Zhang et al., 2005). In similar experiments, Khodadoust, Reddy, and Narla (2006) found significant degradation of 2,4-DNT to occur within kaolinite test samples.

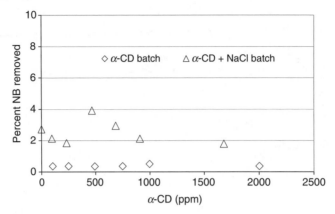

Figure 13.3. Average percentage of nitrobenzene (NB) removed from montmorillonite in batch experiments using washing solutions composed of α-CD and α-CD + NaCl.

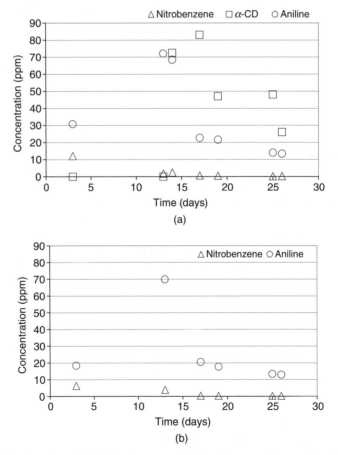

Figure 13.4. Changes in cathode chamber sample chemistry over time with α-CD added to the purging solution (a) and without α-CD (b).

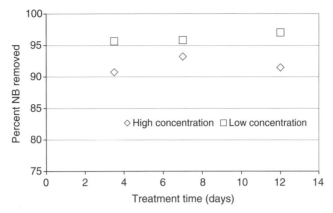

Figure 13.5. Average percentage of nitrobenzene (NB) removed from montmorillonite by electroosmosis for several different treatment times.

The data indicate that the addition of α-CD to the purging solution has no noticeable effect on the amount of nitrobenzene transported into the cathode chamber. Instead, it is surmised that the binding strength of nitrobenzene to the clay layers is sufficiently weak to allow cation exchange with free Na^+ ions in solution. Nitrobenzene liberated in this way could then be transported (up to its solubility limit) along with the electroosmotic flow to the cathode chamber. Figure 13.5 shows the average percentage of nitrobenzene removed from samples treated for 3, 6, and 12 days. The data indicate that over 90% of the nitrobenzene initially sorbed to the montmorillonite was removed for all of the treatment times considered in this study. We reiterate that in batch experiments, it was possible to remove at most 4% of the nitrobenzene, regardless of washing solution (cf. Fig. 13.3). Thus, the ability to remove the model nitroaromatic contaminant, nitrobenzene, from montmorillonite was greatly enhanced by EK treatment. It appears that the treatment had reached its capacity within 3 days. Longer treatment times seemingly did little to increase the percentage of nitrobenzene removed.

The large removal efficiencies found in the nitrobenzene experiments are due in part to the relatively weak binding of that compound with the surfaces of the montmorillonite. As discussed earlier, Haderlein, Weissmahr, and Schwarzenbach (1996) found the adsorption constant of 2,4-DNT with homoionic K^+-montmorillonite to be a factor of 1028 greater than that for nitrobenzene. Hence, it is likely that simple Na^+ ion exchange would not be sufficient to effectively liberate 2,4-DNT from the clay surfaces, an observation supported by the relatively low removal of 2,4-DNT in pumped and EK-treated soils using $NaNO_3$ as the purging solution (Jiradecha, Urgun-Demirtas, and Pagilla, 2006). In an independent set of experiments, we tested the ability of NaCl and NaCl + amino-β-CD purging solutions in conjunction with EK to remove 2,4-DNT from laboratory-spiked montmorillonite. The negatively charged amino-β-CD molecules—introduced at the cathode—move toward the anode via electromigration against the direction of the electroosmotic flow. This effectively increases the length of time for the CD molecules to desorb and form complexes with the 2,4-DNT molecules, a point also made by Jiradecha,

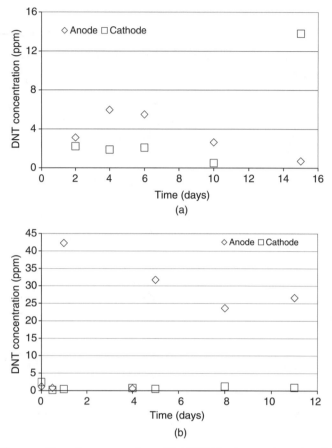

Figure 13.6. Time profiles of concentration of 2,4-DNT in the anode and cathode chambers using NaCl solution (a), and using NaCl + amino-β-CD (b).

Urgun-Demirtas, and Pagilla (2006). The concentrations of 2,4-DNT found in samples drawn at various time intervals from the anode and cathode solutions are plotted for representative experiments using NaCl purging solution (Fig. 13.6a) and NaCl + amino-β-CD purging solution (Fig. 13.6b). Regardless of the type of purging solution, fairly low concentrations of 2,4-DNT were found in the cathode chambers, indicating that in this case the addition of Na$^+$ ions was not sufficient to desorb significant quantities of the contaminant from the clay mineral surfaces. The significant increase in the anode concentration shown in Figure 13.6b suggests that the addition of the amino-β-CD molecules enhanced desorption of 2,4-DNT as hypothesized. Samples were reanalyzed for amino-β-CD content, but in all cases concentrations were below the detection limits, suggesting that the CD molecule was rapidly degraded into unidentified products at the anode. Regardless, the data suggest that the addition of a negatively charged CD molecule into the montmorillonite pore solution enhances the transport of 2,4-DNT out of the clay matrix.

13.5 SUMMARY

In conclusion, we have here covered the more salient points associated with soil and especially clay chemical interactions with energetic compounds, as well as how they affect some of the more promising approaches to remediation for this difficult challenge. In particular, we have discussed the use of solubility-enhancing agents to improve the desorption and transport of hydrophobic organic compounds and have shown that a particular class of agents—CDs—can significantly enhance removal efficiencies for these strongly bound contaminants. Also, the use of *in situ* bioremediation in conjunction with electrokinetics has been introduced and found to be another promising method for the safe and effective removal of explosive compounds from surface and subsurface soils. For further information and details, an interested reader can further investigate the references cited here, as well as the growing literature on the subject. With an efficient, cost-effective, and scalable engineering solution for remediation being the ultimate goal, a number of challenges and open research questions remain. Primary among these might well be (a) EK optimization by soil type and even microenvironment, (b) the adaptation of nanotechnology to this problem, (c) ever-more efficient and directed bioremediation, and (d) how best to combine and/or sequence multiple effects. As work continues in this area, it is expected that further useful insights will ultimately yield a widely used and adaptable suite of remediation technologies.

REFERENCES

Adrian NR, Arnett CM. (2004). Anaerobic biodegradation of hexahydro-1,3,5-trinitro-1,3,5-trazine (RDX) by Acetobacterium malicum strain HAAP-1 isolated from a methanogenic mixed culture. *Current Microbiology* **48**:332–340.

Adrian NR, Lowder A. (1999). Biodegradation of RDX and HMX by a methanogenic enrichment culture. In *Bioremediation of Nitroaromatic and Haloaromatic Compounds* (ed. BC Alleman). Columbus, OH: Battelle, pp. 1–6.

Baizer M. (1973). *Organic Electrochemistry*. New York: Marcel Dekker.

Bardi L, Mattei A, Steffan S, Marzona M. (2000). Hydrocarbon degradation by a soil microbial population with beta-cyclodextrin as surfactant to enhance bioavailability. *Enzyme and Microbial Technology* **27**:709–713.

Barns PW, Heaston MS. (2001). Treatment of explosive-contaminated groundwater by in situ cometabolic reduction. In *Bioremediation of Energetics, Phenolics, and Polycyclic Aromatic Hydrocarbons* (eds. VS Magar, G Johnson, SK Ong, A Leeson). Columbus, OH: Battelle, pp. 25–33.

Beller HR. (2002). Anaerobic biotransformation of RDX (hexahydro-1,3,5-trinitro-1,3,5-triazine) by aquifer bacteria using hydrogen as the sole electron donor. *Water Research* **36**:2533–2540.

Bizzigotti GO, Reynolds DA, Kueper BH. (1997). Enhanced solubilization and destruction of tetrachloroethylene by hydroxypropyl-beta-cyclodextrin and iron. *Environmental Science and Technology* **31**:472–478.

Boving TB, Brusseau ML. (2000). Solubilization and removal of residual trichloroethene from porous media: Comparison of several solubilization agents. *Journal of Contaminant Hydrology* **42**:51–67.

Boving T, Wang XJ, Brusseau ML. (1998). Use of cyclodextrins for remediation of solvent contaminated porous media. *Groundwater Quality: Remediation and Protection. Proceedings of the International Association of Hydrological Sciences*, September 21–25, 1998, Tubingen, Germany. Wallingford, Oxfordshire, UK: International Association of Hydrological Sciences, pp. 437–440.

Brusseau ML, McCray JE, Johnson GR, Wang X, Wood AL, Enfield C. (1999). Field test of cyclodextrin for enhanced in-situ flushing of multiple-component immiscible organic liquid contamination: Project overview and initial results. In *Field Testing of Innovative Subsurface Remediation and Characterization Technologies*. ACS Symposium Series (eds. D Sabatini, MD Annable, ML Brusseau, J Gierke). New York: Oxford University Press, pp. 118–135.

Brusseau ML, McCray J, Wang X. (1996). Using cyclodextrin for in situ remediation of petroleum contamination. *Abstracts of the American Chemical Society* **211**:103–ENVR.

Brusseau ML, Wang XJ, Hu QH. (1994). Enhanced transport of low-polarity organic-compounds through soil by cyclodextrin. *Environmental Science and Technology* **28**:952–956.

Cahill S, Bulusu S. (1993). Molecular-complexes of explosives with cyclodextrins. 1. Characterization of complexes with the nitramines RDX, HMX, and TNAZ in solution by H-1-NMR spin-lattice relaxation-time measurements. *Magnetic Resonance in Chemistry* **31**:731–735.

Cicek N, Govind R. (1997). Micellar enhanced electrokinetic remediation of contaminated soils. *Water Environment Federation (WEFTEC) Conference Proceedings, 70th Annual Conference Exposition*, October 18–22, Chicago, IL. Alexandria, VA: Water Environment Federation, pp. 89–100.

Crini G, Bertini S, Torri G, Naggi A, Sforzini D, Vecchi C, Janus L, Lekchiri Y, Morcellet M. (1998). Sorption of aromatic compounds in water using insoluble cyclodextrin polymers. *Journal of Applied Polymer Science* **68**:1973–1978.

Crini G, Janus L, Morcellet M, Torri G, Morin N. (1999). Sorption properties toward substituted phenolic derivatives in water using macroporous polyamines containing beta-cyclodextrin. *Journal of Applied Polymer Science* **73**:2903–2910.

Croft AP, Bartsch RA. (1983). Synthesis of chemically modified cyclodextrins. *Tetrahedron* **39**:1417–1474.

Dailey OD, Dowler CC, Glaze NC. (1990). Evaluation of cyclodextrin complexes of pesticides for use in minimization of groundwater contamination. In *Pesticide Formulations and Application Systems*, Vol. 10 (eds. LE Bode, JL Hazen, DG Chasin). Philadelphia: ASTM Special Technical Publication 1078-EB, pp. 26–37.

Daun G, Lenke H, Reuss M, Knackmuss HJ. (1998). Biological treatment of TNT-contaminated soil. Anaerobic cometabolic reduction and interaction of TNT and metabolites with soil components. *Environmental Science and Technology* **32**:1956–1963.

Emery EF, Junk T, Ferrell RE, De Hon R, Butler LG. (2001). Solid-state H MAS NMR studies of TNT absorption in soil and clays. *Environmental Science and Technology* **35**:2973–2978.

Eriksson J, Skyllberg U. (2001). Binding of 2,4,6-trinitrotoluene and its degradation products in a soil organic matter two-phase system. *Journal of Environmental Quality* **30**:2053–2061.

Fava F, Di Gioia D, Marchetti L. (1998). Cyclodextrin effects on the ex-situ bioremediation of a chronically polychorobiphenyl-contaminated soil. *Biotechnology and Bioengineering* **58**:345–355.

Fenyvesi E, Szeman J, Szejtli J. (1996). Extraction of PAHs and pesticides from contaminated soils with aqueous CD solutions. *Journal of Inclusion Phenomena and Molecular Recognition in Chemistry* **25**:229–232.

Fikes LE, Winn DT, Sweger RW, Johnson MP, Czarnik AW. (1992). Preassociating alpha-nucleophiles. *Journal of the American Chemical Society* **114**:1493–1495.

Fiorella PD, Spain JC. (1997). Transformation of 2,4,6-trinitrotoluene by Pseudomonas pseudoalcaligenes JS52. *Applied and Environmental Microbiology* **63**:2007–2015.

Funk SB, Roberts DJ, Crawford DL, Crawford RL. (1993). Initial-phase optimization for bioremediation of munition compound-contaminated soils. *Applied and Environmental Microbiology* **59**:2171–2177.

Gao SX, Wang LS, Huang QG, Han SK. (1998). Solubilization of polycyclic aromatic hydrocarbons by beta-cyclodextrin and carboxymethyl-beta-cyclodextrin. *Chemosphere* **37**:1299–1305.

Griffiths DW, Bender ML. (1973). Cycloamyloses as catalysts. *Advances in Catalysis* **23**:209–261.

Haderlein SB, Schwarzenbach RP. (1993). Adsorption of substituted nitrobenzenes and nitrophenols to mineral surfaces. *Environmental Science and Technology* **27**:316–326.

Haderlein SB, Weissmahr KW, Schwarzenbach RP. (1996). Specific adsorption of nitroaromatic explosives and pesticides to clay minerals. *Environmental Science and Technology* **30**:612–622.

Hawari J. (2000). Biodegradation of RDX and HMX: From basic research to field application. In *Biodegradation of Nitroaromatic Compounds and Explosives* (eds. JC Spain, JB Hughes, HJ Knackmuss). Boca Raton, FL: CRC Press, pp. 277–310.

Hawari J, Halasz A, Sheremata T, Beaudet S, Groom C, Paquet L, Rhofir C, Ampleman G, Thiboutot S. (2000). Characterization of metabolites during biodegradation of hexahyro-1,3,5-trinitro-1,3,5-triazine (RDX) with municipal anaerobic sludge. *Applied and Environmental Microbiology* **66**:2652–2657.

Hawari J, Paquet L, Zhou E, Halasz A, Zilber B. (1996). Enhanced recovery of the explosive hexahydro-1,3,5-trinitro-1,3,5-triazine (RDX) from soil: Cyclodextrin versus anionic surfactants. *Chemosphere* **32**(10):1929–1936.

Hodgson JM. (ed.) (1974). *Soil Survey Field Handbook*. Technical Monograph No. 5, Soil Survey. Harpenden, England: Rothamsted Experimental Station.

Huang S, Lindahl PA, Wang C, Bennett GN, Rudolph FB, Hughes JB. (2000). 2,4,6-trinitrotoluene reduction by carbon monoxide dehydrogenase from Clostridium thermoaceticum. *Applied and Environmental Microbiology* **66**:1474–1478.

Jackman SA, Maini G, Sharman AK, Sunderland G, Knowles CJ. (2001). Electrokinetic movement and biodegradation of 2,4-dichlorephenoxyacetic acid in silt soil. *Biotechnology and Bioengineering* **74**:40–48.

Jiradecha C, Kasetsart J. (2000). Removal of naphthalene and 2,4-dinitrotoluene from soils by using carboxymethyl-β-cyclodextrin. *Natural Sciences* **34**:171–178.

Jiradecha C, Urgun-Demirtas M, Pagilla K. (2006). Enhanced electrokinetic dissolution of naphthalene and 2,4-DNT from contaminated soils. *Journal of Hazardous Materials* **136**:61–67.

Kaplan DL, Kaplan AM. (1982). 2,4,6-Trinitrotoluene-surfactant complexes: Decomposition, mutagenicity, and soil leaching studies. *Environmental Science and Technology* **16**:566–571.

Kessler D, Marsh CP, McCormick JJ, Cropek DM, Deguzman AR, Robles R, Gent D. (2004). Investigation of Cyclodextrin-Enhanced Electrokinetic Soil Remediation: Fate and Transport of Nitroaromatic Contaminants and Cyclodextrin Amendments in Expansive Clays. *ERDC-CERL TR-04-3*.

Khodadoust AP, Reddy KR, Narla O. (2006). Cyclodextrin-enhanced electrokinetic remediation of soils contaminated with 2,4-dinitrotoluene. *Journal of Environmental Engineering* **132**:1043–1050.

Kitts CL, Cunningham DP, Unkefer PJ. (1994). Isolation of three hexahydro-1,3,5-trinitro-1,3,5-triazine-degrading species of the family Enterobacteriaceae from nitramine explosive-contaminated soil. *Applied and Environmental Microbiology* **60**:4608–4711.

Ko SO, Schlautman MA, Carraway ER. (1999). Partitioning of hydrophobic organic compounds to hydroxypropyl-beta-cyclodextrin: Experimental studies and model predictions for surfactant-enhanced remediation applications. *Environmental Science and Technology* **33**:2765–2770.

Kokkinidis G., Papoutsis A, Papanastasiou G. (1993). Electrocatalytic reduction of nitrocompounds on gold UPD modified electrodes. I: Reduction of nitrobenzene and 3-nitro-1,2,4-triazole on Au and Au/M(UPD) (M = Pb, Tl). *Journal of Electroanalytical Chemistry* **359**:253–271.

Kopinke F, Remmler M. (1995). Reactions of hydrocarbons during thermodesorption from sediments. *Thermochimica Acta* **263**:123–139.

Lear G, Harbottle MJ, Sills G, Knowles CJ, Semple KT, Thompson IP. (2007). Impact of electrokinetic remediation on microbial communities within PCP contaminated soil. *Environmental Pollution* **146**:139–146.

Lewis TA, Goszczynski S, Crawford RL, Korus RA, Admassu W. (1996). Products of anaerobic 2,4,6-trinitrotoluene (TNT) transformation by Clostridium bifermentans. *Applied and Environmental Microbiology* **62**:4669–4674.

Luo QS, Wang H, Zhang XH, Fan XY, Qian Y. (2006). In situ bioelectrokinetic remediation of phenol-contaminated soil by use of an electrode matrix and a rotational operation mode. *Chemosphere* **64**:415–422.

Maini G, Sharman AK, Sunderland G, Knowles CJ, Jackman SA. (2000). An integrated method incorporating sulfur-oxidizing bacteria and electrokinetics to enhance removal of copper from contaminated soil. *Environmental Science and Technology* **34**:1081–1087.

Marshall TJ. (1947). *Mechanical Composition of Soil in Relation to Field Descriptions of Texture*. Bulletin No. 224. Melbourne, Australia: Council for Scientific and Industrial Research.

McCormick NG, Cornell JH, Kaplan AM. (1981). Biodegradation of hexahydro-1,3,5-trinitro-1,3,5-triazine. *Applied and Environmental Microbiology* **42**:817–823.

McCray JE, Boving TB, Brusseau ML. (2000). Cyclodextrin-enhanced solubilization of organic contaminants with implications for aquifer remediation. *Ground Water Monitoring and Remediation* **20**:94–103.

McCray JE, Brusseau ML. (1998). Cyclodextrin-enhanced in situ flushing of multiple-component immiscible organic liquid contamination at the field scale: Mass removal effectiveness. *Environmental Science and Technology* **32**:1285–1293.

McCray JE, Brusseau ML. (1999). Cyclodextrin enhanced in situ flushing of multiple-component immiscible organic liquid contamination at the field scale: Analysis of dissolution behavior. *Environmental Science and Technology* **33**:89–95.

McCray JE, Bryan KD, Cain RB, Johnson GR, Blanford WJ, Brusseau ML. (1999). Field test of cyclodextrin for enhanced in-situ flushing of immiscible organic liquids: Comparison to water flushing. In *Field Testing of Innovative Subsurface Remediation and Characterization Technologies*. ACS Symposium Series (eds. D Sabatini, MD Annable, ML Brusseau, J Gierke). New York: Oxford University Press, pp. 136–152.

Melton LD, Slessor KN. (1971). Synthesis of monosubstituted cyclohexaamyloses. *Carbohydrate Research* **18**:29–37.

Murakami Y, Kikuchi J, Hisaeda Y, Hayashida O. (1996). Artificial enzymes. *Chemical Reviews* **96**:721–758.

Niqui-Arroyo JL, Bueno-Montes M, Posada-Baquero R, Ortega-Calvo JJ. (2006). Electrokinetic enhancement of phenanthrene biodegradation in creosote-polluted clay soil. *Environmental Pollution* **142**:326–332.

Niqui-Arroyo JL, Ortega-Calvo JJ. (2007). Integrating biodegradation and electroosmosis for the enhanced removal of polycyclic aromatic hydrocarbons from creosote-polluted soils. *Journal of Environmental Quality* **36**:1444–1451.

Nishino SF, Spain JC, He Z. (2000). Strategies for the aerobic degradation of nitroaromatic compounds by bacteria: Process discovery to field application. In *Biodegradation of Nitroaromatic Compounds and Explosives* (eds. JC Spain, JB Hughes, H Knackmuss). Boca Raton, FL: Lewis Publishers, pp. 1–47.

Oliveira MCF. (2003). Study of the hypophosphite effect on the electrochemical reduction of nitrobenzene on Ni. *Electrochimica Acta* **48**:1829–1835.

Ortega-Calvo JJ, Fesch C, Harms H. (1999). Biodegradation of sorbed 2,4-dinitrotoluene in a clay-rich, aggregated porous medium. *Environmental Science and Technology* **33**:3737–3742.

Pasti-Grigsby MB, Lewis TA, Crawford DL, Crawford RL. (1996). Transformation of 2,4,6-trinitrotoluene (TNT) by actinomycetes isolated from TNT-contaminated and uncontaminated environments. *Applied and Environmental Microbiology* **62**:1120–1123.

Petter RC, Salek JS, Sikorsky CT, Kumaravel G, Lin FT. (1990). Cooperative binding by aggregated mono-6-(alkylamino)-beta-cyclodextrins. *Journal of the American Chemical Society* **112**:3860–3868.

Qian Y. (1998). Effect of Cationic Surfactant (CTAB) in the Electrokinetic Remediation of Diesel-Contaminated Soils. MSc Thesis, Department of Biosystems Engineering, University of Manitoba. Winnipeg, Manitoba, Canada.

Reid BJ, Stokes JD, Jones KC, Semple KT. (2000). Nonexhaustive cyclodextrin-based extraction technique for the evaluation of PAH bioavailability. *Environmental Science and Technology* **34**:3174–3179.

Rickert DE. (ed.) (1985). *Toxicity of Nitroaromatic Compounds*. Washington, DC: Hemisphere.

Roberts DJ, Ahmad F, Pendharkar S. (1996). Optimization of an aerobic polishing stage to complete the anaerobic treatment of munitions-contaminated soils. *Environmental Science and Technology* **30**:2021–2026.

Rogers JD, Bunce NJ. (2001). Electrochemical treatment of 2,4,6-trinitrotoluene and related compounds. *Environmental Science and Technology* **35**:406–410.

Saichek RE, Reddy KR. (2005). Electrokinetically enhanced remediation of hydrophobic organic compounds in soils: A review. *Critical Reviews in Environmental Science and Technology* **35**:115–192.

Siegel B. (1979). Preparation and redox properties of a cyclodextrin based ferredoxin model. *Journal of Inorganic and Nuclear Chemistry* **41**:609–610.

Sheremata TW, Hawari J. (2000). Cyclodextrins for desorption and solubilization of 2,4,6-trinitrotoluene and its metabolites from soil. *Environmental Science and Technology* **34**:3462–3468.

Sheremata TW, Thiboutot S, Ampleman G, Paquet L, Halasz A, Hawari J. (1999). Fate of 2,4,6-trinitrotoluene and its metabolites in natural and model soil systems. *Environmental Science and Technology* **33**:4002–4008.

Shi L, Muller S, Harms H, Wick LY. (2008). Effect of electrokinetic transport on the vulnerability of PAH-degrading bacteria in a model aquifer. *Environmental Geochemistry and Health* **30**:177–182.

Spanggord RJ, Spain JC, Nishino SF, Mortelmans KE. (1991). Biodegradation of 2,4-dinitrotoluene by a Pseudomonas sp. *Applied and Environmental Microbiology* **57**:3200–3205.

Stadler-Szoke A, Szejtli J. (1979). Nitroglycerin beta cyclodextrin inclusion complex. *Acta Pharmaceutica Hungarica* **49**:30–34.

Szejtli J. (1998). Introduction and general overview of cyclodextrin chemistry. *Chemical Reviews* **98**:1743–1753.

Tabushi I, Shimizu N. (1979). Cyclodextrin Derivatives, Kosan Co., Ltd., Japan (in Japanese). Japanese Patent No. JP53102985, issued September 7, 1978.

Taha MR. (1996). Micellar Electrokinetic Remediation of TNT from Soil. PhD dissertation, Department of Civil and Environmental Engineering, Louisiana State University and Agricultural and Mechanical College. Baton Rouge, LA.

Taha MR, Gale RJ, Zappi ME. (1994). Surfactant enhanced electrokinetic remediation of NAPLs in soils. *First International Congress on Environmental Geotechnics*, July 10–15, 1994, Edmonton, Alberta, Canada. Richmond, Canada: BiTech Publishers, pp. 373–377.

Taha MR, Soewarto IH, Acar YB, Gale RJ, Zappi ME. (1997). Surfactant enhanced desorption of TNT from soil. *Water Air Soil Pollution* **100**:33–48.

Tsujihara K, Kurita H, Kawazu M. (1977). Highly selective sulfonylation of cycloheptaamylose and syntheses of its pure amino derivatives. *Bulletin of the Chemical Society of Japan* **50**:1567–1571.

Ueno A, Moriwaki F, Osa T, Hamada F, Murai K. (1987). Excimer formation in inclusion complexes of modified cyclodextrins. *Tetrahedron* **43**:1571–1578.

Villiers MA. (1891). Sur la fermentation de la fecule par l'action du ferment butyrique. *Comptes Rendus Academy of Science* **112**:536–538.

Vorbeck C, Lenke H, Fischer P, Spain JC, Knackmuss HJ. (1998). Initial reductive reactions in aerobic microbial metabolism of 2,4,6-trinitrotoluene. *Applied and Environmental Microbiology* **64**:246–252.

Wang XJ, Brusseau ML. (1993). Solubilization of some low-polarity organic compounds by hydroxypropyl-β-cyclodextrin. *Environmental Science and Technology* **27**:2821–2825.

Wang XJ, Brusseau ML. (1995a). Cyclopentanol-enhanced solubilization of polycyclic aromatic hydrocarbons by cyclodextrins. *Environmental Science and Technology* **29**:2346–2351.

Wang XJ, Brusseau ML. (1995b). Simultaneous complexation of organic-compounds and heavy-metals by a modified cyclodextrin. *Environmental Science and Technology* **29**:2632–2635.

Wildman MJ, Alvarez PJJ. (2001). RDX degradation using an integrated Fe(0)-microbial treatment approach. *Water Science Technology* **43**:25–33.

Yamamoto H, Morley MC, Speitel GE, Clausen J. (2004). Fate and transport of high explosives in a sandy soil: adsorption and desorption. *Soil & Sediment Contamination* **13**:459–477.

Yeung AT. (2006). Contaminant extractability by electrokinetics. *Environmental Engineering Science* **23**:202–224.

Young D, Kitts C, Unkefeer P, Ogden K. (1997). Biotransformation of hexahydro-1,3,5-trinitro-1,3,5-triazine (RDX) by a prospective consortium and its most effective isolate Serratia marcescens. *Biotechnology and Bioengineering* **53**:515–522.

Zhang SJ, Yu HQ, Li QR, Yin H. (2005). Effects of an electric or magnetic field on the radiolytic degradation of two biorefractory contaminants. *Journal of Hazardous Materials* **119**:153–158.

PART IV

REMEDIATION OF MIXED CONTAMINANTS

14

ELECTROKINETIC REMEDIATION OF MIXED METAL CONTAMINANTS

Kyoung-Woong Kim, Keun-Young Lee and Soon-Oh Kim

14.1 INTRODUCTION

Heavy metals are of great environmental concern due to their nondegradable characteristics, life cycling in ecosystems, and high toxicity, even in low amount. Soils, sediments, and groundwater can be contaminated with heavy metals derived from various man-made sources (Adriano, 1986). Due to the increasing abundance of metal contaminations, technologies for the decontamination of such sites have been developed, such as electrokinetic remediation. Most hazardous waste sites are contaminated with mixed heavy metals rather than by a single heavy metal. The potential of electrokinetic remediation technology has been successfully demonstrated for the remediation of mixed heavy metal-contaminated soils, sediments, and groundwater (Lageman, 1993; Kim et al., 2002; Suèr, Gitye, and Allard, 2003; Reddy and Chinthamreddy, 2004; Kim, Kim, and Kim 2005a; Nystroem, Ottosen, and Villumsen, 2005). Because electrokinetic remediation is a technology that separates charged contaminants from target areas mainly using electric force, simultaneous treatment of multiple metals at the same site is feasible. In addition, the transport phenomena of multiple heavy metals in an electric field and the factors affecting these phenomena have been studied in many investigations (Vane and Zang, 1997; Wada and Umegaki, 2001; Kim, Kim, and Stüben, 2002; Cherepy and Wildenschild, 2003; Al-Hamdan and Reddy, 2006, 2008). In this chapter, the principles of electrokinetic remediation technology for mixed metals are reviewed through a number of investigations undertaken over the past decades, with the major parameters affecting the final removal efficiency discussed. To support such principles, numbers of representative investigations are described, with several examples introduced in detail. Finally, this chapter deals with some drawbacks that need to be considered

---

Electrochemical Remediation Technologies for Polluted Soils, Sediments and Groundwater,
Edited by Krishna R. Reddy and Claudio Cameselle
Copyright © 2009 John Wiley & Sons, Inc.

only with mixed metal contaminants, and suggests further challenges for overcoming such problems.

14.2 GENERAL PRINCIPLE FOR MIXED METAL CONTAMINANTS

Electrokinetic phenomena are defined as the physicochemical transport of charge, action of charged particles, and the effects of an applied electrical potential on fluid transport in porous media. Movement of charged elements such as heavy metals in soil is caused by an electric field applied to the soil (electromigration). The electric force also leads to the transport of water (electroosmosis), mainly toward the cathode. Movement of charged solid particles such colloids and bacteria is also caused by the imposed electric field (electrophoresis). In addition to such mechanisms, several phenomena affect the electrokinetic transport and removal of contaminants, that is, chemical osmosis and the electrolysis of water. Electromigration and electroosmosis are known to be the most important mechanisms in the electrokinetic removal of contaminants from soils and sediments. Most of these principles were discussed in detail in the previous chapter; therefore, our attention in this chapter will focus only on the principles for multiple metal contaminants. Since electroosmosis is generally affected by controlled artificially conditions, except for specific cases that will be examined further in the next section, electromigration is crucial in the electrokinetic remediation of mixed metals. The main parameters affecting the removal efficiency, as well as transport phenomena, can be summarized as follows:

- Ionic mobility (ionic valence and diffusion coefficient)
- Adsorption capacity and affinity
- Chemical forms in solid matrix
- Inhibitory effect

14.2.1 Theoretical Ionic Mobility and Major Affecting Coefficients

The concepts of ionic mobility (u_i, m^2/s·V) and effective ionic mobility (u_i^*, m^2/s·V) are introduced as representative parameters of electromigration (ionic migration). The effective ionic mobility defines the velocity of the ionic species under the effect of a unit electric field, which can be theoretically estimated using the Nernst–Townsend–Einstein relation (Holmes, 1962). Ionic mobility is related to the ionic valence (z_i) and molecular diffusion coefficient (D_i, m^2/s) of species as follows:

$$u_i = D_i |z_i| F/RT, \qquad (14.1)$$

where F is the Faraday's constant (96.484 C/mol electrons), R the universal gas constant (8.3144 J/K·mol), and T the absolute temperature (K) (Mitchell, 1993; Kim Kim, and Stuben, 2002). The effective mobility u_i^* is defined by

$$u_i^* = n\tau u_i, \qquad (14.2)$$

TABLE 14.1. Diffusion Coefficient, Ionic Mobility at Infinite Dilution and Effective Ionic Mobility in Soil

Metal Species	Molecular Diffusion Coefficient, $D_i \times 10^{10}$ (m²/s)[a]	Ionic Mobility, $u_i \times 10^8$ (m²/s·V)	Effective Ionic Mobility, $u_i^* \times 10^9$ (m²/s·V)[b]
Li^+	10.3	4.01	5.62
Na^+	13.3	5.18	7.25
K^+	19.6	7.63	10.69
Rb^+	20.7	8.06	11.29
Cs^+	20.5	7.98	11.18
Be^{2+}	5.98	4.66	6.52
Mg^{2+}	7.05	5.49	7.69
Ca^{2+}	7.92	6.17	8.64
Sr^{2+}	7.9	6.15	8.61
Ba^{2+}	8.46	6.59	9.22
Pb^{2+}	9.25	7.20	10.09
Cu^{2+}	7.13	5.55	7.77
Fe^{2+}	7.19	5.60	7.84
Cd^{2+}	7.17	5.58	7.82
Zn^{2+}	7.02	5.47	7.65
Ni^{2+}	6.79	5.29	7.40
Fe^{3+}	6.07	7.09	9.93
Cr^{3+}	5.94	6.94	9.72
Al^{3+}	5.95	6.95	9.73

[a] Values from Mitchell (1993).
[b] Values calculated using 0.4 porosity (n) and 0.35 tortuosity (τ).

where n is the porosity of soil and τ the tortuosity. A comparison of the self-diffusion coefficients (Mitchell, 1993) and ionic mobilities for some metal ions are presented in Table 14.1. The effective ionic mobilities of these metals in a soil with a typical porosity of 0.4 and an average tortuosity of 0.35 are used (Kim, Kim, and Stüben, 2002). However, these estimated values cannot reflect the desorption of the metal ions initially existing in a solid matrix. The adsorption capacity and affinity of metal contaminants, which will be discussed further in the next section, are also not considered in the theoretical ionic mobility. Therefore, for practical electrokinetic remediation, it is difficult to predict the transport phenomena and removal properties of mixed metal contaminants using these estimated values. In an ideal environment where the parameters stated above, which has been called a delaying or retardation effect, can be excluded or neglected, the effective ionic mobility value can be the most important factor.

14.2.2 Adsorption Capacity and Affinity of Heavy Metals in Solid Matrix

The state of heavy metal contaminants can generally be classified into two phases: those adsorbed onto solid particles and those dissolved in pore fluid. Heavy metals adsorbed onto solid particles should be transformed by a chemical reaction, that is, desorption, for their removal in an electric field. In electrokinetic remediation processing, the H⁺ generated by the electrolysis of water molecules at the anode compartment is the key element for desorption. This process is referred to as cation exchange reaction, which is introduced in fundamental soil chemistry. It can be illustrated by a simple reaction in which two hydrogen ions (H⁺) displace a metal ion (Me²⁺) from its adsorbed state on a soil surface (Brady, 1990):

$$[\text{Soil}]\text{Me}^{2+} + 2\text{H}^+ \longleftrightarrow [\text{Soil}]{}^{\text{H}^+}_{\text{H}^+} + \text{Me}^{2+}. \quad (14.3)$$

Conversely, dissolved-phase heavy metals in pore fluid are easily removed by the transport mechanisms mentioned above: electromigration and/or electroosmosis. Even though heavy metals in the liquid phase can easily move toward the electrode compartment, there is always the possibility that they will be readsorbed onto the solid particles. As shown by the double arrow in the above illustration, the reaction is reversible. For these reasons, the adsorption capacity (affinity), which is the physicochemical property of each heavy metal, has a high impact on the efficiency of electrokinetic remediation.

Some cations are held much more tightly than others; therefore, they are less likely to be replaced on the solid particle. In general, the higher the charge and the smaller the hydrated radius, the more strongly they will adsorb to the particle. The order of the strength of adsorption for selected cations is as follows (Brady, 1990):

$$\text{Al}^{3+} > \text{Sr}^{2+} > \text{Ca}^{2+} > \text{Mg}^{2+} > \text{Cs}^+ > \text{K}^+ = \text{NH}_4^+ > \text{Na}^+ > \text{Li}^+.$$

For example, considering the relatively loosely held K⁺ ion, if the complementary ions neighboring the K⁺ ion are held loosely, then the H⁺ ions from the acid front are less likely to replace the K⁺ ion (see Fig. 14.1). Conversely, the complementary ions are tightly held, and the K⁺ ion can then be easily replaced. Table 14.2 lists the order of adsorption selectivities of some toxic heavy metals suggested by previous investigations (Schwertman and Taylor, 1989; Yong and Phadungchewit, 1992; Christophi and Axe, 2000). Having a relative high adsorption capacity, lead is a popular heavy metal in electrokinetic experiments. In addition, although lead has a relative high effective ionic mobility value (see Table 14.1), it often shows low removal efficiency compared with other multiple heavy metals in electrokinetic

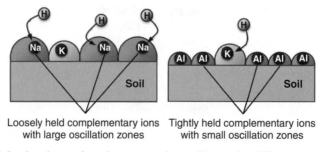

Figure 14.1. Selective desorption phenomena depending on the different adsorption capacities of the cations (modified from Brady, 1990).

TABLE 14.2. Different Selectivity Order of Heavy Metals in Clay Soils (Schwertman and Taylor, 1989; Yong and Phadungchewit, 1992; Christophi and Axe, 2000)

Material	Selectivity Order
Kaolinite clay (pH 3.5–6)	Pb > Ca > Cu > Mg > Zn > Cd
Kaolinite clay (pH 5.5–7.5)	Cd > Zn > Ni
Illite clay (pH 3.6–6)	Pb > Cu > Zn > Ca > Cd > Mg
Montmorillonite clay (pH 3.5–6)	Ca > Pb > Cu > Mg > Cd > Zn
Montmorillonite clay (pH 5.5–7.5)	Cd = Zn > Ni
Goethite	Cu > Pb > Zn > Cd > Co > Ni > Mn
Goethite	Cu > Pb > Cd

treatment (Lageman, 1993; Kim, Moon, and Kim, 2001; Vengris, Binkiené, and Sveikauskaité, 2001; Kim, Kim, and Stüben, 2002; Suèr, Gitye, and Allard, 2003; Kim et al., 2005a; Pedersen, Ottosen, and Villumsen, 2005; Wang et al., 2005), as lead has been found to have a greater affinity for adsorption onto soil surface (Kim, Moon, and Kim, 2001; Kim, Kim, and Stüben, 2002; Kim et al., 2005a). In several recent bench-scale electrokinetic experiments for the decontamination of mixed heavy metals, the removal mechanism of heavy metals was studied considering the adsorption capacity. Yuan et al. (2007) obtained adsorption coefficients for the target heavy metals, copper and cadmium, using the Langmuir equation. The higher adsorption coefficient for copper (1.35) compared with that for cadmium (0.58) suggested a stronger affinity of copper than of cadmium to soil, which was reflected in an enhancement scheme. Al-Hamdan and Reddy (2005) addressed an electrostatic adsorption model for describing the behavior of a kaolin surface for multiple heavy metals, and reported the overall adsorption affinity with kaolin of $Cr^{3+} > Ni^{2+} > Cd^{2+} \gg Cr^{6+}$ in an acidic environment (pH 3–7).

Kim, Moon, and Kim (2001) conducted a study on the electrokinetic recovery of heavy metals from kaolinite and mine tailing soils to explore the feasibility of using this technology on the mine tailing soils in abandoned mine areas, and relate the removal efficiencies to the different behaviors of mixed metal species in soils. The experimental apparatus consisted of a soil cell with a volume of $1215\,cm^3$, a platinum anode, a titanium cathode, electrolyte reservoirs connected on both sides, and a power supply. The soils used in this study were kaolinite soils artificially contaminated with Pb (test 1) and Cd (test 2) and tailing soils (test 3) taken from an abandoned mining area. The tailing soils were contaminated with Pb, Cd, Zn, and Cu. The anode and cathode electrolyte solutions, 0.005 and $0.5\,N\,H_2SO_4$, respectively, were used over a 4-day period. To compare the removal efficiencies of the different contaminants and soil types, all experiments were carried out under the same conditions. Over the period of the experiment, the soil pH decreased in all tests, although the soil pH in the tailing soil test was higher than those in the other tests due to the large pH-buffering capacity of tailing soils. The volume of water transported by electroosmosis also showed similar patterns in all tests, although there were slight differences. However, the voltage profiles and variations of the conductivity in the soil cell showed different trends between soil types, as well as between metal species (Fig. 14.2). The voltage drop in the soil cell was related to the migration of ions in

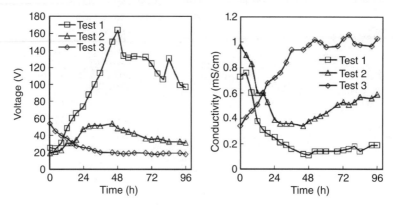

Figure 14.2. Variations in the voltage and conductivity in the soil cell during electrokinetic remediation (Kim, Moon, and Kim, 2001).

the soil cell and to power consumption. Since all tests were conducted under a constant current, the voltage profile directly reflected the conductivity and resistance of the soil bed. The conductivity in the soil cell was calculated using the following equation by Acar et al. (1989):

$$K_a = I_s L / V_s A, \qquad (14.4)$$

where K_a is the apparent conductivity of soils (mS/cm), I_s is the current (mA), L is the length of the soil cell (cm), V_s is the voltage drop in the soils (V), and A is the cross-sectional area of the soil cell (cm^2). Since the current density (I_s/A) and cell length (L) were constant, the conductivity was inversely proportional to the voltage drop in the soil cell. In the initial stages of the kaolinite tests, desorption of heavy metals by hydrogen ions was dominant in the soil system, resulting in decreasing conductivity. In the later stage of the treatments, transport of hydrogen ions and desorbed species were more important than desorption, and the conductivity constantly increased. The difference in the conductivity between Pb and Cd represents the difference in the affinity of soils. In the tailing soils test, hydrogen ions and other species in soils continuously migrated owing to the large void fraction. Therefore, the conductivity in the soil bed increased during the treatment. From the results of the variation in the contaminants in the soil cell, the metal species appear to be gradually transported toward the cathode in the electric field, and the overall amount of contaminants in the soil cell gradually decreased over time. Table 14.3 presents the removal efficiency and energy consumption during each test. When the removal efficiency of the test using kaolinite soils was compared with those using tailing soils, the former showed higher efficiency than the latter. This was explained by the dissolution and desorption of heavy metals on the soil surface being decreased in the tailing soils due to the large pH-buffering capacity, and due to other cationic contaminants (Cu and Zn) competing with the target contaminants (Pb and Cd). In the tests on the removal of Pb, the lower amount of Pb removed was caused by the greater adsorption capacity and the affinity of Pb for the soil surface. This result coincided well with the previously mentioned notion. The energy consumption in test 1 for the removal of Pb was higher than that in test 2 for the removal of Cd,

TABLE 14.3. Removal Efficiency and Energy Consumption of Each Test (Kim, Moon, and Kim, 2001)

Time (day)	Removal Efficiency (%)				Energy Consumption (kWh/ton)		
	Test 1	Test 2	Test 3 (Pb)	Test 3 (Cd)	Test 1	Test 2	Test 3
1	14.5	34.2	12.7	39.2	106.4	66.2	86.9
2	28.9	45.8	20.3	53.6	284.6	169.4	129.9
3	66.1	55.1	34.6	63.6	412.4	265.6	167.2
4	74.9	85.5	49.8	67.6	501.5	328.7	197.3

which was also caused by the strong bonding of Pb to the soil surface. The energy consumption was mainly dependent on the conductivity of the soil cell. As the adsorption and desorption mechanisms predominated in the recovery of heavy metals from soils, the species in soils could not freely migrate through the soil bed; therefore, the conductivity of the soil cell decreased. This research presents a practical example to prove that removal efficiency can be influenced by differences in the affinity or adsorption of target metal species onto soil surfaces, even under identical process conditions.

14.2.3 Chemical Forms of Metal Contaminants in Soils

Heavy metals occur in soils and sediments in various chemical forms, and the variety of heavy metals speciation significantly influences their environmental mobility and bioavailability, which finally determines their potential for environmental contamination (McBride, 1994). Representative speciations of heavy metals in the geoenvironment can be summarized as follows:

1. Dissolved fraction in pore fluid
2. Water soluble or exchangeable fraction
3. Specifically adsorbed fraction
4. Precipitated fraction as insoluble carbonates, sulfides, phosphates, or oxides
5. Organically complexed fraction
6. Crystalline (hydro)oxides fraction
7. Residual fraction

If heavy metals exist as loosely bound fractions, such as (1)–(4), they tend to easily move and disperse. However, metals associated with organics or in crystal lattices, such as (5)–(7), are not easily separated or mobilized. To determine the speciations of heavy metals in any given matrix, sequential extraction analysis has been suggested (Tessier, Campbell, and Bisson, 1979; Gibson and Farmer, 1986), which consists of several extraction steps. Although there are some uncertainties about resorption during the procedures and the selectivity of the extractants, sequential extraction analysis provides useful information to evaluate the potential mobility and fractionations of heavy metals in soils (Suèr, Gitye, and Allard, 2003). Over the

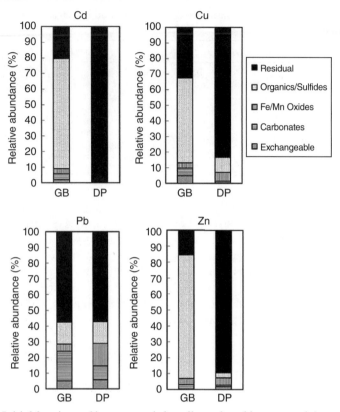

Figure 14.3. Initial fractions of heavy metals in soils analyzed by sequential extraction (Kim, Kim, and Stuben, 2002).

last couple of years, a number of studies have been conducted on the evaluation of electrokinetics, which contemplated the speciation of mixed heavy metals (Kim and Kim, 2001; Kim, Kim, and Stüben, 2002; Kim et al., 2002; Suèr, Gitye, and Allard, 2003; Turer and Genc, 2005; Wang et al., 2005; Zhou et al., 2005; Yuan and Weng, 2006; Traina, Morselli, and Adorno, 2007).

Several studies have emphasized the dependency of removal efficiencies on their speciation, as demonstrated by the different extraction methods used, including sequential extraction, total digestion, and 0.1 N HCl extraction (Kim and Kim, 2001; Kim, Kim, and Stüben, 2002). Contaminated field soils, such as tailing soils, show different physicochemical characteristics, such as initial pH, particle size distribution, and major mineral constituents, and contain high concentrations of target metal contaminants (Cd, Cu, Pb, and Zn) in various forms. Figure 14.3 shows the metal speciation analyzed by sequential extraction, which are summarized in Table 14.4. In the case of the GB tailing soil taken from Gubong abandoned mine area, most metal contaminants existed in the organic and sulfide fractions, with the exception of Pb, which existed predominantly in the residual form. The heavy metals in the DP soil taken from Duckpyeong uraniferous black shale area existed dominantly in the residual fractions, again with the exception of Pb, which was distributed in various fractions. During the electrokinetic process, a constant current

TABLE 14.4. The Sequential Extraction Scheme Suggested by Tessier, Campbell, and Bisson (1979)

Fraction	Chemical Extractant
Exchangeable	0.5 M $MgCl_2$ + NH_4OH/HOAc (pH = 7)
Bound to carbonate or specially adsorbed	1 M NaOAc + NH_4OH/HOAc (pH = 5)
Bound to Fe and Mn oxides	0.04 M NH_2OH HCl in 25% HOAc
Bound to organics and sulfides	0.02 M HNO_3 + 30% H_2O_2 + 3.2 M NH_2OAc in 20% HNO_3 (pH = 2)
Residual	HF/$HClO_4$/HNO_3 (4:2:15)

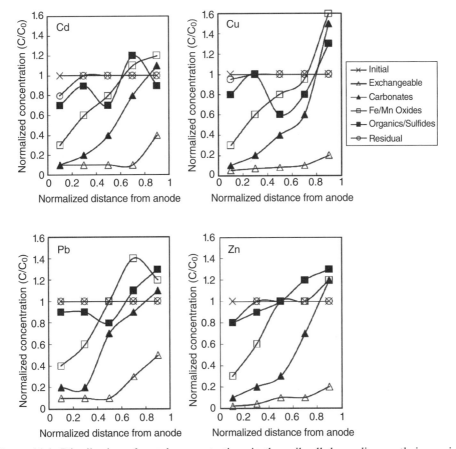

Figure 14.4. Distribution of metal concentrations in the soil cell depending on their speciation after GB tailing soil experiment (Kim, Kim, and Stuben, 2002).

density was maintained, with sulfuric or acetic acid solution used as the cathode electrolyte solution. After the electrokinetic removal treatment, five samples were obtained from the soil bed at equivalent intervals, using a sampler, to analyze the residual concentrations of each fraction of the metal contaminants in the soil bed. Figures 14.4 and 14.5 show the residual metal concentration distributions in the soil

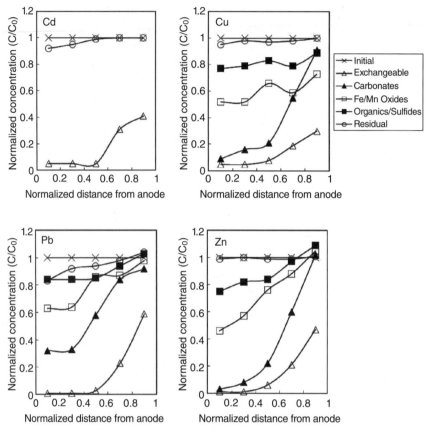

Figure 14.5. Distribution of metal concentrations in the soil cell depending on their speciation after DP soil experiment (Kim, Kim, and Stuben, 2002).

cell for each test. In the electric field, the dissolution and desorption of metal species occurred in the soil cell concurrently with the acid front migration during treatment, and the contaminant species appeared to be gradually transported toward the cathode. The migration of different heavy metals showed similar trends, but these were dependent upon the availability of each species. The overall level of soil contamination gradually decreased with time. Table 14.5 presents the removal efficiencies of the mixed metal contaminants versus their speciation. The removal efficiencies seemed to be influenced by the overall effects of the characteristics of the soils, the specific behaviors of the metals in the soils, and the initial concentrations and speciation of the metal contaminants. The removal efficiencies for Cd, Cu, and Zn in the GB tailing soil, as calculated by total digestion, were higher than those in the DP soil. This directly reflects that the heavy metal contaminants in the DP soil existed more stably than those in the GB tailing soil. Conversely, the Pb in the DP soil, which initially existed in various fractions, with more than 20% in labile fraction, could be efficiently removed compared with the other heavy metals. Therefore, only the Pb removal efficiency from the DP soil was higher than that of the GB tailing soil, as shown from the total digestion. This study

TABLE 14.5. Removal Efficiencies of Metal Contaminants from Soils, As Analyzed by Different Extraction Methods after Electrokinetic Processing (Kim and Kim, 2001; Kim, Kim, and Stüben, 2002)

Extraction Method and Fractions	Removal Efficiency (%)[a]							
	Cd		Cu		Pb		Zn	
	GB	DP	GB	DP	GB	DP	GB	DP
Sequential extraction								
Exchangeable	94.6	92.3	94.8	93.2	90.3	90.3	95.4	92.7
Carbonates	62.1	—	66.0	47.5	47.8	48.0	68.0	76.8
Fe/Mn oxides	21.6	—	31.8	42.6	14.6	25.2	27.7	46.0
Organics/sulfides	11.1	—	20.1	20.6	8.4	13.3	17.1	15.5
Residual	3.1	3.2	2.3	2.4	0.3	2.1	3.4	3.3
Average[b]	13.4	3.5	20.6	7.5	15.6	16.2	17.0	8.0
Total digestion	15.0	3.0	16.0	9.6	13.3	16.4	16.5	4.6
0.1 N HCl	67.6	—	—	—	49.8	—	—	—

[a] The ratio of the initial concentration to that remaining after processing.
[b] Average efficiency = Σ(removal efficiency of each fraction × percentage of initial concentration of each fraction).

definitely supports that the electrokinetic removal efficiencies of heavy metals are significantly influenced by their partitioning in soils. In addition, it was proved that the electrokinetic technique can be effectively applied for the removal of highly mobile and loosely bound metal contaminants in soils, which may cause significant environmental problems due to their high bioavailability and mobility in soils.

14.2.4 Inhibitory Effect of Multiple Metal Contaminants on Electrokinetic Remediation

Wada and Umegaki (2001) made some important findings on the effect of coexisting major cation from their electrokinetic treatment of mixed cationic and anionic contaminants in soil. They monitored the major ion concentrations in pore water, as well as the electrical potential distribution of the soil during treatment, to evaluate the effect of the pore water chemistry on the rate of contaminant removal. They used a surface soil collected from the field, with a constant electric current applied to their soil cell. The water that drained through the cathode compartment was collected, and the pore water at 10 points of the soil body was also collected and analyzed for Ca, Mg, K, Na, and Al as cations, as well as Cl, NO_3, and SO_4 as anions. On application of a voltage, the soil in the vicinity of the anode became acidic due to the generation of hydrogen ions, and the acid front gradually moved toward the cathode. However, the major cation in the acid region was not H^+ ions but Al^{3+} ions, dissolved from soil minerals, which migrated very slowly toward the cathode. Such an anomalously slow migration of Al^{3+} was due to the cationic exchange and precipitation–dissolution reactions in the acidic region, which caused potential flattening and, in turn, diminished the electromigration in this region. In conclusion, the coexistence of high concentrations of major cations can cause a relatively low removal rate of target contaminants in electrokinetic remediation.

If the concentration of nontarget species is higher than that of target species in a contaminated area, the final removal efficiency is unexpectedly decreased (Ottosen, Hansen, and Hansen, 2000; Ribeiro et al., 2000). One reason for this phenomenon can be theoretically explained by the concept of the transference number, as pointed out by Acar and Alshawabkeh (1993). The principle started from the question on how the current would be distributed among a mixture of species in the pore fluid. They assumed the current (I) to be the result of only ion migration in the free pore fluid, where the total current can be related to the migrational mass flux of each species via Faraday's law for equivalence of mass flux and charge flux, as follows:

$$I = \sum_i t_i I = \frac{z_i u_i^* c_i}{\sum_1^n z_j u_j c_j} I, \qquad (14.5)$$

where t_i is the transference number of the ith ion, identifying the contribution of the ith ion to the total effective electric conductivity; z_i is the charge of the ith species; and u_i and u_i^* are the ionic and effective ionic mobilities of the ith ion, respectively. The summation of the transference numbers for all ions in the pore fluid should be equal to one. The transference number is related to the ionic mobility and concentration and therefore increases with the increasing ionic concentration of that specific species. This means that as the concentration of a species decreases relative to the total electrolyte concentration in the pore fluid, its transport under an electric field will be less efficient. This also clarifies why as the concentration of the target species is decreased by transport during electrokinetic remediation, the relative increase in the hydrogen ion concentration in the pore fluid decreases the transference numbers of the other species, thus decreasing their removal efficiency. This is also relevant to the reason the electrokinetic remediation for mixed metal contaminants generally shows lower removal efficiency than that for an individually contaminating metal. In their study, Reddy, Chinthamreddy, and Al-Hamdan (2001) discovered that the migrations of Cr and Ni were higher when present individually compared with when coexisting with Cd in both kaolinite and glacial till soils. Turer and Genc (2005) obtained similar results from their electrokinetic experiments using soils artificially spiked with Pb, Zn, and Cu. When the soil was contaminated by only Pb, the removal efficiency was 48%. However, the Pb removal efficiency decreased to 32% when the soil was contaminated by multiple heavy metals, Pb, Zn, and Cu. Similar results were observed for Zn (from 92% to 37%) and Cu (from 34% to 31%). In conclusion, when soil and sediment highly contaminated with mixed heavy metals, or including significant nontarget ions, are remediated, there is a requirement to find how unexpected low efficiency can occur; therefore, greater time and cost may be required to attain this goal.

14.3 REPRESENTATIVE STUDIES ON ELECTROKINETIC REMEDIATION OF MIXED HEAVY METALS

A number of studies on electrokinetic remediation of mixed heavy metals have been conducted by many researchers using various contaminated sources, enhancement schemes, as well as monitoring and assessment techniques. Table 14.6 lists the representative studies performed over the past decade. It covers the evaluation of process efficiency as well as the transport mechanisms of heavy metals contaminated multiply in soil and sediment, and even solid wastes of environmental concern, such as sludge or ash. Herein, some significant examples from several perspectives are considered.

14.3.1 Electrolyte Conditioning

The most important aspect in the removal of heavy metals from solid matrix is their mobilization. Various enhancing reagents have been tested on electrolyte or pretreatment solutions, and some elucidated different effects and trends depending on the different metal species, even under equivalent conditions. Reddy and his colleagues conducted a number of investigations to evaluate many chemical reagents, such as acetic acid, citric acid, ethylenediaminetetraacetic acid (EDTA), diethylene triamine penta acetic acid (DTPA), potassium iodide (KI), hydroxypropyl-β-cyclodextrin (HPCD), H_2SO_4, NaOH, and NaCl as electrolyte solutions, and humic acid, ferrous iron, and sulfide as reducing agents, to remove mixed heavy metals (Cr, Ni, and Cd) from kaolin and glacial till soils (Reddy, Chinthamreddy, and Al-Hamdan, 2001; Reddy, Xu, and Chinthamreddy, 2001; Reddy and Chinthamreddy, 2003, 2004; Reddy, Danda, and Saichek, 2004; Reddy and Ala, 2005; Al-Hamdan and Reddy, 2006). The results showed that 46%–82% Cr(VI) was removed from the soil, depending on the purging solution used. Although these tests were conducted on high acid-buffering soils such as glacial till soils, the removal efficiencies of Ni(II) and Cd(II) were 48% and 26%, respectively, using acetic acid as the purging solution (Reddy and Chinthamreddy, 2004). Basically, in Cr geochemistry, its speciation is controlled mainly by pH and redox potential. In an oxidizing environment, Cr(VI) exists as $HCrO_4^-$ under very acidic conditions, as $Cr_2O_7^{2-}$ under low to neutral pH conditions, and as CrO_4^{2-} under alkaline conditions, in which species are generally soluble and mobile. On the other hand, Cr(III) species are commonly more immobile than Cr(VI). Cr(III) exists as a soluble Cr^{3+} cation under acidic conditions but precipitates as Cr_2O_3 or $Cr(OH)_3$ under neutral to alkaline conditions. Since Cr was used in the form of Cr(VI) in their test, it migrated toward the anode, contrary to Ni(II) and Cd(II). Accordingly, in a system containing mixed metal contaminants, metal migration may simultaneously be driven in two opposite directions, providing complicated factors considered in the remediation process. In the experiments targeting the clay with low buffering capacity, specifically kaolin soils, 83% of the initial Cr was removed using EDTA as the purging solution in the cathode (Reddy and Chinthamreddy, 2003). The removal of Ni and Cd was relatively low, as also was the case in the experiments that used glacial till soils. A sequentially enhanced approach, accompanying the use of water as an initial purging solution at both electrodes, followed by the use of acetic acid as the cathode purging solution and NaOH solution as the anode purging solution, was tested. This sequential approach resulted in maximum removal of Cr, Ni, and Cd of 68%–71%, 71%–73%, and 87%–94%, respectively. This study has demonstrated that the sequential use of appropriate electrolyte solutions, rather than a single solution, is necessary to remediate soils containing multiple heavy metals. Conversely, a study introducing reducing agents, such as humic acid, ferrous iron, and sulfide, showed no significant

TABLE 14.6. Representative Studies on Electrokinetic Remediation of Mixed Heavy Metals over the Last Decade

Class	Target Material	Heavy Metals	Pollution	Notes	Reference
Soil	Loamy sand (with lime)	Cu, Cr, Hg, Pb, Zn	Polluted	Electrodialytic remediation; relative inefficiency in lime-containing soil	Hansen et al. (1997)
Soil	Clay	Pb, Cu, Ca	Spiked	Evaluation of clay particle zeta potential affecting parameters: clay type, pH, ionic strength of pore fluid	Vane and Zang (1997)
Soil	Kaolinite, glacial till (clay)	Cr, Ni, Cd	Spiked	Evaluation of reducing agent effects; inconsequential results by sulfide: Cr(VI) reduction, pH increase	Reddy and Chinthamreddy (1999)
Soil	Sand, clay	Pb, Zn, Cd	Spiked	pH-dependent removal efficiency: Cd and Zn > Pb, sand > clay	Vengris, Binkienė, and Svekauskaitė (2001)
Soil	Kaolinite, glacial till (clay)	Cr, Ni, Cd	Spiked	Evaluation of the effects of multiple heavy metals; removal efficiency depending on soil pH, polarity of contaminants, type of soil	Reddy, Chinthamreddy, and Al-Hamdan (2001)
Soil	Kaolinite, glacial till (clay)	Cr(VI), Cr(III), Ni, Cd	Spiked	Different migration of initial metal species; efficiency evaluation by sequential extraction	Reddy, Xu, and Chinthamreddy (2001)
Soil	Kaolinite, tailing (clay)	Pb, Cd	Spiked and polluted	Different removal efficiency: Pb < Cd, kaolinite > tailing	Kim, Moon, and Kim (2001)
Soil	Tailing (mainly silt)	Cd, Cu, Pb, Zn	Polluted	Efficiency evaluation depending on heavy metal speciation in soil; main removal mechanism: electromigration > electroosmosis	Kim et al. (2002)
Soil	Various natural clays	Cu, Pb, Zn	Spiked	Heavy metal migration dominated by crystalline clay minerals; lowest efficiency: humic-allophanic and allophonic soils	Darmawan and Wada (2002)
Soil	Kaolinite	Cr, Ni, Cd	Spiked	Sequentially enhanced electrokinetics by electrolyte conditioning	Reddy and Chinthamreddy (2003)
Soil	Factory site (clay)	Ca, Mg, Mn, Fe, Ni, Cu, Zn, Pb, Cd	Polluted	Selective leaching test after electrokinetics; low potential for remobilization of metals after electrokinetics	Suèr, Gitye, and Allard (2003)

Soil	Factory site	Cu, Cd	Polluted	Combination of electrokinetics and phytoremediation	O'Connor et al. (2003)
Soil	Kaolinite	Cr, Ni, Cd	Spiked	EDTA-enhanced electrokinetic; high removal efficiency of only Cr in high pH	Reddy, Danda, and Saichek (2004)
Soil	Glacial till (clay)	Cr, Ni, Cd	Spiked	Enhanced by various electrolyte solutions; simultaneous removal of multiple heavy metals: NaCl/EDTA or acetic acid	Reddy and Chinthamreddy (2004)
Soil	Waste disposal site	Cr, Cd	Polluted	Different removal efficiency depending on scale; bench scale < field scale; field scale: 78% Cr, 70% Cd after 6 months	Gent et al. (2004)
Soil	Kaolinite	Pb, Cd	Spiked	Enhanced by ion exchange membrane; efficiency evaluation depending on electrode configuration	Kim, Kim, and Kim (2005a)
Soil	Red soil	Cu, Zn	Polluted	Catholyte pH controlled by lactic acid and $CaCl_2$; without $CaCl_2$: Zn > Cu; with $CaCl_2$: only enhanced Cu	Zhou et al. (2005)
Soil	Silt clay	Pb, Zn, Cu	Spiked	Comparison of single and multiple heavy metals; two types of electrode geometry	Turer and Genc (2005)
Soil	Field soil (clay)	19 metals (spiked: Pb, Hg)	Polluted and spiked	Enhanced by different extracting solutions; efficient solutions: EDTA, KI	Reddy and Ala (2005)
Soil	Kaolinite	Cr(VI), Cr(III), Ni, Cd	Spiked	Evaluation of pH-dependent adsorption of metals; zeta potential change by heavy metals	Al-Hamdan and Reddy (2005)
Soil	Kaolinite	Cr(VI), Cr(III), Ni, Cd	Spiked	Adsorption and precipitations properties of single and multiple heavy metals	Al-Hamdan and Reddy (2006)

TABLE 14.6. (Continued)

Class	Target Material	Heavy Metals	Pollution	Notes	Reference
Soil	Black cotton soil	Cr, Fe	Spiked	Electroosmotic flow monitoring; different removal trends by heavy metals	Sivapullaiah and Nagendra Prakash, (2007)
Soil	Kaolinite, glacial till (clay)	Cr, Ni, Cd	Spiked	Different removal efficiency; kaolinite > glacial till; geochemical assessment using MINEQL+	Al-Hamdan and Reddy (2008)
Sediment	Harbor sediment	Cu, Zn, Pb Cd	Polluted	Electrodialytic remediation; control pH and liquid-to-solid ratio; maximum removal efficiency: 87% Cu, 98% Cd, 97% Zn, 96% Pb	Nystroem, Ottosen, and Villumsen (2005)
Sediment	Estuarine harbor sediment	As, Cd, Cr, Cu, Ni, Pb, Zn	Polluted	Electrodialytic remediation; average removal efficiency: 75%; dissolved organic carbon (DOC) from sediment: before EDR < after EDR	Gardner, Nystroem, and Aulisio (2007)
Sludge	Wastewater sludge	Cd, Cr, Cu, Pb	Polluted	Pilot-scale study; abiotic and biotic speciation of heavy metals	Kim, Kim, and Stüben (2002)
Sludge	Wastewater sludge	As, Cr, Ni, Pb, Cu, Zn	Polluted	Sludge preacidification and catholyte pH control; sequential extraction after treatment; Cu > Zn > Ni > Cr > As > Pb	Wang et al. (2005)
Sludge	Wastewater sludge	Cr, Cu, Fe, Ni, Pb, Zn	Polluted	Comparison by different electrolyte solutions; highest efficiency: citric acid; Cu > Pb > Ni > Fe > Zn > Cr	Yuan and Weng (2006)
Ash	Incineration fly ash	Cd, Pb, Zn, Cu, Cr	Polluted	Electrodialytic remediation; enhanced by ammonium citrate filling solution; 86% Cd, 20% Pb, 62% Zn, 81% Cu, 44% Cr	Pedersen, Ottosen, and Villumsen (2005)
Ash	Incinerator bottom ash	Pb, Cu, Zn, Cd	Polluted	Comparative experiments by different current density and process duration	Traina, Morselli, and Adorno (2007)

removal of the heavy metal contaminants from the soils (Reddy and Chinthamreddy, 1999). The pH and redox conditions in the soils after the electrokinetic treatment were affected by the presence of reducing agents. In particular, the introduction of sulfides into kaolin caused a significant increase in the soil pH, resulting in low migration of mixed heavy metals. Yuan and Weng (2006) proposed both sodium dodecylsulfate (SDS) and citric acid (CA) as the processing fluids in their electrokinetic experiments on wastewater sludge. Their results showed that the removal efficiency of heavy metals for SDS-enhanced and CA-enhanced systems were 37.2%–76.5% and 43.4%–78.0%, respectively. The priority of the removal of the investigated metals from sludge using the process was Cu > Pb > Ni > Fe > Zn > Cr. Zhou et al. (2005) reported that the application of both lactic acid and $CaCl_2$ as a catholyte pH-conditioning solution improved the removal of Cu and Zn from soil. The removal efficiencies of Cu and Zn reached 63% and 65%, respectively, as this electrolyte conditioning lowered the pH of the soil column and increased the soil electrical conductivity.

14.3.2 Electrodialytic Remediation

The technique using ion exchange membranes for electrokinetics is referred to as electrodialytic remediation. When the electrokinetic treatment is applied without any conditioning, metals may precipitate as hydroxides near the cathode region, where the pH is increased, resulting in a decreased removal efficiency. A popular enhancement scheme to prevent hydroxide precipitates is to apply an ion exchange membrane to the electrode compartments (Hansen et al., 1997; Li, Yu, and Neretnieks, 1998; Ottosen, Hansen, and Hansen, 2000; Ribeiro et al., 2000; Kim, Kim, and Kim, 2005a; Nystroem, Ottosen, Villumsen, 2005; Pedersen, Ottosen, Villumsen, 2005; Gardner, Nystroem, and Aulisio, 2007). When a cation selective membrane is placed in front of the cathode, a low soil pH is maintained. Thus, heavy metals are unable to precipitate and may migrate out of the soil, where they are finally collected in the catholyte. The effects of electrode geometry on removal efficiency are also important in electrokinetic processing. In some studies comparing the electrode geometry of rectangular and cylindrical types, the latter was found to be better (Turer and Genc, 2005). Kim, Kim, and Kim (2005a) evaluated the feasibility of a combination of these two enhancement schemes for the treatment of mixed heavy metals in clay soils. In addition, an auxiliary solution cell (ASC) has been developed to overcome the fouling problem within the cation exchange membrane. Small holes were included into the cathode compartment. A number of hydroxide ions can move from the cathode electrolyte into the ASC through these holes, and as a consequence, metal contaminants are precipitated in the ASC rather than in the cathode electrolytes. The results of each comparative test have proved the applicability of the ion exchange membrane, the cylindrical electrode geometry, and the ASC in the cathode region for the removal of mixed metal contaminants. Another effective study on the removal of heavy metals in harbor sediment has been conducted, where the influence of the changing experimental conditions, such as the applied current and liquid-to-solid ratio, as well as the sample drying conditions were investigated (Nystroem, Ottosen, and Villumsen, 2005). These experiments used HCl as a desorbing agent, which was stirred for 14 days during the electrodialytic remediation. It was found that a liquid-to-solid ratio of 8, along with

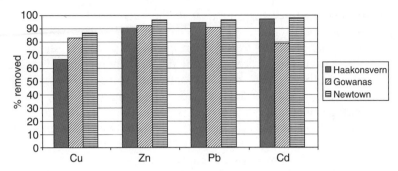

Figure 14.6. The removal of heavy metals from the three sediments at 70 mA and L/S 8, with a remediation time of 14 days and air-dried sediment (Nystroem, Ottosen, and Villumsen, 2005).

a dried sample and a 70-mA current, allowed the maximum removal efficiency of all heavy metals. More severely contaminated sediments were treated using these experimental conditions. In some cases, all the metal concentrations in sediments were reduced below the ecotoxicological assessment criteria (EAC) for heavy metals using this treatment (Fig. 14.6). In this study, however, the ionic strength of sediment pore water was not considered before and after the experiments in determining mineral solubility. Because this parameter can strongly affect the electrokinetic phenomena, especially in an environment conditioned by a high liquid-to-solid ratio such as harbor sediment, the ionic strength of pore water should be taken into account.

14.3.3 Stability Assessment

As discussed previously, the initial speciation of heavy metals in contaminated field soils can potentially be investigated using sequential extraction or selective leaching techniques. They can also be utilized to assess the stability of residual heavy metals in soil after electrokinetic remediation. Suèr, Gitye, and Allard (2003) performed a study on mixed metal and macroelements. They took soil from the site of a chlor-alkali factory, which was predominantly illitic clay. They used NaCl solution in the electrode compartments, with an electric field of 30 V/27 cm applied for 182 days. Soil samples were taken from the electrokinetic cell during and after the experiments. The dried soil samples were subjected to selective leaching as well as to total digestion. As expected, most metals were transported from the acidic part of the soil and accumulated in the high-pH area. From the results of metal selective leaching, however, acidic leaching of the soil in the alkaline zone did not mobilize the metals or macroelements (Fig. 14.7), where the low mobility after electrokinetic remediation may support the low potential for their remobilization. Thus, it was concluded that both mobilization and immobilization using electrokinetics may be feasible methods for the remediation of contaminated soils. We could find a similar result from the experiment conducted by Zhou et al. (2005). In their electrokinetic experiments, without any special electrolyte and just using deionized water, Zn was sequentially extracted after treatment, which showed that Zn existed predominantly as a residual phase. Most of the Zn was not removed from the soil cell but accumu-

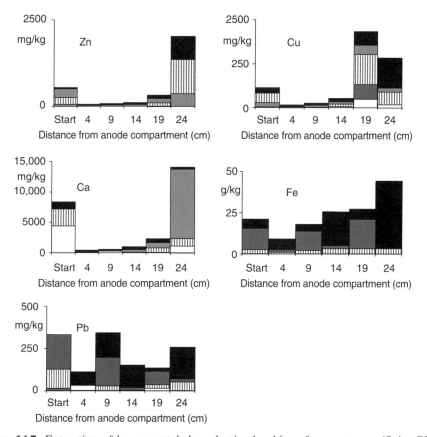

Figure 14.7. Extraction of heavy metals by selective leaching after treatment (Suèr, Gitye, and Allard, 2003).

lated in the area near the cathode electrode; therefore, it can be regarded as being in a chemically stable form that cannot be extracted under normal environmental conditions. These two examples indicate that the electrokinetic remediation may also be feasible for the removal of highly mobile heavy metal contaminants (extractable fractions in the early steps of sequential extraction) in soils.

14.3.4 Zeta Potential and Mixed Heavy Metals

As pointed out in the previous section of this chapter, attention should be paid to electromigration rather than electroosmosis to establish the removal mechanisms of mixed heavy metals. However, high concentrations of multiple metal contaminants, especially divalent cations, can affect the electroosmotic flow in electrokinetic remediation, which is a factor that should be taken into consideration with regard to the removal mechanisms of mixed heavy metals. The electroosmotic velocity, V_e, is given by the Helmholtz–Smoluchowski equation (Acar and Alshawabkeh, 1993; Mitchell, 1993):

$$V_e = -\varepsilon \zeta E / \eta, \tag{14.6}$$

where ε is the permittivity of the liquid, ζ is the zeta potential, E is the electric field strength, and η is the viscosity of the liquid. Electroosmosis is related to several factors, as shown in this equation. Although the permittivity and viscosity are influenced by the concentrations of dissolved ions and temperature, they can generally be considered constant values in electrokinetic processing. The most important parameter is the zeta potential, which determines the direction of flow, and the usual negative values for soils result in a flow toward the cathode. Low pH causes the zeta potential to become less negative, and even reach positive values. The zeta potential also tends to become more positive as the ionic strength or metal concentration increases (Page and Page, 2002). Some studies have focused on this phenomena within heavy metal-contaminated soils. Yukselen and Kaya (2003) showed that when monovalent cations such as Li^+ and Na^+ are present in a system, the zeta potential of kaolinite becomes more negative, even at a concentration of $10^{-2}M$. Conversely, when divalent cations such as Co^{2+}, Cu^{2+}, and Pb^{2+} are present in solution, the zeta potential of kaolinite becomes less negative with increasing cation concentration. Vane and Zang (1997) reported that the effect of divalent cations in clay soils was more influential than the ionic strength of electrolyte solutions. In their experiments, charge reversal was observed for kaolinite with 100 ppm of Pb^{2+} (pH 5) and a background ionic strength of $10^{-2}M$ KCl compared with only 10 ppm of Pb^{2+} and a background of $5 \times 10^{-4}M$ KCl. Al-Hamdan and Reddy (2005) observed that the presence of various heavy metals such as Cr(VI), Cr(III), Ni(II), and Cd(II), either individually or in combination in the system, affected the zeta potential of the kaolin surface, which shifted to a more positive value. In conclusion, the zeta potential and surface charge of soil particles are sensitive to the type and concentration of heavy metals, which in turn, influence the electroosmotic flow and adsorption process.

14.4 SPECIFIC INSIGHT FOR REMOVAL OF MIXED HEAVY METALS, INCLUDING Cr, As, AND Hg

Based on the laboratory experimental and field application results, electrokinetic remediation technology has been shown to be a promising method for simultaneously recovering multiple metal contaminants. However, the process is accompanied by limitations on the removal of several specific heavy metals. In this last section, the previous investigations, which focused on the removal of heavy metals, including Cr, As, and Hg, will be examined in more detail. Heavy metals have some reasonable properties, which should be considered more specifically:

- Mobility and toxicity depending on their oxidation state (Cr, As)
- Anionic species in general (Cr, As) or in particular (Hg)
- Relatively low removal efficiency in electrokinetic treatment of multiple metals (Cr, As, Hg)

14.4.1 Chromium Mixed with Heavy Metals

Many investigations have focused on Cr removal using electrokineitcs, which have shown satisfactory efficiency of the process. Nevertheless, Cr mixed with other heavy metals can be problematic; sometimes, only Cr is successfully removed in multiple metal contaminants (Reddy and Chinthamreddy, 2004; Reddy, Danda, and Saichek, 2004) but on other occasions is the most difficult of mixed metals to remove (Yuan and Weng, 2006). If reducing agents are present in natural soils, Cr(VI) is likely to be reduced to Cr(III), which may significantly affect the electrokinetic migration of Cr, as well as the migration of coexisting heavy metals [e.g. Ni(II) and Cd(II)] (Reddy and Chinthamreddy, 1999). Basically, these results are caused by its redox-sensitive property and various forms in solid and liquid phases. A number of studies on electrokinetic remediation of mixed heavy metals, including Cr, have been conducted by Reddy and his colleagues. Some of their studies verified that the electrokinetic process was capable of removing multiple heavy metal contaminants, even those containing Cr, from a high acid-buffering clay soil (Reddy, Chinthamreddy, and Al-Hamdan, 2001; Reddy and Chinthamreddy, 2003, 2004). In a test using tap water as the anode electrolyte and EDTA as the cathode electrolyte, the removal of Cr from the soil was ineffective compared with other tests using special purging solutions. When EDTA was used, the Ni– and Cd–EDTA complexes migrated toward the anode compartment due to their negative charge. In a test using NaCl combined with EDTA, the removal of Cr was considerably increased to 79%; it is likely that the current and electroosmotic flow were sustained due to the the presence of NaCl, which would be beneficial for the removal of Cr (Fig. 14.8). Ni and Cd significantly migrated toward the anode in this test, and eventually accumulated in the soil near the anode. In conclusion, the NaCl/EDTA combination gave the highest migration of all contaminants toward the regions close to the electrodes. Compared with the previously mentioned results, the application of acetic acid in the cathode reservoir gave a low Cr removal efficiency of 57% (Fig. 14.9). The removal efficiencies of Ni and Cd in this test, however, were highest at approximately 48% and 26%, respectively. Therefore, acetic acid was found to be the most effective solution for simultaneously removing all three contaminants.

14.4.2 Arsenic Mixed with Heavy Metals

In laboratory-scale electrokinetics for treatment of sewage sludge, the mobility and removal efficiency of heavy metals (Ni, Zn, Cu, As, Cr, and Pb) were evaluated (Wang et al., 2005). Acidifying schemes, such as preacidification of the sludge cell and electrolyte pH conditioning, were applied. However, such enhancement treatments, when efficiently applied to multiple heavy metals, show a limitation for the treatment of coexisting As. In this experiment, the mobility and removal efficiency of As (31%) were very low compared with that of other metals (95% for Zn, 96% for Cu, and 90 % for Ni). Major As species in common environments are dominantly in the form of oxyanions: As(V)–$H_2AsO_4^-$ and $HAsO_4^{2-}$, As(III)–$H_2AsO_3^-$; and neutral hydroxy complexes: As(III)–$H_3AsO_3^0$. Due to the specific characteristics of As, its mobility is affected by the pH, as well as by the different valence states compared with other heavy metals; therefore, the electrokinetic remediation of As presents difficulties. As(V) is usually more strongly adsorbed than As(III), but neither are easily adsorbed onto soils under alkaline conditions. Alkaline conditions have been indicated as being favorable for As electromigration (Kim, Kim, and Kim, 2005b), but high-pH conditions reinforces normal electroosmosis toward the cathode, resulting in the retardation of the electromigration of As toward the anode.

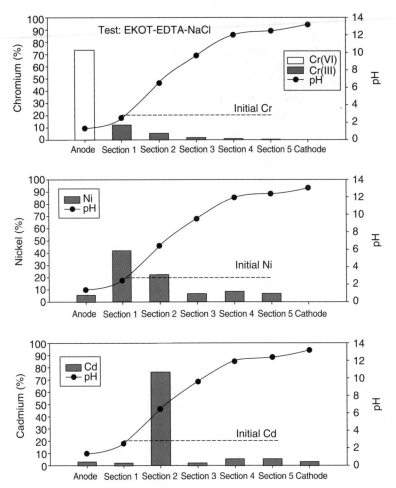

Figure 14.8. Distribution of heavy metals in the soil after electrokinetic treatment: EDTA–NaCl enhancement (Reddy and Chinthamreddy, 2004).

To achieve the desired process efficiency and to improve the system performance, it is necessary to inject an enhancement solution directly into the electrolyte compartements. In order to enhance the electromigration process, sodium carbonate, sodium hydroxide, and sodium hypochlorite have been introduced (Le Hécho, Tellier, and Astruc, 1998). In the electrokinetic processing of both As-contaminated kaolinite and mine tailing soils, high-pH conditions are also preferred for high efficiency (Kim, Kim, and Kim, 2005b). H_2O/KH_2PO_4 and $Na_2CO_3/NaOH$ combinations resulted in maximum removal from kaolinite and tailing soils, respectively. Therefore, it was concluded that the efficiency of the electrokinetic remediation of As from soils was influenced by a number of factors, such as the soil pH, the chemical forms of the As species, the electroosmosis affected due to the zeta potential of the soil, and the electric field intensity.

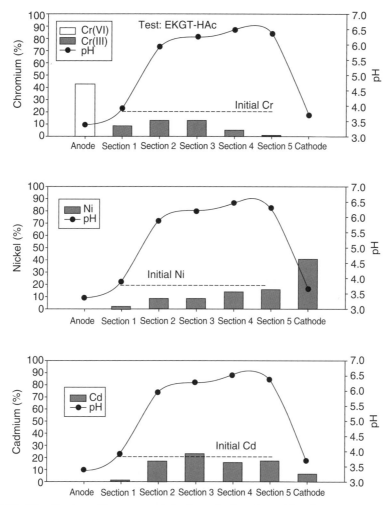

Figure 14.9. Distribution of heavy metals in the soil after electrokinetic treatment: acetic acid enhancement (Reddy and Chinthamreddy, 2004).

14.4.3 Mercury Mixed with Heavy Metals

The electrokinetic remediation of Hg from contaminated soils is notoriously very difficult due to its low solubility, as stated in the previous chapter. Moreover, the electrokinetics of Hg mixed with heavy metals has not been extensively studied. The most efficient removal of Hg in soils was conducted by the oxidation of reduced insoluble Hg(I) to Hg(II) using I_2 (Cox, Shoesmith, and Ghosh, 1996). Here, an anionic complex is formed, where Hg(II) ions are available to migrate through the soil toward the anode. In a recent investigation of the decontamination of mixed heavy metals from contaminated field soils, only Hg was observed to have a different removal property from more than the 10 other metal contaminants (Reddy and Ala, 2005). The system where EDTA solution was applied as the electrolyte was effective for the removal of heavy metals, including Pb, due to the formation of EDTA complexes. Electrokinetic treatment using EDTA has been considered as an efficient scheme for various heavy metals. Hg was not removed in the EDTA system but was efficiently removed as HgI_4^{2-} in the KI system. In their experiments, Hansen *et al.* (1997) also reported that Hg behaved differently from other heavy metals in electrokinetic processing due to the easy reduction of Hg to metallic Hg. Also, the electroosmotic effect was not sufficient to remove metallic Hg, but the anionic form of Hg was removed. The addition of chloride and oxidizing agents to soil has been suggested to mobilize Hg and increase the rate of removal. These investigations support that the choice of strategy for enhancing the removal of metals will depend on the type of metal contamination. In particular, organically complexed Hg species should be taken into great account because it has been reported their transport is thoroughly complicated in the system containing mixed metal contaminants.

14.5 SUMMARY

In order to insure the sufficient efficiency of electrokinetic removal of multiple heavy metals from porous media, it is essential to understand the main parameters affecting the transport and electrokinetic phenomena. Such parameters can be summarized as (a) the theoretical ionic mobility related to the ionic valance and molecular diffusion coefficient of species, (b) the delaying or retardation effect caused by the affinity of heavy metals in solid matrix, and (c) the chemical forms of metal contaminants initially existing in soils. In addition, some unexpected effects especially brought about in the electrokinetic remediation of mixed metal contaminants should be considered. The electrokinetic remediation for mixed metal contaminants generally shows lower removal efficiency than that for individual metal contaminants. High concentrations of multiple metal contaminants can be related to other parameters, for example, transference number, zeta potential, electroosmotic flow, and so on, which are factors that should be taken into consideration with regard to the removal mechanisms.

The potential of electrokinetic remediation technology has been demonstrated for the remediation of mixed metal-contaminated soils, sediments, and groundwater over the past decade. Various enhancement schemes have been developed by the laboratory experimental and field applications, such as electrolyte conditioning and electrodialytic remediation. Despite such advances in the technology, it still has some limitations on the removal of mixed metal contaminants, including several specific heavy metals such as Cr, As, and Hg. Finding a simultaneously applicable process for the remediation of mixed heavy metals will require further study.

REFERENCES

Acar YB, Alshawabkeh AN. (1993). Principles of electrokinetic remediation. *Environmental Science and Technology* **27**(13):2638–2647.

Acar YB, Gale RJ, Putnam G, Hamed J. (1989). Electrochemical processing of soils: Its potential use in environmental geotechnology and significance of pH gradients. *2nd*

International Symposium on Environmental Geotechnology, May 14–17, 1989, Shanghai, China. Bethlehem, PA: Envo Publishing, pp. 25–38.

Adriano DC. (1986). *Trace Elements in Terrestrial Environments*. New York: Springer-Verlag.

Al-Hamdan AZ, Reddy KR. (2005). Surface speciation modeling of heavy metals in kaolin: Implications for electrokinetic soil remediation processes. *Adsorption* **11**:529–546.

Al-Hamdan AZ, Reddy KR. (2006). Geochemical reconnaissance of heavy metals in kaolin after electrokinetic remediation. *Journal of Environmental Science and Health Part A* **41**(1):17–33.

Al-Hamdan AZ, Reddy KR. (2008). Transient behavior of heavy metals in soils during electrokinetic remediation. *Chemosphere* **71**(5):860–871.

Brady NC. (1990). *The Nature and Properties of Soils*, 10th ed. New York: Macmillan.

Cherepy NJ, Wildenschild D. (2003). Electrolyte management for effective long-term electro-osmotic transport in low-permeability soils. *Environmental Science and Technology* **37**(13):3024–3030.

Christophi CA, Axe L. (2000). Competition of Cd, Cu, and Pb adsorption on goethite. *Journal of Environmental Engineering* **126**:66–74.

Cox CD, Shoesmith MA, Ghosh MM. (1996). Electrokinetic remediation of mercury-contaminated soils using iodine/iodide lixiviant. *Environmental Science and Technology* **30**(6):1933–1938.

Darmawan, Wada SI. (2002). Effect of clay mineralogy on the feasibility of electrokinetic soil decontamination technology. *Applied Clay Science* **20**:283–293.

Gardner KH, Nystroem GM, Aulisio DA. (2007). Leaching properties of estuarine harbor sediment before and after electrodialytic remediation. *Environmental Engineering Science* **24**(4):424–433.

Gent DB, Bricka RM, Alshawabkeh An, Larson SL, Fabian G, Granade S. (2004). Bench- and field-scale evaluation of chromium and cadmium extraction by electrokinetics. *Journal of Hazardous Materials* **110**:53–62.

Gibson MJ, Farmer JG. (1986). Multi-step sequential chemical extraction of heavy metals from urban soils. *Environmental Pollution* **11**:117–135.

Hansen HK, Ottosen LM, Kliem BK, Villumsen A. (1997). Electrodialytic remediation of soils polluted with Cu, Cr, Hg, Pb and Zn. *Journal of Chemical Technology and Biotechnology* **70**:67–73.

Holmes PJ. (1962). *The Electrochemistry of Semiconductors*. London: Academic.

Kim SO, Kim KW (2001). Monitoring of electrokinetic removal of heavy metals in tailing-soils using sequential extraction analysis. *Journal of Hazardous Materials* **B85**:195–211.

Kim SO, Kim WS, Kim KW. (2005b). Evaluation of electrokinetic remediation of arsenic-contaminated soils. *Environmental Geochemistry and Health* **27**:443–453.

Kim SO, Kim KW, Stüben D. (2002). Evaluation of electrokinetic removal of heavy metals from tailing soils. *Journal of Environmental Engineering* **128**(8):705–715.

Kim SO, Moon SH, Kim KW. (2001). Removal of heavy metals from soils using enhanced electrokinetic soil processing. *Water, Air, and Soil Pollution* **125**:259–272.

Kim SO, Moon SH, Kim KW, Yun ST. (2002). Pilot scale study on the ex situ electrokinetic removal of heavy metals from municipal wastewater sludges. *Water Research* **36**: 4765–4774.

Kim WS, Kim SO, Kim KW. (2005a). Enhanced electrokinetic extraction of heavy metals from soils assisted by ion exchange membranes. *Journal of Hazardous Materials* **B118**: 93–102.

Lageman R. (1993). Electroreclamation. Applications in the Netherlands. *Environmental Science and Technology* **27**(13):2648–2650.

Le Hécho I, Tellier S, Astruc M. (1998). Industrial site soils contaminated with arsenic or chromium: Evaluation of the electrokinetic method. *Environmental Technology* **19**: 1095–1102.

Li Z, Yu JW, Neretnieks I. (1998). Electroremediation: Removal of heavy metals from soils by using cation selective membrane. *Environmental Science and Technology* **32**(3): 394–397.

McBride M. (1994). *Environmental Chemistry of Soils*. Oxford: Oxford University Press.

Mitchell JK. (1993). *Fundamentals of Soil Behavior*, 2nd ed. New York: John Wiley & Sons.

Nystroem GM, Ottosen LM, Villumsen A. (2005). Electrodialytic removal of Cu, Zn, Pb, and Cd from harbor sediment: Influence of changing experimental conditions. *Environmental Science and Technology* **39**(8):2906–2911.

O'Connor CS, Lepp NW, Edwards R, Sunderland G. (2003). The combined use of electrokinetic remediation and phytoremediation to decontaminate metal-polluted soils: A laboratory-scale feasibility study. *Environmental Monitoring and Assessment* **84**: 141–158.

Ottosen LM, Hansen HK, Hansen CB. (2000). Water splitting at ion-exchange membranes and potential differences in soil during electrodialytic soil remediation. *Journal of Applied Electrochemistry* **30**:1199–1207.

Page MM, Page CL. (2002). Electroremediation of contaminated soils. *Journal of Environmental Engineering* **128**(3):208–219.

Pedersen AJ, Ottosen LM, Villumsen A. (2005). Electrodialytic removal of heavy metals from municipal solid waste incineration fly ash using ammonium citrate as assisting agent. *Journal of Hazardous Materials* **B122**:103–109.

Reddy KR, Ala PR. (2005). Electrokinetic remediation of metal-contaminated field soil. *Separation Science and Technology* **40**:1701–1720.

Reddy KR, Chinthamreddy S. (1999). Electrokinetic remediation of heavy metal-contaminated soils under reducing environments. *Waste Management* **19**:269–282.

Reddy KR, Chinthamreddy S. (2003). Sequentially enhanced electrokinetic remediation of heavy metals in low buffering clayey soils. *Journal of Geotechnical and Geoenvironmental Engineering* **129**(3):263–277.

Reddy KR, Chinthamreddy S. (2004). Enhanced electrokinetic remediation of heavy metals in glacial till soils using different electrolyte solutions. *Journal Environmental Engineering* **130**(4):442–455.

Reddy KR, Chinthamreddy S, Al-Hamdan AZ. (2001). Synergistic effects of multiple metal contaminants on electrokinetic remediation of soils. *Remediation: The Journal of Environmental Cleanup Costs, Technologies & Techniques* **11**(3):85–109.

Reddy KR, Danda S, Saichek RE. (2004). Complicating factors of using ethylenediamine tetraacetic acid to enhance electrokinetic remediation of multiple heavy metals in clayey soils. *Journal Environmental Engineering* **130**(11):1357–1366.

Reddy KR, Xu CY, Chinthamreddy S. (2001). Assessment of electrokinetic removal of heavy metals from soils by sequential extraction analysis. *Journal of Hazardous Materials* **B84**:279–296.

Ribeiro AB, Mateus EP, Ottosen LM, Bech-Nielsen G. (2000). Electrodialytic removal of Cu, Cr and As from chromate copper arsenate-treated timber waste. *Environmental Science and Technology* **34**(5):784–788.

Schwertman U, Taylor RM. (1989). *Minerals in Soil Environments*, 2nd ed. SSSA Book Series No. 1. Madison, WI: Soil Science Society of America.

Sivapullaiah PV, Nagendra Prakash BS. (2007). Electroosmotic flow behavior of metal contaminated expansive soil. *Journal of Hazardous Materials* **143**:682–689.

Suèr P, Gitye K, Allard B. (2003). Speciation and transport of heavy metals and macroelements during electroremediation. *Environmental Science and Technology* **37**(1):177–181.

Tessier A, Campbell RGC, Bisson M. (1979). Sequential extraction procedure for the speciation of particular trace metals. *Analytical Chemistry* **51**:844–851.

Traina G, Morselli L, Adorno GP. (2007). Electrokinetic remediation of bottom ash from municipal solid waste incinerator. *Electrochimica Acta* **52**:3380–3385.

Turer D, Genc A. (2005). Assessing effect of electrode configuration on the efficiency of electrokinetic remediation by sequential extraction analysis. *Journal of Hazardous Materials* **B119**:167–174.

Vane LM, Zang GM. (1997). Effect of aqueous phase properties on clay particle zeta potential and electro-osmotic permeability: Implications for electro-kinetic soil remediation processes. *Journal of Hazardous Materials* **55**:1–22.

Vengris T, Binkienè R, Sveikauskaité A. (2001). Electrokinetic remediation of lead-, zinc- and cadmium-contaminated soil. *Journal of Chemical Technology and Biotechnology* **76**:1165–1170.

Wada SI, Umegaki Y. (2001). Major ion and electrical potential distribution in soil under electrokinetic remediation. *Environmental Science and Technology* **35**(11):2151–2155.

Wang JY, Zhang DS, Stabnikova O, Tay JH. (2005). Evaluation of electrokinetic removal of heavy metals from sewage sludge. *Journal of Hazardous Materials* **B124**:139–146.

Yong RN, Phadungchewit Y. (1992). *Principle of Contaminant Transport in Soils*. Amsterdam, the Netherlands: Elsevier Science.

Yuan C, Weng CH. (2006). Electrokinetic enhancement removal of heavy metals from industrial wastewater sludge. *Chemosphere* **65**:88–96.

Yuan S, Xi Z, Jiang Y, Wan J, Wu C, Zheng Z, Lu X. (2007). Desorption of copper and cadmium from soils enhanced by organic acids. *Chemosphere* **68**:1289–1297.

Yukselen Y, Kaya A. (2003). Zeta potential of kaolinite the presence of alkali, alkaline earth and hydrolysable metal ions. *Water Air Soil Pollution* **145**:155–168.

Zhou DM, Deng CF, Cang L, Alshawabkeh AN. (2005). Electrokinetic remediation of a Cu–Zn contaminated red soil by controlling the voltage and conditioning catholyte pH. *Chemosphere* **61**:519–527.

15

ELECTROKINETIC REMEDIATION OF MIXED METALS AND ORGANIC CONTAMINANTS

MARIA ELEKTOROWICZ

15.1 CHALLENGE IN REMEDIATION OF MIXED CONTAMINATED SOILS

The most well-known chemical groups of organic contaminants found in soils are fuel hydrocarbons, polycyclic aromatic hydrocarbons (PAHs), polychlorinated biphenyls (PCBs), chlorinated aromatic compounds, detergents, and pesticides. Inorganic contaminants mostly include a variety of heavy metals and radionuclides. Among the sources of these contaminants are agricultural runoffs, acidic precipitates, industrial waste materials, and radioactive fallouts (USEPA, 2007). The majority of polluted sites contains more than one pollutant. The presence of both organic and inorganic contaminants forms a specific category of mixed contaminated soils. Organic contaminants and heavy metals often derive from the same sources of contamination. Mixed contamination of soils and groundwater is ubiquitous; following the GERLED (1983) report (the first inventory of contaminated sites in the province of Quebec, Canada), more than 30% of polluted sites fell in the category of mixed contamination.

Many attempts are being made to decontaminate polluted soils, including a range of both *in situ* and off-site techniques. However, technologies that treat mixed contaminants at sites, both inorganic (e.g. heavy metals, radionuclides) and organic contaminants (e.g. PAHs), are very limited. Soil washing/flushing, bioremediation, and solidification/stabilization technologies have been the most common technologies to treat soils at these sites. These technologies, however, fail when low permeability soils are encountered. Soil flushing fails due to a low permeability, stabilization agents are vulnerable in the presence of clayey colloids (Elektorowicz *et al.*, 2008a),

---

Electrochemical Remediation Technologies for Polluted Soils, Sediments and Groundwater,
Edited by Krishna R. Reddy and Claudio Cameselle
Copyright © 2009 John Wiley & Sons, Inc.

and bioavailability of organic xenobiotics is limited (Elektorowicz, Ju, and Oleszkiewicz, 1999).

Mixtures of contaminants that include both petroleum compounds and heavy metals present a particularly difficult challenge for site remediation technology. The methods currently applied for sites polluted with organic contaminants alone or heavy metals alone do not necessarily give satisfactory results in the case of mixed contamination. There are several problems related to remediation on mixed contaminated sites: (a) both groups of contaminants (organic and inorganic) influence reciprocally their properties and behavior in soil, (b) various soil components respond in a different (very often unknown) way in the presence of both groups of contaminants, (c) the method applied for the remediation of one group of contaminants might have an inadequate and even undesirable impact on other groups of contaminants and the environment, (d) remediation techniques commonly used for one group of contaminants might fail in the presence of other group of contaminants, and (e) there are difficulties with an adequate delivery of the reactive agents required for the treatment of each group. Some examples are given below.

The mobility of contaminants and their subsequent removal depend also on their sorption capacity to soil components. Therefore, soil properties, particularly those related to the presence of mineral colloids (e.g. montmorillonite, chlorite, kaolinite) and organic matter, additionally challenge the remediation of mixed contaminated soils.

The quantity of organic matter can have a significant influence on the mobility of heavy metals, particularly if it is in a colloidal form (Scott, 1968; Stevenson, 1994). Humic substances and humic and fulvic acids can form complexes and precipitate; thus, they play an important role in the fixation of heavy metals (Serpaud *et al.*, 1994; Kaschl, Römheld, and Chen, 2002). The mechanism of exchange of ions depends on appropriate conditions, the charge of the ions, and the amount of ions hydrated (Flag, 1974; Stevenson, 1994). Moreover, the concentration of hydrogen ions present in the solution and pH also influence the exchange capacity. Though there are several other possibilities of adsorption of heavy metals by organic matter, in general, there is a lack of information regarding the mechanism of movement and competition of heavy metals in the presence of organic matter (Kaschl, Römheld, and Chen, 2002); but generally, it is assumed that the maximum amount of any given metal that can be bound is found to be approximately equal to the number of carboxyl groups (Alloway, 1990). Humic substances also have an impact on the adsorption process in soil that absorb low molecular weight organic compounds. This adsorption process takes place in soil via different mechanisms in the soil that include (Flaig, Beutelspacher and Rietz, 1975; Sparks *et al.*, 1993) (a) ionic bonding, (b) hydrogen and covalent bonding, (c) electron donor or acceptor mechanisms, (d) cation bridge, (e) van der Waals forces, and (f) water bridging.

Due to various factors (e.g. higher affinity of contaminants to colloids, technological difficulties), the degree of remediation decreases when fine fraction (e.g. montmorillonite, illite) in soil increases. Furthermore, the mixed metals compete for adsorption places in soil depending on the type of soil, type of metals, and their concentrations (Kaoser *et al.*, 2005). These effects might be more visible in mixed contaminated soils. Exchangeable fraction might be occupied by residual organic contaminants leading to a decrease in the adsorption capacity of heavy metals (Kaoser *et al.*, 2004a,b). In addition, the mobility of some metals might be affected by synergistic and antagonistic effects among the metals, particularly in mixed contaminated soil (Reddy et al., 1999). This situation varies from case to case and might lead to a high uncertainty in the design of remediation processes for specific mixed contaminated soils.

Moreover, bioremediation is also affected in the presence of clay colloids due to constraints in the transport of nutrients and electron acceptors (Elektorowicz and Boeva, 1996) and bioavailability limitations (Elektorowicz, Ju, and Oleszkiewicz, 1999). Then, advanced studies might be required to establish microorganisms' response to remediation of soils containing a specific mixture of metals and organic contaminants (before each remediation process). In addition, soils exposed to long-term contamination with a complex mixture of hydrocarbons and heavy metals may present extreme challenges to the maintenance of diverse structural and functional microbial communities (Griffiths et al., 1997; Juck et al., 2000). In the 1990s, the Toronto Harbour Commissioner was involved in cleaning the Lake Ontario coastal area using a chelating agent in an *ex situ* washing system, followed by bioremediation of organic contaminants. Although the separation of fractions and heavy metals elimination worked well, the removal of hydrocarbons was not satisfactory (USEPA, 1993).

In situ remediation of mixed contaminated sites presents a much higher challenge. In this situation, the method applied for the remediation of one group of contaminants might have a very high and even undesirable impact on another group of contaminants. For example, when bioremediation is considered for petroleum contaminants, the supply of an electron acceptor and biological production of CO_2 is expected. However, with the increase of partial pressures of CO_2 in soil, pH decreases by more than one unit. This considerably influences the speciation of heavy metals (Degtiareva and Elektorowicz, 2001). The calculated speciation (using FACT software), which has been performed for zinc, lead, nickel, and cadmium, showed a constant increase of free ions. At partial pressures of CO_2 equal to 10^{-2} atm, almost all zinc in the solution was present as free ions. Furthermore, the concentration of free mobile ions increased by almost 8–10 orders of magnitude when anaerobic conditions ($P_{CO2} = 0.003$ atm) were changed to aerobic conditions. However, buffer capacity of soil due to the presence of calcium and magnesium can attenuate this situation. The creation of suitable aerobic condition for biodegradation can lead to an increase in the concentration of mobile forms of heavy metals in the pore water. A procedure controlling metals' mobility has to be implemented in these cases.

Meanwhile, the mobility of the metals is influenced by pH, redox potential, cation exchange capacity of the solid phase, competition with other metal ions, stability of metal complexes, and their concentration in the soil solution (Merian, 1991). Hence, in many cases, conditioning liquids are applied.

The washing system was used by Zhang and Lo (2007). They investigated the effects of the operating conditions, the initial concentrations of marine diesel fuel (MDF) and the coexisting Pb, and the ethylenediaminetetraacetic acid (EDTA) in a solution on MDF removal by sodium dodecyl sulfate (SDS). They performed the optimization of the process parameters and determined the MDF removal efficiency by SDS under different contamination conditions.

On-site remediation such as phytoremediation was also tested. Palmroth et al. (2007) studied the microbial ecology of weathered hydrocarbon and heavy metal-contaminated soil undergoing phytoremediation. They found that despite the presence of viable hydrocarbon-degrading microbiota, the decomposition of hydrocarbons from weathered hydrocarbon contaminated soil over 4 years, regardless of the presence of vegetation, was low in unfertilized soil.

Elektorowicz, Emon, and Ayadat (2008b) stabilized soil containing phenanthrene (600 mg/kg) and nickel (1000 mg/kg) using silica base grout. Mixed contaminants present in soils ranging from sand to silty sand were entirely stabilized. However, stabilized specimens subjected to water (simulating groundwater conditions) showed various vulnerabilities. While no phenanthrene release from grouted specimens was observed, the nickel release to water reached, in some cases, 4 mg/l.

Very often, due to soil-specific properties (e.g. high specific surface area, low permeability, and high cation exchange capacity), the removal of contaminants from mixed contaminated soil presents an almost impossible task. In these cases, the application of the electrokinetic phenomena can be useful. Electrokinetics is a complex phenomenon induced by an applied electric field, and it refers to the relative motions of charged species. The principle transport phenomena, such as electromigration, electroosmosis, and electrophoresis, are affected by the mixture of organic and inorganic pollutants. The movement of a contaminant could be the result of charged, dispersed species or of the continuous phase.

Several studies were conducted on the application of electrokinetics in different types of soils (see other chapters). They showed that electrokinetic remediation has the potential to remove heavy metals or organic contaminants, notably PAHs from soils. Based on previous experience, the decontamination using the electrokinetic process appears to be the least damaging and most environmentally acceptable technique.

The contaminant removal efficiency generally depends on how great a percentage and how rapidly the contaminant can be transformed into a soluble form, since both inorganics and organics are not easily transported from the solid phase to the aqueous phase. In order to mobilize contaminants for electrokinetic transport, the use of conditioning liquids should be applied. Furthermore, the application of an electric field can uniformly distribute conditioning liquids from injection points and enhance ionic migration and metal removal. This can be difficult if several contaminants coexist simultaneously.

15.2 APPLICATION OF ELECTROKINETIC PHENOMENA TO THE REMOVAL OF ORGANIC AND INORGANIC CONTAMINANTS FROM SOILS

15.2.1 Feasibility Studies on Electrokinetic Remediation of PAHs and Heavy Metal-Contaminated Soils

Despite the fact that the efficiency of electrokinetic remediation was tested when these contaminant groups existed individually, few studies have investigated the feasibility of using electrokinetic remediation for mixed contamination (Elektorowicz, Chifrina, and Konyukhov, 1995; Elektorowicz and Hakimipour, 2001, 2003a,b; Reddy and Kumar, 2002; Khodadoust, Reddy, and Maturi, 2004; Reddy and Maturi, 2005; Maturi and Reddy, 2006, 2008; Reddy and Ala, 2006; Reddy et al., 2006; Reddy and Karri, 2008a). These studies examined the efficiency of the electrokinetic remediation of mixed contaminants (organic pollutants and heavy metals) in low permeability soils (clayey materials).

Khodadoust, Reddy, and Maturi (2004) investigated the feasibility of using different extracting solutions at various concentrations to remove PAHs as well as heavy metals from low permeability clayey soils. Kaolin was selected as a model clayey soil, and it was spiked with phenanthrene as well as nickel at concentrations of 500 mg/kg each to simulate typical field contamination. The extraction solutions selected were surfactants, cosolvents, chelating agents, organic acids, and cyclodextrins. Based on the test results,

1. the extracting solutions, surfactants, and cosolvents were efficient for the removal of phenanthrene, but they were ineffective for the removal of nickel;
2. on the contrary, chelating agents and organic acids were efficient for the removal of nickel but were futile for the removal of phenanthrene;
3. on the other hand, cyclodextrins were found to be inefficient for the removal of both nickel and phenanthrene;
4. to remove both phenanthrene and nickel, it was found that 5% Tween 80 followed by 1 M citric acid or 1 M citric acid followed by 5% Igepal CA-720 was efficient for the subtraction of both nickel and phenanthrene from the kaolin soil;
5. overall, it can be concluded that the sequential use of selected extracting solutions can be efficient for the removal of both heavy metals and PAHs.

Reddy and Maturi (2005) examined the feasibility of using electrokinetic remediation for the removal of mixed contaminants (i.e. mixtures of heavy metals and PAHs) from kaolin (low permeability soil). Likewise, different types of flushing solution were evaluated by a laboratory experimental program, including a cosolvent (n-butylamine), surfactants (3% Tween 80 and 5% Igepal CA-720), and a cyclodextrin (10% hydroxypropyl-β-cyclodextrin or HPCD). It was reported that

1. nickel was found to migrate in the direction of the cathode, and most of it accumulated within the soil close to the cathode due to high pH conditions generated by electrolysis reaction, except with the cosolvent;
2. the experiment with surfactant as a flushing solution resulted in the phenanthrene removal below the detection limit;
3. in the experiment with cyclodextrin, approximately one pore volume of flushing resulted in approximately 50% phenanthrene removal from the soil near the anode;
4. solubilization of the contaminants and the continual electroosmosis flow are the crucial factors that contribute to the removal of both heavy metals and PAHs from low permeability soils.

Further evaluation of the different flushing agents to improve the efficiency of electrokinetic remediation of soil from a manufactured gas plant contaminated with PAHs and heavy metals was performed (Reddy et al., 2006). They concluded that

1. the cosolvent (*n*-butylamine) increased the soil pH, though the surfactants (3% Tween 80 and 5% Igepal CA-720) and cyclodextrin (10% HPCD) did not induce significant change in the soil pH;
2. the density of the current for the tests conducted with the cosolvent extracting solution was higher when compared with the tests performed with the surfactants and the HPCD;
3. it is worthwhile to note that the electroosmotic flow was the maximum with the cosolvent, while the lowest flow was observed with the Tween 80 surfactant;
4. generally, Igepal CA-720 surfactant yielded the highest removal efficiency due to partial solubilization of PAHs, causing some PAHs to travel toward the cathode;
5. however, heavy metals are found to be strongly adsorbed/precipitated and showed negligible migration behavior in all tests.

A comparable laboratory experimental program was conducted on a contaminated sediment at an Indiana harbor. The sediment, having a high organic content of 19% and high moisture content of 78%, was contaminated with PAHs and heavy metals. The same flushing solutions as those used by Reddy and Maturi (2005) were employed. The results obtained showed that

1. 20% of *n*-butylamine and 5% of Igepal CA-720 systems were efficient for the partial solubilization of the PAHs in the sediment;
2. whereas the 3% Tween 80 and 10% HPCD systems were found to be ineffective for desorption/solubilization of PAHs in the sediment;
3. overall, none of the flushing agents were found to be efficient for the removal of heavy metals from the contaminated sediment.

Moreover, the effectiveness of using the cosolvent *n*-butylamine as an extracting solution for the improved removal of an organic contaminant (i.e. PAHs) in the presence of heavy metal from soils having low permeability using electrokinetics was also assessed by Maturi and Reddy (2008) through a bench-scale electrokinetic experimental program. A series of tests was conducted on kaolin soil spiked with phenanthrene and nickel each at 500 mg/kg under a periodic voltage application of 2.0 VDC/cm in cycles of 5 days. It was found that

1. during the initial stages of the experiments when the soil pH was low, nickel existed as a cation and electromigrated toward the cathode. Nevertheless, as the soil pH increased due to hydroxyl ions generated at the cathode and also flushing of high pH *n*-butylamine solution from the anode, nickel precipitated with no further migration;
2. in proportion to the concentration of *n*-butylamine, phenanthrene was found migrating toward the cathode. Both the electroosmotic flow and the concentration of *n*-butylamine governed the extent of phenanthrene removal, but the presence of nickel did not influence the transport and removal of phenanthrene.

The efficiency of the electrokinetic chemical oxidation/reduction for mixed contaminated soil remediation was also investigated (e.g. Reddy and Karri, 2006a; Roach and Reddy, 2006; Reddy and Karri, 2008a,b). In order to assess the effects of voltage gradient on the remedial efficiency of integrated electrochemical remediation (IECR) of mixed contaminants in low permeability soils, a laboratory experimental program was conducted on clayey soil contaminated with nickel and phenanthrene (Reddy and Karri, 2006a). A series of tests was conducted on kaolin soil spiked with phenanthrene and nickel each at 500 mg/kg under two voltage gradients, 1.0 and 2.0 VDC/cm. Two different concentrations (5% and 10%) of the H_2O_2 solution were introduced at the anode, and each experiment was conducted for a total duration of 4 weeks. It was noted from the results that an important migration of nickel occurred from anode to cathode in all tests. However, nickel precipitated near the cathode due to high pH conditions. Moreover, greater migration of nickel toward the cathode was noted when increasing the voltage gradients to 2 VDC/cm. This was due to the diminution of pH throughout the soil as compared with that of the 1 VDC/cm experiment. Nevertheless, nickel precipitated in the close vicinity of cathode even in the 2 VDC/cm experiment, due to high pH conditions. The soil pH distribution and transport of nickel in the soil appeared to be not influenced by the presence of phenanthrene and the use of hydrogen peroxide solution in the anode. Phenanthrene subtraction from the soil was insignificant in all the experiments; however, 28% and 34% of the phenanthrene was oxidized within the soil in the 5% and 10% H_2O_2 experiments, respectively, under 1 VDC/cm. At 2 VDC/cm and under 5% and 10% H_2O_2 concentrations, the phenanthrene oxidation augmented to about 32% and 42%, respectively. Due to a greater migration of the acidic pH front toward the cathode under 2 VDC/cm, nickel migration was slightly higher in the case of 2 VDC/cm than in the case of 1 VDC/cm. In general, the results showed that an increase in the voltage gradient from 1 to 2 VDC/cm improved slightly on the whole remedial performance.

Furthermore, both researchers (Reddy and Karri, 2006b) evaluated the IECR for clayey soil (i.e. kaolin) polluted with mixed contaminants (phenanthrene and nickel). Based on their experimental investigation, the following conclusions were drawn: (a) in batch tests, 76% of phenanthrene was oxidized with 5% H_2O_2, and 87% oxidation of phenanthrene was obtained in the case of 30% H_2O_2 (considerable oxidation of phenanthrene can occur using H_2O_2 as an oxidizer); (b) however, in electrokinetic tests, about 27% oxidation of phenanthrene was achieved with 5% H_2O_2, whereas with 30% H_2O_2, almost 56% oxidation of phenanthrene was obtained; (c) significant migration of nickel toward the cathode occurred; moreover, the mobility of the nickel was slightly increased in the presence of H_2O_2; and (d) the native iron was not a limiting factor for catalytic reaction of H_2O_2 to oxidize phenanthrene.

Chung and Kamon (2005) applied electrokinetics enhanced by ultrasonic devises to flashing soil contaminated with lead (500 mg/kg) and phenanthrene (500 mg/kg). A constant current of 50 mA was applied to graphite electrodes to initiate electrokinetic phenomena. Some test specimens were also subjected to ultrasonic waves at 30-Hz frequency. Tests were continued to a maximum of 15 days. A series of bench tests demonstrated an increase (up to 6%) in the removal of both inorganic and organic pollutants from clayey soil using a combination of electrokinetic and ultrasonic devises.

Elektorowicz and Hakimipour (2001, 2003a) presented a technology that permitted the simultaneous removal of heavy metals and PAHs from natural soil called Simultaneous Electrokinetic Removal of Inorganic and Organic Pollutants (SEKRIOP). This technology used EDTA for metal mobility and zwitterionic surfactants for hydrocarbon mobility. Furthermore, the application of cationic reactive membranes permitted capturing free metallic ions generated by electrokinetic phenomena before their precipitation in the cathode area. The capture of metal–EDTA complexes was done on anionic reactive membranes.

15.2.2 SEKRIOP Technology for the Simultaneous Treatment of Mixed Contaminated Soils

The objective of SEKRIOP was the simultaneous removal of heavy metals and hydrophobic organic carbons from natural soil using the electrokinetic phenomena. Since the main goal of electrokinetic processes is to affect the migration of contaminants in an imposed electric field via electroosmosis, electromigration, and electrophoresis, the experimental approach considered (a) the adequate application of conditioned liquids for the mobilization and transport of organic and inorganic contaminants, (b) the control of the electrokinetic phenomena, particularly anodic and cathodic zones, and (c) the assessment of the effect of timing on contaminant removal and energy consumption. To achieve such objectives, the remediation technology should include (a) choosing adequate conditioning liquids, (b) designing their supply systems, and (c) designing a system for their removal.

Elektorowicz and Hatim (2000) showed that amphoteric surfactants have useful properties for solubilization, desorption, and mobilization of PAH and subsequent transportation within a direct current (DC) electrical field. The use of chelating agents, such as EDTA, to mobilize and recover heavy metals from soil (Choudhury and Elektorowicz, 1997) has also given excellent results when the system included reactive membranes in order to uptake metals before their precipitation in the cathode area (Elektorowicz et al., 1996).

The simultaneous introduction of several conditioning liquids presents a challenge, particularly in clayey soils. Elektorowicz and Hakimipour (2003a) designed a system for the simultaneous desorption of heavy metals and PAHs, their simultaneous transport through the clayey soil, and removal. The system included the simultaneous utilization of a surfactant, a chelating agent, and special electrode supply/removal systems in combination with electrokinetic transport.

To achieve the above-mentioned objective, the following processes have to be considered: (a) formation of H^+ at an oxidized anode area (decrease of pH) and OH^- at a reduced cathode area (increase of pH); (b) dissociation of soluble compounds within an entire length of cell; (c) movement of cations and anions into respective electrodes; (d) displacement of negatively charged and pH-dependent colloidal particles of clays toward the anode; (e) formation of pH-dependent complexes of EDTA–metals and their transport toward the anode; (f) electroosmotic transport of inert particles (phenanthrene compound) toward the cathode; (g) amphoteric surfactant behavior (in the presence of a variable pH within the cell), formation of micelles, and desorption of phenanthrene; (h) displacement of micelles and their transformation; and (i) transport and removal of conditioning liquids.

The various stabilities of complexes, the formation of an acidic front in the soil, and the unknown sorption/desorption processes of phenanthrene in the presence of heavy metals and kinematics of these processes necessitated the creation of eight experimental systems. Therefore, each electrokinetic cell (from C1 to C8), representing a different system, has introduced a new contribution to the entire experiment. Subsequently, each system had a different sequence of conditioning liquids supply and time of contaminant removal.

The laboratory investigations were performed using clayey soil excavated in a Montreal area (Canada). The analysis showed the following characteristics of soil (after the removal of gravel): pH = 7.6, cation exchange capacity (CEC) = 21 meq/100g, organic matter content = 1.3 %, carbonates = 4.5%, sulfates = 0.6 ppm, and the concentration of significant metals (Ca = 250 mg/kg, Fe = 273 mg/kg, K = 176 mg/kg, Ni = 31 mg/kg, Pb = 9 mg/kg). The mineralogical composition of soil showed the following: chlorite (14.1%), illite (49.8%), amphibole (6.0%), kaolinite (1.3%), quartz (4.6%), dolomite (3.2%), and feldspar (21%).

The above-described dried soil was spiked with phenanthrene (600 mg/kg) and with metal solutions (1000 mg/L) of low mobility lead and mobile nickel. The soil was left for the aging process. The contaminated soil in the above-described way was placed in eight electrokinetic cells and compacted in several layers.

Perforated stainless steel electrodes (diameter = 1.0 cm) were fixed to the cell, and two porous zones were placed close to the anode and cathode for the injection of conditioning liquids (Fig. 15.1). The electrode system was enhanced with reactive membranes (ion exchange textile [IET]). In all sets, the electrodes were connected to a DC power supply (TES 6230) to achieve a voltage gradient of 0.3 V/cm. Electrical parameters along the distance between electrodes were monitored during the tests using a silver probe electrode. The cathode design permitted on collecting and measuring the cathode liquid discharge on a daily basis.

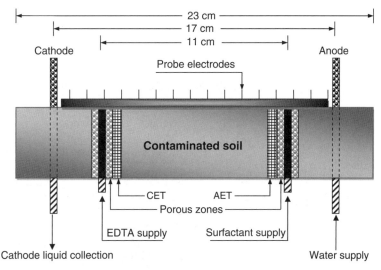

Figure 15.1. Configuration of the electrokinetic system Simultaneous Electrokinetic Removal of Inorganic and Organic Pollutants (SEKRIOP). CET, cation exchange textile; AET, anion exchange textile.

TABLE 15.1. Order of Conditioning Liquids Supply into Eight Electrokinetic Cells

Cell Number	Surfactant Supply Period	EDTA Supply Period	Entire Period of Tests (Days)
C1	No supply	No supply	22
C2	1st–22nd day	1st–22nd day	22
C3	1st–22nd day	11th–22nd day	22
C4	11th–22nd day	1st–22nd day	22
C5	1st–11th day	11th–22nd day	22
C6	11th–22nd day	1st–11th day	22
C7	1st–21st day	21st–42nd day	42
C8	21st–42nd day	1st–21st day	42

In order to optimize the simultaneous removal, eight cell systems (C1–C8) were designed, and each system had a different sequence of conditioning liquids supply (Table 15.1). Zwitterionic surfactant was supplied continuously through the anode porous zone, and the chelating agent EDTA was supplied continuously through the cathode porous zone during periods shown in Table 15.1. The cell C1 was used as a control cell, and during 22 days of test period, only water was injected to both porous zones.

At the end of the experiment, the soil was sliced in an approximately equal thickness around each probe location between electrodes. This sampling protocol allowed the determination of metals, phenanthrene, and other parameter variation with distance between electrodes.

The results showed that the application of the multifunctional method SEKRIOP to clayey soil permitted the simultaneous removal of heavy metals (Pb, Ni) and PAH (phenanthrene) *in situ*. The removal was observed in all cells including the control cell (Table 15.2). However, the removal percentage [normalized calculations expressed as the final concentration C (mg/kg) to the initial concentration C_o (mg/kg)] varied due to the different response of physicochemical processes, which were different in each cell.

TABLE 15.2. Summary of the Results of Mixed Contaminated Soil Treatment Using SEKRIOP

Cell	Surfactant Supply Period	EDTA Supply Period	Average Metal Removal (%)		Average Phenanthrene Removal (%)	Energy Consumption (kWh/m³)
			Lead	Nickel		
C1	No supply	No supply	35.2	75.9	55.6	13.02
C2	1–22	1–22	82.3	73.1	73.6	25.56
C3	1–22	11–22	69.9	57.1	76.3	29.14
C4	11–22	1–22	76.5	58.6	69.6	24.74
C5	1–11	11–22	73.6	82.0	71.6	21.52
C6	11–22	1–11	84.9	84.3	73.6	28.43
C7	1–21	21–42	78.8	86.9	60.1	27.84
C8	21–42	1–21	85.2	73.6	61.6	34.32

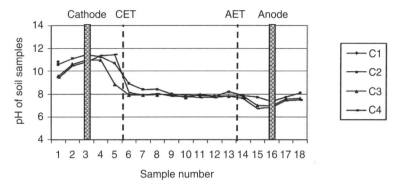

Figure 15.2. pH distribution of soil in cells C1–C4 after the treatment of the mixed contaminated soil. CET, cation exchange textile; AET, anion exchange textile.

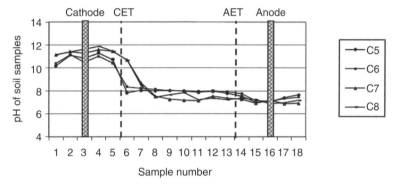

Figure 15.3. pH distribution of soil in cells C5–C8 after the treatment of the mixed contaminated soil. CET, cation exchange textile; AET, anion exchange textile.

Removal higher than 85% was observed for both "low mobile" lead and "mobile" nickel where EDTA was supplied at the beginning of the electrokinetic process (cells C6, C7, and C8). The highest phenanthrene removal of 76% was achieved where surfactant was used at the beginning of the treatment (C3) in the absence of EDTA. Among all experimental sets, cell C6 has had the best results in the simultaneous removal of organic and inorganic contaminants. The system applied to cell C6 was capable of removing an average value of 85% of lead, 84% of nickel, and 74% of phenanthrene. However, when particular soil slices were compared in all cells, the removal was able to achieve as much as 96%, 95%, and 84% for lead, nickel, and phenanthrene, respectively. In this cell, EDTA was supplied during an entire 22-day period, and the surfactant supply system was connected to the cell in the second phase (after 11 days).

Figures 15.2 and 15.3 show the pH distribution between electrodes at the end of the tests for cells C1–C4 and cells C5–C8, respectively. As expected, there was a general increase from anode to cathode. The initial pH of soil was 7.6, and the formation of hydrogen ions at the anode lowered the soil pH. The water supply at the anode attenuated a usual drop of pH in this area. The increase of pH in the cathode region was due to the formation of OH⁻ ions. In this area, the reactive membranes influenced pH distribution and prevented the precipitation of metals before membranes.

The presence of some lead contents close to the anode area was directly related to the effectiveness of EDTA in the mobilization of lead. Lead has a strong affinity for clay soil and has a higher power of exchange with soils than other heavy metals. The highest removal of lead was observed when EDTA was supplied during the first period of treatment in the electrokinetic process. This removal was more than two times higher than those shown in the control cell C1. Due to a low metal mobility, the response of membranes in control cells was much lower than in other cells. Due to a high stability of lead–EDTA complexes, the anion reactive membranes captured more than 30 times lead complexes than cation ones.

The results for cells C3, C4, C5, and C6 showed 57%, 59%, 82%, and 84% average removal of nickel between reactive membranes, respectively. Cells C3–C6 showed a high removal of nickel beside the anode area, leading to an average removal in the range of 57%–84%. Generally, the relocation of nickel was much more efficient than lead. It seems that the nickel responded well for both conditioners as well as to the water washing system in the presence of the DC field. Nickel is easily subjected to complexation in the presence of various compounds including organics, which gives a high mobility in soil. In fact, the anionic reactive membranes captured more than 20 times nickel than cationic ones; which demonstrates once more a high stability of Ni–EDTA complexes.

The distribution of phenanthrene in the soil for cells C3, C4, C5, and C6 showed, 76%, 70%, 72%, and 74% average removal of phenanthrene between reactive membranes, respectively. It can be concluded that coupling the supply of surfactant with electrokinetics enhanced the mobility of phenanthrene. The highest removal of phenanthrene was achieved in cell C3. The surfactant was supplied for a longer period in this cell. No impact of the duration of the selected test on the phenanthrene removal was observed. All systems demonstrated an insignificant capture of phenanthrene on membranes.

The simultaneous transport of organic and inorganic contaminants within the soil in all cells was observed, and the following order of eight systems for various contaminants was assessed.

The order of lead removals was

$$C1 < C3 < C5 < C4 < C7 < C2 < C6 < C8.$$

The order of nickel removals was

$$C3 < C4 < C2 < C8 < C1 < C5 < C6 < C7.$$

The order of phenanthrene removals was

$$C1 < C7 < C8 < C4 < C5 < C2 < C6 < C3.$$

In order to demonstrate the effectiveness of a particular setup, the energy consumption in each cell was estimated (Table 15.2). Generally, the resistance in the electrokinetic cell, filled with soil, increases with time (Fig. 15.4, cell C1). However,

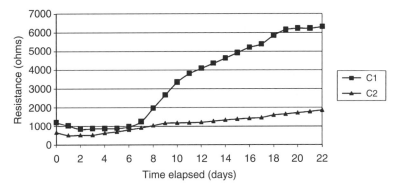

Figure 15.4. Changes of soil resistance in cells C1 (control) and C2 versus time.

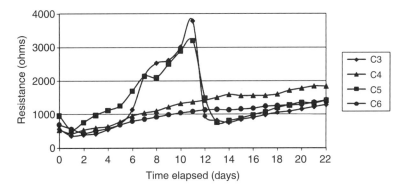

Figure 15.5. Changes of soil resistance in cells C3, C4, C5, and C6 versus time.

the introduction of EDTA prevents a high increase of resistance; for example, cell C2 showed the resistance four times lower than in the control cell. This fact provoked the increase of energy consumption in cell C2. At the same time, the chemical analysis of soil after the tests showed that the lead removal increased 2.3 times and phenanthrene removal by 1.3 times in cell C2.

Figure 15.5 shows the total resistance for cells C3, C4, C5, and C6 versus the elapsed time. As time progressed, the total resistance increased in all cells. However, the injection of EDTA since day 11 decreased the total resistance. Similar patterns were observed in long running cells (C7 and C8). However, doubling the removal time increased the energy consumption by only 1.3 times. Finally, the following order of energy consumption was found:

$$C1 < C5 < C4 < C2 < C7 < C6 < C3 < C8.$$

For this optimal removal with the application of both conditioning liquids, the energy consumption was calculated to be $28.4\,kWh/m^3$ of mixed contaminated soil. Assuming in Canadian dollar (CAD) that CAD 0.03/kWh is the unitary cost, the electrical energy cost for soil remediation would be CAD 0.85 per m^3. The lowest energy cost (beside control cell) of CAD $0.65/m^3$ was observed when a surfactant was applied during the first half period and a chelating agent during the second half period of treatment. In this case, the costs associated with conditioning liquids would also be the lowest. However, the remediation performance in this case was lower than in the above-described cell C6. The laboratory tests also permitted assessing the influence of conditioning liquids on the power consumption. For the optimal removal condition (C6), the following volumes of conditioning liquids were used for contaminant mobilization and moisture increase over 22 days: EDTA: 0.078 L/L of soil; surfactant: 0.16 L/L of soil; water: 0.21 L/L of soil.

The major factors significantly affecting the unit price of mixed contamination treatment are (a) site properties (e.g. soil matrix, initial and target contaminant concentrations, moisture content, depth of contamination, concentration of nontarget ions), (b) electrokinetic system applied (e.g. array of electrodes, voltage, type of electrodes), (c) costs of conditioners and their application time, (d) site preparation requirements, (e) variable operation costs (e.g. electricity, labor, residual waste processing).

Considering the effectiveness of the simultaneous heavy metals and PAH removal, it can be concluded that SEKRIOP might be used for an electrokinetic *in situ* remediation of mixed contaminated soils. The development of the above-described multifunctional method permits remediating the soils, particularly those characterized with low permeability. The results from the research can be applied to various municipal and industrial sites containing petroleum products and heavy metals.

15.3 SUMMARY

1. Numerous sites are contaminated with both inorganic (heavy metals) and organic contaminants (e.g. PAHs), and the technologies that treat such mixed contaminants are very limited. Electrokinetic technologies have the potential to remediate mixed contaminated soils; however, the efficiency of this technology depends on
    i. the multifunctioning electrode system,
    ii. the type of conditioner applied (e.g. surfactants, cosolvents, chelating agents, cyclodextrin),
    iii. the voltage gradient applied, and
    iv. the combination and order of conditioned liquids applied leading to
        a. control of pH and resistance distribution,
        b. priority in complex formation, and
        c. priority in mobility of such complexes subjected to subsequent or simultaneous electroosmotic, electrophoretic, and ion migration motions.
2. Electrokinetics can be applied *in situ* and *ex situ* into a mixed contamination in low permeability soil, mud, sludge, and marine dredging; in fact, soil type does not pose any significant limitation on the technology.
3. The technology uses low-voltage DC and is capable of separating and removing water, heavy metals, radionuclides, and organic contaminants.

4. SEKRIOP technology is capable of the simultaneous removal of heavy metals and organic pollutants using a multifunctional electrode system.
5. The results from the research can be applied to various municipal and industrial sites containing petroleum products and heavy metals.
6. The introduction of a conditioning liquid changes the resistance of the system, increasing the energy costs. Subsequently, the accurate duration and concentration of applied conditioned liquids should be optimized.
7. The costs of mixed contaminated site *in situ* remediation are still comparatively lower than other treatment methods.
8. The remediation of contaminated sites may be efficient when a combination of electrokinetics with other methods is designed.
9. The entire process should be properly controlled, since remediation of one group of contaminants might transform others to more mobile/toxic forms.
10. The technology design might vary with site specifics; therefore, the ultimate decision regarding technological parameters (e.g. type of conditioning liquids, voltage applied) is based on the site's and contaminants' characteristics. Preliminary bench tests are recommended.

REFERENCES

Alloway BJ. (1990). *Heavy Metals in Soils*. New York: John Wiley & Sons.

Choudhury A, Elektorowicz M. (1997). Enhanced electrokinetic methods for lead and nickel removal from contaminated soils. *32nd Central Symposium on Water Pollution Research*, February 4–5, Burlington, Ontario, Canada: CAWQ.

Chung HI, Kamon M. (2005). Ultrasonically enhanced electrokinetic remediation for removal of Pb and phenanthrene in contaminated soils. *Engineering Geology* **77**(3–4):233–242.

Degtiareva A, Elektorowicz M. (2001). A computer simulation of water quality change due to dredging of heavy metals contaminated sediments in the Old Harbor of Montreal. *Water Quality Research Journal of Canada* **36**(1):1–19.

Elektorowicz M, Boeva V. (1996). Electrokinetic supply of nutrients in soil bioremediation. *Environmental Technology* **17**:1339–1349.

Elektorowicz M, Chifrina R, Konyukhov B. (1995). Behaviour of tension-active compounds used to clean up of contaminated manufactured gas plant sites. *Land Contamination & Reclamation* **3**:2–4.

Elektorowicz M, Chifrina R, Kozak M, Hatim G. (1996). The behavior of ion exchange membranes in the process of heavy metals removal from contaminated soil. *Proceedings, CSCE-4th Environmental Engineering Specialty Conference*, June 28–July 1, Edmonton, AB, Canada: CSCE, pp. 229–240.

Elektorowicz M, Emon MH, Ayadat T. (2008b). Stabilization of metals and PAHs in mix contaminated soil. *Proceedings, CSCE Conference*, June 10–13, Quebec, QC, Canada.

Elektorowicz M, Hakimipour M. (2001). Electrical field applied to the simultaneous removal of organic and inorganic contaminants from clayey soil. *18th Eastern Research Symposium on Water Quality*, October 18, Montreal, Canada: CAWQ.

Elektorowicz M, Hakimipour M. (2003a). Practical consideration for electrokinetic remediation of contaminated soil. *CSCE 8th Environmental and Sustainable Engineering Specialty Conference and Offshore Engineering*, June 4–7, Moncton, Canada: CSCE, pp. 689–698.

Elektorowicz M, Hakimipour M. (2003b). Electrokinetic soil remediation method for mixed contaminated soil. *6th Conference on Civil Engineering*, May 5–7, 2003, Isfahan, Iran.

Elektorowicz M, Hasnawie R, Ayadat T, Chifrina R. (2008). Formation of silica grouts curtains or containments in mineralized groundwater. *Journal of Environmental Engineering and Science* **7**:275–287.

Elektorowicz M, Hatim G. (2000). Hydrophobic organic compound removal from clayey soil due to surfactant enhanced electrokinetic system. *CGS 53rd Canadian Geotechnical Conference*, October 15–18, Montreal, Canada: CGS, pp. 617–621.

Elektorowicz M, Ju L, Oleszkiewicz J. (1999). Bioavailability limitations related to the presence of clays. In *Bioavailability of Organic Xenobiotics in the Environment—Practical Consequences for Environment* (eds. Ph Baveye, JC Block, VV Goncharuk). Dordrecht, the Netherlands: Kluwer Academic Publishers, pp. 349–376.

Flag W. (1974). Biochemistry of soil organic matter (a review). Fiche no: 29112. *Soils Bulletin (FAO)* **27**:31–69.

Flaig W, Beutelspacher H, Rietz E (1975). Chemical composition and physical properties of humic substances. In *Soil Components*, Vol. **1** (ed. JE Gieseking). New York: Springer, pp. 1–211.

GERLED. (1983). *Le Groupe d'étude et de restauration des lieux d'élimination de déchets dangereux*. Quebec, Canada: Environment Quebec Publications.

Griffiths BS, Diza-Ravina M, Ritz K, McNicol JW, Ebbelwhite N, Baath E. (1997). Community DNA hybridization and profiles of microbial community from heavy metal polluted soils. *FEMS Microbiology Ecology* **24**:103–112.

Juck D, Charles T, Whyte LG, Greer CW. (2000). Polyphasic microbial community analysis of petroleum hydrocarbon-contaminated soils from two Northern Canadian communities. *FEMS Microbiology Ecology* **33**:241–249.

Kaoser S, Barrington S, Elektorowicz M, Wang L. (2004a). Copper adsorption with Pb and Cd in sand-bentonite liners under various pHs. Part I. Effect on total adsorption. *Journal of Environmental Science and Health* **39**(09):2241–2256.

Kaoser S, Barrington S, Elektorowicz M, Wang L. (2004b). Copper adsorption with Pb and Cd in sand-bentonite liners under various pHs. Part II. Effect on adsorption sites. *Journal of Environmental Science and Health* **39**(09):2257–2274.

Kaoser S, Barrington S, Elektorowicz M, Wang L. (2005). Effect of Pb and Cd on Cu adsorption of bentonite liners. *NRC Canadian Journal of Civil Engineering* **32**(1):241–249.

Kaschl A, Römheld V, Chen Y. (2002). Cadmium binding by fractions of dissolved organic matter and humic substances from municipal solid waste compost. *Journal of Environmental Quality* **31**:1885–1892.

Khodadoust R, Reddy KR, Maturi K. (2004). Removal of nickel and phenanthrene from kaolin soil using different extractants. *Environmental Engineering Science* **21**(6):691–704.

Maturi K, Reddy KR. (2006). Simultaneous removal of organic compounds and heavy metals from soils by electrokinetic remediation with a modified cyclodextrin. *Chemosphere* **63**(6):1022–1031.

Maturi K, Reddy KR. (2008). Cosolvent-enhanced desorption and transport of organic and metal contaminants in soils during electrokinetic remediation. *Water, Air, and Soil Pollution* **189**(1–4):199–211.

Merian E. (1991). *Metals and Their Compounds in the Environment: Occurrence, Analysis, and Biological Relevance*. New York: VCH Publishing.

Palmroth MRT, Koskinen PEP, Kaksonen AH, Münster U, Pichtel J, Puhakka JA. (2007). Metabolic and phylogenetic analysis of microbial communities during phytoremediation of soil contaminated with weathered hydrocarbons and heavy metals. *Biodegradation* **18**(6):769–782.

Reddy KR, Ala PR. (2006). Electrokinetic remediation of contaminated dredged sediment. *Journal of ASTM International* **3**(6):254–267.

Reddy KR, Ala PR, Sharma S, Kumar SN. (2006). Enhanced electrokinetic remediation of contaminated manufactured gas plant soil. *Engineering Geology* **85**(1–2):132–146.

Reddy KR, Donahue M, Saichek RE, Sasaoka R. (1999). Preliminary assessment of electrokinetic remediation of soil and sludge contaminated with mixed waste. *Journal of the Air & Waste Management Association* **49**(7):823–830.

Reddy KR, Karri MR. (2006a). Effect of voltage gradient on integrated electrochemical remediation of contaminated mixtures. *Land Contamination & Reclamation* **14**(3):685–698.

Reddy KR, Karri MR. (2006b). Integrated electrochemical remediation of mixed contaminants in subsurface. *5th International Congress on Environmental Geotechnics*, Cardiff, Wales, UK. London: Thomas Telford Publishing, pp. 271–278.

Reddy KR, Karri MR. (2008a). Effect of oxidant dosage on integrated electrochemical remediation of contaminant mixtures in soils. *Journal of Environmental Science and Health Part A* **43**(8):881–893.

Reddy KR, Karri MR. (2008b). Electrokinetic delivery of nanoiron amended with surfactant and cosolvent in contaminated soil. *International Conference on Waste Engineering and Management*, May 28–30, Hong Kong: HKIE.

Reddy KR, Kumar SN. (2002). Electrokinetic remediation of organic silty sand contaminated with heavy metals and PAHs at a MGP site. *Soil and Sediment Contamination: An International Journal* **11**(3):426.

Reddy KR, Maturi K. (2005). Enhanced electrokinetic remediation of mixed heavy metal and organic contaminants in low permeability soils. *16th International Conference on Soil Mechanics and Geotechnical Engineering*, September 12–16, Osaka, Japan. Rotterdam, the Netherlands: Millpress Science Publishers, pp. 2429–2432.

Roach N, Reddy KR. (2006). Electrokinetic delivery of permanganate into low permeability soils. *International Journal of Environment and Waste Management* **1**(1):4–19.

Scott JE. (1968). Patterns of specificity in the interaction of organic cations with acid mucopolysaccharides. In *The Chemical Physiology of Mucopolysaccharides* (ed. G Quintarelli). Boston: Little, Brown and Co., pp. 171–187.

Serpaud B, Al-Shukry R, Casteignau M, Matejka G. (1994). Adsorption des métaux lourds (Cu, Zn, Cd et Pb) par les sédiments superficiels d'un cours d'eau: rôle du pH, de la température et de la composition du sediment [Heavy metal adsorption (Cu, Zn, Cd and Pb) by superficial stream sediments: Effects of pH, temperature and sediment composition]. *Revue des sciences de l'eau* **7**(4):343–365.

Sparks DL, Fendorf SE, Zhang PC, Tang L. (1993). Kinetics and mechanisms of environmentally important reactions on soil colloidal surface. In *Migration and Fate of Pollutants in Soils and Subsoils*, Vol. 32, NATO ASI Series G (eds. D Petruzzeli, F Helfferich). Berlin, Germany: Springer-Verlag, pp. 141–168.

Stevenson FJ. (1994). *Humus Chemistry: Genesis, Composition, Reactions*. New York: John Wiley & Sons.

USEPA. (1993). Toronto Harbour Commissioners (THC). Soil Recycle Treatment Train Applications. *Analysis Report. EPA/540/AR-93/517*.

USEPA. (2007). *Risk Assessment Guidance for Superfund, Human Health Evaluation Manual*. Washington, DC: Office of Emergency and Remedial Response.

Zhang W, Lo IMC. (2007). Chemical-enhanced washing for remediation of soils contaminated with marine diesel fuel in the presence/absence of Pb. *Journal of Environmental Engineering* **133**(5):548–555.

PART V

ELECTROKINETIC BARRIERS

16

ELECTROKINETIC BARRIERS FOR PREVENTING GROUNDWATER POLLUTION

ROD LYNCH

16.1 INTRODUCTION

Here is a typical example of a relevant problem: an ancient Sardinian hillside mine. The island of Sardinia has been mined for silver, zinc, copper, and lead since Phoenician and Roman times and, until the 1970s, was the largest source of heavy metals in Europe. It has left a legacy of pollution problems associated with mine drainage, carrying pollutants from hillside mines down to valleys below. This is a situation that could well be applicable to treatment by an electrokinetic barrier. The application is as shown in Figure 16.1.

There are a number of situations (e.g. the one outlined above) in which clean uncontaminated land adjacent to a problem area is at risk of contamination. Drainage from hillside mines located above an agricultural land or leakage from old landfill sites is a potential problem. It is in such circumstances that a possible solution to preventing contamination is the use of an electrokinetic barrier (Fig. 16.2). This is different from remediation by electrokinetics in the following ways:

1. There is no attempt to remove the contaminant, only prevention of movement onto adjacent clean land.
2. The accumulation of pollutants around one of the electrodes may cause the pore fluid flow rate to fall, and hence, a partial hydraulic barrier can be formed.
3. In addition, the electrical resistance often rises in an ion-depleted zone, which means that power consumption is reduced. So, the power requirement for the

---

Electrochemical Remediation Technologies for Polluted Soils, Sediments and Groundwater,
Edited by Krishna R. Reddy and Claudio Cameselle
Copyright © 2009 John Wiley & Sons, Inc.

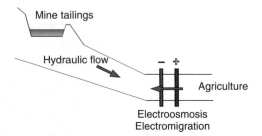

Figure 16.1. Schematic diagram of an electrokinetic barrier.

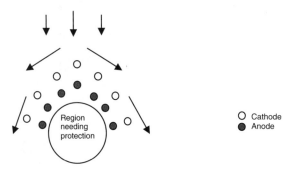

Figure 16.2. Plan view—contaminated groundwater is redirected due to the electrokinetic barrier.

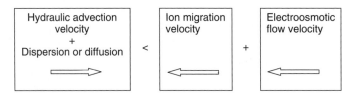

Figure 16.3. The concept of an electrokinetic barrier.

maintenance of an electrokinetic barrier may be significantly less than the initial power condition, or that required for conventional electrokinetic remediation.

16.1.1 Electrokinetic Barrier Concept

The concept of an electrokinetic barrier is extremely simple (Fig. 16.3). The action of an electric field is capable of moving ions in the pore fluid of soil (i.e. electromigration), and this can be used to oppose the spread of pollutants moving in groundwater under a natural hydraulic gradient or diffusion. In addition, an accompanying electroosmotic flow (EOF) aids this process by opposing the natural groundwater flow. The process is outlined in Section 16.1.3. So, the barrier can function if the contaminant ion movement by conventional, hydraulic processes is less than the ion migration velocity plus the EOF velocity.

The above inequality lists the major players in the action of an electrokinetic barrier. Whether diffusion is important will depend very much on the timescale (Fig. 16.3). For the movement of contaminants through sands, it may be that diffusion can be neglected initially. The movement of chemicals in a solution by diffusion can be quite slow compared with other processes. The distance moved is proportional to the square root of twice the diffusion coefficient times the time taken. For example, for a molecule with a free diffusion coefficient of $10^{-9}\,m^2/s$, the distance moved in 1 s is about 40 μm; in 1 day, about 1.3 cm; in 1 week, about 3.4 cm; and in 1 year, about 25 cm. So, for a first approximation, we can perhaps neglect diffusion. However, for long-term protection, for example, of a landfill, diffusion would of course have to be considered. Adsorption, biodegradation, and reduction–oxidation are also not included in this simple approach.

16.1.2 Electromigration

This is the movement of ions in the fluid due to an applied field. In groundwater, the ions are free to move through the water in the pores; however, their progress may be impeded somewhat by the soil particles. So, the mobility of the ions depends on the strength of the applied electric field, the size of the pores (or porosity), and the tortuosity, which allows for the fact that the particles have to weave their way around the soil particles.

Ion velocity due to electromigration can be calculated (Narasimhan and Sri Ranjan, 2000):

$$v_{ion} = u \frac{dV}{dL} n\tau, \qquad (16.1)$$

where τ is tortuosity, n is porosity, and dV/dL is the strength of the electric field, volts (V) across the length of soil (L). u is the ion mobility, which is given in Equation 16.2 (Acar and Alshawabkeh, 1993):

$$u = \frac{D_i z F}{RT}, \qquad (16.2)$$

where D_i is the diffusion coefficient of species i in dilute solution, z is the ionic charge, F is the Faraday constant, R is the gas constant, and T is the absolute temperature (K).

Typical approximate values for ion mobilities in a free solution at 25 °C are Na:5, K:8, H:36, OH:21 ($10^{-8}\,m^2/(V \cdot s)$) (Atkins, 2001). Note that hydrogen ions move almost twice as fast as OH$^-$ ions. So, in a field of 1 V/cm, the electromigration velocity of a sodium ion in a free solution is about 1.8 cm/h. In soil, tortuosity must be taken into account, so it will be somewhat less than this. Acar and Alshawabkeh (1993) reported that the electromigration velocity in soil is typically about 4.7 times less than that in a free solution, so this will reduce the velocity in the above example to about 0.38 cm/h.

16.1.3 Electroosmosis

Many soils are negatively charged on their surface, caused by the presence of oxygen atoms attached to silica or aluminium atoms. The cations in a solution are attracted to this negative surface layer, which leads to an uneven charge distribution in the fluid. Water molecules are polar and tend to associate themselves with ions. Since there are more mobile cations than anions, when the electric field is applied, there is a net movement of water toward the cathode, and this is called EOF. The EOF depends on the surface charge of the soil (which itself may be affected by the chemicals in the water), the concentration of ions in a solution, the strength of the electric field, and the porosity.

EOF velocity can be determined by Equation 16.3 (Casagrande, 1949):

$$v_{EOF} = \frac{Q_{EOF}}{A} = k_{EOF}\frac{dV}{dL}, \qquad (16.3)$$

where Q_{EOF} is the EOF and A is the cross-sectional area of soil, and where

$$k_{EOF} = \frac{\zeta D e n \tau}{\eta}, \qquad (16.4)$$

where D is the dielectric constant, ζ is the zeta potential, e is the permittivity of vacuum, n is porosity, and η is viscosity.

Typical values of k_{EOF} are 5.7 in kaolin, 5 in fine sand, and 5.8 in London clay ($10^{-5}\,cm^2/(V \cdot s)$) (Mitchell and Soga, 2005). So, the typical EOF velocity for a field of 1 V/cm in sand and an area of 1 cm² is about 0.18 cm/h.

16.1.4 Barrier Condition

For the barrier to prevent the movement of heavy metal cations, it follows that the hydraulic flow velocity of groundwater through the soil (v_{Hyd}) should not be greater than the sum of the electroosmotic velocity (v_{EOF}) plus the electromigration velocity (v_{ion}), assuming diffusion is neglected:

$$v_{Hyd} < v_{ion} + v_{EOF}. \qquad (16.5)$$

Combining Equations 16.1 and 16.3 above,

$$k_{perm}\frac{i}{n} < un\tau\frac{dV}{dL} + k_{EOF}\frac{dV}{dL}, \qquad (16.6)$$

where k_{perm} is the permeability (hydraulic conductivity) of the soil, i is the hydraulic gradient, and other terms are as defined above.

So, in the example above of a sodium ion moving through soil under the influence of a 1 V/cm electric field, the hydraulic flow must be less than the sum total of ion migration velocity and EOF velocity, which is $0.38 + 0.18 = 0.56$ cm/h or 1.5×10^{-6} m/s.

In the above simplified explanation, the effects of chemical gradients, principally diffusion, have been neglected. A more rigorous model has been given by Narasimhan and Sri Ranjan (2000).

It may be that this simple approach, which is relevant for a single ionic species, is an oversimplification and is not valid for more complex situations with multiple pollutant ions. For these cases, an irreversible thermodynamic approach has been suggested (Olsen, Gui, and Lu, 2007).

16.2 HISTORY OF ELECTROKINETIC BARRIER DEVELOPMENT

In 1986, a workshop was held at the University of Washington to investigate the electrokinetic remediation of waste sites. As a means of improving containment, Mitchell (1986) mentioned that applying a voltage across a clay liner could counter the gravity flow of liquids through the liner, and similarly, Probstein and Renaud (1986) suggested that a landfill site could be surrounded by a double row of electrodes to oppose hydraulic leakage by a counterflow generated by electroosmosis. In 1989, Lageman, Pool, and Seffinga coined the phrase "electrokinetic fence" in which a line of electrodes of alternating polarity was placed across a plume and contaminants are removed as they approach the fence. This is a combined remediation and barrier technology, because the contaminants are removed from the solutions around the electrodes.

As for a barrier alone, Mitchell and Yeung (1990) suggested that an electrokinetic counterflow may be an effective measure to stop the migration of contaminants under a hydraulic gradient. They proposed that an electrokinetic fluid flow barrier could be created by the continuous or periodic application of an electrical gradient across a compacted clay liner. Yeung and Mitchell (1993) were interested in the combined effects of chemical, electric, and hydraulic gradients on the movement of ions, and how this might help the situation of pollution spreading from leaking landfill sites. They investigated the idea of an electrokinetic barrier in clays and tested it with experiments mobilizing sodium chloride (Yeung, 1990). It appears that little was published about electrokinetic barriers in the latter half of the 1990s. In 2000, Narasimhan and Sri Ranjan published a model for predicting the buildup of ions at the electrodes. They tested their model with the sodium chloride data of Yeung and Mitchell and obtained good agreement between the model and the experiment.

Gregolec *et al.* (2001) showed that experiments combining electric and hydraulic gradients could be used to hinder charged species from moving with the groundwater flow while uncharged species will still be transported in the direction of the hydraulic flow. Experiments were done using sodium chloride solution moving though sand, both without and with an applied hydraulic gradient of 0.001. Without an applied hydraulic gradient, electromigration caused a high resistance zone, depleted of ions to limit the current. When a hydraulic gradient was applied, electromigration dominated the hydraulic flow at high voltage and vice versa at low voltages.

16.2.1 Combined Cleanup and Barrier

In 1989, a related technology of cleanup and barrier combined was described by Lageman, Pool, and Seffinga (Fig. 16.4). A line of electrodes placed across a plume of pollution was used to clean up the pollutant as it moved toward the "electrokinetic fence." The electrodes are operated submerged in an electrode solution, which can be removed for later treatment. Full-scale trials report successful cleanup, approaching 80% (Lageman, 1993). Further details are given in Chapter 19.

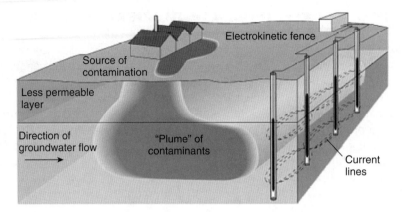

Figure 16.4. Concept of an electrokinetic fence, which is a barrier and cleanup combined (Geokinetics, Inc., Houston, TX) (US EPA, 1997).

16.3 RECENT STUDIES

The following summarizes the studies of electrokinetic barrier tests that have been done in the last 10 years or so.

16.3.1 Sodium Transport

As mentioned earlier, Narasimhan and Sri Ranjan (2000) tested their barrier model with the data produced earlier by Yeung and Mitchell (1993). In Yeung's (1990) experiments, a hydraulic head of 500 cm was applied to a silty clay soil sample of 10-cm length, generating a hydraulic gradient of 50. An electric field of 1 V/cm applied for just 1 h per day successfully opposed the hydraulic flow. After 25 days, the sodium ions were still contained within the soil sample. For further details, the reader is referred to Chapter 30.

16.3.2 Cadmium

In 2002, Ruggeri and Muntoni began to work on the barrier concept at the University of Cagliari in Sardinia to test its applicability to Sardinian pollution problems (Lynch, Muntoni, and Ruggeri, 2003; Muntoni, Ruggeri, and Lynch, 2003; Ruggeri, 2005; Lynch et al., 2007). They began with one-dimensional laboratory experiments in a conventional electrokinetic tube cell, with a length of 20 cm and a diameter of 8 cm, but adapted to supply a large hydraulic head on one side of the cell (Fig. 16.5). The cell contained 50 ppm cadmium-polluted Sardinian soil, an illitic-kaolinitic clay (soil A), which was surrounded on both sides by clean soil (see Fig. 16.6).

Without the aid of an electric field, under the influence of the hydraulic gradient, the cadmium spread quite rapidly to the soil down-gradient, sections 1, 2, and 3 (Fig. 16.7).

However, when a modest electric field of 10 V was applied over 20 cm of soil (0.5 V/cm), the cadmium migration was retarded and had only slightly migrated to section 3, the next section down-gradient of the contamination. At applied voltages

RECENT STUDIES 341

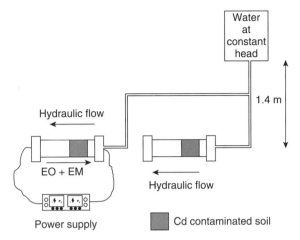

Figure 16.5. Experimental system for measuring cadmium or zinc movement through the barrier, conducted in parallel, with and without electric field (Ruggeri, 2005). Arrows show the direction of the hydraulic flux, and electroosmotic and ion migratory flow. The dark area is the cadmium-contaminated soil. EO, electro-osmosis; EM, electro-migration (ion migration).

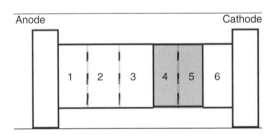

Figure 16.6. Sections 4 and 5 are contaminated before the start of the experiment (Ruggeri, 2005).

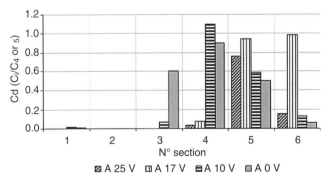

Figure 16.7. Movement of cadmium under the influence of different electric fields (soil A). Hydraulic flow is from right to left (Ruggeri, 2005; Lynch et al., 2007).

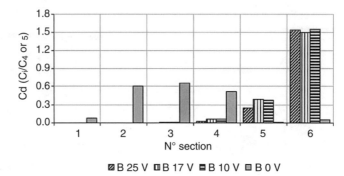

Figure 16.8. Movement of cadmium under the influence of different electric fields (soil B). Hydraulic flow is from right to left (Ruggeri, 2005; Lynch et al., 2007).

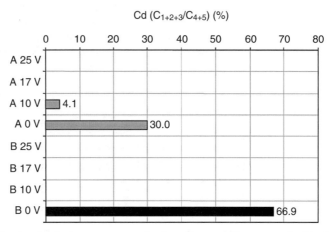

Figure 16.9. Ratio of cadmium concentration in adjacent soil compared with that of contaminated soil.

of 17 and 25 V, the cadmium was prevented from moving, even against a significant hydraulic gradient of 7, leaving the adjacent soil remaining unpolluted. Similar tests were done with soil B, an 80:20 mix of quartz sand and kaolinite clay, which was five times more permeable than soil A. The results are shown in Figure 16.8. With no applied electric field, cadmium spread significantly to sections 3 and 2, and even a little to section 1, the soil farthest away from the incoming pollutant. At applied voltages of 10, 17, and 25 V (i.e. voltage gradients of 0.5, 0.85, and 1.25 V/cm), the clean soil in sections 1, 2, and 3 was uncontaminated, and most cadmium was pushed up-gradient into section 1.

Figure 16.9 represents a measurement of the efficiency of the barrier; it shows the ratio of the mean cadmium concentrations in previously unpolluted and polluted sections of soil. In soil A with no electric field, 30% of cadmium polluted the adjacent soil. With 10 V of applied voltage (0.5 V/cm), still a small amount of contamination occurs, so the limiting voltage for full barrier efficiency lies between 10 and 17 V. In soil B, which is of higher permeability, the increased hydraulic flow results

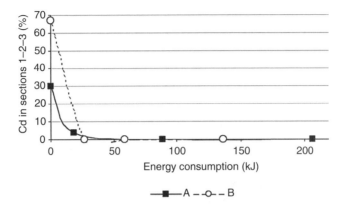

Figure 16.10. Relation between energy expenditure and contamination level in the down-gradient sections 1–2–3.

in more contamination when no electric field is present. Even 10 V is sufficient for the barrier to function.

Figure 16.10 indicates the energy expenditure compared with the degree of contamination that occurred. The optimum efficiency seems to occur at around 25 kJ, which is at a power level of around 0.07 W. Since this is for an area of barrier of 50 cm^2, this indicates that a power level of around 14 W/m^2 would be required.

A word of caution: At very high pH (usually greater than 10), some heavy metals, such as Cd, Pb, and Cr, can form anions, and their solubility can increase with increasing pH. This would mean that under these extreme conditions in an electric field, they would travel back toward the anode, that is, back down-gradient. However, they would eventually find themselves in a lower pH environment and therefore become more insoluble again. In the cadmium experiments of Ruggeri above, the pH levels found were less than 10, and no evidence of reversal of cadmium was found.

16.3.3 Zinc

Similar tests to those described above were also done with zinc pollution of the same soils (Ruggeri, 2005). The apparatus used is as shown in Figure 16.4. Similar to the cadmium experiments (Fig. 16.6), sections 4 and 5 were contaminated with 2000 ppm of zinc. Figure 16.11 shows how the transport of zinc down-gradient into soil sections 1–2–3 is severely reduced when the voltage is applied. At electric fields of 0.5 and 0.85 V/cm (10 and 17 V), there was some pollution down-gradient but very little at 1.25 V/cm (25 V) (Ruggeri, 2005).

16.3.4 Copper

16.3.4.1 Laboratory Two-Dimensional Tests in a Flat Tank Previous work in Cambridge had been underway to try an electrical sweeping method for cleanup (Chan, Lynch, and Ilett, 2001). In this method, a flat tank of depth 5 cm was used to simulate a test area. Two rows of oppositely charged electrodes were first used

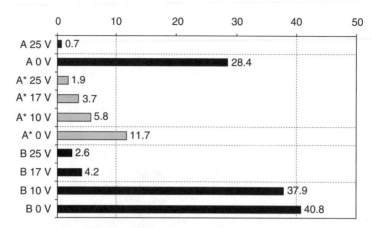

Figure 16.11. Percentage of total initial zinc concentration in sections 1–2–3 of soils at the end of the test (Ruggeri, 2005). [A and A* are two sources from Escalaplano, Sardinia of illite-kaolinite clay. Soil B is 100% kaolinite].

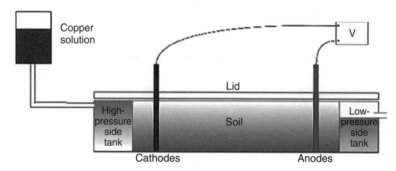

Figure 16.12. Schematic side view of the flat tank experiment (Winfield, 2004).

to mobilize copper ions toward the cathode row. Then, the polarity was changed, that is, only the electrodes at the opposite ends of the original cathode row were left oppositely charged, so that the copper ions could be moved at the right angles, to attempt to sweep the copper into a corner of the square tank. The experiment was only moderately successful in that the first part successfully moved the copper, but the second stage produced only a modest extra movement of copper. Once the copper compounds precipitated in the pores of the soil, it was difficult to move them, which led to the conclusion that maybe the problems of the buildup of pollutants around the electrodes and the subsequent fall in current could perhaps be used to advantage in a barrier technology, where lower current and hence energy costs would be a benefit, not a hindrance.

In a collaboration with Ruggeri and Muntoni in Cagliari, Lynch and Winfield in Cambridge began a series of tests in the same flat tank to test the barrier concept (Lynch et al., 2005). The flat tank was used to simulate a test area, and two side tanks were used to arrange a hydraulic gradient across it (see Figs. 16.12 and 16.13). A lid was fitted so that a significant hydraulic gradient could be used (Lynch, Muntoni, and Ruggeri, 2003; Muntoni, Ruggeri, and Lynch, 2003).

RECENT STUDIES 345

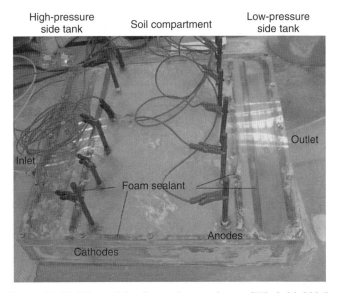

Figure 16.13. View of the flat tank experiment (Winfield, 2004).

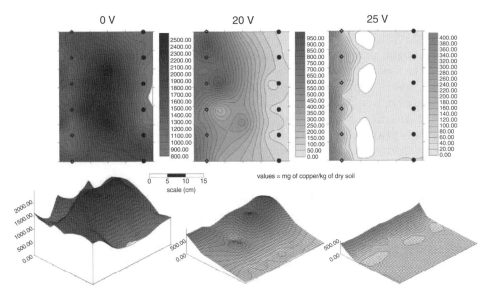

Figure 16.14. Distribution of copper in the soil at 0, 20, and 25 V.

The 40 × 30 × 5 cm flat tank was filled with wet silica silt, then left overnight to consolidate (see Fig. 16.13). Twelve carbon electrodes were placed across the soil, and the tank was sealed with a lid. A container with 1000 mg/L copper solution supplied contaminated water to one side of the soil, to maintain a hydraulic gradient of 1.3. Typical current was 25 mA at 25 V. Samples of soil were taken in a grid across the tank and analyzed for copper content. The resulting copper concentrations are shown in Figure 16.14, for applied voltages of 20 and 25 V, as well as the control case, with no voltage applied.

Without an applied voltage, due to the effect of the hydraulic flow, copper breakthrough was detected throughout the tank in 22 h. However, with the application of 25 V, most of the copper was prevented from entering the soil, even though there was a hydraulic gradient of 1.3. With an applied voltage of 20 V, some incursion of copper had occurred, but it was contained between the electrodes. The energy usage at the end of the test was only 14 W·h. This is equivalent to a power of 31 W/m^2 of barrier.

16.3.4.2 Toward a Solar-Powered Barrier

16.3.4.2.1 Two-Dimensional Copper Tests with Pulsed Voltage. An attractive idea, which is under investigation at Cambridge, is to use only solar power for providing the electric field. In principle, the provision of solar power should be much easier for a barrier than for cleanup because the power requirements should be much less.

In 2004, Nowicki had been working in the Photonics group on the efficient use of solar power and had designed a suitable circuit, which was applied to the electrokinetic barrier experiment (Nowicki, Lynch, and Amaratunga, 2005). Three 12-V, 5-W amorphous silicon panels were used, backed up by two 2.3 A·h lead acid batteries. The operation was as follows: The battery voltage rose as the battery was being charged. At an upper battery limit, it was disconnected from the panel, as otherwise it would overcharge. After disconnection, the duty cycle was then chosen to force the voltage of the load to be 25 V. On the other hand, if there was not enough sunlight, then the system went into shading mode, in which the battery powered the load. The transition was made whenever the input voltage dropped below the shading limit. Whenever the voltage of the panel rose above the shading limit, the system would switch to solar provision of power.

The test was carried out at a far-from-ideal time on December 28–30, almost the shortest days of the year. Two hours of sunlight were received before the panels went into shade and switched to battery power. Nevertheless, power was maintained for 28 h (see Fig. 16.14).

The copper concentration distribution across the soil is shown in Figure 16.15. Approximately 40% of the tank (the lower part of Fig. 16.13) was left without electrodes, to provide a comparison with no field present. The copper concentration across the tank, parallel to the flow, are compared in Figures 16.16 and 16.17. Where there are no electrodes, the copper pollutant has advanced significantly.

Where there was an electric field applied, the advance of the pollutant had been reduced significantly. The power failure due to the lack of light/battery backup meant that the integrity of the barrier was not maintained for 34% of the test.

16.3.4.2.2 One-Dimensional Tests with Pulsed Voltage. In the next series of experiments (Reeve, 2007; Reeve and Lynch, 2007), the object was to investigate the possibility of eliminating a battery-supported solar power supply to provide a current overnight. The philosophy can be explained as follows: At night, when the solar panel is inactive, the pollutant is permitted to advance through the soil between the electrodes. When the solar panel again becomes active, the contaminant is repelled. It is required to push the contaminant far enough against the hydraulic gradient so that when power goes off again, the contaminant does not escape past the anodes (Fig. 16.18).

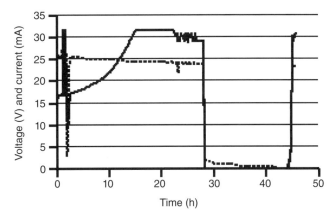

Figure 16.15. The voltage (dashed line) and the current at the electrodes for the duration of the experiment. Duration: from 1:00 p.m. on Tuesday, December 28 to Thursday, December 30, at 10:30 a.m. (Nowicki, 2005).

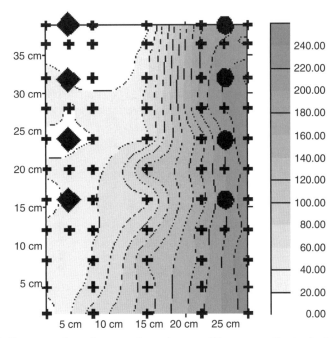

Figure 16.16. Concentration of copper in parts per million across the tank at the end of the experiment. The crosses represent the location of the samples. The diamonds are the anodes, and the circles are the cathodes (Nowicki, 2005).

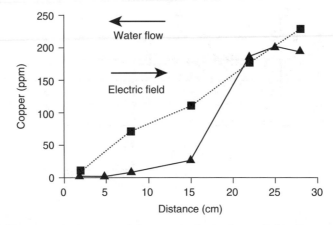

Figure 16.17. Two cross sections were taken across the tank, parallel to the contaminant flow. Dotted line: under no electric field. Continuous line: electric field present (Nowicki, Lynch, and Amaratunga, 2005).

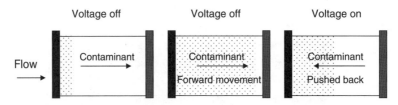

Figure 16.18. Diurnal pulsed power barrier operation (Reeve, 2007).

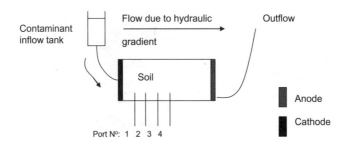

Figure 16.19. Schematic diagram of the barrier test (Reeve, 2007).

The barrier system was investigated in a one-dimensional experiment in a Perspex tube cell (Fig. 16.19). Fine sand of 90–150 μm was used as a test soil, and copper sulfate solution was used as a contaminant, at a level of 1000 mg/l of copper. Sampling tubes fitted with a filter at the end were located so that the pore fluid could be sampled during the test (Fig. 16.19). A 12-h cycle was used, starting with 3 h power off and 9 h power on. This was used for two cycles, followed by 6 h off and 6 h on. A summary of the results are shown in Figure 16.20. The breakthrough of the contaminant occurred when the power remained switched off after 96 h. In

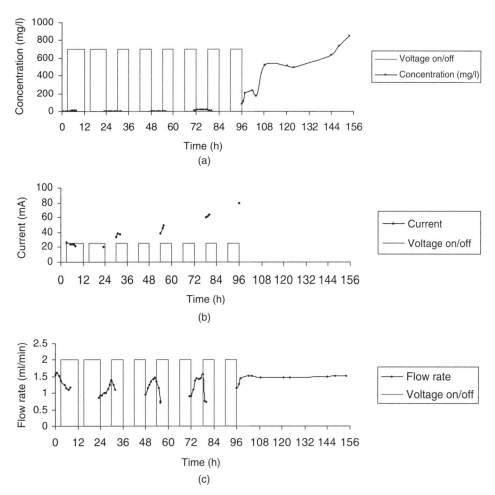

Figure 16.20. (a) Copper concentration at cell outlet, and voltage versus time; (b) current versus time; and (c) flow rate versus time, for a pulsed experiment in fine sand. Cylindrical cell is 11 cm in diameter, and soil length is 20 cm (Reeve, 2007).

this case, the level of power has not been sufficient to maintain the barrier for greater than this time.

Ledden has recently carried out a similar experiment (Ledden, 2008). The experimental setup was very similar except the cell was 40-cm long and 60 V was applied for only 9 h per day. Figure 16.21 shows the variation in flow rate as the electric field is applied, and Figure 16.22 shows the breakthrough curve. It is possible to estimate the relative average copper ion velocities due to EOF and ion migration. Without power supplied to the electrodes, the average flow rate recorded was about 1.24 ml/min. The velocity due to hydraulic flow measured from the flow rate when the power was off is therefore 1.6 cm/h, using a porosity value of 0.476. The cross section of the cylindrical cell was 95 cm^2. When 60 V was applied, the flow initially reduced to zero in about 4 h. Power was supplied for 9 h per day, that is, 38% of the time. When the voltage was applied, the flow rate reduced quickly; the average

350 ELECTROKINETIC BARRIERS

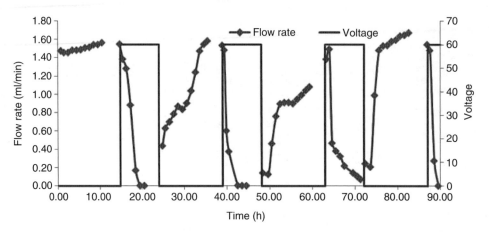

Figure 16.21. Graph of flow rate against time (Ledden, 2008).

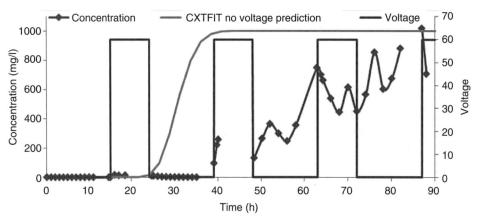

Figure 16.22. Graph of the copper concentration at outflow against time, showing the CXTFIT breakthrough curve prediction for no applied voltage (Ledden, 2008).

value was 0.41 ml/min. So, the net flow rate, taking into account the (negative) EOF contribution, is about 1.24 − 0.41 = 0.83 ml/min or equivalent to a fluid velocity of 1.10 cm/h. This is active for only 38% of the time, so averaged over the day, this amounts to 0.40 cm/h.

The ion mobility of copper in a free solution is given as $5.56 \times 10^{-4} \text{cm}^2/(\text{V} \cdot \text{s})$ (Atkins, 1982). From Equation 16.1, the average copper ion velocity, due to electromigration, is calculated to be 0.50 cm/h, taking porosity as 0.476 and tortuosity as 0.35 (Acar and Alshawabkeh, 1993). The ion mobility velocity is also active for only 38% of the time, so the daily average velocity is 0.19 cm/h.

The daily average velocity of copper is the sum of these three components: 1.6 − 0.40 − 0.19 = 1.0 cm/h. The breakthrough is therefore to be expected through 40 cm of sand at about 40 h, if there is no retardation. However, we expect a retardation factor of 1.5 (Ledden, 2008), so the breakthrough is expected at 60 h, in reasonable agreement with that found experimentally in Figure 16.22. Also shown

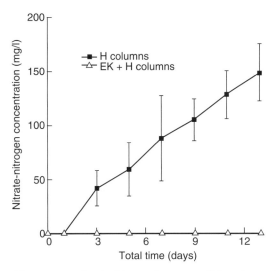

Figure 16.23. Nitrate-N concentration inside a column test (Manokararajah and Sri Ranjan, 2005a). H, Hydraulic; EK, electrokinetic.

in Figure 16.22 is the theoretical breakthrough from the CXTFIT advection–dispersion model, assuming a retardation of 1.5, but no effect from the electric field. Details of the well-known CXTFIT model are given in Toride, Leij, and van Genuchten (1995). Therefore, in this case, the short field, on time was insufficient for the barrier to hold, in this permeable soil.

16.3.5 A Nitrate Barrier

In this case, the electrode position needs to be reversed with the anode in the up-gradient position, because the contaminating species is an anion. Eid *et al.* (2000) reported that applying a current of 6 mA across 30 cm of sandy soil was sufficient to retain significant levels of nitrate against a hydraulic gradient corresponding to a Darcian velocity of 17 cm/h, during a 12-h test. Manokararajah and Sri Ranjan (2005a) recently showed good results with 500 mg/l of nitrate incursion. Laboratory tests were done with a column test similar to those above, with a few refinements.

Figure 16.23 shows the nitrate-nitrogen concentration at a distance of 20% inside the column from the anode end, for both an electrokinetic barrier and no barrier, against a hydraulic gradient of 1.25. After 13 days, the barrier is still seen to be holding the nitrates back (Manokararajah and Sri Ranjan, 2005a).

In similar experiments, Figure 16.24 shows the nitrate concentration within and at the end of column as a function of time (Manokararajah and Sri Ranjan, 2005b). A nice addition is that if the polarity is temporarily reversed, nitrate can be converted to gaseous nitrogen.

16.4 USE WITH OTHER TECHNOLOGIES

Narasimhan (1999) suggested that an electrokinetic barrier can work with the pump and treat technology to improve the cleanup in fine-grained soils. By arranging the

352 ELECTROKINETIC BARRIERS

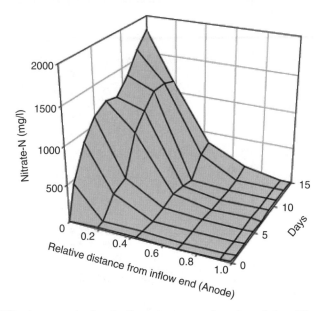

Figure 16.24. Nitrate concentration in the column as a function of time. The anode is at the inlet (Manokararajah and Sri Ranjan, 2005b).

electrodes so that EOF is encouraged toward the well, the pump and treat process could be enhanced. Vapor extraction of volatile pollutants could also be improved by positioning the electrodes to oppose hydraulic flow, and so lowering the water table below the contaminated zone. This can release more vapor-filled pores for extraction (Narasimhan, 1999). Simulations of both of these processes were carried out using the Narasimhan and Sri Ranjan model.

16.4.1 Combined Electrokinetic and Permeable Reactive Barriers (PEREBAR)

An interesting proposal is to combine an electrokinetic barrier with a PEREBAR. Figure 16.25 shows one possible configuration of electrodes and reactive barrier. The electrodes are placed upstream of the barrier. The electric field hinders some groundwater constituents from moving with the groundwater flow into the barrier.

16.4.1.1 Arsenic and Lead, Combined Barrier Lee and Chung (2007) have reported a field experiment in which electrokinetic remediation was coupled with a PEREBAR. Contaminant levels of arsenic and lead fell more than twofold when electrokinetic was used in addition to the permeable barrier of atomizing slag.

16.4.1.2 Organics, Combined Barrier Chung and Lee also tested zeolite and sand, and iron and sand mixtures, with an electrokinetic barrier, and reported that 300 ppm of ethylene glycol and 50 ppm of cadmium were effectively removed (Chung and Lee, 2007a).

In another series of experiments, trichloroethylene (TCE) and cadmium were both 90% removed by an electrokinetic PEREBAR of atomizing slag (Chung and Lee, 2007b).

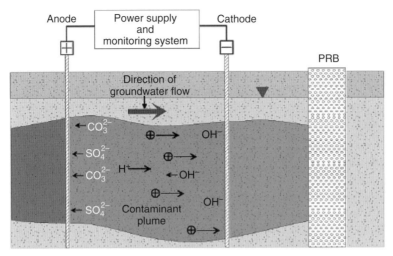

Figure 16.25. Possible configuration of electrodes and reactive barrier for an electrokinetic fence (Gregolec, Roehl, and Czurda, 2002). PRB, permeable reactive barrier.

16.4.1.3 Chromate Red mud, which contains about 35% hematite, has been used effectively in a PEREBAR coupled with electrokinetics (De Gioannis *et al.*, 2007). The red mud is located near the anode so that the acid produced at the anode is partially neutralized, allowing the alkaline front to move through the soil, under a field of 1 V/cm.

16.4.2 A Reactive Iron Barrier

A further method of generating a rather special electrokinetic barrier is that proposed by Faulkner, Hopkinson, and Cundy (2005). Using sacrificial iron electrodes, a low permeability iron-rich layer can be built up between the electrodes. Hydraulic conductivities around 10^{-9} m/s are reported. Further details are given in the reference.

16.5 SUMMARY

This chapter has reviewed recent studies of experimental electrokinetic barriers for both heavy metal and nitrate pollution. The use of such a barrier remains worthy of further work and upscaling. Lageman's work, using a combined barrier and cleanup approach, has been demonstrated in real situations. The use of a simple barrier, without cleanup, is still a possibility under conditions where the natural hydraulic flow is relatively slow, and the application of an electric field is able to establish a band of (relatively) high resistance soil so that low currents can be maintained over considerable periods. The combination of these barriers with PEREBARs and, indeed, the generation of an electrically generated iron reactive barrier are other exciting possibilities of which we shall hear more in the future.

ACKNOWLEDGMENTS

Grateful thanks are due to the following: Professor Aldo Muntoni and Dr. Romano Ruggeri, University of Cagliari, Sardinia, Italy; former students: Dr. May Chan, Marion Nowicki, Katie Winfield, Jo Reeve, Emanuela Contini, and Charlotte Ledden; and also to the reviewers for useful references and suggested improvements.

REFERENCES

Acar YB, Alshawabkeh AN. (1993). Principles of electrokinetic remediation. *Environmental Science & Technology* **27**(13):2638–2647.

Atkins PW. (1982). *Physical Chemistry*. Oxford, UK: Oxford University Press.

Atkins P. (2001). *The Elements of Physical Chemistry*, 3rd Edition. Oxford University Press, p. 190.

Casagrande L. (1949). Electroosmosis in soils. *Geotechnique* **1**(3):159–177.

Chan MSM, Lynch RJ, Ilett DJ. (2001). Use of cation selective membrane and acid addition for pH control in two-dimensional electrokinetic remediation of copper. In *Proceedings of EREM 2001* (eds. C Czurda, R Haus, C Kappeler, R Zorn). Karlsruhe, Germany: Schr. Angew. Geol., pp. 33-1–33-12.

Chung HI, Lee MH. (2007a). Electrokinetic permeable reactive barrier for the removal of heavy metal and organic substance in contaminated soil and groundwater. *6th Symposium on Electrokinetic Remediation (EREM 2007)* (ed. C Cameselle), June 12–15, Vigo, Spain: University of Vigo, pp. 17–18.

Chung HI, Lee MH. (2007b). A new method for remedial treatment of contaminated clayey soils by electrokinetics coupled with permeable reactive barriers. *Electrochimica Acta* **52**:3427–3431.

De Gioannis G, Muntoni A, Ruggeri R, Zilstra H, Floris M. (2007). Chromate adsorption in a transformed red mud permeable reactive barrier using electrokinesis. *6th Symposium on Electrokinetic Remediation* (ed. C Cameselle), June 12–15, Vigo, Spain: University of Vigo, pp. 15–16.

Eid N, Elshorbagy W, Larson D, Slack D. (2000). Electro-migration of nitrate in sandy soil. *Journal of Hazardous Materials* **B79**:133–149.

Faulkner DWS, Hopkinson L, Cundy AB. (2005). Electrokinetic generation of reactive iron-rich barriers in wet sediments: Implications for contaminated land management. *Mineralogical Magazine* **69**(5):749–757.

Gregolec G, Roehl KE, Czurda C. (2002). Proceedings of PEREBAR project. http://www.image-train.net/products/proceedings_first/chapter_24.pdf (accessed June 5, 2009).

Gregolec G, Zorn R, Kurzbach A, Roehl KE, Czurda K. (2001). Coupling of hydraulic and electric gradients in sandy soils. In *Proceedings of EREM 2001* (eds. C Czurda, R Haus, C Kappeler, R Zorn). Karlsruhe, Germany: Schr. Angew. Geol., pp. 41-1–41-15.

Lageman R. (1993). Electroreclamation. *Environmental Science & Technology* **27**(13): 2648–2650.

Lageman R, Pool W, Seffinga GA. (1989). Electro-reclamation: Theory and practice. *Chemistry & Industry* **18**:585–590.

Ledden C. (2008). A Solar-Powered Fence to Keep Out Pollutants in Groundwater. MEng Dissertation, University of Cambridge Engineering Department, Cambridge, UK.

Lee MH, Soo SK, Chung HI. (2007). In-situ electrokinetic permeable reactive barrier: field investigation in the vicinity of unregulated landfillsite. Proceedings of EREM 2007, Vigo, Spain, June, Paper S11.

Lynch RJ, Muntoni A, Ruggeri R. (2003). An electrokinetic barrier against heavy metal contamination in fine-grained soils. In *Remediation of Contaminated Sediments* (eds. M Pellei, A Porta). Second International Conference on Remediation of Contaminated Sediments. September 30–October 3, Venice, Italy. Columbus, OH: Battelle Press.

Lynch RJ, Muntoni A, Ruggeri R, Winfield KC. (2005). Optimization of energy consumption by using a pulsed electrokinetic barrier to prevent Zn and Cd spread over an uncontaminated site. Operation of the barrier in model tank tests. *Fifth Symposium on Electrokinetic Remediation—EREM 2005*, May 22–25, Ferrara, Italy.

Lynch RJ, Muntoni A, Ruggeri R, Winfield KC. (2007). Preliminary tests of an electrokinetic barrier to prevent heavy metal pollution of soils. *Electrochimica Acta* 52:3432–3440.

Manokararajah, KK, Sri Ranjan R. (2005a). Creation of a barrier for nitrate ions using electrokinetic methods in a silty loam soil. *Canadian Biosystems Engineering* 47:1.15–1.21. http://engrwww.usask.ca/oldsite/societies/csae/protectedpapers/c0353.pdf (accessed June 5, 2009).

Manokararajah KK, Sri Ranjan R. (2005b). Electrokinetic retention, migration, and remediation of nitrates in silty loam soil under hydraulic gradient. *Engineering Geology* 77(3–4): 263–272.

Mitchell JK. (1986). *Proceedings of USEPA Workshop on Electrokinetic Treatment and Its Application in Environmental Geotechnical Engineering for Hazardous Waste Remediation*. August 4. Seattle, WA: University of Washington, Department of Civil Engineering, pp. 1–2.

Mitchell JK, Soga K. (2005). *Fundamentals of Soil Behavior*, 3rd ed. New York: Wiley, p. 291.

Mitchell JK, Yeung AT. (1990). Electro-kinetic flow barriers in compacted clay. *Transportation Research Record* (1288):1–9.

Muntoni A, Ruggeri R, Lynch RJ. (2003). Investigation of an electrokinetic fence for protection against heavy metal contamination in soils. *Fourth Symposium on Electrokinetic Remediation—EREM 2003*, May 14–16, 2003, Mol, Belgium: SCK·CEN Belgian Nuclear Research Centre, pp. 51–52.

Narasimhan B. (1999). Electrokinetic Barriers to Contaminant Transport: Numerical Modelling and Laboratory Scale Experimentation. MSc Thesis, Department of Biosystems Engineering, University of Manitoba, Winnipeg, Canada.

Narasimhan B, Sri Ranjan R. (2000). Electrokinetic barrier to prevent subsurface contaminant migration: Theoretical model development and validation. *Journal of Contaminant Hydrology* 42(1):1–17.

Nowicki M. (2005). Photovoltaic Energy Conversion and Storage for Electrokinetic Groundwater Purification. MPhil Dissertation, Cambridge University Engineering Department, Cambridge, UK.

Nowicki MA, Lynch RJ, Amaratunga G. (2005). Stand-alone solar powered electrokinetic fence for preventing heavy metal pollution of groundwater. *Proceedings of the 20th EU Photovoltaic Solar Energy Conference*, June 6–10, Barcelona, Spain.

Olsen HW, Gui S, Lu N. (2007). Critical review of coupled flow theories for clay barriers. *Transportation Research Record* 1714:57–64.

Probstein RF, Renaud PC. (1986). Quantification of fluid and chemical flow in electrokinetics. *Proceedings of workshop on electrokinetic treatment and its application in environmental geotechnical engineering for hazardous waste remediation*. Seattle, WA: University of Washington, Department of Civil Engineering, pp. IV–IV-48.

Reeve J. (2007). A Solar Powered Electrokinetic Barrier. MEng Dissertation, University of Cambridge Engineering Department, Cambridge, UK.

Reeve J, Lynch RJ. (2007). Use of a pulsed electric field for resisting groundwater pollution. *6th Symposium on Electrokinetic Remediation (EREM 2007)* (ed. C Cameselle), June 12–15. Vigo, Spain: University of Vigo, pp. 19–20.

Ruggeri R. (2005). *Applicazione Dell'Elettrocinesi*. PhD Thesis, DIGITA, University of Cagliari, Cagliari, Italy.

Toride N, Leij FJ, van Genuchten MTH. (1995). The CXTFIT Code for Estimating Transport Parameters from Laboratory or Field Tracer Experiments, Version 2.0. August 1995. U.S. Salinity Laboratory, U.S. Department of Agriculture, Riverside, CA. Research Report No. 137.

US Environmental Protection Agency (US EPA). (1997). Recent Developments for In Situ Treatment of Metal Contaminated Soils. March 5, 1997. U.S. Office of Solid Waste and Emergency Response, Technology Innovation Office, Washington, DC. *Work Assignment Number: 011059.* http://www.epa.gov/tio/download/remed/metals2.pdf (accessed June 5, 2009).

Winfield K. (2004). Containment of Soil and Groundwater Pollution by an Electric Barrier. MEng Dissertation, Cambridge University Engineering Department, Cambridge, UK.

Yeung AT. (1990). Electrokinetic Barrier to Contaminant Transport through Compacted Clay. PhD Dissertation, University of California at Berkeley, Berkeley, CA.

Yeung AT, Mitchell JK. (1993). Coupled fluid, electrical and chemical flows in soil. *Geotechnique* **43**(1):121–134.

17

ELECTROKINETIC BIOFENCES

REINOUT LAGEMAN AND WIEBE POOL

17.1 INTRODUCTION

An electrokinetic biofence (EBIS®) is a derivative of an electrokinetic fence (EKIS®). In general, electrokinetic fences can be applied as a passive *in situ* method to fence off, contain, and remediate polluted groundwater plumes. EKIS® is applied in the case of inorganic pollutants, such as heavy metals and other electrically charged contaminants, while EBIS® is used for pollution with organic contamination. Under the influence of the electrical direct current (DC) field, metal ions, transported by the groundwater toward the fence, are deflected toward the electrode filters and captured in the electrolyte solutions circulating the electrodes. Subsequently, they are removed from the electrode solutions above ground. The efficiency or effectiveness of an electrokinetic fence can be defined according to Equation 17.1,

$$Nd = \frac{\text{Number of charged particles caught by the fence}}{\text{Number of charged particles entering the fence area}}$$

or

$$Nd = \frac{C_b - C_e}{C_b} 100, \tag{17.1}$$

where

Nd = efficiency (%),
C_b = concentration in front of the fence (µg/l), and
C_e = concentration behind the fence (µg/l).

Electrochemical Remediation Technologies for Polluted Soils, Sediments and Groundwater,
Edited by Krishna R. Reddy and Claudio Cameselle
Copyright © 2009 John Wiley & Sons, Inc.

TABLE 17.1. Important Parameters for an Electrokinetic Fence

Symbol	Parameter	Dimension
V_{gw}	Groundwater velocity	m/year
K_{ek}	Electrokinetic mobility	m^2/Vs
Φ	Electric potential	V
K_s	Specific electric conductivity of the soil	S/m
L	Length of the electrodes	m
D	Distance between the electrodes	m

The yield can be determined by computer simulations that calculate the velocity of the charged particles coming toward the fence for each value of the relevant physical/chemical and geohydrological parameters. Important parameters in this respect are listed in Table 17.1.

In the case of organic pollutants, the fence acts as an electrobiofence (EBIS®). Such a fence should be envisioned as a more or less elongated zone, wherein biodegradation is enhanced and/or from where downstream biodegradation is being optimized. Nutrients, such as nitrogen, phosphorus, oxygen donors, organic compounds, and spore elements necessary for the biodegradation of organic pollutants dissolved in water, appear almost always as electrically charged compounds and can be dispersed through the soil electrokinetically.

17.2 APPLICATION IN THE FIELD

The basic setup consists of a row of electrodes bordering a high-concentration area or polluted groundwater plume. The row of electrodes is set perpendicular to the prevailing groundwater flow direction, while the depth of the electrodes coincides with the lowest depth where pollutants are found. Depending on the type of pollution (inorganic or organic), the following configurations are possible.

17.2.1 EKIS®

The electric current is induced into the ground by means of alternating anodes and cathodes. Anodes as well as cathodes are integrated into separate closed-loop pump systems, wherein electrolytes are circulated. Via these electrolytes, pH is controlled at a predetermined level, and the pollutants, transported by the groundwater, are captured by the electrodes under the influence of the applied potential (Fig. 17.1). Conditioning of the electrolytes as well as periodical removal of the contaminants from the electrolytes is effectuated in a special containerized installation. If necessary, electricity cables and extraction ducts and pipes can be installed underground.

17.2.2 EBIS®

Depending on site-specific circumstances, an electrokinetic biofence is set up according to the following configuration: A row of alternating anode and cathode

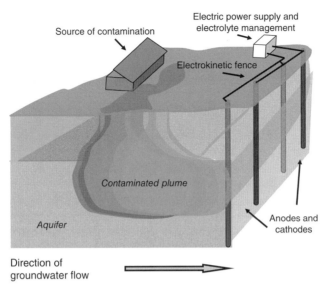

Figure 17.1. Schematic representation of an electrokinetic fence. Heavy metals and other polar contaminants are captured in the electrolytes and periodically removed. There were no contamination downstream of the fence area and no disturbance of the groundwater flow regime.

electrodes is installed perpendicular to the main groundwater flow direction. Anodes as well as cathodes are integrated into separate closed-loop pump systems, wherein electrolytes are circulated. Via these electrolytes, pH is controlled and kept neutral (pH = 7), by mixing anolyte and catholyte above ground. A row of infiltration filters is installed several meters upstream from the electrodes (Fig. 17.2). The infiltration filters are used to bring a nutrient-rich aqueous solution into the ground. DC is applied by which electrically charged nutrients, transported by the groundwater, are deflected and dispersed homogeneously within the zone of the fence. A part of the nutrients are captured by either the anodes or the cathodes, depending on their electrical charge. The neutralized electrolyte mixture, which is returned to the field, induces anionic nutrients into the soil from the anodes and cationic nutrients from the cathodes.

The organic pollutants, transported by the groundwater, are degraded by enhanced microbiological activity, either within or downstream of the zone. Electrokinetic biofences can be installed at relatively great depths and over great distances, avoiding extraction of large volumes of groundwater.

17.3 CASE STUDY

At the site of chemical laundry, a project with an electrokinetic biofence (EBIS®) started in April 2001. This fence has been installed to disperse nutrients in the ground and groundwater in order to enhance reductive dechlorinization of present perchloroethene (PCE), trichloroethene (TCE), *cis*-1,2-dichloroethene (C-DCE),

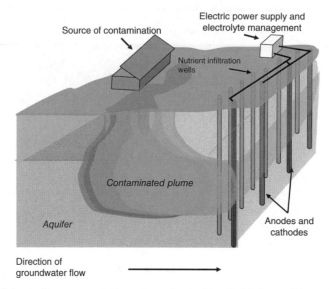

Figure 17.2. Schematic representation of an electrokinetic biofence. The enhancement of biodegradation inside and downstream of the fence area was by inducement of nutrients and homogeneous dispersion of nutrients.

and vinylchloride (VC). The fence acts as a temporary safety measure to avoid further migration of the contamination from the source area underneath the building to the plume area.

17.3.1 Background Information

Previous investigations indicated high grades (10 g/kg) of volatile organic compounds (VOCs) underneath the building. In the groundwater at the front of the building, concentrations of PCE of 180 mg/l in the ground indicated mass transport of free product through the upper aquifer. It consisted of fine to medium fine sand down to the clay layer at approximately 10 m below ground surface (bgs).

The heavily contaminated soil underneath and around the laundry acted and is still acting as a source for the downstream groundwater contamination. In this groundwater plume, C-DCE and VC were found in concentrations above the Dutch intervention value at a distance of 300 m from the laundry and at a depth of 11 m bgs.

17.3.2 Electrokinetic Biofence

An alternative for containment of source areas and remediation of polluted groundwater plumes is to deploy an EBIS®. An EBIS® avoids the necessity to pump and treat hundreds of thousands of cubic meters of groundwater, which is only slightly contaminated. It also means that the groundwater flow regime is not influenced.

Because of the enhanced biodegradation, downstream migration of the contamination will be controlled and eventually halted. The energy necessary to activate the fence depends on the velocity of the groundwater flow. The amount and

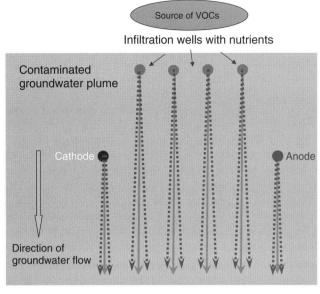

Figure 17.3. Flow paths of nutrients without an electrical field.

composition of the nutrients depend on the amount of VOCs in the groundwater and on the nutrients already present in the groundwater.

The basic setup consists of a row of electrodes (anodes and cathodes) that are installed perpendicular to the main direction of groundwater flow at the boundary of a hot spot or groundwater plume, up to the maximum depth where contaminants are still present.

The infiltration filters contain pouches with nutrients in solid form. These nutrients dissolve slowly and are then dispersed homogeneously through the soil, under the influence of the groundwater flow and the electrical field (Figs. 17.3 and 17.4)

A part of the nutrients are captured in the electrolytes circulating around the electrodes. Nutrients are also added to the electrolytes. Since pH conditions should be neutral, the electrolytes coming from the acidifying anodes and the alkalinizing cathodes are mixed continuously above ground, and the nutrients that are intercepted at the cathodes will be drawn into the soil via the anodes; the nutrients intercepted at the anodes will be drawn into the soil via the cathodes.

The amount of nutrients entering the groundwater per unit of time through the electrode filters is managed electrokinetically, but the amount of nutrients entering the groundwater through the infiltration filters depends among others on the concentration of the nutrients in the infiltration filters and on the construction of the infiltration filters. The extent of the homogeneous dispersal of the nutrients beyond the fence depends on the groundwater flow and the applied voltage over the electrodes. Another important aspect is the stability of the nutrients in the groundwater and the extent of adsorption on soil particles. Although parameters such as adsorption and retardation can be determined under laboratory conditions, the obtained results are often a mismatch with the actual situation in the field.

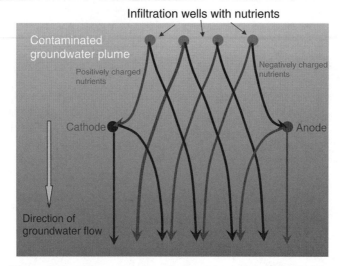

Figure 17.4. Flow paths of nutrients with an electrical DC field.

The organic contaminants migrating via the groundwater are decomposed within the fence zone or downstream thereof, because of the highly increased microbial activity. The "bulk" of induced nutrients will, in principle, be less adsorbed onto the soil particles, and they will hence catch up earlier with the contaminants traveling via the groundwater; this is then followed by bacteriological decomposition of these contaminants. The zone of biological activity will continue to expand over time in the flow direction of the groundwater.

17.3.3 Field Setup

The electrokinetic biofence consists of three cathodes and two anodes. The electrode wells have been placed in a row at a mutual distance of 5 m, downstream of the chemical laundry up to 10 m depth. At a distance of approximately 2 m upstream of the electrode wells, 24 infiltration wells have been installed, 19 of which are in use so far. These infiltration filters consist of polyethylene (PE) filter tubes with a diameter of 50 mm, length of 10 m, and perforation from 1 to 10 m bgs. Each filter can be filled with approximately 13 kg of solid nutrients, consisting of sodium, calcium, nitrogen, phosphorus, and ammonium compounds.

Apart form the 24 filters located directly upstream of the five electrodes wells, seven reference wells have been placed adjacent to, but outside of the zone with the electrode wells. Downstream, 18 monitoring wells have been installed to follow the dispersion of nutrients and the concentration of VOCs.

Cathodes and anodes have been integrated into separate electrolyte circulation systems. Electrolyte management consists of preventing alkalinization at the cathodes and acidification at the anodes by mixing anolyte and catholyte and thus neutralizing both electrolytes to pH 7. An additional advantage of mixing anolyte and catholyte is that anionic nutrients captured in the anolyte end up in the catholyte, and likewise, cationic nutrients end up in the anolyte. Thus, cathodes and

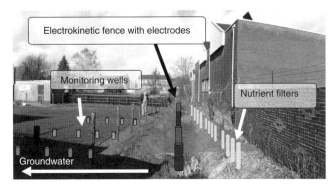

Figure 17.5. Field electrokinetic biofence system.

anodes are acting as new infiltration sources for nutrients. Figure 17.5 gives an overview of the site with the location of the electrodes, nutrient infiltration wells, and monitoring wells.

17.3.4 Project Results

17.3.4.1 Nutrients For the optimal biodegradation of VOCs, electron donors (C) and nutrient nitrogen (N) + phosphorus (P) are required (van Agteren *et al.*, 1998). In Table 17.1, the average concentration of nitrogen-Kjeldahl (N), phosphorus (P as ortho-PO_4^{3-}), and dissolved organic carbon (DOC) are given.

Table 17.2 indicates that sufficient nutrients in the form of a carbon source (DOC), nitrogen (N-Kjeldahl), and phosphorus are present in the groundwater. The concentration of nitrogen increased with nearly a factor 10 during measurement in July 2003 and decreased to a concentration of about 6 mg/l at the end of last year. Phosphate increased 5 times in July 2003 and 15 times at the end of 2005. The concentration of DOC first increased with a factor 2 but decreased in the last period to 52 mg/l. It cannot be derived from the determination of DOC whether all present carbon forms a suitable source for the decomposition of VOCs. Experience and data from literature indicate that the nutrients, administered through the electrode wells and infiltration filters, are quickly decomposed and that they have a positive influence on the decomposition process of VOCs. The data show clearly that the concentration of these nutrients and electron donors increases, caused by the infiltration and dispersion of nutrients by the electrokinetic biofence.

17.3.4.2 VOCs An electrokinetic biofence is designed to remove VOCs in a groundwater plume or to stop their downstream migration. The biofence will stimulate and create optimal conditions for biological activity to dechlorinate VOCs downstream of the fence. In Figures 17.6 and 17.7, the development in concentrations of PCE + TCE and C-DCE + VC is represented. Note that the building of the chemical laundry starts just 1 m south of the nutrient infiltration.

From Figure 17.6, it can be observed that there is a decrease of PCE and TCE as these compounds are being dechlorinated and their degradation products C-DCE and VC are formed. This effect is shown in Figure 17.7, which depicts the

TABLE 17.2. Changes in the Concentration of Nutrients

Concentration of Nutrients	March 2001	December 2004
Nitrogen-Kjeldahl (mg/l)	3.3	5.7
Orthophosphate (mg/l)	1.1	17
DOC (mg/l)	47	52

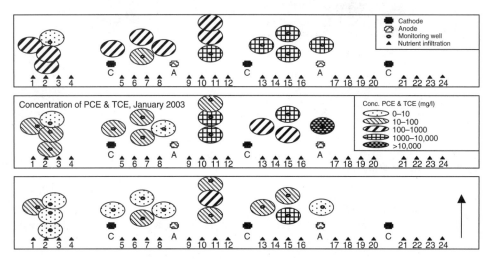

Figure 17.6. Changes in concentration of PCE + TCE. GW, groundwater.

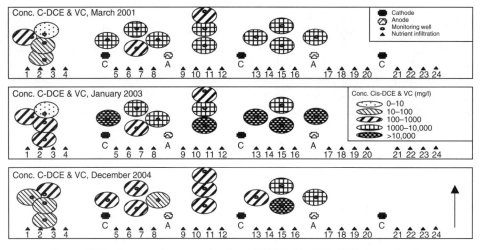

Figure 17.7. Changes in concentration of C-DCE + VC. GW, groundwater.

concentration of C-DCE and VC. In January 2003, the highest concentrations are reached, where after a period of decrease starts.

17.3.4.3 Chloride Index (Cl-Index)
To monitor the dechlorinization of VOCs, the Cl-index is being used. By calculating this index, an indication of biological activity can be obtained, based on the concentrations of VOCs. The Cl-index has been adapted slightly, and it is calculated as follows:

$$\text{Cl-index} = \frac{\{4*[PCE]+3*[TCE]+2*[C\text{-}DCE]+2*[T\text{-}DCE]+1*[VC]\}}{\{[PCE]+[TCE]+[C\text{-}DCE]+[T\text{-}DCE]+[VC]\}} \quad (17.2)$$

wherein

- [PCE] = concentration of perchloroethylene in mol/l,
- [TCE] = concentration of trichloroethene in mol/l,
- [C-DCE] = concentration of *cis*-dichloroethylene in mol/l,
- [T-DCE] = concentration of *trans*-dichloroethylene in mol/l, and
- [VC] = concentration of vinyl chloride in mol/l.

When VOCs are present in groundwater, the Cl-index may vary from 4 (only PCE was dissolved) to 1 (only VC was dissolved). Thus, during biodegradation of PCE to VC, the Cl-index changes from 4 to 1. The Cl-index at the fence changed from an average of 2.9 to 2.0. Some monitoring wells, however, have Cl-indexes >3. The location of these monitoring wells coincides with the area, where an inflow of VOCs was observed from the original soured area underneath the building. In spite of these high concentrations, total Cl-index stays low, which indicates that the biological activity has been enhanced to such an extent that it can cope with high influxes of VOCs.

17.3.5 Energy Requirements and Consumption

The calculations for the electrical power requirements, needed to transport nutrients electrokinetically through the soil, are based on the velocity of the groundwater flow. If the voltage is too high, all nutrients will end up either at the cathode or anode. If the voltage is too low, nutrients will not be dispersed. The optimal voltage therefore is the voltage at which nutrients are dispersed homogeneously through the soil, as shown in Figure 17.3.

The average groundwater velocity at the site amounts to some 3 m/year. Based on this velocity, the EBIS® needs an electrical power input of about 300–350 W. This energy input can be supplied by solar panels. The solar panels installed at the site have a maximum output of 500 W. This is sufficient for sunny days. At night and on cloudy days, the electricity is taken from the grid.

17.4 SUMMARY

Electroreclamation in the form of an electrokinetic fence (EKIS®) can be applied as a passive *in situ* method to fence off, contain, and remediate groundwater plumes polluted with inorganic species like heavy metals, arsenic, and cyanide. In the case of organic pollutants, the fence acts as an electrobiofence (EBIS®). Such a fence should be envisioned as a more or less elongated zone, wherein biodegradation is enhanced and/or from where downstream biodegradation is being optimized by the inducement of nutrients. Nutrients, such as nitrogen, phosphorus, oxygen donors, organic compounds, and spore elements necessary for the biodegradation of organic pollutants dissolved in water, appear as electrically charged compounds and can be dispersed homogeneously through the soil electrokinetically.

A pilot project with an electrokinetic biofence (EBIS®) has been carried out at the beginning of the 2000s at the site of a chemical laundry. The aim of the EBIS® was to enhance the biodegradation of the VOCs in groundwater at the zone of the fence by electrokinetic dispersion of nutrients dissolved in the groundwater upstream of the fence. After a 3-year test period, clear and positive results have been observed. The concentration of nutrients in the zone has increased, the Cl-index has decreased, and VOCs were being dechlorinated by bioactivity. The electrical energy for the EBIS® is being supplied by solar panels in combination with electricity from the grid.

REFERENCE

van Agteren MH, Keuning S, Janssen DB. (1998). *Handbook on Biodegradation and Biological Treatment of Hazardous Organic Compounds*. Dordrecht, the Netherlands: Kluwer Academic Publishers.

PART VI

INTEGRATED (COUPLED) TECHNOLOGIES

18

COUPLING ELECTROKINETICS TO THE BIOREMEDIATION OF ORGANIC CONTAMINANTS: PRINCIPLES AND FUNDAMENTAL INTERACTIONS

Lukas Y. Wick

18.1 INTRODUCTION

18.1.1 Constraints of Bioremediation: A Physical Perspective

Efficient bioremediation of subsurface contaminants depends on both abiotic and microbial factors. The former include the physicochemical properties of the compounds, environmental factors, such as temperature or humidity, the availability of nutrients, terminal electron acceptors (TEAs), and contaminants. The driving force of an efficient soil bioremediation, however, is the presence of a homogeneously dispersed microbial community's catabolic potential. In order to be effective in overcoming the constraints to bioavailability, any environmental biotechnology method therefore has to ensure the rapid transport of microorganisms and/or chemicals at least over the distances typically separating hot spots of pollution (Harms and Wick, 2006). Due to their high solid–water distribution ratios, this is of special relevance for hydrophobic organic contaminants (HOCs), which are typically associated with carbon phases of particles and hence very slowly released from the soil, sediments, and aquifer solids by diffusive transport processes (Semple, Morriss, and Paton, 2003; Johnsen, Wick, and Harms, 2005). Cases of highest initial homogeneity are subsurface contaminations arising from diffuse pollution or resulting from long diffusion processes after point source pollution (Alexander, 2000). Bosma et al. (1997) calculated that the average distance between bacterial microcolonies in soil

---

Electrochemical Remediation Technologies for Polluted Soils, Sediments and Groundwater,
Edited by Krishna R. Reddy and Claudio Cameselle
Copyright © 2009 John Wiley & Sons, Inc.

is in the range of 50–100 μm. Although microcolonies containing particular catabolic capabilities may be further apart, distances to be bridged by bacterial movement to achieve the bioavailability of all molecules of a homogenous pollution are likely to be in the submillimeter to centimeter range. Higher degrees of heterogeneity exist where pollutants have entered the soil or sediment as nonaqueous phase liquids or solids, or where the distribution of nutrients and electron acceptors is highly patchy. In this situation, electrokinetic dispersion will need to transport substantial numbers of indigenous bacteria toward relatively large pockets of the pollutants or into microzones of appropriate chemical conditions, or, inversely, contaminants and nutrients to catabolically active microorganisms. The principal advantage of applying electrokinetics to bioremediation hence is to stimulate biodegradation by effective dispersal of microorganisms and contaminants. Besides this, electric fields and current may induce other biologically relevant physicochemical phenomena in the soil, such as soil heating, desiccation, electrolysis of water and subsequent H_2 and O_2 generation, formation of pH gradients, precipitation of salts, or the decomposition of minerals and organic matter (Virkutyte, Sillanpää, and Latostenmaa, 2002). These phenomena, however, are not central to Chapter 21 and will be discussed in other chapters of this book.

18.1.2 Electrobioremediation

Electrobioremediation is a generic name for a large group of cleanup methods that use both microbiological phenomena for degradation and electrokinetics for the transport of subsurface contaminants, nutrients, TEA, and contaminant-degrading microbes (Chillingar et al., 1997). A wealth of studies has reported on the electrokinetically induced stimulation of bioremediation (Marks, Acar, and Gale, 1994; Ho et al., 1995; Chillingar et al., 1997; Hayes, Flora, and Khan, 1998; Gent and Bricka, 2001; Jackman et al., 2001; Luo et al., 2005; Niqui-Arroyo et al., 2006; Niqui-Arroyo and Ortega-Calvo, 2007; Suni et al., 2007). An overview of successful applications of electrokinetics for the remediation of soil contaminated by organic pollutants *in situ* is given by Lageman and coworkers (Lageman, Clarke, and Pool, 2005). Whereas nutrients, TEA, metals, or water-soluble contaminants can be transported over larger distances, electrokinetic transport of HOC is limited due to their low aqueous solubility (Acar and Alshawabkeh, 1993). For *in situ* electrobioremediation of HOCs, however, "microscale" dispersion rather than "macroscopic" HOC transport (HOC extraction) may be all that is required as bacteria are ubiquitous in soil (Grundmann, 2004). As electrokinetic processes enhance the mobilization of subsurface environment components, contaminants, and/or microorganisms, they have the potential to enhance the contact probability of the bacteria and their substrates (Wick, Shi, and Harms, 2007) and consequently to bridge the distances needed for efficient biotransformation. This is of particular importance in fine-grained subsurface, where the conductivity for pressure-driven (e.g. hydraulic) water flow is extremely small and thus leads to quasi-stagnant water in micro- and nanopores not allowing for adequate mixing of contaminants, nutrients, and TEA and, consequently, for efficient contaminant biodegradation (Page and Page, 2002).

In the following chapter, an outline of the influences of direct current (DC) electric fields on the bioavailability of contaminants and their bioremediation in soil will be developed based on the "electrobioremediation tetrahedron" (Fig. 18.1).

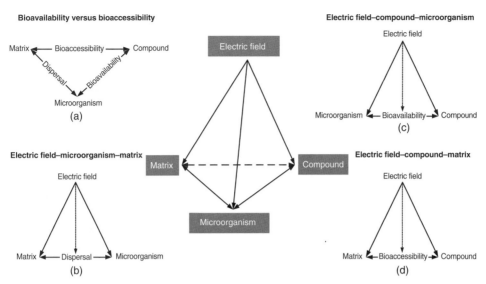

Figure 18.1. Conceptualization of the critical factors influencing the biotransformation of (hydrophobic) organic compounds during subsurface electrokinetic treatment; the "electrobioremediation tetrahedron" in this figure visualizes the influences of electric fields on the bioavailability of subsurface contaminants and presents it as processes at the interface between biological dynamics and physical constraints. The tetrahedron's base triangle (a) represents the subsurface matrix–microbe–compound interactions that govern the chemical's bioavailability and bioaccessibility; (b) relates electric current to the abiotic compound–matrix interactions and conceptualizes the dispersing role of electrokinetics for compounds, nutrients, and terminal electron acceptors (TEAs), whereas (c) and (d) illustrate the impact of electric current on the physics, physiology, and ecology of the organism–compound and organism–matrix interplay influencing both microbial biotransformation and dispersal.

18.2 PRINCIPLES AND FUNDAMENTAL INTERACTIONS OF ELECTROBIOREMEDIATION

18.2.1 The "Electrobioremediation Tetrahedron"

The "electrobioremediation tetrahedron" (Fig. 18.1) attempts to conceptualize the relevant processes for subsurface bioremediation, that is, the impact of an electric field on the microbial subsurface habitat and the mutual interactions involved. The following sections will unfold the four sides of the "electrobioremediation tetrahedron" and discuss the relevant interactions with a special emphasis on the biodegradation of hydrophobic organic compounds in subsurface systems exposed to DC electric fields.

The tetrahedron's base triangle (Fig. 18.1a) represents the subsurface matrix–microbe–compound interactions that govern the chemical's bioavailability and bioaccessibility as outlined in Section 18.2.2. Figure 18.1b relates DC to microbe–matrix interactions and describes the electrokinetic dispersal of bacteria (Section 18.2.3). Figure 18.1c,d represents the impact of electric current on compound biotransformation (Section 18.2.4) and mobilization (Section 18.2.5). Hence, they reflect the DC influence on microorganism–compound and matrix–compound interchanges.

18.2.2 Bioavailability versus Bioaccessibility of Organic Compounds

Active microbes are typically surrounded by water and their interactions with contaminants normally involves the water phase (Wodzinski and Johnson, 1968; Wodzinski and Coyle, 1974; Ogram et al., 1985; Harms and Zehnder, 1995; Harms and Wick, 2004). The hydrophobicity of a chemical is thus often taken as an indicator for its *bioavailability*. The term bioavailability is generally used to refer to *the degree of interaction of chemicals with living organisms* (Semple et al., 2004). A more operational definition of bioavailability, however, is needed before beginning to inspect the various bioavailability processes that govern the interactions of microbes with contaminants in electric fields (Fig. 18.1a). Bosma et al. (1997) defined bioavailability as the ratio of the rate at which a given environmental habitat can provide a chemical to microorganisms inhabiting it to the rate at which these microbes can theoretically degrade the compound. Translated to a practical perspective in bioremediation, this means that the biotransformation rate of a compound is an immediate measure and appropriate dimension of bioavailability. Figure 18.2 specifies the various molecular processes influencing the uptake and transformation of a compound by a single microorganism (Semple et al., 2004). Hydrophobic (i.e. poorly water soluble) chemicals are normally sorbed to organic or inorganic soil matrices. Before they can be taken up by a microorganism, they need to (a) be released to a water-dissolved form (process A) and (b) be transported to the cell's surface (i.e. become bioavailable; process B) before (c) crossing the cell membrane and being transformed by biocatalysis (process D). Processes C and E describe the interactions of the microbes with the soil matrix, such as microbial deposition and dispersal or

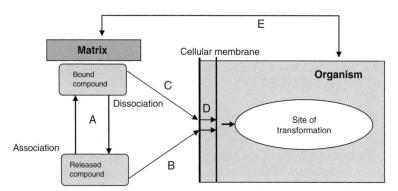

Figure 18.2. Visualization of the microscale processes influencing the bioavailability and the uptake of a compound by a single microorganism. Process A describes the release of a chemical from an organic or inorganic matrix to a better accessible, water-soluble form and, in reverse, its adsorption to a subsurface matrix; processes B, C, and D illustrate the chemical's transport to and uptake through a microorganism's membrane; processes C and E symbolize the interactions of the microbes with the geoparticles, such as microbial attachment and dispersal or the autecological improvement of bioavailability by altering bacterial cell surface properties. The contaminant concentration to which a microbe is exposed to is consequently determined by the actual microbial uptake rate and the environmental mass transfer kinetics and simply depends on how rapidly the habitat can compensate for substrate biodegradation.

the autecological improvement of contaminant bioavailability by altering bacterial cell surface properties. From Figure 18.2, it becomes clear that the capacity of the habitat to deliver a compound will depend on the physical state of the chemical (dissolved, sorbed, separate phase), its physical characteristics (hydrophobicity, effective diffusivity), its spatial distribution relative to the catabolically active microorganism, and the type of organisms involved (Guerin and Boyd, 1992; Calvillo and Alexander, 1996; Reid, Jones, and Semple, 2000). The contaminant concentration to which a microbe is exposed is consequently determined by the actual microbial uptake rate and the environmental mass transfer kinetics. It depends simply on how rapidly the habitat can compensate for substrate biodegradation. Various studies have shown that the sequestration of HOCs in the solid subsurface by sorption and entrapment reduces their bioavailability (Zhang, Bouwer, and Ball, 1998; Semple, Morriss, and Paton, 2003). Three potentially rate-limiting steps of HOC release have been identified: (a) diffusion of the sorbate within the molecular nanoporous network of natural organic matter (Hatzinger and Alexander, 1995), (b) pore or surface diffusion in aggregated minerals constituting natural particles (Alexander, 2000), and (c) diffusion of the sorbate across an aqueous boundary layer at the exterior of soil particles (Luthy et al., 1997).

The capacity of an environment to provide a chemical at a constant rate may become limited due to depletion of labile pools of the chemical. This has inspired the differentiation of an immediate bioavailability [the "freely dissolved" compound immediately ready for uptake (Reichenberg and Mayer, 2006)] from a potential bioavailability, which has been named *bioaccessibility* (the compound that may become bioavailable by, e.g. dissolution, or desorption over time) (Semple et al., 2004). In this concept, the bioaccessible compound is a fraction of the total compound and appears to be an appropriate descriptor of the possible degradation end point.

The principle goal of electrobioremediation is therefore to make bioaccessible compound, nutrient, and TEA fractions bioavailable and consequently increase biotransformation rates (Fig. 18.3). The following chapter will specify the impact of DC fields on bacterial mobilization in and deposition to subsurface matrices (Fig. 18.1b).

18.2.3 Influence of DC on Microbial Dispersal and Deposition

In order to achieve substantial biodegradation, microorganisms must be in contact with bioavailable contaminants, and this may require microbial movement. This is of great importance as soil bacteria are commonly immobilized *in situ*, that is, are attached to soil particles (Costerton and Lappin-Scott, 1989) or form microcolonies rather than releasing single cells into the soil water. Numerous laboratory (Lahlou et al., 2000), field-scale (Dybas et al., 2002; Major et al., 2002), and modeling studies (Schäfer et al., 1998; Shein, Polyanskaya, and Devin, 2002; Kim, 2006) have been performed to understand and predict microbial transport in porous media [for a review, see Harvey and Harms (2001)]. The observed inefficiency of bacterial transport in many soils has been mainly attributed to the extremely low hydraulic conductivity in soil micropores (Li, Loehle, and Malon, 1996; Silliman et al., 2001) and bacterial attachment to the surfaces of soil particles (Baygents et al., 1998). Strategies such as the chemical modification of bacterial surfaces (Gross and Logan, 1995) and

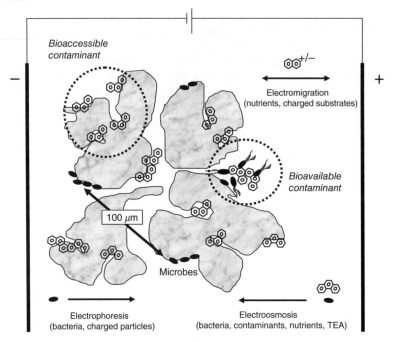

Figure 18.3. Biodegradation of subsurface contaminants depends on the presence of contaminant-degrading bacteria and the provision of optimal environmental conditions for microbial activity. From this figure, it becomes clear that the capacity of the habitat to deliver a compound will depend on the physical state of the chemical (water dissolved vs. solid, liquid, or geosorbent bound), its electrokinetic transport, its spatial distribution relative to the catabolically active microorganism (bioavailable concentration vs. bioaccessible fraction), and the type of organisms involved. As indicated by the arrows, the optimal bioavailability of the pollutants, nutrients, and terminal electron acceptors (TEAs) can be achieved by electrokinetically dispersing the heterogeneously distributed contaminants. The principle goal of electrobioremediation thus can be seen as to efficiently make bioaccessible fractions of chemicals (contaminants, nutrients, TEA) bioavailable and, consequently, increase the biotransformation rates and end points, respectively.

treatment with surfactants (Jackson, Roy, and Breitenbeck, 1994; Brown and Jaffe, 2001) have been proposed to enhance the efficiency of bacterial transport by decreasing bacterial attachment. Centimeter to meter scale electrokinetic transport of bacteria and yeast cells through sand, soil, and aquifer sediments has been reported as a result of the electrophoretic movement of negatively charged microorganisms to the anode (DeFlaun and Condee, 1997; Lee and Lee, 2001) and/or their dispersal with the electroosmotic water flow to the cathode (Suni and Romantschuk, 2004; Wick et al., 2004). Typical electrophoretic transport rates of $0.02\,\text{cm}^2/(\text{V}\cdot\text{h})$ for yeast cells (Lee, Jahng, and Lee, 1999) and up to $4\,\text{cm}^2/(\text{V}\cdot\text{h})$ for bacteria (DeFlaun and Condee, 1997; Wick et al., 2004) have been reported with their extent depending on the type of the subsurface matrix and the physicochemical cell surface properties of the microorganisms studied (Wick et al., 2004). For the transport of weakly charged bacteria, electroosmosis was found to be the predominant transport mechanism, with electrophoresis accounting for less than 20% of the cells dispersed (Wick *et al.*, 2004). This observation is in agreement with the data by Liu and coworkers (Liu, Chen, and Papadopulos, 1999) showing that *Escherichia coli* is exclusively moved by electroosmosis over a wide range of pH values and ionic strengths at electrical field strengths of >0.3 V/cm. In these experiments, at voltages below 0.2 V/cm, however, the undirected movement to both electrodes was observed, most likely due to the superimposed effects of electroosmosis, electrophoresis, galvanotaxis (Shi, Stocker, and Adler, 1996), and random bacterial motility. Differential electrokinetic transport of individual microorganisms was ascribed to intrapopulation heterogeneity with respect to the cell surface charge, with bacteria transported toward the anode being generally more negatively charged than the electroosmotically translocated cells (Shi *et al.*, 2008b). This is not surprising as the role of intrapopulation heterogeneities for bacterial dispersal and attachment has been recognized and discussed previously (Simoni *et al.*, 1998; Bolster *et al.*, 2000; van der Mei and Busscher, 2001; Mailloux *et al.*, 2003), and capillary zone electrophoresis (CZE) is an accepted method for the separation of differently charged bacteria in the laboratory (Rodriguez and Armstrong, 2004). Electroosmotically transported bacteria (transport rate: 0.1–$0.4 \, cm^2 h^{-1} V^{-1}$) were ca. 50% slower in comparison to injected conservative tracers (Wick *et al.*, 2004). Furthermore, poor electrokinetic transport of strongly charged and highly adhesive bacteria (Wick *et al.*, 2002) occurred in a range of different model aquifers pointing to a dominant retardation effect for this type of cells (Wick *et al.*, 2004). The strong affinity for these solid matrices, however, could be partially overcome by treating the bacteria with the nonionic surfactant Brij35, leading to up to 80% enhanced electrokinetic dispersion (Wick *et al.*, 2004).

Little information is available on the mechanisms governing microbial retardation and detachment in subsurface matrices exposed to electric fields. A recent study by Shi and coworkers (Shi *et al.*, 2008b), analyzing the influence of cell size, membrane integrity, chromosome contents, surface charge, and hydrophobicity on the spatial distribution of suspended and matrix-bound bacteria, showed that DC had no influence on the deposition efficiency and consequent retardation of bacteria tested. More knowledge though exists on bacterial attachment to medical devices under the influence of electric potential, electric current, and electric charge transfer on the extent and mechanisms of bacterial deposition and detachment on electrode surfaces (Sjollema, Busscher, and Weerkamp, 1989). These data point to an inhibition of bacterial surface colonization at low electric currents ($17.2 \, \mu A/cm^2$) (Liu *et al.*, 1993), a lack on initial bacterial adhesion (Poortinga, Bos, and Busscher, 2001b), and an increase of initial (Poortinga, Bos, and Busscher, 2001b) yet no influence on the final extents of detachment to glass (van der Borden, van der Mei, and Busscher, 2003). Changes of the electric potential (yet not of the capacitance) of semiconducting indium-tin-oxide electrodes propose the presence of microbially induced charge transfer (about 10^{-14} C per bacterium) during bacterial deposition (Poortinga, Bos, and Buscher, 1999). Hence, it was proposed that the direction of electron transfer may affect the stability of bacterial adhesion, that is, that bacteria adhere more strongly when donating than when accepting electrons from semiconductive surfaces (Poortinga, Bos, and Busscher, 2001a). Electron transfer to conducting surfaces is also known from anodophilic bacteria in microbial fuel cells (Logan and Regan, 2006). These bacteria may be referred to as exoelectrogens, based on their ability to exocellularly release electrons probably due to the involvement of outer-membrane cytochromes in the exogenous electron transfer or the use of soluble electron shuttles that eliminate the need for direct contact between the cell and the electron acceptor.

Overall, it appears that the effects of electric fields on detachment, adhesion, and biofilm formation on electrodes are a result from electric current exchanged rather than the electric potential applied (Fig. 18.1b).

18.2.4 Influence of DC on the Physiology and Ecology of Biotransformation

Efficient biotransformation requires that the application of any form of electric current or electric fields has no negative effect on the biocatalysts, that is, the catabolically active, contaminant-degrading microbial communities (Fig. 18.1c). To date, only limited information on the effect of DC on pollutant-degrading microorganisms is available (Jackman et al., 1999; Lear et al., 2007; Shi et al., 2008c). A recent study has shown that weak DC electric fields, as typically used for electro-bioremediation measures, have no harmful effect on the biodegradation activity and culturability of polycyclic aromatic hydrocarbon (PAH)-degrading soil bacteria (Wick et al., 2004; Shi et al., 2008c). Relative to the control bacteria in the absence of an electric field, DC-exposed cells exhibited up to 60% elevated intracellular adenosine triphosphate (ATP) levels yet remained unaffected on all other levels of cellular integrity and physiological activity tested. This finding was corroborated by the observation that electrokinetically mobilized cells exhibited unchanged physiological fitness as assessed by cell wall integrity analysis on a single cell level (Shi et al., 2008a). Unaffected cell physiology agrees with earlier findings by Lear et al. (2005) reporting that low DC current (0.314mA/cm^2) per se had no influence on the composition and structure of soil microbial communities. Observed community changes close to the electrodes were attributed to electrokinetically induced changes of the soil's pH and physicochemical structure. In other studies, DC-induced pH and eH changes or the production of hydrogen close to the cathode was found to stimulate the biological denitrification of nitrate-contaminated groundwater (Hayes, Flora, and Khan, 1998) or the activity of sulfur-oxidizing bacteria (e.g. during bioleaching processes) (Natarajan, 1992; Jackman et al., 1999).

For a better understanding of the potential effects of electric fields on microbial physiology, one has to consider that the cytoplasm of a cell is electrically conducting, whereas the lipid bilayer of its membrane is dielectric. Hence, the application of electric fields to cells may cause the buildup of electrical charge at the cell membrane, especially where the membrane pointing at the cathode and anode, and thus induces a change in voltage across the membrane (Zimmermann, Schulz, and Pilwat, 1974). For externally applied electric fields (normally above 1 kV/cm) leading to a transmembrane potential difference of about 70–250 mV, the membrane permeability increases and leads to reversible or irreversible "dielectric" breakdown (Zimmermann, Schulz, and Pilwat, 1974) and concomitant cell death (Grahl and Markl, 1996). Local instabilities in the membranes due to electromechanical compression and electric field-induced tension (Ho and Mittal, 1996; Weaver and Chizmadzhev, 1996) have also been proposed as the basis for electropermeabilization (electroporation) of cells exposed to pulses of high electric fields. This property has found extensive application in biotechnology for the electrorelease of cell ingredients and the uptake of extracellular molecules to which the cells are normally not permeable (Turgeon et al., 2006). Pulsed high electrical field gradients, as observed during natural lightning events, to soil have further been found to provoke horizontal gene transfer between soil bacteria (Ceremonie et al., 2006). In biotechnology, permanent cell membrane breakdown has been used to inactivate microbes, for example, for the preservation of food (Elez-Martinez et al., 2004) or disinfection of sludge (Huang, Elektorowicz, and Oleszkiewicz, 2008). Inhibitory effects have been demonstrated by increased propidium iodide uptake (Wouters, Bos, and Ueckert, 2001; Aronsson, Ronner, and Borch, 2005; Garcia et al., 2007), loss of culturability (Yao, Mainelis, and An, 2005), or ATP leakage (Sixou et al., 1991; Wouters, Bos, and Ueckert, 2001; Aronsson, Ronner, and Borch, 2005). Other microbial responses included toxic electrode effects, metabolic stimulation due to enhanced substrate mass transfer and/or electrokinetic removal of toxic metabolites (Pribyl et al., 2001), and sublethal injuries (Guillou et al., 2003) as well as changes in the physicochemical surface properties (Shimada and Shimahara, 1985; Luo et al., 2005). The latter are of special interest in soil bioremediation as they influence bacterial adhesion to surfaces and, concomitantly, their mobility in the subsurface (Redman, Walker, and Elimelech, 2004). Another study has reported that the natural electromagnetic field at the earth's surface may be used by *E. coli* as a supplemental energy source to sustain growth, that is, that some microorganisms may be capable of converting electromagnetic waves into chemical energy (Gusev and Schulze-Makuch, 2005). The conversion of electric energy to physiologically accessible ATP-based energy has also been described by bacteria exposed to oscillating (Tsong, Liu, and Chauvin, 1989), high-voltage pulsed ($X = 1$–$6\,kV/cm$) (Teissie, 1986), or even low-DC electric fields (Zanardini et al., 2002; Shi et al., 2008c). Increased ATP levels have been interpreted as the result of cumulative effects of DC-promoted transmembrane pH gradients and/or membrane potential differences (Shi et al., 2008c).

Adverse electrochemical reactions are the main reasons for the antimicrobial activity of low DC fields, especially in the vicinity of the electrodes (Liu, Brown, and Elliott, 1997). Electrochemical reactions include the formation of toxic substances, such as free chlorine (Davis et al., 1994), metallic ions (Berger et al., 1976), or reactive oxygen species or H_2O_2 (Shimada and Shimahara, 1982; Liu, Brown, and Elliott, 1997; Guillou and Murr, 2002). Not surprisingly, electrokinetic effects, however, depend on the medium's conductivity and osmotic properties, the amperage, treatment time, and the type and the growth phase of the microorganisms studied (Velizarov, 1999).

More relevant for the bioremediation of subsurface systems may, however, be the electrokinetically enhanced mass transfer processes due to the combined effect of electrophoresis and electroosmosis. In laboratory tests, growth stimulation of immobilized cells by 140% due to the electrokinetic removal of inhibitory products and electroosmotically enhanced glucose supply has been reported (Chang, Grodzinsky, and Wang, 1995), suggesting that effective electrochemical biotransformation can be accomplished by selectively coupling microorganisms and suitable conditions. For instance, the enhanced biotransformation of benzonitrile to benzoic acid and ammonia (Mustacchi et al., 2005) electrically enhanced ethanol fermentation (Shin, Zeikus, and Jain, 2002), the reductive carboxylations (Dixon et al., 1989), the reduction of chloropyruvate to chloroacetate catalyzed by a mediator-dependent D-lactate dehydrogenase, or the regeneration of pyridine nucleotides (Schulz et al., 1995) has been reported. Electrokinetic transport may have also been the reason for greatly increased antimicrobial activity of antibiotics against bacterial biofilms by Costeron et al. (1994). Electrokinetically increased soil temperatures have also been found to boost bioremediation of creosote-contaminated soil in cold climate zones (Suni et al., 2007).

18.2.5 Influence of DC on Contaminant and Nutrient Transport

Electrically driven transfer of compounds in the subsurface includes the mobilization of charged species (electromigration), electrophoretic translocation of particle-bound chemicals, and/or electroosmotic solute transport (Pamukcu and Wittle, 1994; Yeung, Hsu, and Menon, 1997; Saichek and Reddy, 2005). By a combination of these factors, it can stimulate microbial activity in the subsurface. Successful electroosmotic flushing has been applied to soil contaminated with phenolic compounds (Yeung, Hsu, and Menon, 1997), hexachlorobenzene (Pamukcu and Wittle, 1994), chlorinated solvents (Ho et al., 1995), low molecular weight fuel components (Bruell, Segall, and Walsh, 1992; Segall and Bruell, 1992; Ho et al., 1995), or polycyclic aromatic hydrocarbons (Electric Power Research Institute, 1994; Maini et al., 2000). However, the electroosmotic transport of HOCs over long distances was generally solely feasible when HOCs were present as droplets, colloidal particles (Mitchell, 1991; Probstein and Hicks, 1993; Kuo and Papadopoulos, 1996), or solubilized by surfactants, cyclodextrins, or chelating agents (Probstein and Hicks, 1993; Kim and Lee, 1999; Ko, Schlautman, and Carraway, 2000; Popov et al., 2001; She et al., 2003; Reddy and Saichek, 2004; Yuan and Weng, 2004). Bruell and coworkers suggested that long-distance electroosmotic removal may depend on the compound's Freundlich constant and its aqueous solubility (Bruell, Segall, and Walsh, 1992). Assumptions based on equilibrium isotherms models, however, may underestimate the release of chemicals as they assume that the dynamic chemical equilibrium in the dissolved phase is quickly reached. This is unlikely to happen with geosorbents when electroosmotic water flow passes along their surfaces. Furthermore, such models do not further account for the possibility that microscale electroosmotic solute transport may stimulate microbial activity (Shi, Harms, and Wick, 2008), which itself may act on macroscopically observed release and transport rates of HOC. At the microscale, intraparticle diffusional mass transfer resistances often impose serious limitations on the rate biotransformation in soil. This is a particular problem in fine-grained soils since their conductivity for pressure-driven (e.g. hydraulic) water flow is extremely small and leads to stagnant water in micro- and nanopores. In principal, electroosmosis is likely to overcome mass-transfer bottlenecks in low permeable soil matrices by (a) increasing the release of sorbates by inducing liquid flow at the immediate exterior of soil particles, (b) creating flow in nanopores in the organic sorbent phase, which are inaccessible to hydraulic flow, and (c) influencing the pore or surface diffusion among aggregated minerals. Little mechanistic information exists on the impact of DC on HOC desorption and microscale transport kinetics of subsurface environments. Indications for possible bioavailability-enhancing effects, however, come from the field of capillary electrochromatography, where electroosmotic flow is used to enhance the transfer of solute between the mobile and stationary phases. As a result, capillary electrochromatography separates uncharged molecules better than high-performance liquid chromatography that relies on pressure-driven flow (Robson et al., 1997). A recent study reported on the stimulated release of the polycyclic aromatic hydrocarbon phenanthrene from a model polymer release system (alginate) in the presence of relevant strengths of DC electric fields (0.5–2 V/cm) (Shi, Harms, and Wick, 2008). Alginate beads exhibit a similar release behavior to natural organic matter (Wells, Wick, and Harms, 2005). In the presence of electroosmotic flow, the phenanthrene release flux from alginate increased up to 1.8-fold and up to 120-fold than under stagnant water conditions (Shi, Harms, and Wick, 2008). The increased release could be explained by the differences between the parabolic and plug-flow profiles (Knox and Grant, 1987) of hydraulic and electroosmotic water flow (Shi, Harms, and Wick, 2008). Electroosmotically induced flow processes therefore result in steepened chemical concentration gradients and concomitantly enhanced transfer of surface-associated contaminants (and nutrients) to the bulk water. Stimulated release as a consequence may increase the local HOC-bioavailability and the HOC-bioremediation efficiency, especially in stagnant water zones often found under hydraulic flow regimes restricted by low permeability (Paillat et al., 2000).

In order to effectively biodegrade contaminants, microorganisms additionally require a sufficient supply of nutrients (e.g. nitrogen phosphorus, potassium) and TEAs (e.g. oxygen, nitrate, sulfate, or Mn^{2+} and Fe^{3+}). A lack of adequate provision of nutrients and TEA to catabolically active communities has often been encountered as additional bottleneck for efficient *in situ* bioremediation. A coupled approach of electroosmotic substrate mobilization and accelerated electrokinetic transfer of charged nutrients and TEA may be useful for efficient subsurface bioremediation. Electromigration rates of nutrients and TEA are up to one order of magnitude higher than electroosmotic solute transport (Virkutyte, Sillanpää, and Latostenmma, 2002). Effective biodegradation of diesel-contaminated soil has, for instance, been shown as a result of effective electrokinetic dispersal of both the externally applied nutrient triethylphosphate and nitrate (Lee, Ro, and Lee, 2007). Other studies reported on successful electrokinetically enhanced injection (Azzam and Oey, 2001) of charged nutrients, cosubstrates, water, and/or TEAs into soil (Marks, Acar, and Gale, 1994; Hayes, Flora, and Khan, 1998; Gent and Bricka, 2001; Lageman, Clarke, and Pool, 2005; Wu et al., 2007) that otherwise would not be achieved by traditional remediation techniques. Hence, tailor-made electrokinetic mobilization of contaminants into defined treatment areas such as reactive barrier zones can lead to optimal biodegradation efficiency (Ho et al., 1995; Davis-Hoover et al., 1999; Luo et al., 2005; Chang and Cheng, 2006). Future work therefore should focus on the stimulating effects of electrokinetic transport phenomena on the ecology of microbial contaminant degradation.

18.3 RESEARCH NEEDS

Conceptually, biotransformation of contaminants in soils appears to be influenced by at least three interacting levels of complexity: (a) the heterogeneous soil pore network, (b) the abiotic environmental conditions, and (c) the network of catabolically interacting microorganisms responding to their environment. Thus, understanding of the interplay of structure and function of this complex (heterogeneous and dynamic) system during electrokinetic regimes is important for controlling the success of electrobioremediation. Special focus should thereby be put to molecular to microscale dispersal processes at the soil-contaminant and microorganism-contaminant interface and their impact on macroscale bioremediation. In order to be universally applicable at field situation, electrobioremediation therefore needs to be based on a quantifiable and scalable understanding of the impact of electric fields on the physical and physiological processes governing the biotransformation of contaminants in highly heterogeneous matrices (Figs. 18.1 and 18.2). More specifically, research may need to include (a) a more advanced understanding of the electroosmosis-driven contaminant release and mobilization in nano- and micropores of heterogeneous soil matrices, (b) an improved knowledge of the impact of low electric fields/currents on the physiology and activity of microorganisms at a single cell level, (c) a better understanding of the (aut)ecological adaptations of contaminant-degrading bacteria and microbial consortia to cope with electrokinetically enhanced fluxes of mixed contaminants containing both toxic compounds and beneficial substrates, and (d) the mathematical prediction of the kinetics governing the contaminant and nutrient transfer fluxes to single cells in the bulk growing phase.

ACKNOWLEDGMENTS

This chapter has been written as a contribution to the German–Argentinean project ELECTRA (D/07/09664 and PA/DA/07/03) supported by the Federal German Ministry of Education and Research (BMBF) and the Secretário de Ciencia y Technica de la Nación (SECYT).

REFERENCES

Acar YB, Alshawabkeh AN. (1993). Principles of electrokinetic remediation. *Environmental Science & Technology* **27**:2638–2647.

Alexander M. (2000). Aging, bioavailability, and overestimation of risk from environmental pollutants. *Environmental Science & Technology* **34**:4259–4265.

Aronsson K, Ronner U, Borch E. (2005). Inactivation of *Escherichia coli*, *Listeria innocua* and *Saccharomyces cerevisiae* in relation to membrane permeabilization and subsequent leakage of intracellular compounds due to pulsed electric field processing. *International Journal of Food Microbiology* **99**:19–32.

Azzam R, Oey W. (2001). The utilization of electrokinetics in geotechnical and environmental engineering. *Transport in Porous Media* **42**:293–314.

Baygents JC, Glynn JR, Albinger O, Biesemeyer BK, Ogden KL, Arnold RG. (1998). Variation of surface charge density in monoclonal bacterial populations: Implications for transport through porous media. *Environmental Science & Technology* **32**:1596–1603.

Berger T, Spadaro J, Chapin S, Becker R. (1976). Electrically generated silver ions: Effects on bacterial and mammalian cells. *Antimicrobial Agents and Chemotherapy* **9**:357–358.

Bolster CH, Mills AL, Hornberger G, Herman J. (2000). Effect of intra-population variability on the long-distance transport of bacteria. *Ground Water* **38**:370–375.

van der Borden AJ, van der Mei HC, Busscher HJ. (2003). Electric-current-induced detachment of *Staphylococcus epidermidis* strains from surgical stainless steel. *Journal of Biomedical Materials Research Part B* **68**:160–164.

Bosma TNP, Middeldorp PJM, Schraa G, Zehnder AJB. (1997). Mass transfer limitation of biotransformation: Quantifying bioavailability. *Environmental Science & Technology* **31**:248–252.

Brown DG, Jaffe PR. (2001). Effects of nonionic surfactants on bacterial transport through porous media. *Environmental Science & Technology* **35**:3877–3883.

Bruell CJ, Segall BA, Walsh MT. (1992). Electroosmotic removal of gasoline hydrocarbons and TCE from clay. *Journal of Environmental Engineering* **118**:62–83.

Calvillo YM, Alexander M. (1996). Mechanisms of microbial utilization of biphenyl sorbed to polyacrylic beads. *Applied Microbiology and Biotechnology* **45**:383–390.

Ceremonie H, Buret F, Simonet P, Vogel TM. (2006). Natural electrotransformation of lightning-competent *Pseudomonas* sp. strain N3 in artificial soil microcosms. *Applied Environmental Microbiology* **72**:2385–2389.

Chang JH, Cheng SF. (2006). The remediation performance of a specific electrokinetics integrated with zero-valent metals for perchloroethylene contaminated soils. *Journal of Hazardous Materials* **131**:153–162.

Chang Y, Grodzinsky A, Wang D. (1995). Augmentation of mass transfer through electrical means for hydrogel-entrapped *Escherichia coli* cultivation. *Biotechnology and Bioengineering* **48**:149–157.

Chillingar GV, Loo WW, Khilyuk LF, Katz SA. (1997). Electrobioremediation of soils contaminated with hydrocarbons and metals: Progress report. *Energy Sources* **19**:129–146.

Costerton J, Ellis B, Lam K, Johnson F, Khoury A. (1994). Mechanism of electrical enhancement of efficacy of antibiotics in killing biofilm bacteria. *Antimicrobial Agents and Chemotherapy* **38**:2803–2809.

Costerton JW, Lappin-Scott HM. (1989). Behaviour of bacteria in biofilms. *ASM News* **55**:650–654.

Davis C, Shirtliff M, Trieff N, Hoskins S, Warren M. (1994). Quantification, qualification, microbial killing efficiencies of antimicrobial chlorine-based substances produced by iontophoresis. *Antimicrobial Agents and Chemotherapy* **38**:2768–2774.

Davis-Hoover WJ, Bryndzia LT, Roulier MH, Murdoch LC, Kemper M, Cluxton P. (1999). In situ bioremediation utilizing horizontal LASAGNA. In *Engineered Approaches for In-Situ Bioremediation of Chlorinated Solvent Contamination* (eds. A Leeson, BC Alleman). Columbus, OH: Battelle Press, pp. 263–277.

DeFlaun MF, Condee CW. (1997). Electrokinetic transport of bacteria. *Journal of Hazardous Materials* **55**:263–277.

Dixon NM, James EW, Lovitt RW, Kell DB. (1989). Electromicrobial transformations using the pyruvate synthase system of *Clostridium sporogenes*. *Bioelectrochemistry and Bioenergetics* **21**:245–259.

Dybas MJ, Hyndman DW, Heine R, Tiedje J, Linning K, Wiggert D. (2002). Development, operation, long-term performance of a full-scale biocurtain utilizing bioaugmentation. *Environmental Science & Technology* **36**:3635–3644.

Electric Power Research Institute. (1994). Electrokinetic Removal of Coal Tar Constituents from Contaminated Sites. Electric Power Research Institute, Palo Alto, CA. Final report of project 2879-21. EPRI TR-103320.

Elez-Martinez P, Escola-Hernandez J, Soliva-Fortuny RC, Martin-Belloso O. (2004). Inactivation of *Saccharomyces cerevisiae* suspended in orange juice using high-intensity pulsed electric fields. *Journal of Food Protection* **67**:2596–2602.

Garcia D, Gomez N, Manas P, Raso J, Pagan R. (2007). Pulsed electric fields cause bacterial envelopes permeabilization depending on the treatment intensity, the treatment medium pH and the microorganism investigated. *International Journal of Food Microbiology* **113**:219–227.

Gent D, Bricka RM. (2001). Electrokinetic movement of biological amendments through natural soils to enhance in-situ bioremediation. In *Bioremediation of Inorganic Compounds* (eds. A Leeson, BM Peyton, JL Means, VS Magar). Richland, WA: Battelle Press, pp. 241–248.

Grahl T, Markl H. (1996). Killing of microorganisms by pulsed electric fields. *Applied Microbiology and Biotechnology* **45**:148–157.

Gross MJ, Logan BE. (1995). Influence of different chemical treatments on transport of *Alcaligenes paradoxus* in porous media. *Applied Environmental Microbiology* **61**:1750–1756.

Grundmann GL. (2004). Spatial scales of soil bacterial diversity—the size of a clone. *FEMS Microbiology Ecology* **48**:119–127.

Guerin WF, Boyd SA. (1992). Differential bioavailability of soil-sorbed naphthalene to two bacterial species. *Applied Environmental Microbiology* **58**:1142–1152.

Guillou S, Besnard V, Murr NE, Federighi M. (2003). Viability of *Saccharomyces cerevisiae* cells exposed to low-amperage electrolysis as assessed by staining procedure and ATP content. *International Journal of Food Microbiology* **88**:85–89.

Guillou S, Murr NE. (2002). Inactivation of *Saccharomyces cerevisiae* in solution by low-amperage electric treatment. *Journal of Applied Microbiology* **92**:860–865.

Gusev VA, Schulze-Makuch D. (2005). Low frequency electromagnetic waves as a supplemental energy source to sustain microbial growth? *Die Naturwissenschaften* **92**:115–120.

Harms H, Wick LY. (2004). Mobilization of organic compounds and iron by microorganisms. In *Physicochemical Kinetics and Transport at Chemical-Biological Interphases* (eds. HP van Leuven, W Koester). Chichester: Wiley, pp. 401–444.

Harms H, Wick LY. (2006). Dispersing pollutant-degrading bacteria in contaminated soil without touching it. *Engineering in Life Sciences* **6**:252–260.

Harms H, Zehnder AJB. (1995). Bioavailability of sorbed 3-chlorodibenzofuran. *Applied Environmental Microbiology* **61**:27–33.

Harvey RW, Harms H. (2001). Transport of microorganisms in the terrestrial subsurface: In situ and laboratory methods. In *Manual of Environmental Microbiology* (eds. CJ Hurst, GR Knudsen, MJ McInerney, LD Stetzenback, RL Crawford). Washington, DV: ASM Press, pp. 753–776.

Hatzinger P, Alexander M. (1995). Effect of aging of chemicals in soil on their biodegradability and extractability. *Environmental Science & Technology* **29**:537–545.

Hayes AM, Flora JRV, Khan J. (1998). Electrolytic stimulation of denitrification in sand columns. *Water Research* **9**:2830–2834.

Ho SY, Mittal GS. (1996). Electroporation of cell membranes: A review. *Critical Reviews in Biotechnology* **16**:349–362.

Ho SV, Sheridan PW, Athmer CJ, Heitkamp MA, Brackin JM, Weber D, Brodsky PH. (1995). Integrated in situ soil remediation technology: The Lasagna process. *Environmental Science & Technology* **29**:2528–2534.

Huang J, Elektorowicz M, Oleszkiewicz JA. (2008). Dewatering and disinfection of aerobic and anaerobic sludge using an electrokinetic (EK) system. *Water Science and Technology* **57**:231–236.

Jackman SA, Maini G, Sharman AK, Knowles CJ. (1999). The effects of direct current on the viability and metabolism of acidophilic bacteria. *Enzyme and Microbial Technology* **24**:316–324.

Jackman SA, Maini G, Sharman AK, Sunderland F, Knowles CJ. (2001). Electrokinetic movement and biodegradation of 2,4-dichlorophenoxyacetic acid in silt soil. *Biotechnology and Bioengineering* **74**:40–48.

Jackson A, Roy D, Breitenbeck G. (1994). Transport of a bacterial suspension through a soil matrix using water and an anionic surfactant. *Water Research* **28**:943–949.

Johnsen AR, Wick LY, Harms H. (2005). Principles of microbial PAH degradation. *Environmental Pollution* **133**:71–84.

Kim J, Lee K. (1999). Effects of electric field directions on surfactant enhanced electrokinetic remediation of diesel-contaminated sand column. *Journal of Environmental Science and Health Part A* **34**:863–877.

Kim SB. (2006). Numerical analysis of bacterial transport in saturated porous media. *Hydrological Processes* **20**:1177–1186.

Knox JH, Grant IH. (1987). Miniaturization in pressure and electroendosmotically driven liquid-chromatography—some theoretical considerations. *Chromatographia* **24**:135–143.

Ko SO, Schlautman MA, Carraway ER. (2000). Cyclodextrin-enhanced electrokinetic removal of phenanthrene from a model clay soil. *Environmental Science & Technology* **34**:1535–1541.

Kuo CC, Papadopoulos KD. (1996). Electrokinetic movement of settled spherical particles in fine capillaries. *Environmental Science & Technology* **30**:1176–1179.

Lageman R, Clarke RL, Pool W. (2005). Electro-reclamation, a versatile soil remediation solution. *Engineering Geology* **77**:191–201.

Lahlou M, Harms H, Springael D, Ortega J-J. (2000). Influence of soil components on the transport of polycyclic aromatic hydrocarbon-degrading bacteria through saturated porous media. *Environmental Science & Technology* **34**:3649–3656.

Lear G, Harbottle MJ, van der Gast CJ, Jackman SA, Knowles CJ, Sills G, Thompson IP. (2005). The effect of electrokinetics on soil microbial communities. *Soil Biology & Biochemistry* **36**:1751–1760.

Lear G, Harbottle MJ, Sills G, Knowles CJ, Semple KT, Thompson IP. (2007). Impact of electrokinetic remediation soil microbial communities within PCP contaminated soil. *Environmental Pollution* **146**:139–146.

Lee GT, Ro HM, Lee SM. (2007). Effects of triethyl phosphate and nitrate on electrokinetically enhanced biodegradation of diesel in low permeability soils. *Environmental Technology* **28**:853–860.

Lee H-S, Lee K. (2001). Bioremediation of diesel-contaminated soil by bacterial cells transported by elctrokinetics. *Journal of Microbiology and Biotechnology* **11**:1038–1045.

Lee HS, Jahng D, Lee K. (1999). Electrokinetic transport of a NAPL-degrading microorganisms through a sandy soil bed. *Biotechnology and Bioprocess Engineering* **4**:151–153.

Li BL, Loehle C, Malon D. (1996). Microbial transport through heterogeneous porous media: Random walk, fractal, and percolation approaches. *Ecological Modelling* **85**:285–302.

Liu WK, Brown RW, Elliott TSJ. (1997). Mechanisms of the bactericidal activity of low amperage current (DC). *The Journal of Antimicrobial Chemotherapy* **39**:687–695.

Liu WK, Tebbs SE, Byrne PO, Elliott TSJ. (1993). The effects of electric current on bacteria colonising intravenous catheters. *The Journal of Infection* **27**:261–269.

Liu Z, Chen W, Papadopulos KD. (1999). Electrokinetic movement of *Escherichia coli* in capillaries. *Environmental Microbiology* **1**:99–102.

Logan BE, Regan JM. (2006). Microbial fuel cells—challenges and applications. *Environmental Science & Technology* **40**:5172–5180.

Luo Q, Zhang X, Wang H, Qian Y. (2005). The use of non-uniform electrokinetics to enhance in situ bioremediation of phenol-contaminated soil. *Journal of Hazardous Materials* **B121**:187–194.

Luthy RG, Aiken GR, Brusseau ML, Cunningham SD, Gschwend PM, Pignatello JJ. (1997). Sequestration of hydrophobic organic contaminants by geosorbents. *Environmental Science & Technology* **31**:3341–3347.

Mailloux BJ, Fuller ME, Onstott TC, Hall J, Dong HL, DeFlaun MF. (2003). The role of physical, chemical, and microbial heterogeneity on the field-scale transport and attachment of bacteria. *Water Resources Research* **39**:1142–1158.

Maini G, Sharman AK, Knowles CJ, Sunderland G, Jackman SA. (2000). Electrokinetic remediation of metals and organics from historically contaminated soil. *Journal of Chemical Technology Biotechnology* **75**:657–664.

Major DW, McMaster ML, Cox EE, Edwards EA, Dworatzek SM, Hendrickson ER. (2002). Field demonstration of successful bioaugmentation to achieve dechlorination of tetrachloroethene to ethene. *Environmental Science & Technology* **36**:5106–5116.

Marks RE, Acar YB, Gale RJ. (1994). In situ remediation of contaminated soils containing hazardous mixed wastes by bioelectric remediation and other competitive technologies. In *Remediation of Hazardous Waste Contaminated Soils* (eds. DL Wise, DJ Trantolo). New York: Marcel Dekker, pp. 405–436.

van der Mei HC, Busscher HJ. (2001). Electrophoretic mobility distributions of single-strain microbial populations. *Applied and Environmental Microbiology* **67**:491–494.

Mitchell JK. (1991). Conduction phenomena: From theory to practice. *Geotechnique* **41**: 299–340.

Mustacchi R, Knowles CJ, Li H, Dalrymple I, Sunderland G, Skibar W, Jackman SA. (2005). Enhanced biotransformation and product recovery in a membrane bioreactor through application of a direct electric current. *Biotechnology and Bioengineering* **89**:18–23.

Natarajan KA. (1992). Bioleaching of sulphides under applied potentials. *Hydrometallurgy* **29**:161–172.

Niqui-Arroyo JL, Bueno-Montes M, Posada-Baquero R, Ortega-Calvo JJ. (2006). Electrokinetic enhancement of phenanthrene biodegradation in creosote-polluted clay soil. *Environmental Pollution* **142**:326–332.

Niqui-Arroyo JL, Ortega-Calvo JJ. (2007). Integrating biodegradation and electroosmosis for the enhanced removal of polycyclic aromatic hydrocarbons from creosote-polluted soils. *Journal of Environmental Quality* **36**:1444–1451.

Ogram AV, Jessup RE, Ou LT, Rao PS. (1985). Effects of sorption on biological degradation rates of (2,4-dichlorophenoxy)acetic acid in soils. *Applied Environmental Microbiology* **49**:582–587.

Page MM, Page CL. (2002). Electroremediation of contaminated soils. *Journal of Environmental Engineering* **128**(3):208–219.

Paillat T, Moreau E, Grimaud PO, Touchard G. (2000). Electrokinetic phenomena in porous media applied to soil decontamination. *IEEE Transactions on Dielectrics and Electrical Insulation* **7**:693–704.

Pamukcu S, Wittle JK. (1994). Electrokinetically enhanced in situ soil decontamination. In *Remediation of Hazardous Waste Contaminated Soils* (eds. DL Wise, DJ Trantolo). New York: Marcel Dekker, pp. 245–298.

Poortinga AT, Bos R, Busscher HJ. (1999). Measurement of charge transfer during bacterial adhesion to an indium tin oxide surface in a parallel plate flow chamber. *Journal of Microbiological Methods* **38**:183–189.

Poortinga AT, Bos R, Busscher HJ. (2001a). Charge transfer during staphylococcal adhesion to TiNOX® coatings with different specific resistivity. *Biophysical Chemistry* **91**:273–279.

Poortinga AT, Bos R, Busscher HJ. (2001b). Lack of an externally applied electric field on bacterial adhesion to glass. *Colloids and Surfaces. B, Biointerfaces* **20**:189–194.

Popov K, Kolosov A, Yachmenev VG, Shabanova N, Artemyeva A, Frid A. (2001). A laboratory-scale study of applied voltage and chelating agent on the electrokinetic separation of phenol from soil. *Separation Science and Technology* **36**:2971–2982.

Pribyl M, Chmelikova R, Hasal P, Marek M. (2001). Modeling of hydrogel immobilized enzyme reactors with mass-transport enhancement by electric field. *Chemical Engineering Science* **56**:433–442.

Probstein RF, Hicks RE. (1993). Removal of contaminants from soils by electric fields. *Science* **260**:499–503.

Reddy KR, Saichek RE. (2004). Enhanced electrokinetic removal of phenanthrene from clay soil by periodic electric potential application. *Journal of Environmental Science and Health Part A* **39**:1189–1212.

Redman J, Walker S, Elimelech M. (2004). Bacterial adhesion and transport in porous media: Role of the secondary energy minimum. *Environmental Science & Technology* **38**:1777–1785.

Reichenberg F, Mayer P. (2006). Two complementary sides of bioavailability: Accessibility and chemical activity of organic contaminants in sediments and soils. *Environmental Toxicology and Chemistry* **25**:1239–1245.

Reid B, Jones K, Semple K. (2000). Bioavailability of persistent organic pollutants in soils and sediments—a perspective on mechanisms, consequences and assessment. *Environmental Pollution* **108**:103–112.

Robson MM, Cikalo MG, Myers P, Euerby MR, Bartle KD. (1997). Capillary electrochromatography: A review. *The Journal of Microcolumn Separations* **9**:357–372.

Rodriguez MA, Armstrong DW. (2004). Separation and analysis of colloidal/nano-particles including microorganisms by capillary electrophoresis: A fundamental review. *Journal of Chromatography B* **800**(1-2):7–25.

Saichek RE, Reddy KR. (2005). Electrokinetically enhanced remediation of hydrophobic organic compounds in soils: A review. *Critical Reviews in Environmental Science and Technology* **35**:115–192.

Schäfer A, Ustohal P, Harms H, Stauffer F, Dracos T, Zehnder AJB. (1998). Transport of bacteria in unsaturated porous media. *Journal of Contaminant Hydrology* **33**:149–169.

Schulz M, Leichmann H, Gunther H, Simon H. (1995). Electromicrobial regeneration of pyridine-nucleotides and other preparative redox transformations with clostridium-thermoaceticum. *Applied Microbiology and Biotechnology* **42**:916–922.

Segall BA, Bruell CJ. (1992). Electroosmotic contaminant-removal processes. *Journal of Environmental Engineering* **118**:84–100.

Semple K, Doick K, Jones K, Burauel P, Craven A, Harms H. (2004). Defining bioavailability and bioaccessibility of contaminated soil and sediment is complicated. *Environmental Science & Technology* **15**:229A–231A.

Semple KT, Morriss AWJ, Paton GI. (2003). Bioavailability of hydrophobic organic contaminants in soils: Fundamental concepts and techniques for analysis. *European Journal of Soil Science* **54**:809–818.

She P, Liu Z, Ding FX, Yang JG, Liu X. (2003). Surfactant enhanced electroremediation of phenanthrene. *Chinese Journal of Chemical Engineering* **11**:73–78.

Shein EV, Polyanskaya LM, Devin BA. (2002). Transport of microorganisms in soils: Physicochemical approach and mathematical modelling. *Eurasian Soil Science* **35**:500–508.

Shi L, Harms H, Wick LY. (2008). Electroosmotic flow stimulates the release of alginate-bound phenanthrene. *Environmental Science & Technology* **42**:2105–2110.

Shi L, Müller S, Harms H, Wick LY. (2008a). Effect of electrokinetic transport on the vulnerability of PAH-degrading bacteria in a model aquifer. *Environmental Geochemistry and Health* **30**:177–182.

Shi L, Müller S, Harms H, Wick LY. (2008b). Factors influencing the electrokinetic dispersion of PAH-degrading bacteria in a laboratory model aquifer. *Applied Microbiology and Biotechnology* **80**(3):507–515.

Shi L, Müller S, Loffhagen N, Harms H, Wick LY. (2008c). Activity and viability of PAH-degrading *Sphingomonas* sp. LB126 in a DC-electrical field typical for electro-bioremediation measures. *Applied Microbiology and Biotechnology* **1**:53–61.

Shi W, Stocker BAD, Adler J. (1996). Effect of the surface composition of motile *Escherichia coli* and motile *Salmonella* species on the direction of galvanotaxis. *Journal of Bacteriology* **178**:1113–1119.

Shimada K, Shimahara K. (1982). Responsibility of hydrogen peroxide for the lethality of resting *Escherichia coli* B exposed to alternating current in phosphate buffer solution. *Agricultural and Biological Chemistry* **46**:1329–1337.

Shimada K, Shimahara K. (1985). Changes in surface charge, respiratory rate and stainability with crystal violet of resting *Escherichia coli* B cells anaerobically exposed to alternating current. *Agricultural and Biological Chemistry* **49**:405–411.

Shin HS, Zeikus JG, Jain MK. (2002). Electrically enhanced ethanol fermentation by *Clostridium thermocellum* and *Saccharomyces cerevisiae*. *Applied Microbiology and Biotechnology* **58**:476–481.

Silliman SE, Dunlap R, Fletcher M, Schneegurt MA. (2001). Bacterial transport in heterogeneous porous media: Observations from laboratory experiments. *Water Resources Research* **37**:2699–2707.

Simoni SF, Harms H, Bosma TNP, Zehnder AJB. (1998). Population heterogeneity affects transport of bacteria through sand columns at low flow rates. *Environmental Science & Technology* **32**:2100–2105.

Sixou S, Eynard N, Escoubas JM, Werner E, Teissie J. (1991). Optimized conditions for electrotransformation of bacteria are related to the extent of electropermeabilization. *Biochimica et Biophysica Acta* **1088**:135–138.

Sjollema J, Busscher HJ, Weerkamp AH. (1989). Real-time enumeration of adhering microorganisms in a parallel plate flow cell using automated image analysis. *Journal of Microbiological Methods* **9**:73–78.

Suni S, Malinen E, Kosonen J, Silvennoinen H, Romantschuk M. (2007). Electrokinetically enhanced bioremediation of creosote-contaminated soil: Laboratory and field studies. *Journal of Environmental Science and Health Part A* **42**:277–287.

Suni S, Romantschuk M. (2004). Mobilisation of bacteria in soils by electro-osmosis. *FEMS Microbiology Ecology* **49**:51–57.

Teissie J. (1986). Adenosine 5′-triphosphate synthesis in *Escherichia coli* submitted to a microsecond electric pulse. *Biochemistry* **25**:368–373.

Tsong TY, Liu D-S, Chauvin F. (1989). Electroconformational coupling (ECC): An electric field induced enzyme oscillation for cellular energy and signal transductions. *Bioelectrochemistry and Bioenergetics* **21**:319–331.

Turgeon N, Laflamme C, Ho J, Duchaine C. (2006). Elaboration of an electroporation protocol for *Bacillus cereus* ATCC 14579. *Journal of Microbiological Methods* **67**:543–548.

Velizarov S. (1999). Electric and magnetic fields in microbial biotechnology: Possibilities, limitations and perspectives. *Electro- and Magnetobiology* **18**:185–212.

Virkutyte J, Sillanpää M, Latostenmaa P. (2002). Electrokinetic soil remediation—critical overview. *The Science of the Total Environment* **289**:97–121.

Weaver JC, Chizmadzhev YA. (1996). Theory of electroporation: A review. *Bioelectrochemistry and Bioenergetics* **41**:135–160.

Wells MC, Wick LY, Harms H. (2005). A model polymer release system study of PAH bioaccessibility: The relationship between 'rapid' release and bioaccessibility. *Environmental Science & Technology* **39**:1055–1063.

Wick LY, Mattle PM, Wattiau P, Harms H. (2004). Electrokinetic transport of PAH-degrading bacteria in model aquifers and soil. *Environmental Science & Technology* **38**:4596–4602.

Wick LY, de Munain AR, Springael D, Harms H. (2002). Responses of *Mycobacterium* sp. LB501T to the low bioavailability of solid anthracene. *Applied Microbiology and Biotechnology* **58**:378–385.

Wick LY, Shi L, Harms H. (2007). Electro-bioremediation of hydrophobic organic soil contaminants: A review of fundamental interactions. *Electrochimica Acta* **52**:3441–3448.

Wodzinski RS, Coyle JE. (1974). Physical state of phenanthrene for utilization by bacteria. *Applied Microbiology* **27**:1081–1084.

Wodzinski RS, Johnson MJ. (1968). Yields of bacterial cells from hydrocarbons. *Applied Microbiology* **16**:1886–1891.

Wouters PC, Bos AP, Ueckert J. (2001). Membrane permeabilization in relation to inactivation kinetics of *Lactobacillus* species due to pulsed electric fields. *Applied Environmental Microbiology* **67**:3092–3101.

Wu X, Alshawabkeh AN, Gent DB, Larson SL, Davis JL. (2007). Lactate transport in soil by DC fields. *Journal of Geotechnical and Geoenvironmental Engineering* **133**:1587–1596.

Yao M, Mainelis G, An HR. (2005). Inactivation of microorganisms using electrostatic fields. *Environmental Science & Technology* **39**:3338–3344.

Yeung AT, Hsu CN, Menon RM. (1997). Physicochemical soil-contaminant interactions during electrokinetic extraction. *Journal of Hazardous Materials* **55**:221–237.

Yuan C, Weng CH. (2004). Remediation ethylbenzene-contaminated clayey soil by a surfactant-aided electokinetic (SAEK) process. *Chemosphere* **57**:225–232.

Zanardini E, Valle A, Gigliotti C, Papagno G, Ranalli G, Sorlini C. (2002). Laboratory-scale trials of electrolytic treatment on industrial wastewaters: Microbiological aspects. *Journal of Environmental Science and Health Part A* **37**:1463–1481.

Zhang WX, Bouwer EJ, Ball WP. (1998). Bioavailability of hydrophobic organic contaminants: Effects and implications of sorption-related mass transfer on bioremediation. *Ground Water Monitoring & Remediation* **18**:126–138.

Zimmermann U, Schulz J, Pilwat G. (1974). Dielectric breakdown of cell membranes. *Biophysical Journal* **14**:881–899.

19

COUPLED ELECTROKINETIC–BIOREMEDIATION: APPLIED ASPECTS

SVENJA T. LOHNER, ANDREAS TIEHM, SIMON A. JACKMAN, AND PENNY CARTER

19.1 BIOREMEDIATION OF SOILS

Microbes are able to biodegrade almost all organic chemicals that have been discovered or produced to date. Indeed, microbial reactions are only generally limited by the thermodynamics of any given chemical reaction. Furthermore, enzymatic degradation is frequently more efficient and requires less harsh conditions than chemical breakdown. Biodegradation is therefore the approach of choice for many aqueous systems comprising complex organic mixtures. When moving into the soil matrix, the challenges begin. Despite the great promise of reactions in liquid systems, there is frequently a tighter constraint on the system in soil, which leads to partial or very slow degradation (Thompson, 2002). These constraints may be physical, chemical, or biological. Overcoming them within an appropriately managed risk-based framework is the challenge for the bioremediation industry.

At present, bioremediation is receiving a lot of interest as it appears sustainable, low cost, and low energy and therefore meets the needs of a remediation solution that will both benefit the environment directly and also have a relatively low carbon footprint. In practice, it often needs to be combined with other approaches to achieve an effective solution. This requirement is frequently driven by the need to remediate over a short timescale, in the case of *ex situ* bioremediation, or the need to promote long-term effectiveness under a regime such as monitored natural attenuation or permeable reactive barriers. The engineering of bioremediation to enhance its effectiveness has been studied for the last few decades with considerable success. It has become a mainstream part of the toolbox for contaminated sites.

Electrochemical Remediation Technologies for Polluted Soils, Sediments and Groundwater,
Edited by Krishna R. Reddy and Claudio Cameselle
Copyright © 2009 John Wiley & Sons, Inc.

19.1.1 Relevant Biodegradable Contaminants

In looking at the prevalence of contaminants in sites both within Europe and globally, there are two classes of contaminant that stand out, namely hydrocarbons and chlorinated solvents. Hydrocarbons are derived principally from oil- and gas-related activities and are therefore located both at factory sites and oil refineries and also in commercial and domestic environments where oil is supplied to homes and offices. Common contaminants include saturates (e.g. *n*-alkanes and cycloalkanes, C6–C16), aromatics [e.g. polycyclic aromatic hydrocarbons (PAHs) and benzene, toluene, ethylbenzene, and xylenes (BTEX)], resins (e.g. amides), and asphaltenes (e.g. ketones) (Pollard, Hrudey, and Fredorak, 1994; Potter and Simmons, 1998). Contamination plumes in the soil resulting from domestic kerosene spills frequently contain concentrations of total petroleum hydrocarbons (TPHs) in the range between 1000 and 100,000 mg/kg (RAW Group, pers. comm.). They constitute a potential risk to human health (Nathanail, Bardos, and Nathanail, 2002). The individual constituents of fuel oils have varied physicochemical properties resulting in heterogeneous migration through a soil system (Dror, Gerstl, and Yaron, 2001). When oil contacts with the soil, large areas can become contaminated through the lateral spread of the contamination, influenced by the environmental conditions, and through the presence of water and a hydraulic gradient. The hydrophobic nature of hydrocarbons gives them a high affinity to bind to the soil resulting in reduced bioavailability. Bioavailability describes the extent of interactions between the bacteria and the target compound. Low bioavailability indicates that there is limited contact between the bacteria and the target compound, leading to a reduction in degradation rate (Amellal, Portal, and Berthelin, 2001; Tiehm and Stieber, 2001).

The biodegradative pathways of most hydrocarbon contaminants have been previously characterized. Of the aromatic compounds, benzene, toluene, and xylene are all degradable in the soil under optimum conditions. The principal pathways of degradation include dioxygenase attack on the aromatic ring and monoxygenase attack on the methyl substituents (Tsao, Song, and Bartha, 1998). The biodegradation of alkanes (cyclic and straight chain) involves the oxidation of the chemical to the corresponding alcohol, aldehyde, and fatty acid, with the latter undergoing β-oxidation to form acetate and propionate (Rosenberg and Ron, 1996).

Chlorinated solvents have been used over the years for degreasing and lubricating activities. Due to their extensive use, toxicity, and their persistence in the environment, chloroethenes are among the most problematic contaminants worldwide. They form persistent plumes of contaminated groundwater reaching lengths of several kilometers. For a long time, it was postulated that chloroethenes are nonbiodegradable, recalcitrant substances. Generally, in the field, only slow and partial degradation of chloroethenes is observed, since microbial degradation in the field is limited by a lack of electron donors and electron acceptors (Bradley, 2003; Aulenta, Majone, and Tandoi, 2006). Today, there is a variety of microorganisms known to be capable of degrading chlorinated compounds. During microbial reductive dechlorination, chloroethenes are sequentially degraded from perchloroethene (PCE), via trichloroethene (TCE), *cis*-dichlorethene (*cis*-DCE), and vinyl chloride (VC), to ethene via a reductive dechlorination pathway (Fig. 19.1). The lower chlorinated metabolites, such as VC and *cis*-DCE, can additionally undergo oxidative

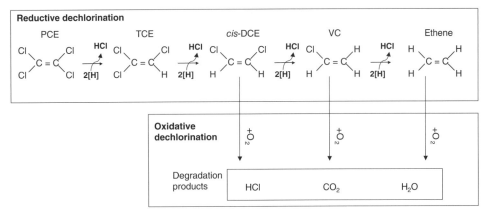

Figure 19.1. Schematic view of the reductive and oxidative biodegradation pathway for chloroethenes.

degradation with oxygen as the electron acceptor (Verce, Ulrich, and Freedman, 2000; Coleman et al., 2002a,b; Tiehm et al., 2008) (Fig. 19.1).

19.1.2 Biotransformation of Contaminants

The biotransformation of contaminants is driven by the gain of energy for microorganism growth and maintenance. The energy-producing process is the physiological coupling of oxidation and reduction reactions (redox process) in which electrons are transferred from one compound (electron donor) to an electron-accepting substrate (electron acceptor). Within the cell, electrons are moved through an electron transport system, and the electrical energy is converted to chemical energy [adenosine triphosphate (ATP)] by a process called chemiosmosis (Chapelle, 2001). In general, the compound acting as electron donor is characterized by a relatively reduced state and is oxidized in a redox process. Common examples are natural organic materials, dissolved hydrogen, and many contaminants such as aromatic hydrocarbons. Conversely, electron acceptors occur in relatively oxidized states and are reduced during microbial metabolism (Wiedemeier et al., 1999; Tiehm and Schmidt, 2007). The most important terminal electron acceptors for microbial metabolism are dissolved oxygen, nitrate, sulfate, iron(III) and manganese(IV) minerals, and carbon dioxide, which occur naturally in the subsurface.

Transformation processes, in which the contaminant transfers electrons to an electron acceptor, are called oxidative degradation. Reductive degradation occurs when the organic compound acts as an electron acceptor in presence of an electron donor. In Figure 19.2, both processes are illustrated. Additionally, fermentation reactions can occur, in which the organic compound serves as both electron donor and acceptor.

The maximum usable energy for the microorganisms that is produced from the redox process can be quantified by the Gibbs free energy of the reaction (ΔG_r^0). Negative values indicate that the reaction is energy producing (exothermic), while positive values indicate that the reaction consumes energy (endothermic).

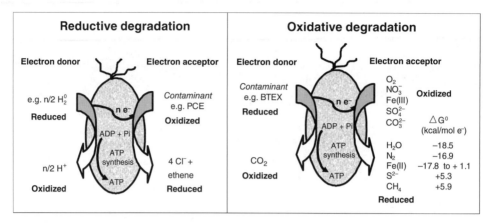

Figure 19.2. Schematic view of microbial degradation processes in the subsurface. ΔG^0, standard Gibbs energy [modified from Löffler et al. (2003)].

Considering both the electron-consuming and the electron-donating half reactions, microorganisms can only facilitate those redox processes that will yield energy (i.e. ΔG_r^0 of complete redox reaction <0), and reactions with the most energy release are preferred over reactions resulting in less energy. In Figure 19.2, the Gibbs free energy of different electron acceptor half reactions are given. The data show that some electron acceptors such as oxygen are energetically more favorable compared with others such as sulfate. According to their energetic data, a sequential utilization of electron donors in the order of oxygen (aerobic degradation), nitrate, iron, sulfate, and finally, carbonate (anaerobic oxidation) is observed in the field (Christensen, Bjerg, and Kjeldsen, 2001).

For reductive anaerobic transformation processes, such as reductive dechlorination, hydrogen was identified to be the key electron donor (DiStefano, Gossett, and Zinder, 1992; Ballapragada et al., 1997; Löffler, Tiedje, and Sanford, 1999; He et al., 2002). Depending on the microbial community, more complex substrates such as acetate, butyrate, or lactate can also serve as initial electron donors. However, the fermentation of these substrates results in hydrogen and formate/acetate, which act as ultimate electron donors (Smatlak, Gossett, and Zinder, 1996; Ballapragada et al., 1997; He et al., 2002). In general, biodegradation is only possible when all necessary components such as contaminants, electron donors, and acceptors as well as nutrients are bioavailable for the degrading microorganisms (Bosma et al., 1997; Xu and Obbard, 2004).

19.1.3 Limitations of Bioremediation

In practice, physical, chemical, and microbial constraints limit the efficiency of bioremediation and lead to slow or incomplete biodegradation of contaminants.

Physical constraints may be considered in terms of temperature or moisture levels in a soil environment or, alternatively, entrapment of contaminants in a location where the microbes are unable to gain access to them, such as in soil micropores. Furthermore, contaminants may bind to soil, thereby reducing their bioavailability (Hatzinger and Alexander, 1995; Allard and Neilson, 1997;

Carmichael, Christman, and Pfaendeer, 1997). This is further exacerbated by poor contaminant solubility and pollutant "aging." With temperature and moisture levels, there may be a reduction in rate of reaction, whereas in the case of inaccessibility of contaminants, there may be a level of residual contaminant remaining at the end of the biodegradation process.

Chemical constraints to biodegradation can take a number of forms. An example of the different chemistries affecting the potential for biodegradation is that of polychlorinated biphenyls (PCBs), which contain a biphenyl nucleus that is susceptible to biodegradation. However, the attachment of chlorine atoms to the nucleus results in electron withdrawal and a reduced susceptibility to biodegradation. Depending on the number and placement of chlorine atoms, biodegradation can be significantly reduced and may not be able to occur under aerobic conditions, relying on anaerobic dechlorination, which is a fortuitous and less predictable process in its timescale and outcome. Further chemical constraints relate to the controls on the bioavailability of the organic compounds as a function of solubility. In this instance, the supply of surfactants may facilitate solubilization and biodegradation (Tiehm, 1994; Tiehm et al., 1997; Cho et al., 2002).

Microbiological constraints to biodegradation may take a number of forms but frequently relate to the availability of nutrients and electron acceptors to support biodegradation. In some cases, where sites are contaminated with a mixture, there may be issues with organisms being unable to degrade or tolerate organic wastes (Huesemann, Hausmann, and Fortman, 2003). High concentrations of contaminant may be inhibitory or toxic to bacteria or toxic transformation products may build up (Pollard, Hrudey, and Fredorak, 1994; Lorton et al., 2001; Wackett and Hershberger, 2001). Microbes are well known to require nitrogen, phosphorus, and potassium for growth. They may also require other metals such as iron. In addition, for anaerobic degradation processes, there is a requirement for electron donors to enable the microbial dechlorination of, for example, chlorinated solvents.

Experiments in contaminated groundwater demonstrate that in a well mixed, homogeneous environment, contaminants such as BTEX and PAH can be successfully degraded under nitrate-, iron(III)- and sulfate-reducing conditions (Fig. 19.3). The microbial degradation of toluene in the presence of nitrate as electron acceptor is shown in more detail, illustrating that during toluene degradation, the electron acceptor (nitrate) becomes depleted. However, in natural systems, a heterogeneous distribution of microorganisms, electron acceptors, and nutrients is observed in the subsurface (Harms and Bosma, 1997; Gonod, Chenu, and Soulas, 2003; Nambi et al., 2003; Pallud et al., 2004). Additionally, in the aquifer, only a very limited transversal mixing of compounds by dispersion and diffusion occurs (Thornton et al., 2001; Thullner et al., 2002; Rahman et al., 2005). Especially in contaminant plumes, bioavailable electron acceptors and nutrients are quickly depleted due to microbial activity (Fig. 19.4). Consequently, the limited bioavailability of electron acceptors such as oxygen, nitrate, or sulfate, and nutrients, such as phosphorus and ammonium, is most pronounced under *in situ* conditions and is considered to be an important limiting factor for a successful microbiological cleanup of polluted sites.

Concerning nutrient availability, from bioremediation practice, it was observed that a ratio of approximately C:N:P of 250:10:3 is suitable for microbial growth (Geller, 1991). However, this ratio is mostly not found at contaminated field sites.

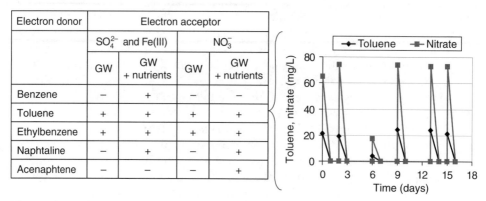

Figure 19.3. Microbial contaminant degradation in groundwater with nitrate, sulphate, and iron(III) as electron acceptors and toluene biodegradation under nitrate-reducing conditions [modified from Schulze (2004)]. GW, groundwater; +, biodegration observed; –, no biodegration observed.

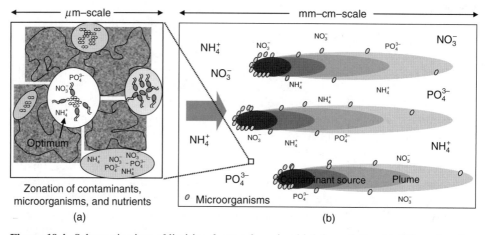

Figure 19.4. Schematic view of limiting factors for microbial degradation in different scales. Heterogeneous distribution of contaminants, electron acceptors, nutrients, and microorganisms on the μm-scale (a) and depletion of electron acceptors and nutrients in a contaminant plume on a mm-scale (b).

In Figures 19.3 (table) and 19.5, microbial degradation activity in BTEX- and PAH-contaminated groundwater is illustrated under nutrient-limited and nonlimited conditions. When nutrients such as phosphate and ammonium are added to the groundwater, a more efficient biodegradation takes place. In contrast, with no nutrients available, either no or less microbial contaminant degradation was observed. These results demonstrate how important the bioavailability of necessary components is for microbial degradation processes.

Different approaches exist to enhance bioavailability and bioremediation in the field, which include:

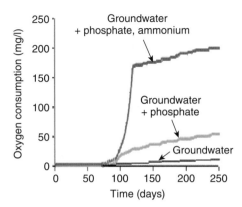

Figure 19.5. Nutrient-limited and -stimulated biodegradation of contaminants (BTEX, PAH) in groundwater [from Schulze (2004)].

- bioaugmentation with organisms either appropriate for biodegradation or especially acclimatized to conduct biodegradation under site conditions;
- biostimulation through the addition of chemicals such as nutrients, electron donors, and electron acceptors;
- enhanced bioremediation through the addition of other compounds such as surfactants to mobilize contaminants; and
- "pump and treat" approaches to driving contaminants through the subsurface through hydraulic pressures.

19.2 COMBINATION OF ELECTROKINETICS AND BIOREMEDIATION

One of the greatest challenges with bioremediation is that of being able to predict the performance of the microorganisms in biodegrading the contaminants. This is central to both *ex situ* and *in situ* remediations. Any complementary techniques that can direct and control the bioremediation process are therefore of interest to practitioners. In principle, electrokinetics, with its ability to control the flow of ions and water through the soil and to manipulate pH and redox, is a strong candidate for complementing bioremediation. Its efficacy in clay and other low porosity soils gives it potential advantages over hydraulic techniques that are severely constrained in these environments. Furthermore, electrodes can contribute to stimulate biodegradation by increasing the bioavailability of nutrients and electron donors and acceptors in the field. In general, the potential benefits of electrokinetic and electrochemical processes coupled with bioremediation include:

1. enhancement of contaminant bioavailability through electrokinetic mobilization,
2. increase of restricted soil bacteria mobility by electrokinetic transport processes,
3. electrokinetic-induced mass transfer and transport of ionic electron acceptors and nutrients, and

4. electrochemical production of limited electron donors (H_2) and acceptors (O_2).

The application of electrodes represents a new approach for distributing and producing stimulating agents in polluted soil and groundwater. Especially in recent years, the interest arose in coupled bio-electro technologies to enhance biodegradation efficiency. If the constraints to biodegradation and bioavailability can be tackled using electrokinetics, then there is the potential for bioremediation to become more widespread in its application to contaminated sites and sediments.

In the following sections, an outline of each electrobioremediation approach is given. No general statements about operation conditions, removal rates, or effectiveness of the coupled technology can be provided at this stage as the current knowledge of applied electrobioremediation is still patchy and therefore limited. Instead, the current knowledge of electrobioremediation is reviewed based on bench- and field-scale studies described in the literature.

19.2.1 Electrobioremediation: Mobilization of Hydrocarbons and Increase of Bacteria Mobility

Bioremediation has been applied to hydrocarbon contamination in a range of contexts, from water systems to oil spills and land contamination. For land remediation, most *in situ* and *ex situ* technologies are difficult to use when there is a complex infrastructure present in the vicinity of the contamination (Vik and Bardos, 2002). For this reason, there is a demand for a reliable *in situ* technology that is capable of remediating hydrocarbons by overcoming problems of bioavailability, nutrient delivery, and the ability to optimize biodegradation without having direct access to the plume (McCarty, 1991). This therefore represents a potential niche for the application of electrokinetics that may provide a cost structure that promotes its uptake as a technology (Fig. 19.6).

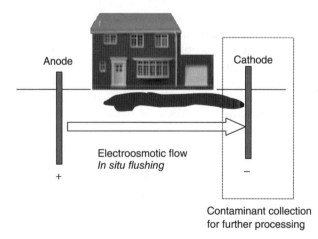

Figure 19.6. Schematic representation of electrokinetic treatment under a house.

The application of a low-voltage direct current in a soil system can induce a number of physicochemical processes including electroosmosis, electromigration, electrophoresis, and electrolysis. Electroosmosis can be used as an *in situ* flushing technique for the mobilization of the noncharged hydrocarbon or chloroethene contaminants from the subsurface. Once accessible, contaminants may be degraded *in situ* or could be moved further through the soil into an accessible area and then treated *ex situ* by physical or biological treatment processes without the requirement to excavate subsurface soils. For electroosmosis to be the driving force for the removal of contaminants, the hydraulic conductivity of the soil should be less than 10^{-5} mm/s (Bruell, Burton, and Walsh, 1992). If the opposing hydraulic gradient is greater than 10^{-5} mm/s the movement of contaminants toward the cathode of the circuit would be negated by water movement in the opposite direction. It is reported that electroosmotic conductivity is generally between 1 and 10×10^{-9} m²/(V·s) for a variety of soil types (Casagrande, 1947).

To date, a number of studies have been undertaken to investigate the use of electroosmosis as a process to remediate contaminated soils (Saichek and Reddy, 2005). However, as the use of electroosmotic flow for *in situ* soil flushing and the removal of organic contaminants are discussed in other chapters of this book, here, only a brief overview of some successful examples is provided. The electroosmotic movement of phenol and acetic acid was achieved in laboratory-scale experiments (Shapiro and Probstein, 1993), and *para* (*p*)-nitrophenol (Ho *et al.*, 1995), TCE, and benzene, toluene, and *m*-xylene (BTX) (Bruell, Burton, and Walsh, 1992; Segall and Bruell, 1992) have been investigated. Bruell, Burton, and Walsh (1992) compared conventional contaminant-transport equations to contaminant breakthrough graphs and additionally used glass column experiments to analyze the electroosmotic movement of BTX and TCE. This provides some supporting evidence that the electroosmotic removal of contaminants with high solubility and low distribution coefficients is possible. Renaud and Probstein (1987) identified that it may be feasible to use electroosmosis to control groundwater flow in contaminated soils and landfill sites, although their conclusions were based predominantly on theoretical models.

Studies conducted by authors Jackman and Carter demonstrated that electrokinetics alone has the potential to enhance the biodegradation of kerosene constituents in soil. They also showed that the efficiency is highly dependent on the chemical nature of the constituents and upon soil type (Table 19.1). Besides contaminants, microorganisms are also subject to electrokinetic transport processes. Microorganisms, especially pure cultures, have been shown to be transported in an electric field (DeFlaun and Condee, 1997; Wick *et al.*, 2004; Suni and Romantschuk, 2005; Wick, 2005). Depending on the microorganism species, at neutral pH, the negative or positive charges are predominant on the surface of the cell, leading to an electrokinetic transport toward the anode or cathode, respectively. Detailed information on microorganism interaction with the electric field is provided in Wick, Shi, and Harms (2007) or in the corresponding chapter in this book.

In spite of promising studies confirming the electrokinetic potential to mobilize contaminants and microorganisms, only a few laboratory studies and even less field studies actually showed the combination of electrokinetics and biodegradation and confirm the stimulating potential of the coupled process. A review of the electrobioremediation potential and some successful applications is given by Chilingar

TABLE 19.1. Mass Balance Analysis for Contaminants in Different Soils During Electrokinetic Treatment and in Controls without Electric Field

Soil	Test	Contaminant	Initial Concentration (mg/kg)	Final Concentration (mg/kg)	Extractable Contaminant of Initial Spike (%)
Reading soil (sandy loam)	EK	Dodecane	6941.0	6690.0	96.4
	Control	Dodecane	6941.0	7877.0	101.1
	EK	p-Xylene	1054.0	217.4	20.6
	Control	p-Xylene	1054.0	932.1	88.4
	EK	m-Xylene	3253.9	687.8	21.1
	Control	m-Xylene	3253.9	2696.5	82.9
	EK	o-Xylene	920.7	191.0	20.7
	Control	o-Xylene	920.7	776.2	84.3
	EK	Tetradecane	3769.5	3286.8	87.2
	Control	Tetradecane	3769.5	3905.5	103.6
Wytham soil (silty clay)	EK	Dodecane	8229.4	1683.6	20.5
	Control	Dodecane	8229.4	4474.2	54.4
	EK	p-Xylene	768.3	83.2	10.8
	Control	p-Xylene	768.3	233.3	30.4
	EK	m-Xylene	2573.8	274.4	10.7
	Control	m-Xylene	2573.8	724.1	28.1
	EK	o-Xylene	727.7	85.6	11.8
	Control	o-Xylene	727.7	191.2	26.3
	EK	Tetradecane	4805.7	560.1	11.7
	Control	Tetradecane	4805.7	1911.4	39.8

Analytical start and final concentration data are means with $n = 11$ (start) and $n = 6$ (final).
EK, Electrokinetic treatment.

et al. (1997). Early fieldwork combining both technologies has been conducted by Loo (1994) who treated gasoline- and diesel-contaminated clayey mud of Hayward, California. The remediation process, including passive biotreatment and electrokinetic transport of amendments and contaminants, was successful in removing concentrations of 100–3900 ppm TPH to less than 100 ppm TPH in soil and 10 ppm in groundwater within 4 weeks. In another approach, Loo, Wang, and Fan (1994) combined biodegradation and electrokinetic transport with a hot air venting system and ultraviolet light biocontrol system to remediate an area of about 2400 ft^2 contaminated with gasoline in concentrations ranging from 100 to 2200 ppm. Within 90 days of treatment, concentrations were less than the proposed cleanup level of 100 ppm, and the remediation costs were calculated to be about $50 per ton of soil treated.

More recent laboratory studies from Niqui-Arroyo et al. (2006) and Niqui-Arroyo and Ortega-Calvo (2007) demonstrated an up to 10-fold increase of PAH degradation in experiments with applied electric field compared with control experiments without electric current. Additionally, they showed that a periodic change of electrode polarity resulted in a more stable and better degradation efficiency. Also Luo et al. (2005) observed a stimulated microbial degradation of phenol in the presence of an electric field. In their experiments, bioremediation rates could be increased

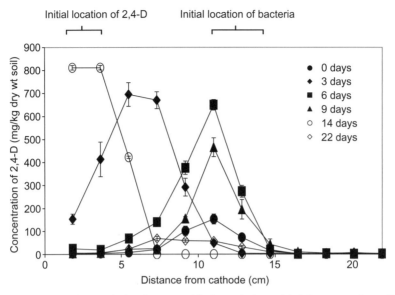

Figure 19.7. Distribution and simultaneous biodegradation of 2,4-D across the soil during electrokinetic treatment. Results are shown for different treatment times. The positions of 2,4-D and bacterial cells at the beginning of the experiment are shown [from Jackman et al. (2001)].

about five times by using a bench-scale, nonuniform electrokinetic system, in which the polarity of the electrodes was periodically reversed. In experiments from Mustacchi et al. (2005) to increase the yield of biosynthesis by electrokinetic separation of the synthesis product, they observed a significantly enhanced microbial transformation of benzonitrile to benzoic acid when an electric field was applied. The same working group demonstrated a simultaneous electrokinetic transport of the contaminant 2,4-dichlorophenoxyacidic acid (2,4-D) and the degrading microorganisms in silt soil (Jackman et al., 2001). With current densities of $3.72 A/m^2$, 2,4-D was transported at rates of approximately 4 cm/day toward the anode. At the same time, 2,4-D degrading bacteria (*Burkholderia* spp. RASC) were injected into the soil close to the anode. During the electrokinetic transport of 2,4-D, its simultaneous biodegradation was observed in the soil compartment in which the contaminant and the microorganisms met (Fig. 19.7).

Olszanowski and Piechowiak (2006) used *Pseudonomonas putida*, *Bacillus subtilis*, and *Klebsiella pneumoniae* to study stimulated cell migration and subsequent biodegradation of crude oil. In the soil samples, they observed facilitated migration of bacteria and showed larger biodegradation of crude oil under the influence of the electric field. A more field-related study was conducted by Lee and Lee (2001), who introduced *Pseudonomas* species into a diesel-contaminated aquifer and used electrodes for a homogeneous distribution of the microorganisms by electrophoresis and electroosmosis. They observed up to 60% degradation of diesel within 8 days by electrokinetically distributed *Pseudonomas* cells.

A large full-scale field application of electrobioremediation represents the so-called Lasagna technology. This *in situ* soil remediation technology was developed by a consortium of industry partners (Monsanto, DuPont, and General Electric), the Department of Energy (DOE), and the Environmental Protection Agency (EPA). It was especially designed for cleaning up contaminations in heterogeneous or low-permeability soils. The technology uses electrokinetic transport processes to drive contaminants into special treatment zones, which are installed directly in the contaminated area. The treatment zones are typically placed in between the electrodes at close spacing (5 ft apart) and generally consist of adsorptive or reactive materials such as activated carbon or iron. Once the contaminants reach the treatment zones, they can either be biodegraded in the case of activated carbon fillings or abiotically transformed by reaction with the iron filling. The successful application of this technology has been shown in a small field test at a TCE-contaminated site in which over 98% of TCE could be removed (Ho et al., 1999a). Upscaling to a large field test at a TCE-contaminated site in Paducah, Kentucky, demonstrated a TCE removal ranging from 95% to 99%. Challenges to face were very high contamination levels and a complex hydrogeology in the subsurface. On the basis of the field results, the calculated treatment costs for a typical one-acre site with contamination from 15 to 45 ft deep were $45–$80 per cubic yard (Ho et al., 1999b).

19.2.2 Electrobioremediation: Induced Mass Transfer and Transport of Microbial Nutrients and Electron Acceptors

The idea of this strategy is that in the electric field, ionic contaminants, and electron acceptors such as sulfate, nitrate, and nutrients, such as ammonium and phosphate, are transported by electromigration. At contaminated sites, especially downgradient the contaminant source, electron acceptors and nutrients are depleted in many areas due to microbial processes. The application of an electric field perpendicular to the flow direction enables the transport of sulfate, nitrate, and nutrients from the surrounding environment, still enriched in necessary compounds, into the plume. At the same time, organic contaminants are moved by electroosmosis and microorganisms by electrophoresis. On a macroscopic scale, the electrokinetic processes result in an increase of the dispersion coefficient, which subsequently results in better bioremediation. In this way, a better mixing of heterogeneously distributed compounds for microbial degradation can be achieved.

In the literature, single studies in the field of groundwater remediation reported stimulated biodegradation of contaminants in combination with the use of electric fields. However, in the beginning it could not be clarified which processes led to these observations (Lageman and Pool, 2001). In this context, other authors postulated that electrokinetic processes result in a more homogeneous distribution of nutrients in the soil (Acar and Alshawabkeh, 1996; Elektorowicz and Boeva, 1996), and a potential combination of electrokinetics and bioremediation was considered to be promising for future remediation strategies (Rahner et al., 2001). Several investigations were conducted to prove the successful electrokinetic transport of nutrients and electron acceptors under different experimental conditions (Runnels and Wahli, 1993; Elektorowicz and Boeva, 1996; Acar, Rabbi, and Ozsu, 1997; Budhu et al., 1997; Thevanayagam and Rishindran, 1998; Eid et al., 2000; Børresen, 2001). In recent studies (Lohner, Katzoreck, and Tiehm, 2008a,b), which focused on the determination of comparable ion migration rates due to a standardized experimental design, an efficient movement of nitrate, sulfate, phosphate, and ammonium in

TABLE 19.2. Average Electrokinetic Transport Parameters for Sulfate, Nitrate, Phosphate, and Ammonium in Sandy Soil Saturated with Demineralized Water

20 V/10 cm	Average Transport Rate (cm/h)	Mass Transported within Approximately 48 h (c_0 = 1 g/l) (mg)
Nitrate	1.34	210
Sulfate	1.91	321.5
Phosphate	0.48	37.7
Ammonium	0.40	6

Experiments were conducted in a 10-cm electrokinetic cell with a voltage gradient of 2 V/cm (Lohner, Katzoreck, and Tiehm, 2008a).

sandy soils saturated with demineralized water was demonstrated, and specific transport rates and mass transport for the different ion species could be determined (Table 19.2).

Sulfate shows the highest transport rate followed by nitrate, phosphate, and ammonium. The determining factors for successful electromigration are the conductivity of the soil, soil porosity, pH gradient, applied voltage gradient, initial concentration of the specific ions, and the presence of competitive ions. The electromigrative velocity v_{em} of an ion is proportional to the ion charge and the local electric field strength:

$$v_{em} = u_i \cdot z_i \cdot n \cdot \tau \cdot F \cdot E, \tag{19.1}$$

where u_i is the ion mobililty [m²/(V·s)], z_i is the valence, n is porosity, τ is tortuosity, F is the Faraday constant (96485 C/mol), and E is the electric field strength (V/m).

Figure 19.8 demonstrates the effect of voltage gradient and initial ion concentration on the electrokinetic transport parameters. As expected from Equation 19.1, a doubled electric field strength results in a doubled transport rate of the ions. Also, the mass transport increases significantly with higher voltage. However, for the mass transport, no linear relationship is observed. Changes in initial ion concentrations do not affect the rate of electromigration but the amount of ions is considerably higher with an electric gradient of 2 V/cm compared with 1 V/cm. When ion mixtures are introduced into the electric field, both the transport rate and the mass transport decrease compared with the single ion transport due to competitive transport of other species, the need for charge balance, and electrical neutrality when multispecies are present.

In soils that have a low buffering capacity, the effect of the pH change on ion transport has to be considered. Due to electrode reactions, forming H^+ at the anode and OH^- at the cathode, a distinct pH gradient develops throughout the soil. Solute pH close to the cathode can reach values of about 7–12.5, while the pH near the anode decreases to 1.5. Where H^+ and OH^- species meet in the soil, they form H_2O, and a waterfront is generated in the soil (Dzenitis, 1997). The waterfront acts as an insulation layer, and corresponding to the pH gradient, a steep local voltage gradient develops. As indicated in Equation 19.1, transport rates are high where steep local voltage gradients develop. Thus, ions entering the region with a high electric field strength will migrate fast, whereas the migration decreases in the region with low-voltage gradient, and an accumulation of ions in the soil occurs. A high buffer

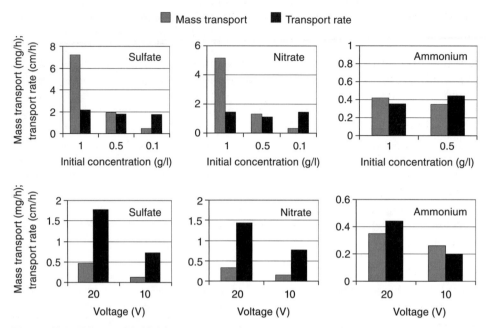

Figure 19.8. Effects of initial ion concentration and voltage gradient on the electrokinetic transport parameters for sulfate, nitrate, and ammonium [from Lohner, Katzoreck, and Tiehm (2008a)].

TABLE 19.3. Comparison of Electromigrative Transport Parameters of Nitrate in Demineralized Water and Groundwater (Lohner, Katzoreck, and Tiehm, 2008b)

Nitrate (20V, 0.5 g/l)	Transport Rate (cm/h)	Mass Transported within 53 h (mg)
Groundwater	1.64	85.6
Distilled H$_2$O, single ion	1.13	53.4
Distilled H$_2$O, mixed ions	0.61	36.9

Experiments were conducted in a 10-cm electrokinetic cell with a voltage gradient of 2 V/cm. The initial nitrate concentration was 0.5 g/l.

capacity of the treated soil is therefore beneficial for electromigration. Experiments in contaminated groundwater (Lohner, Katzoreck, and Tiehm, 2008b) confirm that electromigrative nitrate transport is enhanced compared with experiments in demineralized water due to a higher buffer capacity (Table 19.3).

In addition to the pH effects on electrokinetics, it has to be considered that many microorganism species are only active within a specific pH range. If the pH in the soil falls below or above these values due to electrode reactions, microbial activity will be inhibited and biodegradation will cease in the aquifer. Thus, the pH is one of the most important parameters influencing electrokinetic-enhanced biodegradation.

The actual application of electrokinetic cometabolites injection or nutrient amendment into contaminated groundwater to stimulate bioremediation has been demonstrated only in a few studies. One approach, the injection of cometabolites into contaminated groundwater, was done by Rabbi *et al.* (2000). In this study, sulfate and the cometabolic substrate benzoic acid were distributed in the soil by means of electrokinetics. The results showed that both the electrokinetic injection of sulfate and benzoic acid facilitated the biodegradation of TCE. In a 1-m soil column study, the determined TCE first-order degradation rates were approximately 0.039 per day. Other bench-scale experiments were conducted by Lee, Ro, and Lee (2007) who investigated the electrokinetically enhanced bioremediation of diesel in low-permeability soils. Diesel (59.4%) was anaerobically biodegraded with nitrate as electron acceptor when triethyl phosphate (TEP) was electrokinetically provided compared with 18.7% in the control without electrokinetic treatment. Nutrient amendment not only stimulates the biodegradation of organic compounds. A study by Reddy *et al.* (2003) showed that the bioremediation of chromium-contaminated soil can also be stimulated by electrokinetics. Acetate, phosphate, and ammonium were successfully delivered into the soil by electrokinetics and, moreover, improved Cr(VI) reduction when microorganism cultures were employed.

Godschalk and Lageman (2005) introduced an electrobioremediation technology, which they called the electrokinetic biofence (EKB). The basic setup of their technology consists of a row of alternating electrodes (anodes and cathodes), which are installed perpendicular to the groundwater flow direction. Nutrients such as nitrogen and phosphorous or electron acceptors, organic compounds, and micronutrients needed for bioremediation processes can be injected into infiltration wells upstream of the electrodes and into the electrode wells. When an electric field is applied, the injected compounds are electrokinetically dispersed homogeneously throughout the contaminated soil. The electric field strength, and hence the energy for the EKB, is dependent on the velocity of groundwater flow. Nutrient dispersion efficiency is additionally dependent on the nutrient stability in the groundwater (e.g. precipitation reactions) and the extent of adsorption on soil particles. This technology was especially designed for the removal of volatile organic carbons (VOCs) in a groundwater plume to stop downstream migration of the contaminants. With VOCs, in addition to the beneficial effects of electrokinetic nutrient injection and dispersion, the effect of soil heating during electrokinetic treatment can also be used to create optimal conditions for the biological activity for VOC biodegradation.

A field-scale test to demonstrate the successful application of the EKB was conducted at a VOC-contaminated site of a chemical laundry at Wildervank in the Netherlands (Godschalk, 2005; Godschalk and Lageman, 2005). After 2 years of operation, a significant biodegradation of the chlorinated compounds and an increase of nutrient concentrations in the contaminated area were observed, which was due to the electrokinetic dispersion of dissolved nutrients in the groundwater. The EKB is one of the first technologies implying electrobioremediation, which is commercially available to date.

19.2.3 Electrobioremediation: Electrochemical Production of Electron Donors and Acceptors

In this approach, the electrode reactions are used to provide the necessary electron acceptors and donors to soil microorganisms. In aqueous solutions, after applying an electric current, besides electrokinetic processes, water electrolysis takes places, generating hydrogen and oxygen at the electrodes according to the following equations:

$$\text{Cathode}(-): \quad 2e^- + 2H_2O \rightarrow H_2(g) + 2OH^-,$$

$$\text{Anode}(+): \quad 2H_2O \rightarrow O_2(g) + 4H^+ + 4e^-.$$

Down-gradient the cathode, hydrogen is available as an electron donor for reductive degradation processes, whereas at the anode, oxygen is provided for oxidative biodegradation.

Only a few studies concentrate on the coupling of electrolysis and microbial degradation, mainly focusing on stimulated nitrification and denitrification. Szekeres *et al.* (2001) used electrolytically generated hydrogen for microbial denitrification, and other studies demonstrated the successful coupling of electrolysis and denitrification in biofilm-electro reactors (BERs) (Hayes, Flora, and Khan, 1998; Islam and Suidan, 1998; Sakakibara and Nakayama, 2001; Feleke and Sakakibara, 2002). The electrogenerated hydrogen serves as the electron donor for denitrification to proceed:

$$2NO_3^- + 5H_2 + 2H^+ \rightarrow N_2 + 6H_2O.$$

The degrading microorganisms were immobilized on the cathode and utilized the produced hydrogen with denitrification efficiencies of up to 98% (Flora *et al.*, 1994). A stimulated denitrification was also observed when microorganisms were immobilized on the filling material of the reactor (quartz sand) instead of the electrodes (Hayes, Flora, and Khan, 1998). In similar experiments, nitrification was stimulated by generated oxygen at the anode:

$$NH_4^+ + 2O_2 \rightarrow NO_3^- + 2H^+ + H_2O.$$

Degradation efficiencies reached 84%, and control experiments showed that no abiotic transformation of ammonium occurred. Watanabe, Hashimoto, and Kuroda (2002) managed to simultaneously stimulate nitrification at the anode and denitrification at the cathode in a bio-electro reactor system.

In experiments by Lohner and Tiehm (2008), it was demonstrated that chloroethene biodegradation can also be stimulated by electrolysis. The effect of electrogenerated electron donors and acceptors on microbial chloroethene degradation is shown in Figure 19.9. During electrolysis, much more chloride is released due to enhanced microbial degradation than without electrolysis. Control experiments without microorganisms did not result in significant chloride increase, indicating that no electrochemical, abiotic transformation of the chloroethenes occurred.

Due to the sequential biodegradation pathway (Fig. 19.1), a sequential electrolytic production of hydrogen and oxygen holds the opportunity to stimulate the reductive dechlorination of the higher chlorinated compounds (PCE, TCE) down-gradient the cathode in the first step and subsequently the oxidative degradation of the lower chlorinated metabolites (*cis*-DCE, VC) down-gradient the anode. The

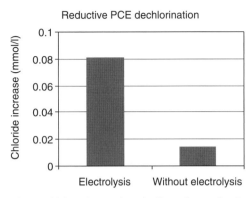
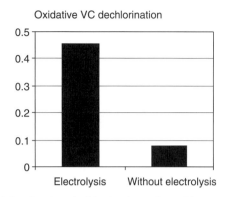

Figure 19.9. Electrochemically enhanced microbial reductive dechlorination of perchloroethene (PCE) and oxidative degradation of vinyl chloride (VC). Stimulated biodegradation is indicated by a chloride increase during electrochemical treatment.

principle of this process is shown in Figure 19.10a. Experiments proved the successful electrolytic production of hydrogen and oxygen in laboratory experiments, which is shown in Figure 19.10b. Maximal values of dissolved hydrogen downgradient the cathode were 0.592 mg/l, and dissolved oxygen was measured in concentrations up to 11.1 mg/l behind the anode. In a flow-through quartz sand column system, the coupled bio-electro process was simulated. PCE was used as the model contaminant. Figure 19.10c demonstrates the contaminant distribution in the active column system during electrolysis. The diagram shows that PCE was reductively transformed to *cis*-DCE, VC, and ethene in the cathodic column with the electrogenerated hydrogen as electron donor. In the second step, anodic oxygen was used as the electron acceptor for complete mineralization of the lower chlorinated metabolites.

One major advantage of this process is that the degradation efficiency can be controlled by the applied current. The correlation of degradation efficiency and electric current is given by Faraday's law, which defines that the amount of the produced substance at the electrodes is proportional to the electric current:

$$I \cdot t = z \cdot F \cdot n, \qquad (19.2)$$

where I is the current intensity (A), t is the electrolysis time (s), z is the valence, F is the Faraday constant (96485 C/mol), and n is the amount of the substance produced at the electrode (mol).

Faraday's law indicates that with higher electric current, more hydrogen and oxygen are produced at the electrodes for microbial degradation. Consequently, the experimentally determined degradation efficiency in the column system increases with increasing current intensity, which is shown in Lohner and Tiehm (2008).

Until now, no studies were conducted to show the coupling of electrolysis and biodegradation in the field. However, recent laboratory experiments in original groundwater from a chloroethene-contaminated site indicate that electrolytic stimulation can be successfully supplied under hydrochemical field conditions (S.T. Lohner and A. Tiehm, unpublished data).

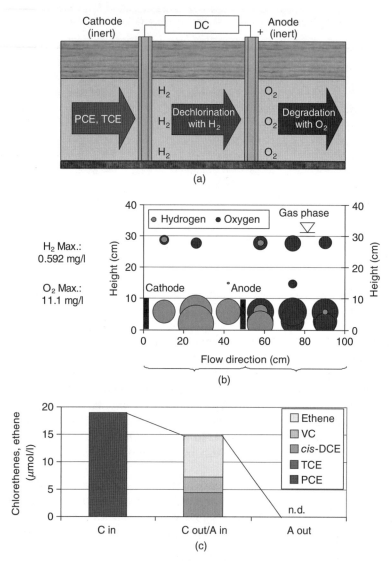

Figure 19.10. (a) Principle of electrolytical stimulation of chloroethene biodegradation. (b) Electrochemical production of hydrogen and oxygen down-gradient the electrodes in a model aquifer. (c) Electrolytically stimulated biodegradation of PCE in a flow-through column system. DC, direct current; C, cathodic column; A, anodic column; in, influent; out, effluent; n.d., not detected.

19.3 PRACTICAL CONSIDERATIONS AND LIMITATIONS FOR COUPLED BIO-ELECTRO PROCESSES

In the previous sections, the promising potential for coupled bio-electro processes was shown. This paragraph wants to emphasize some points that have to be considered for the most efficient application of coupled bio-electro processes.

19.3.1 Electrode Reactions

In technologies in which electrolysis is coupled with biodegradation, specific electrode reactions are desired to generate compounds such as hydrogen and oxygen for the degrading microorganisms. For these products to be bioavailable, a direct electrode contact between groundwater and electrodes is required. The main point that has to be kept in mind is that, in complex media such as groundwater, during electrolysis not only hydrogen and oxygen generation takes place. Depending on the electrochemical potential and the groundwater composition, side reactions can occur, which can significantly affect the process efficiency. In the following, the most relevant competition reactions are given:

$$\text{Anode}(+): \quad 2Cl^- \rightarrow 2e^- + Cl_2,$$
$$\text{Anode}(+): \quad Me^0 \rightarrow Me^{n+} + ne^-,$$
$$\text{Cathode}(-): \quad O_2 + 2H^+ + 2e^- \rightarrow H_2O_2,$$
$$\text{Cathode}(-): \quad NO_3^- + 2H^+ + 2e^- \rightarrow NO_2^- + H_2O,$$
$$\text{Cathode surrounding}: \quad Ca^{2+} + HCO_3^- + OH^- \rightarrow CaCO_3 + H_2O,$$
$$Mg^{2+} + 2OH^- \rightarrow Mg(OH)_2.$$

Anodic side reactions include chlorine evolution that is especially relevant in groundwater with high chloride concentrations, which can be the case at landfills on which salts were disposed. Depending on the electrode material, in corrosive groundwater, the anode can corrode, which reduces the oxygen production efficiency. When using stainless steel materials, depending on the electrode composition, secondary contaminants such as heavy metals (nickel, chromium, and iron) can be introduced into the groundwater due to anodic corrosion.

Potential cathodic side reactions are hydrogen peroxide evolution in oxygen-enriched groundwaters or nitrate reduction, leading to the production of toxic nitrite. In addition, pH changes at the electrodes can result in precipitation reactions of metals or calcium carbonate at the cathode, which reduces the porosity of the aquifer and might, in the worst case, lead to a blocking of the groundwater flow.

Some literature exists addressing these electrochemical side reactions mainly focusing on minimizing or optimizing a desired reaction at a special electrode material. In laboratory column studies using stainless steel electrodes, Franz *et al.* (2002) found light to heavy cathode coating of calcium carbonate that, however, did not affect the oxygen generation efficiency. This group also investigated the production of H_2O_2 and Cl_2 and could only find minimal hydrogen peroxide production. In electrolytes with high chloride concentrations (355 mg/l), significant chlorine evolution increasing with increasing current densities was observed. Kraft *et al.* (1999a,b) investigated electrochemical hypochlorite production showing that hypochlorite production rates were dependent on temperature, chloride concentrations, and applied current densities. Deposits at the cathode were also detected in this study consisting of $CaCO_3$ and $Mg(OH)_2$. Hydrogen peroxide formation with a faradic yield of 0.21 was measured by Drogui *et al.* (2001) at a vitreous carbon cathode. The rate of peroxide production was shown to be proportional to the current density.

Not only do these reactions reduce the efficiency of hydrogen and oxygen production, but electrolysis products such as chlorine can also have detrimental effects on degrading microorganisms, which is demonstrated in the following paragraph.

19.3.2 Microbial Activity

One of the most important requirements for coupled bio-electro processes is a sustained microbial activity in the electric field or during electrokinetic or electrolytical treatment. In studies focusing on the behavior of bacteria in intense electric fields, the viability, metabolism, and morphology of microorganisms were assessed showing that within very short treatment times, viable organisms such as *Escherichia coli* can be significantly reduced (Katsuki et al., 2000; Schoenbach, Joshi, and Stark, 2000). Electroporation, the irreversible permeabilization of cell membranes, oxidative stress, and cell death due to electrochemically generated oxidants and electrochemical oxidation of vital cellular constituents have been identified to be responsible for cell death during electrochemical exposure (Drees, Abbaszadegan, and Maier, 2003). Only recently, such investigations have also attracted attention in environmental studies with the objective to assess microbial activity of degrading microorganisms in coupled bio-electro processes.

Only very few studies exist that deal with the effects of weak electric fields on contaminant-degrading soil microorganisms (Liu, Brown, and Elliott, 1997; Jackman et al., 1999; Drees, Abbaszadegan, and Maier, 2003; Shi et al., 2008a,b; Tiehm, Lohner, and Augenstein, 2008). It has been shown that, depending on the electrode material, microbial activity can be inhibited by electric current densities in the range of $4\,mA/cm^2$ when electrodes are not shielded from the microorganisms (Tiehm, Lohner, and Augenstein, 2008). Figure 19.11 shows the total viable numbers during electrolysis. At lower current intensities that are applicable for bio-electro processes ($100\,mA$ corresponding to $4\,mA/cm^2$), no significant effects on microbial numbers are observed. However, increasing the current to $350\,mA$ ($14\,mA/cm^2$) results in

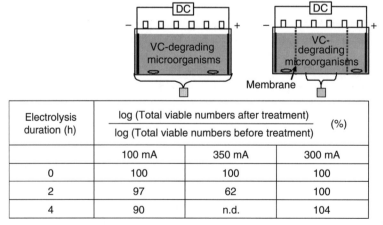

Figure 19.11. Effect of current intensity on microbial numbers during electrolysis with and without electrode shielding. DC, direct current; n.d., not detectable (<detection limit).

microbial numbers below the detection limit within 4 h of electrolysis. When the electrodes are separated from the microorganisms at similar currents (300 mA), no effects on microbial activity are observed. These results indicate that electrogenerated products are responsible for the inhibition of the microbial numbers. Many studies from other authors confirm the killing potential of electrogenerated products such as chlorine, hydrogen peroxide, or different radical species (Patermarakis and Fountoukidis, 1990; Davis et al., 1994; Liu, Brown, and Elliott, 1997; Hunt and Marinas, 1999; Drogui et al., 2001; Khadre and Yousef, 2001; Tanner et al., 2004; Vencel et al., 2004; Bergmann and Koparal, 2005).

Consequently, considerations about the microbial activity are especially important for applications in which electrochemical processes are applied in combination with microbial processes without shielding the electrolysis products from the microorganisms. A shielding is recommended for applications involving electrokinetics as for these processes, the generated electric field is important and not the electrogenerated electrode products.

19.4 SUMMARY

This chapter has provided an overview of potential innovative approaches to enhance biodegradation through the application of electrodes and electrokinetics, respectively. We summarized the current knowledge on electrobioremediation of contaminants based on the available literature and have outlined several approaches of coupling electrokinetics or electrolysis to bioremediation. The potential to overcome limitations of bioremediation by the use of electrobioremediation has been confirmed in numerous studies. However, data are mostly limited to bench-scale studies, and only a few field studies demonstrate the potential for the upscaling of the coupled technology. Besides the mentioned Lasagna technology and the EKB approach, the field-scale application of electrobioremediation is still in its infancy. However, the examples of applied electrobioremediation demonstrated in the literature, so far illustrate the breadth of what can be achieved and also highlight the significant opportunities and challenges associated with applying the coupled technique to real contaminated sites.

It is worth concluding that, while there is further development work to be done to enable these techniques to be applied in the field, there is significant potential for their application. Remarkably, electrokinetics has applicability to most of the common constraints associated with bioremediation, whether it be the bioavailability of hydrophobic contaminants or the lack of nutrients or electron acceptors/donors.

ACKNOWLEDGMENTS

We gratefully acknowledge the financial support by the European COST Action D32, stimulating knowledge exchange and fruitful discussions. A. Tiehm and S. Lohner acknowledge the financial support by the Federal Ministry of Economics and Technology and the German Federation of Industrial Research Associations "Otto von Guericke" e.V. (AiF) (project no. 15131 N and 150 ZN).

REFERENCES

Acar YB, Alshawabkeh AN. (1996). Electrokinetic remediation. I: Pilot-scale test with lead-spiked kaolinite. *Journal of Geotechnical Engineering* **122**:173–185.

Acar YB, Rabbi MF, Ozsu E. (1997). Electrokinetic injection of ammonium and sulfate ions into sand and kolinite beds. *Journal of Geotechnical and Geoenvironmental Engineering* **123**(3):239–249.

Allard A-S, Neilson AH. (1997). Bioremediation of organic waste sites: A critical review of microbiological aspects. *International Biodeterioration & Biodegradation* **39**(4):253–285.

Amellal N, Portal J-M, Berthelin J. (2001). Effect of soil structure on the bioavailability of polycyclic aromatic hydrocarbons within aggregates of a contaminated soil. *Applied Geochemistry* **16**:1611–1619.

Aulenta F, Majone M, Tandoi V. 2006. Enhanced anaerobic bioremediation of chlorinated solvents: Environmental factors influencing microbial activity and their relevance under field conditions. *Journal of Chemical Technology and Biotechnology* **81**(9):1463–1474.

Ballapragada BS, Stensel HD, Puhakka JA, Ferguson JF. (1997). Effect of hydrogen on reductive dechlorination of chlorinated ethenes. *Environmental Science & Technology* **31**(6):1728–1734.

Bergmann H, Koparal S. (2005). The formation of chlorine dioxide in the electrochemical treatment of drinking water for disinfection. *Electrochimica Acta* **50**:5218–5228.

Børresen, M. (2001). Increased mobility for nutrients in fine grained soils using electrokinetics. In *3. Symposium on Electrokinetic Remediation* (eds. K Czurda, R Haus, C Kappeler, R Zorn), April 18–20, 2001, Karlsruhe, Germany. Karlsruhe, Germany: Schr. Angew. Geol., pp. 38.1–38.13.

Bosma TNP, Middeldorp PJM, Schraa G, Zehnder AJB. (1997). Mass transfer limitation of biotransormation: Quantifying bioavailability. *Environmental Science & Technology* **31**: 248–252.

Bradley PM. (2003). History and ecology of chlorethene biodegradation: A review. *Bioremediation Journal* **7**(2):81–109.

Bruell CJ, Burton AS, Walsh MT. (1992). Electroosmotic removal of gasoline hydrocarbons and TCE from clay. *Journal of Environmental Engineering* **118**(1):68–83.

Budhu M, Rutherford M, Sills G, Rasmussen W. (1997). Transport of nitrate through clay using electrokinetics. *Journal of Environmental Engineering* **123**:1251–1253.

Carmichael LM, Christman RF, Pfaendeer FK. (1997). Desorption and mineralization kinetics of phenanthrene and chrysene in soils. *Environmental Science & Technology* **31**: 126–132.

Casagrande L. (1947). The Application of Electroosmosis to Practical Problems in Foundations and Earthworks. Department of Scientific and Industrial Research, Building Research, London, UK. *Technical Paper 30*.

Chapelle FH. (2001). *Ground-Water Microbiology and Geochemistry*, 2nd ed. New York: John Wiley & Sons.

Chilingar GV, Loo WW, Khilyuk LF, Katz SA. (1997). Electro-bioremediation of soils contaminated with hydrocarbons and metals: Progress report. *Energy Sources* **19**:129–146.

Cho H-H, Choi J, Goltz MN, Park JW. (2002). Combined effect of natural organic matter and surfactants on the apparent solubility of polycyclic aromatic hydrocarbons. *Journal of Environmental Quality* **31**:275–280.

Christensen TH, Bjerg PL, Kjeldsen P. (2001). Natural attenuation as an approach to remediation of groundwater pollution at landfills. In *Treatment of Contaminated Soil* (eds. R Stegmann, G Brunner, W Calmano, G Matz). Berlin, Germany: Springer, pp. 587–602.

Coleman NV, Mattes TE, Gossett JM, Spain JC. (2002a). Biodegradation of *cis*-dichlorethene as the sole carbon source by a β-proteobacterium. *Applied Environmental Microbiology* **68**:2726–2730.

Coleman NV, Mattes TE, Gossett JM, Spain JC. (2002b). Phylogenetic and kinetic diversity of aerobic vinyl chloride-assimilating bacteria from contaminated sites. *Applied Environmental Microbiology* **68**:6162–6171.

Davis CP, Shirtliff ME, Trieff NM, Hoskins SL, Warren MM. (1994). Quantification, qualification, and microbial killing efficiencies of antimicrobial chlorine-based substances produced by ionophoresis. *Antimicrobial Agents and Chemotherapy* **38**(12):2768–2774.

DeFlaun MF, Condee CW. (1997). Electrokinetic transport of bacteria. *Journal of Hazardous Materials* **55**:263–277.

DiStefano TD, Gossett JM, Zinder SH. (1992). Hydrogen as an electron donor for dechlorination of tetrachloroethene by an anaerobic mixed culture. *Applied Environmental Microbiology* **58**:3622–3629.

Drees KP, Abbaszadegan M, Maier RM. (2003). Comparative electrochemical inactivation of bacteria and bacteriophage. *Water Research* **37**:2291–2300.

Drogui P, Elmaleh S, Rumeau M, Bernard C, Rambaud A. (2001). Hydrogen peroxide production by water electrolysis: Application to disinfection. *Journal of Applied Electrochemistry* **31**:877–882.

Dror I, Gerstl Z, Yaron B. (2001). Temporal changes in kerosene content and composition in field soil as a result of leaching. *Journal of Contaminant Hydrology* **48**:305–323.

Dzenitis JM. (1997). Steady state and limiting current in electroremediation of soil. *Journal of the Electrochemical Society* **144**(4):1317–1322.

Eid N, Eishorbagy W, Larson D, Slack D. (2000). Electro-migration of nitrate in sandy soil. *Journal of Hazardous Materials* **B79**:133–149.

Elektorowicz M, Boeva V. (1996). Electrokinetic supply of nutrients in soil bioremediation. *Environmental Technology* **17**:1339–1349.

Feleke Z, Sakakibara Y. (2002). A bio-electrochemical reactor coupled with adsorber for the removal of nitrate and inhibitory pesticide. *Water Research* **36**:3092–3102.

Flora JRV, Suidan MT, Islam S, Biswas P, Sakakibara Y. (1994). Numerical modeling of a biofilm-electrode reactor used for enhanced denitrification. *Water Science Technology* **29**(10–11):517–524.

Franz JA, Williams RJ, Flora JRV, Meadows ME, Irwin WG. (2002). Electrolytic oxygen generation for subsurface delivery: Effects of precipitation at the cathode and an assessment of side reactions. *Water Research* **36**:2243–2254.

Geller A. (1991). *Handbuch Mikrobiologische Bodenreinigung Materialien zur Altlastenbearbeitung*. Karlsruhe: Landesanstalt für Umweltschutz Baden-Württemberg.

Godschalk B. (2005). Electro-bioreclamation of PCE and TCE contaminated soil at a site of a former silver factory. *Symposium on Electrokinetic Remediation*, May 22–25, Ferrara, Italy.

Godschalk MS, Lageman R. (2005). Electrokinetic biofence, remediation of VOCs with solar energy and bacteria. *Engineering Geology* **77**(3–4):225–231.

Gonod LV, Chenu C, Soulas G. (2003). Spatial variability of 2,4-dichlorophenoxyacetic acid (2,4-D) mineralisation potential at a millimetre scale in soil. *Soil Biology & Biochemistry* **35**(3):373–382.

Harms H, Bosma TNP. (1997). Mass transfer limitation of microbial growth and pollutant degradation. *Journal of Industrial Microbiology and Biotechnology* **18**:97–105.

Hatzinger PB, Alexander M. (1995). Effect of aging of chemicals in soil on their biodegradability and extractability. *Environmental Science & Technology* **29**:537–545.

Hayes AM, Flora JRV, Khan J. (1998). Electrolytic stimulation of denitrification in sand columns. *Water Research* **32**(9):2830–2834.

He J, Sung Y, Dollhopf ME, Fathepure BZ, Tiedje JM, Löffler FE. (2002). Acetate versus hydrogen as direct electron donor to stimulate the microbial reductive dechlorination process at chloroethene-contaminated sites. *Environmental Science & Technology* **36**:3945–3952.

Ho SV, Athmer C, Sheridan PW, Hughes BM, Orth R, McKenzie D, Brodsky PH, Shapiro A, Sivavec TM, Salvo J, Schultz D, Landis R, Griffith R, Shoemaker S. (1999b). The Lasagna technology for in situ soil remediation. 2. Large field test. *Environmental Science & Technology* **33**(7):1092–1099.

Ho SV, Athmer C, Sheridan PW, Hughes BM, Orth R, McKenzie D, Brodsky PH, Shapiro A, Thornton R, Salvo J, Schultz D, Landis R, Griffith R, Shoemaker S. (1999a). The Lasagna technology for in situ soil remediation. 1. Small field test. *Environmental Science & Technology* **33**(7):1086–1091.

Ho SV, Sheridan PW, Athmer CJ, Heitkamp MA, Brackin JM, Weber D, Brodsky PH. (1995). Integrated in situ soil remediation technology: The Lasagna process *Environmental Science & Technology* **29**:2528–2534.

Huesemann MH, Hausmann TS, Fortman TJ. (2003). Assessment of bioavailability limitations during slurry biodegradation of petroleum hydrocarbons in aged soils. *Environmental Toxicology and Chemistry* **22**(12):2853–2860.

Hunt NK, Marinas BJ. (1999). Inactivation of *Escherichia coli* with ozone: Chemical and inactivation kinetics. *Water Research* **33**(11):2633–2641.

Islam S, Suidan MT. (1998). Electrolytic denitrification: Long term performance and effects of current intensity. *Water Research* **32**(2):528–536.

Jackman SA, Maini G, Sharman AK, Knowles CJ. (1999). The effect of direct electric current on the viability and metabolism of acidophilic bacteria. *Enzyme and Microbial Technology* **24**:316–324.

Jackman SA, Maini G, Sharman AK, Sunderland G, Knowles CJ. (2001). Electrokinetic movement and biodegradation of 2,4-dichlorephenoxyacetic acid in silt soil. *Biotechnology and Bioengineering* **74**(1):40–48.

Katsuki S, Majima T, Nagata K, Lisitsyn I, Akiyama H, Furuta M, Hayashi T, Takahashi K, Wirkner S. (2000). Inactivation of *Bacillus stearothermophilus* by pulsed electric field. *IEEE Transactions on Plasma Science* **28**(1):155–160.

Khadre MA, Yousef AE. (2001). Sporicidal action of ozone and hydrogen peroxide: A comparative study. *International Journal of Food Microbiology* **71**:131–138.

Kraft A, Blaschke M, Kreysig D, Sandt B, Schröder F, Rennau J. (1999b). Electrochemical water disinfection. Part II: Hypochlorite production from potable water, chlorine consumption and the problem of calcareous deposits. *Journal of Applied Electrochemistry* **29**:895–902.

Kraft A, Stadelmann M, Blaschke M, Kreysig D, Sandt B, Schröder F, Rennau J. (1999a). Electrochemical water disinfection. Part I: Hypochlorite production from very dilute chloride solutions. *Journal of Applied Electrochemistry* **29**:861–868.

Lageman R, Pool W. (2001). Thirteen years electro-reclamation in the Netherlands. In *3. Symposium on Electrokinetic Remediation* (eds. K Czurda, R Haus, C Kappeler, R Zorn). April 18–20, Karlsruhe, Germany. Karlsruhe, Germany: Schr. Angew. Geol., pp. 1.1–1.17.

Lee GT, Ro HM, Lee SM. (2007). Effects of triethyl phosphate and nitrate on electrokinetically enhanced biodegradation of diesel in low permeability soils. *Environmental Technology* **28**(8):853–860.

Lee HS, Lee K. (2001). Bioremediation of diesel-contaminated soil by bacterial cells transported by electrokinetics. *Journal of Microbiology and Biotechnology* **11**(6):1038–1045.

Liu W-K, Brown MRW, Elliott TSJ. (1997). Mechanisms of the bacterial activity of low amperage electric current (DC). *Journal of Antimicrobial Chemotherapy* **39**:687–695.

Löffler FE, Cole JR, Ritalathi KM, Tiedje JM. (2003). Diversity of dechlorinating bacteria. In *Dehalogenation. Microbial Processes and Environmental Applications* (eds. MM Häggblom, ID Bossert). Boston: Kluwer Academic Publishers, pp. 53–89.

Löffler FE, Tiedje JM, Sanford RA. (1999). Fraction of electrons consumed in electron acceptor reduction and hydrogen thresholds as indicators of halorespiratory physiology. *Applied Environmental Microbiology* **65**:4049–4056.

Lohner ST, Katzoreck D, Tiehm A. (2008a). Electromigration of microbial electron acceptors and nutrients: (I) transport in synthetic media. *Journal of Environmental Science and Health Part A* **43**(8):913–921.

Lohner ST, Katzoreck D, Tiehm A. (2008b). Electromigration of microbial electron acceptors and nutrients: (II) transport in groundwater. *Journal of Environmental Science and Health Part A* **43**(8):922–925.

Lohner ST, Tiehm A. (2008). Application of electrolysis to stimulate microbial reductive PCE dechlorination and oxidative VC biodegradation. *Environmental Science & Technology* (in revision).

Loo WW. (1994). Electrokinetic enhanced passive in situ bioremediation of soil and groundwater containing gasoline, diesel and kerosene. Paper presented at the HAZMACON Conference, March 29–31, San Jose, CA.

Loo WW, Wang IS, Fang J. (1994). Electrokinetic enhanced bioventing of gasoline in clayey soil: A case history. Paper presented at the Superfund XV Conference, November 29–December 1, Washington, DC.

Lorton DM, Jones ADG, Smith S, Mason JR. (2001). The study of polycyclic hydrocarbon bioavailability in soils. *The Sixth International Symposium—In Situ and On-Site Bioremediation*, Platform abstracts, San Diego, California, June 4–7, 2001.

Luo QS, Zhang XH, Wang H, Qian Y. (2005). The use of non-uniform electrokinetics to enhance in situ bioremediation of phenol-contaminated soil. *Journal of Hazardous Materials* **121**(1–3):187–194.

McCarty PL. (1991). Engineering concepts for in situ bioremediation. *Journal of Hazardous Materials* **28**:1–10.

Mustacchi R, Knowles CJ, Li H, Dalrymple I, Sunderland G, Skibar W, Jackman SA. (2005). Enhanced biotransformations and product recovery in a membrane bioreactor through application of a direct electric current. *Biotechnology and Bioengineering* **89**(1):18–23.

Nambi IM, Werth CJ, Sanford RA, Valocchi AJ. (2003). Pore-scale analysis of anaerobic growth along the transverse mixing zone of an etched silicon pore network. *Environmental Science & Technology* **37**:5617–5624.

Nathanail J, Bardos RP, Nathanail P. (2002). *Contaminated Land Management: Ready Reference*. EPP Publications and Land Quality Press in association with Environmental Technology Ltd. and Land Quality Management Ltd. at the University of Nottingham. Richmond, UK: EPP Publications.

Niqui-Arroyo JL, Bueno-Montes M, Posada-Baquero R, Ortega-Calvo JJ. (2006). Electrokinetic enhancement of phenantrene biodegradation in creosote-polluted clay soil. *Environmental Pollution* **142**(2):326–332.

Niqui-Arroyo JL, Ortega-Calvo JJ. (2007). Integrating biodegradation and electroosmosis for the enhanced removal of polycyclic aromatic hydrocarbons from creosote-polluted soils. *Journal of Environmental Quality* **36**(5):1444–1451.

Olszanowski A, Piechowiak K. (2006). The use of an electric field to enhance bacterial movement and hydrocarbon biodegradation in soils. *Polish Journal of Environmental Studies* **15**(2):303–309.

Pallud C, Dechesne A, Gaudet JP, Debouzie D, Grundmann GL. (2004). Modification of spatial distribution of 2,4-dichlorophenoxyacetic acid degrader microhabitats during growth in soil columns. *Applied Environmental Microbiology* **70**(5):2709–2716.

Patermarakis G, Fountoukidis E. (1990). Disinfection of water by electrochemical treatment. *Water Research* **24**(12):1491–1496.

Pollard SJT, Hrudey SE, Fredorak PM. (1994). Bioremediation of petroleum and creosote-contaminated soils: A review of constraints. *Waste Management & Research* **12**:173–194.

Potter TL, Simmons KE. (1998). Composition of petroleum mixtures. In *Total Petroleum Hydrocarbon Criteria Working Group Series*, Vol. **2**. Amherst, MA: Scientific Publishers Amherst.

Rabbi MF, Clark B, Gale RJ, Ozsu-Acar E, Pardue J, Jackson A. (2000). In situ TCE bioremediation study using electrokinetic cometabolite injection. *Waste Management* **20**(4):279–286.

Rahman MA, Jose SC, Nowak W, Cirpka OA. (2005). Experiments on vertical transverse mixing in a large-scale heterogeneous model aquifer. *Journal of Contaminant Hydrology* **80**(3–4):130–148.

Rahner D, Ludwig G, Röhrs J, Neumann V, Nitsche C, Guderitz I. (2001). Electrochemically induced reactions in soils—a new approach to the in-situ remediation of contaminated soils? In *3. Symposium on Electrokinetic Remediation* (eds. K Czurda, R Haus, C Kappeler, R Zorn). April 18–20, 2001, Karlsruhe, Germany. Karlsruhe, Germany: Schr. Angew. Geol., pp. 2.1–2.20.

Reddy KR, Chinthamreddy S, Saichek RE, Cutright TJ. (2003). Nutrient amendment for the bioremediation of a chromium-contaminated soil by electrokinetics. *Energy Sources* **25**(9):931–943.

Renaud PC, Probstein RF. (1987). Electroosmotic control of hazardous wastes. *Physicochemical Hydrodynamics* **9**(1–2):345–360.

Rosenberg R, Ron EZ. (1996). Bioremediation of petroleum contamination. In *Bioremediation: Principles and Applications* (eds. RL Crawford, DL Crawford). Cambridge, UK: Cambridge University Press, pp.100–124.

Runnels DD, Wahli C. (1993). In situ electromigration as a method for removing sulfate, metals and other contaminants from ground water. *Ground Water Monitoring and Research* **13**(1):121–129.

Saichek, RE, Reddy KR. (2005). Electrokinetically enhanced remediation of hydrophobic compounds in soils: A review. *Critical Reviews in Environmental Science and Technology* **35**:115–192.

Sakakibara Y, Nakayama T. (2001). A novel multi-electrode system for electrolytic and biological water treatments: Electric charge transfer and application to denitrification. *Water Research* **35**(3):768–778.

Schoenbach KH, Joshi RP, Stark RH. (2000). Bacterial decontamination of liquids with pulsed electric field. *IEEE Transactions on Dielectrics and Electrical Insulation* **7**(5): 637–645.

Schulze S. (2004). Mikrobieller Abbau und Redoxzonierung im Abstrom einer teerölkontaminierten Altablagerung. Dissertation, University of Dresden, Dresden, Germany.

Segall BA, Bruell CJ. (1992). Electroosmotic contaminant-removal processes. *Journal of Environmental Engineering ASCE* **118**(1):84–100.

Shapiro AP, Probstein RF. (1993). Removal of contaminants from saturated clay by electro-osmosis. *Environmental Science & Technology* **27**:283–291.

Shi L, Muller S, Harms H, Wick LY. (2008a). Factors influencing the electrokinetic dispersion of PAH-degrading bacteria in a laboratory model aquifer. *Applied Microbiology and Biotechnology* **80**(3):507–515.

Shi L, Muller S, Harms H, Wick LY. (2008b). Effect of electrokinetic transport on the vulnerability of PAH-degrading bacteria in a model aquifer. *Environmental Geochemistry and Health* **30**(2):177–182.

Smatlak CR, Gossett JM, Zinder SH. (1996). Comparative kinetics of hydrogen utilization for reductive dechlorination of tetrachloroethene and methanogenesis in an anaerobic enrichment culture. *Environmental Science & Technology* **30**:2850–2858.

Suni S, Romantschuk M. (2005). Mobilization of bacteria in soils by electro-osmosis. *FEMS Microbiology Ecology* **49**:51–57.

Szekeres S, Kiss I, Bejerano TT, Soares MIM. (2001). Hydrogen-dependent denitrification in a two-reactor bio-electrochemical system. *Water Research* **35**(3):715–719.

Tanner BD, Kuwahara S, Gerba CP, Reynolds KA. (2004). Evaluation of electrochemically generated ozone for the disinfection of water and wastewater. *Water Science and Technology* **50**(1):19–25.

Thevanayagam S, Rishindran T. (1998). Injection of nutrients and TEAs in clayey soils using electrokinetics. *Journal of Geotechnical and Geoenvironmental Engineering* **124**(4):330–338.

Thompson IP. (2002). It's a bug's life. *Chemistry in Britain* **4**:32–33.

Thornton SF, Quigley S, Spence MJ, Banwart SA, Bottrell S, Lerner DN. (2001). Processes controlling the distribution and natural attenuation of dissolved phenolic compounds in a deep sandstone aquifer. *Journal of Contaminant Hydrology* **53**(3–4):233–267.

Thullner M, Mauclaire L, Schroth MH, Kinzelbach W, Zeyer J. (2002). Interaction between water flow and spatial distribution of microbial growth in a two-dimensional flow field in saturated porous media. *Journal of Contaminant Hydrology* **58**(3–4):169–189.

Tiehm A. (1994). Degradation of polycyclic aromatic hydrocarbons in the presence of synthetic surfactants. *Applied Environmental Microbiology* **60**:258–263.

Tiehm A, Lohner ST, Augenstein T. (2008). Effects of direct electric current and electrode reactions on the viability and activity of VC degrading microorganisms. *Electrochimica Acta* (in press).

Tiehm A, Schmidt KR. (2007). Methods to evaluate biodegradation at contaminated sites. In *Environmental Geology. Handbook of Field Methods and Case Studies*, 1st ed. (eds. K Knödel, G Lange, H-J Voigt). Berlin: Springer, pp. 876–940.

Tiehm A, Schmidt KR, Pfeiffer B, Heidinger M, Ertls S. (2008). Growth kinetics and stable carbon isotope fractionation during aerobic degradation of cis-1,2-dichloroethene and vinyl chloride. *Water Research* **42**:2431–2438.

Tiehm A, Stieber M. (2001). Strategies to improve PAH bioavailability: Addition of surfactants, ozonation and application of ultrasound. In *Treatment of Contaminated Soil—Fundamentals, Analysis, Applications* (eds. R Stegmann, W Brunner, W Calmano, G Matz). Berlin: Springer-Verlag, pp. 299–324.

Tiehm A, Stieber M, Werner P, Frimmel FH. (1997). Surfactant-enhanced mobilization and biodegradation of polycyclic aromatic hydrocarbons in manufactured gas plant soil. *Environmental Science & Technology* **31**(9):2570–2576.

Tsao C-W, Song H-G, Bartha, R. (1998). Metabolism of benzene, toluene and xylene hydrocarbons in soil. *Applied Environmental Microbiology* **64**(12):4924–4929.

Vencel LV, Likirdopulos CA, Robinson CE, Sobsey MD. (2004). Inactivation of enteric microbes in water by electro-chemical oxidant from brine (NaCl) and free chlorine. *Water Science and Technology* **50**(1):141–146.

Verce MF, Ulrich RL, Freedman DL. (2000). Characterization of an isolate that uses vinyl chloride as a growth substrate under aerobic conditions. *Applied Environmental Microbiology* **66**:3535–3542.

Vik EA, Bardos P. (2002). Remediation of Contaminated Land Technology Implementation in Europe. A report from the Contaminated Land Rehabilitation Network for Environmental Technologies. http://www.umweltbundesamt.at/en/umweltschutz/altlasten/projekte1/international1/clarinet/clarinet_results (accessed May 26, 2009).

Wackett LP, Hershberger CD. (2001). *Biocatalysis and Biodegradation: Microbial Transformation of Organic Compounds*. Washington, DC: ASM Press.

Watanabe T, Hashimoto S, Kuroda M. (2002). Simultaneous nitrification and denitrification in a single reactor using bio-electrochemical process. *Water Science and Technology* **46**(4–5):163–169.

Wick LY. (2005). Electrokinetic transport of PAH-degrading bacteria in model aquifers and soil. *Symposium on Electrokinetic Remediation*, May 22–25, Ferrara, Italy.

Wick LY, Mattle PA, Wattiau P, Harms H. (2004). Electrokinetic transport of PAH-degrading bacteria in model aquifers and soil. *Environmental Science & Technology* **38**:4596–4602.

Wick LY, Shi L, Harms H. (2007). Electro-bioremediation of hydrophobic organic soil-contaminants: A review of fundamental interactions. *Electrochimica Acta* **52**(10): 3441–3448.

Wiedemeier TH, Rifai HS, Newell CJ, Wilson JT. (1999). *Natural Attenuation of Fuels and Chlorinated Solvents in the Subsurface*. New York: John Wiley & Sons.

Xu R, Obbard JP. (2004). Bioremediation and biodegradation. Biodegradation of polycyclic aromatic hydrocarbons in oil-contaminated beach sediments treated with nutrient amendments. *Journal of Environmental Quality* **33**:861–867.

20

INFLUENCE OF COUPLED ELECTROKINETIC–PHYTOREMEDIATION ON SOIL REMEDIATION

M.C. Lobo Bedmar, A. Pérez-Sanz, M.J. Martínez-Iñigo, and A. Plaza Benito

20.1 SOIL CONTAMINATION: LEGISLATION

Soil is considered the superficial layer of the earth's crust. Its main constituents are mineral particles, organic matter, water, air, and living organisms. Thus, soil is a live system and a nonrenewable resource due to the slow speed of the formation process. It plays an important role due to its capacity to store, filter, and transform many substances, including water, nutrients, and carbon. It is the biggest carbon store in the world (1.5 gigatons). These functions must be protected because of their socioeconomic and environmental importance (Commission of the European Communities, 2006). The soil matrix is extremely complex and variable. In Europe more than 300 major soil types have been identified, and within each there are enormous variations in physical, chemical, and biological properties. For this reason, the response of each soil type to a degradation process will be different.

Soil is subject to a series of degradation processes or threats. These include erosion, loss of organic matter, contamination (local and diffuse), sealing, compaction, decline in biodiversity, salinization, floods, and landslides. Since the 1970s, the increase in industrial processes and in waste treatment systems has led to the appearance of different types of substances—including heavy metals and organic compounds—that constitute contamination sources in the soil and prevent the soil from performing its functions. Soil has natural functions that are essential for life and as a habitat for humans, animals, plants, and soil organisms through its water

---

Electrochemical Remediation Technologies for Polluted Soils, Sediments and Groundwater,
Edited by Krishna R. Reddy and Claudio Cameselle
Copyright © 2009 John Wiley & Sons, Inc.

and nutrient cycles. It is a medium for decomposition, balance, and restoration as a result of its filtering, buffering, and substance-converting properties, especially for groundwater protection (Miehlich, 2001). As consequence, any damage produced in soil by the effect of pollutants leads to damage to other environmental media and ecosystems. Even so, every remediation method influences natural soil functions and soil's function as an archive of natural and cultural history.

Nowadays, soil contamination is a global problem. Several countries, including the United States, Japan, Canada, Australia, Brazil, have established soil protection policies that include legislation, guidance documents, monitoring systems, and identification of risk areas, inventories, remediation programs, and funding mechanisms for contaminated sites for which no responsible party can be found. Such policies ensure a comparable level of soil protection. In 1980, the US Congress enacted the Comprehensive Environmental Response, Compensation, and Liability Act (CERCLA), commonly known as Superfund, to address the danger of abandoned or uncontrolled hazardous waste sites. CERCLA provides the US Environmental Protection Agency (EPA) and other federal agencies with the authority to respond to a release or a substantial threat of a release of "any pollutant or contaminant which may present an immediate and substantial danger to public health or welfare" (EPA, 2007).

Soil degradation is also a serious problem in Europe. Although soil degradation processes vary considerably from member state to member state, with different threats having different degrees of severity, soil degradation is an issue all over the European Union (EU). It is driven or exacerbated by human activity such as inadequate agricultural and forestry practices, industrial activities, tourism, and industrial and urban expansion. These activities have a negative impact, preventing the soil from performing its functions and services to humans and ecosystems. This results in loss of soil fertility, carbon and nutrient content, biodiversity, changes in physical properties, and reduced degradation of contaminants.

European governments have considered that soil degradation has a direct impact on water and air quality, biodiversity, and climate change, and therefore consequences on human health. Against this background, the Commission of the European Communities contends that a comprehensive EU strategy for soil protection is necessary. This strategy should take into account all the different functions that soils can perform, their variability and complexity, and the range of different degradation processes to which they can be subjected, while also considering socioeconomic aspects. On September, 25, 2006, the European Commission was presented to the European Parliament, which defined new laws to prevent industries and agriculture from contaminating the soil (COM, 2006). The EU's executive proposed new rules to preserve soil quality in Europe and to ask the governments of member states to address the problem of soil contamination immediately, which would otherwise cost Europe millions of euros every year. The proposal submitted by the European Commission would require all the member states to identify areas that are at risk of soil contamination, landslides, erosion, and other related threats, and to set targets for reducing these risk factors. Soil contamination can be caused by fuel leaking from tanks in factories, use of heavy machinery in farming, and the building of roads and other infrastructure; in general, due to the industrial activities. Moreover, agricultural activities can lead to contamination due to the excessive use of pesticides and organic residues. A chemical is considered phytotoxic when its presence reduces the growth or alters the normal development of plants (Fletcher, 1997). Thousand of organic chemicals termed "phytotoxins" have been demonstrated to have such properties. Phytotoxins enter the soil components of ecosystems as a result of naturally occurring, ecological events such as when one plant species releases a chemical that influences the growth or another species (allelopathy) or from human activities when man-made compounds are released into the environment (e.g., from industry or agriculture)

Soil degradation due to contamination represents an increasing hazard that requires the development of improved methods for soil remediation with two main objectives: (1) remove the contaminants from the soil and (2) allow the soil to recover its functionality.

20.2 WHAT IS THE LIMIT OF THE REMEDIATION? SOIL RECOVERY

In general, the limits of a remediation process are conditioned by the soil's future use and the time required for the cleanup. Over the past decades, different technologies have been applied to soil remediation, e.g., incineration, inertization, soil washing, soil venting, etc., in an effort to remove pollutants according to economic criteria and time. With regard to the variability of soil types and the degree of disturbance of soil functions, there are considerable differences among the remediation methods. Methods without soil excavation, such as soil–air stripping, soil flushing, or hydraulic treatments, may influence natural soil functions adversely by changing soil structure, porosity, and water balance. Other methods based on soil excavation, such as thermal treatment, soil washing, etc., influence soil functions to a greater extent. At the contaminated site, the excavation of topsoil causes a reduction of soil organic matter, nutrients, and cation exchange capacity with the consequences that some properties as filtering, buffering, and biodegradation are reduced. Thus, the soil will lose its function as an archive of natural and cultural history. The decontaminated soil materials change their properties to different degrees depending on treatment, which also leads to change in the soil.

Nowadays, it is possible to restore natural soil functions with the redeposition of soil materials after remediation. Some of the natural soil functions can be reestablished very quickly after recultivation, such as soil characteristics, whereas others take tens to hundred of years for regeneration, such as soil structure, soil type, and biodiversity. There are many experiences in relation to the recultivation of open mining sites and degraded soils (Alcañiz et al., 1996).The results show that the restoration of chemical soil properties, such as pH, organic matter, and nutrients is much easier than the restoration of physical soil properties such as pore volume, pore size distribution, and soil structures. The selection of the appropriate soil texture is of great importance in order to create a suitable habitat for flora and fauna, favoring water supply, root penetrability, biomass for decomposers, and also to establish maximum capacity for filtering, buffering, and substance transformation. Soil recovery has to be carried out using suitable techniques and machines in order to avoid compaction that leads to air shortage and conditions of low redox potential in soil.

The materials used in restoration vary in their suitability for recultivation depending on the decontamination procedure applied. Thermal treatment leads to such a change in properties that these materials may only be used as filling material in deeper soil layers. In the case of soil washing, washed materials are suitable for dry and nutrient-poor soils only, due to their coarse texture. However, when bioremediation has been applied, biological decontaminated materials are suitable for a wide range of applications. Furthermore, it has to be considered that the decontamination procedures themselves may affect soils or other compartments of the ecosystem beyond the actual remediation site, such as the effects of emissions on soil, water, the atmosphere, and the biosphere by the decontamination treatment plan.

In recent years, there has been increased interest in finding new and innovative solutions for the removal of contaminants from soil and consequently solving the problem of groundwater and soil contamination. Management of contaminated soils is a global problem with regard to human health. Pollutants reach the soil through failures in industrial processes or waste disposal and even through agricultural practices. With the aim of avoiding or minimizing adverse impacts, it is necessary to consider the extent and the magnitude of the risk and efficient cleanup goals. The most important goal in managing contaminated soil is to render the soil environmentally acceptable through management or remediation so the soil can recover its functionality. This goal applies to situations where there has been recent waste disposal, where the contamination has occurred over time, or where the pollutants have been "weathering" at a specific site for many years. Thus a distinction must be made between current and historical contamination. An important question about the management of contaminated soils is: How clean is clean? What determines an acceptable environmental endpoint in the remediation technique? This question is asked at the national level as new laws or changes to existing environmental (Loehr, 1996). Traditionally, soil cleanup criteria have been set using concentration-based standards that often require the application of a remediation technology to reach background levels or to other specified levels that are considered acceptable according to specific regulation at the regional or national level. Recently however, soil cleanup criteria now includes risk-based criteria. In this situation, the approach used recognizes that each pollutant or contaminant poses a risk to human health due to its capacity to be mobilized and transported to the receptor. The risk-based approach considers the need to establish the concentration of each pollutant that is acceptable in each soil. In this situation, it is necessary to define the degree of availability of the chemical substances. Such an approach recognizes that for a chemical in a soil to pose a risk, it must first be made available to a receptor (such as a human being) through mobilization and transport, and must then elicit an adverse response from the release due to that exposure. Under a risk-based approach, answering "How clean is clean?" in a meaningful way requires determining what concentration of a chemical, such as a hydrocarbon or a heavy metal, is environmentally acceptable at a specific site.

In determining the relative risk posed by pollutants in soils, it is not enough to simply measure the chemical concentration. It is also necessary to address the risk-assessment paradigm, the pathways by which human health and the environment can be affected, and the availability of the released chemical for transport and adverse impact (Loehr, 1996). Environmental risk is commonly defined as the adverse impact resulting from human or environmental exposure (real or potential) to contaminants under site-specific circumstances. But while decisions on the suitability of soil-remediation processes commonly focus on chemical-concentration

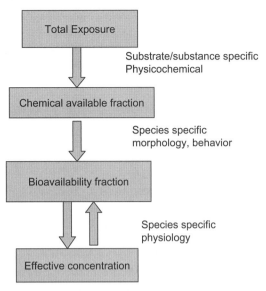

Figure 20.1. Influence variables on the biologically effective concentration of a contaminant in the soil/test substrates (Kördel and Hund-Rinke, 2001).

reduction, other parameters important to risk-evaluation decisions—such as the mobility and relative toxicity of the residues from the remediation process—need to be determined and evaluated.

A harmful soil change is an impairment of the soil functions that can lead to increased risk. The existence and extent of a single contaminant or the combination of different contaminants in a soil leads to an impairment of soil functions. This depends on the type of substance as well as on its concentration, the combination of different substances, soil properties, and the respective use of the soil. In this case, it is necessary to define concepts as "exposure" and "bioavailability." Exposure to pollutants is commonly divided into a chemically available fraction, the fraction available to the organisms (bioavailable fraction), and the fraction that is taken up by the organism (effective concentration). Figure 20.1 shows the important terms and influencing factors for the biologically effective concentration of contaminants in soils. Nowadays, there are regulations in different countries in relation to environmental risk assessment for new and existing chemicals, pesticides, medicines, biocides, etc., (Williamson, Loehr, and Kimura, 1998) that facilitate the application of measurement to soil protection.

20.3 INFLUENCE OF THE ELECTROKINETIC TECHNOLOGY ON SOIL PROPERTIES

Electrokinetic remediation is a developing technique that has a significant potential for *in situ* remediation of low-permeability soils contaminated with heavy metals and/or organic compounds. The technique involves applying a low direct current or a low potential gradient to electrodes that are inserted into the ground (Acar and Alshawabkeh, 1993; Ugaz *et al.*, 1994). Due to electromigration and electro-osmosis processes, the contaminants are transported to either the cathode or the anode and collected near the electrodes. Then they are removed by pumping the water in the electrodes zone, or by using other different methods such as electroplating, precipitation or coprecipitation at the electrodes, or complexing with ion-exchange resins. This technology seems to be an adequate alternative to removing pollutants from the soil. The advantages of electrokinetic remediation are the low cost, the nonintrusive character, the applicability to a wide range of contaminants, and the insensitivity to the pore size of the soil, which makes it suitable for fine-grained soils (Virkutyte, Sillanpaa, and Latostenmaa, 2002). However, different assays have shown a direct relation between yield removal and soil characteristics, such as particle size, pH, and organic matter and carbonate content that can be limiting factors when trying to obtain good results. Ionic mobility is dependent on pH. A purging solution with different ionic strengths is necessary in order to remove pollutants from different soils (Saicheck and Reddy, 2003). Water and citric acid are able to mobilize the pollutants in some soils, but when pollutants are strongly fixed to the soil constituent, it is necessary to apply other acids or chelating agents (Kim, Moon, and Kim, 2001; Zhou, Zorn, and Kurt, 2003) that increase the mobilization of the pollutant but that can affect the soil dynamic. As part of the electrokinetic technology application, electro-osmosis and electromigration can cause electrochemical electrode reactions, as well as solution, precipitation, redox reactions; the changing of the surface potential of soil particles, pH, electrolyte concentration, thickness of the diffuse double layer and water content; destroying clay mineral under acidic conditions, as well as exchanging of ions. Under these variable conditions, a changing of the soil structure, texture, and mineralogical composition is possible, and, consequently, the physicochemical soil parameter can be modified as well (Mitchell, 1993; Yeung, Hsu, and Menon, 1997; Steger *et al.*, 2005). The impact that electrokinetic extraction will have on the physical chemical and biological soil characteristics has not been adequately quantified yet.

Soils are complex microbial habitats because they are dominated by a soil phase surrounded by aqueous and gaseous phases that fluctuate markedly with time and space. The soil phase is a three-dimensional system composed of minerals of various particle sizes and intermingled with plant, animal, and microbial residues in various stages of decay of living and metabolizing microbiota. (Haider, 1995). Organic and inorganic components are closely associated, and the presence of these metal and clay-organic matter complexes affects soil quality (Schnitzer, 1995) and likely exerts significant influence on the physical, chemical, and biological properties of soil:

- Physical effects concern soil aggregation, erosion, drainage, tilth, aeration, water-holding capacity, bulk density, permeability, mechanical properties.
- Chemical effects relate to exchange capacity, buffering capacity, supply and availability of macro- and micronutrients, adsorption of pesticides and other chemicals, etc.
- Biological effects are involved with activities in the soil by bacteria, fungi, actinomycetes, earthworths, roots, root hairs, soil biomass.

Soil structure is influenced by the association of soil organic matter with minerals to form aggregates. Aggregate formation improves soil structure and water infiltration and increases water-holding capacity. These changes improve root growth and provide habitats for diverse soil organisms. The pumping of the pollutant led to the

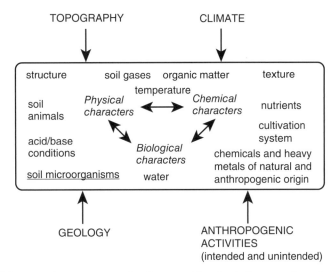

Figure 20.2. The complex structure of soil created by the influence of geology, topography, and climate as well as anthropogenic activities (Tortessen, 1997).

movement of soil particles and further removal from the soil. Changes in the soil structure can occur and can have a negative effect on soil equilibrium. Soil is composed of three-dimensional arrangements of solid particles and pores. Soil structure is determined by the distribution and the size of these soil aggregates and pore spaces. A good soil structure is vital, as it can affect the availability of air, water, and nutrients for plant growth. The application of the current content could significantly alter soil structure and, as a consequence, reduce plant growth, making it difficult for plants to obtain water, air, and nutrients.

To evaluate the impact of the technology on soil quality, it is necessary to consider soil as a complex structure with a considerable heterogeneity in its physical, chemical, and biological composition (Fig. 20.2) (Tortesson, 1997). The importance of biological processes for the function of a soil ecosystem is unequivocal, and a large number of soil processes are, on a superficial level, dependent on biological activity. During the electrokinetic process, the removal of macroelements is produced by the pumping of pollutants. This fact affects the cation exchange capacity and the stability of organomineral complexes. However, some studies suggest a low availability of metals and macroelements after electrokinetic remediation, due to a strong attachment to the soil after the application of the electric field (Suer, Gitye, Allard, 2003). In this case, this technology could be used to immobilize metals and macroelements in contaminated soils.

Soil organisms are exposed to changing environmental conditions (Voroney, 2006). Extremes are marked by stimulation, depression, or death. Domsch, Jagnow, and Anderson (1983) have considered the following naturally occurring stress situations on soil microbial populations and their activities. Some of them can occur in soil treated by electrokinetic technology:

- Temperature fluctuations: Conditions of constant temperature are a rare exception in soil. Damages to the microbial population are produced by cold shock or warming up.

- Water potential extremes: Changes in the availability of water by drying, flooding, irrigation, or elevated soil salinity are reflected in considerable changes in population densities.
- Soil pH extremes: In a soil, great differences in pH may occur in microhabitats. A change in pH, e.g., by liming, causes negative effects on certain microorganisms while others may increase.
- Decreased supply of nutrients: The primary production of green plants is affected.

All the above factors have to be accounted for in the application of electrokinetic technology. Further research is needed to make a diagnosis about the real effect on the soil depending on these properties. Due to the importance of mitigating the impact of the technical application from the point of view of future use of the soil, the application of combined technologies (bio- or phytoremediation) constitutes a suitable strategy in soil remediation (Lobo, 2007).

20.4 PHYTOREMEDIATION

Phytoremediation is defined as the use of living plants and their associated microorganisms to remove, degrade, or sequester inorganic and organic pollutants from *in situ* remediation of contaminated soil, sludges, sediments, and groundwater. It has been reported to be an effective, nonintrusive, inexpensive, aesthetically pleasing, and socially accepted technology (Alkorta et al., 2004). Phytoremediation is widely viewed as the ecologically responsible alternative to the environmentally destructive physical remediation methods currently practiced. Table 20.1 shows the cost of different technologies in comparison with phytoremediation. The advantages of phytoremediation over other methods of remediation are well known and have been reviewed by many authors (Salt *et al.*, 1995; Salt, Smith, and Raskin, 1998; Chaney *et al.*, 1997; Raskin, Smith, and Salt, 1997; Chaudhry *et al.*, 1998; Wenzel *et al.*, 1999; Meagher, 2000; Navari-Izzo and Quartacci, 2001; Salt and Baker, 2001; Lasat, 2002; McGrath, Zhao, and Lombi, 2002; Wolfe and Bjornstad, 2002; McGrath and Zhao, 2003). Originally, the technique was applicable to heavy metal contaminants, but it also suitable to organic pollutants. Phytoremediation can be applied *in situ*. Hence, it reduces soil disturbance, the spread of contaminants, and the amount of waste going to landfills. Early estimates of the costs of phytoremediation

TABLE 20.1. Cost of Different Remediation Technologies

Process	Cost (US $/ton)	Other factors
Vitrification	75–425	Long-term monitoring
Land filling	100–500	Transport/excavation/monitoring
Chemical treatment	100–500	Recycling of contaminants
Electrokinetics	20–200	Monitoring
Phytoextraction	5–40	Disposal of phytomas

Source: Glass (1999).

indicate that the technique is cheaper than conventional remediation methods. It is also easy to implement and maintain.

However, phytoremediation does have certain disadvantages and limitations. This technology is limited by depth (roots) and also by the solubility and the availability of the pollutant. Although it is faster than natural attenuation, phytoremediation requires long time periods and is restricted to sites with low contaminant concentrations. The plant biomass obtained from phytoextraction requires proper disposal as hazardous waste. Phytoremediation depends on the climate and season. It can also lose its effectiveness when damage occurs to vegetation from disease or pests. The introduction of inappropriate or invasive plant species should be avoided (non-native species may affect biodiversity). Contaminants may be transferred to another medium, the environment, and/or the food chain. Amendments and cultivation practices may have negative consequences on contaminant mobility.

20.4.1 Categories of Phytoremediation

Depending on the contaminants, the site conditions, the level of cleanup required, and the types of plants, phytoremediation technology can be used for containment (phytoimmobilization and phytostabilization) or removal (phytoextraction and phytovolatilization) purposes (Padmavathiamma and Li, 2007).

20.4.1.1 Phytostabilization Phytostabilization uses certain plant species to immobilize contaminants in soil through absorption and accumulation by roots, adsorption onto roots or precipitation within the root zone, and physical stabilization of soils. The aim of phytostabilization is to prevent secondary contamination and exposure by reducing the mobility of the contaminants and preventing migration to ground, water, or air. Phytostabilization can provide a dense vegetative ground cover, which can greatly reduce soil erosion and human exposure to contaminants via dermal contact and inhalation (Cunningham *et al.*, 1997). Phytostabilization is not intended to remove contaminants from a site; for that reason, it is applied in situations where there are potential human health impacts, and exposure to substances of concern can be reduced to acceptable levels by contaminant. The disruption to site activities may be less than with more intrusive soil remediation technologies. Characteristics of plants that are appropriate for phytostabilization include tolerance to high levels of contaminants; high production of root biomass able to immobilize these contaminants through uptake, precipitation, or reduction; and retention of applicable contaminants in roots, as opposed to a transfer to shoots, which avoids the need for special handling or disposal of the shoots. The lack of appreciable metals in shoot tissue also eliminates the need to treat harvested shoot residue as a hazardous waste. Phytostabilization involves soil amendments to promote the formation of insoluble metal complexes that reduce biological availability and plant uptake, thus preventing metals from entering the food chain. One way to facilitate such immobilization is by altering the physicochemical properties of metal–soil complex by introducing a multipurpose anion such as phosphate (Bolan, Adriano, and Naidu, 2003); by addition of humified organic matter such compost with lime to raise the pH (Kuo, Jellum, and Baker, 1985); or soil acidification.

20.4.1.2 Phytofiltration Phytofiltration is the use of plant roots (rizhofiltration) or seedlings (blastofiltration) to absorb or adsorb pollutants, mainly metals, from water and aqueous waste streams (Prasad and Freitas, 2003). Mechanisms involved in bioabsorption include chemisorption, complexation, ion exchange, microprecipitation, hydroxide condensation onto the biosurface, and surface absorption (Gardea-Torresdey, de la Rosa, and Peralta-Videa, 2004). Rhizofiltration uses terrestrial plants instead of aquatic plants because the former feature much larger fibrous root systems covered with root hairs with extremely large surfaces areas. The process involves raising plants hydroponically and transplanting them into metal-polluted waters where the plants absorb and concentrate the metals in their roots and shoots.

Wetlands phytoremediation is a good example of phytofiltration. Wetlands are identified as areas covered by water or that have waterlogged soils for significant periods during the growing season. The expansive rhizosphere of wetland herbaceous shrub and tree species provides an enriched culture zone for microbes involved in degradation. Redox conditions in most wetland soil/sediment zones enhance degradation pathways requiring reducing conditions. However, assessing the phytoremediation potential of natural wetlands is complex due to variable conditions of hydrology, soil/sediment types, plant species diversity, growing season, and water chemistry (Williams, 2002). The situations where constructed wetlands will be most valuable are when plumes are inaccessible to the rizhosphere. In theses cases, phytoremediation could be used to pump deeper plume waters into surface wetlands, which enhances contaminant removal. Contaminated plant biomass following accumulation must be addressed because hydrophytes decomposition is more rapid than the terrestrial species.

20.4.1.3 Phytovolatilization Phytovolatilization is the use of plants to volatilizate pollutants (Terry *et al.*, 1992). Plants are capable of absorbing not only elemental forms of metals such as mercury or selenium, but also water-soluble organic contaminants from soil, and biologically converting them to gaseous species within the plant and releasing them into the atmosphere. This is the most controversial of phytoremediation technologies, because there are doubts about whether the volatilization of these elements into the atmosphere is desirable or safe. Once contaminants have been removed via volatilization, there is a loss of control over their migration to other areas. Some authors suggest that the addition of gases to the atmosphere through phytovolatilization would not contribute significantly to the atmospheric pool, since the contaminants are likely to be subjected to more effective or rapid natural degradation process such as photodegradation. However, phytovolatilization should be avoided for sites near population centers and at places with unique meteorological conditions that promote the rapid deposition of volatile compounds. Hence, the consequences of releasing metals into the atmosphere need to be considered carefully before adopting this method as a remediation tool.

20.4.1.4 Phytoextraction Phytoextraction, also called phytoaccumulation, is the use of plants for the uptake and transport of metals or organic contaminants from the soil into the roots and aboveground plant biomass, which can subsequently be harvested with conventional agricultural methods. Phytoextraction can be used instead of phytoremediation, but phytoremediation is a concept, whereas phytoextraction is a specific cleanup technology. Most plants that grow from metal-contaminated soil can accumulate and store heavy metals in their roots and shoots without visible symptoms. Some of them have developed strategies to exclude pollutants, mainly metals from uptake, and are thus not suitable for phytoextraction, although they can be used for revegetation purposes. There are also certain plants called hyperaccumulators that can actively accumulate and concentrate heavy metals in their shoots to levels of shoot metal concentration of 1% (Zn, Mn), 0.1% (Ni, Co, Cr, Cu, Pb, and Al), 0.01% (Cd and Se) or 0.001% (Hg) of the dry weight shoot biomass (Baker and Brooks, 1989). The main advantage of phytoextraction over other remediation methods is that it removes hazardous compounds from soil without destroying the soil structure and has a limited impact on soil fertility. However, the time required for remediation depends on the type and the extent of metal contamination, the duration of the growing season, and the efficiency of the metal removal by plants, but it normally ranges between 1 and 20 years. This technique is suitable for remediating large areas of land contaminated at shallow depths with low to moderate levels of metal contaminants. There two basic strategies for improving the effectiveness of phytoextraction: chelated-assisted phytoextraction and long-term continuous phytoextraction. Chelates or acidifying agents can be added to the soil to increase the metal solubility and thus metal uptake into plants (Blaylock *et al.*, 1997; Huang *et al.*, 1997; Lombi *et al.*, 2001). Several chelate agent such as ethylenediamine tetracetic acid (EDTA), etilene glycol-bis-(2-aminoethylether-N,N,N,N, tetracetic (EGTA), SS-ethylene diaminedisuccinic acid (EDDS), and citric acid have been found to enhance phytoextraction by mobilizing metals and increasing metal accumulation (Tandy, Schulin, and Nowack, 2006; Cooper *et al.*, 1999). While occasionally effective (Meers *et al.*, 2004), there is the potential risk of metals leaching into groundwater and a lack of reported detailed studies regarding the persistence of metal-chelating agent complexes in contaminated soils (Angle and Linacre, 2005). Moreover, the effects of the addition of the chelate to the soil microbial community are usually disregarded.

20.4.2 Plants Used for Phytoremediation

According to Kärenlampi *et al.* (2000), a plant suitable for phytoremediation should have ability to accumulate the metal(s) intended to be extracted, preferably in its aboveground parts; tolerance to the metal concentrations accumulated; fast growth and highly "effective" (i.e., metal-accumulating) biomass; and easily harvestable. Table 20.2 shows the most effective plants to be used in phytoremediation. Those few plants can accumulate metals to exceptionally high concentrations in their shoots, the so-called hyperaccumulators (Baker and Brooks, 1989), are a particularly valuable resource. The most recent published inventory listed 418 species of vascular plants as being metal hyperaccumulators (Reeves and Baker, 2000). They came from a wide range of taxonomic groups and geographic areas, and as such, have a wide diversity of morphological, physiological, and ecological characteristics (Pollard *et al.*, 2002). Some examples are *Thlaspi caerulescens*; *T. goesingense*; *T montanum*, and *Arabiopsis halleri* L. *T. caerulescens* has been recorded as hyperaccumulating zinc, nickel, cadmium, and possibly lead, and appears to have the capacity to hyperaccumulate at least two others—manganese and cobalt—under laboratory conditions. However, the majority of hyperaccumulators usually accumulate only a single metal. The great of majority of plants are sensitive to metal

TABLE 20.2. Examples of Hyperaccumulators and Their Bioaccumulation Potential

Plant species	Metal	Content (mg kg^{-1})	Reference
T. caerulescens	Zn	39,600 (shorts)	Reeves and Brooks (1983)
T. caerulescens	Cd	1,800	Baker and Walter (1989)
Ipomea alpine	Cu	12,300	Baker and Walter (1989)
Severita acuminate	Ni	25% by wt. dried sap	Jaffre et al. (1976)
Humaniastrum robertii	Co	10,200	Brooks (1998)
A. racemosus	Se	14,900	Beath et al. (1937)
P. vittata	As	27,000	Wang et al. (2002)
Berkheya coddii	Ni	5,500	Robinson et al. (1997)
Iberis intermedia	Ti	3,070	Leblanc et al. (1999)

Source: Padmavathiamma and Li (2007).

toxicity, so they are restricted to areas with low metal concentrations. There are species, however, which are grown exclusively in soils with high metal concentrations. Examples are from lead/zinc mine spoil in Europe and include *Viola calaminaria* (DC). On the other hand, there are plants that have populations on both normal and metalliferous substrates. They represent probably the most interesting category of metal-tolerant plant, not only in a genetic context (Pollard et al., 2002), but also for remediation. Well-known examples of these include *Agrostis capillaries* L., *Plantago lanceolata* L., *Mimulus guttatus* DC, and *Silene vulgaris* (Moench) Garcke. They are interested both in phytoextraction and phytostbilization, while the hyperaccumulators have shown good results in phytoextraction.

Aquatic species were also found to be hyperaccumulators in wetland ecosystems (Williams, 2002). *Ceratophyllum demersum* is described as an arsenic accumulator (Kalbitz and Wenrich, 1998), whereas water hyacinths, *Eichhornia crassipes*, have been found to be effective in accumulating cadmium, lead, and mercury, and in the uptake of pesticides residues. Duckweed (*Lemna minor*) and water velvet (*Azolla pinnata*) were both found to effectively remove iron and copper at low concentrations in laboratory experiments, and also cadmium. The yellow water lily (*Nuphar variegatum*) accumulated copper and zinc.

20.5 USE OF THE ELECTROKINETIC PROCESS TO IMPROVE PHYTOREMEDIATION

The efficient removal or permanent immobilization of metals achieved by current techniques such as physical extraction, chemical leaching, and vitrification results in both soil structure and fertility detriment. On the other hand, remediation technologies based on biological systems (bioremediation and phytoremediation) seem to be sound processes that enhance the natural response of soil to heavy metal contamination. In these cases, the slow rates of heavy metal removal in soil are strong drawbacks to scaling up the processes for technological applications.

In past decades, the development of phytoremediation techniques has faced several limitations because of hyperaccumulators' low rate of growth and biomass production. Also, soil conditions do not always favor metal extraction by plant roots because of the inadequate speciation and reduced availability of metals. In many cases, the aid of chelating agents is required, which involves an additional impact to soil. Although few investigations have been reported on phytoremediation assisted by the electromobilization of heavy metals, the combined technology seems to be an improved implement for soil remediation that increases the benefits and overcomes the limitations of each principle or strategy of remediation.

Electrokinetic remediation is a physical–chemical extractive technology specifically convenient for the removal of heavy metals regardless the clay content of soils. Electrokinetic remediation and phytoremediation alike have been reported to be efficient in the remediation of moderate to low contaminated soils. The fact that the techniques can be carried out *in situ* without complex devices enables the simultaneous growth of crops on the contaminated site that is to be treated. Many questions still arise from the attempt to successfully combine the two remediation processes, most of them dealing with the behavior of plants under the artificial influence of direct electric current. According to the literature, plants grown in between electrodes are not directly affected by the electric current at the laboratory scale assay. But the effect of current on other parameters such as pH, metal availability, and chelating agents can modify plant responses.

Lim *et al.* (2004) optimized the time and electric field applied to each Indian mustard (*Brassica juncea*) plant in a remediation experiment performed in 1.2-kg pots of lead- and arsenic-contaminated soil. The electric potential of 30–40 V applied 1 h per day were the optimal conditions to reach maximum phytoextraction after 9-day treatment. In this case, the increase of EDTA availability by the electrical current caused a rapid toxicity response of plants that limited the remediation process.

Alternative chelators and purging solutions of reduced toxicity toward plants have to be investigated. Ethylenediaminedisuccinate is recently reported as a green chelating agent that is found in roots and shoots at concentrations equal to metal uptake without causing plant damage (Tandy, Schulin, and Nowack, 2006). In another approach, the biological acidification of soil as a result of the release of protons in roots exudates to replace ammonium ion uptake by roots and/or microorganisms is considered. In this case, soil is supplied with either urea, ammonium sulfate, or organic amendments with the aim to increase soluble metal contents. Broadly, plant roots play a specific role in assisting remediation processes. Exudates released from plant roots to soil in the rhizosphere give rise to favorable conditions to promote bacterial diversity and activity. It has been reported in the literature that nitrifying bacteria in the rhizosphere contribute to the solubilization of heavy metals. A controlled process can be attempted by inoculating nitrifying bacteria on soil rhizospheres that prevents the use of acidic purging solutions in a phytoelectrokinetic remediation. Other microorganisms of interest in plant rhizospheres are those that produce biosurfactants (Tran, Kruijt, and Raaijmakers, 2007). Frequently these strains are involved in metal mobilization in the rhizosphere so the colonization of soil rhizospheres by selected bacteria can significantly increase phytoextraction rates of solubilized metals, the mobility of which is improved under the electrical field (Sheng *et al.*, 2008).

Since the electric field efficiently drives increased amounts of soluble heavy metals toward plant roots, which results in stress conditions for plants, hyperaccumulator plants with a rapid growth period are candidates for use in combination with electrokinetic techniques. Heavy metals removal to clean up soils seems feasible just with successive crops.

In another experimental study (O'Connor et al., 2003), a constant voltage of 30 V was continuously applied to a contaminated soil that supported the growth of perennial ryegrass (*Lolium perenne* cv Elka). In this case, the routine harvesting of shoots was the preferred means of ensuring prolonged phytoremediation. A different redistribution of metals and biomass production in soil among electrodes was observed. Furthermore, the difference in the electromobility of copper as compared to cadmium was related with enhancement of copper plant uptake, which in turn, led to a number of harvests that decreased in the cooper-contaminated soil. This study also showed that soil close to the electrodes achieved extreme pH values as a consequence of the metal electrokinetics, which caused a considerable impact that prevented the growth of plants. The benefits of the combined techniques were not widely extended over the contaminated soil.

In light of the results obtained from this preliminary assay to the combined technology, several aspects still need to be adjusted. Changes in the polarity of electrodes during the process can avoid fixed redistribution of heavy metals and soil pH values that are associated with different rates of plant biomass and phytoextractions to the proximity to electrodes. The combined technique is homogenously applied to the contaminated site.

Ongoing research is encouraged to evaluate how effectively the combination of phyto- and electrokinetic remediation techniques accomplishes soil remediation.

20.6 PHYTOREMEDIATION AFTER ELECTROKINETIC PROCESS

The application of the electrokinetic technology to achieve a remediated soil can produce significant changes in soil properties, thereby preventing subsequent restoration. Different studies have demonstrated the efficiency of the electric field in the mobilization and removal of pollutants (Reddy et al., 2001, 2002; Page and Page, 2002; Pazos, Sanroman, and Cameselle, 2006), with a pollutant removal rate of approximately 90%. Most of the experiments have been realized using small samples (5–20 g soil). Due to soil characteristics and heterogeneous composition, it is difficult to extrapolate the effect and efficiency of the technology in field conditions.

Some experiments have been carried out in containers with 1 ton of soil, applying this technology to remove heavy metals from soil (Lobo et al., 2003, 2007). Soils with different pH and permeability were tested with the application of an electric potential of 100 V (100 cm between electrodes) and citric acid as a purging solution. In these conditions, the results suggested that pollutant migration is strongly dependent on soil pH and carbonate content rather than on pore size. The pollutant removal efficiency also depended on water–soil content and electro-osmotic purging solutions that also controlled the suitable pH soil. In alkaline soil (pH>8), the removal of heavy metals is limited due to their strong link to soil carbonates. In these cases, the technology efficiency is lower than 50%. However, in the same conditions, in neutral pH soil, the effect of the electric field led to removal percentages between 65% and 70%. For both assays, the residual concentration was more difficult to remove and even increased process time. In general, the migration of the pollutants was highly related to soil mineralogy, and the overall contaminant

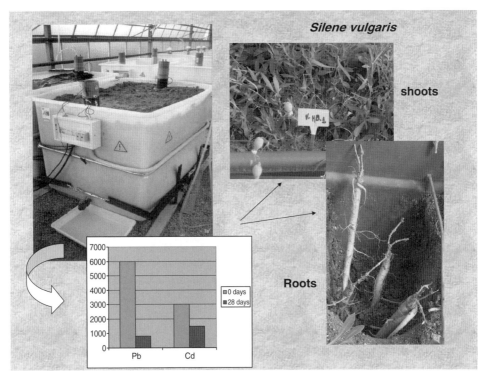

Figure 20.3. Soil remediation assay applying electrokinetic and phytoremediation (Lobo et al., 2007). (Different bars represent heavy metal remaining in the soil after electrokinetic as initial conditions to phytoremediation.)

removal efficiency was controlled by different geochemical reactions that included oxidation reduction, adsorption and precipitation, and led to metal immobilization after direct current application in some soils. To improve the efficiency of the mobilization and removal of the pollutants, it may be necessary to use more aggressive agents (acids or chelants) and disturbing chemical and biological soil properties that prevent the further recovery.

The use of hyperaccumulator or tolerant plants to remove residual concentration and to achieve clean soil after the electrokinetic process can be an alternative that contributes to the recovery of soil properties, improving soil structure through the influence of the root system. Before the establishment of vegetal cover, the application of organic matter or compost could be an useful alternative to restore the nutrients removed with the metals during the electroremediation process. *Silene vulgaris*, a wild plant, has been used to remove residual concentration of heavy metals in soil previously treated by electrokinetic (Fig. 20.3) (Lobo et al., 2007). Heavy metals were absorbed by the plant and significant concentrations were accumulated both in roots and leaves. Higher yields were observed when soils were amended with compost with the aim to incorporate organic matter and nutrients to improve soil structure (Pérez-Sanz et al., 2007) (Fig. 20.4). In general, a few studies have been carried out evaluating the influence of the direct current on soil microbial activity, although it is known that the loss of nutrients and structure can affect the microbial life.

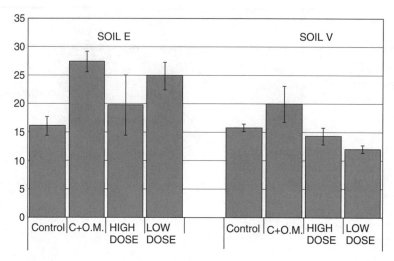

Figure 20.4. Biomass (mg/pot) of *Silene vulgaris* in soils treated by electrokinetic remediation. E: soil pH alkaline; V: soil pH neutral. C+O.M.: control soil+compost. High dose and low dose: Different metal concentration in soil after electrokinetic application (Perez-Sanz et al., 2007).

In neutral soils treated by electrokinetic technology to remove heavy metals, Lobo *et al.* (2008) did not find significant differences in microbial activity (soil respiration, enzymatic activities) in relation to the control soil. Assays during 30 days applying 100V (100cm between electrodes) led to changes in the soil structure due to losses in the clay content; however, no effect on microbial activity was observed. As a consequence, further studies in field experiments are necessary to evaluate the impact of the technology on soil biology (enzymatic activities, population biodiversity) that could constitute a limiting factor in achieving a soil restoration process that preserves soil functionality.

20.7 SUMMARY

Soil is a living system and a nonrenewable resource due to the slow speed of the formation process. The soil matrix is extremely complex, and for this reason, the response of each soil to a degradation process will be different. Even so, every remediation method influences the natural soil functions and its function as an archive of natural and cultural history. Soil degradation due to contamination represents an increasing hazard that requires the development of improved methods for soil remediation with two main objectives: removal of the contaminants from the soil, and allowing the soil to recover its functionality. Electrokinetic technology seems to be an adequate alternative for the removal of pollutants from the soil. The advantages of this technology are its low cost, nonintrusive character, and applicability to a wide range of contaminants. The impact that electrokinetic extraction will have on the physical, chemical, and biological characteristics of the soil is not adequately quantified yet. Further research is necessary to make a diagnosis about the real effect on the soil depending on its properties. Due to the importance in the mitigation of the impact of the application of this technology from the point of view of the future use of the soil, the application of combined technologies (bio- or phytoremediation) constitutes a suitable strategy in soil remediation.

Phytoremediation is defined as the use of living plants and their associated microorganisms to remove, degrade, or sequester inorganic and organic pollutants from *in situ* remediation of contaminated soil, sludges, sediments, and groundwater. This technology is limited by depth (roots) and also by the solubility and the availability of pollutants. Although it is faster than natural attenuation, phytoremediation requires long time periods and is restricted to sites with low contaminant concentration. Although few investigations have been made on phytoremediation assisted by the electromobilization of heavy metals, the combined technology seems to be an improved tool for soil remediation under certain conditions, low dose of heavy metals, and the need to restore soil functions.

Additional research is encouraged to scale up to field experiments in order to evaluate the level of accomplishment of the combination of phytoremediation and electrokinetic techniques.

REFERENCES

Acar YB, Alshawabkeh AN. (1993). Principles of electrokinetic remediation. *Environmental Science and Technology* **27**(13):2638–2647.

Alcañiz JM, Comellas, L, Pujolá, M. (1996). *Manual de Restauració d`activitats extractives amb fangs de depuradora. Recuperació de terrenys marginals.* Junta de Sanejament. Departament de Medi Ambent, Generalitat de Catalunya, Barcelona, p. 69.

Alkorta I, Hernandez-Allica J, Becerril JM, Amezaga I, Albizu I, Garbisu C. (2004). Recent findings on the phytoremediation of soils contaminated with environmentally toxic heavy metals and metalloids such as zinc, cadmium, lead and arsenic. *Reviews in Environmental Science and Biotechnology* **3**:71–90.

Angle JS, Linacre NA. (2005). Metal phytoextraction: A survey of potential risks. *International Journal of Phytoremediation* **7**:241–257.

Baker AJM, Brooks RR. (1989). Terrestrial higher plants which hyper accumulate metallic elements. A review of their distribution, ecology, and phytochemistry. *Biorecovery* **1**:81–126.

Baker AJM, Walker PL. (1989). Ecophysiology of metal uptake by tolerant plants. In *Heavy Metal Tolerance in Plants: Evolutionary Aspects*. Boca Raton, FL: CRC, pp. 155–177.

Beath OA, Eppsom HF, Gilbert GS. (1937). Selenium distribution in seasonal variation of vegetation type occurring on seleniferous soils. *Journal of the American Pharmaceutical Association* **26**:394–405.

Blaylock MJ, Salt DE, Dushenkov S, Zakharova O, Gussman C, Kapulnik Y. (1997). Enhanced accumulation of Pb in Indian mustard by soil-applied chelating agents. *Environmental Science and Technology* **31**(3):860–865.

Bolan NS, Adriano DC, Naidu R. (2003). Role of phosphorus in immobilization and bioavailability of heavy metals in the soil–plant system. *Reviews of Environmental Contamination and Toxicology* **177**:1–44.

Brooks RR, Chambers MF, Nicks LJ, Robinson BH. (1998). Phytomining. *Trends in Plant Science* **3**(9):359–362.

Chaney RL, Malik M, Li YM, Brown SL, Brewer EP, Angle JS, Baker AJ. (1997). Phytoremediation of metals. *Current Opinion in Biotechnology* **8**:279–284.

Chaudhry TM, Hayes WJ, Khan AG, and Khoo CS. (1998). Phytoremediation—Focusing on accumulator plants that remediate metal-contaminated soils. *Australasian Journal of Ecotoxicology* **4**:37–51.

Commission of the European Communities. COM (2006) 232. Proposal for a Directive of the European Parliament and of the Council. http://ec.europa.eu/comm/environment/soil/index.htm

Cooper EM, Sims JT, Cunningham SD, Huang JW, Berti WR. (1999). Chelate-assisted phytoextraction of lead from contaminated soil. *Journal of Environmental Quality* **28**:1709–1719.

Cunningham SD, Shann JR, Crowley DE, Anderson TA. (1997). Phytoremediation of contaminated water and soil. In *Phytoremediation of Soil and Water Contaminants*, ACS Symposium Series, Vol. **664** (eds. EL Kruger, TA Anderson, JR Coats). Washington, DC: American Chemical Society, pp.2–19.

Domsch KH, Jagnow G, Anderson TH. (1983). An ecological concept for assessment of side-effects of agrochemicals on soil microorganims. *Residue Reviews* **86**:65–105.

Environmental Protection Agency. (2007). Treatment Technologies for Site Cleanup: Annual Status Report, 12th ed. EPA-542-R-07-012. Washington, DC, USA.

Fletcher JS. (1997). Testing and knowledge of organic phytotoxins. In *Soil Ecotoxicology* (eds. J Tarradellas, G Bitton, D Rossel). Boca Raton, FL: CRC Press, pp. 271–289.

Gardea-Torresdey JL, de la Rosa G, and Peralta-Videa JR. (2004). Use of phytoremediation technologies in the removal of heave metals: A review. *Pure and Applied Chemistry* **76**(4):801–813.

Glass, DJ. (1999). *US and International Markets for Phytoremediation, 1999–2000*. Needham, MA: D. Glass Associates, p. 266.

Haider K. (1995). Sorption phenomena between inorganic and organic compounds in soils: Impact on transformation processes. In *Environmental Impact of Soil Component Interactions. Natural and Anthropogenic Organics* (eds. PM Huang, J Berthelin, JM Bollag, WB McGill, AL Page). Boca Raton, FL: CRC Press Inc.

Huang JW, Chen J, Berti WR, Cunningham SD. (1997). Phytoremediation of lead contaminated soil: Role of synthetic chelates in lead phytoextraction. *Environmental Science and Technology* **31**(3):800–805.

Jaffre T, Brooks RR, Lee J, Reeves RD. (1976). *Sebertia acumip*. A nickel-accumulating plant from New Caledonia. *Science* **193**:579–580.

Kalbitz K, Wenrich R. (1998). Mobilization of heavy metals and arsenic in polluted wetland spills and its dependence of dissolved organic matter. *The Science of the Total Environment* **209**:27–39.

Kärenlampi S, Schat H, Vangronsveld J, Verkleij JAC, van der Lelie D, Mergeay M, Tervahauta AI. (2000). Genetic engineering in the improvement of plants for phytoremediation of metal polluted soils. *Environmental Pollution* **107**:225–231.

Kim S, Moon SH, Kim KW. (2001). Removal of heavy metals from soils using enhanced electrokinetic soil processing. *Water, Air, Soil Pollution* **125**(1–4):259–272.

Kördel W and Hund-Rinke K. (2001). Ecotoxicological assessment of soils-bioavailavility from an ecotoxicological point of view. In *Treatment of Contaminated Soil. Fundamentals, Analysis, Applications* (eds. R Stegmann, G Brunner, W Calamano, G Matz). Berlin: Springer-Verlag.

Kuo S, Jellum EJ, Baker AS. (1985). Effects of soil type, liming, and sludge application on zinc and cadmium availability to Swiss chard. *Soil Science* **139**:122–130.

Lasat MM. (2002). Phytoextraction of toxic metals—A review of biological mechanisms. *Journal of Environmental Quality* **31**:109–120.

Leblanc M, Petit D, Deram A, Robinson B, Brooks RR. (1999). The phytomining and environmental significance of hyperaccumulation of thallium by *Iberis intermedia* from southern France. *Economic Geology* **94**(1):109–113.

Lim JM, Salido AL, Butcher DJ. (2004). Phytoremediation of lead using Indian mustard (*Brasica juncea*) with EDTA and electrodics. *Microchemical Journal* **76**:3–9.

Lobo MC. (2007). Sustainable soil remediation. The use of combined technologies. *Proceedings of the Sixth Symposium on Electrokinetic Remediation*, Vigo, Spain.

Lobo MC, Martínez-Iñigo J, Alonso J, Perucha C, De Fresno A, Laguna J. (2003). Application of electrokinetic remediation techniques to soil contaminated with heavy metals and organic compounds. *Proceedings of the Fourth European Congress of Chemical Engineering*, Granada, Spain.

Lobo MC, Martínez-Iñigo MJ, Pérez-Sanz A, Plaza A, Alonso J, Perucha C. (2007). Electrokinetic remediation of a soil contaminated with heavy metals in a scale pilot experiment. *Proceedings of the Sixth Symposium on Electrokinetic remediation*, Vigo, Spain.

Lobo MC, Perez-Sanz A, Martínez-Iñigo MJ, Vicente MA, Plaza A. (2008). Influence of electrokinetic process on soil properties in a pilot scale assay. *Third International Meeting on Environmental Biotechnology and Engineering (3IMEBE)*, Palma de Mallorca, Spain.

Loehr RC. (1996). The environmental impact of soil contamination: Bioavailability, Risk Assessment, and Policy Implications. Policy Study No. 211. The Reason Foundation. Los Angeles, CA 90034.

Lombi E, Zhao FJ, Dunham SJ, MacGrath SP. (2001). Phytoremediation of heavy metal-contaminated soils: Natural hyperaccumulation versus chemically enhanced phytoextraction. *Journal of Environmental Quality* **30**:1919–1926.

McGrath SP, Zhao FJ. (2003). Phytoextraction of metals and metalloids. *Current Opinion in Biotechnology* **14**:277–282.

McGrath SP, Zhao FJ, Lombi E. (2002). Phytoremediation of metals, metalloids, and radionuclides. *Advances in Agronomy* **75**:1–56.

Meagher RB. (2000). Phytoremediation of toxic elemental and organic pollutants. *Current Opinion in Plant Biology* **3**:153–162.

Meers E, Hopgood M, Lesage E, Vervaeke P, Tack FMG, Verloo MG. (2004). Enhanced phytoextraction: In search of EDTA alternatives. *International Journal of Phytoremediation* **6**:95–109.

Miehlich G. (2001). Do contaminated soils have to be decontaminated? In *Treatment of Contaminated Soil. Fundamentals, Analysis, Applications* (eds. R Stegmann, G Brunner, W Calamano, G Matz). Berlin: Springer-Verlag.

Navari-Izzo F, Quartacci MF. (2001). Phytoremediation of metals. Tolerance mechanisms against oxidative stress. *Minerva Biotecnologica* **13**:73–83.

O'Connor CS, Lepp NW, Edwards SR, Sunderland G. (2003). The combined use of electrokinetic remediation and phytoremediation to decontaminate metal-polluted soils: A laboratory-scale feasibility study. *Environmental Monitoring and Assessment* **84**(1–2): 141–148.

Padmavathiamma PK, Li LY. (2007). Phytoremediation technology: Hyperaccumulation metals in plants. *Water, Air, Soil Pollution* **184**:105–126.

Page MM, Page CL. (2002). Electroremediation of contaminated soil. *Journal of Environmental Engineering* **128**(3):208–219.

Pazos M, Sanroman MA, Cameselle C. (2006). Improvement in the electrokinetic remediation of heavy metals spiked kaolin with the polarity inversion technique. *Chemosphere* **62**(5):819–822.

Pérez-Sanz A, Alonso J, Alarcón R, García Gonzalo P, Lobo MC. (2007). Preliminary test to evaluate metal accumulation in *Silene vulgaris* grown in polluted soils treated previously with electrokinetic technologies. *Proceedings of the Ninth International Conference on the Biogeochemistry of Trace Elements (ICOBTE)*, Beijing, China.

Pollard AJ, Powell KD, Harper FA, Andrew J, Smith C. (2002). The genetic basis of metal hyperaccumulation in plants. *Critical Reviews in Plant Sciences* **21**(6):539–566.

Prasad MNV, Freitas H. (2003). Metal hyperaccumulation in plants—Biodiversity prospecting for phytoremediation technology. *Electronic Journal of Biotechnology* **6**: 275–321.

Raskin I, Smith RD, Salt DE. (1997). Phytoremediation of metals: using plants to remove pollutants from the environment. *Current Opinion in Biotechnology* **8**:221–226.

Reddy KR, Saichek RE, Maturi K, Ala P. (2002). Effects of soil moisture and heavy metals concentrations on electrokinetic remediation. *Indian Geotechnical Journal* **32**(3):258–288.

Reddy KR, Chinthamreddy S, Al-Hamdan A. (2001). Synergist effects of multiple metal contaminants on electrokinetic remediation of soils. *Remediation: The Journal of Environmental Cleanup Costs, Technologies & Techniques* **11**(3):85–109.

Reeves RD, Baker AJM. (2000). Metal-accumulating plants. In *Phytoremediation of Toxic Metals* (eds. I Raskin and BD Ensley). New York: Wiley, pp. 193–229.

Reeves RD, Brooks RR. (1983). Hyperaccumulation of lead and zinc by two metallophytes from a mining area of Central Europe. *Environmental Pollution Series A*, **31**:277–278.

Robinson BH, Brooks RR, Howes AW, Kirkman JH, Gregg PEH. (1997). The potential of the high-biomass nickel hyperaccumulator *Berkheya coddii* for phytoremediation and phytomining. *Journal of Geochemical Exploration* **60**:115–126.

Saichek RE, Reddy KR. (2003). Effect of pH control at the anode or the electrokinetic removal of phenanthrene from kaolin. *Chemosphere* **51**:273–287.

Salt DE, Baker AJM. (2001). Phytoremediation of metals. In *Biotechnology, Vol. 11b: Environmental Processes II, Soil Decontamination* (eds. HJ Rehm, G Reed). New York: Wiley VCH, pp. 386–397.

Salt DE, Blaylock M, Kumar PBAN, Dushenkov V, Ensley BD, Chet I, Raskin I. (1995). Phytoremediation: a novel strategy for the removal of toxic metals from the environment using plants. *Biotechnology* **13**:468–475.

Salt DE, Smith RD, Raskin I. (1998). Phytoremediation. *Annual Review of Plant Physiology and Plant Molecular Biology* **49**:643–668.

Schnitzer M. (1995). Organic–inorganic interactions in soils and their effects on soil quality. In *Environmental Impact of Soil Component Interactions. Natural and Anthropogenic Organics* (eds. PM Huang, J Berthelin, JM Bollag, WB McGill, AL Page). Boca Raton, FL: CRC Press Inc.

Sheng X, He L, Wang Q, Ye H, Jiang C. (2008). Effects of inoculation of biosurfactant-producing Bacillus sp. J119 on plant growth and cadmium uptake in a cadmium-amended soil. *Journal of Hazardous Materials* **55**(1–2):17–22.

Steger H, Zorn R, Gregolec G, Czurda K, Borst M. (2005). Soil structure changing caused by electrokinetic process. *Proceedings of the Fifth Symposium on Electrokinetic Remediation*, Ferrara, Italy.

Suer P, Gitye K, Allard B. (2003). Speciation and transport of heavy metals and macroelements during electroremediation. *Environmental Science and Technology* **37**:177–181.

Tandy S, Schulin R, Nowack B. (2006). The influence of EDDS on the uptake of heavy metals in hydroponically grown sunflowers. *Chemosphere* **62**(9):1454–1463.

Terry N, Carlson C, Raab TK, Zayed A. (1992). Rates of selenium volatilization among crop species. *Journal of Environmental Quality* **21**:341–344.

Tortensson L. (1997). Microbial assays in soils. In *Soil Ecotoxicology* (eds. J Tarradellas, G Bitton, D Rossel). Boca Raton, FL: CRC Press Inc.

Tran H, Kruijt M, Raaijmakers JM. (2007). Diversity and activity of biosurfactant-producing Pseudomonas in the rhizosphere of black pepper in Vietnam. *Journal of Applied Microbiology* **104**(3):839–851.

Ugaz A, Puppah S, Gale RJ, Acar YB. (1994). Electrokinetic soil processing. Complicating features of electrokinetic remediation of soils and slurries: saturation effects and the role of the cathode electrolysis. *Chemical Engineering Communications* **129**:183–200.

Virkutyte J, Sillanpaa M, Latostenmaa P. (2002). Electrokinetic soil remediation—critical overview. *The Science of the Total Environment* **289**(1–3):97–121.

Voroney RP. (2006). The soil habitat. In *Soil Microbiology, Ecology, and Biochemistry* (ed. EA Paul). Burlington, MA: Academic Press.

Wang J, Zhao F, Meharg AA, Raab A, Feldman J, McGrath SP. (2002). Mechanism of arsenic hyperaccumulation in *Pteris vittata*. Uptake kinetics, interactions with phosphate, and arsenic speciation. *Plant Physiology* **130**:1552–1561.

Wenzel WW, Adriano DC, Salt D, Smith R. (1999). Phytoremediation: A plant–microbe-based remediation system. In *Bioremediation of Contaminated Soils. Agronomy Monograph No. 37*, Madison, WI: SSSA, pp. 457–508.

Williams JB. (2002). Phytoremediation in wetland ecosystems: Progress, problems and potential. *Critical Reviews in Plant Sciences* **21**(6):607–635.

Williamson D, Loehr RC, Kimura Y. (1998). Release of chemical from contaminated soils. *Soil and Sediment Contamination* **7**(5):543–558.

Wolfe AK, Bjornstad DJ. (2002). Why would anyone object? An exploration of social aspects of phytoremediation acceptability. *Critical Reviews in Plant Sciences* **21**(5):429–438.

Yeung AT, Hsu CH, Menon RM. (1997). Physicochemical soil-contaminant interaction during electrokinetic extraction. *Journal of Hazardous Materials* **55**(1–3):221–237.

Zhou DH, Zorn R, Kurt C. (2003). Electrochemical remediation of copper contaminated kaolinite boy conditioning analyte and catholyte pH simultaneously. *Journal of Environmental Science-China* **15**(3):396–400.

21

ELECTROKINETIC–CHEMICAL OXIDATION/REDUCTION

GORDON C. C. YANG

21.1 INTRODUCTION

Since the 1990s, remediation by the electrokinetic (EK) process has received much attention by many researchers worldwide. Some research has focused on the treatment of inorganic pollutants (e.g., heavy metals and nitrates) and organic pollutants (e.g., aromatic compounds, chlorinated compounds, and gasoline hydrocarbons), and/or relevant mechanisms (Probstein and Hicks, 1993). However, the EK process per se has its limits for practical applications. The question of how to overcome those problems and enhance the performance of the EK process is of significance. Many researchers have focused on the development of new treatment technologies to combine the EK process with one or more other technologies such as the Lasagna™ process (Ho et al., 1995; US DOE, 2002), the EK–Fenton process (Yang and Long, 1998), the EK–nanoiron wall (Yang, Li, Hung, 2004), the EK–ultrasound process (Chung and Kamon, 2005), the EK–IEM (ion exchange membrane) process (Kim, Kim, and Kim, 2005), nanoiron injection–EK (Yang and Chang, 2006), the EK–reagent flushing process (Reddy et al., 2006), the electro-bioremediation process (Wick, Shi, and Harms, 2007), EK–Fenton–biodegradation (Chen, 2000), and the EK–Fenton–catalytic iron wall (Hung, 2002).

21.2 GENERAL PRINCIPLES

In addition to early works by Acar and Alshawabkeh (1993) and Probstein and Hicks (1993), Virkutyte, Sillanpää, and Latostenmaa (2002) also reviewed the principles of the EK process and several integrated technologies of the EK and other

---

Electrochemical Remediation Technologies for Polluted Soils, Sediments and Groundwater,
Edited by Krishna R. Reddy and Claudio Cameselle
Copyright © 2009 John Wiley & Sons, Inc.

processes. That review paper, however, is mainly concerned with heavy metal remediation. In this chapter the fundamental principles relevant to coupled EK–chemical redox (reduction/oxidation) processes will be briefly reviewed and summarized.

21.2.1 Fundamentals of the EK Process

It is well known that the EK process relies on several interacting mechanisms, including (1) advection resulting from electro-osmotic flow and externally applied hydraulic gradients, (2) diffusion of the acid front to the cathode, and (3) the migration of cations and anions toward the respective electrode. The electrolysis of water is the dominant and most important electron transfer reaction that occurs at the electrodes during the EK process.

Electrokinetic transport offers a unique and effective treatment method when dealing with low-permeability soils. The main goal of EK remediation is to effect the migration of subsurface contaminants in an imposed electric field via electro-osmosis, electromigration, and electrophoresis. These three phenomena in a soil matrix subjected to a low-voltage direct current can be briefly described as follows: (1) electro-osmosis, the movement of soil water or groundwater from the anode to the cathode of an electrochemical system; (2) electromigration (also known as ionic migration), the transport of ions and ion complexes to the electrode of opposite charge; and (3) electrophoresis, the transport of charged particles of colloidal size and bound contaminants under the influence of an electric field.

In many instances, electro-osmotic flow plays the most important role in the removal of contaminants within the system. Electromigration takes place when highly soluble ionized inorganic species (e.g., metal cations, chlorides, nitrates, and phosphates) are present in moist soil environments. To enhance the performance of the treatment, an integration of the EK process with another treatment technology (e.g., the Fenton process) could be necessary. In some cases, coupling the EK process with more than one technology could also be considered.

21.2.2 Fundamentals of Chemical Oxidation Processes

21.2.2.1 Fenton and Fenton-like Processes The Fenton process has been proven to be a very effective technology for the degradation of various organics in the presence of wastewater and soil (Barbeni *et al.*, 1987; Bowers *et al.*, 1989; Watts *et al.*, 1990; Yang and Lai, 1997). Conventionally, H_2O_2 and Fe^{2+} are used in the Fenton process for generating hydroxyl radicals to chemically destruct organic pollutants. In addition, a Fenton-like reaction using zero-valent iron or iron minerals instead of Fe^{2+} has been found to be effective in the degradation of organic pollutants in both wastewater and soil (Tang and Chen, 1996; Greenberg *et al.*, 1998). Because of abundant iron species in the subsurface environment, a Fenton-like reaction might play a significant role in subsurface remediation. The basics of the Fenton process and the Fenton-like process are similar. The former uses a transition metal to activate hydrogen peroxide for the formation of hydroxyl radicals (·OH or HO·), which further would attack the C–H bonds of organics in the neighborhood. Hydroxyl radicals have a very strong oxidizing capability next to fluorine. Thus, the Fenton process could be used for treating many biorefractory organic compounds.

The Fenton's reagents (i.e., H_2O_2 and Fe^{2+}) per se are not stable. As they are in contact, several reactions would occur simultaneously, resulting in the formation of HO·, HO_2·, Fe^{3+}, and O_2 as shown in the following reaction equations (Chen et al., 2001):

$$Fe^{2+} + H_2O_2 \rightarrow Fe^{3+} + OH^- + HO· \qquad (21.1)$$

$$Fe^{2+} + HO· \rightarrow Fe^{3+} + OH^- \qquad (21.2)$$

$$H_2O_2 + HO· \rightarrow H_2O + HO_2· \qquad (21.3)$$

$$Fe^{2+} + HO_2· \rightarrow Fe^{3+} + HO_2^- \leftrightarrow H_2O_2 \qquad (21.4)$$

$$Fe^{3+} + HO_2· \rightarrow Fe^{2+} + H^+ + O_2 \qquad (21.5)$$

$$Fe^{3+} + H_2O_2 \rightarrow Fe^{2+} + HO_2· + H^+ \qquad (21.6)$$

Although HO·, HO_2·, and O_2 are all oxidants, hydroxyl radicals have the strongest oxidizing ability for a variety of organic compounds. When organic compounds encounter the Fenton's reagents, chain reactions would take place as shown below:

$$RH + HO· \rightarrow R· + H_2O \quad \text{(Chain propagation)} \qquad (21.7)$$

$$R· + H_2O_2 \rightarrow ROH + HO· \quad \text{(Chain propagation)} \qquad (21.8)$$

$$R· + HO· \rightarrow ROH \quad \text{(Nonchain termination)} \qquad (21.9)$$

$$2R· \rightarrow \text{Product (Dimer)} \quad \text{(Nonchain termination)} \qquad (21.10)$$

$$R· + Fe^{3+} \rightarrow Fe^{2+} + \text{product} \quad \text{(Regenerate } Fe^{2+} \text{ for chain initiation)} \qquad (21.11)$$

$$R· + Fe^{2+} \rightarrow Fe^{3+} + RH \quad \text{(Chain termination)} \qquad (21.12)$$

In solution, hydroxyl radicals generally would react vigorously with the dissolved organics, H_2O_2, and Fe^{2+}. In fact, hydroxyl radicals would react among themselves as well. Due to their high reaction rate as compared with their generation rate in the reaction system, hydroxyl radicals generally have a very short life cycle (a few nanoseconds). Reportedly, the performance of the Fenton process would be affected by the following reaction parameters: pH, type of inorganic salts, natural organic matter, molecular structures of contaminants, dissolved oxygen, dose ratio of hydrogen peroxide to ferrous ion, and iron species and dosage.

21.2.2.2 Ultrasonic Oxidation Process Sonochemical degradation methods are relatively new and involve exposing aqueous solutions containing the organic pollutant to ultrasound. In the past decade, considerable interest has been shown on the application of ultrasonic degradation for the removal of hazardous contaminants from water (Vinodgopal et al., 1998; Kim, Huang, and Chiu, 2001; Manousaki et al., 2004; Kidak and Ince, 2006). The chemical effects of high intensity ultrasound have long been known to arise from acoustic cavitation of liquids. This rapid formation, growth, and implosive collapse of gas vacuoles generates short-lived (~ ns); localized "hot spots," whose peak temperatures and pressures have been measured at ~3000 K and ~300 atm. As a result, a variety of high energy species (e.g., H_2O_2 and HO·) would be produced in solution due to water sonolysis. This is the so-called cavitation phenomenon in the ultrasonic oxidation (Suslick, Schubert, and Goodale, 1981; Suslick *et al.*, 1983; Suslick, Hammerton, and Cline Jr., 1986). The cavitation phenomenon leads to subsequent chemical reactions in three phases: internal cavity, interface boundary layer, and liquid bulk. In the reaction system, those substances with high hydrophobicity and low boiling points will enter the cavities and be completely decomposed by combustion or pyrolyzation therein. In addition, the reactions with HO· would occur in the interfacial boundary layer regions or in the surrounding bulk liquid at normal temperature and pressure. A simple scheme showing radical formation and depletion during water sonolysis is given below (Ince *et al.*, 2001):

$$H_2O +))) \rightarrow HO\cdot + H\cdot \quad \text{(pyrolysis)} \quad (21.13)$$

$$HO\cdot + H\cdot \rightarrow H_2O \quad (21.14)$$

$$2HO\cdot \rightarrow H_2O + O\cdot \quad (21.15)$$

$$2HO\cdot \rightarrow H_2O_2 \quad (21.16)$$

The optimization of ultrasonic irradiation as an advanced oxidation technology can be achieved by adjusting a variety of parameters. In general, the influencing factors on ultrasonic degradation of organic compounds can be categorized into two groups: (1) system operating parameters such as ultrasonic frequency, ultrasonic intensity, and ultrasonic density; and (2) system characteristics such as type of saturating gas, viscosity, vapor pressure, interfacial tension, hydrostatic pressure, solution temperature, and solution pH. A review on background theories and environmental applications of sonochemical degradation can be found elsewhere (Ince *et al.*, 2001).

21.2.3 Fundamentals of Chemical Reduction Processes

21.2.3.1 Acid-driven Surface-mediated Process Among other substances, the oxidation of zero-valent metals (e.g., Fe^0) and magnetite (Fe_3O_4) would release electrons to the system. It is well understood that some species such as NO_3^- and Cr^{6+} would be reduced to less toxic compounds if they accepted electrons in the system. Several studies have indicated the final products of the chemical reduction of nitrate by zero-valent iron could be N_2 or NH_3 depending on the experimental conditions (Flis, 1991; Agrawal and Tratnyek, 1996; Huang, Wang, and Chiu, 1998; Choe *et al.*, 2000). In the aqueous system, the nitrate ion would accept the electrons released from the surface of iron resulting in chemical reduction of nitrate, as shown in Eqs. 21.17–21.19 (Choe *et al.*, 2000):

$$Fe^0 + 2H_2O \leftrightarrow Fe^{2+} + H_2 + 2OH^- \quad (21.17)$$

$$5Fe^0 + 2NO_3^- + 6H_2O \leftrightarrow 5Fe^{2+} + N_2 + 12OH^- \quad (21.18)$$

$$NO_3^- + 6H_2O + 8e^- \leftrightarrow NH_3 + 9OH^- \quad (21.19)$$

It was also reported (Huang *et al.*, 1998; Su and Puls, 2004; Yang and Lee, 2005) that the system pH would greatly affect the reduction efficiency of nitrate. In general, chemical reduction of nitrate would drop when the system pH was greater than 5, whereas the reduction efficiency would be increased under acidic conditions (e.g., pH ≈ 3). Based on the findings obtained using iron nanoparticles for nitrate reduction, Yang and Lee (2005) proposed the following two possible reaction pathways for the acid-driven surface-mediated process with the first reaction pathway probably being the predominant one:

$$NO_3^- + 4Fe^0 + 10H^+ \rightarrow 4Fe^{2+} + NH_4^+ + 3H_2O \tag{21.20}$$

and

$$2NO_3^- + 5Fe^0 + 12H^+ \rightarrow 5Fe^{2+} + N_{2(g)} + 6H_2O \tag{21.21}$$

Very recently, the present author's group (Chen, 2008) further investigated the reaction behaviors among laboratory-prepared nanomaterials (e.g., Fe_3O_4) against NO_3^- and Cr^{6+} in simulated groundwater. Experimental results illustrated the occurrence of reductive adsorption of the said nanomaterial toward target contaminants under acidic conditions. Part of the NO_3^- was reduced to NO_2^- and NH_4^+, whereas Cr^{3+} was formed for the case of Cr^{6+}. Therefore, acid-driven chemical reduction of some inorganic contaminants would take place if an acidic condition prevails.

21.2.3.2 Reductive Dechlorination and Hydrodechlorination Reductive dechlorination is a common biological process that is used to describe certain types of biodegradation of chlorinated solvents in groundwater. Nevertheless, reductive dechlorination can be referred to as the degradation of chlorinated compounds or inorganic pollutants via the acceptance of electrons that have originated from any organic electron donors (e.g., molasses, whey, or vegetable oil) or inorganic electron donors (e.g., zero-valent metals or ferrous chloride) in the system. In this chapter, reductive dechlorination and hydrodechlorination by nanoiron and its bimetals will be explored.

Due to its small size and high reactivity, the use of nanoiron (i.e., nanoscale zerovalent iron) for *in situ* remediation of groundwater contaminants has received much attention in the past decade. Environmental contaminants that are rapidly reduced by nanoiron or bimetallic nanoparticles (e.g., Pd/Fe nanoparticles) include chlorinated organics (Zhang and Wang, 1997; Masciangioli and Zhang, 2003; Liu *et al.*, 2005; Quinn *et al.*, 2005; Li, Elliott, and Zhang, 2006; Saleh *et al.*, 2007), heavy metals (Ponder, Darab, and Mallouk, 2000; Kanel *et al.*, 2005), inorganic anions (Choe *et al.*, 2000; Mondal, Jegadeesan, and Lalvani, 2004; Yang and Lee, 2005), and other contaminants (Zhang, 2003; Joo *et al.*, 2005; Nurmi *et al.*, 2005). Basically, the reduction reactions promoted by zero-valent iron (ZVI) are surface mediated. The high reactivity of nanoiron has been attributed to its large specific surface area.

Primarily, reductive dechlorination occurs on the surface of ZVI. According to Matheson and Tratnyek (1994), the reaction mechanisms and pathways of reductive dehalogenation of chlorinated compounds by iron metal can be summarized as follows: First, chlorinated compound(s) (designated RCl) transport to the surface of ZVI via mass transfer. Owing to the oxidation of ZVI, electrons are released to the system. By accepting the electrons, dechlorination of RCl takes place. During this reaction process, Fe^{2+} is generated as a result of Fe^0 oxidation. Meanwhile, Cl^- forms due to dechlorination. Second, further oxidation of Fe^{2+} to Fe^{3+} releases one more electron. This leads to a further formation of Cl⁻ through dechlorination. Third, the corrosion of ZVI in aqueous solution will generate hydrogen gas. Upon the catalyzation by a suitable microorganism or precious metal, hydrogenolysis would take place between $H_{2(g)}$ and RCl. Here $H_{2(g)}$ plays the role of electron donor, which results in further dechlorination of RCl. The relevant reactions are shown below:

$$Fe^0 \rightarrow Fe^{2+} + 2e^- \qquad (21.22)$$

$$RCl + H^+ + 2e^- \rightarrow RH + Cl^- \qquad (21.23)$$

$$Fe^0 + RCl + H^+ \rightarrow Fe^{2+} + RH + Cl^- \qquad (21.24)$$

Hydrodechlorination is the reaction between $H_{2(g)}$ and an organic compound that contains C-Cl bond(s). During the reaction, chlorine is removed from the chlorinated compound as the HCl product, and the original C-Cl bond is replaced with a new C-H bond as shown below:

$$R\text{-}Cl + H_{2(g)} \leftrightarrow R\text{-}H + HCl \qquad (21.25)$$

Usually, hydrodechlorination is performed in the presence of noble metals (e.g., Pd, Pt, and Rh). Different metals will lead to different hydrodechlorination reaction activity and selectivity. In many instances, Pd is found to have the highest hydrodechlorination reaction activity.

Graham and Jovanovic (1999) pointed out that the degradation of chlorinated compounds by reacting with Pd/Fe bimetal includes the following reactions: (i) surface reactions, (ii) solution reactions, and (iii) a dechlorination reaction. The relevant schematic diagram (see Fig. 21.1) and reaction equations are given as follows:

I. Surface reactions

$$Fe^0 \longrightarrow Fe^{2+} + 2e^- \qquad (21.26)$$

$$2H^+ + 2e^- \xrightarrow{Fe} H_{2(g)} \qquad (21.27)$$

$$H^+ + e^- \xrightarrow{Pd} H^* \qquad (21.28)$$

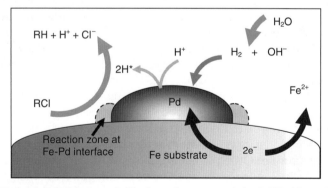

Figure 21.1. Hydrodechlorination of chlorinated compounds by Pd/Fe bimetal (Yang, Hung, and Tu, 2005).

II. Solution reactions

$$2H_2O \rightarrow 2H^+ + 2OH^- \qquad (21.29)$$

$$2HCl \rightarrow 2H^+ + 2Cl^- \qquad (21.30)$$

III. Dechlorination reaction

$$2H^* + R\text{-}Cl \rightarrow R\text{-}H + HCl \qquad (21.31)$$

IV. Overall dechlorination reaction

$$Fe^0 + R\text{-}Cl + H^+ \rightarrow R\text{-}H + Fe^{2+} + Cl^- \qquad (21.32)$$

21.2.4 Fundamentals of Electrochemical Oxidation/Reduction Processes

In recent years, electrochemistry applications for environmental pollution abatement have received much attention (Jüttner, Galla, and Schmieder, 2000; Chen, 2004). In general, the removal and destruction of pollutant species can be carried out directly or indirectly by electrochemical oxidation/reduction processes in an electrochemical cell without a continuous feed of redox chemicals. Besides, the high selectivity of many electrochemical processes helps to prevent the production of unwanted by-products, which in many cases have to be properly treated before disposal. Jüttner, Galla, and Schmieder (2000) noted that the application of electrochemical technologies for wastewater treatment generally can provide the following advantages: versatility, energy efficiency, amenability to automation, and cost effectiveness. Despite these advantages, electrochemical technologies also have limitations (e.g., electrochemical processes are heterogeneous in nature; the long-term stability and activity of the electrode material and cell components).

21.2.4.1 The Classical Electrochemical Principle

21.2.4.1.1 Electrochemical Oxidation Process. An electrochemical oxidation/destruction treatment of various pollutants can be subdivided into two important categories: direct oxidation at the anode, and indirect oxidation using appropriate anodically formed oxidants (Jüttner, Galla, and Schmieder, 2000). A critical review on electrochemical oxidation of organic pollutants for the wastewater treatment can be found elsewhere (Martínez-Huitle and Ferro, 2006).

Indirect electro-oxidation of pollutants in aqueous solution can be fulfilled in different ways: (i) through the use of anodically generated chlorine and hypochlorite (Allen, Khader, and Bino, 1995; Abuzaid *et al.*, 1999); (ii) through the use of the electrochemically generated hydrogen peroxide (Matsue, Fujihira, and Osa, 1981; Brillas, Sauleda, and Casado, 1998); and (iii) through the use of electrically generated ozone (El-Shal *et al.*, 1991). Mediated electro-oxidation is another kind of electro-oxidation proposed by Farmer *et al.* (1992) using Ag^{2+}, Co^{3+}, Fe^{3+}, Ce^{4+} and Ni^{2+} as mediators.

Direct anodic oxidation, the other type of electro-oxidation of pollutants, is to generate physically adsorbed active oxygen (i.e., adsorbed ·OH) or chemisorbed active oxygen (i.e., oxygen in the oxide lattice, MO_{x+1}) directly on anodes (Comninellis, 1994). Such physically adsorbed active oxygen would cause the complete combustion of organic compounds, whereas the chemisorbed active oxygen would participate in the formation of selective oxidation products within electrocatalytic O-transfer mechanisms (Feng and Johnson, 1990). In general, HO· is more effective for pollutant oxidation than O in MO_{x+1}.

21.2.4.1.2 Electrochemical Reduction Process. Electrochemical reduction has long been practiced for the removal of metal ions in aqueous solution through metal deposition onto the surface of the cathode. Electroreduction for the transformation of chlorinated compounds to their parent compounds had been verified more than three decades ago by Farwell, Beland, and Geer (1975). Electrochemical reduction can also be categorized into direct electroreduction and indirect electroreduction.

As direct electro-oxidation, direct electroreduction can be fulfilled through direct electrolysis of a substrate (e.g., chlorobenzenes or phenols) in a solvent-supporting electrolyte (e.g., tetraethylammounium bromide in methanol) at an electrode (e.g., graphite) surface (Farwell, Beland, Geer, 1975).

During the period 1996–2000, the Voss group at the University of Hamburg, Germany successfully established indirect electrolysis with mediators (e.g., nickel complexes Ni(cyclam)Cl$_2$ and Ni(bipy)Cl$_2$) as a method of detoxification (http://www.chemie.uni-hamburg.de/oc/voss/e_elek.htm; Nünnecke, 2000). The scheme and principle of indirect electroreduction proposed by the Voss group is given in Figure 21.2. Chlorinated naphthalenes, dibenzofurans and dibenzo-*p*-dioxins with a low degree of chlorination in contaminated soils could be dehalogenated to the unsubstituted compounds by applying indirect electroreduction (Voss *et al.*, 2001).

21.2.4.2 The Microconductor Principle Aside from the conventional concept of EK phenomena, a German research group adapted the concept of an electro-

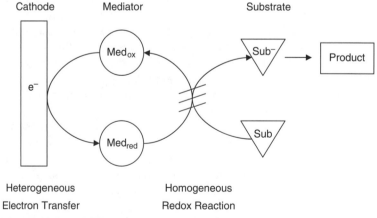

Figure 21.2. Principle of indirect electroreduction (redrawn from the figure given at http://www.chemie.uni-hamburg.de/oc/voss/mem_dn2.htm).

chemical solid bed reactor to the soil matrix in order to suggest a new reaction principle for the electrochemically induced remediation of contaminated soils (Rahner, Ludwig, and Röhrs, 2002; Röhrs, Ludwig, and Rahner, 2002). In this principle, the wet soil matrix acts in its sum as an ionic conductor. Only in the presence of such "microconductors" will an electrochemical reaction inside the soil matrix be induced by an external electric field. These reactions may occur parallel to the EK phenomena. The microconductor should itself be a redox system suitable for an electron transfer. Microconductors possess large numbers of electrochemical microreaction sites. They may act as monopolar or bipolar microelectrodes that allow for the occurrence of redox processes. Through the application of an electric field, mobile species (e.g., cations, anions, and dissolved contaminants) in solution are transported by electromigration and electro-osmosis to the electrodes (i.e., the principle of EK remediation). On the other hand, inactive particles in the system remain unchanged with respect to any electrochemical activity. Their surface charge may be changed by the recharging of their double layer capacity. Particles with electronic conductivity (microconductors, conducting films, or inactive particles covered by conducting films) are polarized in the electric field and may act as microelectrodes for electrochemical induced reactions. The nature and the chemical composition of the microconductor, together with redox systems in the vicinity of the microconductor particle, determine the electrochemical and catalytic activity of the proposed microreaction sites.

21.3 REPRESENTATIVE STUDIES

Many studies had been conducted using the integration of the EK process with several other technologies. However, only some selected ones will be introduced here.

21.3.1 EK–Fenton Process and EK–Fenton–Another Process

Since 1995, the present author's group has conducted research using a technological combination of the EK process with other processes (e.g., the Fenton process). The very first document in this regard was released by Long (1997). That research was to evaluate the feasibility of using a combined technology of EK soil processing and Fenton process for *in situ* remediation of a sandy loam (i.e., Soil No. 1) and a silt loam (i.e., Soil No. 2) contaminated by phenol and 4-chlorophenol, respectively. An electric potential gradient of 1 V/cm and deionized water in the cathode reservoir were employed in all experiments with EKs. In the Fenton process, scrap iron powder and $FeSO_4$ were used as catalysts. When scrap iron powder was placed in the soil column at a distance of 5 cm from the anode, 0.3% H_2O_2 was added to the anode reservoir. The treatment time was 10 days. When $FeSO_4$ solution was used, on the other hand, different concentrations of $FeSO_4$ were provided in the anode reservoir. After a 3-day EK reaction, 0.3% H_2O_2 was used to replace $FeSO_4$ solution in the anode reservoir. The treatment continued further for additional 7–10 days. In that study, the directions of all electro-osmotic (EO) flows were found to be positive (i.e., from the anode toward the cathode). Based on the mass balances for target pollutants, different reaction mechanisms were determined. When scrap iron powder was used as a catalyst, "removal" and "destruction" were found to be the dominant reaction mechanisms for Soils No. 1 and No. 2, respectively. On the other hand, when the $FeSO_4$ solution was employed, mechanisms of "destruction" and "removal" were found to be associated with concentrations higher and less than 0.0196 M, respectively. The results of that study also showed that by adding 1.05 g of scrap iron powder as a catalyst, the highest (i.e., 99.7%) destruction and removal efficiency (DRE) of phenol was obtained for Soil No. 1. However, under the same treatment conditions, a DRE of only 36% was obtained for Soil No. 1 contaminated by 4-chlorophenol. In the case of Soil No. 1 contaminated by phenol, it was found that an increase of the amount of scrap iron powder placed in the soil column would delay the arrival of hydrogen peroxide or hydroxyl radicals to the cathode reservoir. This can be verified by comparing the cumulative increased mass of phenol in the cathode reservoir for various amounts of scrap iron powder used. When 0.0196 M $FeSO_4$ was used as a catalyst, the highest DREs for Soil No. 1 contaminated by phenol and 4-chlorophenol were found to be 99.5% and 89.0%, respectively. It was also found that the cumulative increased mass of pollutants in the cathode reservoir decreased after 3–4 days of reaction. When a concentration of $FeSO_4$ less than 0.0196 M was used, the cumulative increased mass of pollutants in the cathode reservoir was found to gradually increase throughout the treatment period. By using the same operating conditions, a much lower DRE was obtained for Soil No. 2 regardless of the pollutant type. It was postulated that a much higher content of the organic matter associated with Soil No. 2 would be responsible for the finding. It was also found that the capabilities of removing organic pollutants in soil would be about the same for the EK process alone and the technology combined EK and Fenton processes. However, the latter was found to be superior to the former in terms of chemical destruction of organic pollutants. Parts of the findings of that work can be found elsewhere (Yang and Long, 1998, 1999a, 1999b).

Since the appearance of the very first journal paper on the EK–Fenton process by Yang and Long (1999b), many researchers around the world have devoted their efforts to studying the different aspects of this hybrid process, including the treatment of various contaminants. A few selected studies are presented in the following paragraphs for the reader's reference.

Yang and Liu (2001) reported a study on remediation of trichloroethylene (TCE) contaminated soils by an *in situ* EK–Fenton process. The treatment performance and cost analysis of the *in situ* EK–Fenton process for the oxidation of TCE in soils were evaluated in that work. In all experiments, an electric potential gradient of 1 V/cm, deionized water as the cathode reservoir fluid and a treatment time of 10 days were employed. Treatment efficiencies of TCE were evaluated in terms of the electrode material, soil type, catalyst type, and catalyst dosage and granular size if applicable. Test results showed that graphite electrodes were superior to stainless steel electrodes. It was found that the soil with a higher content of organic matter would result in lower treatment efficiency (e.g., a sandy loam is less efficient than a loamy sand). Experimental results showed that the type of catalyst and its dosage would markedly affect the reaction mechanisms (i.e., "destruction" and "removal") and the treatment efficiency. Aside from $FeSO_4$, scrap iron powder (SIP) in the form of a permeable reactive barrier was also found to be an effective catalyst for a Fenton-like reaction to oxidize TCE. In general, the smaller the granular size of the SIP, the lower the "overall treatment efficiency (= removal efficiency +

destruction efficiency)" and the greater the destruction efficiency. When a greater quantity of SIP was used, a decrease of the overall treatment efficiency and an increase of the percent destruction of TCE were found. Experimental results have shown that the quantity of the EO flow decreased as the quantity of SIP increased. It has been verified that the treatment performances are closely related to the corresponding EO permeability. Results of the cost analysis have indicated that the EK–Fenton process employed in this work is very cost effective with respect to TCE destruction.

A further study in a sand box for the treatment of phenol-contaminated soils by a combined EK–Fenton process had also been conducted by the present author's group (Chen, 2002). The purpose of that study was to evaluate the treatment efficiency of phenol-contaminated soils by the EK process conducted in sand boxes (60 cm × 30 cm × 30 cm; L × W × H). The electric potential gradient, electrode polarity reverse, and Fenton's reagent were employed as the experimental factors in that study to assess the variations of soil characteristics, potential difference, and residual phenol concentration distribution during a treatment period of 20 days and after the treatment. It was found that the anode reservoir pH decreased to around 2 and the cathode reservoir pH increased to approximately 12 after 2–3 days of treatment in the no-electrode-polarity-reverse system. However, the variation of pH in the anode and cathode reservoirs was less obvious in the case with electrode polarity reverse. A general trend of a lower pH in the anode reservoir and a higher pH in the cathode reservoir was found regardless of whether a constant potential system or a constant current system was employed. The acid front generated at the anode reservoir flushed across the soil specimen toward the cathode and the base front advanced toward the anode. However, in the central region of the sand box, unsaturated and saturated soil specimens maintained neutrality. For EK or EK–Fenton experiments, under constant potential conditions, the potential difference relative to the cathode versus the distance from the anode was found to have a linear relationship at the beginning of the electrical potential application. As the treatment time elapsed, the potential gradient became nonlinear (see Fig. 21.3). Nevertheless,

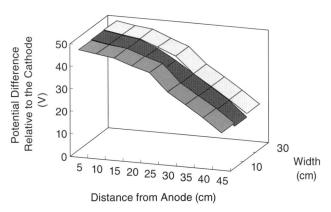

Figure 21.3. The distribution of electric potential difference across the soil matrix in a sand-box scale remediation of phenol-contaminated soil by the EK–Fenton process after 20 days of reaction.

there was no remarkable potential gradient change in the case with electrode polarity reverse. Although capillarity has resulted in an increase of the moisture content of unsaturated soil (from 25.34% to 30% after 20 days), EO flow was not obvious in the unsaturated zone. Experiments with electrode polarity reverse had a much greater EO flow quantity. The corresponding EO permeability coefficients for the constant potential and constant current systems were $6.42 \times 10^{-6}\,cm^2/V \cdot s$ and $9.47 \times 10^{-6}\,cm^2/V \cdot s$, respectively. It was also found that the existence of contaminants did reduce the EO flow quantity. Regardless of the employment of a constant potential or constant current system, the destruction efficiency of phenol was less obvious than the removal efficiency in the electrode-polarity-reverse system. In addition, a frequent reverse of electrode polarity also resulted in a frequent change of EO flow direction. Thus, a flow hysteresis of phenol in the soil compartment was found.

In fact, even earlier, the present author's group (Chen, 2000) had already released findings of a study on *in situ* treatment of pentachlorophenol- (PCP) contaminated soils by the EK–Fenton process and the EK–Fenton process combined with biodegradation. That research was to evaluate the treatment efficiency for *in situ* treatment of PCP-contaminated soil by the EK–Fenton process combined with biodegradation. An electric potential gradient of 1 V/cm and graphite electrodes were employed in all experiments. Soil types, catalyst types and dosage, hydrogen peroxide concentration, cathode reservoir liquid species, and reaction time were employed as the experimental factors. In that study, a prolonged reaction time of the EK–Fenton process could promote the DRE of target pollutant from soil. By using 0.0196 M $FeSO_4$ with 3.5% H_2O_2, the DRE was only 2% lower than 0.098 M $FeSO_4$ with 3.5% H_2O_2. It showed that using 0.0196 M $FeSO_4$ could provide enough Fe^{2+} to react with H_2O_2. By increasing H_2O_2 concentration from 0.35% to 3.50%, a DRE rose from 68.34% to 79.77%. When iron powder was used as the catalyst, the residual PCP concentration in soil near the anode reservoir was lower than the case using 0.0196 M $FeSO_4$. The corresponding DRE of PCP was only 56.58% for the case using iron powder as compared with 68.34% for the case using 0.0196 M $FeSO_4$. As to the influence of soil types on the EK–Fenton process, the residual concentration of pollutants for Soil No. 2 (i.e., sandy loam) was higher than Soil No. 1 (i.e., loamy sand). A DRE of only 59.22% was obtained. A lower treatment efficiency was ascribed to a higher content of organic matter in the said soil that would consume a greater magnitude of hydroxyl radicals in the system. To determine the influence of different reservoir liquid species, in that study a 0.1 M acetic buffer solution was used as the cathode reservoir liquid with the goal of enhancing removal efficiency. However, the experimental results did not support this goal. From the experiment of the EK process combined with cometabolism, a treatment efficiency of only 25.67% was obtained, but by using an EK–Fenton process to first pretreat the pollutant within soil, an increased efficiency of biodegradation was found. Further, a prolonged treatment time could completely eliminate the target pollutant from soil. If the EK–Fenton process proceeded only with intrinsic iron minerals in Soil No. 1, a DRE of only 20% to 30% was obtained. In comparison, the DRE of Soil No. 2 was found to be higher because it had a higher content of iron in soil to produce more hydroxyl radicals for the destruction of the target pollutant.

The present author's group has also conducted a pilot study on the *in situ* treatment of a chlorinated-hydrocarbons-contaminated site by using the combined technologies of EK processing–Fenton process–catalytic iron wall (Hung, 2002). The L_9 orthogonal arrays of an experimental design were utilized to investigate the effects of four experimental factors (i.e., H_2O_2 concentration, size fraction of iron particles, mass of iron particles, and elapsed time) on the treatment efficiency. The experimental results were further subjected to the analysis of variance (ANOVA) and regular analysis. According to the ANOVA of the results of nine experiments conducted under an electric gradient of 1 V/cm, the H_2O_2 concentration, mass of iron particles, and treatment time were determined to be very significant parameters for the DRE of 1,1,2,2-tetrachloroethane (TeCA). In that system, the optimal conditions with respect to the DRE of TeCA would be 2% H_2O_2, 50–100 mesh iron, 0.2 wt % iron and a 20-day treatment time. Under optimal conditions, it was possible to obtain a DRE of 69.56%, and the total cost of scrap iron powder, hydrogen peroxide, and energy consumption was determined to be about US\$21.71/m^3. In addition to the constant voltage operation, a constant current operation was also employed in this study. The latter was found to be superior to the former in terms of EO flow quantity and DRE. Experimental results of soil column tests showed that TeCA was transformed into trichloroethylene (TCE); TCE could be regarded as a daughter product of TeCA degradation. Test results further showed that "destruction" dominated the DRE of TeCA, whereas "removal" played a much more important role in the DRE of another test. Based on the lab-scale test results, a 9-day pilot test using the same combined treatment technologies (i.e., EK–Fenton–catalytic iron wall) was carried out in a petrochemicals-contaminated site. Figure 21.4 shows the cross-sectional view of the setup used in the pilot test. The pilot test results were found to be very satisfactory. DREs of vinyl chloride, dichloroethane, and TCE were found to be >96%, >96%, and >94%, respectively, in the anode and cathode wells. The concentrations of TCE in both the anode and cathode wells were found to be lower than the regulatory threshold (i.e., 2 μg/l). Table 21.1

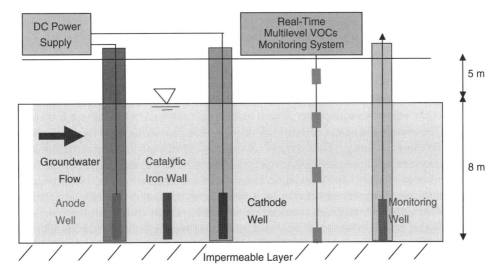

Figure 21.4. The cross-sectional view of the setup used in the pilot test for *in situ* remediation of a chlorinated-hydrocarbons-contaminated site by the combined technologies of EK processing–Fenton process–catalytic iron wall.

TABLE 21.1. Concentration Variations of Major Contaminants Detected in the Monitoring Well in an *In Situ* Pilot Test Using EK–Fenton–Catalytic Iron Wall Technology

Elapsed time of test	Beginning	1 day	2 days	3 days	8 days	9 days
VC (µg/l)	105.0	36.2	40.8	26.8	14.1	25.4
TeCA (µg/l)	33.0	15.1	15.0	12.3	6.6	9.6
TCE (µg/l)	97.0	19.2	21.2	14.4	7.3	12.9

Note: VC: vinyl chloride; TeCA: 1,1,2,2-tetrachloroethane; TCE: trichloroethylene.

shows the concentration variations of major contaminants detected in the monitoring well. Surprisingly, in that pilot test, the operating cost, limited to scrap iron powder, hydrogen peroxide, and electric power consumption, was determined to be as low as US$1.85/m^3.

Research was conducted by Kim, Kim, and Han (2005) to investigate the phenomenon and applicability of the EK–Fenton process for the remediation of low-permeability soil contaminated with polynuclear aromatic hydrocarbons (PAHs). The experimental results suggested that an increased H_2O_2 addition in the anode solution would result in the increase of anions (HO_2^- and $O_2 \cdot^-$) for subsequent chain reactions. The results of that research showed that the residual concentration of H_2O_2 was proportional to the transfer rate of the acid front and increase of electrolysis rate along with the transport of ionic composition in the soil. Furthermore, the phenanthrene degradation yield was proportional to the transfer rate of the acid front and H_2O_2 stability. It was concluded that an injection of acid is necessary (but to an extent that does not decrease the EO flow rate) to effectively treat the sorbed contaminants on soils during the EK–Fenton process.

Isosaari *et al.* (2007) conducted a study on the integration of electrokinetics and chemical oxidation for the remediation of creosote-contaminated clay. The benefits of this integrated process were evaluated in lab-scale experiments lasting 8 weeks. A voltage gradient of 48 V/m of direct current, 4.7 V/m of alternating current, and periodic additions of chemical oxidants were applied to creosote-contaminated soil. Electrokinetically enhanced oxidation with sodium persulfate resulted in better PAHs removal (35%) than either the EK process (24%) or persulfate oxidation (12%) alone. However, the improvement was shown only within one-third (5 cm) of the soil compartment. Isosaari and coworkers found that EKs did not improve the performance of Fenton oxidation. Experimental results showed that the addition of chemical oxidants (H_2O_2 and $Na_2S_2O_8$) resulted in more positive oxidation-reduction potential than the EK process alone. On the other hand, persulfate treatment impaired the EO flow rate. It was concluded that further optimization of an integrated remediation technology combining the beneficial effects of EKs, persulfate oxidation, and Fenton oxidation would be worth conducting.

Reddy and Karri (2008b) also reported research investigating Fenton-like oxidation coupled with EK remediation of low-permeability soils contaminated with both heavy metals and PAHs. That study examined the simultaneous oxidation of organic contaminants and removal of heavy metals. Kaolin spiked with nickel and phenanthrene each at a concentration of 500 mg/kg of dry soil to represent typical heavy metal and PAH contaminants found at contaminated sites. Four-week lab-scale EK experiments were conducted using a voltage gradient of 1 V/cm and rather high concentrations of H_2O_2 solution (i.e., 5%, 10%, 20%, and 30%) as the anolyte, with the use of deionized (DI) water as a control. Batch tests showed that phenanthrene oxidation increased from 76% to 87% when the H_2O_2 concentration increased from 5% to 30%. The EK experiments showed substantial EO flow in all the tests. Approximately 1 pore volume of flow was generated in the DI baseline test, whereas about 1.2–1.6 pore volumes were generated in case of H_2O_2 tests. Phenanthrene was partially oxidized in the H_2O_2 tests, and its removal from the soil was insignificant. The oxidation of phenanthrene increased with increasing concentrations of H_2O_2; a maximum of 56% oxidation was observed with 30% H_2O_2. Nickel migrated from the anode toward the cathode. This migration was more pronounced in the H_2O_2 tests as compared with the DI baseline test. As expected, nickel precipitated in all the tests near the cathode due to high pH conditions. These results showed that it is important to optimize H_2O_2/catalyst concentration and voltage gradient, as well as to control soil pH to enhance the removal of Ni and the oxidation of phenanthrene.

A study on the feasibility of the enhanced EK–Fenton process for the remediation of hexachlorobenzene (HCB) in a low-permeability soil was conducted by Oonnittan, Shrestha, and Sillanpää (2008). In that work, kaolin was spiked with HCB and treated by the EK and EK–Fenton processes. β-cyclodextrin was used to enhance the solubility of HCB in pore fluid. Two EK experiments were conducted to observe the suitability of β-cyclodextrin as a flushing solution for these processes. Another test conducted was the EK–Fenton test using β-cyclodextrin as an enhancing agent. Results showed that the type of flushing solution, system pH, electric current, and EO flow are of significance for HCB removal.

A Spanish research group also reported their research findings in remediation of phenanthrene-contaminated kaolinite by EK–Fenton technology (Alcántara et al., 2008b). The objective of that study was to evaluate the capability of using a combination of the EK and Fenton processes to decontaminate PAH-polluted soils using phenanthrene as the target contaminant. Kaolinite was spiked with phenanthrene at a concentration of 500 mg/kg. The EK process alone resulted in a negligible removal of phenanthrene from the kaolinite sample. Faster and more efficient degradation of this compound was enhanced by the introduction of a strong oxidant (e.g., HO·) into the soil. For this reason, further tests were carried out using H_2O_2 (10%) as the anolyte and catholyte, kaolinite polluted with iron as the model sample (pH ≈ 3.5), and a voltage gradient of 3 V/cm. An overall removal and destruction efficiency of 99% for phenanthrene was obtained in 14 days. It was concluded that the combined technology of the EK process and the Fenton reaction is capable of simultaneously removing and degrading of PAHs in polluted kaolinite.

21.3.2 EK–Electrochemical Processes

A new wastewater treatment technology using the integrated technology of EK and electro-oxidation (i.e., EK–EO process) was developed by Wang et al. (2004). The EK–EO process was found to take advantage of both electro-oxidation on the anode surface and the EK process of anionic impurities under an electric field, resulting in the enhancement of the total organic carbon (TOC) removal in the electrolysis process. Experimental results had shown that under an electric field an anionic azo dye Acid Red 14 (AR14) was transported into the anode compartment and then efficiently mineralized. After 360 min electrolysis of 100 mg/l AR14 solutions at 4.5 mA/cm^2, complete discoloration was observed in both the cathode and anode compartments. About 60% TOC was electromigrated from the cathode compartment to the anode compartment, in which more than 25 mg/l TOC was abated. The test results suggested that both electro-oxidation–electroreduction and electromigration pathways coexisted in EK–EO process: (i) electro-oxidation resulted in mineralization of organic pollutant; (ii) electromigration of AR14 might lead to comparatively high specific abatement efficiencies; and (iii) electroreduction partly contributed to decoloration of AR14 in the cathode compartment.

In an EK–electrochemical study by Cong, Ye, and Wu (2005), it was found that the system pH significantly affected the migration of contaminants in soil. The authors found that it was easier to remove the weak organic acids such as chlorophenols at higher pH values. However, partial degradation of chlorophenols occurred close to the electrodes. The contaminated soil can be remediated by the combination of EK process and electrochemical reaction. After 140 min of treatment, the removal efficiencies of chlorophenols could reach 85% at a voltage gradient of 1200 V/m, current of 10 mA, and pH equal to 9.8.

Mikkola et al. (2008) employed an approach of generating the necessary oxidizing agents (i.e., HO· and $S_2O_8^{2-}$) electrochemically within a treatment system. That study aimed at evaluating the feasibility of an innovative EK process for an *in situ* application against fuel-contaminated soil. However, no experimental data regarding the remediation of contaminants were reported in that preliminary work.

A rather different approach of combining both EK and electrochemical oxidation processes for the treatment purposes was studied by others. Pazos et al. (2007) reported a study on evaluating the feasibility of the combination of EK remediation and electrochemical oxidation for the remediation of polluted soil with organic compounds. The model kaolinite was spiked with an azo dye [i.e., Reactive Black 5 (RB5)]. The process consisted of two phenomena for the remediation of RB5-spiked kaolinite: soil treatment by the EK process and liquid treatment by electrochemical oxidation. RB5 (0.39 g dye/kg) could be effectively removed from the kaolinite matrix by the EK process under optimal operating conditions. Complete removal of RB5 was achieved using K_2SO_4 as the processing fluid (for enhanced desorption of RB5 from the kaolinite matrix) and a constant pH of 7 in the anode chamber. This would favor the alkalinization of the system and the ionization of RB5. It would further lead to its migration toward the anode chamber where RB5 could be oxidized electrochemically. It was also pointed out that under the optimal conditions the electric power consumption (56 kW/mg of removed dye) was 10 times lower as compared with the unenhanced EK process having no pH control in the electrolytes. Separate electrochemical decolorization tests of RB5 confirmed the effectiveness of K_2SO_4 in the efficiency of the process. A linear relationship between K_2SO_4 concentration and the decolorization rate was found. A nearly complete decolorization was achieved after 2 and 3 h of electrochemical treatment when the electrolyte concentrations were 0.1 and 0.01 M of K_2SO_4, respectively.

A similar study for the dye-contaminated kaolinite was also reported by Pazos, Cameselle, and Sanromán (2008). The selected target dye stuff was Lissamine Green B (LGB). Experimental results showed that unenhanced EK treatment did not result in any significant removal of LGB dye from the kaolinite sample. However, the use of Na_2HPO_4 in the processing fluid did increase the EO flux, improve desorption of dye from the surface of kaolinite particles, and prevent the acidification of the model sample. After 2 days, 94% of the dye was transported to the cathode chamber. The use of Na_2SO_4 in the processing fluid also improved the results as compared with the unenhanced experiment. Nevertheless, the removal of dye only reached 75% due to the lower EO flux. After the LGB dye was removed from kaolinite sample, the dye was decolorized through electrochemical treatment. Based on the dye concentration determined, Na_2HPO_4 was found to be capable of reaching high decolorization in a short treatment time.

Further, in a work by the same research group in Spain, a two-stage process combining soil EK remediation and liquid electrochemical oxidation for the remediation of polluted soil with organic compounds was developed and evaluated using phenanthrene-spiked kaolinite (Alcántara *et al.*, 2008a). Application of an unenhanced EK process resulted in negligible removal of phenanthrene from the kaolinite sample. The addition of cosolvents (e.g., ethanol) and electrolytes (e.g., Na_2SO_4) to the processing fluid used in the electrode chambers enhanced phenanthrene desorption from the kaolinite matrix and favored the EO flow. Near-complete removal of phenanthrene was achieved using ethanol and Na_2SO_4 in the processing fluid. Phenanthrene was transported to the cathode chamber where it was collected and later treated in a separate electrochemical cell by electrochemical oxidation. Complete degradation of phenanthrene was achieved after 9h using Na_2SO_4 as the anolyte.

21.3.3 EK–PRB Process and the Lasagna™ Process

The Lasagna™ process is an *in situ* remedial technology being developed by a consortium of Monsanto, DuPont, and General Electric in conjunction with the U.S. Environmental Protection Agency (EPA) to treat organic contamination in low-permeability soils (Ho *et al.*, 1995; Athmer, 2004). This process uses an applied direct current electric field to drive solvent-contaminated soil pore water by electro-osmosis through treatment zones installed in the contaminated soil between electrodes. Typically, the treatment zones are vertical zones composed of iron filings and kaolin clay. Actually, the Lasagna patent claims any use of the EK process with *in situ* treatment between the electrodes, including any permeable reactive barrier (PRB) placed between the electrodes or an injection of oxidants or reductants between electrodes. This patented process has been demonstrated for *in situ* remediation of solvent contaminated sites with the most well-known being the Paducah Gaseous Diffusion Plant in Paducah, Kentucky (US DOE, 2002).

Aside from the Lasagna™ process, other research was also conducted using the coupled technology of EKs and PRBs. The feasibility of using EKs coupled with a zero-valent iron treatment wall ("iron wall" for short) to abiotically remediate nitrate-contaminated soils was investigated by Chew and Zhang (1998). Upon completion of each test run, the contaminated soil specimen was sliced into five sections and analyzed for NO_3^-–N, NH_3–N, and NO_2^-–N. Nitrogen mass balance was used to determine the major transformation products. In control experiments where only EKs were used at various constant voltages, 25% to 37% of the nitrate–nitrogen was transformed. The amount of NO_3^-–N transformed improved when an iron wall (20g or about 8–10 wt %) was placed near the anode. For test runs at various constant voltages, the amount of nitrate–nitrogen transformed ranged from 54% to

87%. By switching to the operating mode of constant current, the amount of NO_3^-–N transformed was about 84% to 88%. The major transformation products were NH_3–N and NH_3. NO_2^-–N was less than 1% in all experimental runs. As expected, two localized pH conditions existed in the system: a low pH region near the anode and a high pH region near the cathode. Placing an iron wall near the anode increased the pH in that area as time elapsed. Migration of the acid front, however, did not flush across the cathode in the test runs. This study demonstrated that the EK–iron wall process could be used to remediate nitrate-contaminated groundwater. Similar research was also reported elsewhere by the same authors (Chew and Zhang, 1999).

Instead of using microsized zero-valent iron, Yang, Li, and Hung (2004) reported an investigation on the treatment of nitrates in the subsurface environment by a nanosized zero-valent iron wall enhanced by the EK process. In that study, a lab-scale treatment system that coupled EKs and PRBs made of nanosized (50–80 nm) or microsized (75–150 μm) iron was used to evaluate its capacity to remove nitrate from a soil matrix. Experimental results have indicated that the treatment capability of the nanosized iron is much superior to that of microsized one. More specifically, 20 g of microsized iron would be needed to obtain a nitrate degradation of 95.08%. However, only 2.5 g of nanosized iron was needed to attain a nitrate degradation of 98.06% for a treatment time of 6 days. Using the present experimental setup, it was found that the most suitable location for PRB installation was 5 cm from the anode reservoir. The treatment efficiency was found to increase as the mass of nanosized iron increased to an optimal mass; after that, efficiency began to drop. Experimental results also showed that nitrate degradation increased with the application of increasing electric potential gradient. A practice of polarity reversal in the course of EK–PRB processing has been proved to be beneficial to nitrate degradation. Table 21.2 summarizes all the test conditions and results.

Chang and Cheng (2006) reported a study on the remediation performance of a specific EK process integrated with zero-valent metals (ZVMs) for perchloroethylene- (PCE) contaminated soils. Various experimental conditions were controlled such as different voltage gradients, the position of the ZVMs, and the ZVM species (i.e., iron and zinc). The appropriate operational parameters were concluded as follows: (1) 0.01 M Na_2CO_3 as the working solution; (2) the voltage gradient of 1.0 V/cm; and (3) the ZVM wall near the anode. Based on the above operating conditions, the pH value of the working solution could maintain a neutral range so that the soil acidification was avoided. Neutral pH also resulted in a stable consumption of electricity during the test period. The removal efficiencies of PCE reached 99% and 90% for the pore-water and soil, respectively, after a 10-day treatment. It was found that Zn^0 outperformed Fe^0 in PCE degradation. Moreover, the soils treated by the EK process and ZVMs were found to maintain their original properties.

Yuan (2006) reported a study on the effect of Fe^0 on EK remediation of clay contaminated with PCE. That work investigated the effect of iron wall position and Fe^0 quantity on the remediation efficiency and EK performance of PCE-contaminated clay under an electric potential gradient of 2 V/cm for 5 days. The iron wall was composed of 2–16 g of Fe^0 mixed with Ottawa sand at a ratio of 1:2. Its positions were located at the anode, the middle, and the cathode end of the EK cell, respectively. Test results showed that a relatively higher remediation of 66% of PCE was found as the iron wall located at the cathode side, which corresponded to a factor

TABLE 21.2. A summary of Experimental Results for Nitrate-Contaminated Soil Treated by EK–Iron Wall

Test No.	Distance of iron wall (PRB) from the anode reservoir (cm)	Applied electric gradient (V/cm)	Treatment time (day)	Mass of nanosized iron in PRB (g)	Removal (%)	Degradation (%)	Overall treatment efficiency (%)
1	0	1.0	6	2.0	18.80	69.13	87.93
2	5	1.0	6	2.0	7.21	81.13	88.34
3	10	1.0	6	2.0	48.46	43.57	92.03
4	15	1.0	6	2.0	52.15	45.87	98.02
5	5	1.0	6	2.5	1.53	98.06	99.58
6	5	1.0	6	4.0	6.83	87.72	94.55
7	5	1.0	6	2.0	6.57	85.81	92.38
8	5	1.0	3	2.0	1.36	89.20	90.56
9	5	1.0	10	2.0	0.99	98.66	99.65
10	5	1.0	6	20.0 (microsized)	1.97	95.08	97.05
11	5	0.5	6	2.0	3.52	72.83	76.35
12	5	1.5	6	2.0	1.84	82.03	83.87
13	N/A	1.0	6	N/A	50.65	0.05	50.70
14	5	1.0	6	10.0 (microsized)	11.43	78.54	89.97
15	5	1.0	4	2.0	2.38	94.41	96.79
16	5	1.0	5	2.0	2.62	95.40	98.02

2.4 times greater than that in the EK system alone (27.0%). As the Fe^0 quantity increased to 16 g, the highest remediation efficiency of 90.7% was obtained. It was concluded that the PCE removal in the EK/Fe^0 system was dominated by the Fe^0 quantity rather than the iron wall position.

A study for remedial treatment of contaminated clayey soils by the EK process coupled with PRBs was also reported by Chung and Lee (2007). The objective of that research was to investigate the potential use of atomizing slag (a product cutoff converter slag generated by the atomizing method) as an inexpensive PRB material coupled with an EK process to remediate contaminated ground of low-permeability soils. Lab-scale remediation experiments using the EK process and the EK process with PRB remediation were conducted to evaluate the treatment capacity of atomizing slag. Experimental results showed the TCE concentrations of the effluent through the PRB material were much lower than those of EK remediation without atomizing slag. In general, the removal efficiencies for both TCE and Cd were about 90%. The removal rate of Cd from the soil specimen was higher than that of TCE owing to the additional effects of electromigration for its positive charge. Based on the preliminary test results, the coupled technology of EKs with PRBs could be effective in treating contaminated groundwater and soil *in situ*. Applying atomizing slag as a PRB reactive material was very promising for the sorption of Cd and the dechlorination of TCE, although the investigators claimed that some of the TCE passing through the PRB material would have been dechlorinated by the atomizing slag based on the increase of chloride concentrations. However, no relevant mechanism was provided to support that.

The EK process coupled with iron-wall technology for the treatment of soils contaminated by heavy metals was also reported by other researchers. Weng *et al.* (2007) investigated the effectiveness of incorporating ZVI into the EK process to remediate hyper-Cr(VI) contaminated clay (2497 mg/kg). A ZVI wall was installed in the center of the soil specimen and was filled with 1:1 (w/w) ratio of granular ZVI and sand. Results showed that the transport of H^+ was greatly retarded by the strong migration of anionic chromate ions in the opposite direction. As a result, a reversed EO flow was observed. Chromium removal was characterized by a high Cr(VI) concentration that ended up in the anolyte and the presence of Cr(III) precipitates in the catholyte. The Cr(VI) reduction efficiencies for the process without ZVI wall were 68.1% and 79.2% for 1 and 2 V/cm, respectively. As the ZVI wall was installed, the corresponding reduction efficiencies increased to 85.8% and 92.5%. The costs for energy and ZVI utilized in this process were calculated to be US\$41.0/cm^3 and US\$57.5/cm^3 for the system with electric potential gradient of 1 and 2 V/cm, respectively.

Yuan and Chiang (2007) reported a study on arsenic removal from soil by EK–PRB hybrid technology. Batch tests with PRB media of Fe^0 and FeOOH under an electric potential gradient of 2 V/cm for 5 days' duration were conducted to evaluate the removal mechanisms of arsenic. An enhancement of 1.6–2.2 times of arsenic removal was achieved when a PRB system was installed in the EK system. The best performance was found for the system with the FeOOH wall located in the middle of the EK cell. This was due largely to a higher surface area of FeOOH, and the migration of $HAsO_4^{2-}$ toward the anode (electromigration effect) was retarded by the EO flow. SEM-EDS (scanning electron microscopy coupled with energy dispersive X-ray spectrometry) results further confirmed that arsenic existed on the passive layer surface of PRB media. In comparison with the EK process alone, a greater removal of arsenic in EK–PRB systems might be ascribed to surface adsorption/precipitation of arsenic species on PRB media. As for the removal mechanisms, electromigration was considered to be predominant as compared with the EO flow. Test results also showed that surface adsorption and precipitation, respectively, were the principal removal mechanisms for acid environment near the anode and basic environment near the cathode.

21.3.4 EK–Ultrasonic Process

A combination of the EK process and ultrasonic technology has been tested for environmental treatment by Chung and Kamon (2005). The objective of that laboratory-scale investigation was to evaluate the coupled effect of EK and ultrasonic techniques for the extraction of ionic and nonionic matter from contaminated soils. A series of tests were conducted for the EK process alone and coupled EK–ultrasonic process. The main findings of that investigation were given as follows: (1) Water and the contaminant in porous soil media were allowed to flow and migrate under the actions of EO flow and electromigration by electric power and acoustic flow by ultrasonic waves for the coupled EK–ultrasonic process. (2) The accumulated outflow and contaminant removal rate were higher with the addition of vibration, cavitation, and sonication effects in the case of coupled EK–ultrasonic process than in the case of the EK process alone. (3) The accumulated quantity of outflow was 120 ml/h for EK process alone and 143 ml/h for coupled EK–ultrasonic test after 360 h, so the quantity of outflow increased with 19% due to the coupled effects of the EK and ultrasonic phenomena. (4) The final pH value across the soil specimen slightly decreased due to an increase of outflow and further advance of an acid front in the case of the enhancement test by the introduction of an ultrasonic wave to the electric field. (5) The removal rates of lead and phenanthrene were about 88% and 85% for EK alone test, respectively, and about 91% and 90% for the coupled EK–ultrasonic test. Finally, the authors even claimed that the new remedial technique combining EKs with the ultrasonic process could be effectively applied for the removal of ionic and nonionic contaminants in the subsurface.

A series of laboratory experiments involving simple, ultrasonic, EK, and EK–ultrasonic flushing tests for the treatment and removal of heavy metals and hydrocarbons from contaminated groundwater in sandy layers under a river bank were also carried out by Chung (2007). The test results showed that the EK–ultrasonic flushing technique is the most effective one for the removal of heavy metals and hydrocarbon from contaminated sandy layers. It was also found that the EK process is the most effective one to enhance the removal efficiency of heavy metals (e.g., Cd) from contaminated sandy soil under the river bank. On the other hand, the ultrasonic technique is the most effective one to enhance the removal efficiency of hydrocarbon contaminants (e.g., diesel fuel) from contaminated soil.

21.4 ELECTROKINETIC TREATMENT COUPLED WITH INJECTION OF NANOMATERIALS

21.4.1 Electrokinetic Delivery of Nanomaterials in Porous Media

The use of the delivery vehicle concept for environmental remediation has received much attention in recent years (Schrick *et al.*, 2004; Kanel and Choi, 2007;

Saleh et al., 2007). This concept employs a carrier that strongly binds and transports reactive reagents (e.g., nanoscale ZVI) to the location where chemical reactions needed. This technique can be particularly useful for contamination sites (e.g., under buildings) that cannot easily be reached by hydrofracturing techniques, or in urban areas where conventional excavation methods or PRB technology are not possible.

In recent years, several attempts have been made by researchers trying to study the feasibility of electrokinetically enhanced chemical delivery of nanoparticles in various porous matrixes including the subsurface environment. Of these, only limited studies used bare nanoiron, while the rest used surface modified nanoparticles to investigate the delivery behavior and relevant outcome.

Adams (2006) reported that the electrokinetically induced movement of bare nanoiron (i.e., nanoscale ZVI) was ineffective in penetrating the porous matrix under a hydraulic gradient, probably due to the nanoiron agglomerating to form particles that were too large to effectively migrate through the porous matrix.

Cardenas and Struble (2006) reported a study on the EK nanoparticle treatment of hardened cement paste for reduction of permeability. In that research, colloidal nanoparticles were electrokinetically transported into hardened cement paste pores where they underwent chemical reactions resulting in reduced permeability. Nanoparticles of SiO_2 (20 nm) and Al_2O_3 (2 nm) were combined with simulated pore fluids to assess precipitate production. One precipitate formed was C–S–H, the binder material native to portland cement paste. Permeability tests were conducted to study the effect of these processes on hardened cement pastes of high water/cement ratio and of high and low alkali contents. It was observed that a 5-min treatment using 5 V of potential applied over a span of 0.15 m (i.e., about 0.03 V/cm) was found to be sufficient to drive the concerned nanoparticles into the pore system, enabling them to react and form precipitates. The permeability coefficient for each paste was reduced by 1–3 orders of magnitude. It was concluded that reactive nanoparticles could be electrokinetically inserted to reduce the permeability of hardened cement paste, even in the presence of an opposing hydraulic flow.

Using the same concept, Cardenas et al. (2007) also reported some research findings on EK transport and the processing of bone repair agents in a porous substrate. In that investigation, calcium phosphate cements (CPCs) were used extensively for bone replacement because of their similarity to the mineral component of bone. The progression of changes in morphology of the CPCs observed by SEM indicated the evidence of bioactivity resulting from exposure to simulated body fluid. Teflon Millipore filters were used to capture suspended CPC particles as they were undergoing EK transport. The permeability of the filters was reduced by approximately a factor of 100 after 6 h of EK treatment. Weight measurement results showed that the treatment increased the mass of the filters by approximately 64%. That work demonstrated that bioceramic suspensions could be electrokinetically delivered onto a porous substrate, rapidly increasing the mass and reducing the permeability of the structure.

A study conducted by Pamukcu, Hannum, and Wittle (2008) demonstrated that the transport of polymer-coated dispersed nanoiron could be electrokinetically enhanced for subsurface remediation of tight clay soils. The polymer-coated iron nanoparticles possessed a positive zeta potential below pH 8.3 and remained suspended in solution rather than settling or agglomerating over time. The injection of nanoiron particles in the lab tests showed a positive shift in the oxidation–reduction potential (ORP) where the effectiveness of nanoiron as an environmental catalyst was demonstrated. The presence of nanoiron and an applied electric field together pushed the system ORP to higher positive values than the EK effects or nanoiron alone. The diffusion of nanoiron without the electrical field showed no activation of the iron, as indicated by little or no change in the ORP. These results showed that nanoiron was both transported and activated by the applied electric field in the test system.

In a study conducted by Yang, Tu, and Hung (2007), it was found that the transport distance of the 1 vol % PAA (polyacrylic acid)-modified nanoiron under typical groundwater conditions in the vertical loamy sand soil column was estimated to be about 0.25 m. For the horizontal column packed with simulated groundwater-saturated similar soil, the transport distance of the PAA-modified nanoiron under the influence of an external electric field was found to be 10 times greater regardless of potential adverse effects due to a high ionic strength of groundwater in the reaction system.

21.4.2 Injection of Nanosized Iron Slurry Coupled With the EK Process for Subsurface Remediation

Recently, EK delivery of nanoparticles in the subsurface for remediation purposes has received much attention from academia and industrial sectors. A US patent was issued to Yang (2008b) that details a method for treating a body of a polluted porous medium as follows: (1) prepare a reactive solution containing nanoparticles; (2) inject the reactive solution into the body of the polluted porous medium so as to decompose pollutants in the polluted porous medium by the reaction of the nanoparticles with the pollutants; and (3) apply an electric field to the body of the polluted porous medium so as to enhance the transporting effect of the nanoparticles in the body of the polluted porous medium.

In a study, Reddy and Karri (2007) investigated the potential *in situ* remediation of low-permeability soils contaminated with PCP by EK delivery of nanoscale iron particles (NIPs), which consist of an elemental iron core and a magnetite shell in approximately the same amounts by weight. Kaolin soil was artificially spiked with PCP (1000 mg/kg of dry soil). Lab-scale EK experiments were conducted using deionized water or NIPs 50–300 nm in size at slurry concentrations of 5 g/l and 10 g/l at the anode. All experiments were conducted for 427 h at a constant voltage gradient of 1 V/cm. Experimental results showed that substantial EO flow was induced initially and then decreased. The test results also showed the EO flow was not hindered by the NIPs. The total iron in the soil increased from the anode to the cathode, indicating that NIPs might have been transported toward the cathode. However, the transport of NIPs in the soil was limited by the aggregation and settling of NIPs in the anode. In addition, the NIPs might have transformed into Fe^{3+} ions under the oxygenated and low pH conditions that existed at the anode. Therefore, NIPs might not have contributed to PCP degradation. Instead, 47% to 55% of the PCP was degraded in the cathode chamber by reductive dechlorination in all tests. Complete PCP degradation did not occur in any of the tests because of limited transport of PCP into the cathode. It was concluded that for an EK system without NIPs to be effective for the remediation of PCP, enhanced transport of PCP into the cathode chamber and high EO flow are needed.

Reddy and Karri (2008a) further examined the enhanced delivery of nanoiron amended with surfactants or cosolvents under an electric field for the remediation of a low-permeable kaolin soil spiked with PCP (1000 mg PCP per kg of dry soil). Bench-scale EK experiments were conducted first by using bare nanoiron suspension (50–300 nm particle size; 5 g/l concentration) in the anode chamber and then applying an electric potential of 1 V/cm. Results revealed that only a limited amount of nanoiron could be transported into the soil due to aggregation and settlement as well as partial oxidation of nanoiron within the anode chamber. Additional experiments were conducted using nanoiron amended with a nonionic surfactant (5% Igepal CA720) or a cosolvent (5% ethanol) in the anode chamber and applying the same electric potential of 1 V/cm. Results showed that the EO flow was not significantly influenced by the amendment; however, the transport of nanoiron was limited similarly to the case of bare nanoiron. PCP was partially degraded in all of the experiments not because of the nanoiron, but mainly due to reductive dechlorination at the cathode. The extent of PCP reduction was slightly greater in the cosolvent system, possibly due to the enhanced solubilization and transport of PCP into the cathode chamber. It was concluded that prevention of aggregation, settlement, and oxidation of nanoiron is needed for effective EK delivery of nanoiron and remediation of PCP in low-permeability soils.

Very recently, the present author's group (Yang, 2008a) carried out similar research to simulate the treatment of subsurface TCE. First, nanoiron was prepared by the borohydride reduction method. During the growth of iron nuclei in the solution, a soluble starch (3 wt %) among other dispersants was determined to be the best to form the stabilized nanoiron slurry for later uses. A fluffy substance over the surface of nanoiron [as shown in a transmission electron microscopy (TEM) image] was confirmed to be starch by SEM mapping of elements on the surface of the nanoiron. Based on the preliminary test results, nanoiron slurry (nanorion dose: 1.25 g/l; soluble starch: 3 wt %) was verified for its capability to degrade TCE in soil and groundwater. In this series of tests, lab-prepared, saturated TCE-bearing soil was firmly packed in horizontal soil columns to simulate the groundwater flow in the subsurface. Soil specimens of sandy clay loam and sand were selected in order to compare the transport behaviors of nanoiron slurry in the simulated subsurface environment as driven by EKs. In the EK tests, 20 ml of nanoiron slurry was injected to the selected electrode (anode or cathode) reservoir daily with a constant applied electric potential gradient of 1 V/cm and for a reaction time of 7 days. The initial TCE concentration in the sandy clay loam was approximately 285 mg/kg, and about 68 mg/kg in sand. Experimental results showed that the anode reservoir is a better injection spot as compared with its counterpart. Under the circumstances, a TCE degradation efficiency of about 99% could be obtained. The EK tests were found to be capable of transporting nanoiron slurry in the simulated groundwater system, which resulted in a greater TCE removal. Moreover, the number of microorganism colonies in the soil was found to increase as a result of the injection of the nanoiron slurry. This might be ascribed to the fact that starch molecules surrounding iron nanoparticles provide a good carbon source for the growth of microorganism colonies in the neighborhood.

21.4.3 Injection of Nanosized Iron Emulsion Coupled With the EK Process for Subsurface Remediation

Recently, the present author's group (Chang, 2007) also conducted research to evaluate the treatment efficiency of a TCE-contaminated aqueous solution and soil by the combined technologies of the injection of emulsified nanoscale zero-valent iron slurry (ENZVIS) and the EK process. Nanoiron (30–50 nm) was synthesized using the chemical reduction method by industrial grade chemicals. The nanoiron emulsion was prepared by mixing two surfactants (Span 80 and Tween 40) with soybean oil to yield the ENZVIS. The degradation of TCE by ENZVIS under various operating parameters was carried out in batch experiments. These experimental results indicated that emulsified nanoiron outperformed nanoiron in TCE dechlorination rates. ENZVIS (0.75 g-Fe^0/l) degraded TCE (initial concentration of 10 mg/l) down to 45%. An increase of the oil dosage could improve the stability of the emulsion but yield a negative influence on the degradation of TCE. Experimental results also showed that ENZVIS could remove TCE up to 94% when pH equaled 6, and that a higher initial concentration TCE would result in a higher TCE removal efficiency. In addition, using ENZVIS to degrade TCE-contaminated artificial groundwater has indicated that nitrate and carbonate of groundwater will suppress nanoiron reaction with TCE. Especially, a high concentration of carbonate in the reaction system might form a passive film or precipitates on the nanoiron surface. This study further evaluated the treatment efficiency of coupled ENZVIS injection and the EK process in treating TCE-contaminated soil. Experimental conditions were given as follows: (1) initial TCE concentration in the range of 98~118 mg/kg; (2) an electric potential gradient of 1 V/cm; (3) a daily addition of 20 ml ENZVIS; and (4) a reaction time of 10 days. Experimental results have shown that the addition of ENZVIS to the anode reservoir of strongly acidic and oxidative environment would cause the nanoiron to corrode rapidly and decrease TCE removal efficiency. On the other hand, the addition of ENZVIS to the cathode reservoir would enhance the degradation of TCE therein. In summary, an addition of ENZVIS to the cathode reservoir would yield the best TCE removal efficiency.

21.4.4 Injection of Nanosized Palladium/Iron Slurry Coupled With the EK Process for Subsurface Remediation

21.4.4.1 Removal and Degradation of Trichloroethylene Yang and Chang (2006) carried out research to evaluate the treatment efficiency of a TCE-contaminated soil with the combined technologies of the injection of palladized nanoiron slurry (size: 50–80 nm; specific surface area: 100.61 m^2/g) and the EK process. An addition of 1 wt % of PAA during the nanoiron preparation stage was found to yield a good stabilization of palladized nanoiron. Previous studies have shown that the application of an electric field in the subsurface environment would yield an EO flow that enhanced the groundwater movement. Thus, the final stage of this study was to evaluate the treatment efficiency of combined technologies of the injection of the slurry of palladized nanoiron and the EK process in treating TCE-contaminated soil. Test conditions used were given as follows: (1) initial TCE concentration in soil: 160–181 mg/kg; (2) electric potential gradient: 1 V/cm; (3) daily addition of 20 ml of the palladized nanoiron (2.5 g/l) slurry to the electrode reservoir(s); and (4) reaction time: 6 days. Test results showed that the addition of palladized nanoiron slurry to the anode reservoir yielded the lowest residual TCE concentration in soil, about 92.5% removal of TCE from soil. On the other hand, the addition of palladized nanoiron slurry to the cathode reservoir enhanced the degradation of TCE therein. Based on the above findings, the treatment method employed in this work was proved to be a novel and efficient one for *in situ* remediation of TCE-contaminated soil.

21.4.4.2 Removal and Degradation of Nitrate In addition to the treatment of TCE in the subsurface using nanosized Pd/Fe slurry coupled with the EK process as indicated above, Yang, Hung, and Tu (2008) also used the same combined technology to remediate nitrate in a simulated system. To mimic the subsurface environment, a bench-scale horizontal column packed with nitrate-contaminated soil was treated by the EK process coupled with the injection of nanosized Pd/Fe slurry. The estimated transport distance of the Pd/Fe nanoparticles in soil based on the calculated sticking coefficient was about 9 m by the enhancement of electrokinetics, primarily by EO flow. The injection of the nanosized Pd/Fe slurry into the anode reservoir coupled with the application of an electric field would result in the greatest overall removal efficiency of nitrate in soil and reservoir fluids as compared with other injection positions, the cathode reservoir being the worst injection spot. Figure 21.5 illustrates the total mass of residual nitrate in different positions of the EK soil column for various tests. In summary, over 99% of nitrate in the whole system could be removed and degraded even if only 0.05 wt % of nanoscale Pd/Fe bimetal was injected into the anode reservoir of an EK system. Presumably, the driving force of the moving nitrate ion (i.e., ionic migration) and negatively charged PAA-modified nanoparticles (i.e., electrophoresis) toward the anode must be the predominant migration mechanisms. Chemical reduction of nitrate occurred mostly in the anode reservoir where nanosized Pd/Fe bimetal existed.

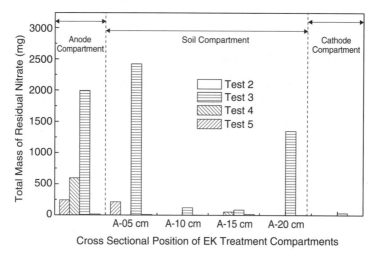

Figure 21.5. Distributions of the total mass of residual nitrate in the electrokinetic soil columns for different tests with different injection positions of nanosized Pd/Fe slurry.

21.5 PROSPECTIVE

This chapter reviews the principles of electrokinetic remediation (the EK process) and various chemical redox-related processes and introduces some selected studies on a variety of environmental contaminants treated by the EK process alone or by EK-coupled technologies. As indicated above, the EK process is a versatile technology. It has many advantages over other physicochemical technologies for *in situ* remediation of a variety of environmental contaminants. In particular, the EK process could be used in low hydraulic conductivity media and below buildings. Inevitably, the EK process per se has limitations.

An in-depth understanding of the concepts behind these EK-coupled chemical redox processes will help us to realize the merits and defects of these hybrid technologies as well as the EK process itself. How to further take advantages of the EK phenomena and turn their negative effects into positive ones through the integration with other technologies is worth exploring. As stated in a very famous lecture given by physicist Richard Feynman in 1959, this chapter's author also believes "There's plenty of room at the bottom" for EK research and applications.

REFERENCES

Abuzaid NS, Al-Hamouz Z, Bukhari AA, Essa MH. (1999). Electrochemical treatment of nitrite using stainless steel electrodes. *Water Air Soil Pollution* **109**:429–442.

Acar YB, Alshawabkeh AN. (1993). Principles of electrokinetic remediation. *Environmental Science and Technology* **27**(13):2638–2647.

Adams A. (2006). *Transport* of nanoscale zero valent iron using electrokinetic phenomena. http://www.sese.uwa.edu.au/__data/page/96395/Adams_2007.pdf. Accessed January 16, 2008.

Agrawal A, Tratnyek PG. (1996). Reduction of nitro aromatic compounds by zero-valent iron metal. *Environmental Science and Technology* **30**:153–160.

Alcántara T, Pazos M, Cameselle C, Sanromán MA. (2008a). Electrochemical remediation of phenanthrene from contaminated kaolinite. *Environmental Geochemistry and Health* **30**:89–94.

Alcántara T, Pazos M, Gouveia S, Cameselle C, Sanromán MA. (2008b). Remediation of phenanthrene from contaminated kaolinite by electroremediation-Fenton technology. *Journal of Environmental Science and Health Part A* **43**(8):901-906.

Allen SJ, Khader KYH, Bino M. (1995). Electrooxidation of dyestuffs in waste waters. *Journal of Chemical Technology and Biotechnology* **62**:111–117.

Anonymous. Electrolysis of chlorinated and polycyclic aromatics. http://www.chemie.uni-hamburg.de/oc/voss/e_elek.htm. Accessed May 12, 2008.

Athmer C. (2004). In-situ remediation of TCE in clayey soils. *Soil Sediment Contamination* **13**:381–390.

Barbeni M, Minero C, Plizzetti E, Borgarello E, Serpone N. (1987). Chemical degradation of chlorophenols with Fenton's reagent. *Chemosphere* **16**(10–12):2225–2237.

Bowers AR, Eckenfelder WW, Gaddipati P, Mosen RM. (1989). Treatment of toxic or refractory wastewater with hydrogen peroxide. *Water Science and Technology* **21**:447–486.

Brillas E, Sauleda R, Casado J. (1998). Degradation of 4-chlorophenol by anodic oxidation, electro-Fenton, photoelectro-Fenton, and peroxi-coagulation processes. *Journal of the Electrochemical Society* **145**:759–765.

Cardenas HE, Struble, LJ. (2006). Electrokinetic nanoparticle treatment of hardened cement paste for reduction of permeability. *Journal of Materials in Civil Engineering* Jul/Aug: 546–560.

Cardenas HE, Vasam SS, Zhao Y, Morishetti D. (2007). Electrokinetic transport and processing of bone repair agents. *Sixth Symposium on Electrokinetic Remediation (EREM 2007), Book of Abstracts*, June 12–15, 2007, Vigo, Spain, p. 55.

Chang YI. (2007). Treatment of Trichloroethylene in Aqueous Solution Using Nanoscale Zero-Valent Iron Emulsion. MS Thesis, National Sun Yat-Sen University, Kaohsiung, Taiwan (in Chinese).

Chang JH, Cheng SF. (2006). The remediation performance of a specific electrokinetics integrated with zero-valent metals for perchloroethylene contaminated soils. *Journal of Hazardous Materials* **131**(1–3):153–162.

Chen G. (2004). Electrochemical technologies in wastewater treatment. *Separation and Purification Technology* **38**(1):11–41.

Chen CD. (2000). A Study on In-Situ Treatment of PCP Contaminated Soils by Electrokinetics-Fenton Process Combined With Biodegradation. MS Thesis, National Sun Yat-Sen University, Kaohsiung, Taiwan (in Chinese).

Chen YH. (2008). Reaction Behavior of Nanoscale Fe_3O_4 and $[Fe_3O_4]MgO$ with Different Inorganic Pollutants (NO_3^-, Cd^{2+} and Cr^{6+}) in Simulated Groundwater. MS Thesis, National Sun Yat-Sen University, Kaohsiung, Taiwan (in Chinese).

Chen YS. (2002). Treatment of Phenol-Contaminated Soils by Combined Electrokinetic-Fenton Process. MS Thesis, National Sun Yat-Sen University, Kaohsiung, Taiwan (in Chinese).

Chen G, Hoag GE, Chedda P, Nadimb F, Woody BA, Dobbs GM. (2001). The mechanism and applicability of in situ oxidation of trichloroethylene with Fenton's reagent. *Journal of Hazardous Materials* **87**(1–3):171–186.

Chew CF, Zhang TC. (1998). In-situ remediation of nitrate-contaminated ground water by electrokinetics/iron wall processes. *Water Science and Technology* **38**(7):135–142.

Chew CF, Zhang TC. (1999). Abiotic degradation of nitrates using zero-valent iron and electrokinetic processes. *Environmental Engineering Science* **16**(5):389–401.

Choe S, Chang YY, Hwang Y, Khim J. (2000). Kinetics of reduction denitrification by nanoscale zero-valent iron. *Chemosphere* **41**:1307–1311.

Chung HI. (2007). Treatment of contaminated groundwater in sandy layer under river bank by electrokinetic and ultrasonic technology. *Water Science and Technology* **55**(1–2): 329–338.

Chung HI, Kamon M. (2005). Ultrasonically enhanced electrokinetic remediation for removal of Pb and phenanthrene in contaminated soils. *Engineering Geology* **77**(3–4): 233–242.

Chung HI, Lee MH. (2007). A new method for remedial treatment of contaminated clayey soils by electrokinetics coupled with permeable reactive barriers. *Electrochimica Acta* **52**(10):3427–3431.

Comninellis C. (1994). Electrocatalysis in the electrochemical conversion/combustion of organic pollutants for waste water treatment. *Electrochimica Acta* **39**:1857–1862.

Cong Y, Ye Q, Wu Z. (2005). Electrokinetic behaviour of chlorinated phenols in soil and their electrochemical degradation. *Process Safety and Environmental Protection* **83**(B2): 178–183.

El-Shal W, Khordagui H, El-Sebaie O, El-Sharkawi F, Sedahmed GH. (1991). Electrochemical generation of ozone for water treatment using a cell operating under natural convection. *Desalination* **99**:149–157.

Farmer JC, Wang FT, Hawley-Fedder RA, Lewis PR, Summers LJ, Foiles L. (1992). Electrochemical treatment of mixed and hazardous wastes: oxidation of ethylene glycol and benzene by silver (II). *Journal of the Electrochemical Society* **139**:654–662.

Farwell SO, Beland FA, Geer RD. (1975). Reduction pathways of organohalogen compounds. Part I. Chlorinated benzenes. *Journal of Electroanalytical Chemistry* **61**(3): 3303–313.

Feng J, Johnson DC. (1990). Electrocatalysis of anodic oxygen-transfer reactions: Fe-doped beta-lead dioxide electrodeposited on noble metals. *Journal of the Electrochemical Society* **137**(2):507–510.

Flis J. (1991). Stress corrosion cracking of structural steels in nitrate solutions. In *Corrosion of Metals and Hydrogen-Related Phenomena*. Materials Science Monograph, Vol. **59** (ed. J Flis). Amsterdam: Elsevier, pp. 57–94.

Graham LJ, Jovanovic G. (1999). Dechlorination of p-chlorophenol on a Pd/Fe catalyst in a magnetically stabilized fluidized bed; implications for sludge and liquid remediation. *Chemical Engineering Science* **54**:3085–3093.

Greenberg RS, Andrews T, Kakarla PKC, Watts RJ. (1998). In-situ Fenton-like oxidation of volatile organics: laboratory, pilot, and full-scale demonstration. *Remediation* **8**(2): 29–42.

Ho SV, Sheridan PW, Athmer CJ, Heitkamp MA, Brackin JM, Weber D, Brodsky PH. (1995). Integrated in situ soil remediation technology: the Lasagna process. *Environmental Science and Technology* **29**(10):2528–2534.

Huang CP, Wang HW, Chiu PC. (1998). Nitrate reduction by metallic iron. *Water Research* **32**:2257–2264.

Hung YC. (2002). Pilot-Scale in Situ Treatment of a Chlorinated Hydrocarbons Contaminated Site by Combined Technologies of Electrokinetic Processing-Fenton Process-Catalytic Iron Wall. MS Thesis, National Sun Yat-Sen University, Kaohsiung, Taiwan.

Ince NH, Tezcanli G, Belen RK, Apikyan IG. (2001). Ultrasound as a catalyzer of aqueous reaction systems-the state of the art and environmental applications. *Applied Catalysis B: Environmental* **29**:167–176.

Isosaari P, Piskonen R, Ojala P, Voipio S, Eilola K, Lehmus E, Itävaara M. (2007). Integration of electrokinetics and chemical oxidation for the remediation of creosote-contaminated clay. *Journal of Hazardous Materials* **144**:538–548.

Joo SH, Feitz AJ, Sedlak DL, Waite TD. (2005). Quantification of the oxidizing capacity of nanoparticulate zero-valent iron. *Environmental Science and Technology* **39**(5): 1263–1268.

Jüttner K, Galla U, Schmieder H. (2000). Electrochemical approaches to environmental problems in the process industry. *Electrochimica Acta* **45**(15–16):2575–2594.

Kanel SR, Choi H. (2007). Transport characteristics of surface-modified nanoscale zero-valent iron in porous media. *Water Science and Technology* **55**(1–2):157–162.

Kanel SR, Manning B, Charlet L, Choi H. (2005). Removal of arsenic(III) from groundwater by nano scale zero-valent iron. *Environmental Science and Technology* **39**:1291–1298.

Kidak R, Ince NH. (2006). Effects of operating parameters on sonochemical decomposition of phenol. *Journal of Hazardous Materials* **137**:1453–1457.

Kim ILK, Huang CP, Chiu PC. (2001). Sonochemical decomposition of dibenzothiophene in aqueous solution. *Water Research* **35**:4370–4378.

Kim SS, Kim JH, Han SJ. (2005). Application of the electrokinetic-Fenton process for the remediation of kaolinite contaminated with phenanthrene. *Journal of Hazardous Materials* **118**:121–131.

Kim WS, Kim SO, Kim KW. (2005). Enhanced electrokinetic extraction of heavy metals from soils assisted by ion exchange membranes. *Journal of Hazardous Materials* **118**(1–3): 93–102.

Li XQ, Elliott DW, Zhang WX. (2006), Zero-valent iron nanoparticles for abatement of environmental pollutants: materials and engineering aspects. *Critical Reviews in Solid State and Materials Sciences* **31**:111–122.

Liu Y, Majetich, SA, Tilton RD, Sholl DS, Lowry GV. (2005). TCE dechlorination rates, pathways, and efficiency of nanoscale iron particles with different properties. *Environmental Science and Technology* **39**(5):1338–1345.

Long YW. (1997). A Study on Treatment of Phenol and 4-Chlorophenol Contaminated Soils by Electrokinetics-Fenton Process. MS Thesis, National Sun Yat-Sen University, Kaohsiung, Taiwan (in Chinese).

Manousaki E, Psillakis E, Kalogerakis N, Mantzavinos D. (2004). Degradation of sodium dodecylbenzene sulfonate in water by ultrasonic irradiation. *Water Research* **38**:3751–3759.

Martínez-Huitle CA, Ferro S. (2006). Electrochemical oxidation of organic pollutants for the wastewater treatment: direct and indirect processes. *Chemical Society Reviews* **35**: 1324–1340.

Masciangioli T, Zhang WX. (2003). Environmental technology at the nanoscale. *Environmental Science and Technology* **37**:102A–108A.

Matheson LJ, Tratnyek PG. (1994). Reductive dehalogenation of chlorinated methanes by iron metal. *Environmental Science and Technology* **28**(12):2045–2053.

Matsue T, Fujihira M, Osa T. (1981). Oxidation of alkylbenzenes by electrogenerated hydroxyl radical. *Journal of the Electrochemical Society* **128**:2565–2569.

Mikkola H, Schmale JY, Wesner W, Petkovska S. (2008). Laboratory pre-assays for soil remediation by electro synthesis of oxidants and their electrokinetic distribution. *Journal of Environmental Science and Health Part A* **43**(8):907–912.

Mondal K, Jegadeesan G, Lalvani SB. (2004). Removal of selenate by Fe and NiFe ultrafine particles. *Industrial & Engineering Chemistry* **43**(16):4922–4934.

Nünnecke D. (2000). Indirect Electroreduction of Chloroarenes in Methanol Mediated by Nickel Complexes. PhD Dissertation, University of Hamburg, Hamburg, Germany (in German).

Nurmi JT, Tratnyek PG, Sarathy V, Baer DR, Amonette JE, Pecher K, Wang C, Linehan JC, Matson DW, Penn RL, Driessen MD. (2005). Characterization and properties of metallic iron nanoparticles: spectroscopy, electrochemistry, and kinetics. *Environmental Science and Technology* **39**(5):1221–1230.

Oonnittan A, Shrestha RA, Sillanpää M. (2008). Remediation of hexachlorobenzene in soil by enhanced electrokinetic Fenton process. *Journal of Environmental Science and Health Part A* **43**(8):894–900.

Pamukcu S, Hannum L, Wittle JK. (2008). Delivery and activation of nano-iron by DC electric field. *Journal of Environmental Science and Health Part A* **43**(8):934–944.

Pazos M, Ricart MT, Sanroman MA, Cameselle C. (2007). Enhanced electrokinetic remediation of polluted kaolinite with an azo dye. *Electrochimica Acta* **52**(10):3393–3398.

Pazos M, Cameselle C, Sanroman MA. (2008). Remediation of dye-polluted kaolinite by combination of electrokinetic remediation and electrochemical treatment. *Environmental Engineering Science* **25**(3):419–428.

Ponder SM, Darab JG, Mallouk TE. (2000). Remediation of Cr(VI) and Pb(II) aqueous solutions using supported nanoscale zero-valent iron. *Environmental Engineering Science* **34**:2564–2569.

Probstein RF, Hicks RE. (1993). Removal of contaminants from soils by electric fields. *Science* **260**(5107):498–503.

Quinn J, Geiger C, Lausen C, Brooks K, Coon C, O'Hara S, Krug T, Major D, Yoon WS, Gavaskar A, Holdsworth T. (2005). Field demonstration of DNAPL dehalogenation using emulsified zero-valent iron. *Environmental Science and Technology* **39**:1309–1318.

Rahner D, Ludwig G, Röhrs J. (2002). Electrochemically induced reactions in soils—a new approach to the in-situ remediation of contaminated soils? Part 1: The microconductor principle. *Electrochimicca Acta* **47**(9):1395–1403.

Reddy KR, Ala PR, Sharma S, Kumar SN. (2006). Enhanced electrokinetic remediation of contaminated manufactured gas plant soil. *Engineering Geology* **85**(1–2):132–146.

Reddy KR, Karri MR. (2007). Electrokinetic delivery of nanoscale iron particles for in-situ remediation of pentachlorophenol-contaminated soils. *Proceedings of International Symposium on Geo-Environmental Engineering for Sustainable Development* (eds. BP Han, LM Hu, HH Liu, XQ Zhu), October 22–24, 2007, Xuzhou, China, pp. 74–79.

Reddy KR, Karri MR. (2008a). Electrokinetic delivery of nanoiron amended with surfactant and cosolvent in contaminated soil. *Proceedings of the International Conference on Waste Engineering and Management*, May 28–30, 2008, Hong Kong, pp. 1–11.

Reddy KR, Karri MR. (2008b). Effect of oxidant dosage on integrated electrochemical remediation of contaminant mixtures in soils. *Journal of Environmental Science and Health Part A* **43**(8):881–893.

Röhrs J, Ludwig G, Rahner D. (2002). Electrochemically induced reactions in soils—a new approach to the in-situ remediation of contaminated soils? Part 2: Remediation experiments with a natural soil containing highly chlorinated hydrocarbons. *Electrochimica Acta* **47**:1405–1414.

Saleh N, Sirk K, Liu Y, Phenrat T, Dufour B, Matyjaszewski K, Tilton R, Lowry GV. (2007). Surface modifications enhance nanoiron transport and DNAPL targeting in saturated porous media. *Environmental Engineering Science* **24**(1):45–57.

Schrick B, Hydutsky BW, Blough JL, Mallouk TE. (2004). Delivery vehicles for zerovalent metal nanoparticles in soil and groundwater. *Chemistry of Materials* **16**:2187–2193.

Su C, Puls RW. (2004). Nitrate reduction by zerovalent iron: effects of formate, oxalate, citrate, chloride, sulfate, borate, and phosphate. *Environmental Science and Technology* **38**:2715–2720.

Suslick KS, Schubert PF, Goodale JW. (1981). Sonochemistry and sonocatalysis of iron carbonyls. *Journal of the American Chemical Society* **103**:7342–7344.

Suslick KS, Goodale JW, Wang HH, Schubert PF. (1983). Sonochemistry and sonocatalysis of metal carbonyls. *Journal of the American Chemical Society* **105**:5781–5785.

Suslick KS, Hammerton DA, Cline, Jr. RE. (1986). The sonochemical hot spot. *Journal of the American Chemical Society* **108**:5641–5642.

Tang WZ, Chen RZ. (1996). Decolorization kinetics and mechanisms of commercial dyes by H_2O_2/iron powder system. *Chemosphere* **32**(5):947–958.

U.S. Department of Energy. (2002). Final Remedial Action Report for Lasagna™ Phase IIb In-Situ Remediation of Solid Waste Management Unit 91 at the Paducah Gaseous Diffusion Plant, Paducah, Kentucky. http://www.rtdf.org/public/lasagna/lastechp.htm.

Vinodgopal K, Peller J, Makogon O, Kamat PV. (1998). Ultrasonic mineralization of reactive, textile azo dye, Remazol Black B. *Water Research* **32**:3646–3650.

Virkutyte J, Sillanpää M, Latostenmaa P. (2002). Electrokinetic soil remediation—critical overview. *Science of the Total Environment* **289**:97–121.

Voss J, Altrogge M, Golinske D, Kranz O, Nünnecke D, Petersen D, Waller E. (2001). Degradation of chlorinated arenes by electroreduction. In *Treatment of Contaminated*

Soil: Fundamentals, Analysis, Applications (eds. R Stegmann, G Brunner, W Calmano, G Matz). Berlin: Springer-Verlag, Chapter 34.

Wang A, Qu J, Liu H, Ge J. (2004). Degradation of azo dye Acid Red 14 in aqueous solution by electrokinetic and electrooxidation process. *Chemosphere* **55**(9):1189–1196.

Watts RJ, Udell MD, Rauch PA, Leung SW. (1990). Treatment of pentachlorophenol-contaminated soils using Fenton's reagents. *Hazardous Waste and Hazardous Materials* **7**(4):335–345.

Weng CH, Lin YT, Lin TY, Kao CM. (2007). Enhancement of electrokinetic remediation of hyper-Cr(VI) contaminated clay by zero-valent iron. *Journal of Hazardous Materials* **149**(2):292–302.

Wick LY, Shi L, Harms H. (2007). Electro-bioremediation of hydrophobic organic soil-contaminants: A review of fundamental interactions. *Electrochimica Acta* **52**(10): 3441–3448

Yang GCC. (2008a). Development and Application of Green Nanotechnology: Technology Development and Application of Environmentally Benign Nanoscale Zero-Valent Iron for In Situ Remediation of Simulated Soil/Groundwater Pollution. Final report nr EPA 08496007 (in Chinese).

Yang GCC. (2008b). "Method for treating a body of a polluted porous medium," U.S. Patent 7,334,965 (February 26, 2008). National Sun Yat-Sen University, Kaohsiung, Taiwan, assignee.

Yang GCC, Chang DG. (2006). Degradation of trichloroethylene in the subsurface by nanoscale bimetallic Pd/Fe slurry under an electric field. *EnviroNano 2006* (ed. GCC Yang), July 6, 2006, Kaohsiung, Taiwan, pp. 40–47.

Yang GCC, Hung CH, Tu HC. (2005). A preliminary study on kinetics of TCE degradation by nanoscale Pd/Fe in aqueous solution. *The Third Conference on Soil and Groundwater*, November 18–19, 2005, Chungli, Taiwan. Proceedings on CD-ROM (in Chinese).

Yang GCC, Hung CH, Tu HC. (2008). Electrokinetically enhanced removal and degradation of nitrate in the subsurface using nanosized Pd/Fe slurry. *Journal of Environmental Science and Health Part A* **43**(8):945–951.

Yang GCC, Lai WH. (1997). Chemical oxidation treatment of phenol-contaminated soil by Fenton process. *1997 Extended Abstracts for the ACS Special Symposium on Emerging Technologies in Hazardous Waste Management IX*, September 15–17, 1997, Pittsburgh, PA, p. 107.

Yang GCC, Lee HL. (2005). Chemical reduction of nitrate by nanosized iron: Kinetics and pathways. *Water Research* **39**:884–894.

Yang GCC, Li HL, Hung CH. (2004). Treatment of nitrates in the subsurface environment by nanosized zero-valent iron wall enhanced by electrokinetic remediation. *Journal of the Chinese Institute of Environmental Engineers* **14**(4):255–260.

Yang GCC, Liu CY. (2001). Remediation of TCE contaminated soils by in situ EK-Fenton process. *Journal of Hazardous Materials* **85**:317–331.

Yang GCC, Long YW. (1998). Treatment of phenol-contaminated soil by electrokinetics-Fenton process. *ACS Special Symposium on Emerging Technologies in Hazardous Waste Management X, Book of Abstracts, 216th ACS National Meeting*, August 23–27, 1998, Boston, MA.

Yang GCC, Long YW. (1999a). Treatment of phenol contaminated soil by in-situ electrokinetic remediation and Fenton-like process. *Proceedings of the Second Symposium on Heavy Metals in the Environment and Electromigration Applied to Soil Remediation*, July 7–9, 1999, Lyngby, Denmark, pp. 65–70.

Yang GCC, Long YW. (1999b). Removal and degradation of phenol in a saturated flow by in-situ electrokinetic remediation and Fenton-like process. *Journal of Hazardous Materials* **69**:259–271.

Yang GCC, Tu HC, Hung CH. (2007). Stability of nanoiron slurries and their transport in the subsurface environment. *Separation and Purification Technology* **58**(1):166–172.

Yuan C. (2006). The effect of Fe(0) on electrokinetic remediation of clay contaminated with perchloroethylene. *Water Science and Technology* **53**(6):91–98.

Yuan C, Chiang TS. (2007). The mechanisms of arsenic removal from soil by electrokinetic process coupled with iron permeable reaction barrier. *Chemosphere* **67**:1533–1542.

Zhang W, Wang C. (1997). Nanoscale metal particles for dechlorination of PCE and PCBs. *Environmental Science and Technology* **31**:2154–2156.

Zhang WX. (2003). Nanoscale iron particles for environmental remediation: An overview. *Journal of Nanoparticles Research* **5**:323–332.

22

ELECTROSYNTHESIS OF OXIDANTS AND THEIR ELECTROKINETIC DISTRIBUTION

W. Wesner, Andrea Diamant, B. Schrammel, and M. Unterberger

22.1 OXIDANTS FOR SOIL REMEDIATION

The basic idea of providing electrochemical oxidation equivalents on site by simply sticking anodes in the soil was promising, but in experiments, the limitations of this arrangement became apparent. Anode-generated radicals for direct oxidation steps are limited in respect to the lifetime of the formed species. OH radicals generated in water using dimension stable anodes (DSAs) are known to have lifetimes in the range of nanoseconds to milliseconds. Oxidants with higher lifetimes can be produced from components of the soil solution, either at the anode or from secondary reactions of the OH radicals with given molecules. The direct oxidation by the electrode only works in the area surrounding the anode. In summary, these approaches are known to show a poor efficiency and only very localized effects in soil (Gerzabek and Reichenauer, 2006).

One new approach separates the production of tailored oxidants, with defined potentials and optimized lifetimes, and the electrokinetic transport of the oxidants to the contamination. Most of the oxidants are negatively charged, but they have to migrate in the electric field starting near the negatively charged cathode in order to be attracted by the positive anode. For that reason it is not possible to use the same anode for both the production and transport of oxidants. For an appropriate efficiency, it is useful to separate the production unit from the transport unit. Anodic production of oxidants works in the potential range of 2–5 V. Larger spaces between the electrodes are counterproductive; these just produce waste heat. The amount of the produced oxidants is a function of the current (ampere) applied

---

Electrochemical Remediation Technologies for Polluted Soils, Sediments and Groundwater,
Edited by Krishna R. Reddy and Claudio Cameselle
Copyright © 2009 John Wiley & Sons, Inc.

to the electrode; increasing the potential does not result in more oxidation equivalents.

Oxidants used to transport oxidation equivalents from the electrode to the reaction point are called mediators. A variety of electrochemical production technologies for oxidants are available due to high-tech electrode materials, and nearly any species of oxidants can be produced on site.

Electrokinetic soil remediation is based on the technology of the transportation of substances through the soil by an applied electric field (Acar *et al.*, 1995; Virkutyte, Silanpää, Latostenmaa, 2002). This technology is very useful in catching charged ions such as heavy metals that are soluble in soil–water phases, or even for transporting charged organic ions such as surfactants, which emulate organic contaminations. To treat hydrophobic organic contaminations, which are not mobile in soil–water phases, it is useful to transport the oxidants through the soil (Wieczorek *et al.*, 2005).

For the oxidation of organic contaminants in soil, there are a number of techniques that have been in use for decades, which simply introduce oxidants such as permanganate or hydrogen peroxide by pouring them out over the surface of the contaminated soil or pumping them into the soil. These methods, collectively called *in situ* chemical oxidation (ISCO), work well in gravel or in soil where particles are large. Fine, loamy soil contaminated with hydrophobic hydrocarbons cannot be treated with ISCO because there is no efficient penetration through simple diffusion or convection. With these types of soil, the electrokinetic approach is the only one that can effect an effective transport of oxidants directly to the contamination.

Classical ISCO uses ready-made chemicals, which can be used only once (Acar *et al.*, 1995). With on-site electrochemical production of oxidants, chemicals with short lifetimes (in the range of a few hours) can be used. No stabilization of peroxides is necessary, the danger of a high quantity of chemicals is avoided, and because of the constant production and consumption process, logistics are easier to handle. The best case is a closed loop of production to transportation to reaction to transportation to production of the mediators (see Fig. 22.1).

In reality, a partial loss of the mediator has to be considered. One important restriction that must be accounted for is that the oxidant used be environmentally

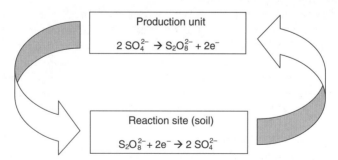

Figure 22.1. Use of oxidants as mediators—closed loop. The mediator is completely recycled, electrons (oxidation units) are transported from the organic pollutants in soil to the electrodes. The sulfate/persulfate system is one possible mediator; many others are to be considered.

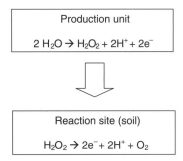

Figure 22.2. Use of oxidants as mediators—one-way. The mediator will be used once, and the electrons for producing oxidants are taken from water or other substances that are part of soil and may remain there. Oxidants are transported into the soil; the organic pollutants are decomposed by setting free water or other environmentally friendly substances.

compatible and not cause secondary pollution in the soil. In some cases (for example, hydrogen peroxide or ozone), the oxidants are produced from components of the soil, so there is no problem with their transportation because it is a one-way direction only to the reaction site, where the oxidants are decomposed again. The electrolyte is part of the soil and can be introduced in any quantity (see Fig. 22.2).

To decide which oxidant is the best mediator for a given remediation process, the following attributes have to be taken into account:

1. The oxidant must be stable enough to be delivered to the place of reaction.
2. The oxidant should perform a fast reaction in the soil.
3. The oxidant should have the potential to be able to oxidize a wide variety of pollutants.
4. The oxidant is not allowed to cause secondary pollutions in the soil either by itself or by persistent by-products that have been generated in reactions with the contamination or with the soil.

General points to consider when using oxidants to oxidize organic pollutants in soil are that an elevated temperature always results in a higher reaction rate and that most reactions that use oxidants for the production of radicals need a special minimum temperature to work. In some cases, temperatures that are too low can be compensated for by using catalysts such iron salts for the production of radicals.

For many oxidants, there is an optimum reaction at a pH below 7. In an electrokinetically treated soil, the pH is not uniform, and a gradient occurs, decreasing the pH from cathode to anode. By measuring this gradient, the region of reaction can be defined. Oxidants are brought in at the cathode in alkaline pH conditions, and then they are transported toward the anode while pH decreases. Once it reaches a pH of 7, the activation of an oxidant such as persulfate starts. The rate of radical generation depends on the temperature and the concentration of dissolved iron ions in the soil solution. Due to the enrichment of anions at the anode and cations at the cathode, the pH gradient is not uniform. Fast changes can be observed near the electrodes, while in between there is an approximately continuous slow rise at medium pH. Using electrode distances of about 1 to 5 m, all transportation and reactions take place in the range of moderate pH changes.

The amount of oxidizable metals and organic components in the soil plays an important role for side reactions, which are responsible for destroying the oxidants. Equivalents used for the oxidation of soil components can be further used by biological remediation processes by reducing metals after the chemical oxidation. The whole process lasts longer than just the chemical treatment alone. The oxidation of organic particles and humus is not desired, but as contaminants are often more stable against oxidation, selective oxidation of contaminants only is not possible. Therefore, it is easier to treat soils with low organic content.

22.1.1 Hydrogen Peroxide

Hydrogen peroxide has been well known as an oxidant for soil remediation for more than a century. It reacts with the evolution of heat and gas:

$$H_2O_2 + 2H^+ + 2e^- \rightarrow 4H_2O \quad E_0 = +1.78\,V \tag{22.1}$$

or with Fe^{2+} activation as Fenton's reagents:

$$H_2O_2 + Fe^{2+} \rightarrow OH^- + \cdot OH + Fe^{3+} \quad E_0 = 2.8\,V \tag{22.2}$$

The optimal pH for the reaction is 3 to 5; a pH higher than 6 can lead to the catalytic degradation of hydrogen peroxide into water and oxygen. Carbonates in the soil solution may act as radical scavengers. Nevertheless, it is possible to apply Fenton processes in neutral soil solutions. In this case, special attention has to be paid to the addition of the iron salt (Yang and Long, 1999; Yang and Liu, 2001). Chelated iron may be used to control the reaction. In some cases there is enough iron in the soil to find appropriate conditions for a Fenton reaction without the need for additional iron.

The classical electrochemical production of hydrogen peroxide uses the electrolytic production of persulfate from sulfuric acid and hydrolyzes the persulfate in a second step into hydrogen peroxide and sulfuric acid. Hydrogen peroxide is always a by-product for the anodic production of oxidants in dilute solutions (Fig. 22.3).

An alternative method of electrochemical on-site production uses the cathodic reduction of oxygen at a graphite or gas diffusion electrode. Da Pozzo et al. (2005) studied the production of hydrogen peroxide through the cathodic reduction of oxygen in an acidic medium by comparing the results obtained by using commercially available graphite and a gas diffusion electrode. A low pH was required to allow the application of hydrogen peroxide generation to an electro-Fenton process. The influence of applied potential and the gas flow composition were investigated. The gas diffusion electrode demonstrates a higher selectivity for hydrogen peroxide production without significantly compromising the iron regeneration, thus making its successful application to a cathodic Fenton-like treatment possible. Unlike the graphite cathode, the gas diffusion cathode also proved to be effective in the air flow.

The cathodic production of hydrogen peroxide from air in combination with anodic production processes at boron-doped diamond electrodes could be the key

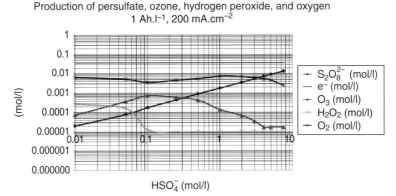

Figure 22.3. Production of oxidants on boron-doped diamond electrodes with different concentrations of sulfate.

to an optimized level of efficiency, so that the cathodic hydrogen production is eliminated for the benefit of hydrogen peroxide production. Further work in this field of research is needed.

22.1.2 Persulfates

Persulfates have a high potential without any activation:

$$S_2O_8^{2-} + 2e^- \rightarrow 2SO_4^{2-} \quad E_0 = 2.0\,V \tag{22.3}$$

that can even be higher in case of activation to sulfate radicals:

$$S_2O_8^{2-} + 2Fe^{2+} \rightarrow 2SO_4^- \cdot + 2Fe_3^+ \tag{22.4}$$

The sulfate radicals can be Fenton-like produced with iron ions or at elevated temperatures above 40 °C (Neta *et al.*, 1977). The optimal pH for this application is 3 to 7. Persulfate is industrially produced by the electrolysis of potassium sulfate on platinum or glassy carbon electrodes and delivered as potassium salt. Potassium persulfate has the lowest solubility in water (50 g/l, 20 °C). Therefore, sodium persulfate (545 g/l, 20 °C) or ammonium persulfate (620 g/l, 20 °C) are favored raw materials for the on-site production of oxidants and application in soil remediation (Park *et al.*, 2005).

The oxidation of hydrocarbons is enhanced in combination with permanganate. The transport of these oxidants in the soil is easy to achieve due to a high stability of 1–2 weeks in average soil. An important factor controlling the system's reactions is the distribution of iron ions. Some iron ions are naturally in the soil; therefore, a slow reaction takes place anyway. Iron ions can be added to the oxidant or can be separately distributed in the soil. Another method is to use iron electrodes as anodes to set up the electric field in order to establish migration. These electrodes dissolve when current is applied. The positively charged iron ions migrate toward the cathode; on their way they meet the persulfate ions moving to the anode, and the Fenton-like reaction can take place.

22.1.3 Oxygen/Mild Oxidants CaO_2, MgO_2, Percarbonates

Oxygen is the major waste product in the electrosynthesis of oxidants. At the same time, it is a very useful tool to aid biological remediation. In the electrolysis process, pure oxygen is formed, and its solubility in water is five times higher than that of atmospheric oxygen (21% O_2 in atmosphere). If possible, the formed oxygen should be used to aerate the unsaturated zones of the soil to optimize the overall oxidation. Using tap water or soil–water-based electrolytes, calcium and magnesium peroxides, as well as percarbonates, are also formed during electrolysis. These substances are weak oxidants but able to augment oxygen delivery to enhance biological remediation. These mechanisms are used following the treatment of the main contamination.

22.1.4 Ozone

Ozone is a very strong oxidizer with a high potential of 2.07 V. It has a high reactivity and a short stability, and its solubility in water is exceedingly modest. Therefore, it is more practicable to inject ozone as gas into the unsaturated zone of the soil than to try to create a solution with ozone. Transport of dissolved ozone in the soil–water phase is only possible by electro-osmosis, and due to the short lifetimes, only short distances are possible. Ozone is a by-product in many anodic processes, and the amount of its production is directly proportional to the applied current density. The more efficient way to produce ozone uses classical corona discharge ozone generators. Ozone is not the optimal mediator for electrokinetic remediation, but an alternative or additional method for a remediation concept using aeration techniques. Ozone works as a direct oxidant and as well as a free radical:

$$O_3 + 2H^+ + 2e^- \rightarrow O_2 + H_2O \tag{22.5}$$

$$O_3 + OH^- \rightarrow O_2^- + HO_2 \cdot \tag{22.6}$$

The combination of ozone with hydrogen peroxide is a known technology.

22.1.5 Permanganate

Permanganate is a quite stable oxidant that is easy to distribute. Due to its color, it is easy to detect, and it reacts in a wide range of pH. The reaction follows:

$$MnO_4^{4-} + 4H^+ + 3e^- \rightarrow MnO_2 + 2H_2O \quad E_0 = 1.68\,V \tag{22.7}$$

In some soil types the permeability can be decreased with the formation of MnO2, which may cause some problems for remediation. Sodium permanganate has a higher solubility than potassium permanganate and is therefore preferred in field applications.

The production of permanganate requires anodic oxidation following oxidation with air. Due to the insolubility of MnO_2, it is not possible to recycle the manganese by anodic oxidation.

22.1.6 Chlorine

To avoid the formation of halogenated hydrocarbons, chlorides are not used in high concentrations. Low concentrations occur when using tap water or a soil solution as an anolyte due to dissolved chlorides.

22.2 PRODUCTION OF OXIDANTS

22.2.1 Electrodes

While cathode materials can theoretically be chosen from a wide range of noncorroding metals such as stainless steel, anode materials are restricted to DSAs. Due to the higher pH and the reactions at the cathode, precipitations take place and isolate the electrode surface. Precipitations do not occur in absolutely clean solutions only, without calcium, magnesium, and metal ions (e.g., working with clean sodium sulfate in distilled water). If the electrolyte is based on a soil solution or tap water, it is useful to change the polarity of the electrodes in time intervals—changing cathode to anode and back again. The low pH at the anode can clean the precipitations of the electrodes' cathodic phase. In this case, both electrodes have to be dimensionally stable so they resist cathodic and anodic treatment. The reversing of the polarity restricts the possible electrode materials to some titanium-based mixed oxide electrodes, platinum electrodes, and boron-doped diamond electrodes. Steel electrodes would dissolve during the anodic cycle. Lead electrodes, longtime favorite anodes because of their high overpotential for oxygen production, dissolve in the cathodic phase. The lifetimes of mixed oxide electrodes will be reduced due to changing polarity, even if there are some types that can withstand this reversal of polarity up to 50,000 times. Boron-doped diamond electrodes show no changes due to the polarity change, but metal-based boron-doped diamond can have problems due to the instability of the metal–diamond junction, especially at higher current densities (Wesner, 2007, unpublished data).

Classical DSAs based on titanium mixed oxides show higher chlorine and oxygen productions in various chloride-containing electrolytes, while all kinds of boron-doped diamond electrodes oxidize, preferably, sulfate and water for persulfate and hydrogen peroxide (Michaud *et al.*, 2000); at higher current densities, ozone is produced (Haenni *et al.*, 1999). The best results regarding the durability of the electrode material were obtained using bipolar diamond electrodes made of single boron-doped diamonds fixed in a fluorinated polymer. As a bipolar arrangement is necessary, it is not useful to operate these electrodes *in situ*, but on site is possible.

22.2.2 Electrolyte

In anodic oxidation processes, the composition of the electrolyte is, besides the type of electrode employed, the most important factor defining the kind and quantity of the oxidants produced. Using clean water, the production of oxygen, hydrogen peroxide, and ozone, according to the applied current densities, is possible. In the presence of chlorides, chlorine is also formed. This is not acceptable in many applications because of the secondary reaction of chlorine with organic substances, which leads to adsorbable organic halides (AOX) formation. As the production of chlorine is a function of the chloride concentration and the concentration of other oxidizable substrates at the anode, in many cases the reaction can be kept at low production rates by designing the composition of the treated solution. With defined conditions it is possible to generate chlorine dioxide, which is much more stable and therefore more suitable to be transported in soil. In the presence of sulfates, peroxodisulfate is produced—one of the most powerful oxidants for soil remediation. Especially in combination with iron ions in the soil, highly active OH radicals will be formed *in situ*. If carbonates and hydrogen carbonates are present in the electrolyte, percarbonates, well known for laundry detergent, will also be formed. For special applications, manganese, complexed iron, and silver can transfer oxidant equivalents to contaminants in the soil under high potential conditions. Organic substances in the electrolyte can be directly oxidized at the electrodes, which can provide a high overpotential for the evolution of oxygen.

The composition of the produced oxidants at the electrode is directly dependent on the mixture of oxidizable molecules in the soil solution. The composition of these oxidants can be approximated by modelling the whole reactions, based on the substance specific production rates of the single reactions. The rate of direct destruction of organic substances at the anode can also be shown to be dependent on the concentration of oxidants at the electrode and dependent on the temperature at the surface of the electrode.

For economic reasons, one possible way to optimize the reactions for on-site production of oxidants is to create highly concentrated solutions of pure substrates. An alternative is to mix different oxidant-forming agents to optimize the production in a way that maximizes the amount of oxidation equivalents. A number of mixtures containing sulfates, carbonates, and other ions have been tested successfully, giving higher yields of equivalents than concentrated solutions of single components.

22.2.3 Reactions

At mixed oxide electrodes, free orbitals of the metals from the platinum group (Pt, Ir, Ru) are able to catalyze direct reactions such as the formation of chlorine. First, chloride is bound to the surface of the electrode; oxidation happens in the second step.

With platinum, water is oxidized to oxygen at low potentials, but only at higher current densities can different species of oxygen be formed. Using boron-doped diamond electrodes, the reaction is more unspecific. Due to the important characteristics of boron-doped diamond electrodes and the high overpotential for oxygen production in aqueous solutions, a wide range of reactions is possible. Low and high energetic oxidants are produced at the same time. It is assumed that production is happening to some extent directly at the electrode and that the main parts of the oxidations are generated as secondary reactions of the produced OH^- radical.

To produce 1 mol of the OH^- radical of water,

$$H_2O \rightarrow OH\cdot + H^+ + e^- \qquad (22.8)$$

an amount of 26.8 Ah is theoretically necessary.

The OH· radical is extremely reactive; therefore it has a short lifetime. It reacts with all kinds of reaction partners surrounding the radical. Some important processes in oxidant production are listed below:

$$\cdot OH + \cdot OH \rightarrow H_2O_2 \tag{22.9}$$

$$\cdot OH \rightarrow \cdot O + H^+ + e^- \tag{22.10}$$

$$\cdot O + \cdot O \rightarrow O_2 \tag{22.11}$$

$$\cdot O + O_2 \rightarrow O_3 \tag{22.12}$$

$$\cdot OH + HSO_4^- \rightarrow SO_4^- \cdot + H_2O \tag{22.13}$$

$$SO_4^- \cdot + SO_4^- \cdot \rightarrow S_2O_8^{2-} \tag{22.14}$$

$$\cdot OH + Cl^- + H^+ \rightarrow Cl \cdot + H_2O \tag{22.15}$$

$$Cl \cdot + Cl \cdot \rightarrow Cl_2 \tag{22.16}$$

$$Cl \cdot + \cdot OH \rightarrow HOCl \tag{22.17}$$

Depending on the concentrations of the different species surrounding the ·OH radicals, mixtures of reactions happen, and different product results are generated.

Figure 22.3 shows that only at high sulfate concentrations does the persulfate production become greater than the production of oxygen. A difference of 40% to 60% compared to theoretical calculations should be taken into account for the production of any oxidant. If it is possible to bring the by-product oxygen in a dissolved form into the soil system, a nearly 100% efficiency can be established. While the persulfate will be transported by electromigration to the anode, oxygen and ozone may be transported with the electro-osmotic water flow toward the cathode. The distribution of the anolyte from the production cell should be in the area between the anode and cathode; the method of distribution is dependent on the transport velocities of both directions.

Lead-oxide electrodes react in a comparable way, and they also show a high overpotential for oxygen production. In the literature, the use of lead-oxide electrodes for the applications discussed is common, due to their instability as cathodes and the danger of releasing lead molecules. In modern production units they have been replaced by boron-doped diamond electrodes.

22.3 DISTRIBUTION OF OXIDANTS

Most oxidants are anions. The electrokinetic transport accelerates the distribution of anionic oxidants toward the anode. At the same time, the electro-osmotic flow carries water to the cathode. Depending on the soil parameters, the electrokinetic transport can be faster than the electro-osmotic flow. In very dense ground with a high amount of clay, the electro-osmotic flow can be even faster than the electrokinetic transport. In any case, it is important to test the velocity of transport in both directions, to find the optimal distances for dosing the oxidant, and to plan the arrangement of the electrodes depending on the transport velocity and the expected duration of remediation. The velocity of the electrokinetic transport increases linearly with the applied voltage. In average soil types and an electric field of 4V/cm, up to 2cm per hour of transport velocity could be observed.

Another possible on-site application for the use of electrochemically produced oxidants in soil remediation is to treat the water of a "pump and treat application" in an electrochemical cell. Water is pumped into the soil and contaminants transported by the water flow can be oxidized in order to reuse the water to wash out further contaminants (Yang and Long, 1999).

As in most cases where the soil is not uniform and regions of high density alternate with regions of low hydrodynamic resistance, pumping is always a good way to support electrokinetic distribution. Low-density regions are served fast, and high-density regions are served with electrokinetic transport.

REFERENCES

Acar Y, Gale R, Alshawabkeh A, Marks R, Puppala S, Bricka M, Parker R. (1995). Electrokinetic remediation: Basics and technology status. *Journal of Hazardous Materials* **40**:117–137.

Da Pozzo A, Di Palma L, Merli C, Petrucci E. (2005). An experimental comparison of a graphite electrode and a gas diffusion electrode for the cathodic production of hydrogen peroxide. *Journal of Applied Electrochemistry* **35**(4):413–419.

Gerzabek MH, Reichenauer TG. (2006). *Innovative In-situ Methoden zur Sicherung und Sanierung von Altablagerungen und Altstandorten*. Wien: Facultas.

Haenni W, Borel M, Perret A, Correa B, Michaud P-A, Comninellis C. (1999). Production of Oxidants on Diamond Electrodes. CSEM Scientific Report.

Michaud P-A, Mahe E, Haenni W, Perret A, Cominellis C. (2000). Preparation of peroxidisulfuric acid using boron-doped diamond thin film electrodes. *Electrochemical and Solid-State Letters* **3**(2):77–79.

Neta P, Madhavan V, Zemel H, Fessenden RW. (1977). Rate constants and mechanism of reaction of SO_4^{2-} with aromatic compounds. *Journal of the American Chemical Society* **99**(1):163–164.

Park J, Kim S, Lee Y, Baek K, Yang J. (2005). EK-Fenton process for removal of phenantrene in a two-dimensional soil system. *Engineering Geology* **77**:217–224.

Virkutyte J, Sillanpää M, Latostenmaa P. (2002). Electrokinetic soil remediation—critical overview. *The Science of the Total Environment* **289**:97–121.

Wieczorek S, Weigand H, Schmid M, Marb C. (2005). Electrokinetic remediation of an electroplating site: design and scale-up for an in-situ application in the unsaturated zone. *Engineering Geology* **77**:203–215.

Yang G, Liu C. (2001). Remediation of TCE contaminated soils by in situ EK-Fenton process. *Journal of Hazardous Materials* **85**:317–331.

Yang G, Long Y. (1999). Removal and degradation of phenol in a saturated flow by in-situ electrokinetic remediation and Fenton-like process. *Journal of Hazardous Materials* **69**:259–271.

23

COUPLED ELECTROKINETIC–PERMEABLE REACTIVE BARRIERS

CHIH-HUANG WENG

23.1 INTRODUCTION

Subsurface contamination at hazardous waste disposal and brownfield sites is a serious environmental problem. A majority of these sites contain high concentrations of heavy metals, radionuclides, and nonbiodegradable organic pollutants that would be difficult to remediate with traditional technologies, particularly when they reside in clayey or silty soils. *Ex situ* remediation of such sites by the excavation of contaminated soil is usually followed by soil washing. Although soil washing is a promising process, the post-treatment of the extracted solution and the recovery of the extracted soil after treatment are the two major concerns when using this costly process. The effectiveness of this method is also affected by the depth of the contaminant underneath the subsurface, soil properties, extracting agents, and contaminant types, which may increase the treatment difficulties.

In situ or on-site cleanup methods could be applied to reduce environmental impacts and the cost of remediation. Recently, a proven technology that has drawn attention by site governmental officials and remediation engineers is electrokinetic (EK) remediation. This emerging *in situ* technology can separate the contaminants from low-permeability soils, i.e., clays and silts, by applying a low intensity direct current to the soil. The EK process involved in soil remediation can be based on three key mechanisms: (1) the advection of electro-osmotic (EO) flow (forced water movement) driven by electric current, (2) the movement of H^+ ions produced from water electrolysis at the anode advancing through the soil toward the cathode, and (3) the electromigration of charged ions toward the opposite electrodes (Alshawabkeh, Yeung, and Bricka, 1999). Dissolution, adsorption/desorption, reduction-oxidation, and precipitation are always accompanied during the process. The

---

Electrochemical Remediation Technologies for Polluted Soils, Sediments and Groundwater,
Edited by Krishna R. Reddy and Claudio Cameselle
Copyright © 2009 John Wiley & Sons, Inc.

EK process has been reported to be successful and cost effective to treat both organic (Yalcin, Li, and Gale, 1992; Ho, 1999; Yang and Liu 2001; Weng, Yuan, and Tu, 2003; Yuan and Weng 2004) and inorganic contaminants (Yalcin et al., 1992; Coletta et al., 1997; Sah and Lin, 2000; Weng and Yuan, 2001; Reddy, Xu, and Chinthamreddy, 2001; Reddy and Chinthamreddy, 2003; Weng and Huang, 2004) from low-permeability soils. This technology has been commercialized for years under such names as Lasagna™, Electro-Klean™, and Electrokinetic Ltd. The remediation limitation of the EK process by itself is that it can only remove the contaminants from the soil phase to the solution phase or it can degrade partially the organic contaminants in the system. As such, when considering a total remediation case, further treatment of the extracted solution may be needed in conjunction with the existing EK system.

The permeable reactive barrier (PRB) is a process that contains a reactive zone in groundwater for the destruction of halogenated organic compounds and for the removal of specific inorganic compounds. The PRB is applied *in situ* groundwater remediation via three configurations: continuous trench, funnel and gate, and reactive vessel. In the reactive zone, various materials can be used as a fill. The barrier, when filled with selected microorganisms, can act like a biological reactor to degrade the target contaminants. Other materials such as limestone, hydroxyapatite, activated carbon, and zeolite, which act as chemical precipitators or adsorbents to retain the contaminants within the barrier, are also applicable. Generally, the use of adsorptive materials in barriers to collect the intended contaminants would not cause problems related to permeability loss of the PRB system because the adsorption process does not involve decreasing the pore size of the adsorbent. However, adsorption of indifferent compounds and saturation of the adsorbent could cause a low removal action (Rocca, Belgiorno, and Meriç, 2007). Aside from the abovementioned materials, Fe^0 is the most popular reactive materials used in PRBs to remove certain persistent organic and inorganic contaminants such as dense nonaqueous phase liquids (DNAPLs) and Cr(VI) (Blowes, Ptacek, and Jambor, 1997; Puls, Paul, and Powell, 1999; Blowes et al., 2000; Simon et al., 2001; Mayer, Blowes, and Frind, 2001; Morrison, Metzler, and Carpenter, 2001; Phillips et al., 2003; Kamolpornwijit et al., 2004; Wilkin et al., 2005; Ebert et al., 2006; Lai and Lo, 2008). The degradation process involves the corrosion of Fe^0, which provides the electrons necessary for the reduction of compounds such as Cr(VI) and trichloroethylene (TCE). The main reasons to adapt Fe^0 in PRB groundwater remediation are outlined as follows:

- It is a good reducing agent for the destruction of many recalcitrant contaminants.
- It is readily available and relatively inexpensive.
- It has *in situ* applicability.
- It requires little maintenance and has a long-life reactivity.
- It creates nontoxic products.
- It offers simultaneous removal of many priority pollutants.
- It has been used in field application for two decades.

However, some potential users still worry about the long-term performance of *in situ* Fe^0 PRBs. When groundwater contains a high concentration of dissolved inorganic species, the long-term performance of Fe^0-based PRBs would be limited due to porosity losses caused by mineral precipitation (Mayer, Blowes, and Frind, 2001; Parbs, Ebert, Dahmke, 2007). The corrosion of Fe^0 produces mineral precipitates that may alter the system's hydraulic integrity (Liang *et al.*, 2000). Because zero-valent iron (ZVI) in the PRBs can be oxidized into ferrous or ferric iron, resulting in decreased reactivity with time, the contaminant degradation rates will decrease with time in the long-term performance of the Fe0 PRB (Phillips *et al.*, 2000). The gradual deterioration of the Fe^0 barrier, loss of the barrier reactivity, and system clogging are problems to be overcome. When iron-reducing bacteria is added to the Fe^0 PRB, it can reconvert Fe(III) to Fe(II) and therefore will prolong the reductive removal of chlorinated organic contaminants in groundwater. A study by Shin, Singhal, and Park (2007) has verified that iron-reducing bacteria in the oxidized Fe^0 barriers could enhance the removal rate of chlorinated organic compounds and influence on the long-term performance of Fe^0 reactive barriers. Based on an 8-year operation in a Fe^0-PRB site at the US Coast Guard Support Center near Elizabeth City, NC, evidence indicated that secondary iron-bearing mineral products may enhance the capacity of Fe^0 barriers to remediate Cr in groundwater, either through redox reactions at the mineral–water interface or by the release of Fe(II) to solution via mineral dissolution and/or metal corrosion (Wilkin *et al.*, 2005).

In view of the advantages of the EK process (capable of removing contaminants from low-permeable media) and PRBs (ability to adsorb/reduce/degrade contaminants in the barrier), this chapter describes the EK process coupled with PRBs (EK–PRB) that has been developing (see Fig. 23.1). The problems of clogging in the PRB system due to the precipitation of ferric oxide or ferric hydroxide would be overcome by EK operation. Since the remediation period of the EK process is much shorter than that of PRB for groundwater remediation, the long-term performance of the loss of barrier reactivity in the EK–PRB process would not be a serious problem if there is sufficient Fe^0 in the barrier. Reactive heavy metals and

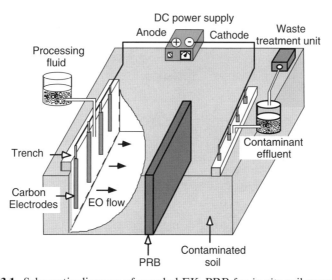

Figure 23.1. Schematic diagram of coupled EK–PRB for *in situ* soil remediation.

chlorinated organic compounds are known to be the dominant pollutants at many hazardous waste sites in many developing countries. The information presented in this chapter will assist in implementing this integrated electrochemical process at contaminated sites.

23.2 DESIGN OF REACTIVE BARRIER IN THE EK–PRB PROCESS

23.2.1 Iron Fillings

Depending on the contaminants in the soil, the fillings in the barrier will be different. Cr(VI)-contaminated soil is used as an example. To create a reactive zone for Cr(VI) reduction, the ZVI aggregate and silica sand can be used as filling material in the barrier. Commercial iron aggregate (Fig. 23.2), such as Connelly–GPM Inc.,

Figure 23.2. Fe^0 aggregate (from Connelly–GPM Inc., USA) before use. (a) Light micrograph (original size), (b) scanning electron microscope image (100× magnification).

USA, with a particle size ranging from 0.30 to 2.36 mm and a weight of approximately 2240–2560 kg/m^3 is stuffed in the barrier. To insure that the barrier is permeable, a 1:1 weight ratio of ZVI aggregate to silica sand mixture is used. This mix allows sufficient reaction time for Cr(VI) reduction and for the avoidance of possible clogging due to the precipitation of chromium and iron hydroxide/carbonate precipitates. The likely precipitates are either initiated from the reaction products in the barrier, such as $Cr(OH)_3(s)$, $Fe(OH)_3(s)$, and $Cr_xFe_{1-x}(OH)_3(s)$, or the metal salts precipitates resulting from the processing fluids containing ligands, such as SO_4^{2-} and CO_3^{2-}. Formation of the precipitate $Cr_xFe_{1-x}(OH)_3(s)$, even under acidic conditions, has been documented (Eary and Rai, 1988; Weng et al., 1994). Recent study has shown that Cr(VI) reduction by ZVI would produce precipitates of FeOOH, Fe_2O_3, and $(Fe-Cr)_2O_3$ (Yang et al., 2007).

In the design of the barrier, two key parameters ought to be considered: the amount of ZVI in the barrier and thickness of the barrier. Depending upon the Cr(VI) species (Fig. 23.3) involved in the reaction, the amount of Fe^0 in the PRB was determined based on Cr(VI) and ZVI reactions:

$$2Fe^\circ + Cr_2O_7^{2-} + 14H^+ \rightarrow 2Fe^{3+} + 2Cr^{3+} + 7H_2O \tag{23.1}$$

$$Fe^\circ + CrO_4^{2-} + 8H^+ \rightarrow Fe^{3+} + Cr^{3+} + 4H_2O \tag{23.2}$$

$$Fe^\circ + HCrO_4^- + 7H^+ \rightarrow Fe^{3+} + Cr^{3+} + 4H_2O \tag{23.3}$$

The chromate species are closely related to the pH and the total Cr(V) concentration. The dominant species in the solution are $HCrO_4^-$ and CrO_4^{2-} for low Cr(V) concentration, while $Cr_2O_7^{2-}$ appears at an elevated Cr(VI) concentration (Fig. 23.3). For the EO solution having a high concentration of Cr(VI), each one of chromate ions, CrO_4^{2-}, $HCrO_4^-$, and $Cr_2O_7^{2-}$, needs to be accounted for for ZVI estimation. If the concentration of Cr(VI) in the EO flow is less than 50 mg/l, then $HCrO_4^-$ and CrO_4^{2-} are the dominant species at all pHs. Based on Eqs. 23.1–23.3, 1 molar concentration of ZVI can reduce 1 molar of Cr(VI). To ensure all chromate species are completely reduced to Cr(III), an excess amount of ZVI is needed. Thus, the minimum ZVI requirement (W_{ZVI} in kilograms) in the barrier can be estimated as follows:

$$W_{ZVI} = [Cr(VI)]_{initial} \times W_{soil} \div Cr_{MW} \times M_{Cr(VI)}/M_{ZVI}, \tag{23.4}$$

where $[Cr(VI)]_{initial}$ (mg/kg) is the initial concentration of Cr(VI) in soil, W_{soil} (kg) is the mass of soil, Cr_{MW} is atomic weight of Cr (52 g/mol) and $M_{Cr(VI)}/M_{ZVI}$ is the molar ratio of chromates and ZVI.

By accounting for only 90% of iron content in the ZVI aggregate sample with 10% impurities, the minimum requirement of ZVI in the barrier to treat 100 tons of soil with a Cr(VI) concentration of 2000 mg/kg was estimated as follows:

$$\begin{aligned} W_{ZVI} &= [Cr(VI)]_{initial} \times W_{soil} \div Cr_{MW} \times M_{Cr}/M_{Fe} \\ &= 2000\,mg/kg \times 100\,tons \times 1000\,kg/ton \div \\ &\quad 52(g/mol) \times 56(g/mol) \div 1000(g/mol) \div 9 \\ &= 239,316\,g \cong 240\,kg \text{ (or } \cong 0.1\,m^3 \text{ ZVI aggregate)} \end{aligned} \tag{23.5}$$

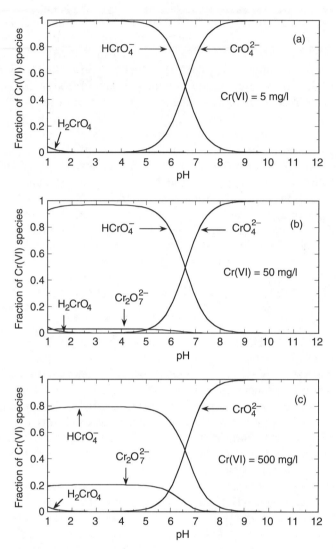

Figure 23.3. The Cr(VI) speciation diagram as function of pH for Cr(VI) concentration of (a) 5 mg/l, (b) 50 mg/l, and (c) 500 mg/l.

In this case, at least 240 kg ZVI have to be applied to provide enough ZVI for Cr(VI) reduction. If the price of this commercial ZVI is US$500/ton, then the minimum cost of ZVI alone in the installation of the PRB is only US$120. Other installation fees, including construction and sand are not estimated here.

23.2.2 Evaluation of the Thickness of the Barrier

Major reactions, including adsorption/desorption, reduction-oxidation (redox), precipitation, and dissolution have to be considered when the EO flow passes through the barrier. For the target contaminant in solution, Cr(VI), the dominant reaction

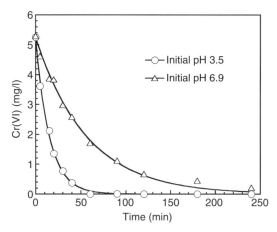

Figure 23.4. Cr(VI) reduction by Fe^0 as a function of time. Solid lines represent the best fit of the first-order kinetic equation. Experimental conditions: Cr(VI) concentration 5.2 mg/l; Fe^0 5 g/l; $NaClO_4$ 1×10^{-2} M; 27 °C.

between Cr(VI) and ZVI will be redox reaction. In order to design the thickness of ZVI barrier the kinetic information has to be given. The reduction rate of Cr(VI) by ZVI aggregates is influenced by factors such as pH, temperature, constituents of soil, and dissolved oxygen in EO solution. Among these, pH is usually to be considered the key factor affecting the reduction rate. Figure 23.4 shows an example of Cr(VI) reduction by ZVI. As shown, the reduction rate is fast at a lower pH, and it becomes slower at a higher pH. The reduction results clearly indicate that the ZVI can effectively reduce Cr(VI) to Cr(III) regardless of the solution pH and follows the first-order kinetic equation:

$$C = C_o e^{-kt}, \quad (23.6)$$

where C_o is the initial Cr(VI) concentration (mg/l) in the EO solution; k is the first-order rate constant (1/min); t is the reaction time (min). For the example shown, the rate constants are 0.064 and 0.019 1/min, respectively, for solution pH 3.5 and 6.9. The thickness of wall (T_{barrier}, cm) can be calculated as follows:

$$T_{\text{barrier}} = K_{eo} \times i_e \times t_{Cr}, \quad (24.7)$$

where K_{eo} (cm^2/V-s) is electro-osmotic permeability; i_e (V/cm) is the applied potential gradient to the soil matrix; and t_{Cr} (sec) is the reaction time in the barrier for Cr(VI) to be completely reduced to Cr(III) by ZVI. The value of t_{Cr} varies by the source of ZVI and the experimental conditions conducted (Table 23.1). The design of a PRB depends to a great extent on the experimental conditions. From Table 23.1, it is seen that a faster reduction occurred for the barrier that contained ZVI only. Since the barrier has to maintain its permeable condition to allow the EO flow to pass through, sand or other high porous media are usually used and mixed with the ZVI aggregate; the rate will be less than with the ZVI only. As stated previously,

TABLE 23.1. Reduction of Cr(VI) by ZVI

Initial Cr(VI) (mg/L)	PRB fillings (ZVI/Sand)	pH	k (1/min)	t_{Cr} (h)	Cr(VI) reduction (%)	Source
5–30	1/0	6.3–8.5	—	<0.1	99	Liu et al., 2003
5	1/1	3.5	0.0198	4	99	Weng et al., 2007
5	1/1	6.9	0.0097	8	99	Weng et al., 2007
5	1/0	3.5	0.064	1	99	Fig. 23.4
5	1/0	6.9	0.019	4	99	Fig. 23.4

solution pH is another key factor governing the reaction kinetics. An acidic environment usually promotes the reduction rate.

By adopting the K_{eo} value of $1 \times 10^{-5}\,cm^2/V\text{-}s$ and t_{Cr} of 8 h with potential gradient, i_e, of 2 V/cm, for instance, the thickness of barrier, $T_{barrier}$, is estimated as follows:

$$T_{barrier} = 1 \times 10^{-5}\,cm^2/V\text{-}s \times 2\,V/cm \times 8\,h \times 3600\,s/h \approx 0.6\,cm \tag{24.8}$$

To confirm that the Cr(VI) solution passing through the barrier will be completely reduced to Cr(III), a minimum length of the reaction zone (path) greater than 0.6 cm needs to be constructed. For a site with the dimensions of 2 m deep, 5 m wide, and 5 m long, a total of 240 kg (or $\cong 0.10\,m^3$ ZVI aggregate) and 240 kg sand (or $\cong 0.133\,m^3$ sand with a bulk density of 1800 kg/m^3) are needed for a barrier filled with 1:1 Fe^0 to sand weight ratio to completely reduce a Cr(VI) concentration of 2000 mg/kg. If a continuous trench type of barrier is constructed (Fig. 23.1), the thickness of the barrier proving for the filling is

$$\begin{aligned}\text{Dimension of barrier} &= \text{Depth} \times \text{Width} \times \text{Thickness} \\ &= 2\,m \times 5\,m \times \text{thickness} \\ &= (0.1+0.133)\,m^3 = 0.233\,m^3\end{aligned} \tag{23.9}$$

$$\text{Thickness of barrier} = 0.0233\,m \cong 24\,cm$$

Hence, since the thickness of the barrier (24 cm) is greater than the minimum length of reaction zone (0.6 cm), the thickness of barrier would provide a sufficient time for Cr(VI) reduction. The above estimation does not account for solution ligands such as chloride, carbonate, sulphate, citrate, oxalate, nitrate, and phosphate that can complex with Fe^0 to decrease chromate reduction rate. In addition, the overall design of the barrier should consider the possible existence of dissolved oxygen in the EO flow or the O_2 gas bubble produced from the electrolysis reaction at the anode that can oxidize the Fe^0 and decrease the efficiency of the ZVI reduction performance.

23.3 IMPLEMENTATION OF EK–PRB TO POLLUTED SOIL

A number of recent studies have been carried by integrating PRBs into the EK process for decontamination purposes. The performance recorded is summarized in Table 23.2. The research results on this approach are outlined in this section.

TABLE 23.2. Comparison of the Performance of Coupled EK–PRB in Various Applications

Soil			EK				PRB		Removal	Reference
Texture	Contaminant	Electrodes	Processing fluid	Electric gradient	Duration	Material	Number/position			
Natural clay	Cr(VI): 1150 mg/kg	Graphite rods	GW	2 V/cm	6 days	ZVI/sand w/w 1:2	1/close to anode		100%	Weng et al., 2006
Natural clay	Cr(VI): 1150 mg/kg	Graphite rods	GW	2 V/cm	6 days	ZVI/sand w/w 1:2	1/middle		100%	Weng et al., 2006
Natural clay	Cr(VI): 1150 mg/kg	Graphite rods	GW	2 V/cm	6 days	ZVI/sand w/w 1:2	1/close to cathode		100%	Weng et al., 2006
Natural clay	Cr(VI): 2497 mg/kg	Graphite rods	GW	1 V/cm	6 days	ZVI/sand w/w 1:1	1/middle		85.8 %	Weng et al., 2007
Natural clay	Cr(VI): 2497 mg/kg	Graphite rods	GW	2 V/cm	6 days	ZVI/sand w/w 1:1	1/middle		92.5%	Weng et al., 2007
Silt loam	As(V): 966 mg/kg	Graphite rods	GW	2 V/cm	5 days	ZVI/sand w/w 1:2	1/middle		51%	Yuan and Chiang, 2007
Silt loam	As(V): 966 mg/kg	Graphite rods	GW	2 V/cm	5 days	FeOOH/sand w/w 1:2	1/middle		60%	Yuan and Chiang, 2007
Sandy loam	PCE: 60 mg/kg	Graphite plate	0.01 M $NaCO_3$	1 V/cm	10 days	ZVI	1/close to anode		99% pore water 85% soil	Chung and Lee, 2007
Sandy loam	PCE: 60 mg/kg	Graphite plate	0.01 M $NaCO_3$	2 V/cm	10 days	ZVZ	1/close to anode		99% pore water 80% soil	Chung and Lee, 2007
Natural clay	TCE: 116 mg/kg	Graphite rods	Deionized water	0.2 V/cm	8.3 days	Atomizing slag	2/in the electrode compartments		85%	Chung and Lee, 2007
Natural clay	Cd: 116 mg/kg	Graphite rods	de-ionized water	0.2 V/cm	200 h	Atomizing slag	2/in the electrode compartments		90%	Chung and Lee, 2007

Note: ZVI: zero-valent iron, ZVZ: zero-valent zinc, GW: groundwater.

23.3.1 Remediation of Cr(VI)-Contaminated Soil

In situ chemical chemical/organic reduction (He and Traina, 2005; James, 2001; Ludwig *et al.*, 2007) and bioreduction (Smith, 2001) can be applied to remediate Cr-contaminated soils. However, these processes cannot effectively remove Cr from a low-permeability soil such as clayey soil. A laboratory scale of the EK process coupled with PRBs to remediate Cr(VI)-contaminated clay was first introduced by Weng *et al.* (2006). The effectiveness of this process was tested using a 12-cm-long cylindrical cell (4.2-cm diameter) consisting of approximately 300 g of the artificially Cr(VI)-contaminated clay (1150 mg/kg) with a moisture content of $32 \pm 1\%$. Simulated groundwater was used as processing fluid (conductive medium). A reactive barrier filled with granular Fe^0 and sand (1:2 w/w) was installed either near the cathode or anode or in the middle of the soil specimen. Weng *et al.* reported that the position of the barrier had an impact on the EO permeability but not on the Cr(VI) reduction. When the barrier was installed near the anode, a relatively lower EO permeability was found than at the barrier installed in the middle or near the cathode. This phenomenon was probably associated with dissolution of Fe at the acidic region, which resulted in a greater electric resistance caused by more Fe hydroxide precipitates along the cell. Installation of a ZVI barrier in the EK process could increase both Cr(VI) reduction and total Cr removal. Experimental results showed that the removal of Cr from clay was characterized by the high Cr(VI) concentration that occurred in the anode reservoir and the presence of Cr(III) precipitates in the cathode reservoir (Fig. 23.5). Weng *et al.* pointed out that this process could achieve a nearly 100% reduction of the Cr(VI) in the clayey soil. Up to 60 to 71% total Cr removal was achieved for the process with an Fe^0 barrier with respect to only a 28% removal efficiency for the process without barrier. They indicated that the electro-osmotic advection and ionic migration are the dominant transport mechanisms responsible for the removal of Cr from clay. Based on their study, Cr(VI) reduction was significantly enhanced by coupling Fe^0 to the barrier in the EK process. The total costs, counting the energy consumption, graphite rod

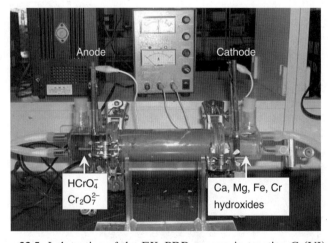

Figure 23.5. Lab testing of the EK–PRB process in treating Cr(VI) soil.

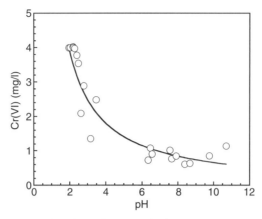

Figure 23.6. Effect of pH on the adsorption of Cr(VI) onto natural clay. Experimental conditions: Cr(VI) concentration 4.03 mg/l; natural clay 1 g/l; $NaClO_4$ 1×10^{-2} M; 27 °C.

electrodes, and Fe^0 being used were US\$37.9 /m^3 for the EK process coupled with an Fe^0 barrier.

Since this coupled EK–PRB process can reduce all Cr(VI) in the soil to the less toxic form of Cr(III), Weng *et al.* (2007) continued to study this subject and further tested its effectiveness to remediate a clay sample contaminated with a hyper-Cr(VI) concentration of 2497 mg/kg. In their tests, the barrier was installed in the center of the soil specimen and was filled with 1:1 w/w ratio of granular ZVI and sand. They revealed that transport of the acid front (H^+) was greatly retarded by the strong opposite migration of anionic chromate ions, whereupon a reverse electro-osmosis flux with K_e of 0.5×10^{-5} to 1.0×10^{-5} cm^2/V-s resulted, and an alkaline zone across the specimen was developed that promoted the release of Cr(VI) from the clay. Figure 23.6 gives the effect of pH on the amount of Cr(VI) adsorbed. The marked adsorption was observed under acidic conditions, and the amount adsorbed decreased with increasing pH. Note that the desorption of Cr(VI) from soil will favor the alkaline condition.

This pronounced change in the direction of EO flow was mainly attributed to the strong migration of anionic chromate ions toward the anode. The impact of this reversed EO flow is that it caused an alkaline condition across the soil specimen and created a favorable condition for Cr(VI) desorption from soil. Therefore, reverse EO flow would promote migration of Cr(VI) toward the anode and consequently accumulate in the anodic reservoir and/or by reduced by the ZVI barrier. They stated that the removal of Cr(VI) was accomplished by either EO flow or by the reduction of ZVI as the flow passed through the barrier. A large portion of Cr ions migrated into the anode reservoir as they were dragged by the reverse EO flow, and they presented mainly as Cr(VI) with a small amount of Cr(III). As such, chromium removal was characterized by a high Cr(VI) concentration that occurred in the anolyte and by the presence of Cr(III) precipitates in the catholyte. The Cr(VI) reduction efficiencies for the process without a ZVI wall were only 68.1% and 79.2% for 1 and 2 V/cm, respectively. As the ZVI barrier was installed, the corresponding reduction efficiencies increased to 85.8% and 92.5%. The costs for

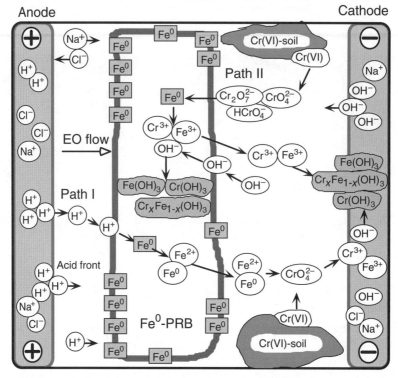

Figure 23.7. Cr(VI) reduction pathway in contaminated soil by coupled EK–PRB.

energy and ZVI utilized in this process are US$41.0/m³ and US$57.5/m³ for the system with an electric gradient of 1 and 2 V/cm, respectively. The major advantage of the EK–PRB process is that the ZVI barrier can effectively reduce Cr(VI) contamination.

In the EK–PRB treatment system, the removal mechanism for Cr(VI) can be illustrated by Fig. 23.7. Since the removal is believed to involve the reduction of Cr(VI) to Cr(III) by the oxidation of Fe^0 to Fe(II)/Fe(III) and the accumulation of Cr(VI) in the reservoirs, two simple removal paths are sketched. The first path is reduction of Cr(VI) due to the corrosion of the acid front passing through the barrier and bringing the dissolved Fe^0/Fe^{2+} ions into the EO flow. The second path is the migration of Cr(VI) ions directly into a PBR reacting with the ZVI. The reaction scenarios of this reductive precipitation mechanism are described as below.

The electrolysis reaction produces both H^+ and OH^- ions in the anode and cathode, respectively. The movement of H^+ ions (acid front) advancing through soil toward the cathode can result in the release of Fe^0/Fe^{2+} ions to the EO flow via mineral dissolution (corrosion of the ZVI). At the same time, the OH^- ion move toward the anode creates a favorable alkaline zone for Cr(VI) desorption from soil. Thus, the chromates are reduced either at the Fe^0–water interface in the barrier or by the Fe^0/Fe^{2+} ions in the EO flow path. The existence of Cr(VI) in the cathode reservoir due to the diffusion of soluble Cr(VI) caused by the concentration gradient near the cathode will also be reduced to Cr(III) by Fe^{2+} ions. In addition, the occurrence of a secondary electrolysis reaction could reduce Cr(VI) to Cr(III)

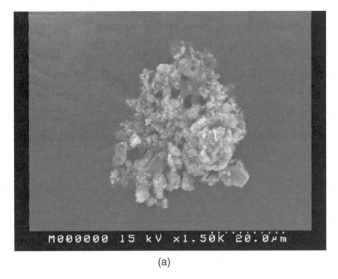

Figure 23.8. Fe^0 aggregates (from Connelly–GPM Inc., USA) after 6 days of use for treating Cr(VI). (a) SEM photomicrograph of Fe^0 aggregate with surface precipitates (1500× magnification). (b) Energy dispersive X-ray spectroscopy (EDS) spectra shown for the entire used Fe^0 aggregate. The semiquantitative results indicate Si 5.28 wt %, Fe 79.86 wt %, O 6.65 wt %, Cr 8.22 wt %.

under a highly reductive environment in the catholyte. Because a basic medium was generated in the catholyte, all the cationic metal species, such as the reducing products Fe(III) and Cr(III) would precipitate as sparingly soluble Fe(III)-Cr(III) oxyhydroxides $[Cr_xFe_{1-x}(OH)_3(s)]$ or hydroxides $[Cr(OH)_3(s)$ and $Fe(OH)_3(s)]$ in the catholyte when the concentration reached its solubility limit.

While a portion of chromates (CrO_4^{2-}, $HCrO_4^-$, and $Cr_2O_7^{2-}$) is reduced by the advection of the EO flow that contains Fe^0/Fe^{2+} ions from the corrosion of ZVI in the barrier, most chromates do directly migrate into the PRB and react with the ZVI. The subsequent reaction products, Cr(III) and Fe(III), will migrate toward the cathode or reside in the soil pore solution or in the barrier to form precipitates of Cr(III)-Fe(III) oxyhydroxides or hydroxides. Figure 23.8 shows the Fe^0 filings

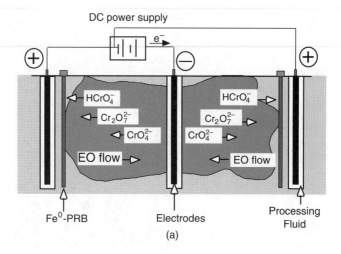

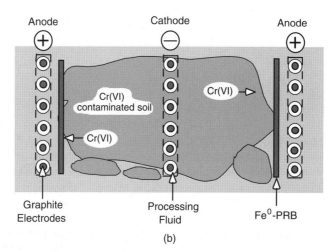

Figure 23.9. (a) Conceptual diagram and (b) top view of the coupled EK–PRB process for remediation of hyper Cr(VI)-contaminated soil.

after being used in EK–PRB process. The scanning electron microscope (SEM) image showed that the hollows were attributed to the corrosion of the Fe aggregate and the reaction products—ferric oxide or ferric hydroxide—attached on the surface of the filings.

A conceptual diagram of the EB–PRB approach to remediate Cr(VI)-soil is proposed (Fig. 23.9). This series EK process is operated with two ZVI barriers, which are placed near the electrodes allowing Cr(VI) reduction in the barrier. The design of the layout of this process is site specific. Depending on the contamination level, distribution of contaminants, soil properties, mineralogical and geochemical data, other PRB configurations, including the use of the funnel and the gate may applicable in order to complete capture of the contaminants.

23.3.2 Remediation of Arsenic-Contaminated Soil

The EK–PRB process is also applicable to treat arsenic-contaminated soil. Yuan and Chiang (2007) tested the effectiveness of this process using soil preloaded with arsenic concentrations of 423 and 966 mg/kg under a potential gradient of 2 V/cm. The PRB media they used were ZVI powder and FeOOH (goethite). In their experiments, the ZVI powder was produced by the reduction of Fe^{3+} (Eq. 23.10), and FeOOH was formed by reacting FeCl with NaOH (Eq. 23.11).

$$4Fe^{3+} + 3BH_4^- + 9H_2O \rightarrow 4Fe^0(s) + 3H_2BO_3^- + 12H^+ + 6H_2 \quad (23.10)$$

$$Fe^{3+} + 3OH^- + 9H_2O \rightarrow FeOOH + H_2O \quad (23.11)$$

Yuan and Chiang (2007) reported that the installation of a barrier in the EK system did enhance arsenic removal from the soil matrix. Arsenic removal efficiency was achieved that was 1.6–2.2 times higher than that of the systems without a barrier. In particular, a better performance of arsenic removal was shown in a barrier filled with FeOOH than with ZVI due to the higher surface area of FeOOH adsorbing more anionic As(V), i.e., $HAsO_4^{2-}$ and AsO_4^{3-} ions. The best performance in removing arsenic was achieved when the barrier was placed in the middle of the soil specimen because the electromigration of $HAsO_4^{2-}$ ion was inhibited by the EO flux, and the sorption of arsenic was much less at the alkaline region near the cathode side. Unlike a reverse EO flow (moving from cathode toward anode) that was found in the study by Weng's group (2007), a similar reverse flow was not observed in the study by Yuan and Chiang (2007), although the anionic As(V) is likely move toward anode by electromigration effect.

Yuan and Chiang reported that the overall As(V) removal efficiency for the process without ZVI barrier was 27% for 966 mg/kg at a 5-day treatment. As barriers were installed at the center of the soil specimen, the corresponding As(V) removal efficiencies increased to 51% and 60%, respectively for the ZVI and FeOOH fillings. Although ZVI in the barrier can serve as a good reagent for As(V) reduction, they reported that during the test periods, no As(III) was found in the system. They concluded that the removal of As was largely attributed to the surface adsorption and precipitation on PRB media (ZVI and FeOOH). Relevant studies have shown that the use of ZVI in PRB could remediate arsenic-contaminated groundwater by promoting adsorption of arsenic onto the iron surface (Farrell et al., 2001; Melitas, Conklin, and Farrell, 2002). The formation of arsenic precipitates under alkaline conditions (pH 12) can be described as follows (Bothe and Brown, 1999):

$$HAsO_4^{2-} + Ca^{2+} + nH_2O \rightarrow CaHAsO_4 \cdot nH_2O \quad (23.12)$$

$$2AsO_4^{3-} + 3Ca^{2+} + nH_2O \rightarrow Ca_3(HAsO_4)_2 \cdot nH_2O \quad (23.13)$$

23.3.3 Remediation of DNAPLs-Contaminated Soil

Dense nonaqueous phase liquids (DNAPLs) are the chlorinated organic compounds that have a density higher than water. Because of this unique property, DNAPLs will pass through the vadose zone and contaminate groundwater. Some reside in clay lenses, while others remain in the vadose zone of the soil layer and become a contamination source for groundwater. Due to their low aqueous solubility, broad distribution, and persistence in soil, remediation of the subsurface contaminated with DNAPLs can pose a great challenge. Among the technologies developed for targeting DNAPLs decontamination, the ZVI destruction technique has become one of the priority solutions for subsurface remediation because of its advantages. It is effective, commercially inexpensive and readily available, and *in situ* applicable (Shin, Singhal, and Park, 2007; Plagentz, Ebert, and Dahmke, 2006; Cho *et al.*, 2005; Parbs, Ebert, and Dahmke, 2007). This section outlines recent studies of EK–PRB process, which is considered a feasible *in situ* technique for remediation of DNAPLs-contaminated soil.

Chang and Cheng (2006) demonstrated that the EK–PRB process could be used to remediate perchloroethylene- (PCE) contaminated soils. In their study, the barrier was positioned close to the anode. The experiments were conducted at a constant voltage gradient of 1.0 V/cm, and 0.01 M sodium carbonate served as the processing fluid to maintain at neutral range so as to avoid soil acidification. In this process, the ZVI and ZVZ (zero-valent zinc) served as dechlorination reaction agents. The reductive dechlorination reaction in the aqueous solution for ZVI and chlorinated organic compounds (RCl) is expressed as follows (Johnson, Scherer, Tratnyek, 1996):

$$Fe^0 + RCl + H^+ \rightarrow Fe^{2+} + RH + Cl^- \qquad (23.14)$$

$$2Fe^{2+} + RCl + H^+ \rightarrow 3Fe^{3+} + RH + Cl^- \qquad (23.15)$$

The formation of reaction products such as ferric oxide or ferric hydroxides on the ZVI surface will decrease the activity of the ZVI, therefore limiting the operational life of ZVI (Li and Farrell, 2000). As shown in Eqs. 23.14 and 23.15, the proton (H^+) is not only beneficial to the reductive dechlorination reactions but it can also eliminate the formation of precipitates (ferric oxide or ferric hydroxide) to maintain the ZVI activity. Chang and Cheng (2006) have made an important note that the electrolysis reaction in the EK system would act as a proton (H^+) provider when incorporating ZVI into the system. Therefore, the higher the voltage gradient that was applied to the process, the more H^+ production at the anode enhanced PCE degradation on the process. It was proved that the 2.0 V/cm operation performs a better PCE soil remediation efficiency than the lower potential gradients, i.e., 1.0 and 0.5 V/cm. The best removal efficiency tested could reach 99% and 90% for decontaminating the PCE in the pore water and the soil, respectively, after a 10-day treatment. They also found that as the barrier filled with ZVZ, the process achieved a better PCE degradation efficiency than that of ZVI. The soils treated by this process still roughly maintained their original properties.

By the same token, the concept of EK–PRB can also be applied to remediate soil contaminated with TCE. A new approach was proposed by Chung and Lee (2007). They used atomizing slag (commercial name: PS Ball, worldwide patent) and sand as the filter media in the barrier for the decontamination of soil/groundwater contaminated with TCE (300 mg/l) under an applied voltage gradient of 0.2 V/cm. The atomizing slag is a porous material with a major component of Fe_2O_3 (28.2% to 44.2%), CaO (17.4% to 40.4%), SiO_2 (11.7% to 20.0%), Al_2O_3 (3.0% to

7.6%), and others (4.6% to 10.2%) (SAT and PS Ball Ecomaster, 2008). The main application of this slag is as construction material. Chung and Lee chose this slag as the barrier material in their study is because it is less expensive than other PRB materials such as ZVI aggregate and Zeolite.

They reported that the effluent TCE concentrations through the PRB process were much lower than those of EK remediation without atomizing slag. In addition, the effluent chloride concentrations were greater under the EK process with the PRB system than those under EK remediation only, which was caused by the dechlorination of TCE from the reaction between TCE and the reactive material. Although both TCE and chloride concentrations were measured during the experiments, the daughter products such as dichloroethylene (DCE) and vinyl chloride (VC) were not monitored in their study. This information, however, is necessary in order to verify the effectiveness of the reductive dechlorination reaction using this patented commercial material (atomizing slag). Overall, a 90% TCE removal efficiency was achieved. Based on these findings, Chung and Lee concluded that the coupled EK-PRB could effectively remediate contaminated groundwater and soil *in situ*. The use of atomizing slag as the PRB reactive material has shown promise for the dechlorination of TCE. Figure 23.10 presents the conceptual diagram of the EB–PRB approach to remediate TCE-contaminated soil. Because the movement of TCE accompanies the EO flow, the electrical polarity is switched with respect to the layout of the remediation of Cr(VI) shown previously (Fig. 23.9).

23.4 PERSPECTIVES

The *in situ* application of the EK process has become popular in soil remediation. PRBs have long been used in the remediation of groundwater that contains chlorinated organics, concentrated heavy metals, and nitrates. Fe^0 is the most popular media used in PRB fillings to destroy the target contaminants that pass through the barriers. It also allow the simultaneous removal of concerned priority pollutants from groundwater. By adapting the advantages of PRBs into the EK process, this chapter gives particular relevance to the potential *in-situ* application of this innovative EK–PRB process and its ability remove those reactive contaminants from low-permeability soil. A brief review of the current research subjects on this matter shows that the optimistic field implementation of incorporating the Fe^0-barrier technology into the EK process.

Since all the tests presented were performed in the laboratory, before field application approaches are undertaken, scaled-up research and further on-site testing experiences are highly recommended. The heterogeneity nature of the soil textures and mixed contaminants that may coexist in the site are the major concerns in the development of this EK–PRB process. In addition, highly contaminated hazardous waste sites can have extreme pH and E_h (redox potential) values that may accelerate the corrosion of Fe^0 in the barrier or change the barrier reactivity. More research is needed to address such influences as to ultimate the performance of EK–PRB.

This chapter has highlighted the unique features of this coupled technique, but the reader should not assume that this technique can do it all or that it can offer a complete solution in achieving cleanup criteria. Indeed, in most site remediation

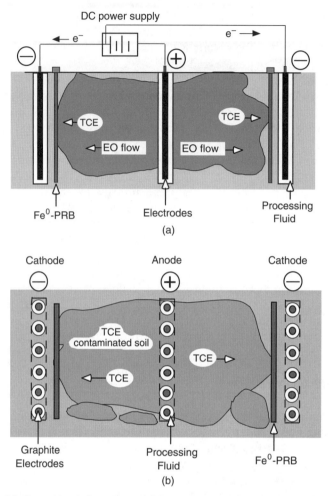

Figure 23.10. (a) Conceptual diagram and (b) top view of the coupled EK–PRB process for remediation of TCE-contaminated soil.

cases, a variety of remedial technologies often needs to be applied (also known as the treatment train) to meet the remediation goals.

REFERENCES

Alshawabkeh AN, Yeung AT, Bricka MR. (1999). Practical aspects of in-situ electrokinetic extraction. *Journal of Environmental Engineering* **125**(1):27–35.

Blowes DW, Ptacek CJ, Jambor JL. (1997). In-situ remediation of Cr(VI)-contaminated groundwater using permeable reactive walls: Laboratory studies. *Environmental Science and Technology* **31**(12):3348–3357.

Blowes DW, Ptacek CJ, Benner SG, McRae CWT, Bennett TA, Puls RW. (2000). Treatment of inorganic contaminants using permeable reactive barriers. *Journal of Contaminant Hydrology* **45**:123–137.

Bothe, Jr. JV, Brown PW. (1999). Arsenic immobilization by calcium at arsenate formation. *Environmental Science and Technology* **33**:3806–3811.

Coletta TF, Bruell CF, Ryan DK, Inyang HIJ. (1997). Cation-enhance removal of lead from kaolinite by electrokinetics. *Environmental Engineering* **123**(12):1227–1233.

Chang JH, Cheng SF. (2006). The remediation performance of a specific electrokinetics integrated with zero-valent metals for perchloroethylene contaminated soils. *Journal of Hazardous Materials* **131**:153–162.

Cho HH, Lee T, Hwang SJ, Park JW. (2005). Iron and organo-bentonite for the reduction and sorption of trichloroethylene. *Chemosphere* **58**:103–108.

Chung HI, Lee MH. (2007). A new method for remedial treatment of contaminated clayey soils by electrokinetics coupled with permeable reactive barriers. *Electrochimica Acta* **52**:3427–3431.

Eary LE, Rai D. (1988). Chromate removal from aqueous wastes by reduction with ferrous ion. *Environmental Science and Technology* **22**:972–977.

Ebert M, Kober R, Parbs A, Plagentz V, Schafer D, Dahmke A. (2006). Assessing degradation rates of chlorinated ethylenes in column experiments with commercial iron materials used in permeable reactive barriers. *Environmental Science and Technology* **40**:2004–2010.

Farrell J, Wang J, O'Day P, Conklin M. (2001). Electrochemical and spectroscopic study of arsenate removal from water using zero-valent iron media. *Environmental Science and Technology* **35**(10): 2026–2032.

He YT, Traina SJ. (2005). Cr(VI) reduction and immobilization by magnetite under alkaline pH conditions: The role of passivation. *Environmental Science and Technology* **39**(12): 4499–4504.

Ho VS. (1999). The lasagna technology for in situ soil remediation. I: Small field test. *Environmental Science and Technology* **33**(7):1086–1091.

James B. (2001). Remediation-by-reduction strategies for chromate contaminated soils. *Environmental Geochemistry and Health* **23**(3):175–179.

Johnson TL, Scherer MM, Tratnyek PG. (1996). Kinetics of halogenated organic compound degradation by iron metal. *Environmental Science and Technology* **30**: 2634–2640.

Kamolpornwijit W, Liang L, Moline GR, Hart T, West OR. (2004). Identification and quantification of mineral precipitation in Fe0 fillings from a column study. *Environmental Science and Technology* **38**(21):5757–5765.

Lai KCK, Lo IC. (2008). Removal of chromium (VI) by acid-washed zero-valent iron under various groundwater geochemistry conditions. *Environmental Science and Technology* **42**:1238–1244.

Li T, Farrell J. (2000). Reductive dechlorination of trichloroethylene and carbon tetrachloride using iron and palladized iron cathodes. *Environmental Science and Technology* **34**:173–179.

Liang L, Korte N, Gu B, Puls R, Reeter C. (2000). Geochemical and microbial reactions affecting the long-term performance of in situ "iron barriers." *Advances in Environmental Research* **4**:273–286.

Liu F, Lu Y, Chen H, Liu Y. (2003). Removal of Cr^{6+} from groundwater using zero valence iron in the laboratory. *Chemical Speciation and Bioavailability* **14**:75–77.

Ludwig R, Su C, Lee TR, Wilkin RT, Acree RT, Ross RR, Keeley A. (2007). In situ chemical reduction of Cr(VI) in groundwater using a combination of ferrous sulfate and sodium dithionite: A field investigation. *Environmental Science and Technology* **41**(15):5299–5305.

Mayer KU, Blowes DW, Frind EO. (2001). Reactive transport modeling of an in situ reactive barrier for the treatment of hexavalent chromium and trichloroethylene in groundwater. *Water Resources Research* **37**(12):3091–3103.

Melitas N, Conklin M, Farrell J. (2002). Electrochemical study of arsenate and water reduction on iron media used for arsenic removal from potable water. *Environmental Science and Technology* **36**(14):3188–3193.

Morrison SJ, Metzler DR, Carpenter CE. (2001). Uranium precipitation in a permeable reactive barrier by progressive irreversible dissolution of zerovalent iron. *Environmental Science and Technology* **35**(2):385–390.

Parbs A, Ebert M, Dahmke A. (2007). Influence of mineral precipitation on the performance and long-term stability of Fe-0-permeable reactive barriers: A review on the basis of 19 Fe-0-reactive barrier sites. *Groundwater* **12**(4):267–281.

Phillips DH, Gu B, Watson DB, Roh Y, Liang L, Lee SY. (2000). Performance evaluation of a zero valent iron reactive barrier: Mineralogical characteristics. *Environmental Science and Technology* **34**:4169–4176.

Phillips DH, Watson DB, Roh Y, Gu B. (2003). Mineralogical characteristics and transformations during long-term operation of a zero valent iron reactive barrier. *Journal of Environmental Quality* **32**:2033–2045.

Plagentz V, Ebert M, Dahmke A. (2006). Remediation of ground water containing chlorinated and brominated hydrocarbons benzene and chromate by sequential treatment using ZVI and GAC. *Environmental Geology* **49**:684–695.

Puls RW, Paul CJ, Powell RM. (1999). The application of in situ permeable reactive (zero-valent iron) barrier technology for the remediation of chromate-contaminated groundwater: A field test. *Applied Geochemistry* **14**(8):989–1000.

Reddy KR, Chinthamreddy S (2003). Effects of initial form of chromium on electrokinetic remediation in clays. *Advances in Environmental Research* **7**:353–365.

Reddy KR, Xu CY, Chinthamreddy S. (2001). Assessment of electrokinetic removal of heavy metals from soils by sequential extraction analysis. *Journal of Hazardous Materials* **84**:279–296.

Rocca CD, Belgiorno V, Meriç S. (2007). Overview of *in-situ* applicable nitrate removal processes. *Desalination* **204**:46–62.

Sah JG, Lin LY. (2000). Electrokinetic study on copper contaminated soils. *Journal of Environmental Science and Health Part A* **35**(7):1117–1139.

SAT and PS Ball Ecomaster. (2008). http://www.ecomaister.com/renewal/eng/business/business120.php. Accessed November 24, 2008.

Shin HY, Singhal N, Park JW. (2007). Regeneration of iron for trichloroethylene reduction by Shewanella alga BrY. *Chemosphere* **68**:1129–1134.

Simon FG, Meggyes TT, Czurda K, Roehl KE. (2001). Long-term behaviour of permeable reactive barriers used for the remediation of contaminated groundwater. *Proceedings of the 8th International Conference on Radioactive Waste Management and Environmental Remediation*, Vol. 1, ICEM 2001, pp. 637–641.

Smith WL. (2001). Hexavalent chromium reduction and precipitation by sulphate-reducing bacterial biofilms. *Environmental Geochemistry Health* **23**(3):297–300.

Weng CH, Huang CP. (2004). Preliminary study on treatment of soil enriched in COPR by electrokinetics. *Practice Periodical of Hazardous, Toxic, and Radioactive Waste Management* **8**(2):67–72.

Weng CH, Huang CP, Allen HE, Cheng AHD, Sanders PF. (1994) Chromium leaching behavior in soil derived from chromite ore processing residue. *Science of the Total Environment* **154**:71–86.

Weng CH, Lin TY, Chu SH, Yuan C. (2006). Laboratory-scale evaluation of Cr(VI) removal from clay by electrokinetics incorporated with Fe(0) barrier. *Practice Periodical of Hazardous, Toxic, and Radioactive Waste Management* **10**(3):171–178.

Weng CH, Lin YT, Lin TY, Kao CM. (2007). Enhancement of electrokinetic remediation of hyper-Cr(VI) contaminated clay by zero-valent iron. *Journal of Hazardous Materials* **149**(2):292–302.

Weng CH, Yuan C. (2001). Removal of Cr(III) from clay soils by electrokinetics. *Environmental Geochemistry Health* **23**(3):281–285.

Weng CH, Yuan C, Tu HH. (2003). Removal of trichloroethylene from clay soil by series-electrokinetic process. *Practice Periodical of Hazardous Toxic and Radioactive Waste Management* **7**(1):25–30.

Wilkin RT, Su C, Ford RG, Paul CJ. (2005). Chromium removal processes during groundwater remediation by a zero valent iron permeable reactive barrier. *Environmental Science and Technology* **39**(12):4599–4605.

Yalcin B, Li H, Gale RJ. (1992). Phenol removal from kaolinite by electrokinetics. *Journal of Geotechnical Engineering* **118**(11):1837–1852.

Yang GCC, Liu CY. (2001). Remediation of TCE contaminated soil by in situ EK-Fenton process. *Journal of Hazardous Materials* **85**:317–3331.

Yang JE, Kim JS, Ok YS, Yoo KR. (2007). Mechanistic evidence and efficiency of the Cr(VI) reduction in water by different sources of zerovalent irons. *Water Science and Technology* **55**(1–2):197–202

Yuan C, Chiang TS. (2007). The mechanisms of arsenic removal from soil by electrokinetic process coupled with iron permeable reaction barrier. *Chemosphere* **67**:1533–1542.

Yuan C, Weng CH. (2004). Remediating ethylbene-contaminated clayey soil by a surfactant aided electrokinetic (SAEK) process. *Chemosphere* **57**(3):225–232.

24

COUPLED ELECTROKINETIC–THERMAL DESORPTION

GREGORY J. SMITH

24.1 FUNDAMENTAL PRINCIPLES

Where electrical energy (either direct or alternating current) is applied to soil, resistive heating occurs. This process can be beneficial in the treatment of volatile organic contaminants when harnessed properly. The heating can be used to accelerate many chemical and biological reactions, and modify physical properties. Heating can be used to increase the desorption of many organic compounds and to remove dense nonaqueous phase liquids (DNAPLs). Thermal effects during electrokinetic treatment methods have not been well studied, and electrolytic effects have not been thoroughly evaluated with respect to *in situ* thermal treatment. Therefore, to evaluate the thermal desorption effects during treatment using electrokinetic methods, we will be relying on the body of knowledge developed from *in situ* thermal treatment. *In situ* thermal methods have been used for site remediation since the mid- to late 1980s, beginning in the form of steam injection and thermal conductive heating, with *in situ* electrical resistance heating (ERH) being patented in 1996 and commercialized for implementation shortly thereafter.

This chapter will present the principles governing thermal desorption and from this, the changes in chemical and biological reactions that occur during *in situ* thermal treatment.

24.2 THERMAL PRINCIPLES

Heating resulting from electrokinetic techniques in the subsurface involves the resistance to the passage of electrical current through soil moisture. It is this

---

Electrochemical Remediation Technologies for Polluted Soils, Sediments and Groundwater,
Edited by Krishna R. Reddy and Claudio Cameselle
Copyright © 2009 John Wiley & Sons, Inc.

24.2.1 Heat Transport

Energy transport in the subsurface is attributable to heat conduction in the porous matrix as well as heat transport by fluid motion (advection). In the absence of fluid movement, energy flow by conduction only is described by the relationship

$$q_x = -k(dT/dx), \qquad (24.1)$$

where:

q_x = heat energy flux in the x direction (W·m²),
k = thermal conductivity (W·m·°K), and
dT/dx = temperature gradient in the x direction (°K·m).

The terms "convection" and "advection" are often used interchangeably to describe energy transport due to movement of heated fluids. Convection includes heat diffusion as well as advection; however, heat transport by fluid diffusion tends to be minor compared with advection and conduction. In general, heat convection is the most effective transport mechanism in materials with moderate to high permeability, such as sand and gravel, while conduction is the dominant process in materials with lower permeability such as silt and clay (Fig. 24.1).

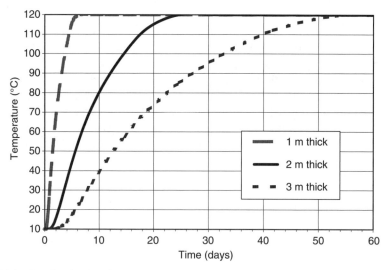

Figure 24.1. Conductive heating of 1-, 2-, and 3-m-thick impermeable layers from both sides. Steam temperature above and below the layer is 120°C, ambient temperature is 10°C, and soil diffusivity is 5.82×10^{-7} m²/s.

TABLE 24.1. Thermal Properties of Representative Materials and Fluids

Material	Thermal Conductivity (W/m·K)	Heat Capacity (kJ/m³·K)	Diffusivity (m²/s)	Density (g/cm³)
Quartz	8.79	2008	4.38×10^{-6}	2.66
Clay minerals	2.93	2008	1.46×10^{-6}	2.65
Organic matter	0.25	2510	9.96×10^{-8}	1.30
Silty sand (dry)	1.23	1906	9.76×10^{-7}	1.52
Silty sand (wet)	1.41	4359	5.82×10^{-7}	1.80
Silt (dry)	0.96	1078	1.29×10^{-6}	1.44
Silt (wet)	1.26	5030	4.77×10^{-7}	1.90
Water	0.57	4184	1.36×10^{-7}	1.00
Air	0.0218	1.3	1.68×10^{-5}	0.0013
Engine oil	0.15	1669	7.71×10^{-8}	0.89

Source: From Sundberg 1988; Wijk 1963.

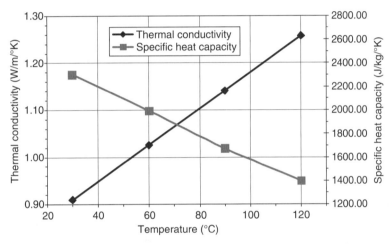

Figure 24.2. Thermal conductivity and specific heat capacity of dry silty sand versus temperature (U.S. EPA, 2002a).

24.2.2 Thermal Conduction and Thermal Diffusivity

Thermal conductivity is a measure of the ability of a material to conduct heat. Thermal diffusivity is a measure of a material's ability to conduct heat relative to its ability to store heat (defined as thermal conductivity divided by heat capacity; see next section). Heat can migrate relatively quickly through a material with high thermal conductivity, while heat flow into a material with high thermal diffusivity will result in a relatively rapid temperature increase. Typical values of thermal conductivity are presented in Table 24.1.

The thermal conductivity of solids and fluids varies with temperature, and representative values should be selected for the temperature range of interest (Fig. 24.2). Measured thermal conductivity of porous materials is a combination of conduction and convection occurring within the pore fluids. The change in measured thermal conductivity may be influenced by increased convection with temperature. Additional thermal conductivity values are available in the published literature for soil materials (Johansen, 1977; Sandberg, 1988) and for fluids (Poling, Prausnitz, and O'Connell, 2001). Note that the bulk thermal conductivity (k_{bulk}) for a soil is the combined thermal conductivity of the soil particles and the fluid contained in the pore space:

$$k_{bulk} = k_{particle}(1-n) + k_{fluid}(n), \tag{24.2}$$

where n is the soil porosity (unitless).

24.2.3 Heat Capacity

The specific heat of a material ($kJ \cdot kg^{-1} \cdot °K^{-1}$) is the energy required to raise the temperature of a unit mass by one degree (Fig. 24.2). The product of the specific heat and density is referred to as the heat capacity ($kJ \cdot m^{-3} \cdot °K^{-1}$) and provides a measure of the material's ability to store heat. The heat capacity of soils and fluids change with temperature; however, the range of variation for heat capacity of different soils is generally small compared with the variability of other parameters such as permeability. Similar to thermal conductivity, the bulk heat capacity of a soil is the combined heat capacity of the soil particles and the fluid in the pore space.

24.2.4 Heat of Vaporization

The heat of vaporization ($kJ \cdot kg^{-1}$) is the amount of energy required to vaporize a unit mass of material, that is, to boil a kilogram of liquid until it is entirely converted to gas. For example, 1 cal of energy is required to increase 1 cm^3 of water 1 °C in temperature, but it takes 80 cal to convert 1 cm^3 of water at 100 °C to steam. At the liquid boiling point, the energy input does not result in a temperature increase but rather a phase change. The total heat required to vaporize a liquid that is originally at ambient temperature will be the sum of the heat required to raise the liquid to its boiling point (product of specific heat, temperature increase, and mass) and the total heat required to vaporize the liquid (product of heat of vaporization and mass).

24.2.5 The Arrhenius Equation

The relationship between the rate at which a reaction proceeds and its temperature can be defined by the Arrhenius equation. A useful generalization supported by the Arrhenius equation is that, for many common chemical reactions at room temperature, the reaction rate doubles for every 10 °C increase in temperature. The Arrhenius equation is presented below:

$$k = A \cdot e^{(-E_a/RT)}, \tag{24.3}$$

where:

k = rate constant,
A = a constant,

E_a = activation energy,
R = universal gas constant $(8.314 \times 10^{-3} \, kJ \cdot mol^{-1} \cdot °K^{-1})$, and
T = temperature (°K).

24.2.6 Thermodynamics of Nonaqueous Phase Liquid (NAPL)/Water/Dissolved Gas Ebullition

According to Amos *et al.* (2005), ebullition in saturated porous media occurs when the sum of the vapor pressures exceeds the hydrostatic and capillary pressures. For pure water, this represents a phase change from liquid to gas (i.e. boiling). Since the total vapor pressure is the sum of partial pressures of all of the components of the mixture, the boiling point of the mixture (eutectic point of the azeotropic mixture) can be achieved at a lower temperature than the boiling points of any of the separate components (Lupis, 1983). This phenomenon has also been called co-distillation or steam distillation (Davis, 1998). The implication for *in situ* thermal remediation is that many contaminants can be removed as vapor at temperatures below that of the boiling point of water, or at steam temperatures where their boiling temperatures are greater than 100 °C. Table 24.2 provides eutectic points for select binary NAPL azeotropic systems.

24.2.7 Gas Mixtures

Four fundamental laws of thermodynamics describe the composition of a gas mixture:

- Dalton's law: the pressure exerted by a mixture of gases is the sum of the pressures that could be exerted by each individual gas occupying the same volume alone. This pressure is referred to as the partial pressure.
- Raoult's law: the equilibrium gas-phase partial pressure of a compound is equal to the product of the vapor pressure of the pure compound and the mole fraction of the compound in the solution phase.
- Henry's law: the equilibrium gas-phase partial pressure of a compound is equal to the mole fraction of the compound in the aqueous phase multiplied by a constant known as Henry's law constant (see Fig. 24.3).
- The ideal gas law: the mole fraction of a compound in the gas phase is equal to the ratio of its partial pressure and the total pressure.

24.2.8 Carbon Dioxide

Groundwater contains many dissolved gases, with one of the most significant being carbon dioxide (CO_2), which is at saturation in natural groundwater (Stumm and Morgan, 1981). Heating natural groundwater results in creating supersaturated conditions with respect to carbon dioxide, creating bubbles and, potentially, ebullition if the vapor pressures exceed the hydrostatic and capillary pressures (as mentioned above). Carbon dioxide results from the dissolution of carbonate minerals into groundwater. The equilibrium expression for the dissolution of calcium carbonate at neutral pH can be represented by (Stumm and Morgan, 1981)

TABLE 24.2. Select NAPL Compounds and Ebullition (Eutectic) Points

NAPL Mixture	Component Boiling Points (°C)	Ebullition Point (°C) (Eutectic Point)
Benzene	80.1	69.4
Water	100	
Carbon tetrachloride	76.8	66.8
Water	100	
Chlorobenzene	132	56.3
Water	100	
Chloroform	61.2	56.3
Water	100	
1,2-Dichloroethane	83.5	72.0
Water	100	
Dichloromethane	40.1	<39.9
Water	100	
1,4-Dioxane	83.5	87.8
Water	100	
Etylbenzene	136.2	92.0
Water	100	
Hexane	69.0	61.6
Water	100	
Styrene	145.2	93.9
Water	100	
Tetrachloroethene	121	88.5
Water	100	
Toluene	110.6	85.0
Water	100	
1,1,2-Trichloroethane	113.7	86.0
Water	100	
Trichloroethene	87.1	73.1
Water	100	
Xylene	139.1	94.5
Water	100	

Source: From Lide, 1999.

$$CaCO_3(s) + CO_2 + H_2O = Ca^{2+} + 2HCO_3^-; \quad {}^+K_{psO}. \tag{24.4}$$

This can be further represented by

$$\frac{[Ca^{2+}][HCO_3^-]^2}{p_{CO_2}} = {}^+K_{psO}. \tag{24.5}$$

Because of electroneutrality, $2[Ca^{2+}] \approx [HCO_3^-]$, the above equation can be rewritten as

$$[Ca^{2+}] \approx 0.63 \cdot {}^+K_{psO}^{1/3} \cdot p_{CO_2}^{1/3}, \tag{24.6}$$

$$\therefore p_{CO_2}^{1/3} \cong \frac{[Ca^{2+}]}{0.63 \cdot {}^+K_{psO}^{1/3}}, \tag{24.7}$$

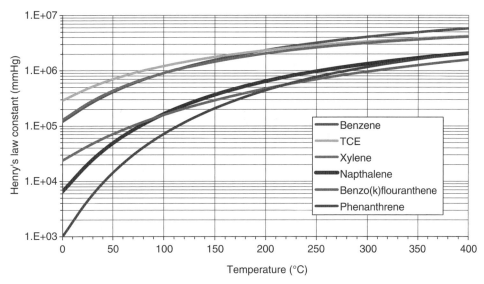

Figure 24.3. Henry's law constants versus temperature for various compounds (U.S. EPA, 1984).

where:

$^+K_{psO} = K_{so} \times K_1 \times K_H \times K$,

K_{so} = solubility constant,
K_1 = first acidity constant,
K_2 = second acidity constant, and
K_H = Henry constant.

Therefore, one can estimate the partial pressure of carbon dioxide by knowing the concentration of calcium.

Figure 24.4 presents photomicrographs of carbon dioxide bubble growth in saturated porous media. Globules of trichloroethene (TCE) as DNAPL are also present. In photomicrograph 1, we can see the saturated porous media and the TCE globules. Where carbon dioxide is at supersaturation in groundwater, bubbles form and grow. Carbon dioxide is a nonpolar and organic gas, which facilitates its ability to dissolve organic liquids and gases. Carbon dioxide being an organic gas, and since like dissolves like, provides for the dissolution of organic compounds into this gas in preference to water or steam. In photomicrograph 2, we see the CO_2 bubble grows and comes into contact with the TCE DNAPL globule. Note that the contact between the TCE globule and the CO_2 bubble is maximized; this facilitates the dissolution of TCE into CO_2. In photomicrographs 3 and 4, the CO_2 bubble remains relatively static in size, yet the TCE globule is reduced, showing that the TCE is being dissolved into the CO_2 bubble. In photomicrograph 5, we see the CO_2 bubble continues to grow, as would occur during heating and, in photomicrograph 6, comes into contact with another TCE globule, again with the contact between the two organic materials maximized.

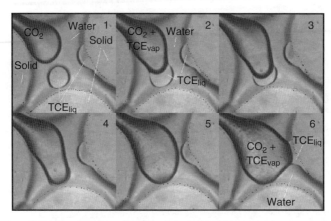

Figure 24.4. Photomicrograph of carbon dioxide bubble growth in saturated porous media. (Courtesy of Dr. Marios Ioannidis.)

The dehalogenation of chlorinated compounds in groundwater results in increases in chloride ions. Increases in anions result in increases in cations through what is known as the "common ion effect." Where calcium is the dominant cation, increases in chloride result in increase in calcium and hence carbon dioxide. This, in turn, increases the partial pressure of CO_2, facilitating the removal of chlorinated DNAPL compounds.

The "React" module of the Geochemists Workbench® (GWB) was used to model the changes in groundwater chemistry that occurred during ERH at a confidential site in the Chicago area by simulating the increase in calcium concentrations and CO_2 fugacity* presented in Table 24.3 (Smith, 2008). This was done by simultaneously increasing chloride concentrations and the temperature of groundwater. Chloride concentrations were increased linearly by 75.08 mg/l per day to mimic the increases in chloride concentrations observed from the dehalogenation of the perchloroethene (PCE) and TCE, as well as the associated daughter compounds, while increases in temperature were also simulated, increasing linearly from a baseline of 18 to 82°C over a period of 37 days. The modeled calcium and field data are presented in Table 24.3, with the corresponding partial pressures.

It can be seen that the model underpredicts the actual calcium concentrations. The model assumes instantaneous equilibrium, and the nature of the program is to ignore reaction kinetics, which in this case results in predictions of lower concentrations of Ca than were observed. The baseline conditions at the Chicago site showed Ca at 393 mg/l. The GWB "Spec8" module evaluated equilibrium conditions and determined that 372.9 mg/l represented the initial equilibrium free Ca concentration. According to GWB Spec8, $CaCl^+$ would be found at 22.49 mg/l (Ca equivalent at 11.93 mg/l), $CaHCO_3^+$ at 15.92 mg/l (Ca equivalent at 7.16 mg/l), $CaSO_4$ at 4.458 mg/l (Ca equivalent at 0.11 mg/l), and $CaCO_3$ at 0.03344 mg/l (Ca equivalent at 0.013 mg/l). These minerals had calculated activity coefficients of 0.8542, 0.85531, 1.000, and 1.000, respectively. As such, $CaSO_4$ and $CaCO_3$ would be precipitating, while $CaCl^+$ and $CaHCO_3^+$ would be in solution. This would raise the concentrations of Ca to

---

* Assuming ideal gas behavior, fugacity is equivalent to partial pressure.

TABLE 24.3. Modeled Calcium Concentrations and Fugacity of CO_2

Day	Temperature (°C)	Modeled Calcium Concentration (mg/l)	Actual Calcium Field Data (mg/l)	Modeled Fugacity of CO_2 (Bars)	Water Vapor Pressure (Bars)	PCE Vapor Pressure (Bars)	TCE Vapor Pressure (Bars)	Total Vapor Pressure (Bars)
0	18	204.6	393	0.6735	0.02023	0.03	0.06	0.78373
4.6	24.4	356.8	—	0.792	0.02993	0.03	0.09	0.94193
8.2	30.8	508.9	—	0.9213	0.04351	0.04	0.12	1.12481
11.8	37.2	660.8	—	1.061	0.0622	0.04	0.17	1.3332
15.4	43.6	812.7	—	1.209	0.08749	0.06	0.22	1.57649
19	50.0	964.4	—	1.366	0.1212	0.08	0.29	1.8572
22.6	56.4	1116	—	1.528	0.1655	0.11	0.38	2.1835
26.2	62.8	1267	—	1.696	0.2229	0.15	0.48	2.5489
29.8	69.2	1419	—	1.866	0.2964	0.19	0.60	2.9524
33.4	75.6	1570	—	2.038	0.3891	0.25	0.73	3.4071
37.0	82	1721	1870	2.208	0.5048	0.32	0.89	3.9228

392.13, 99.77% of the actual laboratory analysis, providing good agreement between modeled and field data. At 84 °C, GWB Spec8 predicted calcium at 2790 mg/l versus a laboratory measurement of 3279 mg/l. Calcium compounds with activity coefficients of less than one include $CaCl^+$ (619.7 mg/l; Ca equivalent of 329.1 mg/l), $CaHCO_3^-$ (204 mg/l; Ca equivalent of 81.69 mg/l), $CaSO_4$ (121.5 mg/l; Ca equivalent of 35.73 mg/l), and $CaCO_3$ (0.9691 mg/l; Ca equivalent of 0.388 mg/l), giving a total calcium concentration of 3236.91 mg/l, 98.7% of the actual laboratory analysis, again providing good agreement between field and modeled data. Therefore, it is believed that GWB Spec8 and React modules provide very good predictions of the calcium concentrations resulting from the common ion effect with chloride. However, because it assumes instantaneous reactions, it underpredicts the free calcium in the water and hence may underpredict the fugacity of CO_2. From Table 24.3, the sum of the partial pressures is at 1 atm at approximately 30 °C, a far lower temperature than one would expect by simply evaluating the contribution of the partial pressures of water, PCE, and TCE.

24.2.9 Other Gases

Groundwater contains many dissolved gases besides CO_2, including nitrogen and trace amounts of argon (Amos *et al.*, 2005). With the release of organic compounds to the subsurface, additional gases are generated, including nitrogen and ammonia from the reduction of nitrates and nitrites in groundwater, hydrogen sulfide from the reduction of sulfate in groundwater, and methane from the reduction of water and bicarbonate.

Quantification of various processes that produce these gases, including the rate of methane production, consumption, and transport in the saturated and unsaturated zones, as well as mass transfer between the zones, is difficult (Amos *et al.*, 2005). Methane production from below the water table may be significant where highly reducing conditions are found, and because methane is fairly insoluble, formation of gas bubbles and ebullition may be possible (Reeburgh, 1972; Kipphut and Martens, 1982; van Breukelen *et al.*, 2003). Methane may also be produced in the vadose zone but may be oxidized by the flux of atmospheric oxygen from the ground surface as part of the normal flow of air and percolation of water (Hers *et al.*, 2000; Chaplin *et al.*, 2002).

24.2.10 Thermodynamics of Steam Stripping via *In Situ* Steam Generation

Steam stripping is defined as a process where contaminants partition from soil, water, or NAPL phases into the vapor (steam) phase and are removed (stripped) from their source areas by vapor flow. The removal of the vaporized contaminants prevents the liquid or solid phases from reaching equilibrium with the vapor phase, allowing vaporization to continue at maximum rates. The stripped contaminants are condensed along with the steam at the edge of the steam zone (condensation front). When the concentration of a condensing contaminant exceeds its solubility, NAPL or solid compounds are deposited near the steam front, forming a "contaminant bank." This contaminant bank has the ability to migrate as the heating front propagates, which can be used to direct NAPL toward recovery wells for collection.

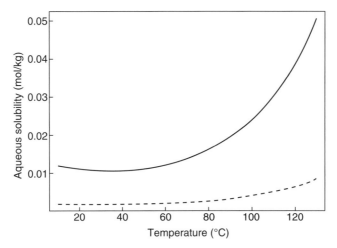

Figure 24.5. Aqueous solubilities of trichloroethylene (solid line) and tetrachloroethylene (dashed line) as a function of temperature. Reprinted from Knauss *et al.* (2000). Copyright (2000), with permission of Elsevier.

24.3 PHYSICAL AND CHEMICAL PRINCIPLES

24.3.1 Aqueous Solubility

For hydrocarbons, data on solubility at the temperatures attained during *in situ* thermal remediation are scarce. The aqueous solubilities of TCE and PCE as functions of temperature are presented in Figure 24.5. While the solubilities of the two compounds depicted increase exponentially with temperature, they do so only at temperatures above the conventional boiling point of water. The solubility actually achieves a minimum value around 30–50 °C as observed in experimental studies for both TCE and PCE (Imhoff, Frizzel, and Miller, 1997; Heron, Christensen, and Enfield, 1998; Heron *et al.*, 1998; Knauss *et al.*, 2000).

The temperature dependency for the aqueous solubility of large molecules such as naphthalene is relatively strong. Figure 24.6 shows the aqueous solubility of solid naphthalene in the temperature range from 20 to 140 °C. The solubility increases about 45 times (from 31 to 1350 mg/l) by heating from 25 to 100 °C. The solubility of naphthalene, when present in a mixture such as creosote, increases by a factor of 5–10 at temperatures typically seen in *in situ* thermal remediation (Davis, 2002). This helps to explain the dramatic increases seen in dissolved polycyclic aromatic hydrocarbon (PAH) concentrations during field-scale steam remediation at wood-treating sites such as Visalia Pole Yard (G. Heron, pers. comm.) and Wyckoff-Eagle Harbor (M.K. LeProwse and G. Heron, pers. comm.). An interesting result in investigating the temperature effect on the dissolution rate for NAPLs into water was that, even if the aqueous solubility of PCE did not change dramatically with temperature, the dissolution rate for the DNAPL increases about fivefold (Imhoff, Frizzel, and Miller, 1997). As a result, for systems with mass-transfer limitations, where the extracted water is below saturation, heating may increase the mass removal rate in the dissolved phase substantially.

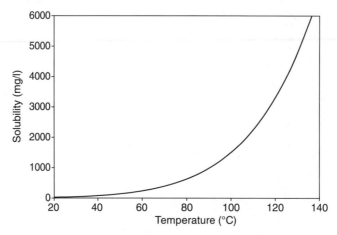

Figure 24.6. Calculated aqueous solubility of naphthalene as a function of temperature. [Based on the equations of Reid, Prausnitz, and Poling (1987).]

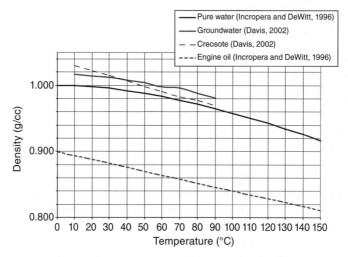

Figure 24.7. Density versus temperature for various liquids (Incropera and DeWitt, 1996; Davis, 2002).

Overall, the increased solubility of the contaminants can lead to enhanced dissolution, and for PAHs with low volatility, the aqueous phase removal can become a substantial component of the remedy.

24.3.2 Density

Figure 24.7 shows the variation of density with temperature for different fluids. In general, hydrocarbon compound densities will decrease about 10% for a temperature change of 100 °C (Davis, 1997). The density of water decreases about 4% over the temperature range from 0 to 100 °C. Although these changes are small, they can affect contaminant migration because of the more rapid change of NAPL density

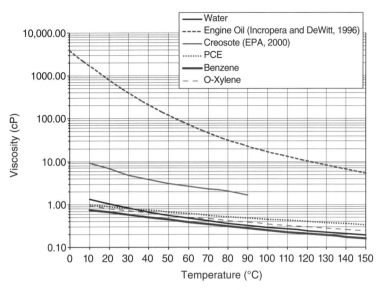

Figure 24.8. Viscosity versus temperature for liquids (Incropera and DeWitt, 1996; Poling, Prausnitz, and O'Connell, 2001; U.S. EPA, 2002a).

relative to groundwater density. DNAPLs with densities close to the density of water, such as creosote or halogenated hydrocarbon and oil and grease mixtures, can become light nonaqueous phase liquids (LNAPLs) at elevated temperatures.

24.4 FLUID AND ENERGY TRANSPORT

24.4.1 Viscosity

Figure 24.8 shows the temperature dependence of viscosity for different fluids. In general, the viscosity of most liquid organic chemicals decreases by about 1% for a temperature increase of 1 °C (Davis, 1997; Poling, Prausnitz, and O'Connell, 2001). Gas viscosities tend to be one to two orders of magnitude less than liquid viscosities but increase proportionally with temperature. Typically, the viscosity of a gas will increase about 30% with a temperature increase of 100 °C (Davis, 1997), facilitating transport through porous media.

24.4.2 Sorption

Soil–water sorption coefficients affect only the aqueous-phase transport of compounds. In general, sorption coefficients will decrease as a function of temperature, increasing the ability of hot fluids to remove contaminants from the soil. It is typically observed that during *in situ* heating, concentrations of volatile organic compounds in groundwater will increase during the early stages of heating, before a decline takes place with the associated removal of mass. The effect of temperature is specific to the contaminant, soil type, and water content. The soil–water sorption coefficient (saturated conditions) for TCE can theoretically decrease by 50% from 20 to 90 °C, while the soil–gas sorption (dry soil) coefficient for TCE can decrease

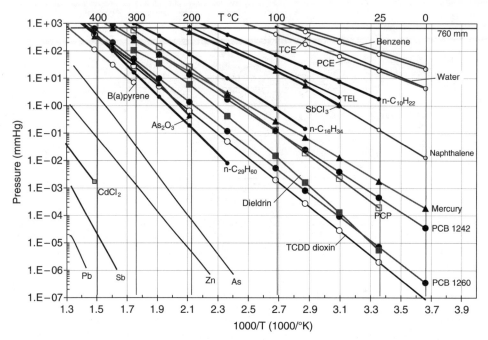

Figure 24.9. Vapor pressure of selected compounds versus temperature. (Note: Atmospheric pressure is 760 mm Hg.) (Copyright, 2001, CRC Press, used with permission) (Stegemeier and Vinegar, 2001). TEL, tetra-ethyl lead; PCP, pentachlorophenol; TCDD, 2,3,7,8-tetrachlorodibenzo-p-dioxin; PCB, polychlorinated biphenyls.

by an order of magnitude in the same temperature range (Heron, Christensen, and Enfield, 1998).

Based on these principles, the concentrations of hydrocarbon contaminants in vapor extracted from a thermal treatment zone will increase with temperature. Further, the relative proportion of each component in the vapor phase is dependent on its volatility (Henry's law constant) and its concentration in the liquid phase.

24.4.3 Vapor Transport

Volatilization of a chemical compound is controlled by its vapor pressure, ambient pressures (which vary with depth), and the partial pressures of associated chemicals and dissolved gases in groundwater. Vapor pressure increases with temperature (see Fig. 24.9), and ebullition occurs when the sum of the partial pressures of the chemical compound, water, and associated dissolved gases exceeds hydrostatic plus capillary pressures (Amos et al., 2005). That is, for ebullition to occur, bubbles must form under sufficient vapor pressure to rise to the water table and then overcome the capillary pressures to enter the vadose zone. As a result, evaporation or vaporization may occur below the pure-component boiling point.

Because the gas-phase concentration of a hydrocarbon compound depends on its Henry's law constant, a temperature increase will improve mass transport by increasing the concentration in the vapor phase, even if the temperature is below the boiling point. Figure 24.9 shows the vapor pressure of a variety of compounds

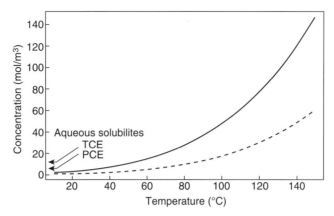

Figure 24.10. Vapor-phase concentrations of trichloroethylene (solid line) and tetrachloroethylene (dashed line) as functions of temperature. (Note: Calculated with the Clapeyron equation assuming ideal gas behavior). The arrows to the ordinate axis indicate the aqueous solubilities of the two compounds at 25 °C.

as a function of temperature. Figure 24.10 presents the predicted vapor-phase volumetric concentration of TCE and PCE as a function of temperature; the volumetric aqueous solubilities at room temperature are noted.

24.4.4 Hydrolysis

Hydrolysis involves the chemical reaction with water, without regard to redox conditions, dissolved minerals, or the presence of soil microbes. For the reaction to take place, the organic compound needs to be dissolved in water, and the reaction can be modified by pH and/or temperature. The general hydrolysis reaction, involves the exchange of some functional group X (e.g. chloride, Cl$^-$) with the hydroxide group in water:

$$Rx + H_2O \rightarrow ROH + XH. \tag{24.8}$$

While pH influences the rate of some hydrolysis reactions, there are few environmental situations where it can be used, resulting in temperature adjustment to be the most practical means to modify the reaction. Heating the subsurface from ambient temperatures to 80 °C will increase the hydrolysis rates of most organic compounds by a factor of greater than 1000. Heating to 100 °C will increase the rate of hydrolysis by a factor of over 10,000.

Thermally enhanced hydrolysis is generally the most cost-effective remediation method for halogenated alkanes, and many fumigants and pesticides. A listing of common compounds with their hydrolysis half-lives at 100 °C is shown in Table 24.4. *In situ* thermal methods have been successfully used to hydrolyze 1,1,1-trichloroethane (TCA), 1,1,2,2-tetrachloroethane (TeCA), dichloromethane (methylene chloride), and ethylene dibromide to remediate groundwater.

For compounds with relatively low volatility and nonhazardous hydrolysis progeny, it may be possible to rely on hydrolysis alone to treat the site, with

TABLE 24.4. Hydrolysis Half-Lives at pH 7

Compound	Half-Life at 15°C (Years)	Half-Life at 100°C (Days)	pH Effect	Hydrolysis Product	Reference
Chloroform	8019	117	7.1+	Mineralizes	Jeffers et al. (1989)
Methylene chloride	3282	35	–	Mineralizes	Mabey and Mill (1978)
Bromoform	3089	43	pH++	Mineralizes	Washington (1995)
Dichlorofluoromethane	973	22	6.0+	Mineralizes	Jeffers and Wolfe (1996)
1,1,2-TCA	479	39	pH++	DCE	Jeffers et al. (1989)
1,2-DCA	386	6	–	Ethylene glycol	Jeffers et al., (1989)
1,1-DCA	285	3	–	Acetaldehyde	Jeffers et al. (1989)
Carbon tetrachloride	206	1	–	Mineralizes	Jeffers et al. (1989)
1,2,3-Trichloropropane	192	3	–	Glycerol	Ellington et al. (1987)
1,1,1,2-TeCA	178	0.3	pH++	TCE	Jeffers et al. (1989)
1,2-Dichloropropane	78	0.8	–	Propylene glycol	Washington (1995)
Ethylene dibromide	9	0.5	–	Mineralizes	Weintraub et al. (1986)
1,3-Dichloropropane	8	0.4	–	Propylene glycol	Jeffers et al. (1989)
1,1,1-TCA	6	0.03	–	1,1-DCE	Jeffers et al. (1989)
1,1,2,2-TeCA	1	0.1	pH++	TCE	Jeffers et al. (1989)

+ = the neutral and alkaline reactions are equal at the stated pH. Higher pH accelerates hydrolysis.
pH++ = alkaline dominant, a unit increase in pH will increase the hydrolysis rate by a factor of 10.
– = change in pH in the range of pH 5–9 does not significantly change the rate of hydrolysis.
DCA, dichloroethane; DCE, dichloroethene.

potentially no vapor recovery and treatment. Examples include the fumigants dichloropropane and trichloropropane, which have the progenies propylene glycol and glycerol, respectively. Both of these progenies are relatively easy to biodegrade.

As can be seen from Table 24.4, the hydrolysis rates for halogenated alkenes (e.g. TCE and PCE) tend to be very slow, even at steam temperatures. However, the hydrolysis rates for halogenated alkanes (e.g. TCA and carbon tetrachloride) tend to be very fast.

24.5 HYDRAULIC PRINCIPLES

Hydraulic conductivity in liquid-saturated porous media is dependent on liquid density, liquid viscosity, and soil characteristics including grain size (Hubbert, 1956). As noted above, liquid density and viscosity are modified by temperature. That is, hydraulic conductivity (K) is separable into distinct contributions due to the fluid properties and porous media permeability:

$$K = k\rho g/\eta, \qquad (24.9)$$

where:

ρ = density of the liquid ($M \cdot l^{-3}$),
η = dynamic viscosity of the liquid ($M \cdot l^{-1} \cdot T$),
g = gravitational constant ($l \cdot T^{-2}$), and
k = intrinsic permeability of the porous matrix (l^2).

Laboratory testing has not shown significant variations of intrinsic permeability of sands with temperature (Sageev et al., 1980). Saturated hydraulic conductivity values are not applicable under partially saturated conditions, that is, in the vadose zone, or when various combinations of air, steam, liquid water, or NAPL are present. In general, the permeability of a granular soil matrix is not significantly affected by the application of heat. However, *in situ* thermal techniques can modify the permeability of fine-grained materials, either by hydraulic fracturing from *in situ* steam generation, consolidation induced by the application of an electrical field, the tendency of calcium bentonite clays to flow with heating, or by desiccation from drying at temperatures above the boiling point of groundwater. Smectite and montmorillonite clays in the subsurface consolidate with electrokinetic forces. These factors (with the exception of calcium bentonite clays) are expected to result in a reduction of primary porosity but may enlarge joints and fractures, resulting in an increase in secondary porosity. Calcium bentonite clays, on the other hand, may tend to flow and, depending on the setting, could result in a sealing of joints or fractures.

24.5.1 Multiphase Flow

The performance of *in situ* thermal remediation systems are strongly affected by the simultaneous flow of two (gas and water) or three (gas, water, and NAPL)

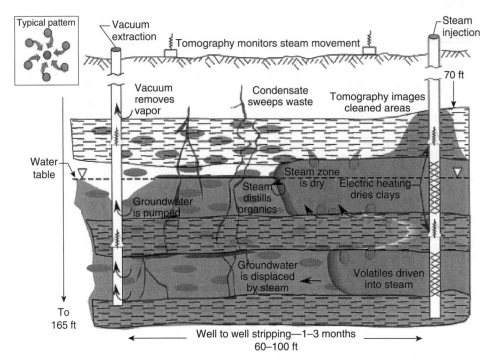

Figure 24.11. Conceptual diagram of the multiphase flow at a dynamic underground stripping project (Lawrence Livermore National Laboratory, 1994).

phases through a porous matrix at elevated temperatures. The following paragraphs summarize the principles that control the interactions of immiscible fluids in the subsurface. Figure 24.11 shows a conceptual diagram of fluid movement during steam injection.

24.5.2 Fluid and Energy Transport

24.5.2.1 Effect of Fluid Transport on Energy Transport Energy may be conveyed in the subsurface through conduction, electrical or electromagnetic stimulation, convective or advective flow, or by fluid-phase changes (e.g. condensation of steam) in the porous matrix. Hot fluid transfers heat to soil particles, raising the temperature of the soil according to its heat capacity. When hot gases condense in soil, however, the heat of vaporization is also transferred to the soil particles, resulting in an additional temperature increase. Since the heat of vaporization of a liquid is many times larger than its heat capacity, the condensation accompanying fluid flow has the potential to deliver much more energy to the subsurface than fluid flow alone.

24.6 BIOLOGICAL PROCESSES AT ELEVATED TEMPERATURES

McNab and Narasimhan (1994) presented a framework that describes the reactions occurring in the fate and transport of the organic compounds in groundwater

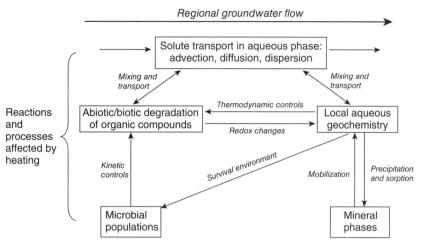

Figure 24.12. Coupling between chemical transport, geochemistry, and degradation reactions in groundwater (modified from McNab and Narasimhan [1994]).

(Fig. 24.12). These reactions are modified as the soil and groundwater contaminated with organic compounds are heated.

Four main abiotic and biotic reactions can result in the breakdown of halogenated aliphatic compounds (Vogel, Criddle, and McCarty, 1987). These are substitution (such as hydrolysis), dehydrohalogenation, oxidation, and reduction. Hydrolysis has already been discussed above. Dehydrohalogenation occurs in mammalian digestive systems where complex enzymes are found; there are no corresponding reactions that occur in aqueous systems. Since the act of heating results in lowering the oxidation–reduction potential (discussed further below) in water, oxidation is not considered a significant reaction during heating. Reduction mechanisms include hydrogenolysis, dehalo-elimination, and coupling. Dehalo-elimination occurs both abiotically and biotically, such that without performing some diagnostic testing, it may be difficult to differentiate the actual biotic or abiotic mechanism.

For dehalo-elimination reactions, it is helpful to simplify the reactions that are occurring as involving terminal electron acceptors (TEAs) and terminal electron donors (TEDs). This is also helpful, for in environmental remediation, there are two broad classes of organic compounds that are commonly dealt with: (a) halogenated organic compounds (TEAs) and (b) fuels and residue from fuel production (TEDs).

TEAs can be both natural and man-made. Natural TEAs include oxygen, nitrate, sulfate, ferric iron compounds, and manganese minerals such as pyrolusite (MnO_2), carbohydrates, and carbon dioxide. In the biodegradation process, halogenated compounds can also be considered to be TEAs.

TEDs include hydrogen, natural organic carbon, humic material, and nonhalogenated fuel compounds. For reductive dehalogenation to occur, there must be both a TEA and a TED. For the biotic pathway, organisms must derive energy from the reaction.

TABLE 24.5. Some Redox Processes That Consume Organic Matter and Reduce Inorganic Compounds in Groundwater

Process	Equation
Denitrification	$CH_2O + \frac{4}{5}NO_3^- = \frac{2}{5}N_2(g) + HCO_3^- + \frac{1}{5}H^+ + \frac{2}{5}H_2O$
Manganese reduction	$CH_2O + 2MnO_2(s) + 3H^+ = 2Mn^{2+} + HCO_3^- + 2H_2O$
Iron (III) reduction	$CH_2O + 4Fe(OH)_3(s) + 7H^+ = 4Fe^{2+} + HCO_3^- + 10H_2O$
Sulfate reduction	$CH_2O + \frac{1}{2}SO_4^{2-} = \frac{1}{2}HS^- + HCO_3^- + \frac{1}{2}H^+$
Methane fermentation	$CH_2O + \frac{1}{2}H_2O = \frac{1}{2}CH_4 + \frac{1}{2}HCO_3^- + \frac{1}{2}H^+$

Source: From Freeze and Cherry, 1979.
CH_2O represents organic matter; other organic compounds can also be oxidized and substituted in the equations.

24.6.1 Redox Theory of Biodegradation

Under natural conditions, water interacts with soil organics and minerals, oxidizing the organic material and, in turn, reducing the minerals. When oxygen in the groundwater is consumed, oxidation can still occur, but the oxidizing agents are NO_3^-, MnO_2, $Fe(OH)_3$, and SO_4^{2-}. As these oxidizing agents (TEAs) are consumed, the groundwater environment becomes more and more reduced to the point where methanogenic conditions may result if there are (a) sufficient oxidizable organics, (b) sufficient nutrients, and (c) temperature conditions conducive to bacterial growth. The water is said to progress from nitrate-reducing, to manganese- or iron-reducing, to sulfate-reducing conditions and may eventually progress to methanogenic conditions, resulting in the production of methane. These represent the five reducing zones and can be expressed by the following half reactions (Table 24.5).

Nitrate reduction will result in the reduction of nitrate to nitrite relative to background, also forming nitrogen gas, and may result in the generation of ammonia. Manganese and iron reduction results in the generation of more soluble forms of these elements and results in elevated concentrations in the impacted groundwater. Sulfate reduction results in the reduction of sulfate concentrations and the formation of sulfides including hydrogen sulfide gas. There are no distinct boundaries between these conditions, and the indicator compounds will show overlap (i.e. iron reduction may take place while nitrate reduction is occurring). These processes are the reduction/oxidation (redox) processes that occur in groundwater and are used to define conditions for natural attenuation.

As noted above, TEDs can be in the form of hydrogen, natural organic carbon, or fuels. The presence of TEAs creates a TED demand. According to Wiedemeier *et al.* (1996), each milligram of dissolved organic carbon oxidized via reductive dechlorination consumes 5.65 mg of organic chloride. This ratio can be used to determine organic carbon demand and the change in this organic carbon demand as the compounds degrade. Organic carbon demand is compared with concentrations of organic carbon. Where organic carbon exceeds the organic carbon demand and appropriate redox conditions are present, reductive dehalogenation is occurring. Conversely, where organic carbon demand exceeds available organic carbon, the dehalogenation may be inhibited.

The oxidation–reduction (redox) potential of the groundwater is a good indication of the type of electron activity present. Anaerobic microbial processes, such as denitrification and sulfate reduction, occur at and result in negative redox potentials. Redox potentials under these conditions have been measured in the field and reported primarily as negative. The negative redox potentials indicate that available dissolved oxygen has been consumed and that the bacteria are scavenging mineral sources of oxygen.

Several guidance documents published by the Environmental Protection Agency (EPA) and U.S. Department of Defense (DoD) are available (U.S. Department of the Navy, 1998; U.S. EPA, 1999, 2002b), describing in greater detail the physical, chemical, and biological process involved in *in situ* biodegradation and monitored natural attenuation. These documents should be consulted during the design and review phases of an electrokinetic project and incorporated into the project's monitoring program.

24.6.2 Gibbs Free Energy and Redox Potential

For the biodegradation reaction to occur, energy must be derived by the microorganisms as part of the process. The Gibbs free energy for the process defines the likelihood of a reaction to occur as well as determines if the energy derived from the reaction is favorable to microorganisms. Gibbs free energy of reaction (an indicator of the spontaneity of the reaction) and the redox potential are related to the Nernst equation:

$$E_H^o = \frac{\Delta G^0}{nF} = \frac{RT}{nF} \ln K = 2.3 \frac{RT}{nF} \log K, \qquad (24.10)$$

where:

E_H^o = redox potential,
ΔG^0 = change in Gibbs free energy,
 where:
 $\Delta G^0 = 0$, the process is reversible (endergonic) and
 $\Delta G^0 < 0$, the process is irreversible (exergonic)
n = number of moles,
F = Faraday constant,
R = universal gas constant, and
K = reaction constant between the oxidant and the reductant.

The change in Gibbs free energy may also be defined as

$$dG = dH - TdS - SdT, \qquad (24.11)$$

where:

H = enthalpy,
S = entropy, and
T = temperature (°K).

Under standard conditions, the temperature is considered to be constant (room temperature) and the $-SdT$ term is usually neglected. Since *in situ* thermal treatment increases the temperature from 75 to 80 °C above room temperature, in general, the Gibbs free energy (and correspondingly, the redox potential) becomes more negative because the magnitude of the $-TdS$ and $-SdT$ terms become greater and hence more negative. The more negative the Gibbs free energy value, the more likely the reaction is to occur (i.e. the reaction is exergonic).

Table 24.6 lists the Gibbs free energy and associated redox potentials of a variety of common chlorinated aliphatic compounds and natural TEAs as presented in Dolfing, van Eckert, and Mueller (2006). Other authors have provided estimates of Gibbs free energies and redox potentials (Dolfing and Jansen, 1994; Alberty, 1998; Haas and Shock, 1999; Plyasunov and Shock, 2000).

Three mechanisms have been identified in the metabolic dechlorination of chlorinated aliphatic hydrocarbon. Metabolic processes are coupled to energy conservation and, as a result, are interpreted to be beneficial to the dechlorinating microorganisms. These include

- oxidative conversions,
- chlorinated aliphatic hydrocarbons serving as TEAs, and
- fermentation.

Under the oxidative pathway, chlorinated aliphatic hydrocarbons act as an electron donor. The presence of an electron acceptor, such as Fe^{3+}, NO_3^-, SO_4^{2-}, or carbon dioxide, is required. According to Dolfing *et al.* (2006), only dichloromethane was known to have been dechlorinated via this pathway under anaerobic conditions; therefore, this will not be evaluated further.

Where the chlorinated aliphatic hydrocarbon acts as a TEA, a TED is required, and this reaction may be coupled to growth (Dolfing and Jansen, 1994). There is some disagreement on this point, in that methanogenic bacteria have been observed to use the chlorinated aliphatic hydrocarbons to respirate, producing ethene but no methane, which may be interpreted as no growth of these bacteria (Y. Dong, pers. comm.), yet the bacteria are deriving energy from the reaction. However, this is believed to be the dominant mechanism during ERH remediation.

In the case of fermentation, the chlorinated aliphatic hydrocarbons act as TED, TEA, and carbon source. This may be occurring under *in situ* thermal treatment were yeastlike odors have been noted, but the degree to which this is occurring is uncertain.

Cometabolic processes derive no apparent benefit from the energy derived from the reactions, and as a result, it is interpreted that there is no driving force. From a microbiological standpoint under normal groundwater temperatures, cometabolic transformations present a disadvantage because the input of reducing equivalents costs energy and toxic transformation products are formed. From an *in situ* thermal or electrokinetic perspective, cometabolic reactions may be considered almost serendipitous; since energy is input to heat the system to remove NAPL, the corresponding cometabolic dechlorination reactions appear as a benefit during *in situ* thermal treatment.

Other authors cited above developed their own Gibbs free energies and redox potentials. There is a disagreement among various authors and among publications.

TABLE 24.6. Gibbs Free Energy Changes and Standard Reduction Potentials at 5°C and pH 7 of Selected Organic and Inorganic Redox Couples[a] [from Dolfing, van Eckert, and Mueller (2006)]

Electron Acceptor	Half-Reaction of Reductive Transformations			$\Delta G^{o\prime}$ (kJ/electron)	$E^{o\prime}$ (mV)
Pd^{2+}	$Pd^{2+} + 2e^-$	$\rightarrow$	$Pd(0)$	−88.3	915
1,1,2-TCA	$CHCl_2\text{-}CH_2Cl + 2e^-$	$\rightarrow$	$C_2H_3Cl + 2Cl^-$	−79.8	827
O_2	$O_2 + 4H^+ + 4e^-$	$\rightarrow$	$2H_2O$	−78.7	816
1,2-DCA	$CH_2Cl\text{-}CH_2Cl + 2e^-$	$\rightarrow$	$C_2H_4 + 2Cl^-$	−71.3	738
Fe^{3+}	$Fe^{3+} + e^-$	$\rightarrow$	Fe^{2+}	−74.4	771
CT	$CCl_4 + H^+ + 2e^-$	$\rightarrow$	$CHCl_3 + Cl^-$	−65.0	673
MnO_2	$MnO_2 + HCO_3^- + 3H^+ + 2e^-$	$\rightarrow$	$MnCO_3 + 4H_2O$	−58.9	610
PCE	$C_2Cl_4 + H^+ + 2e^-$	$\rightarrow$	$C_2HCl_3 + Cl^-$	−55.4	574
1,1,1-TCA	$CCl_3\text{-}CH_3 + H^+ + 2e^-$	$\rightarrow$	$CHCl_2\text{-}CH_3 + Cl^-$	−54.1	561
CF	$CHCl_3 + H^+ + 2e^-$	$\rightarrow$	$CH_2Cl_2 + Cl^-$	−54.1	560
TCE	$C_2HCl_3 + H^+ + 2e^-$	$\rightarrow$	$C_2H_2Cl_2 + Cl^-$	−53.1 to −50.9[b]	550–527[b]
1,1,2-TCA	$CHCl_2\text{-}CH_2Cl + H^+ + 2e^-$	$\rightarrow$	$C_2H_4Cl_2 + Cl^-$	−51.9 to −49.8[c]	538–516[c]
DCM	$CH_2Cl_2 + H^+ + 2e^-$	$\rightarrow$	$CH_3Cl + Cl^-$	−47.6	493
CM	$CH_3Cl + H^+ + 2e^-$	$\rightarrow$	$CH_4 + Cl^-$	−44.8	464
CA	$C_2H_5Cl + H^+ + 2e^-$	$\rightarrow$	$C_2H_6 + Cl^-$	−44.6	462
VC	$C_2H_3Cl + H^+ + 2e^-$	$\rightarrow$	$C_2H_4 + Cl^-$	−43.4	450
NO_3^-	$NO_3^- + 2H^+ + 2e^-$	$\rightarrow$	$NO_2^- + H_2O$	−41.7	432
DCE	$C_2H_2Cl_2 + H^+ + 2e^-$	$\rightarrow$	$C_2H_3Cl + Cl^-$	−40.6 to −38.3[b]	420–397[b]
1,1-DCA	$C_2H_4Cl_2 + H^+ + 2e^-$	$\rightarrow$	$C_2H_5Cl + Cl^-$	−38.3	397
1,2-DCA	$CH_2Cl\text{-}CH_2Cl + H^+ + 2e^-$	$\rightarrow$	$C_2H_5Cl + Cl^-$	−36.2	375
$Fe(OH)_3$	$Fe(OH)_3 + 3H^+ + e^-$	$\rightarrow$	$Fe^{2+} + 3H_2O$	−11.4	118
SO_4^{2-}	$SO_4^{2-} + 9H^+ + 8e^-$	$\rightarrow$	$HS^- + 4H_2O$	+20.9	−217
HCO_3^-	$HCO_3^- + 9H^+ + 8e^-$	$\rightarrow$	$CH_4 + 3H_2O$	+23.0	−238
Ni^{2+}	$Ni^{2+} + 2e^-$	$\rightarrow$	$Ni(0)$	+27.0	−280
Proton (pH 7)	$2H^+ + 2e^-$	$\rightarrow$	H_2	+40.5	−420
Fe^{2+}	$Fe^{2+} + 2e^-$	$\rightarrow$	$Fe(0)$	+42.5	−440
Ti-citrate	Ti^{4+} citrate $+ e^-$	$\rightarrow$	Ti^{4+} citrate	+46.3	−480
Zn^{2+}	$Zn^{2+} + 2e^-$	$\rightarrow$	$Zn(0)$	+73.6	−763
Al^{3+}	$Al^{3+} + 3e^-$	$\rightarrow$	$Al(0)$	+160.4	−1662

[a] Calculated on the basis of data from Bard et al. (1985), Zehnder and Wuhrmann (1976), Dolfing and Harrison (1992), Dolfing and Jansen (1994), and Dolfing et al. (2006) for (Cl^-) = 1 mM.
[b] Depending on the isomer formed or dechlorinated (i.e. cis-$C_2H_2Cl_2$, trans-$C_2H_2Cl_2$, or 1,1-$C_2H_2Cl_2$).
[c] Depending on the isomer formed (i.e. 1,1-$C_2H_4Cl_2$ or 1,2-$C_2H_4Cl_2$).

For instance, Dolfing and Jansen (1994) gave a redox potential of 0.402 V for the dehalogenation of dichloromethane to chloromethane at 5°C, and Dolfing, van Eckert, and Mueller (2006) gave a redox potential of 0.493 V for the same reaction at 5°C. So, it is difficult to assign absolute values for these reactions, and for the

TABLE 24.7. Standard Reduction Potentials for Temperatures of 0–100 °C

Reaction	Standard Reduction Potentials	
	E° at 0 °C (V)	E° at 100 °C (V)
O_2 reduction	1.27	1.277
NO_3 reduction	1.221	1.25
MnO_2 reduction	0.956	0.968
Fe^{3+} reduction	0.726	0.907
SO_4 reduction	0.289	0.346
$S°$ reduction	0.133	0.187
CH_4 reduction	0.104	0.115
PCE → TCE	0.704	0.72
TCE → cis-DCE	0.659	0.681
cis-DCE → vinyl chloride	0.549	0.521
Vinyl chloride → ethene	0.522	0.573

cis-DCE, cis-dichlorethene.

purpose of this analysis, we can only look at the relative values presented from a given literature source. Table 24.7 presents the redox values for the dehalogenation of chloroethenes at 0 and at 100 °C from Haas and Shock (1999).

It can be seen from Table 24.7 that the reductive dehalogenation reactions for the chloroethenes remain in the iron-reducing range after heating at 100 °C. However, it can also be seen that the redox ranges for the various reducing conditions change with heating. Manganese reduction occurs over a much narrower range of redox conditions at higher temperatures, while iron- and sulfide-reducing conditions occur over a wider range of redox values. Therefore, from Haas and Shock (1999), dehalogenation of chloroethenes occurs under iron-reducing conditions and lower. It becomes more energetically favorable under sulfate-reducing conditions as the ferric iron in the soils and groundwater becomes reduced and no longer available to act as a TEA. Under sulfate-reducing conditions, reductive dehalogenation is more energetically favorable to the microorganisms than sulfate, as long as the chloroethenes are available to the bacteria.

The lower the redox potential, the more energetically favorable the reductive dehalogenation becomes (Vogel, Criddle, and McCarty, 1987). However, there are exceptions to this. The kinetics of the reductive dechlorination of cis-1,2-dichloroethene, vinyl chloride, and chloroethane actually becomes slower under methanogenic conditions compared with sulfate-reducing conditions (Bradley and Chapelle, 1997).

24.6.3 Electron Donors

All bacterial reductive dehalogenation depends on the presence of an electron donor, unless fermentation is the responsible mechanism. Competition for available electrons can occur between dechlorinating and nondechlorinating species (Dolfing and Jansen, 1994). Whether dechlorination occurs depends on the affinity of the dechlorinating population for the available electron donor. For metabolic dechlorination, the dechlorinating microorganisms can only thrive if the dechlorination provides sufficient energy for maintenance and growth. For cometabolic processes, there must be surplus electron donors present. As noted previously, electron donor demand for dechlorinated compounds can be determined via the relationship presented in Wiedemeier et al. (1996).

As previously mentioned, for the degradation of fuels, natural TEAs are consumed in the biodegradation of these compounds, such that at a majority of fuel release sites, and sulfate-reducing and methanogenic conditions are typically found in the source area (Newell, McLeod, and Gonzales, 1996).

From Newell, McLeod, and Gonzales (1996), the biodegradation capacity (mg/l) of an aquifer is defined as

Biodegradation capacity (mg/l) = {(average up-gradient oxygen concentration) − (minimum source zone oxygen concentration)}/3.14 + {(average up-gradient nitrate concentration) − (minimum source zone nitrate concentration)}/4.9 + {(average up-gradient sulfate concentration) − (minimum source zone sulfate concentration)}/4.7 + {(average observed ferrous iron conc. in source area)}/ 21.8 + {(average up-gradient methane concentration in source area)}/0.78.

In the above equation, each of the TEA concentrations is divided by what are called utilization factors. Since this represents a mathematical division operation, the lower the factor, the greater each of the TEAs affects the biodegradation capacity. From this, methane production would have the greatest influence on biodegradation capacity, but the kinetics of biodegradation may slow down under methanogenic conditions (Bradley and Chapelle, 1997). Oxygen has the next lowest, but concentrations of oxygen do not typically exceed 8 mg/l in natural water and is typically found at much lower concentrations in groundwater. Sulfate is typically found in natural groundwater at concentrations ranging from 50 to 300 mg/l and as high as 3000 mg/l in parts of Montana. Therefore, sulfate is the most significant TEA in determining biodegradation capacity due to its relatively high concentrations and low utilization factor.

26.6.4 Temperature Tolerance of Microorganisms

The occurrence and abundance of microorganisms in a particular environment is controlled by the complex interaction of nutrients with physical and chemical factors (temperature, redox, pH, etc.) present in the environment, which may evolve over time. The presence and success of a specific organism or consortia of microorganisms responsible for degrading hydrocarbons in a contaminated subsurface ecosystem depends on both nutrient requirements and tolerance for the range of physical and chemical conditions.

Temperature is considered to be one of the most important physical factors controlling the distribution and abundance of organisms. Bacteria adapt to a wide variety of environments, including habitats characterized by extreme temperatures, and can be sorted into descriptive groups based on the temperatures at which optimum growth has been observed (Table 24.8).

Thermophiles and extreme thermophiles have characteristic membrane and enzyme systems that allow them to function at temperatures that would otherwise inhibit cellular transport and metabolic activity. These adaptations include high proportions of saturated lipids in cell membranes to prevent melting, enzyme

TABLE 24.8. Generally Recognized Optimum Growth Range for Various Groups of Microorganisms

Bacterial Classification	Optimum Growth Temperature
Psychrophiles	<0 to <20 °C (<32 to 68 °F)
Mesophiles	20 to <45 °C (<68 to <113 °F)
Thermophiles	45–90 °C (<113 to <194 °F)
Extreme thermophiles	90–110 °C (<194 to <230 °F)

systems that remain stable at high temperatures, and high proportions of amino acids guanine and cytosine in nucleic acids to raise the melting point of DNA (Atlas and Bartha, 1993).

In general, elevated temperatures that do not kill microbes or exceed the temperature tolerance of the microbial consortia will result in higher metabolic activity. The increased metabolic activity of enzymatic systems with temperature continues up to the temperature where the enzymes denature or lose the structural stability that enables them to function. Mesophiles are more efficient at degrading hydrocarbons at temperatures from 30 to 40 °C (86–104 °F; Bossert and Bartha, 1984). Thermophiles actively degrade hydrocarbons and recalcitrant NAPL constituents (PAHs and high-molecular-weight hydrocarbons) at temperatures up to 70 °C (158 °F; Huesemann et al., 2002). Even though biocatalytic reactions proceed faster at higher temperatures, the growth rate of thermophiles is often slower than mesophiles at their optimum growth temperatures. Consequently, degradation reactions at elevated temperatures found at an *in situ* thermal treatment site may progress more rapidly as thermophiles are capable of mediating degradation reactions at a faster pace without diverting energy to increasing biomass.

Consequently, increased subsurface temperatures should increase the concentration of contaminants in the dissolved phase and increase the availability of these compounds to degrading microorganisms.

As discussed above, the solubility of PAH constituents generally increases at increased temperatures. At elevated temperatures, the dissolution rate for TCE and PCE increases. This results in greater mass of the contaminant in the dissolved phase and, as a result, greater potential availability to degrading microorganisms. Preliminary data have shown that the increased solubility of selected PAHs at temperatures up to 60 °C enables thermophiles to degrade the PAHs at a rate of up to eight times faster than mesophiles at lower temperatures (Peyton, pers. comm.).

24.6.5 Microorganism Population Changes and Ability to Degrade Contaminants

The combination of increased metabolic activity and greater bioavailability allows for enhancing *in situ* biodegradation at *in situ* thermal remediation sites. During active thermal treatment, *in situ* degradation may be temporarily inhibited once subsurface temperatures increase above the tolerance range of indigenous mesophiles. Microorganisms have several mechanisms for surviving unfavorable conditions, including the formation of nonvegetative structures (i.e. spores or cysts) that are metabolically less active and allow the organism to survive until environmental conditions become more optimal. Bacterial cells in bench-scale tests of creosote-contaminated soil have demonstrated the ability to rapidly become metabolically active following steam injection (Richardson, 2000).

Further studies have also demonstrated that thermophilic degradation of PAHs and nonvolatile hydrocarbons increased at temperatures likely to occur adjacent to the active treatment zone of an *in situ* thermal remediation site (Huesemann *et al.*, 2002). This observation is likely the result of shifts in the population of microorganisms from predominantly mesophilic to predominantly thermophilic. This type of community shift is usually associated with a reduction in the diversity of microorganisms. As subsurface temperatures cool after active thermal treatment, the consortia within the heated zone will again shift as conditions become less favorable for thermophiles and return to the optimum temperatures for mesophiles. The EPA Technology Innovation Office has published a more detailed review of this topic, available at the website www.clu-in.org.

24.6.6 Bioavailability at Elevated Temperatures

In addition to increased metabolic activity that can be attributed to increased temperature, biodegradation at an electrokinetic treatment site can also be enhanced by a temperature-induced increase in bioavailability. For example, the persistence of PAHs in the environment is attributed to the hydrophobicity of these compounds and their tendency to strongly sorb onto soil or sediment particles. Generally, NAPL constituents not in the dissolved phase are not available for microbial degradation, although there is some recent evidence that NAPL contaminants may be degraded by specially adapted bacteria (Wattiau, 2002). However, increases in temperature have been shown to increase the solubility of both PAHs and the rate of dissolution of chlorinated compounds (Hulscher and Cornelissen, 1996; Imhoff, Frizzel, and Miller, 1997; Bonten, Grotenhuis, and Rulkens, 1999; Jayaweera, Marti-Perez, and Diaz-Ferrero, 2002).

24.7 SUMMARY

Thermal desorption at sites undergoing electrokinetic treatment has not been well studied. To evaluate the thermal effects during electrokinetic treatment, information developed from *in situ* thermal treatment was relied upon, with specific focus on *in situ* ERH. Thermal principles provide the basis from which changes in soil and groundwater chemistry as well as chemical and biological effects are evaluated. In general, increases in temperature increase the kinetics of reactions at ambient temperatures. Increases in temperature result to increases in bioactivity in groundwater. As temperature increases, the Gibbs free energy of the reaction becomes more negative, facilitating the breakdown of organic chemicals.

The increases in bioactivity can result in increases in dissolved gases. Anaerobic biodegradation results in the production of hydrogen sulfide, methane, and increases in chloride resulting from the reductive dehalogenation of chlorinated aliphaic compounds in groundwater. Where calcium is the dominant cation in groundwater, this results in increases in calcium in groundwater through the common ion effect, which in turn results in increases in the fugacity of carbon dioxide. Dissolved gases contribute significantly to the volatilization of NAPL compounds, and with increases in temperature, modeling has shown that the groundwater system can achieve vapor pressures in excess of 3.9 bars (Smith, 2008).

Physical properties of various organic chemicals are also observed to be modified by heating. Aqueous solubility, in general, increases, but for compounds such as PCE and TCE, increases are significant only above the boiling point of water. Whereas by heating from 25 to 100 °C, the aqueous solubility of naphthalene increases by a factor of approximately 45.

Viscosity, density, and sorption coefficients all decrease with temperature, facilitating removal from the subsurface.

Rates of hydrolysis increase significantly for halogenated alkanes and pesticides. Heating the subsurface from ambient temperatures to 80 °C increases the hydrolysis rate by a factor of 1000, and increasing temperatures to 100 °C increases the rates by a factor of 10,000. This provides a mechanism for *in situ* destruction that is not normally considered in environmental restoration projects. Increasing the rate of hydrolysis through temperature increases has been successfully used to treat TCA, TeCA, dichloromethane (methylene chloride), and ethylene dibromide.

REFERENCES

Alberty RA. (1998). Calculation of standard transformed Gibbs free energies and standard transformed enthalpies of biochemical reactants. *Archives of Biochemistry and Biophysics* **353**(1):116–130.

Amos RT, Mayer U, Belkins BA, Delin GN, Williams RC. (2005). Use of dissolved and vapor phase gases to investigate methanogenic degradation of petroleum hydrocarbon contamination in the subsurface. *Water Resources Research* **41**(2):1–15.

Atlas RM, Bartha R. (1993). *Microbial Ecology, Fundamentals and Applications*, 3rd ed. Redwood City, CA: The Benjamin/Cummings Publishing.

Bard AJ, Parsons R, Jordan J. (1985). *Standard Potentials in Aqueous Solution*. New York: Marcel Dekker.

Bonten LTC, Grotenhuis TC, Rulkens WH. (1999). Enhancement of PAH biodegradation in soil by physiochemical pretreatment. *Chemosphere* **38**:3627–3636.

Bossert I, Bartha R. (1984). The fate of petroleum in soil ecosystems. In *Petroleum Microbiology* (ed. RM Atlas). New York: Macmillan, pp. 435–473.

Bradley PM, Chapelle FH. (1997). Kinetics of DCE and VC mineralization under methanogenic and Fe(III)-reducing conditions. *Environmental Science & Technology* **31**(9):2692–2696.

van Breukelen BM, Roling WFM, Groen J, Griffioen J, van Verseveld HW. (2003). Biogeochemistry and isotope geochemistry of a landfill leachate plume. *Journal of Contaminant Hydrology* **65**:245–268.

Chaplin BP, Delin GN, Baker RJ, Lahvis MK. (2002). Long term evolution of biodegradation and volatilization rates in a crude oil contaminated aquifer. *Bioremediation Journal* **6**:237–255.

Davis EL. (1997). How Heat Can Enhance In Situ Soil and Aquifer Remediation: Important Chemical Properties and Guidance on Choosing the Appropriate Technique. *EPA/540/S-97/502*.

Davis EL. (1998). Steam Injection for Soil and Aquifer Remediation. *EPA/540/S-97/505*.

Davis EL. (2002). *Steam Injection Treatability Study, Wyckoff/Eagle Harbor Superfund Site*. Robert S. Kerr Research Laboratory, Ada, OK.

Dolfing J. (2003). Thermodynamic considerations for dehalogenation. In *Dehalogenation: Microbial Processes and Environmental Applications* (eds. MM Häggblom, ID Bossert). Boston: Kluwer Academic Publishers, pp. 89–113.

Dolfing J, van Eckert M, Mueller J. (2006). Thermodynamics of low Eh reactions. *Proceedings of the Fifth International Conference on Remediation of Chlorinated and Recalcitrant Compounds*, May 22–26, Monterey, CA.

Dolfing J, Harrison BK. (1992). Gibbs free energy of formation of halogenated aromatic compounds and their potential role as electron acceptors in anaerobic environments. *Environmental Science and Technology* **26**(11):2213–2218.

Dolfing J, Jansen DB. (1994). Estimates of Gibbs free energies of formation of chlorinated aliphatic compounds. *Biodegradation* **5**:21–28.

Ellington JJ, Stancil FE, Jr, Payne WD, Trusty CD. (1987). Measurement of hydrolysis rate constants for evaluation of hazardous waste land disposal: Volume II. Data on 54 chemicals. EPA-600/3-87-019. Athens, GA: US Environmental Protection Agency.

Freeze RA, Cherry JA. (1979). *Groundwater*. Englewood Cliffs, NJ: Prentice Hall.

Haas JR, Shock EL. (1999). Halocarbons in the environment: Estimates of thermodynamic properties for aqueous chloroethylene species and their stabilities in natural settings. *Geochimica et Cosmochimica Acta* **63**(19/20):3429–3411.

Heron G, Christensen TH, Enfield CG. (1998). Henry's law constant for trichloroethylene between 10 and 95°C. *Environmental Science & Technology* **32**(10):1433–1437.

Heron G, Christensen TH, Heron T, Larsen T. (1998). Thermally enhanced remediation at DNAPL sites: The competition between downward mobilization and upward volatilization. *Proceedings of the First International Conference on Remediation of Chlorinated and Recalcitrant Compounds*. Columbus, OH: Battelle Press, pp. 193–198.

Hers I, Atwater J, Li L, Zapf-Gilje R. (2000). Evaluation of vadose zone biodegradation of BTX vapors. *Journal of Contaminant Hydrology* **46**:233–264.

Hubbert MK. (1956). Darcy's law and the field equations of the flow of underground fluids. *Transactions AIME* **207**:222–239.

Huesemann MH, Hausmann TS, Fortman TJ, Truex MJ. (2002). Evidence of thermophilic biodegradation for PAHs and diesel in soil. *Proceedings of the Third International Conference on Remediation of Chlorinated and Recalcitrant Compounds*.

Hulscher EM, Cornelissen G. (1996). Effect of temperature on sorption equilibrium and sorption kinetics of organic micropollutants—a review. *Chemosphere* **32**:609–626.

Imhoff PT, Frizzel A, Miller CT. (1997). Evaluation of thermal effects on the dissolution of a nonaqueous phase liquid in porous media. *Environmental Science & Technology* **31**(6):1615–1622.

Incropera PF, DeWitt DP. (1996). *Introduction to Heat Transfer*. New York: John Wiley & Sons.

Jayaweera IS, Marti-Perez M, Diaz-Ferrero J. (2002). Solubility of polycyclic aromatic hydrocarbons under hydrothermal conditions. *Proceedings of the Third International Conference on Remediation of Chlorinated and Recalcitrant Compounds*.

Jeffers PM, Ward LM, Woytowitch LM, Wolfe NL. (1989). Homogeneous hydrolysis rate constants for selected chlorinated methanes, ethanes, ethenes, and propanes. *Environmental Science & Technology* **23**(8):965–969.

Jeffers PM, Wolfe NL. (1996). Homogeneous hydrolysis rate constants—Part II: Additions, corrections and halogen effects. *Environmental Toxicology and Chemistry* **15**(7):1066–1070.

Johansen Ø. (1977). *Thermal Conductivity of Soils*. U.S. Army Cold Regions Res. Eng. Laboratory, Translation 637.

Kipphut GW, Martens CS. (1982). Biogeochemical cycling in an organic-rich coastal marine basin, part 3: Dissolved gas transport in methane-saturated sediments. *Geochimica et Cosmochimica Acta* **46**:2049–2060.

Knauss KG, Dibley MJ, Leif RN, Mew DA, Aines RD. (2000). The aqueous solubility of trichloroethylene (TCE) and tetrachloroethylene (PCE) as a function of temperature. *Applied Geochemistry* **15**:501–512.

Lawrence Livermore National Laboratory. (1994). *Demonstration of Dynamic Underground Stripping at the LLNL Gasoline Spill Demonstration Site*. Lawrence Livermore National Laboratory, Livermore, CA. *Final Report No. UCRL-ID-116964*, Vol. 1–4.

Lide DR. (1999). *CRC Handbook of Chemistry and Physics*. Boca Raton, FL: CRC Press.

Lupis CHP. (1983). *Chemical Thermodynamics of Materials*. New York: Elsevier.

Mabey W, Mill T. (1978). Critical review of hydrolysis of organic compounds in water under environmental conditions. *J. Phys. Chem. Ref. Data* **7**:383–415.

McNab WW, Narasimhan TN. (1994). Modeling reactive transport of organic compounds in groundwater using a partial redox disequilibrium approach. *Water Resources Research* **30**(9):2619–2635.

Newell CJ, McLeod RK, Gonzales JR. (1996). BIOSCREEN Natural Attenuation Decision Support System User's Manual V 1.3. *EPA/600/R-96/087*.

Plyasunov AV, Shock EL. (2000). Standard state Gibbs energies of hydration of hydrocarbons at elevated temperatures as evaluated from experimental phase equilibria studies. *Geochimica et Cosmochimica Acta* **64**(16):2811–2833.

Poling BE, Prausnitz JM, O'Connell JP. (2001). *The Properties of Gases and Liquids*. New York: McGraw-Hill.

Reeburgh WS. (1972). Processes affecting gas distribution in estuarine sediments. *Geological Society of America Bulletin* **133**:383–389.

Reid RC, Prausnitz JM, Poling BE. (1987). *The Properties of Liquids and Gases*. New York: McGraw-Hill.

Richardson RE. (2000). Final Report on Post-Steam Microbial Experiments Performed at U.C. Berkeley for the Wyckoff/Eagle Harbor Superfund Site (submitted for publication).

Sageev A, Gobran BD, Birgham WE, Ramey HJ Jr. (1980). The effect of temperature on absolute permeability to distilled water of unconsolidated sand cores. *Proceedings, 6th Workshop on Geothermal Reservoir Engineering*, SGP-TR-50, Stanford University, Stanford, CA, pp. 297–300.

Smith GJ. (2008). Symbiotic effects of biodegradation during electrical resistance heating. *Proceedings of the Sixth International Conference on Remediation of Chlorinated and Recalcitrant Compounds*, May 19–22, 2006, Monterey, CA.

Stegemeier GL, Vinegar HJ. (2001). Thermal Conduction heating for in situ desorption of soils. In *Hazardous & Radioactive Waste Treatment Technologies Handbook*. Boca Raton, FL: CRC Press.

Stumm W, Morgan JJ. (1981). *Aquatic Chemistry. An Introduction Emphasizing Chemical Equilibria in Natural Waters*, 2nd Edition. New York: John Wiley & Sons.

Sundberg J. (1988). Thermal Properties of Soils and Rocks. Swedish Geotechnical Institute, Linkoping, Sweden. *Report Number 35*.

U.S. Department of the Navy. (1998). Technical Guidelines for Evaluation Monitored Natural Attenuation of Petroleum Hydrocarbons and Chlorinated Solvents in Ground

Water at Naval and Marine Corps Facilities. Prepared by TH Wiedemeier and FH Chapelle for the Naval Facilities Engineering Command.

U.S. EPA. (1984). Process Design Manual for Stripping of Organics. *Report number 600/2/84/139*.

U.S. EPA. (1999). Microbial Processes Affecting Monitored Natural Attenuation of Contaminants in the Subsurface. Office of Solid Waste and Emergency Response, National Risk Management Research Laboratory, Subsurface Protection and Remediation Division, Robert S. Kerr Environmental Research Center, Ada, OK. *EPA/540/S-99/001*.

U.S. EPA. (2002a). Final Design Analysis, Thermal Remediation Pilot Study, PN C1871, Soil and Groundwater Operable Units. Wyckoff/Eagle Harbor Superfund Site, Bainbridge Island, WA. Report to EPA by U.S. Army Corps of Engineers.

U.S. EPA. (2002b). A Discussion of the Effects of Thermal Remediation Treatments on Microbial Degradation Processes. Office of Solid Waste and Emergency Response, Technology Innovation Office, Washington, DC. www.clu-in.org.

Vogel TM, Criddle CS, McCarty PL. (1987). Transformations of halogenated aliphatic compounds: Oxidation, reduction, substitution and dehydrohalogenation reactions occur abiotically or in microbial and mammalian systems. *Environmental Science & Technology* **21**(8):722–736.

Washington JW. (1995). Hydrolysis rates of dissolved volatile organic compounds: Principles, temperature effects and literature review. *Ground Water* **33**:415–424.

Wattiau P. (2002). Microbial aspects in bioremediation of soils polluted by polyaromatic hydrocarbons. In *Biotechnology for the Environment, Strategy and Fundamentals* (eds. SN Agathos, W Reineke). Dordrecht, the Netherlands: Kluwer Academic Publishers, pp. 69–89.

Weintraub RA, Jex GW, Moye HA (1986). Chemical and microbial degradation of 1,2-dibromoethane (EDB) in Florida ground water, soil, and sludge. In *Evaluation of Pesticides in Ground Water* (eds. WY Garner, RC Honeycutt, HN Nigg). Washington, DC: American Chemical Society, pp. 294–310.

Wiedemeier TH, Swanson MA, Moutoux DE, Gordon EK, Wilson JT, Wilson BH, Kampbell DH, Hansen JE, Haas P, Chapelle FH. (1996). *Technical Protocol for Evaluating Natural Attenuation of Chlorinated Solvents in Groundwater*. San Antonio, TX: Air Force Center for Environmental Excellence.

Wijk WR van. (1963). *Physics of Plant Environment*. Amsterdam: North Holland.

Zehnder AJB, Wuhrmann K. (1976). Titanium(III) citrate as a nontoxic oxidation reduction buffering system for the culture of obligate anaerobes. *Science* **194**(4270):1165–1166.

PART VII

MATHEMATICAL MODELING

25

ELECTROKINETIC MODELING OF HEAVY METALS

José Miguel Rodríguez-Maroto and Carlos Vereda-Alonso

25.1 INTRODUCTION

As presented in the chapters of the previous parts (II and III), the electrokinetic soil remediation system uses electric current to extract ionic and nonionic species of contaminants from soils. This chapter is focused on the modeling of the electrokinetic remediation (EKR) of soils contaminated by heavy metals in which electrically charged species are mainly involved. The electrical potential difference applied between the electrodes inserted in the soil causes an electrical direct current between these electrodes that flows through the moist soil. This current is carried by the existing ions in the pore water (ionic current) and drives the contaminant species toward one of the electrodes where they can be recovered. If the contaminant species are charged, they are moved by ionic migration toward one of the electrodes, depending on the sign of their electric charge. In addition, regardless of their electrical charge, the contaminants can move toward one of the electrodes due to the movement of pore water by electroosmosis. The third transport mechanism related to this electrical driving force is the electrophoresis, whose contribution to the global transport process, compared with those of the above two mechanisms, is not significant in the soil system and therefore can be neglected. Besides the above transport processes, diffusion and water advection are normally present as a result of the concentration and pressure gradients, respectively.

Logically, the ionic current that flows through the soil has to be converted to an electronic current carried by the electrons through the electrodes and the external wires that connect this electrical circuit to the power supply. This change occurs at the electrode/electrolyte interface in the electrode wells. An electrochemical reduction reaction takes place at the cathode, which accepts electrons from the external

---

Electrochemical Remediation Technologies for Polluted Soils, Sediments and Groundwater,
Edited by Krishna R. Reddy and Claudio Cameselle
Copyright © 2009 John Wiley & Sons, Inc.

circuit, and a simultaneous electrochemical oxidation reaction occurs at the anode by transferring electrons into the external circuit, thereby completing the electrical circuit. Normally, water is electrochemically oxidized to oxygen and protons at the anode; this produces an acid front that advances across the soil toward the cathode and desorbs heavy metals from the surface of soil particles. Simultaneously, the electrochemical reduction of water usually occurs at the cathode and produces a basic front that migrates through the soil toward the anode. This basic front can originate the precipitation of the heavy metal in the soil or can create a low electrical conductivity zone when it meets the acid front that, if not avoided, makes this cleanup technique ineffective. There are several enhancements of this technique that prevent these inconveniences, such as the addition of acid at the cathode or the use of ion exchange membranes in the electrode compartments (Ottosen et al., 1997) as indicated in previous sections.

As suggested by the above paragraph, other types of phenomena affect the previously described transport mechanisms of the contaminants toward the electrodes. These are the physical–chemical interactions, both between different compounds in the aqueous phase and between these aqueous species and the solid phases of the soil system. Some of these interactions are precipitation, acid–base, complex formation and redox reactions, adsorption, and ion exchange and surface complexation reactions.

Thus, as expected, the design, operation, and performance of the EKR system are not easy. Mathematical models are necessary in order to gain a better understanding of the processes that occurs in the EKR and to allow predictions for the field-scale remediation. Generally, it is a good policy to keep the mathematical model as simple as possible while adequately describing the behavior of the main parameters of the system (principle of parsimony). Thus, models with relatively simple transport equations and few equilibrium equations are able to predict the evolution of parameters such as the rate of recovery of the toxic ion, the maximum recovery, the rate of acid addition, and the energy requirements. The equation of mass conservation for a pore water solute species (e.g. an ion) in an EKR system can be expressed as follows:

$$\frac{\partial C_k}{\partial t} = -\vec{\nabla} \cdot \vec{J}_k + R_k, \tag{25.1}$$

where J_k is the flux of the kth ion, and C_k and R_k are the concentration and the rate of production of the kth ion per unit of volume, respectively. This flux includes electromigration, electroosmosis, diffusion, and advection contributions, each one of which can be expressed as a function of their corresponding gradients:

$$\frac{\partial C_k}{\partial t} = -\vec{\nabla} \cdot \left(-C_k u_k \vec{\nabla} V - C_k k_e \vec{\nabla} V - \vec{\vec{D}}_k \vec{\nabla} C_k + C_k \vec{v} \right) + R_k, \tag{25.2}$$

where the terms in the parenthesis in the right-hand expression represent, respectively, the electromigration, electroosmotic, diffusive, and advective flux, u_k and D_k are the ionic mobility and the diffusion coefficients of the kth ion, k_e is the electroosmotic permeability of the soil, v is the advective velocity of the pore water, and ∇V is the electrical voltage gradient. The values of these parameters correspond with the effective values that would include the porosity and tortuosity factors when needed.

When the EKR is applied to a heavy metal-contaminated soil or when the ionic conductivity of the pore fluid is high, the electromigration becomes the most important transport mechanism since the electroosmotic transport is from 10 to 300 times smaller than the electromigration transport. Regarding diffusive transport, the estimation of the ionic mobilities from the diffusion coefficients using the Nernst–Einstein–Townsend relation indicates that ionic mobility of a charged species is much higher than the diffusion coefficient (about 40 times the product of the charge of the ion multiplied by the electrical potential gradient):

$$u = \frac{D\,z\,F}{R_g\,T}. \tag{25.3}$$

Considering all the aforementioned aspects, a simple EKR model is proposed that assumes electromigration as the most significant transport process, neglecting the electroosmotic and diffusive transports and assuming there is no advection. The mass conservation equation is simplified to

$$\frac{\partial C_k}{\partial t} = -\vec{\nabla}\cdot\left(-C_k u_k\,\vec{\nabla}V\right) + R_k \quad k = 1, 2 \ldots \text{total number of species}. \tag{25.4}$$

The electrical potential distribution is a function of the electrical resistance of the soil and therefore depends on the instantaneous local concentration and mobility of all ions existing in the pore water of the soil, which ultimately determine the electrical resistance evolution. Therefore, a set of conservation equation, one for each ion, has to be simultaneously integrated, with the difficulty that all these differential equations are strongly coupled through the electrical potential distribution. Therefore, an analytical solution would be very difficult to obtain, if at all possible. So, a numerical solution must be used instead.

25.2 ONE-DIMENSIONAL EKR SIMPLE MODEL

As indicated in the previous section, the modeling of an electrokinetic soil remediation system requires a numerical integration of the mass conservation equation for each ion existing in the pore water. In a first approximation, a one-dimensional model is presented here where an electrical circuit analogy is used (Wilson, Rodríguez-Maroto, and Gómez-Lahoz, 1995a).

As shown in Figure 25.1, the numerical solution divides the one-dimensional system into $N_i + 2$ volume elements or cells. The compartments of the electrodes are located at both ends of the system (0 and $N_i + 1$), and the inner cells contain a soil contaminated by heavy metals. Since conventional current is defined in electrical science as a flow of positive charge, the electrical current always flows from the positive potential end of the conductor toward the negative potential end, independently of the actual direction of motion of the differently charged current carriers (cations, anions, or electrons).

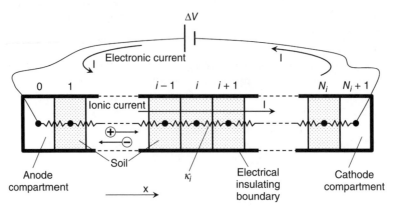

Figure 25.1. One-dimensional model scheme. I, electric current; ΔV, electrical potential drop; κ_i, electrical conductivity at the boundary.

The numerical integration of the mass conservation equations consists basically in a time-iterative method with two sequential steps for each increment of time. In the first one, the transport of ions between each cell is calculated, yielding transient concentrations that are used in the second one, where the chemical equilibria and charge and mass balance equations are solved within each volume element. The first corresponds to the transport phenomena term in the differential conservation equation of each species, and the second corresponds to the rates of production of ions in those same equations. In this way, the model obtains new values that are used to perform the integration forward in time.

The hypotheses of the model are:

- The soil is completely saturated with water.
- The electromigration is the only significant transport process. Electroosmosis and diffusion are neglected, and advection is not present. Models including electroosmotic transport can be found in the literature (Eykholt and Daniel, 1994; Yu and Neretnieks, 1996; Lemaire, Moyne, and Stemmelen, 2007).
- The potential drop in the compartments of the electrodes is usually very low if they are compared with the potential drop in the soil. Since the electrode compartments are well stirred and the electrolytic solutions usually have a high ionic concentration, the potential drop due to ionic current in these compartments is considered negligible.
- The electrode reactions are the water oxidation at the anode and the water reduction at the cathode.
- The chemical reaction rates are high enough to consider that the local equilibrium is reached (i.e. chemical reaction rates are much faster than ion movement).
- The boundary of the EKR system is electrically insulated; the current only flows in the soil regions between the electrodes.
- The current density (i) is kept constant throughout the entire process. Nevertheless, the model can be easily modified if the EKR system operates at constant electrical potential drop.

Therefore, the model is subject to the following constraints:

- The synergistic effects of multiple chemical species are not considered (water ionic strength, activities, etc.).
- The effect of the temperature on the chemical equilibrium constants is not considered.
- The effect of pH and ionic strength on the adsorption equilibria is neglected.

25.2.1 Calculation of the Electrical Voltage Drop between the Electrodes

An EKR system has two voltage drops in series: one due to the electrical resistance of the soil to the ionic current flow and the other due to the electrode processes. Ohm's law and the electrical resistance of the volume elements into which the soil was divided give the first one. The total electrical resistance R is the sum of the resistance of the N_i volume elements, R_i:

$$R = \sum_{i=1}^{N_i} R_i. \tag{25.5}$$

And the resistance of the ith volume element is given by

$$R_i = \frac{\Delta x\, \tau}{A\, \omega} \frac{1}{\sum_{k=1}^{N_k} \lambda_k C_{i\,k}}, \tag{25.6}$$

where Δx is the thickness of the volume element, τ is the tortuosity of the soil, A is the cross-sectional area, ω is the porosity of the water-saturated soil, N_k is the number of ions existing in the pore water, λ_k is the molar conductivity of the kth ion ($cm^2/(\Omega\, mol)$), and $C_{i\,k}$ is the concentration of the kth ion in the ith volume element. Values for the limiting molar conductivity of ions (molar conductivity at infinite dilution) can be found in literature, and the relation between molar conductivity and ionic mobility for an ion at infinite dilution is $\lambda_k = z_k\, u_k\, F$, where z_k is the ion charge and F is the Faraday constant.

The voltage drop resulting from the electrode processes is obtained from both the standard reduction potentials of the half reactions at the electrodes and the Nernst equation, assuming that the partial pressure of the gases is 1 atm (bubbles of gas in the electrolytic solutions) and the temperature is 25 °C:

Cathode: $\quad 4H_2O + 4e^- \rightleftarrows 4OH^- (aq) + 2H_2 (gas) \quad E^0 = -0.828$ V,

Anode: $\quad 2H_2O \rightleftarrows 4H^+ (aq) + O_2 (gas) + 4e^- \quad E^0 = -1.229$ V,

$$E = (-0.828\text{ V} - 1.229\text{ V}) - \frac{RT}{F}(\ln[OH^-]_{\text{cathode}} + \ln[H^+]_{\text{anode}}), \tag{25.7}$$

$$E = -1.23 - 0.059\,(pH_{\text{cathode}} - pH_{\text{anode}}). \tag{25.8}$$

And therefore, the electrical potential difference between the electrodes is the sum of these two voltage drops:

$$\Delta V = |E| + IR, \tag{25.9}$$

$$\Delta V = |-1.23 - 0.059(\text{pH}_{\text{cathode}} - \text{pH}_{\text{anode}})| + I \sum_{i=1}^{N_i} \frac{\Delta x \, \tau}{A \, \omega} \frac{1}{\sum_{k=1}^{N_k} \lambda_k \, C_{ik}}. \tag{25.10}$$

As indicated above, the electric current that flow through the soil can be calculated from this equation when the EKR system works at a constant electrical potential drop. If the electroosmosis is not neglected, these latter equations would be used in the calculation of the electrical potential gradient for electroosmotic flow. Anyway, the system energy requirements are calculated from the values of the electrical potential drop and of the electrical current.

25.2.2 First Sequential Step: Transport Step

The first of the two sequential steps of the iterative numerical integration is described. As indicated above, the electromigration is the only transport phenomena considered in this stage that changes the concentration of the ions. Therefore, the mass conservation equation for a volume element of the soil ($1 \leq i \leq N_i$) is given by

$$(\Delta x \, A \, \omega) \frac{dC_{ik}}{dt} = \dot{n}_{ik} - \dot{n}_{i-1\,k} \quad \text{for } k = 1, 2, 3 \ldots N_k, \tag{25.11}$$

where $\dot{n}_{ik}$ is the molar flow of the kth ions from the $(i+1)$th volume element toward the inside of the ith volume element, which is proportional to the electric current according to the following expression:

$$\dot{n}_{ik} = \frac{-I}{z_k F} t_{ik}, \tag{25.12}$$

where the product of z_k and F is a factor related to Faraday's law, which converts the electrical units to molar units, and t_{ik} is the transport number of the kth ions, which is the current density due to the kth ions divided by the sum of the current densities of all the ions in the electrolyte. The negative sign indicates that the flow of cations through the right face of the ith cell is leaving this cell, while anions are moving in the opposite direction (Fig. 25.2).

Thus, according to Figure 25.2, the total conductivity between the ith and $(i+1)$th volume elements, κ_i, is given by the sum of the contribution to the conductivity of the cations moving rightward and the anions moving leftward. This is represented by

$$\kappa_i = \sum_{k=1}^{N_k} \lambda_k [C_{ik} S(z_k) + C_{i+1\,k} S(-z_k)], \tag{25.13}$$

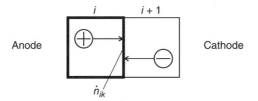

Figure 25.2. Molar flows of cations and anions through the right face of the ith volume element.

where $S(x)$ is a function introduced to take into account the different movements of cations and anions. Its value is 1 if x is higher than 0, and 0 if x is lower or equal to 0:

$$S(x) = \begin{cases} 0 & x \leq 0 \\ 1 & x > 0 \end{cases}. \quad (25.14)$$

Thus, the fraction of the total current carried by the kth ions between those two cells, $t_{i\,k}$, is calculated as the ratio between the contribution of these ions to the conductivity and the total conductivity at the boundary face; this is

$$t_{i\,k} = \frac{\lambda_k \left[C_{i\,k}\, S(z_k) + C_{i+1\,k}\, S(-z_k) \right]}{\kappa_i}, \quad (25.15)$$

where the S function is again introduced to consider the opposite direction of the movement of cations and anions. Therefore, the molar flow of the kth ions between the ith and $(i+1)$th volume elements is

$$\dot{n}_{i\,k} = \frac{-I}{z_k\, F} \frac{\lambda_k \left[C_{i\,k}\, S(z_k) + C_{i+1\,k}\, S(-z_k) \right]}{\kappa_i} \quad \text{for } k = 1, 2, 3 \ldots N_k \text{ and for } 1 \leq i \leq N_i. \quad (25.16)$$

Substituting this expression into the mass conservation equation,

$$(\Delta x\, A\, \omega) \frac{dC_{i\,k}}{dt} = \frac{-I\, \lambda_k}{z_k\, F} \left(\frac{[C_{i\,k}\, S(z_k) + C_{i+1\,k}\, S(-z_k)]}{\kappa_i} - \frac{[C_{i-1\,k}\, S(z_k) + C_{i\,k}\, S(-z_k)]}{\kappa_{i-1}} \right) \quad (25.17)$$

And finally, the new values of the concentrations at the time $t + \Delta t$ can be calculated using a numerical solution of this latter ordinary differential equation. In this case, Euler's method (Perry, Green, and Malone, 1984) is used due to its simplicity, although errors are proportional to Δt. Other method of high order, as Runge-Kutta (Perry, Green, and Malone, 1984), can be used if needed:

$$C_{i\,k}^{*}(t + \Delta t) = C_{i\,k}(t) + \left. \frac{dC_{i\,k}}{dt} \right|_{t} \Delta t, \quad (25.18)$$

$$C^*_{ik}(t+\Delta t) = C_{ik}(t) + \frac{-I\lambda_k}{(\Delta x\, A\, \omega)z_k F}\left(\frac{[C_{ik}(t)S(z_k)+C_{i+1\,k}(t)S(-z_k)]}{\kappa_i(t)} - \frac{C_{i-1\,k}(t)S(z_k)+C_{ik}(t)S(-z_k)}{\kappa_{i-1}(t)}\right)\Delta t. \quad (25.19)$$

These values of the concentrations obtained in this first step, $C^*_{ik}(t+\Delta t)$, will be used in the second step, in which the chemical equilibria among all species considered are achieved. Nevertheless, up to this point, only the inner volume elements have been considered; nothing was said about the electrode compartments, which are discussed in the following paragraphs before beginning the second step, and whose concentrations are needed to calculate the above concentration values in the inner volume elements.

According to the previous hypotheses, the molar flow through the right-hand side of the cathode compartment ($i = N_i + 1$) is not possible as a result of the electrical insulation hypothesis. Thus, only the molar flow between the N_ith cell and the cathode compartment has to be considered, which was already calculated above. Otherwise, the reaction of the reduction of water at the cathode takes place in this compartment, and then the generation of the hydroxyl ions must be considered in the mass conservation equation (one mol of hydroxyl ions per one Faraday of circulating charge). In addition, this same equation must consider the neutralization of a percentage of the hydroxyl ions generated at the cathode if an acid enhancement of the EKR system is used. Both circumstances are considered as follows.

For all ions except hydroxyl ions and anions of the acid used in the enhanced EKR,

$$V_{N_i+1}\frac{dC_{N_i+1\,k}}{dt} = -\dot{n}_{N_i\,k}, \quad (25.20)$$

where V_{N_i+1} is the volume of the cathode compartment:

$$V_{N_i+1}\frac{dC_{N_i+1\,k}}{dt} = \frac{I\lambda_k}{z_k F}\left(\frac{[C_{N_i\,k}S(z_k)+C_{N_i+1\,k}S(-z_k)]}{\kappa_{N_i\,k}}\right). \quad (25.21)$$

For k = hydroxyl ions,

$$V_{N_i+1}\frac{dC_{N_i+1\,k}}{dt} = -\dot{n}_{N_i\,k} + (1-nr)\frac{I}{F}, \quad (25.22)$$

$$V_{N_i+1}\frac{dC_{N_i+1\,k}}{dt} = \frac{I\lambda_k}{z_k F}\left(\frac{[C_{N_i\,k}S(z_k)+C_{N_i+1\,k}S(-z_k)]}{\kappa_{N_i\,k}}\right) + (1-nr)\frac{I}{F}, \quad (25.23)$$

where nr is the ratio of hydroxyl ions neutralized.

For k = anions of the acid used in the enhanced EKR,

$$V_{N_i+1}\frac{dC_{N_i+1\,k}}{dt} = -\dot{n}_{N_i\,k} + nr\frac{I}{F}, \quad (25.24)$$

$$V_{N_i+1}\frac{dC_{N_i+1\,k}}{dt} = \frac{I\lambda_k}{z_k F}\left(\frac{[C_{N_i\,k}S(z_k)+C_{N_i+1\,k}S(-z_k)]}{\kappa_{N_i\,k}}\right) + nr\frac{I}{F}. \quad (25.25)$$

The corresponding value of the concentrations at the time $t + \Delta t$, $C^*_{N_i+1\,k}(t+\Delta t)$, is obtained by the same procedure that was used for the inner compartments (e.g. Euler's method).

Analogously, the left-hand side of the anode compartment ($i = 0$) is electrically insulated, too; therefore, ionic flow only occurs through its right-hand face:

$$\dot{n}_{0\,k} = \frac{-I}{z_k F}\frac{\lambda_k[C_{0\,k}S(z_k)+C_{1\,k}S(-z_k)]}{\kappa_0} \quad \text{for } k = 1, 2, 3\ldots N_k. \quad (25.26)$$

In addition, the generation of the protons at the anode must also be considered in the mass conservation equation for this compartment as follows.

For all ions except protons,

$$V_0\frac{dC_{0\,k}}{dt} = \dot{n}_{0\,k}, \quad (25.27)$$

where V_0 is the volume of the anode compartment:

$$V_0\frac{dC_{0\,k}}{dt} = \frac{-I\lambda_k}{z_k F}\left(\frac{[C_{0\,k}S(z_k)+C_{1\,k}S(-z_k)]}{\kappa_0}\right). \quad (25.28)$$

For protons,

$$V_0\frac{dC_{0\,k}}{dt} = \dot{n}_{0\,k} + \frac{I}{F}, \quad (25.29)$$

$$V_0\frac{dC_{0\,k}}{dt} = \frac{-I\lambda_k}{z_k F}\left(\frac{[C_{0\,k}S(z_k)+C_{1\,k}S(-z_k)]}{\kappa_0}\right) + \frac{I}{F}. \quad (25.30)$$

As previously done, the values of the preliminary concentrations at the time $t + \Delta t$, $C^*_{0\,k}(t+\Delta t)$, are obtained in the same manner. Up to this point, only the transport step has been described. Now the next step corresponding to the sink and source term of the differential mass conservation equation is presented.

25.2.3 Second Sequential Step: Chemical Equilibria

The rates of production/consumption of ions have to be calculated in this step as independent phenomena from the previous step. According to the proposed hypotheses, these rates are high enough to consider that the local equilibrium is reached. In the present case, this means that concentration changes of the ions and species considered in this model can be calculated using the action mass equations of the involved chemical reactions. After the transport step, a set of preliminary values of concentrations, which are not at chemical equilibrium, is known for each volume element ($C^*_{i\,k}$). The solution in each cell of the mass conservation equations, together with the electroneutrality condition of the aqueous phase, subject to the constraints of the set of the nonlinear equilibrium mass action equations, provides the concentrations of all the species at equilibrium (C_{ik}), as sought in this step. For simplicity, the subscripts referred to the compartments are not used in this section of the model, so C_{ik}^* is denoted by C_k^* and the equilibrium concentration C_{ik} by C_k.

The calculation of species concentration in equilibrium is a common problem in environmental modeling, which can be solved using many available computer programs (Visual MINTEQ, PHREEQC, etc.). Unfortunately, these stand-alone programs cannot be easily used as a subroutine in the computer programs written by other engineers and scientists, as in this case. Although, the mathematical procedure described in the full technical documentations of these programs provides a useful guide for writing a source code suitable for each particular problem (Allison, Brown, and Novo-Gradac, 1991; Parkhurst and Appelo, 1999; Carrayrou, Mose, and Behra, 2002).

In all these computer programs, the mass action equations and mass balances are rewritten as a function of a set of species that is composed of only one species for each of the element in the chemical system. This set is known as "master species." This is, if M is the number of master species in the present model, which for programming simplicity is denoted by the subscripts ranging from 1 to M, then (N_k–M) formation reactions, one for each of the nonmaster species, can be found as a function of only the set of master species:

$$\sum_{m=1}^{M} v_{jm} X_m \rightleftarrows Y_j \quad \text{for } j = M+1, M+2 \ldots N_k, \tag{25.31}$$

where v_{jm} is the stoichiometric coefficient of the master species X_m ($1 \leq m \leq M$) in the formation reaction of the nonmaster species Y_j ($M+1 \leq j \leq N_k$). The mass action equation for each one of the above reactions would be written as

$$K_j = C_j \prod_{m=1}^{M} C_m^{-v_{jm}}. \tag{25.32}$$

And therefore, the concentration of the nonmaster species can be expressed as a function of only the master species concentrations:

$$C_j = K_j \prod_{m=1}^{M} C_m^{v_{jm}}. \tag{25.33}$$

The logarithmic form of this equation is

$$\log C_j = \log K_j + \sum_{m=1}^{M} v_{jm} \log C_m. \tag{25.34}$$

With regard to mass balance, there exists an M conservation equation, one for each element, which can be expressed by

$$T_m = \sum_{j=1}^{N_k} v_{jm} C_j, \qquad (25.35)$$

where T_m is the total concentration of the element represented by the mth master species, whose value is usually known. In the present EKR model, the known value of T_m is T_m^*, which can be calculated from the preliminary values of concentrations obtained in the transport step (C_k^*).

The mathematical solution of this nonlinear equation system is normally obtained by an iterative numerical procedure. In this case, the globally convergent Newton–Raphson method, with line searches and backtracking described by Press *et al.* (1992), works well. This iterative procedure is as follows: The concentration of the nonmaster species (C_j) is calculated from an initial guess set of values for the concentration of master species (C_m) using the logarithmic equation given above. Then, the total concentrations of each master species (T_m) is obtained by means of the mass balances. The differences between these calculated total concentrations of the master species (T_m) and the corresponding known values from the transport step (T_m^*) are known as the mass balance residues. If these residues are higher than a tolerance value, the Newton–Raphson method will indicate the new set of values for the concentration of master species that have to be used in the next iteration. This iterative procedure is continued until all the mass balance residues are below the desired tolerance value.

The equilibrium equations that normally have to be considered in the EKR modeling of a soil contaminated by heavy metals can be classified into one of the following categories: complex formation reactions, precipitation of the metal hydroxides or of other species, ion exchange reactions, surface complexation reactions, etc. Anyway, the autoionization of water always has to be considered and the precipitation of carbonates, together with the carbonate–bicarbonate equilibrium, should normally also be considered. However, the above equations have only considered the species in aqueous phase, so if a species precipitates, a new master species has to be included in this equilibrium system, whose "concentration" would be the amount of the precipitated species per unit volume of water. This additional degree of freedom is constrained by the solubility product constant of the precipitate (K^s), because the new solid phase is in equilibrium with the aqueous phase. If there exists N_p precipitated species, the pure-phase equilibria can be represented with the following equation:

$$K_p^s = \prod_{m=1}^{M} C_m^{v_{pm}}, \qquad (25.36)$$

where v_{pm} is the stoichiometric coefficient of the master species m in the precipitation reaction. And the saturation index of the pth species (SI_p) is defined as

$$SI_p = \log \frac{\prod_{m=1}^{M} C_m^{v_{pm}}}{K_p^s} = -\log K_p^s + \sum_{m=1}^{M} v_{pm} \log C_m. \qquad (25.37)$$

$SI_p = 0$ when the pth precipitate is in equilibrium with the aqueous phase; if the saturation index is negative, the system is undersaturated with respect to that species, and if the index is positive, the solution is supersaturated. The Newton–Raphson method uses the logarithmic expressions for the saturation indexes in a similar manner to the mass balance residue expressions.

If a precipitate is present, the mass balance equations should be

$$T_m = \sum_{j=1}^{N_k} v_{j\,m} C_j + \sum_{p=1}^{N_p} v_{p\,m} C_p. \qquad (25.38)$$

Of course, the ion exchange and surface complexation reactions can be easily included in this step of the model in a similar manner to the precipitation reactions. Likewise, the concentration of electron as a master species (pe value) is needed to calculate the concentration of the aqueous species involved in a redox couple. The mass action equation for each of these redox couples is combined to remove the concentration of the electron from the equations, as if a set of redox couple was in a redox equilibrium.

The definitive values of concentrations obtained after this equilibrium step are used to iterate these two sequential steps for the next increment of time. In this way, the numerical integration is performed forward in time.

All these equations can be rearranged in a matrix form that can be easily represented in a tableau similar to Morel's tableau (Morel and Morgan, 1972; Morel, 1983), as shown in Figure 25.3.

The calculation steps are:

1. The aqueous concentrations of species (C_a) are calculated from the tentative values of the "master species" (X_a). The logarithm of concentration of the species at one row is obtained as the sum over that row of the product of the

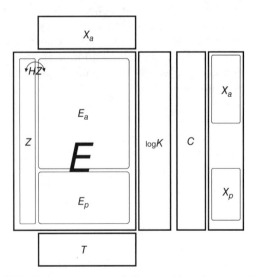

Figure 25.3. Rearrangement of the equations in a matrix form.

stoichiometric coefficients (E_a) and the logarithm of the corresponding values of the concentrations of the "master species" plus the logarithm of the corresponding equilibrium constant:

$$\log(C_a) = \log K_a + E_a \cdot \log(X_a). \tag{25.39}$$

2. The mass balance for each "master species" corresponds to the sum over one column of the product of the stoichiometric coefficients (E) and the concentration values of the species (C). In this step, the concentrations of the "precipitated species" are also considered. The mass balance of protons (normally located at the first column) is exchanged with the electroneutrality condition, and then the column corresponding to the electrical charge is exchanged with the column corresponding to the protons. The mass balance residues (f_{vec}) are calculated for the Newton–Raphson method:

$$E_{aHZ}^T \times C_a + E_{pHZ}^T \times X_p - T = 0 = f_{vec}, \tag{25.40}$$

where $E_{aHZ} = E_a$, exchanging the columns corresponding to protons and electrical charge.

3. Finally, the saturation indexes of each precipitated species are calculated. When a precipitated species is present, the corresponding value of saturation index is appended to the matrix of the residues for the Newton–Raphson method. The saturation index values are calculated by the sum of the product of the stoichiometric coefficients corresponding to the precipitates (E_p) and the logarithm of the corresponding values of the concentrations of the "master species" minus the logarithm of the corresponding solubility product constant:

$$SI = E_p \times \log(X_a) - \log K_p = f_{vec} \ (= 0 \text{ if precipitate is present}). \tag{25.41}$$

25.2.4 Initial Chemical Conditions

Logically, before any integration step, the chemical conditions of the heavy metals in the soil system should be defined. In this sense, the common environmental modeling programs (Visual MINTEQ, PHREEQC, etc.) provide the most probable distribution of the metals between the different chemical species (speciation) according to the initial conditions of the soil (pH, initial concentration of metals, redox potential, etc.). The main species and equilibria to be included in the chemical reaction step described in the above subsection are selected in this manner. An example of the utility of these geochemical models can be found in Al-Hamdan and Reddy (2008a), where the geochemical speciation and distribution of heavy metals during electrokinetics in the soils were assessed by the geochemical model MINEQL⁺.

25.3 TWO-DIMENSIONAL MODEL

Up to this time, numerous one-dimensional models for electrokinetic soil remediation have been presented (Alshawabkeh and Acar, 1992; Choi and Lui, 1995;

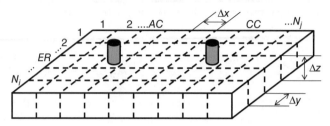

Figure 25.4. Scheme of the numerical model.

Wilson, Rodríguez-Maroto, and Gómez-Lahoz, 1995a; Yu and Neretnieks, 1996; Haran et al., 1997; Mascia et al., 2007; Al-Hamdan and Reddy, 2008b). This two-step sequential strategy was previously presented in electrokinetic models such as TRANQL (Cederberg, Street, and Leckie, 1985) and DYNAMIX (Liu and Narasimhan, 1989). As presented in Chapter 30, similar models have been successfully used to describe the electrokinetic barrier to contaminant transport through compacted clay (Narasimhan and Ranjan, 2000). These models establish which are the factors that control an electrokinetic process in the soil (pH, equilibrium chemistry in the pore water, interactions between contaminants and the solid surface, electrical current, electrode reactions, etc.). Enhancements of the EKR system have been studied using this one-dimensional model, too (Wilson, Rodríguez-Maroto, and Gómez-Lahoz, 1995b; Garcia-Gutierrez et al., 2007). Nevertheless, theoretical studies about the performance of the enhanced method of this technique in a two-dimensional arrangement are not so frequent.

Of course, if an analytical solution is not feasible for a one-dimensional model, there will be little chance of finding an analytical solution for a two-dimensional arrangement. Therefore, a numerical solution is developed following the same procedures as outlined in the latter subsection. In this case, a horizontal surface of saturated soil with a depth of Δz is considered (Fig. 25.4). The area is divided into a two-dimensional grid of N_i rows and N_j columns resulting in $(N_i \times N_j)$ volume elements with horizontal dimensions of Δx by Δy. The concentration of the kth ion in the cell (i, j) is denoted by C_{ijk}, where i and j are row and column indices into the grid. The electrodes are placed perpendicular to the horizontal surface of the soil, the anode compartment is located in the (ER, AC) cell, and the cathode compartment in the (ER, CC) cell.

Again, the numerical integration of the mass conservation equations for this two-dimensional arrangement consists in the same time-iterative method with two sequential steps for each increment of time just described in the previous section. The procedure followed in the current equilibria step is identical to that outlined for the one-dimensional model; therefore, only the transport step is presented here. The transient concentrations obtained in this transport step would be used in the equilibrium step.

Analogously, the hypotheses of the two-dimensional model are essentially the same as those used in the one-dimensional model: saturated soil, electromigration as the only significant transport, negligible potential drop in the electrode compartments, water hydrolysis at the electrodes, local equilibrium assumption, electrically insulated EKR system, and constant current density (i).

TWO-DIMENSIONAL MODEL   553

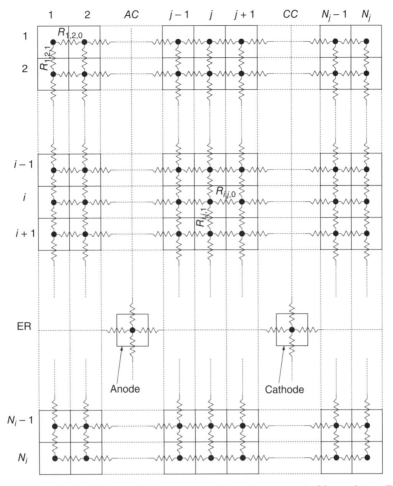

Figure 25.5. Electrical resistances between compartments represented by resistors (Vereda-Alonso *et al.*, 2004).

25.3.1 Calculation of the Distribution of the Current in the Soil

As shown in Figure 25.5, the electrical resistance between each two adjacent cells is represented by a single resistor whose resistance is obtained from the molar conductivities and the concentrations of the ions that flow through the shared face between both cells. The electrical resistance between the cells (i, j) and $(i, j + 1)$ is denoted by R_{ij0} and the one between (i, j) and $(i + 1, j)$ by R_{ij1}. The last subscript indicates if the resistor is in the **x** direction (0) or in the **y** direction (1). The values of these resistors are obtained at each time step as the sum of the electrical resistance of the adjacent half cells:

$$R_{ij0} = \frac{\Delta x \, \tau}{\Delta y \, \Delta z \, \omega} \left(\frac{1/2}{\sum_{k=1}^{N_k} \lambda_k \, C_{ijk}} + \frac{1/2}{\sum_{k=1}^{N_k} \lambda_k \, C_{ij+1k}} \right), \qquad (25.42)$$

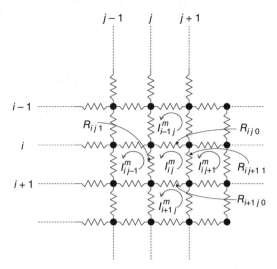

Figure 25.6. Circuit with mesh current labeled.

$$R_{ij1} = \frac{\Delta y \, \tau}{\Delta x \, \Delta z \, \omega} \left(\frac{1/2}{\sum_{k=1}^{N_k} \lambda_k \, C_{ijk}} + \frac{1/2}{\sum_{k=1}^{N_k} \lambda_k \, C_{i+1jk}} \right). \quad (25.43)$$

Once all these electrical resistances are known, Kirchhoff's current and voltage laws (Nilsson, 1990) are used to calculate the currents that are flowing through each one of these resistors when either a voltage drop, ΔV, is applied between the electrodes or a constant current, I, flows between them. Kirchhoff's current law simply states that the sum of currents flowing into a junction equals the sum of currents flowing away from the junction. Kirchhoff's voltage law states that the sum of voltage drops around a closed circuit is equal to the sum of voltage sources.

As shown in Figure 25.6, the mesh current method assumes that a mesh current, I_{ij}^m, is flowing in each closed circuit (essential mesh) in the indicated direction. The system contains $((N_i - 1) \times (N_j - 1) + 1)$ essential meshes whose mesh currents are unknown, and therefore, the same number of equations is required. The additional mesh (not shown in the Fig. 25.5) is that which connect the power supply to the electrodes. The voltage law asserts in this particular case that the sum of the voltage across the four resistors in each essential mesh have to be 0; the voltage drops are positive if the mesh current flows in the indicated direction and negative if it flows in the opposite direction. So, the following equation results for the mesh whose upper-left corner is the center of the cell (i, j) and does not contain a resistor in the horizontal line connecting the electrodes:

$$(R_{ij0} + R_{ij+11} + R_{i+1j0} + R_{ij1})I_{ij}^m - R_{ij0}I_{i-1j}^m - R_{ij+11}\,I_{ij+1}^m - R_{i+1j0}\,I_{i+1j}^m - R_{ij1}\,I_{ij-1}^m = 0. \quad (25.44)$$

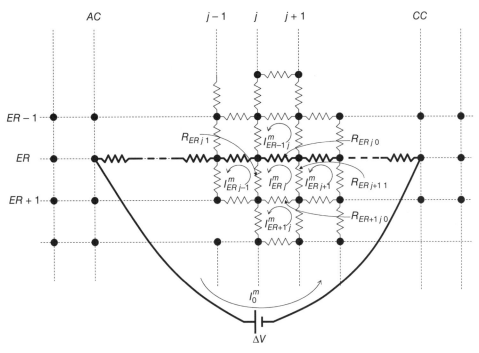

Figure 25.7. Essential mesh including the power supply.

If the mesh includes one of the resistors located at the electrodes row, between the anode and cathode, the latter expression of the voltage law has to include the term corresponding to the essential mesh that connects the power supply to the electrodes (Fig. 25.7), where $|I_0^m|$ is the total current flowing through the soil.

If $i = ER-1$ and $AC \leq j < CC$,

$$(R_{i\,j\,0} + R_{i\,j+1\,1} + R_{i+1\,j\,0} + R_{i\,j\,1})I_{i\,j}^m - R_{i,\,j\,0}\,I_{i-1\,j}^m - R_{i\,j+1\,1}\,I_{i\,j+1}^m - R_{i+1\,j\,0}\,I_{i+1\,j}^m - R_{i\,j\,1}\,I_{i\,j-1}^m - R_{ER\,j\,0}\,I_0^m = 0. \quad (25.45)$$

If $i = ER$ and $AC \leq j < CC$,

$$(R_{i\,j\,0} + R_{i\,j+1\,1} + R_{i+1\,j\,0} + R_{i\,j\,1})I_{i\,j}^m - R_{i\,j\,0}\,I_{i-1\,j}^m - R_{i\,j+1\,1}\,I_{i\,j+1}^m - R_{i+1\,j\,0}\,I_{i+1\,j}^m - R_{i\,j\,1}\,I_{i\,j-1}^m + R_{ER\,j\,0}\,I_0^m = 0. \quad (25.46)$$

The last equation of this system corresponds to the essential mesh that connects the power supply to the electrodes (Fig. 25.7). Again, the current that flows across this mesh is the total current that flows through the soil:

$$\sum_{j=AC}^{CC-1}\left(-R_{ER\,j\,0}I_{ER-1\,j}^m + R_{ER\,j\,0}I_{ER\,j}^m\right) + I_0^m \sum_{j=AC}^{CC-1} R_{ER\,j\,0} + \Delta V = 0. \quad (25.47)$$

Depending on if the EKR system works either at constant voltage or at constant current density, either the voltage or the total current between the electrodes is fixed, and therefore, the other one will be defined by this system of equations. Anyway, the voltage drop indicated in this latter equation does not include the voltage drop resulting from the electrode processes, which will have to be considered if the energy requirement has to be calculated.

The solution of this system of $[(N_j - 1) \times (N_i - 1) + 1]$ linear equations at each time step gives the unknown values of the mesh currents. These values are used to calculate the currents that flow through all the resistors in the circuit, namely the values of the current moving from one cell to the adjacent one. The notation for these currents is similar to the notation used for the resistors. So, the current flowing between the (i,j) and $(i,j+1)$ cells is denoted by I_{ij0} and the current flowing between the (i,j) and $(i+1,j)$ cells by I_{ij1}. The sign convention is as follows: Currents entering a junction are positive and currents leaving a junction are negative, as usual in Kirrchhoff's current law. Therefore, the current flowing into the cell (i,j) is positive, and the current flowing out from the cell (i,j) is negative. Otherwise, $I_0^m = -1$, since conventional current always flows from the positive potential end of the conductor toward the negative potential end. Thus, the values of these currents are calculated using the following equations.

If $i = ER$ and $AC \leq j < CC$ (at the electrode row and between them),

$$I_{ERj0} = I_{ERj}^m - I_{ER-1j}^m + I_0^m. \tag{25.48}$$

In the rest of the cases,

$$I_{ij0} = I_{ij}^m - I_{i-1j}^m, \tag{25.49}$$

$$I_{ij1} = -I_{ij}^m + I_{ij-1}^m. \tag{25.50}$$

As was sought, this step provides the distribution of the current in the soil, which will be used for the calculation of the molar flows of the ions in the next step. Anyway, if it was needed, the electrical potential distribution in the soil can be easily obtained from these results.

25.3.2 Transport Step

The molar flow of each ion between two adjacent cells is obtained from the current flowing between these two cells and the transport number of the considered ion through the shared face between both cells. According to the conventional definition of a positive current, cations flow in the direction of the current and anions flow in the opposite direction. Therefore, if I_{ij0} is higher than 0, cations will move from the cell $(i,j+1)$ toward the (i,j), while anions will move to the opposite direction. If I_{ij0} is negative, the cell (i,j) will lose cations that will be earned by the cell $(i,j+1)$ (inversely for anions). In the y direction, if I_{ij1} is positive, cations will move from the cell $(i+1,j)$ toward the (i,j), while anions will flow from the cell (i,j) toward the $(i+1,j)$. And finally, if I_{ij1} is negative, the cell (i,j) will lose cations that will be earned by the cell $(i+1,j)$ and vice versa for anions.

The total conductivities between cells (i,j) and $(i,j+1)$ and between (i,j) and $(i+1,j)$, which are denoted by κ_{ij0} and κ_{Ij1}, respectively, are obtained by the following expressions:

$$\kappa_{ij0} = \sum_{k=1}^{N_k} \lambda_k [C_{ijk}S(z_k)S(-I_{ij0}) + C_{ij+1k}S(z_k)S(I_{ij0}) + C_{ij+1k}S(-z_k)S(-I_{ij0}) + C_{ijk}S(-z_k)S(I_{ij0})], \quad (25.51)$$

$$\kappa_{ij1} = \sum_{k=1}^{N_k} \lambda_k [C_{ijk}S(z_k)S(-I_{ij1}) + C_{i+1jk}S(z_k)S(I_{ij1}) + C_{i+1jk}S(-z_k)S(-I_{ij1}) + C_{ijk}S(-z_k)S(I_{ij1})]. \quad (25.52)$$

where the S function, which was defined in Section 25.2, is again introduced to take into account the different movement of cations and anions as described in the above paragraph.

Analogously, the transport number of each ion between the cells (i, j) and $(i, j + 1)$, and between (i, j) and $(i + 1, j)$, which are denoted by t_{ijk0} and t_{ijk1}, respectively, are obtained by

$$t_{ijk0} = \frac{\lambda_k [C_{ijk}S(z_k)S(-I_{ij0}) + C_{ij+1k}S(z_k)S(I_{ij0}) + C_{ij+1k}S(-z_k)S(-I_{ij0}) + C_{ijk}S(-z_k)S(I_{ij0})]}{\kappa_{ij0}}, \quad (25.53)$$

$$t_{ijk1} = \frac{\lambda_k [C_{ijk}S(z_k)S(-I_{ij1}) + C_{i+1jk}S(z_k)S(I_{ij1}) + C_{i+1jk}S(-z_k)S(-I_{ij1}) + C_{ijk}S(-z_k)S(I_{ij1})]}{\kappa_{ij1}}. \quad (25.54)$$

Of course, it can be demonstrated that the summation over all ions of the transport number in either the horizontal or vertical direction should be equal to 1.

Therefore, the molar flows of the kth ion into the cell (i, j) from $(i, j + 1)$ and from $(i + 1, j)$, which are denoted by $\dot{n}_{ijk0}$ and $\dot{n}_{ijk1}$, respectively, are obtained from the corresponding current and the corresponding transport number:

$$\dot{n}_{ijk0} = \frac{I_{ij0}}{z_k F} t_{ijk0} \text{ for } k = 1, 2, 3 \ldots N_k \text{ and for } 1 \leq i < N_i \text{ and } 1 \leq j < N_j, \quad (25.55)$$

$$\dot{n}_{ijk1} = \frac{I_{ij1}}{z_k F} t_{ijk1} \text{ for } k = 1, 2, 3 \ldots N_k \text{ and for } 1 \leq i < N_i \text{ and } 1 \leq j < N_j. \quad (25.56)$$

All these definitions assure that the electroneutrality in the compartment (i, j) should be 0 as demonstrated below:

$$\sum_{k=1}^{N_k} z_k (\dot{n}_{ijk0} - \dot{n}_{ij-1k0} + \dot{n}_{ijk1} - \dot{n}_{i-1jk1})$$
$$= \frac{1}{F} \sum_{k=1}^{N_k} (I_{ij0} t_{ijk0} - I_{ij-10} t_{ij-1k0} + I_{ij1} t_{ijk1} - I_{i-1j1} t_{i-1jk1})$$
$$= \frac{1}{F} \left(I_{ij0} \sum_{k=1}^{N_k} t_{ijk0} - I_{ij-10} \sum_{k=1}^{N_k} t_{ij-1k0} + I_{ij1} \sum_{k=1}^{N_k} t_{ijk1} - I_{i-1j1} \sum_{k=1}^{N_k} t_{i-1jk1} \right)$$
$$= \left(\text{by definition the } \sum_{k=1}^{N_k} t_{rskt} = 1 \right)$$
$$= \frac{1}{F} (I_{ij0} - I_{ij-10} + I_{ij1} - I_{i-1j1}) = 0 \text{ (by the Kirchhoff's current law).} \quad (25.57)$$

The set of mass conservation equations for the cell (i, j) in this transport step, in which the electromigration is considered the only transport phenomena, is given by

$$(\Delta x\, \Delta y\, \Delta z\, \omega)\frac{dC_{ijk}}{dt} = \dot{n}_{ijk0} - \dot{n}_{ij-1k0} + \dot{n}_{ijk1} - \dot{n}_{i-1jk1} \quad \text{for } k = 1, 2, 3 \ldots N_k. \quad (25.58)$$

Substituting the expressions of the four flows into the mass conservation equation,

$$(\Delta x\, \Delta y\, \Delta z\, \omega)\frac{dC_{ijk}}{dt} = \frac{1}{z_k F}(I_{ij0}\, t_{ijk0} - I_{ij-10}\, t_{ij-1k0} + I_{ij1}\, t_{ijk1} - I_{i-1j1}\, t_{i-1jk1}). \quad (25.59)$$

These latter equations should be conveniently modified for the anode and cathode compartments, where water oxidation and reduction reactions are respectively taking place in agreement with the hypotheses. Therefore, the generation of protons at the anode and hydroxyl ions at the cathode should be included in the expressions of the corresponding mass balances.

For k = protons at the anode,

$$(\Delta x\, \Delta y\, \Delta z\, \omega)\frac{dC_{ER\,AC\,k}}{dt} = \dot{n}_{ER\,AC\,k0} - \dot{n}_{ER\,AC-1k0} + \dot{n}_{ER\,AC\,k1} - \dot{n}_{ER-1\,AC\,k1} + \frac{I}{F}. \quad (25.60)$$

And for k = hydroxyl ions at the cathode,

$$(\Delta x\, \Delta y\, \Delta z\, \omega)\frac{dC_{ER\,CC\,k}}{dt} = \dot{n}_{ER\,CC\,k0} - \dot{n}_{ER\,CC-1k0} + \dot{n}_{ER\,CC\,k1} - \dot{n}_{ER-1\,CC\,k1} + \frac{I}{F}. \quad (25.61)$$

When the EKR system is enhanced by the addition of an acid, these equations should be adequately modified in the same manner as was done for the one-dimensional model. Finally, the new transient values of the concentrations at the time $t + \Delta t$ can be calculated using a numerical solution of differential equations like Euler's method:

$$C^*_{ijk}(t + \Delta t) = C_{ijk}(t) + \left.\frac{dC_{ijk}}{dt}\right|_t \Delta t. \quad (25.62)$$

The convergence of this method only depends on the Δt value and on the number of nodes used, because Kirchhoff's laws provide an analytical solution of the currents distribution (or the electrical potentials distribution) in the soil.

The transient concentrations obtained in this transport step will be used in the equilibrium step, which is not described here because it is identical to that outlined for the one-dimensional model. The definitive values of concentrations that are obtained after the equilibrium step are then used to iterate these two sequential steps for the next time step. In this way, the numerical integration is performed forward in time.

Examples of the application of this two-dimensional model can be found in Vereda-Alonso et al. (2004, 2007).

NOTATION

A	cross-sectional area (m^2)
AC	anode column in the two-dimensional model
C^*_{ik}	concentration of the kth ion in the ith volume element obtained in this first step of the model calculations (mol/m^3)
C^*_k	concentration of the kth ion in an unspecified volume element obtained in this first step of the model calculations (mol/m^3)
CC	cathode column in the two-dimensional model
C_{ijk}	concentration of the kth ion in the (i, j) volume element, used in the two-dimensional model (mol/m^3)
C_{ik}	concentration of the kth ion in the ith volume element, used in the one-dimensional model (mol/m^3)
C_k	concentration of the kth ion (mol/m^3)
D_k	diffusion coefficient of the kth ion (m^2/s)
E	reduction potentials of the half reactions at the electrodes (V)
E^0	standard reduction potentials of the half reactions at the electrodes (V)
ER	electrodes row in the two-dimensional model
F	Faraday's constant, 96,485 C/mol e$^-$
I	electrical current (A)
I^m_0	total current flowing through the soil in the two-dimensional model (A)
$I^m_{i,j}$	mesh current for the mesh whose upper-left corner is the center of the cell (i, j) in the two-dimensional model (A)
I_{ij0}	current flowing between the (i, j) and $(i, j + 1)$ cells in the two-dimensional model (A)
I_{ij1}	current flowing between the (i, j) and $(i + 1, j)$ cells in the two-dimensional model (A)
I	electrical current density (A/m^2)
J_k	flux of the kth ion (mol/(m^2·s))
K_j	equilibrium constant corresponding to the formation reaction of the non-master species Y_j
K^s_p	solubility product constant of the pth precipitated species
k_e	electroosmotic permeability of the soil (m^2/(V·s))
M	number of master species
N_i	number of volume elements in the one-dimensional model or number of rows in the two-dimensional model
N_j	number of columns in the two-dimensional model
N_k	number of ions (species)
N_p	number of precipitated species
$\dot{n}_{ijk0}$	molar flow of the kth ion into the cell (i, j) from $(i, j + 1)$ in the two-dimensional model (mol/s)
$\dot{n}_{ijk1}$	molar flow of the kth ion into the cell (i, j) from $(i + 1, j)$ in the two-dimensional model (mol/s)

Symbol	Description

$\dot{n}_{ik}$ molar flow of the kth ions from the $(i + 1)$th volume element toward the inside of the ith volume element in the one-dimensional model (mol/s)

nr ratio of hydroxyl ions neutralized at the cathode compartment (dimensionless)

R total electrical resistance of the soil (Ω)

R_g ideal gas constant (J/(mol·k))

R_i electrical resistance of the ith volume element in the one-dimensional model (Ω)

R_{ij0} electrical resistance between the cells (i, j) and $(i, j + 1)$ corresponding to a resistor in the **x** direction (Ω)

R_{ij1} electrical resistance between the cells (i, j) and $(i + 1, j)$ corresponding to a resistor in the **y** direction (Ω)

R_k rate of production of the kth ion (mol/(m^3·s))

$S(x)$ function introduced to take into account the different movements of cations and anions

SI_p saturation index of the pth species

T temperature (°C)

t time (s)

t_{ijk0} transport number of the kth ion between the cells (i, j) and $(i, j + 1)$ in the two-dimensional model (dimensionless)

t_{ijk1} transport number of the kth ion between the cells (i, j) and $(i + 1, j)$ in the two-dimensional model (dimensionless)

t_{ik} transport number of the kth ions between the ith and $(i + 1)$th volume elements (dimensionless)

T_m total concentration of the mth master species

u_k effective ionic mobility of the kth ion (m^{2}s^{-1}V^{-1})

v advective velocity of the pore water (m/s)

V_0 volume of the anode compartment in the one-dimensional model (m^3)

V_{N_i+1} volume of the cathode compartment in the one-dimensional model (m^3)

X_m master species mth

Y_j nonmaster species jth

z_k the electrical charge of the kth ion (mole of electrons per mole of kth ion)

ΔV electrical potential difference between the cathode and the anode (V)

Δx thickness of the volume element in the one-dimensional model or length of the volume element in the two-dimensional model (m)

Δy width of the volume element in the two-dimensional model (m)

Δz depth of the volume element in the two-dimensional model (m)

κ_i electrical conductivity at the boundary between the ith and $(i + 1)$th cells (Ω/m)

κ_{ij0} total conductivity between cells (i, j) and $(i, j + 1)$ in the two-dimensional model (Ω/m)

κ_{ij1} total conductivity between cells (i, j) and $(i + 1, j)$ in the two-dimensional model (Ω/m)

λ_k molar conductivity of the kth ion (m^2/(Ω mol))

ν_{jm} the stoichiometric coefficient of the master species X_m in the formation reaction of the nonmaster species Y_j

ν_{pm} stoichiometric coefficient of the mth master species in the precipitation reaction of the pth species

τ tortuosity of the soil (dimensionless)

ω porosity of the soil assumed saturated (dimensionless)

∇V electrical voltage gradient (V/m)

REFERENCES

Al-Hamdan AZ, Reddy KR. (2008a). Transient behavior of heavy metals in soils during electrokinetic remediation. *Chemosphere* **71**(5):860–871.

Al-Hamdan AZ, Reddy KR. (2008b). Electrokinetic remediation modeling incorporating geochemical effects. *Journal of Geotechnical and Geoenvironmental Engineering* **134**(1): 91–105.

Allison JD, Brown DS, Novo-Gradac KJ. (1991). MINTEQA2/PRODEFA2, a Geochemical Assessment Model for Environmental Systems: Version 3.0 User's Manual. Environmental Research Laboratory, Office of Research and Development, U.S. Environmental Protection Agency, Athens, GA. *Report No.* EPA/600/3-91/021.

Alshawabkeh AN, Acar YB. (1992). Removal of contaminants from soils by electrokinetics: A theoretical treatise. *Journal of Environmental Science and Health Part A* **27**(7): 1835–1861.

Carrayrou J, Mose R, Behra P. (2002). New efficient algorithm for solving thermodynamic chemistry. *AIChE Journal* **48**(4):894–904.

Cederberg GA, Street RL, Leckie JO. (1985). A groundwater mass transport and equilibrium chemistry model for multicomponent systems. *Water Resources Research* **21**(8): 1095–1104.

Choi YS, Lui R. (1995). A mathematical model for the electrokinetic remediation of a contaminated soil. *Journal of Hazardous Materials* **44**(1):61–75.

Eykholt GR, Daniel DE. (1994). Impact of system chemistry on electroosmosis in contaminated soil. *Journal of Geotechnical Engineering* **120**(5):797–815.

Garcia-Gutierrez MD, Gomez-Lahoz C, Rodriguez-Maroto JM, Vereda-Alonso C, Garcia-Herruzo F. (2007). Electrokinetic remediation of a soil contaminated by the pyritic sludge spill of Aznalcollar (SW, Spain). *Electrochimica Acta* **52**(10):3372–3379.

Haran BS, Popov BN, Zheng GH, White RE. (1997). Mathematical modeling of hexavalent chromium decontamination from low surface charged soils. *Journal of Hazardous Materials* **55**(1-3):93–107.

Lemaire T, Moyne C, Stemmelen D. (2007). Modelling of electro-osmosis in clayey materials including pH effects. *Physics and Chemistry of the Earth* **32**(1-7):441–452.

Liu CW, Narasimhan TN. (1989). Redox-controlled multiple-species reactive chemical transport 1. Model development. *Water Resources Research* **25**(5):869–882.

Mascia M, Palmas S, Polcaro AM, Vacca A, Muntoni A. (2007). Experimental study and mathematical model on remediation of Cd spiked kaolinite by electrokinetics. *Electrochimica Acta* **52**(10):3360–3365.

Morel F, Morgan J. (1972). A numerical method for computing equilibria in aqueous chemical systems. *Environmental Science & Technology* **6**(1):58–67.

Morel FMM. (1983). *Principles of Aquatic Chemistry*. New York: Wiley Interscience.

Narasimhan B, Ranjan RS. (2000). Electrokinetic barrier to prevent subsurface contaminant migration: Theoretical model development and validation. *Journal of Contaminant Hydrology* **42**:1–17.

Nilsson JW. (1990). *Electric Circuits*, 3rd ed. Reading, MA: Addison-Wesley cop.

Ottosen LM, Hansen HK, Laursen S, Villumsen A. (1997). Electrodialytic remediation of soil polluted with copper from wood preservation industry. *Environmental Science & Technology* **31**(6):1711–1715.

Parkhurst DL, Appelo CAJ. (1999). User's Guide to PHREEQC (Version 2)—A Computer Program for Speciation, Batch-Reaction, One-Dimensional Transport, and Inverse Geochemical Calculations. U.S. Department of the Interior, U.S. Geological Survey. *Water-Resources Investigations Report No. 99-4259*.

Perry RH, Green DW, Malone JM. (eds.). (1984). *Perry's Chemical Engineers' Handbook*, 6th ed. New York: McGraw-Hill.

Press WH, Flannery BP, Teukolsky SA, Vetterling WT. (1992). *Numerical Recipes in C: The Art of Scientific Computing*. Cambridge: Cambridge University Press.

Vereda-Alonso C, Heras-Lois C, Gomez-Lahoz C, Garcia-Herruzo F, Rodriguez-Maroto JM. (2007). Ammonia enhanced two-dimensional electrokinetic remediation of copper spiked kaolin. *Electrochimica Acta* **52**(10):3366–3371.

Vereda-Alonso C, Rodríguez-Maroto JM, García-Delgado RA, Gómez-Lahoz C, García-Herruzo F. (2004). Two-dimensional model for soil electrokinetic remediation of heavy metals. Application to a copper spiked kaolin. *Chemosphere* **54**(7):895–903.

Wilson DJ, Rodríguez-Maroto JM, Gómez-Lahoz C. (1995a). Electrokinetic remediation. I. Modeling of simple systems. *Separation Science and Technology* **30**(15):2937–2961.

Wilson DJ, Rodríguez-Maroto JM, Gómez-Lahoz C. (1995b). Electrokinetic remediation. II. Amphoteric metals and enhancement with weak acid. *Separation Science and Technology* **30**(16):3111–3125.

Yu JW, Neretnieks I. (1996). Modelling of transport and reaction processes in a porous medium in an electrical field. *Chemical Engineering Science* **51**(19):4355–4368.

26

ELECTROKINETIC BARRIERS: MODELING AND VALIDATION

R. SRI RANJAN

26.1 INTRODUCTION

26.1.1 Extent of Contamination

The increase in petroleum fuel-based transportation and the increased use of chemicals in the agricultural and manufacturing sectors have lead to the contamination of soils and groundwater during the past century. The National Pollutant Release Inventory of Canada has reported the release of 225,000 tons of pollutants to the subsurface from 9500 facilities (Environment Canada, 2006). In 1996, the same report showed a pollutant release of 125,000 tons reported by 1800 facilities. As the population expands, the competition for land and potable water is also increasing. In addition, the advances in medicine have traced some causes of increased cancer exposure to contaminants in soil and groundwater. As a result, the interest in remediation of contaminated soils and groundwater has increased during the past few decades. The competition for potable water from the increasing population has lead to a greater interest in maintaining soil and groundwater quality. The prevention of contamination is an economical method to protect the environment because treating contaminated soil and groundwater is very costly. Traditionally, physical barriers have been used to contain the spread of contaminants.

Proper design of barriers to contaminant movement requires an understanding of the nature of the pathways by which contaminants spread in soils and groundwater. When an accidental spill occurs at a particular location, the contaminant will infiltrate the soil under the influence of gravitational and capillary forces. The deep percolating contaminant may reach the groundwater table, and depending on the density of the contaminant, it will migrate in different ways. If the contaminant is denser than water [dense nonaqueous phase liquid (DNAPL)], it will continue to

---

Electrochemical Remediation Technologies for Polluted Soils, Sediments and Groundwater, Edited by Krishna R. Reddy and Claudio Cameselle
Copyright © 2009 John Wiley & Sons, Inc.

percolate deep below the water table, leaving behind trace contaminants in the soil along its pathway, spanning both the saturated and unsaturated zones above. The trace contaminants left behind in the pathway will amount to at least the residual saturation of the contaminant in the soil within the pathway. The groundwater passing through this contaminated zone will dissolve the residual contaminants and carry it further downstream in the direction of groundwater flow.

If the contaminant is lighter than water [light nonaqueous phase liquid (LNAPL)], it will float on the water table and migrate down-gradient on top of the water table. The residual DNAPLs and LNAPLs remaining in the unsaturated zone will further contaminate the water, flowing through this zone as interflow. Therefore, the initial task in any spill is to quickly assess the extent of contamination and deploy methods to contain the contamination to the spill site. Since a major part of the cost of cleanup is proportional to the quantity of contaminated soil and groundwater, early containment can reduce the cost of cleanup by minimizing the extent of the contamination of the soil and groundwater. Stopping the movement of groundwater through the contaminated zone is an effective way to reduce the spread of the contaminant.

26.1.2 Traditional Containment Methods

Traditionally, containment has been achieved by the construction of slurry trenches where a slurry of fine-grained material such as bentonite/clay is placed in the trench to act as a barrier to groundwater flow and contaminant spread. The slurry has very low hydraulic conductivity causing the groundwater flow to bypass the contaminated site. This is a physical barrier that is usually constructed soon after the spill to minimize the spread of contaminants. The access of heavy equipment to the spill site to accomplish this task may cause delays in the establishment of a barrier. In addition, the slurry trenches are not easy to construct adjacent to existing buildings because they may pose a danger of causing catastrophic structural failure of nearby buildings.

Another physical barrier that is used is the cutoff wall installed into the ground surrounding the contaminated site to prevent the contaminants from moving outside the walls while preventing fresh groundwater from mixing with the contaminated groundwater within this zone. The cutoff walls could be physical barriers that usually consist of sheet piles inserted vertically into the ground to isolate the contaminated zone. The insertion of the sheet pile often requires the use of heavy equipment.

The traditional methods described above require the use of heavy machinery, which may be difficult to access in a built-up environment. Since time is of essence in arresting the spread of contaminants to minimize the cleanup cost, novel methods that can overcome these difficulties are needed. The creation of an electrokinetic barrier is one such method that can address these concerns. In this method, an electrical field generated within/outside the contaminated zone causes the formation of a counter-gradient that can control the flow of water and contaminants. The relative ease of implementing electrokinetic barriers has generated a lot of interest in this technique as a viable alternative to traditional methods. The principles governing the creation of electrokinetic barriers are described in the next section.

26.2 ELECTROKINETIC PHENOMENA

Flow of water through soils occurs under the influence of different driving gradients, of which the hydraulic gradient is the most common one. The hydraulic gradient is generated by the creation of hydraulic pressure differences between two different points, causing the water to flow from a high-pressure location to a lower one. The other driving gradients are electrical potential difference, temperature difference, and chemical concentration difference. Under isothermal conditions, the fluid flow due to thermal gradients can be negligible.

The flow due to an applied electrical potential gradient can be attributed to the movement of cations, found within the diffuse double layer near the negatively charged clay particle surface, which are surrounded by loosely bound water molecules. The negative charge of the clay particle is due to the isomorphous substitution of silicon atoms by aluminum atoms in the silica crystal lattice. This leads to a net negative charge on the surface and attract positively charged cations to the surface. The presence of negatively charged clay particles in the pore space leads to the formation of the diffuse double layer. When an electrical potential gradient is applied, the cations in the pore water move toward the cathode, and the anions move toward the anode. However, the cations, being larger, tend to drag more pore water along with them causing a net movement of pore water toward the cathode. This fluid flow induced by electrical potential gradient is called electroosmosis. Previous studies by Lageman, Pool, and Seffinga (1989), Eykholt and Daniel (1991), and Acar (1992) have also shown the successful application of electroosmotic flow under experimental conditions.

The electroosmotic flow is different from the movement of water due to chemical gradients, which is called normal osmosis. The migration of charged particles toward the electrode of the opposite polarity due to electrical potential gradient is called electrophoresis. The flow of fluid due to gradients other than hydraulic is called coupled flow. The flow is directly proportional to the gradient and the total flow can be calculated by summing the direct flow with the coupled flow while paying attention to the direction of the respective driving gradients. If the electroosmotic flow is induced in the direction opposite to the hydraulic gradient, the resulting net flow could be reduced or even stopped. Therefore, the application of an electrical potential gradient to generate a counter-gradient to the hydraulic gradient is the underlying mechanism that can retard the flow of water within the pore spaces. Thus, the electrokinetic barriers are formed by the strategic placement of electrodes to generate a counter-gradient to the hydraulic gradient.

26.3 DIRECT AND COUPLED FLOW AND TRANSPORT OF IONS

26.3.1 Flow due to Hydraulic Gradient

The water flow through clay soils under a hydraulic gradient is governed by Darcy's law as shown in Equation 26.1:

$$q_h = K_h \cdot i_h, \qquad (26.1)$$

where:

q_h = flux (m/s),
K_h = hydraulic conductivity (m/s), and
i_h = negative of the hydraulic gradient (m/m).

The hydraulic conductivity of soils varies over several orders of magnitude ranging from 1×10^{-3} m/s for coarse sand to 1×10^{-12} m/s for fine-grained dense clays. In fine-grained soils, very high hydraulic gradients need to be generated to cause even a small flux. Therefore, the pump and treat method of cleanup or bioremediation is not effective in clayey soils.

26.3.2 Flow due to Electrical Potential Gradient

The electroosmotic flux is analogous to Darcy's law and can be formulated as shown in Equation 26.2:

$$q_e = K_e \cdot i_e, \quad (26.2)$$

where:

q_e = electroosmotic flux (m/s),
K_e = electroosmotic conductivity [m²/(V·s)], and
i_e = negative of the electrical potential gradient (V/m).

The electroosmotic conductivity does not vary much between different soils and ranges between 1×10^{-9} and 1×10^{-10} m²/(V·s) (Mitchell, 1993). The electroosmotic conductivity is dependent on many different factors. Casagrande (1949) derived the following relationship:

$$K_e = \frac{-\zeta D \varepsilon}{\eta} n\tau, \quad (26.3)$$

where:

ζ = zeta potential (V),
D = dielectric constant of the water (dimensionless),
ε = permittivity of the vacuum (8.85E-12 F/m),
τ = tortuosity,
η = viscosity of the water (1E-3 N s/m²), and
n = effective porosity.

Zeta potential is defined as the potential difference between the shearing surface in the diffuse double layer and the pore water that is moving and ranges from +50 to −50 mV. The sign indicates the relative abundance of cations and anions. If the cation concentration is high relative to anion concentration, the sign will be negative and the flow will be preferentially toward the cathode. Yeung (1994) found that the zeta potential ranged from 0 to −50 mV for most clay soils with zero values being seen in highly acidic soils. Under highly acidic conditions, the abundance of anions makes the sign positive, and the electroosmotic flow has been shown to preferentially occur toward the anode (Eykholt and Daniel, 1994). Eykholt and Daniel (1994) showed the zeta potential to change from +10 mV at a pH of 3 near the anode reservoir to −40 mV at a pH of 11 near the cathode reservoir. When zeta potential is zero at a pH of 4, the isoelectric point is reached and the electroosmotic flow will also be zero. When an electrical potential is applied to the soil, the formation of acidic and basic conditions near the electrodes due to hydrolysis reactions is inevitable. Therefore, the pH changes have to be taken into account by calculating the anion and cation concentrations to properly model the direction of the electroosmotic flow.

Probstein and Hicks (1993) found that the electroosmotic conductivity mainly depended on zeta potential and effective porosity only. The size of the pores does not affect the electroosmotic conductivity, and therefore, the magnitude only varies by an order of magnitude for soils with many orders of magnitude difference in pore size ranges. The zeta potential is dependent on the pH of the pore water. Therefore, the pH can influence the electroosmotic conductivity indirectly.

When an electrical potential gradient is applied in soils, electrolysis reactions occur at the electrodes as shown below:

$$2H_2O - 4e^- \rightarrow 4H^+ + O_2 \uparrow \text{ (anode)},$$

$$4H_2O + 4e^- \rightarrow 4OH^- + 2H_2 \uparrow \text{ (cathode)}.$$

The accumulation of hydrogen ions at the anode decreases the pH and moves toward the cathode as an acid front. Similarly, the hydroxyl ions accumulating at the cathode increases the pH and begins to move toward the anode as a base front. However, the acid front moves much faster than the base front because the mobility of hydrogen ions is 1.75 times that of hydroxyl ions. The movement of these fronts should be considered in the overall formulation of the model. Lemaire, Moyne, and Stemmelen (2007) expressed the need to account for the negative pore pressure generated by the electrolysis of water and included the pH effects to account for this.

In fine-grained soils, the application of a small voltage gradient can induce a large electroosmotic flux. Contrasting this range with hydraulic flux, only a small electrical potential gradient needs to be applied to counter a large hydraulic gradient. However, in the natural environment, the hydraulic gradients are seldom over 0.1 m/m, making the electrokinetic method a promising alternative for creating a barrier to flow. This can be illustrated by an example scenario. A clay soil with an electroosmotic conductivity of 1×10^{-9} m^2/(V·s) and hydraulic conductivity of 1×10^{-10} m/s exists in a location where the natural groundwater gradient is 0.02 m/m. The applied electrical potential gradient needed to create a counter-gradient will only be

$$i_e = \frac{K_h \times i_h}{K_e} = \frac{10^{-10} \times 0.02}{10^{-9}} = 0.002 \text{ V/m}. \tag{26.4}$$

If the hydraulic conductivity of soil is 1×10^{-7} m/s, then the electrical potential gradient needed to counter the hydraulic flux is only 2 V/m for a similar hydraulic gradient. Lageman, Pool, and Seffinga (1989) demonstrated the use of electroosmotic flow to retard the migration of lead, copper, zinc, and cadmium during the remediation of a contaminated site at a paint factory. Laboratory column experiments carried out by Sri Ranjan, Qian, and Krishnapillai (2006), Manokararajah and Sri Ranjan (2005), and Narasimhan and Sri Ranjan (2000) have shown the successful creation of electrokinetic barriers against hydraulic gradients as large as 1.25. Lynch et al. (2007) demonstrated the ability to generate an electrokinetic gradient capable of preventing the contaminant spread, resulting from a hydraulic gradient of 7 for a one-dimensional experiment. They also used an electric field of 125 V/m to prevent the contaminant spread at a gradient of 1.3 in a two-dimensional experiment.

26.3.3 Flow due to Chemical Concentration Gradient

The accumulation of ions near the electrodes creates a chemical potential gradient. This concentration gradient induces the osmotic flow of water through the clay due to natural osmotic potential. The concentration difference also causes the ions to diffuse through the clay sample governed by Fick's law of diffusion. This velocity of ions within the pore fluid induces an electrical potential gradient, called membrane potential, which induces an electroosmotic counterflow (Heister, Kleingeld, and Loch, 2006). The electrically induced water flux was found to be 98% of the chemically induced water flux, and this effect was found to be larger as the salt concentration gradient increased.

26.3.4 Electromigration of Ions

The movement of ions toward the oppositely charged electrode is called electromigration, which is quantified by the effective ionic mobility. The effective ionic mobility (U_j^*) is defined as the velocity of ion within the pore space under the influence of a unit electrical potential gradient. The Nernst–Einstein equation is used to relate the ionic mobility to the diffusion coefficient of the ion in a dilute solution (Koryta, 1982) as follows:

$$U_j^* = \frac{D_j z_j F}{RT} n\tau, \quad (26.5)$$

where:

D_j = diffusion coefficient of species j in a dilute solution (m²/s),
z_j = charge of the chemical species j,
F = Faraday's constant (96,487 C/mol),
R = universal gas constant (8.314 J/mol/K),
T = absolute temperature (K),
τ = tortuosity, and
n = effective porosity.

26.3.5 Streaming Potential

The streaming potential refers to the creation of an electrical potential gradient resulting from the physical movement of ions toward the oppositely charged electrodes. This is the reverse of the electroosmotic flow. Heister *et al.* (2005) showed that the increase of streaming potential due to hydraulic gradient was proportional to the salt concentration of the permeating solution.

26.4 MODEL DEVELOPMENT

Lewis and Garner (1972) developed a finite element model to predict the changes in pore water pressure by considering the coupling of electroosmotic flow with hydraulic flow. They considered the electroosmotic flow and hydraulic flow in two different directions along with streaming potential and electrical conductance. Later, Renaud and Probstein (1987) developed a finite element model of electroosmotic flow and hydraulic flow of groundwater flow. Since this was used to model the diversion of groundwater flow away from a hypothetical hazardous waste site, it did not consider the ionic mobilities. It predicted the voltage gradients, electroosmotic flow, and hydraulic head distribution around the landfill site.

Eykholt and Daniel (1994) investigated the impact of system chemistry on electroosmosis in a contaminated soil. Electroosmotic treatment of copper-spiked kaolinite led to the removal of copper at the acidic anode region and precipitation of copper at the cathode due to the high pH. Acidic conditions were found to favor the mobility and hence the removal of metal ions such as lead and copper. Alkaline conditions were found to favor the mobility of chromate. The pH changes may reverse the sign of the zeta potential and consequently reverse the direction of electroosmotic flow. Therefore, the modeling of the migration of metal ions, because of its pH dependence on ionic mobility, needs to consider the changes in pH of the pore water at different distances from the electrode. Eykholt and Daniel (1994) calculated the electroosmotic conductivity based on the changes in zeta potential based on the pH at a particular location and time within the soil column. This approach is different from the constant electroosmotic conductivity that was used by previous researchers (Hamed, Acar, and Gale, 1991; Bruell, Segall, and Walsh, 1992). Lemaire, Moyne, and Stemmelen (2007) found this approach to be limited by the need to obtain the zeta potential versus pH relationship for a specific soil by experiment. Lynch *et al.* (2007) modeled the prevention of cadmium transport due to a hydraulic gradient by the combined effect of electroosmosis and electromigration, which took into account the soil pH, cadmium ion concentration, and current intensity.

Narasimhan and Sri Ranjan (2000) made simplifying assumptions to model the electrokinetic flow phenomena. The structure of the soil was assumed to be uniform with the net electroosmotic flow taking place toward the cathode. The applied voltage was assumed to be useful in transporting the water and ions. The flux due to hydraulic, electrical, and chemical gradients were superimposed linearly while neglecting the effect of electrophoresis in the fine-grained media. The hydraulic conductivity and porosity were assumed to be constant in time.

Combining Equations 26.1 and 26.2, the coupled flow equation modeling the fluid flux is written as

$$J_w = -K_h \nabla h - K_e \nabla E, \qquad (26.6)$$

where:

J_w = fluid flux/area (m/s),
K_h = hydraulic conductivity (m/s),
K_e = electroosmotic conductivity [m²/(V·s)], and
∇ = gradient operator for hydraulic head (h) and electrical potential difference (E).

The charge flux occurs due to the chemical gradient as well as electrical potential gradient and only the latter was modeled by Narasimhan and Sri Ranjan (2000) as follows:

$$I = F \sum_{j=1}^{N} z_j C_j u_j^* \nabla E \qquad (26.7)$$

where:

I = current density (A/m²),
z_j = charge of the chemical species j,
u_j^* = effective ionic mobility of species j [m²/(V·s)], and
C_j = concentration of the species j (mol/l).

The charge flux due to the chemical potential gradient was neglected because of its small magnitude in comparison to the electrical potential gradient. The mass flux under coupled chemical, hydraulic, and electrical gradients is given by Narasimhan and Sri Ranjan (2000) as

$$J_j = D_j^* \nabla C_j - \left(J_w + u_j^* \nabla E\right) C_j, \qquad (26.8)$$

where:

D_j^* = effective diffusion coefficient (m²/s).

The terms on the right represent diffusion and advection, respectively. Under saturated conditions, the flow equation is given as

$$K_h \frac{\partial^2 h}{\partial x^2} + K_e \frac{\partial^2 E}{\partial x^2} = m_v \frac{\partial h}{\partial t}, \qquad (26.9)$$

where m_v is the coefficient of volume compressibility. The advection–dispersion equation for the mass transport of the j species is given as

$$D_j^* \frac{\partial^2 C_j}{\partial x^2} - J_w \frac{\partial C_j}{\partial x} \pm \frac{r_j}{n} = \frac{\partial C_j}{\partial t}, \qquad (26.10)$$

where r_j is the production rate of species j (per unit pore volume) arising from chemical reactions representing sorption, precipitation, dissolution, and oxidation/reduction. Assuming that the electrical capacitance is zero, the conservation of charge is maintained:

$$F\sum_{j=1}^{N} z_i D_j^* \frac{\partial^2 C_j}{\partial x^2} + F\sum_{j=1}^{N} z_i u_j^* C_j \frac{\partial^2 E}{\partial x^2} = 0. \qquad (26.11)$$

A finite element model was developed based on the above relationship and was validated using experimental data presented by Yeung (1990) and in Yeung and Mitchell (1993).

26.5 MODEL VALIDATION

Narasimhan and Sri Ranjan (2000) validated their model described in the previous section against the experimental data presented in Yeung (1990) and in Yeung and Mitchell (1993). The changes in pH, pore water pressure, and voltage gradient at different distances and time were simulated by the model well. The cation transport was found to be retarded because the electroosmotic flow was counter-gradient to the ionic migration. The application of an electrical potential gradient retarded the migration of cations, while the transport of anions in the opposite direction increased.

Narasimhan (1999) conducted a series of one-dimensional column experiments to verify the model predictions using potassium ions as a tracer. A solution of 0.02 M KCl was used as a permeant at the upstream end subjected to a hydraulic gradient of 3.75. A 100 V/m electrical potential gradient was applied to counter the hydraulic flow on a 2-h on to 22-h off cycle for 32 days. The soil columns were sectioned and analyzed for potassium ion concentration. The symbols in Figure 26.1 present the results of the experiment with and without an electrokinetic barrier at different

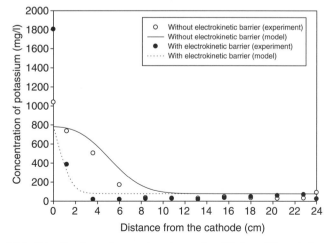

Figure 26.1. Distribution of potassium ions after 32 days (Narasimhan, 1999).

distances from the inflow end. The continuous lines on Figure 26.1 represents the model results with and without electrokinetic treatment. The modeled graph closely matches the experimental results. The migration of potassium ions was retarded by more than 20 times by the application of the electroosmotic counter-gradient.

Park *et al.* (2003) considered diffusion, electromigration, electroosmosis, and hydraulic flow for modeling the movement of phenol by numerical analysis and found reasonable agreement with one-dimensional experimental results. In their approach, the electrode reservoirs were chemically conditioned.

26.6 FIELD APPLICATION SCENARIOS

26.6.1 Soil Vapor Extraction

When volatile contaminants move down through a clay-loam soil profile, a part of it remains in the unsaturated zone as residual amount. During soil vapor extraction, a large volume of air is pumped through the unsaturated contaminated zone and the contaminated air is extracted with a vacuum pump to be scrubbed aboveground. The soil vapor extraction system is most effective when the contaminated zone remains unsaturated. Electrokinetic barriers can be used to lower the water table within the contaminated zone to facilitate this process. Figure 26.2 presents a possible electrode configuration that might be used to lower the water table below the

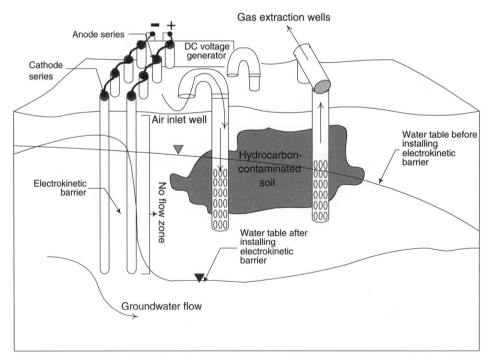

Figure 26.2. Potential electrode configurations for the vapor extraction system (Narasimhan and Sri Ranjan, 1999). DC, direct current.

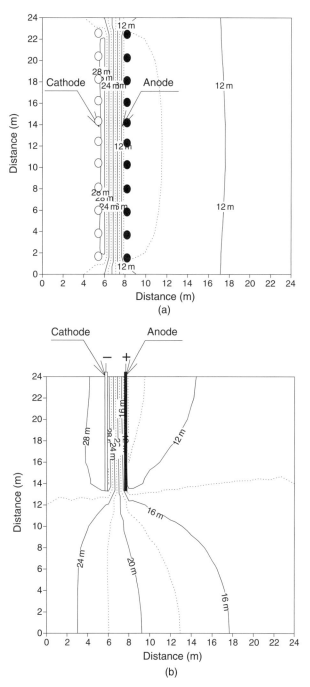

Figure 26.3. Configuration of electrodes and hydraulic head contours: (a) plan view and (b) elevation view (Narasimhan and Sri Ranjan, 1999).

574   ELECTROKINETIC BARRIERS: MODELING AND VALIDATION

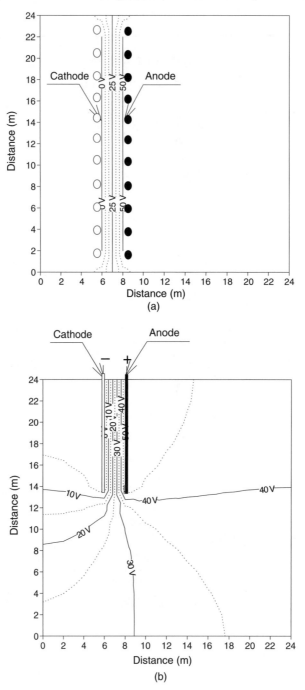

Figure 26.4. Configuration of electrodes and voltage potential contours: (a) plan view and (b) elevation view (Narasimhan and Sri Ranjan, 1999).

contaminated zone. Narasimhan and Sri Ranjan (1999) used their model to simulate the hydraulic head near an array of electrodes placed transversely on the upgradient side of the contaminated zone. An electrical potential gradient of 25 V/m was applied to a series of electrodes that were placed about 2 m apart with an anode–cathode separation of 2 m. Model simulations indicated the creation of a hydraulic head difference of 14 m across the electrodes, which were embedded 11 m into the ground. The hydraulic head shown in Figure 26.3 is when the lowering of the water table has reached a steady state. Figure 26.4 shows the voltage potential at different locations from the electrodes. The drop in voltage due to the formation of gases at the electrodes and current leakage through a conductive soil are some of the possible obstacles that the model did not simulate. However, the simulations show a general trend one might expect in this type of electrode configuration.

26.6.2 Pump and Treat Method

Contaminants that are highly soluble in water are ideal candidates for pump and treat remediation methods. However, access to adequate water at the contaminated site is a major limitation for many spill emergencies. Figure 26.5 shows dual rows of electrodes placed in an arrangement that will raise the water table to facilitate the dissolving of the contaminants. The use of existing water at the site to wash the contaminated site above is a very attractive option mainly because the water may have already been contaminated. The model simulation of the water table elevation shows a marked rise. The electrical potential gradient applied across the arrays of electrodes acts like a pump to move the water through the contaminated zone

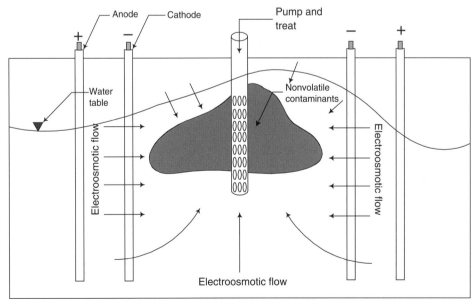

Figure 26.5. Electrode arrangement for the pump and treat method (Narasimhan and Sri Ranjan, 1999).

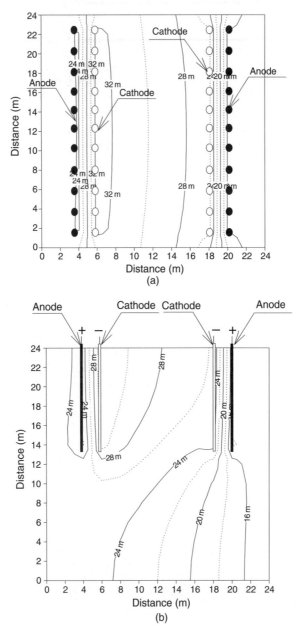

Figure 26.6. Configuration of electrodes and hydraulic head contours: (a) plan view and (b) elevation view (Narasimhan and Sri Ranjan, 1999).

toward the extraction well. In the simulations, an electrical potential gradient of 12 V/m was applied continuously until a steady-state distribution of hydraulic head was attained (Narasimhan and Sri Ranjan, 1999). Figures 26.6 and 26.7 show the plan and elevation views of hydraulic head distribution and electrical potential gradient, respectively. The distribution of the hydraulic head shows the formation of a water table mound. Thus, the use of electroosmotic flow to flood the

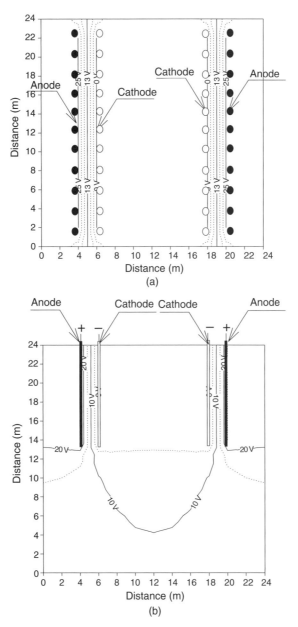

Figure 26.7. Configuration of electrodes and voltage potential contours: (a) plan view and (b) elevation view (Narasimhan and Sri Ranjan, 1999).

contaminated zone well above the prevailing water table elevation can be a very useful feature in pump and treat remediation.

26.7 SUMMARY

The use of counter-gradients generated by electrokinetic methods to create a hydraulic flow barrier has been experimentally shown to be successful. The pH of the pore water plays a major role in determining the direction of the eletroosmotic gradient. Therefore, the numerical model developed to simulate the coupled flow considered the formation of H^+ and OH^- ions at the electrodes and their transport as well. Many factors may alter the process and produce the opposite results to what was expected. The numerical method discussed in this chapter is comprehensive in including the different processes. Given the uncertainty in input parameters, an exact match to experimental results is not easily attained. The validation of the model was carried out with data obtained by two different researchers (Yeung, 1990; Narasimhan, 1999). The simulated results were found to be in close agreement with the experimental data.

The model was then used to simulate two field application scenarios, that is, soil vapor extraction and the pump and treat method. The model predictions seem to be reasonable based on what is expected under the conditions applied in the field.

ACKNOWLEDGMENTS

As part of their theses research, Balaji Narasimhan, Mano Krishnapillai, YuWei Qian, and Steven Thomas contributed to the Electrokinetic Research program at the Soil and Water Engineering Laboratory of the University of Manitoba.

REFERENCES

Acar YB. (1992). Electrokinetic cleanups. *Civil Engineering* **62**(10):58–60.

Bruell CJ, Segall BA, Walsh MT. (1992). Electroosmotic removal of gasoline hydrocarbons and TCE from clay. *Journal of Environmental Engineering* **118**(1):68-83.

Casagrande L. (1949). Electroosmosis in soils. *Geotechnique* **1**(3):159-177.

Environment Canada. (2006). National Pollutant Release Inventory of Canada. http://www.ec.gc.ca/pdb/npri/npri_factsheet06_e.cfm (accessed April 9, 2008).

Eykholt GR, Daniel DE. (1991). Electrokinetic decontamination of soils. *Journal of Hazardous Materials* **28**(1–2):208-209.

Eykholt GR, Daniel DE. (1994). Impact of system chemistry on electroosmosis in contaminates soil. *Journal of Geotechnical Engineering* **120**(5):797–815.

Hamed J, Acar YB, Gale RJ. (1991). Pb(II) removal from kaolinite by electrokinetics. *Journal of Geotechnical Engineering* **117**(2):241-271.

Heister K, Kleingeld PJ, Keijzer TJS, Loch JPG. (2005). A new laboratory set-up for measurement of electrical, hydraulic, and osmotic fluxes in clays. *Engineering Geology* **77**(3–4):295–303.

Heister K, Kleingeld PJ, Loch JPG. (2006). Induced membrane potentials in chemical osmosis across clay membranes. *Geoderma* **136**(1–2):1–10.

Koryta J. (1982). *Ions, Electrodes, and Membranes*. New York: John Wiley and Sons.

Lageman R, Pool W, Seffinga G. (1989). Electro-reclamation: Theory and practice. *Chemistry & Industry* **18**:585-590.

Lemaire T, Moyne C, Stemmelen D. (2007). Modelling of electro-osmosis in clayey materials including pH effects. *Physics and Chemistry of the Earth* **32**:441–452.

Lewis RW, Garner RW. (1972). A finite element solution of coupled electrokinetic and hydrodynamic flow in porous media. *International Journal for Numerical Methods in Engineering* **5**(1):41-55.

Lynch RJ, Muntoni A, Ruggeri R, Winfield KC. (2007). Preliminary test of an electrokinetic barrier to prevent heavy metal pollution of soil. *Electrochimica Acta* **52**(10): 3432–3440.

Manokararajah K, Sri Ranjan R. (2005). Electrokinetic retention, migration and remediation of nitrates in silty loam soil under hydraulic gradients. *Engineering Geology* **77**(3–4): 263–272.

Mitchell JK. (1993). *Fundamentals of Soil Behaviour*, 2nd ed. New York: John Wiley & Sons.

Narasimhan B. (1999). Electrokinetic Barriers to Contaminant Transport: Numerical Modelling and Laboratory Scale Experimentation. MSc Thesis, University of Manitoba, Winnipeg, MB, Canada.

Narasimhan B, Sri Ranjan R. (1999). Creation of a subsurface barrier to contaminant transport using electrical potential gradients. Presented at the *ASAE/CSAE-SCGR Annual International Meeting*, July 18–21, 1999, Paper No. 992124. St. Joseph, MI: ASAE.

Narasimhan B, Sri Ranjan R. (2000). Electrokinetic barrier to prevent subsurface contaminant migration: Theoretical model development and validation. *Journal of Contaminant Hydrology* **42**(1):1–17.

Park J-S, Kim S-O, Kim K-W, Kim BR, Moon S-H. (2003). Numerical analysis for electrokinetic soil processing enhanced by chemical conditioning of the electrode reservoirs. *Journal of Hazardous Materials* **B99**(1):71–88.

Probstein RF, Hicks RE. (1993). Removal of contaminants from soils by electric fields. *Science* **260**:498–503.

Renaud PC, Probstein RF. (1987). Electroosmotic control of hazardous wastes. *PCH PhysicoChemical Hydrodynamics* **9**(1/2):345-360.

Sri Ranjan R, Qian Y, Krishnapillai M. (2006). Effects of electrokinetics and cationic surfactant cetyltrimethylammonium bromide (CTAB) on the hydrocarbon removal and retention from contaminated soils. *Environmental Technology* **27**(7):767–776.

Yeung AT. (1990). Electrokinetic Barrier to Contaminant Transport through Compacted Clay. PhD Thesis, University of California, Berkeley, CA.

Yeung AT. (1994). Electrokinetic flow processes in porous media and their applications. In: *Advances in Porous Media*, Vol. **2**, 2nd ed. (ed. MY Corapcioglu). Amsterdam, The Netherlands: Elsevier, pp. 309–395.

Yeung AT, Mitchell JK. (1993). Coupled fluid, electrical and chemical flows in soil. *Geotechnique* **43**(1):121–134.

PART VIII

ECONOMIC AND REGULATORY CONSIDERATIONS

27

COST ESTIMATES FOR ELECTROKINETIC REMEDIATION

CHRISTOPHER J. ATHMER

27.1 INTRODUCTION

The cost of electrokinetic remediation, like any other remediation technology, is highly site specific. There have been relatively few full-scale installations of electrokinetic processes when compared with conventional remediation processes. There have been more applications of electrokinetics in Europe than in the USA. Since 1987, more than 75 electrokinetic projects have been implemented in Europe (Holland Environment, pers. comm.), while in the USA, very few full-scale applications have been successfully completed.

Cost information available for US projects is scarce and tends to be based on pilot-scale, research studies or demonstration sites (USEPA, 1995, 1999, 2003; DOE, 1996; ITRC, 1997; United States Army Environmental Center, 2000). There has been multiple pilot-scale and full-scale demonstrations completed, but costs for these type of projects are significantly higher than full-scale commercial installations due to the small-scale, intensive monitoring requirements and multiple layers of oversight. Other than Lasagna, there are no recent successful commercial applications in the USA from which to draw accurate cost information not based on research, pilot, or demonstration sites.

Some review articles cited above present cost estimates for full-scale systems based on pilot trial extrapolation with little or no full-scale site data. However, these estimates appear to be fairly accurate, considering inflation, and fall within the range of current costs outlined in Table 27.1. Table 27.1 summarizes the present costs of commercial, full-scale electrokinetic remediation for typical installations to treat organic and inorganic contamination. These data are based on information from Holland Environment BV/RL and Terran Corporation. There may be other

---

Electrochemical Remediation Technologies for Polluted Soils, Sediments and Groundwater,
Edited by Krishna R. Reddy and Claudio Cameselle
Copyright © 2009 John Wiley & Sons, Inc.

TABLE 27.1. Cost Summary for Electrokinetic Remediation

Contaminant Class	Range $/m³ ($/yd³)	Average $/m³ ($/yd³)
Inorganic	115–400 (90–300)	200 (150)
Organic	90–275 (75–200)	200 (150)

sources for electrokinetic remediation services that may have costs outside the presented range.

The primary competition at contaminated sites is rarely other electrokinetic-based remedies but between electrokinetic and conventional remedy choices. Many times, the choice is between electrokinetics and excavation. The economics of electrokinetics compare favorably to excavation and disposal of large amounts of hazardous wastes. The average cost of $200 per cubic meter equates to roughly $90 per ton of saturated loamy soil.

There is still a lot of resistance to using this relatively unknown process. Due to the nature of electrokinetic remediation, it is frequently a remediation of last resort at difficult sites and is provided by very few organizations in Europe and the USA. As the technology gets applied more frequently, future innovations will most likely lower the cost.

Even though rarely considered, the cost of treating organic contamination using electrokinetic techniques is very competitive with thermal technologies. As with all innovative technologies, however, there are few cases to draw upon for cost data to use as leverage with traditional technologies. As more installations are employed, the confidence in cost estimates will increase.

27.2 COST FACTORS

Table 27.2 outlines the major and minor factors that influence the cost of electrochemical remediation installations. Most of the cost factors are common to all types of remediation systems. The largest and most significant cost factor is the size of area to be treated.

Power requirements and materials are directly related to the size of the treatment area. One of the largest unit cost factors is the depth of the contamination. This variable may be more sensitive to Lasagna-type installations and direct-push installations than borehole or well-based installations. Project costs tend to rise more than linearly with depth. Power and materials rise linearly with depth, but the installation labor costs and, possibly, equipment cost may rise more than linearly as time required to install and horsepower can increase significantly as depth increases, especially beyond 15–25 m. It may take longer to drill the boreholes or drive mandrel systems to depth because of soil changes and the increase in soil densities.

Another major cost driver is any time requirement placed on the cleanup process. Time constraints generally require the system to operate more aggressively driving up the cost of the remedy. Aggressive electrokinetic operation often requires greater electrode placement densities (i.e. smaller distances between electrodes) and consequently increased current draw and power consumption. If time is not a primary

TABLE 27.2. Cost Variables for Electrokinetic Remediation Installations

Inorganic	Organic
Major Cost Variables	
Size of area	Size of area
Depth of contamination	Depth of contamination
Time	Time
pH management	Site preparations
Site preparations	Electrical conductivity
Cleanup target	Cleanup target
Electrical conductivity	
Minor Cost Variables	
Soil buffering	Partitioning
Sand content/hydraulic velocity	Electricity cost
Electricity cost	Stray current
Stray current	
Contaminant solubility and EH/pH	

EH, reduction potential (millivolts); pH, hydrogen potential (unitless).

concern, the overall cost can be optimized by balancing materials and installation costs with operating costs.

Site preparation is also a major cost factor. The site may require building or footer removal, utility relocating, electrical isolation, or some other steps before the electrokinetic process can be installed and operated. Any cost for site preparation is added directly to the overall cost before any contamination is addressed or removed.

For inorganic treatment systems, pH control at the electrodes can significantly add to the treatment costs. Specially designed electrodes, material specifications, pump maintenance, and waste management all contribute to the costs of pH management. Electrode pH management is not a concern for most organic applications, including Lasagna installations.

As with any remediation process, the cleanup target can be a major cost driver. Numerical criteria like the generic cleanup number (GCN), maximum concentration limit (MCL), alternative concentration limit (ACL), or other regulatory limit imposed on the remediation performance can significantly increase the treatment costs. Regulatory numerical limits are very hard to attain in most source areas as soil partitioning and binding become important. Bench-scale and pilot-scale tests are normally needed to assure that the electrokinetic process can achieve the remediation goals. More electrodes and/or longer operating time may be required to meet the remediation goals. Less stringent, mass-based removal targets may not require such intensive trials.

Some cost factors of lesser impact include the price of electricity, soil conditions, and stray current control. These factors typically have less than 5% impact on the overall treatment costs.

TABLE 27.3. Cost Breakdown for Typical Electrokinetic Remediation Systems

Item	Typical Range (%)	Average (%)
Electricity	7–25	15
Site preparations	5–25	10
Installation (labor, equipment, materials)	10–60	40
Operation, less electricity (labor, expendables)	15–50	25
Waste management, permits, oversight	5–20	10

Site conditions, such as highly saline environments, can preclude using electrokinetics due to the high current consumption leading to low-voltage gradients and overheating of the soil. Buried metal objects and utility lines can also make electrokinetics impractical. Accessibility to the site can also determine the practicality of electrokinetics. Installing an electrokinetic-based system under an operating facility with metal utilities under the floor may not be safe or feasible.

27.3 COST BREAKDOWN

It is interesting to note that for electrokinetic remediation processes, electricity costs are only about 15% of the total cost. Table 27.3 shows the cost breakdown for typical electrokinetic installations.

Site installation and operations have the largest variability. Lasagna systems tend to have higher installation costs and relatively low operation cost due to the fully *in situ* remediation with no aboveground extraction or treatment systems. There are no pumps or valves to control or maintain and no waste to manage. Standard electrokinetic metal treatment systems cost little to install but require higher operating costs to manage pH and extracted metal-laden wastes.

Site preparations can vary substantially at sites. Some sites require extensive site preparations, while others are fully accessible and ready for installation. Sites with abandoned buildings require the removal of the buildings and footers to allow the installation of the planar electrodes. However, in many circumstances, the buildings need to be removed eventually.

Installing direct current (DC) electrical-based remediation systems in urban areas also requires containment of stray voltage and current. DC systems can cause corrosion of buried gas and water lines or wreak havoc on cathodic protection systems. A good design can minimize the impacts, but sometimes, extra sacrificial anodes need to be installed to contain the electric field, adding to the cost of installation.

27.4 SUMMARY

Electrokinetic remediation of soil and groundwater is cost-effective and averages $200 per cubic meter. At $90 per ton, treatment by electrokinetics is generally less than excavation and disposal as hazardous waste. Electrokinetic treatment of organic contaminated soil is roughly equal in cost to some thermal technologies. As the technology matures, confidence in electrokinetic remediation should rise and costs should drop.

REFERENCES

DOE. (1996). Lasagna™ Soil Remediation Innovative Technology Summary Report. U.S. Department of Energy Office of Environmental Management. report number DOE/EM-0308, Oak Ridge, TN.

ITRC. (1997). Emerging Technologies for the Remediation of Metals in Soils—Electrokinetic Remediation. Interstate Technology Regulatory Cooperation. document number MIS-4, Washington, DC.

United States Army Environmental Center. (2000). In-Situ Electrokinetic Remediation of Metal Contaminated Soils. *Technology Status Report, SFIM-AEC-ET-CR-9902*.

USEPA. (1995). In-Situ Remediation Technology: Electrokinetics. *EPA/542/K-94/007*.

USEPA. (1999). Sandia National Laboratories In-Situ Electrokinetic Extraction Technology Innovative Technology Evaluation Report. *EPA/540/R-97/509*.

USEPA. (2003). Innovative Technology Evaluation Report: Electrochemical Design Associates (Formerly Geokinetics International, Inc.) Lead Recovery Technology Evaluation at the Building 394 Battery Shop. Pearl Harbor Naval Shipyard and Intermediate Maintenance Facility, Honolulu, HI. *EPA/540/R-04/506*.

28

REGULATORY ASPECTS OF IMPLEMENTING ELECTROKINETIC REMEDIATION*

RANDY A. PARKER

28.1 INTRODUCTION

A better understanding of the environmental impact of hazardous waste management practices has led to new environmental laws and a comprehensive regulatory program. This program is designed to address remediation of past waste management practices and to ensure that the hazardous wastes generated today do not become expensive and complex cleanup problems in the future.

During the 1970s and the early 1980s, toxic chemicals and other hazardous wastes were treated by land disposal or by incineration. The Hazardous and Solid Waste Amendments (HSWA) of 1984 to the Resource Conservation and Recovery Act (RCRA) severely restricted disposal of wastes in untreated landfills, surface impoundments, and land treatment units. Additionally, negative public opinion was focused on the disproportionate risk borne by those living near incinerators and other hazardous waste treatment and disposal facilities. Environmental laws passed in the 1970s clearly articulated public and congressional dissatisfaction with early site remedies that wholly consisted of containment, off-site disposal, and incineration.

Given land disposal restrictions, trends against incineration, and restrictions on storage of hazardous wastes, alternative treatment methods such as electrokinetics are increasingly given consideration for the treatment of hazardous waste site

---

* The views expressed in this chapter are those of the author and do not necessarily reflect the views or policies of the United States Environmental Protection Agency. No official endorsement should be inferred.

Electrochemical Remediation Technologies for Polluted Soils, Sediments and Groundwater,
Edited by Krishna R. Reddy and Claudio Cameselle
Copyright © 2009 John Wiley & Sons, Inc.

contamination. Electrokinetic remediation shows potential for treatment of hazardous wastes in ways that are more effective than traditional treatment methods.

Electrokinetic remediation is part of a group of technologies that include electrochemical remediation, electromigration, and electroreclamation. Electrokinetic transport mechanisms can be employed for the remediation of organic and inorganic contaminants, for the injection of nutrients, electron acceptors, and other process additives applicable to *in situ* bioremediation, for the injection of surfactants that enhance solubilization and transport, for the injection of grouts for soil stabilization and waste containment, for precipitation of contaminants in migrating plumes and diversion systems for contaminant plumes, or for dewatering soils, slurries, and sediments (Acar *et al.*, 1995).

The various applications of electrokinetic technologies must adhere to the regulatory framework of environmental laws designed to assure the protection of human health, the natural environment, sensitive habitats, and endangered species. The most relevant regulations addressing the remediation of contamination in the USA are the RCRA (1976) and the Comprehensive Environmental Response, Compensation, and Liability Act (CERCLA).

28.2 OVERVIEW OF ENVIRONMENTAL REGULATION IN THE USA

In 1970, the US Congress, driven by rising and widespread concerns over environmental problems, began to enact a body of legislation that would have a major impact on citizens' lives and the nation's way of conducting business. It began with the signing on January 1, 1970 of the National Environmental Policy Act (NEPA). This law made it national policy to "encourage productive and enjoyable harmony between man and his environment" (NEPA, 1970). NEPA also created the environmental impact review process and the Council on Environmental Quality. NEPA not only gave high visibility to growing environmental concerns but it also started a decade in which environmental concern was transmitted into environmental law.

Throughout the 1970s, Congress reacted to environmental concern by passing several major pieces of environmental law: the Clean Air Act (CAA, 1970) and its amendments (1977), the Clean Water Act (CWA, 1970) and its amendments (1977), the Federal Insecticide, Fungicide, and Rodenticide Act (1972) and it amendments (1975 and 1978), the Safe Drinking Water Act (SDWA, 1974) and its amendments (1977), the RCRA (1976) and its amendments (the HSWA, 1980), the Toxic Substances Control Act (TSCA, 1976), and the CERCLA (1980). Amendments to CERCLA were added to federal law in 1986 (Battle, 1986).

In addition to these explicitly environmental laws, Congress passed additional laws addressing occupational safety (1978), coastal and fisheries management, endangered species, and subsurface mining. By 1980, the USA had passed comprehensive national legislation intended to protect almost every aspect of environmental quality.

The most relevant laws dealing with contaminated sites in the USA are covered in this chapter as well as their relevance to electrochemical remediation technologies in general and, specifically, electrokinetic remediation. It is important to note that in addition to federal laws covered in this chapter, state and local laws also apply to electrochemical remediation technologies deployed in the USA.

28.2.1 CERCLA

The CERCLA is a central part of legislative framework for environmental protection. It is designed to remedy threats to human health and the environment from unexpected spills and other releases of pollutants and contaminants into the environment, and to remedy historical mistakes in hazardous waste management. The law provided the federal government with new response authority and created a US$1.6 billion trust fund for cleaning up abandoned or uncontrolled hazardous waste sites. Because of the trust fund provision, CERCLA came to be known as Superfund. The fund was established primarily by taxes on oil and designated chemicals [United States Environmental Protection Agency (EPA), 2008a].

CERCLA also established prohibitions and requirements related to closed and abandoned waste sites and provided liability for persons responsible for releases of hazardous wastes at those sites. CERCLA amended the National Oil and Hazardous Substances Contingency Plan [National Contingency Plan (NCP)] to provide a regulatory blueprint for federal response to releases of hazardous substances, pollutants, and contaminants.

During the 1980s, as more sites were identified, it became apparent that the problem of abandoned hazardous waste sites was far more extensive than what was initially recognized. Solutions to the problem of abandoned hazardous waste sites would be more complex and time-consuming than originally envisioned. Because the environmental threats were from historical hazardous waste management activities, the responsible parties may be unknown, may no longer be in existence (i.e. a defunct company), or may be unable to pay. To address these problems, Congress passed the Superfund Amendments and Reauthorization Act (SARA) in 1986. SARA directs EPA to (a) use remedial alternatives that permanently and significantly reduce the volume, toxicity, or mobility of hazardous substances, pollutants, or contaminants; (b) select remedial actions that protect human health and the environment, are cost-effective, and involve permanent solutions and alternative treatment or resource recovery technologies to the maximum extent possible; and (c) avoid off-site transport and disposal of untreated hazardous substances or contaminated materials when practicable treatment technologies exist. SARA established new standards and schedules for site cleanup and increased the size of the trust fund from US$1.6 billion to US$8.5 billion. The law also created new programs for informing the public of the risks of hazardous substances in their community and helped prepare communities for hazardous substance emergencies (SARA, 1986).

The Superfund cleanup process is outlined in Table 28.1. After EPA becomes aware that a site may contain contamination, CERCLA requires a nine-step process that begins with a preliminary assessment of the site. If the preliminary assessment reveals that remedial action is necessary, EPA will conduct a more involved study of the site during the site inspection phase. The site is evaluated using the Hazard Ranking System, a scoring system that determines the relative risk to human health and the environment posed by the hazardous substances in groundwater, surface water, air, and soil. Sites most in need of cleanup are placed on the National Priorities List (NPL), EPA's list of priority sites for cleanup. Superfund moneys are available only to nonfederal hazardous waste sites on the NPL. As of March 2008, there were over 1300 sites on the NPL or proposed for inclusion (EPA, 2008b). The

TABLE 28.1. The Superfund Cleanup Process

Site discovery	Discovery or notification to EPA of possible releases of hazardous substances
Preliminary assessment/ site inspection	Investigations of site conditions. If the release of hazardous substances requires immediate or short-term response actions, these are addressed under the Emergency Response Program of Superfund.
National Priorities List (NPL) listing	A list of the most serious sites identified for long-term cleanup
Remedial investigation/ feasibility study	Determines the nature and extent of contamination. Assesses the treatability of site contamination and evaluates the potential performance and cost of treatment technologies
Records of decision (ROD)	Explains which cleanup alternatives will be used at NPL sites. When remedies exceed $25 million, they are reviewed by the National Remedy Review Board.
Remedial design/remedial action (RD/RA)	Preparation and implementation of plans and specifications for applying site remedies. The bulk of the cleanup usually occurs during this phase. All new fund-financed remedies are reviewed by the National Priorities Panel.
Construction completion	Identifies completion of the physical cleanup construction, although this does not necessarily indicate whether final cleanup levels have been achieved
Postconstruction completion	Ensures that Superfund response actions provide for the long-term protection of human health and the environment. Included here are long-term response actions (LTRA), operation and maintenance, institutional controls, 5-year reviews, and remedy optimization.
NPL delete	Removes a site from the NPL once all response actions are complete and all cleanup goals have been achieved
Site reuse/redevelopment	Information on how the Superfund program is working with communities and other partners to return hazardous waste sites to safe and productive use without adversely affecting the remedy

Source: www.epa.gov/superfund/cleanup/index.htm.

process proceeds through the "remedy phases," which include remedial investigation, remedy selection, remedy design, and remedial action. The postconstruction phases assure that the implemented remedy provides for the long-term protection of human health and the environment.

28.2.2 RCRA

The RCRA gives EPA the authority to regulate hazardous waste from its generation to its disposal. This includes transportation, treatment, and storage of hazardous waste. RCRA also sets the standards for the management of nonhazardous solid wastes. RCRA was written to accomplish three primary goals: (a) protection of human health and the environment, (b) reduction of waste and conservation of energy and natural resources, and (c) reduction or elimination of the generation of hazardous waste as expeditiously as possible. RCRA prohibits dumping of solid and hazardous wastes, requires the proper management of hazardous waste, and directs EPA to take steps to minimize the generation of hazardous wastes and to develop alternative land disposal options.

The HSWA to RCRA, promulgated in 1984, focused on waste minimization and phasing out land disposal of hazardous waste as well as corrective action for releases (EPA, 2008c). Although HSWA's primary purpose was to amend the RCRA program, Congress also added a provision under the authority of the SDWA. This provision requires that no hazardous wastes may be disposed of by underground injection into any underground drinking water source or any formation located above such a source. This provision does not apply to injection into the aquifer from which it was withdrawn if the injection is a response action under CERCLA or RCRA, and if such groundwater is treated to substantially reduce hazardous constituents.

RCRA was amended by the Federal Facilities Compliance Act of 1992 and the Land Disposal Program Flexibility Act of 1996. The Federal Facilities Compliance Act amends RCRA to state clearly that federal facilities are part of the regulated community and are subject to the wide range of enforcement actions including fines and penalties. The Land Disposal Program Flexibility Act mandated changes to the RCRA land disposal restriction regulatory program and the nonhazardous landfill groundwater monitoring program.

RCRA regulates how wastes should be managed to avoid potential threats to human health and the environment, while CERCLA comes into consideration when mismanagement of wastes has occurred (i.e. when there has been a release of a pollutant or a contaminant that poses a significant risk of danger to human health).

The act is divided into 10 subtitles (A through J) that provide EPA with the framework and authority to achieve the goals of RCRA. Provisions applicable to electrokinetic remediation are found in RCRA subtitle C, which establishes criteria for controlling hazardous waste from its point of generation to its final disposal (RCRA, 1976).

28.2.3 CAA

The CAA is the comprehensive federal law that regulates air emissions from stationary and mobile sources. The law authorizes EPA to set National Ambient Air Quality Standards (NAAQS) to protect human health and public welfare and to regulate emission of hazardous air pollutants (EPA, 2008d). In addition to setting other goals, the act set goals for achieving NAAQS in every state. The setting of these pollutant standards was coupled with directing the states to develop State Implementation Plans (SIPs) applicable to the appropriate industrial sources in the state, in order to achieve these standards. The Act was amended in 1977 and in 1990 to set new goals for the attainment of NAAQS. The 1990 CAA Amendments, Section 112, requires issuance of standards for major sources and certain area sources. Major sources are defined as a stationary source or a group of stationary sources that emit or have the potential to emit 10 tons or more per year of a single hazardous air pollutant, or 25 tons or more per year of a combination of hazardous air pollutants. An "area source" is any stationary source that is not a major source. The list of hazardous air pollutants is listed in CAA Section 112.

The air emissions of concern to the CAA are hazardous or toxic air pollutants. The CAA regulations not only apply to major sources of pollution but also to remediation units, process vents, and equipment leaks. The CAA requirements apply at facilities at which a site remediation meets the following conditions: (a) The site remediation is colocated at the facility with one or more stationary sources that emit hazardous air pollutants and the remediation itself has potential to emit hazardous air pollutants, and (b) the facility is a major source of hazardous air pollutants. All emission sources at the facility, as well as the site remediation activity, must be included when determining the total amount of hazardous air pollutants. Site remediations to which air emission regulations do not apply are

- site remediations that clean up only material that does not contain hazardous air pollutants;
- site remediations performed under authority of CERCLA as a remedial action or as a non-time-critical removal action;
- site remediations performed under RCRA corrective action conducted at a treatment, storage, and disposal facility that is required by the facility permit issued by EPA or by a state program;
- site remediations performed at a gasoline station to clean remediation material from a leaking underground storage tank;
- site remediations conducted at a residential site; and
- remediations conducted at a research and development facility that meets requirements under the CAA.

28.2.4 CWA

The CWA establishes the basic structure for regulating discharges of pollutants into waters of the USA and for regulating quality standards for surface waters (CWA, 1977). The statute employs a variety of regulatory tools to reduce pollutant discharges into waterways and to manage polluted runoff. Under the CWA, it is unlawful to discharge any pollutant from a point source into navigable waters unless a permit is obtained. The National Pollutant Discharge Elimination System (NPDES) permit program controls discharges. First, water quality standards consistent with statutory goals must be established. If a water body is not meeting water quality standards, strategies for obtaining those standards must be developed. After implementation of these strategies, ambient conditions are again measured and compared to water quality standards. If standards are now being met, only occasional monitoring is needed. If standards are still not being met, a revised strategy is developed and implemented. This iterative process must be repeated until standards are met (EPA, 2008e).

28.2.5 SDWA

The SDWA was originally passed to protect public drinking water supplies. SDWA was amended in 1986 and 1996 and requires many actions to protect both drinking water and its sources; rivers, streams, lakes, reservoirs, springs, and groundwater wells. SDWA authorizes EPA to set standards for drinking water to protect it from naturally occurring and man-made contaminants. Most of the direct oversight of the 160,000 drinking water systems in the USA is performed by state drinking water programs.

EPA sets primary standards for drinking water through a three-step process: identification of contaminants, determination of contaminant risk, and setting maximum contaminant levels. First, EPA identifies contaminants that may adversely affect drinking water with a frequency and with levels that pose a threat to human health. EPA decides which contaminants to potentially regulate. EPA then sets a maximum contaminant level for the contaminants that it decides to regulate. The goal of the regulation is to limit the contaminant to a level to which there is no known or expected health risk. These goals allow for a margin of safety. Lastly, EPA specifies a maximum contaminant level, the maximum permissible level of contaminant in drinking water that is delivered to any user of a public water system. There are cases when it may not be feasible to set a maximum contaminant level, for example, when the technology to treat the contaminant to the required level does not exist or in cases where there is no reliable or economic method to detect the contaminants in drinking water. In these cases, EPA sets a required treatment technique that specifies a way to treat and to remove the contaminant.

28.3 REGULATORY CONSIDERATIONS FOR IMPLEMENTING ELECTROKINETIC REMEDIATION

Electrokinetic remediation is a separation and removal technique that has been applied to a variety of applications for remediation of contaminated media. The technology has been used to treat radionuclides, heavy metals, certain organic compounds, or mixed inorganic contaminants in soils, sediments, and groundwater. The technology uses electricity to affect chemical concentrations and groundwater flow. A number of electrochemical effects may result when a voltage difference is applied across a saturated or partially saturated subsurface zone. Such electrochemical effects include ion diffusion, ion exchange, development of osmotic and pH gradients, dessication due to heat generation at the electrodes, mineral decomposition, precipitation of salts and secondary minerals, electrolysis, hydrolysis, oxidation, reduction, physical and chemical adsorption, and soil fabric change (Acar *et al.*, 1995).

Electrokinetics may conceivably be used in six general ways for hazardous waste site remediation: (a) concentration and dewatering of waste sludges; (b) extraction of pollutants from soils, sediments, and groundwater; (c) creation of hydraulic flow barriers; (d) injection for delivery of nutrients or electron acceptors to enhance bioremediation; (e) injection of grouts or cleanup chemicals; or (f) improvement of the effective permeability of a soil mass (Parker, 1992).

Because of the broad range of contaminants and media that have applicability to electrokinetic remediation, there is a wide range of federal, state, and local regulations that may apply. These regulations apply to transport, treatment, storage, and disposal of wastes and treatment residuals. This section discusses specific federal environmental regulations pertinent to the operation of electrokinetic technologies in the USA. State and local requirements, which may be more stringent, must also be addressed when deploying electrokinetic technologies.

28.3.1 CERCLA Relevant Regulatory Requirements

Rather than establish individual cleanup standards, CERCLA assures that remedies are based on cleanup standards established by other laws (e.g. CAA, CWA, and RCRA). In conjunction with site-specific risk factors, CERCLA requires that remedies attain any legally applicable or relevant and appropriate requirements (ARARs). ARARs are standards, criteria, or limitations under federal and state environmental laws. For example, if electrokinetic remediation involves the on-site treatment, storage, or disposal of hazardous wastes, the remediation activity must meet RCRA standards for such treatment, storage, and disposal. ARARs relative to electrokinetic remediation include (a) the CERCLA, (b) the RCRA, (c) the CAA, (d) the CWA, (e) the SDWA, and (f) Occupational Safety and Health Administration (OSHA) regulations (EPA SITE Program, 2003). These six general requirements are discussed in the sections that follow.

As part of CERCLA, EPA has prepared the NCP for hazardous substance response. The NCP delineates the methods and criteria used to determine the appropriate extent of removal and cleanup for hazardous waste contamination.

28.3.2 Feasibility Criteria for the Use of Electrokinetic Technologies

Two types of remediation activities are possible under CERCLA: removal and remedial actions. Remedial actions are longer-term permanent solutions to hazardous waste contamination. Removal actions are short-term cleanup actions that address immediate threats at a site. They are conducted in response to an emergency situation (e.g. to avert an explosion, to clean up a hazardous waste spill, or to stabilize a site until a permanent solution can be found). Electrokinetic remediation technologies are likely to be part of a CERCLA remedial action. There are nine general criteria that must be addressed by those wishing to implement electrokinetic technologies as part of CERCLA remedial actions:

- overall protection of human health and the environment;
- compliance with ARARs;
- long-term effectiveness and permanence;
- reduction of toxicity, mobility, or volume through treatment;
- short-term effectiveness;
- implementability;
- cost;
- state acceptance; and
- community acceptance.

These criteria are used to determine the feasibility of using the electrokinetic technology at CERCLA sites. Electrokinetic technologies are expected to protect human health and the environment by concentration, removal, or destruction of contaminants in soils, sediments, and groundwater. Overall reduction of human health risk should be evaluated on a site-specific basis because contaminants may change speciation or mobility. For example, volatile organic compounds (VOCs) could be stripped from the soil during treatment and could increase soil vapor concentrations and VOC migration in the soil. Also, the system effluent (e.g. analytes or extracted contaminants) may contain hazardous constituents and may need to be characterized, labeled, and disposed of as hazardous waste.

Specific requirements that may be ARARs for electrokinetic remediation are summarized in Table 28.2. All regulations are listed in Title 40 of the Code of Federal Regulations (CFR). ARARs are determined on a site-by-site basis considering the type of contaminants present, the actions taken, the waste streams generated, and the location of the site relative to sensitive environments. The waste streams generated relate to the material to be treated, the material after treatment, and used personnel protective equipment (PPE) requiring disposal. Other regulations may be appropriate depending on site, contaminant, or location-specific factors. The electrokinetic technology has the potential to comply with federal, state, and local ARARs (EPA SITE Program, 1995).

With respect to long-term effectiveness, electrokinetics has the potential to reduce human health risk by treating contaminated media to an acceptable lifetime excess cancer risk level. Long-term effectiveness is achieved by extraction, containment, or concentration and removal of contaminants. While the time needed to achieve cleanup goals depends primarily on the characteristics of the contaminated media, electrokinetic technologies may take longer than conventional technologies due to the fact that as target levels become low (<100mg/kg in soils), the operational efficiency of the technology may decrease (ITRC, 1997). Periodic review of treatment system performance may be needed to ascertain that treatment effectiveness and efficiency is maintained. Also, regulators may be concerned that soils and sediments may not be able to sustain plant and microbial life due to chemical, physiological, and biological impacts from the application of electric currents. Postremediation monitoring may be necessary to determine the long-term impacts of the treatment.

The criterion of reduction of toxicity, mobility, or volume through treatment refers to anticipated performance of the treatment technology potentially employed in a Superfund remediation. The technology reduces the volume and mobility of contaminants by the concentration or extraction of contaminants for subsequent disposal. Minimal waste is generated during the contaminant treatment process.

Short-term effectiveness addresses the period of time needed to achieve lasting protection of human health and the environment as well as any adverse impact that may be posed during the construction and implementation period before the cleanup goals are achieved. Electrokinetics is expected to present minimal short-term risks to workers and the nearby community, including noise and exposure to airborne contaminants.

The implementability criterion considers the technical and administrative feasibility of a remedial action, including the availability of materials and services to implement a particular option. Electrokinetic remediation can theoretically be deployed anywhere an electrode array can be installed. Prior to implementing electrokinetic remediation at a specific site, field and laboratory screening tests may need to be conducted to determine if the site is amenable to this technology. In the case of deploying the technology in soils, buried metallic objects and utility lines could short-circuit the current path, thereby influencing the voltage gradient and affecting the contaminant extraction rate.

TABLE 28.2. Appropriate and Relevant Requirements (ARARs) for Electrokinetic Remediation

Criteria	ARAR	Regulation Description	General Applicability	Specific Applicability to Electrochemical Remediation Processes
Overall protection of human health and the environment	CERCLA: 40 CFR part 300	Authorizes and regulates the cleanup of environmental contamination	CERCLA site cleanups require that other environmental laws be considered as appropriate to protect human health and the environment.	Applicable to remediation at Superfund sites
Waste characterization	RCRA: 40 CFR part 261 (or the state equivalent)	Standards apply to the identification and characterization of hazardous wastes.	Properties of the waste determine its suitability for treatment by the types of contaminants present and the grain size of the soil/sediment to determine suitability.	Chemical and physical analyses must be performed to determine if waste/contaminants are suitable for electrokinetic treatment.
Waste processing	RCRA: 40 CFR part 264	Standards apply to the treatment of wastes in a treatment facility.	Standards apply to the treatment of wastes in a treatment facility (i.e. there are requirements for operations, record keeping, and contingency planning).	Not likely applicable to electrokinetic remediaton, since the process is not normally conducted at treatment facilities
	CAA: 40 CFR parts 50 and 70 (or the state equivalent)	Regulations govern the toxic pollutants, visible emissions, and particulates.	Any off-gas venting (from buildup of VOCs, etc.) must not exceed limits for the air district of the site.	When treating semi-volatile organic compound (SVOCs) and metals, particulate emissions may contain regulated substances. In such cases, standard monitoring and record keeping apply.
Storage of auxiliary wastes	RCRA: 40 CFR part 264 subpart J or the state equivalent	Regulation governs the standards for tanks at treatment facilities.	Storage tanks for liquid wastes (e.g. decontamination waste) must be placarded appropriately, must have secondary containment, and must be inspected daily.	If storing non-RCRA wastes, RCRA requirements may still be relevant and appropriate.
	RCRA 40 CFR part 264, subpart 1	Regulation covers the storage of waste materials generated.	Potentially hazardous wastes remaining after treatment (i.e. contaminated electrodes) must be labeled as hazardous waste and must be stored in containers that are in good condition. Containers should be stored in a designated storage area. Storage should not exceed 90 days unless a permit is obtained.	Applicable to RCRA wastes; relevant and appropriate for non-RCRA wastes.

Determination of cleanup standards	Local	Standards apply for treatment of soils and groundwater.	Remedial actions for soils, sediments, and groundwater are required to meet local requirements.	The technology will be required to meet specific goals and cleanup standards.
Waste disposal	RCRA: 40 CFR part 262	Standards that pertain to generators of hazardous wastes	Potential hazardous waste generated by electrochemical processes is limited to drill cuttings, well purge water, personal protective equipment, and decontamination wastes.	Generators must dispose of wastes at facilities permitted to handle the waste. Generators must obtain an EPA identification number prior to disposal.
	CWA: 40 CFR parts 403 and/or 122 and 125	Standards for discharge of wastewater to a POTW or to a navigable waterway.	Applicable and appropriate for any decontamination and wastewater generated from the process. Discharge of wastewater to a POTW must meet pretreatment standards; discharges to a navigable waterway must be permitted under NPDES.	Applicable to electrokinetics if groundwater treatment is specified as part of the cleanup criteria. Standards may apply to wastewater generated from decontaminating soils and sediment cores and from electrodes that are removed at the end of treatment.
	RCRA: 40 CFR part 268	Standards regarding the disposal of hazardous wastes	Applicable for the off-site disposal of auxiliary waste (e.g. excess soil and sediment samples)	Hazardous wastes must meet specific treatment standards prior to land disposal or treated using specific technologies.
	SDWA: parts 144 and 145	Standards regarding underground injection of wastes	Applicable to subsurface injection of biological nutrients, surfactants, oxidants, and so on that were prepared aboveground	State-issued Underground Injection Control permits may be required to inject process additives into the subsurface.
	Atomic Energy Act (AEA), 10 CFR and RCRA	Standards regarding the disposal of both radioactive and hazardous wastes	Applicable to the treatment, storage, and disposal of mixed wastes containing both hazardous and radioactive components	Both RCRA and AEA regulations apply to electrokinetic treatment of mixed wastes. EPA and the Depart of Energy directives provide guidance that addresses mixed wastes.
Personnel protection	OSHA: 29 CFR parts 1900 through 1926	OSHA regulates on-site construction activities and health and safety of workers at hazardous waste sites.	Installation and operation of electrochemical systems at Superfund and RCRA sites must meet OSHA requirements.	OSHA standards for personnel protection and medical monitoring apply to electrochemical remediation site activities.

SVOC, semi-volatile organic compound.

The cost criterion addresses capital, operation, and maintenance costs. Treatment costs for the technology vary greatly depending on the size of the treatment system used, the contaminant characteristics and concentrations, the cleanup goals, the volume of contaminant to be treated, and the length of treatment.

State acceptance of the electrokinetic technology at hazardous waste sites will likely depend on expected residual contamination and on how wastes from the remediation processes are handled and disposed. The potential for emissions during drilling and installation of the electrokinetic technology is substantially lower than excavation. Acceptance of the technology may involve consideration of performance data from applications such as field- and pilot-scale demonstrations using actual wastes that will be treated during later, full-scale remediation.

Community acceptance is usually assessed after the public has had an opportunity to review and comment on the proposed remedial activity. No loud or heavy equipment is associated with the electrokinetic technology, and no substantial quantities of wastes are generated as a result of treatment. When proper operational and safety procedures are followed, hazards to personnel are minimized and the potential for a chemical spill to the environment that might endanger the community is minimal. It is often prudent to involve state regulators, permitting authorities, and the community early in the process to identify and to address their concerns.

28.3.3 RCRA Regulatory Requirements

The law referred to as RCRA is actually a combination of the first federal solid waste statute (the Solid Waste Disposal Act of 1965) and all subsequent amendments. RCRA regulations define hazardous wastes and regulate their transport, treatment, storage, and disposal. RCRA-defined hazardous wastes that may be present during electrokinetic soil remediation activities include contaminated soil cuttings and purge water generated during well installation and development, residual wastes generated from any groundwater sampling activities, and used PPE. If wastes are determined to be hazardous wastes (either because of a characteristic or a listing carried by the waste), essentially all RCRA requirements regarding the management and disposal of the hazardous waste need to be addressed by remedial managers. Treatment, storage, or disposal of hazardous waste typically require issuance of an RCRA part B treatment, storage, or disposal permit.

Wastes defined as hazardous under RCRA include characteristic and listed wastes. An RCRA-defined hazardous waste is a waste that appears on one of four lists (F list, K list, U list, or P list) or exhibits at least one of four characteristics (ignitability, corrosivity, reactivity, or toxicity). Criteria for identifying characteristic hazardous wastes are included in Title 40, CFR part 261, subpart C. Listed wastes from industrial sources are itemized in 40 CFR part 261, subpart D. Other regulations that are relevant to the technology include the requirement to characterize the waste for a hazardous waste generator (40 CFR part 262.11), the requirement to determine if the hazardous waste is restricted from land disposal [40 CFR 268.7(a)], requirements for on-site storage of waste for up to 90 days [40 CFR 262.34(a)], or 40 CFR 264.553 for storage of waste in a temporary unit for up to 1 year prior to disposal.

Other RCRA requirements may include a Uniform Hazardous Waste Manifest (or its state counterpart) for off-site shipping of hazardous waste. Transport of hazardous wastes must comply with the United States Department of Transportation waste packaging, labeling, and transportation regulations (4 CFR part 172).

28.3.3.1 RCRA Research, Development, and Demonstration (RD&D) Permits

RCRA also established a mechanism to allow for demonstration of small-scale, state-of-the art technologies and processes and for modifications of existing technologies or processes. This mechanism is the RCRA RD&D permit. It allows for RD&D of technologies that are not yet commercially available, units that require refined or improved performance, or innovative commercial units that require demonstration of performance and cost-effectiveness.

The scale of operation for an RD&D treatment unit is limited to quantities necessary to determine the cost-effectiveness, efficiency, and performance capabilities of the process. Treatment units are not statutorily limited in size, but these units are usually well below the scale of commercial operations. There may also be limits on the amount of hazardous material that can be processed or stored on-site. By statute, the demonstration period for RCRA RD&D permitted units is limited to 1 year. The permit may be renewed up to three times, and each renewal is valid for up to 1 year. Electrokinetic remediation and other types of electrochemical remediation technologies are still considered emerging technologies by the regulatory community. Because of their status as an emerging technology, some electrokinetic remediation processes have been field tested through the RCRA RD&D permitting process. A list of RD&D permit areas for consideration by regulators and applicants, with associated RCRA regulatory citations, is shown in Table 28.3.

28.3.4 CAA Requirements

The CAA establishes national primary and secondary air quality standards for sulfur oxides, particulate matter, carbon monoxide, ozone, nitrogen dioxide, and lead. It also limits the emission of 189 listed hazardous waste pollutants such as vinyl chloride, arsenic, asbestos, and benzene (CAA, 1977). States are responsible for enforcement of the CAA. To assist in this effort, Air Quality Control Regions (AQCRs) were established. Allowable emission limits are determined by the AQCR or its subunit, the Air Quality Management District. These emission limits are based on whether or not the region is currently within attainment of National Air Quality Standards.

The CAA requires that treatment, storage, and disposal facilities comply with primary and secondary air quality standards. Requirements relevant to electrokinetic treatment address toxic pollutants, visible emissions, and particulates (40 CFR part 50, and 40 CFR part 70). Air emissions from electrokinetic treatment systems may result from fugitive emissions such as drilling activities related to system installation (VOC or dust emissions), periodic sampling efforts, and the staging and storing of contaminated drill cuttings. VOC emission control equipment should be provided to reduce emissions if the treatment system is applied to contaminated media that contain VOCs. State air quality standards may require additional measures to prevent fugitive emissions.

28.3.5 CWA Requirements

The objective of the CWA is to restore and to maintain the chemical, physical, and biological integrity of the nation's waters by establishing federal, state, and local

TABLE 28.3. RCRA RD&D Permit Regulations Reference

RD&D Permit Considerations	RCRA Regulatory Citation, Title 40 Code of Federal Regulations (CFR)
A. Facility description	
Description of project	270.13(a), 270.13(m), 270.14(b), 270.65(a)
Description of facility	264.11, 270.13, 270.14
Location information	264.18, 270.14
Seismic standard	264.18
Floodplain standard	264.18
Groundwater	—
B. Waste description	264.13, 270.13, 271.14,
C. RD&D process information	
Description of the technology	—
Description of electrokinetic equipment	270.14
Description proposed remediation zone	—
Sampling and analysis plan	264.13, 270.13, 270.30, 264.73
Equipment installation	264.13
Equipment operation	270.30
Performance monitoring	270.30
Demonstration schedule	270.65
Expected results	—
D. Procedures to prevent hazards	
Security	264.14
General inspection	264.15, 264.174
Emergency preparedness	264.50-56, 270.14, 264.30-37
Prevention of explosion and fires	264.31, 270.14
Prevention of releases	264.31, 270.14
Prevention of ignition, reaction, or explosions of reactive or incompatible materials	270.14, 270.17
Prevention of general hazards	270.14
Contingency plan	264 subpart D
E. Groundwater monitoring	264.9, 264.101, 270.14
F. Personnel qualifications	264.16
G. Closure plan	
Maximum quantity of waste	254.112
Disposal of equipment, structures, and soils	262.34, 264.112
Decontamination of equipment and structures	264.112
Closure schedule	264.112, 264.113
Certification of closure	254.115
Postclosure	264.1116-120
H. Other federal laws	
I. Records	264.13, 264.16, 264.53, 264.73
J. Financial responsibility	264.140

discharge standards. If treated water is discharged to surface water bodies or publicly owned treatment works (POTW), CWA regulations will apply. A facility desiring to discharge water to a navigable waterway must apply for a permit under the NPDES. The permitting program is outlined in 40 CFR parts 122 and 125. When an NPDES permit is issued, it includes waste discharge requirements. Discharges to a POTW must also comply with general pretreatment regulations outlined in 40 CFR part 403.

28.3.6 SDWA Requirements

The SDWA requires EPA to establish regulations to protect human health from contaminants in water designed for drinking use. This includes both aboveground or underground sources. The most direct oversight of water systems is conducted by state drinking water programs. States can apply for "primacy," the authority to implement the SWDA within their jurisdictions, if they can show that they will adopt standards at least as stringent as EPA's and can make sure that water systems meet these standards. All states and territories, except Wyoming, and the District of Columbia have received primacy.

EPA is also required to establish minimum standards for state programs to protect sources of drinking water from endangerment from underground injection of fluids. SWDA sets a framework for the Underground Injection Control (UIC) program to control the injection of wastes into groundwater. EPA and states implement the UIC program, which sets standards for safe waste injection practices and bans certain types of injection altogether. The National Primary Drinking Water Standards are found in 40 CFR parts 141 through 149. Parts 144 and 145 discuss requirements associated with the underground injection of contaminated water. If underground injection of wastewater, biological nutrients, oxidants, and so on is selected as a means of disposal, the approval of EPA or the delegated state for the construction and operation of a new underground injection well may be required.

28.3.7 OSHA Requirements

The OSHA ensures worker and workplace safety. The goal is to make the workplace free of recognized hazards, such as exposure to toxic chemicals, excessive noise levels, mechanical dangers, heat and cold stress, and unsanitary conditions. Both CERCLA remediation activities and RCRA corrective actions must meet OSHA requirements detailed in 20 CFR parts 1900 through 1926, particularly part 1910.120, which provides for the health and safety of workers at hazardous waste sites. Part 1910.129, "Hazardous Waste Operations and Emergency Response," and part 1926, "Safety and Health Regulations for Construction," apply to any on-site construction activities. If working at a hazardous waste site, all personnel involved in the installation and implementation of a treatment process are required to have an OSHA training course and must be familiar with all OSHA requirements relevant to hazardous waste sites. Workers on hazardous waste sites must be enrolled in a medical monitoring program. During construction activities, noise levels should be monitored to ensure that workers are not exposed to noise levels above a time-weighted average of 85 dB over an 8-h day. If noise levels increase above this limit, workers will be required to wear hearing protection. The levels of noise anticipated from electrokinetic treatment systems are not expected to adversely affect the community.

Process chemicals must be managed according to OSHA controls (OSHA part 1926, subpart D, "Occupational Health and Environmental Controls," and subpart

H, "Materials Handling, Storage, and Disposal"). Electric utility hookups must comply with part 1926 subpart K, "Electrical." Interaction with local utility companies is important because high-voltage direct current (DC) underground can interfere with cathodic protection systems and underground natural gas and water lines.

Health and safety plans for site remediation activities should address chemicals of concern and should include monitoring practices to ensure that worker health and safety are maintained. State OSHA requirements, which may be significantly stricter than federal standards, also must be met. These standards for workplace safety have been developed and are enforced throughout the USA.

28.3.8 Other Statutes Applicable to Electrokinetic Remediation

Other laws and statutes that need to be considered when deploying electrokinetic remediation or other electrochemical remediation technologies include the TSCA, the Atomic Energy Act (AEA), and state and local regulatory requirements. While these statutes may not directly affect decisions to deploy electrokinetic remediation, they should be considered in some cases where the technology is used.

The TSCA (1976) was enacted by Congress to give EPA the ability to track the 75,000 industrial chemicals currently produced or imported into the USA. It also gives EPA the ability to track the new chemicals that the industry develops each year with either unknown or possibly dangerous characteristics. EPA can then control these chemicals to protect the environment and human health. The disposal of polychlorinated biphenyls (PCBs) is regulated under Section 6(e) of TSCA. PCBs at concentrations of 50–500 ppm can either be disposed of in TSCA-permitted landfills or destroyed by incineration at a TSCA-approved incinerator. Although electrokinetic technology is not usually used to treat PCBs, the technology may be used at sites (particularly contaminated sediment sites) where PCBs are a part of the contaminated material.

EPA has the authority to issue generally applicable environmental radiation standards. Other federal and state organizations must follow these standards when developing requirements for their areas of radiation protection. Mixed waste (wastes containing both radioactive and hazardous components) is subject to both the AEA and the RCRA. However, when the application of both AEA and RCRA regulations results in a situation inconsistent with AEA (e.g. increased likelihood of radioactive exposure), AEA requirements supercede RCRA (EPA, 1995). The use of electrokinetic remediation at sites with radioactive contamination might involve the treatment or generation of mixed waste. EPA, in conjunction with the Nuclear Regulatory Commission, has issued several directives to assist in the identification, treatment, and disposal of low-level radioactive mixed waste. Various EPA directives include guidance on defining, identifying, and disposing commercial, mixed, low-level radioactive hazardous wastes. If electrokinetics is used to treat low-level mixed wastes, these directives should be considered.

State and local regulations may also apply to electrokinetic remediation. The regulatory requirements are site specific or location specific and may be more stringent than federal standards. Also, other regulations may apply when working in sensitive areas, such as historic sites or sites with endangered species habitats.

28.4 SUMMARY

The principal regulatory considerations for the deployment of electrokinetics at hazardous waste sites are CERCLA and RCRA. CERCLA hazardous substances encompass RCRA hazardous wastes as well as other toxic pollutants regulated by the CAA, CWA, and the TSCA. Electrokinetic treatment is most likely to be part of a remedial action and must address the criteria associated with such cleanups.

Legal, policy, and regulatory concerns are only one facet of the decision to deploy a particular technology. Other areas include stakeholder values and input, financial constraints and opportunities, political constraints and opportunities, technology performance and cost, site-specific information and characterization data, and future land use. To facilitate the acceptance of a technology, it is important to consider all of these areas and to include all stakeholders in the decision-making process.

REFERENCES

Acar YB, Gale RJ, Alshawabkeh AN, Marks RE, Puppala S, Bricka M, Parker RA. (1995). Electrokinetic remediation: basics and technology status. *Journal of Hazardous Materials* **40**:117–137.

Battle JB. (1986). *Environmental Law: Environmental Decision Making and NEPA*, Vol. 1. Cincinnati, OH: Anderson.

Clean Air Act (CAA) (1970). 42 USC 7401 et seq. Public Law 91-604, 84 stat 1676 (December 31, 1970).

Clean Air Act, Amended (1977). 42 USC 7401 et seq. Public Law 95-95, 91 stat 685 (August 17, 1977).

Clean Water Act (CWA) (1970). 33 USC 1251 et seq. Public Law 91-244, 80 stat 1246 (April 3, 1970).

Clean Water Act, Amended (1977). 33 USC 1251 et seq. Public Law 95-217, 90 stat 2139 (December 27, 1977).

Comprehensive Environmental Response, Compensation, and Liability Act (CERCLA). (1980). Public Law 96-510, 42 USC 9601 et seq, 94 stat 2767 (December 11, 1980).

Federal Insecticide, Fungicide and Rodenticide Act (1972). 7 USC 136 et seq. Public Law 92-516, 86 stat 973 (October 21, 1972).

Federal Insecticide, Fungicide and Rodenticide Act, Amended (1975). 7 USC 136 et seq. Public Law 94-140, 89 stat 751 (December 28, 1975).

Federal Insecticide, Fungicide and Rodenticide Act, Amended (1978). 7 USC 136 et seq. Public Law 95-376 (September 30, 1978).

Hazardous and Solid Waste Amendment (1984). 42 USC 6901 et seq. Public Law 98-616 (November 8, 1984).

Interstate Technology and Regulatory Cooperation (ITRC). (1997). *Emerging Technologies for the Remediation of Metals in Soils: Electrokinetics*. Metals in Soils Work Team Final Report. http://www.itrcweb.org/Documents/MIS-4.pdf.

National Environmental Policy Act (NEPA). (1970). Public Law 91-190, 42 USC 4321et seq, sec 2, 83 stat 852 (January 1, 1970).

Parker, RA. (1992). Research and development projects in the U.S. Environmental Protection Agency's Superfund Innovative Technology Evaluation (SITE) Program and their impact on hazardous waste remediation. In *Environmental Geotechnology* (eds. MA Usmen, YB Acar). Rotterdam: Balkema, pp. 559–562.

Resource Conservation and Recovery Act (RCRA). (1976). Public Law 94-580, 42 USC 6901 et seq, 90 stat 2795 (October 21, 1976).

Safe Drinking Water Act (1974). 42 USC 300 et seq. Public Law 93-523, 88 stat 1660 (December 12, 1974).

Safe Drinking Water Act, Amended (1977). 42 USC 300 et seq. Public Law 95-190, 91 stat 1387 (November 16, 1977).

Solid Waste Disposal Act (1965). 42 USC 6901-6991k. Public Law 89-272, 79 Stat. 992 (October 20, 1965).

Superfund Amendments and Reauthorization Act (SARA). (1986). Public Law 99-499, 100 stat 1615 (October 17, 1986).

Toxic Substances Control Act (TSCA). (1976). Public Law 19-469, 15 USC 2601 et seq (October 10, 1976).

United States Environmental Protection Agency (EPA). (2008a). *CERCLA: The hazardous waste cleanup program.* http://www.epa.gov/oswer/ (accessed April 4, 2008).

United States Environmental Protection Agency (EPA). (2008b). *National Priorities List.* http://www.epa.gov/superfund/sites/npl/ (accessed March 19, 2008).

United States Environmental Protection Agency (EPA). (2008c). *Summary of the Resource Conservation and Recovery Act.* http://www.epa.gov/lawsregs/laws/rcra.html (accessed April 10, 2008).

United States Environmental Protection Agency (EPA). (2008d). *Summary of the Clean Air Act.* http://www.epa.gov/lawsregs/laws/caa.html (accessed April 1, 2008).

United States Environmental Protection Agency (EPA). (2008e). *Summary of the Clean Water Act.* http://www.epa.gov/lawsregs/laws/cwa.html (accessed April 10, 2008).

United States Environmental Protection Agency (EPA) SITE Program. (1995). *Sandia National Laboratories in-situ electrokinetic extraction technology.* Innovative Technology Evaluation Report. http://www.epa.gov/ORD/SITE/, EPA/540/12-97/509 (accessed April 14, 2008).

United States Environmental Protection Agency (EPA) SITE Program. (2003). *Electrochemical Design Associates, lead recovery evaluation, building 394 battery shop; Pearl Harbor Naval Shipyard and Immediate Maintenance Facility.* Innovative Technology Evaluation Report. http://www.epa.gov/ORD/SITE/, EPA/540/R-04/506 (accessed April 14, 2008).

PART IX

FIELD APPLICATIONS AND PERFORMANCE ASSESSMENT

29

FIELD APPLICATIONS OF ELECTROKINETIC REMEDIATION OF SOILS CONTAMINATED WITH HEAVY METALS

Anshy Oonnittan, Mika Sillanpaa, Claudio Cameselle, and Krishna R. Reddy

29.1 INTRODUCTION

Electrokinetic remediation is an environmental technique that has been studied extensively at laboratory scale since the late 1980s and has demonstrated several successes in the elimination of heavy metals from soils, sludges, and solid wastes. The removal of heavy metals such as Cu, Cr, Hg, Pb, Zn, Mn, Cd, and others (Hansen et al., 1997; Ottosen et al., 2001; Reddy, Xu, and Chinthamreddy, 2001; Ricart et al., 2004), nonmetallic species such as As (Le-Hécho, Tellier, and Astruc, 1998; Kim, Kim, and Kim, 2005), and inorganic anions (Pomes et al., 1999; Yang, Hung, and Tu, 2008) from soils, sediments, and solid wastes has been the objective of many research studies. In the very beginning, kaolinite was used to model low-permeability soils, and several studies were realized with heavy metal-spiked kaolinite. Those studies were useful to assess the electromigration of heavy metals in soils and to optimize the operating conditions for their removal. However, the data obtained in the laboratory with spiked kaolinite could not always be extrapolated and used for real polluted soil and in field applications, and researchers started to test at bench scale soils, sediments, and industrial wastes from polluted sites. These bench scale studies established the feasibility of electrokinetics for the remediation of polluted soil and the operating conditions obtained at bench scale were the starting point for a large scale treatment.

The first field-scale application of electrokinetics for soil remediation was carried out by Geokinetics in 1987 (Lageman, 1993), although it was previously reported

Electrochemical Remediation Technologies for Polluted Soils, Sediments and Groundwater,
Edited by Krishna R. Reddy and Claudio Cameselle
Copyright © 2009 John Wiley & Sons, Inc.

that electrokinetics was used in the former Soviet Union since early 1970s to concentrate metals and to explore for minerals in deep soils (USEPA, 1997). The subsequent works carried out by Geokinetics were a big leap into the practical field-scale implementation of electrokinetics for soil remediation. Apart from these, the field-scale feasibility study of electrokinetic remediation by Banerjee et al. (1989) undertaken at a Superfund site at Corvallis, Oregon, was one among the other early field-scale studies. Thereafter, extensive laboratory-scale studies and research paved the way to understand the principles and parameter effects on the process (Acar and Alshawabkeh, 1993; Reddy et al., 1997; Alshawabkeh, Yeung, and Bricka, 1999a; Chung and Kang, 1999; Ottosen et al., 2001; Reddy and Chinthamreddy, 2003; Zhou et al., 2005).

Laboratory research during the last two decades clearly demonstrated the feasibility of electrokinetics in removing a wide range of heavy metals from lightly as well as heavily contaminated soils (Reddy et al., 1997; Chung and Kang, 1999; Ottosen et al., 2001; Reddy and Chinthamreddy, 2003; Zhou et al., 2005). Thus, electrokinetics was established as a promising technology for soil remediation of, and especially fine-grained ones, where other remediating techniques failed. However, the field-scale trials were not that successful as the bench-scale test and resulted in unpredictable performance due to the heterogeneities in the soil, presence of metallic objects, low moisture content, aging of the contaminants, and also due to the lack of experience in the implementation of the technology in the field (Ottosen et al., 2002; Virkutyte et al., 2002). The major focus of laboratory-scale experiments was aspects such as feasibility of the technology with different metals, simultaneous presence of metals, soil types, enhancement of existing technologies, and so on. The complex nature of the field conditions and the inability to predict the field performance using available laboratory data led to the unexpected performance in the field (Lageman, Clarke, and Pool, 2005).

This chapter reviews the field-scale applications of electrokinetics for metal removal from soil. Focus is given on some of the major completed field demonstrations including actual cleanup projects, the variations in the processes used, the outcome of field performances, and the conclusions based on these.

29.2 DESCRIPTION OF PROCESSES INVOLVED IN FIELD APPLICATIONS

Electrokinetics involves the application of low-level direct current (DC) between electrodes placed in a contaminated area. Different variations of this process were developed to suit the needs of each case. The processes adopted at each site differ with each other in one or many aspects. Basically, two approaches are defined depending on the type of contaminant. The first approach is the enhanced removal in which the contaminants are transported by electromigration and/or by electroosmosis toward the electrodes for subsequent removal, and the second approach is the treatment without removal, which involves the electroosmotic transport of contaminants through the treatment zones and may also include the frequent reversal of polarity of electrodes to control the direction of contaminant movement (USEPA, 1997). The first approach is applicable for the removal of heavy metals, whereas the second approach was developed for the removal of organic species from contaminated land. Table 29.1 shows a list of some of the major completed field-scale projects. A brief description of the processes adopted for these field projects is given below.

29.2.1 Electroreclamation, Geokinetics International, Inc.

The application of electric field for soil remediation known under different names was first implemented in field scale in 1987, and the technology, then called electroreclamation, was developed by Geokinetics (the Netherlands). This process, which was used for some field trials and actual cleanup projects, was based on the basic electrokinetic phenomena without any amendments or enhancements by the addition of other chemicals. Control and management of electrolyte conditions and pH were the key features of this technology to improve the transportation and elimination of heavy metals from the polluted soil (Lageman, Clarke, and Pool, 2005). The electrode configuration used in these systems was also simple with vertically or horizontally installed anodes and cathodes in the polluted sites, drilling wells or tunnels or trenches around the polluted area. The process was used to remediate soil contaminated with a wide range of metals including copper, zinc, arsenic, cadmium, chromium, lead, and nickel. This process can also be used to fence off hazardous waste sites or hazardous industrial sites, avoiding the uncontrolled spread of the pollutants to the surrounding areas.

29.2.2 *In Situ* Electrokinetic Extraction (ISEE) System, Sandia National Laboratories (SNL)

In 1996, as part of the Superfund Innovative Technology Evaluation (SITE) Program, the United States Environmental Protection Agency (USEPA) demonstrated the ISEE system at the SNL chemical waste landfill site in Albuquerque, New Mexico (USEPA, 1998). The ISEE system was developed by SNL for removing hexavalent chromium from unsaturated soil. This was a cutting edge, since most of the laboratory-scale studies were carried out in saturated soil samples. In a saturated sample, the contact of the interstitial fluid with the solid particles is more effective and aids in the extraction and transportation of pollutants. The two primary transport mechanisms in electrokinetics (electromigration and electroosmosis) require a liquid medium (water), but in the unsaturated soil zone, the lack of water in the interstices makes the solubilization and transportation of the heavy metals precipitated or adsorbed on the solid particles surface more difficult.

The technology involves the *in situ* application of DC to soil through a row of anodes surrounded by two or four rows of cathodes. The distance between the electrodes was 3 ft (about 1 m) and the active zone was up to a depth of about 4.3 m. The anode fluid was circulated for chemical conditioning and pH control. The main feature of the ISEE technology was that the lysimeter technology was used in the construction of electrodes to create a continuum between the electrolyte and pore water. The electrode fluid is held inside the electrode by an applied vacuum, keeping the fluid from saturating the soil. This allowed the removal of chromium from unsaturated soil without significantly altering the soil moisture content.

The ISEE technology is applicable for treating unsaturated soil contaminated with hexavalent chromium. According to SNL, this technology can be modified to

TABLE 29.1. Reported Field Application of Electrokinetic Remediation of Heavy Metal Contaminated Soils

Site Location	Treatment Objective	Treatment Scheme	Contaminants and Concentration Range	Reference
Naval Air Weapons Station (NAWS), Point Mugu	Field demonstration	ERDC treatability study predictions of electrokinetic extraction performance	Chromium:180–1100 mg/kg Cadmium:5–20 mg/kg	Gent et al.(2004)
Honolulu, Hawaii Phase I–2001 Phase II–2002	Site demonstration in two phases using lead-contaminated soil from the location	EDA's lead recovery technology evaluation	Lead: 82,300–8270 mg/kg	USEPA (2003)
NAWS, Point Mugu, 1998	Field demonstration	Validation of ERDC treatability study predictions of electrokinetic extraction performance	Chromium: 25,100 mg/kg Cadmium: 1810 mg/kg	USAEC (2000)
Albuquerque, New Mexico, 1996	Site demonstration to remove hexavalent chromium from unsaturated soil	In situ electrokinetic extraction (ISEE) system	Chromium	USEPA (1998)

The Netherlands	Project 1–Groningen, 1987	Pilot test to remove heavy metals such as lead and copper from peat soil	Electroreclamation process	Lead: 300–5000 ppm Copper: 500–1000 ppm	Lageman (1993)
	Project 2–Delft	Pilot test to remove zinc from a site of galvanizing plant	Electroreclamation process	Zinc: 7010-ppm maximum concentration, with an average of 2410 ppm	
	Project 3–Loppersum, 1989	Actual cleanup of the site to remove arsenic from contaminated heavy clay soil	Electroreclamation process	Arsenic: 400–500 ppm	
	Project 4–Stadskanal, 1990–1992	Cleanup of temporary landfill of soil and sludge contaminated with cadmium and other heavy metals	Electroreclamation process	Cadmium: 2–3400 mg/kg	
	Project 5–Woensdrecht, 1993	Cleanup of temporary depot filled with clay and peat contaminated with heavy metals	Electroreclamation process	Cadmium: 660 mg/kg Lead: 730 mg/kg Copper: 770 mg/kg Nickel: 860 mg/kg Chromium: 7300 mg/kg Zinc: 2600 mg/kg	
Corvallis, Oregon, 1989		Field study to evaluate the technical feasibility of electrokinetic remediation technique to treat a site contaminated with chrome plating wastes	Electrokinetic remediation technique	Chromium	Banerjee et al. (1989)

treat saturated contaminated soil and to remove contaminants dissolved in the pore water. One major shortcoming of this process was that it did not have a system for contaminant recovery and disposal. Therefore, the waste had to be disposed of as hazardous waste; however, the waste treatment system can be designed and installed separately for each application considering the physicochemical properties of the heavy metals in these sites.

29.2.3 Electrokinetic Remediation, United States Army Environmental Center (USAEC) and Engineer Research and Development Center (ERDC)

A field demonstration of electrokinetic remediation was conducted by the USAEC and the ERDC at the Naval Air Weapons Station (NAWS), Point Mugu, California, in 2000 (USAEC, 2000). Major contaminants in this area were chromium and cadmium, and the system was developed with the aim to reduce the contaminant concentrations below the regulatory action levels for metal concentration and toxicity criteria. The developed system was basically the same as other existing electrokinetic remediation processes. It involved the use of citric acid as an amendment to control the formation of the pH front in the treatment area. The design included a system for the recovery of the electrolyte after separating the metal contaminants.

29.2.4 Lead Recovery Technology, Electrochemical Design Associates (EDA)

Another endeavor of the EDA (formerly Geokinetics International, Inc.), the lead recovery technology, was evaluated by the USEPA SITE Program to assess the ability of electrokinetics to remove lead from soil. The demonstration was held at Building 394 Battery Shop, Pearl Harbor Naval Shipyard and Intermediate Maintenance Facility, Honolulu, Hawaii, in 2001 and 2002 (USEPA, 2003).

This technology uses a rectangular tank that allows the flux of the injecting solution vertically and horizontally. This type of operation (*ex situ*) is similar to the *in situ* treatment of polluted soil with the advantage that it eliminates the problems caused by loss of fluids and potential mobilization of the contaminants (in this case Pb) beyond the treatment area. The innovative feature in this technology was that ethylenediaminetetraacetic acid (EDTA) was used in the electrolyte solution as a chelating agent to remove lead as soluble Pb–EDTA complex from the soil. The system was operated as a batch closed-loop process, and the soluble Pb–EDTA complex formed in the electrolyte was removed from the treatment tank. The electrolyte solution was then treated to remove the Pb and to recover the EDTA that was reused after lead recovery. So, the advantage of this system is the effective removal of lead from polluted soils with no generation of harmful wastes, and with the recovery and reuse of both: the extracting/complexing agent (EDTA) and the contaminant itself.

29.2.5 Electrosorb Process, Isotron Corporation

A pilot-scale demonstration of electrokinetic extraction was conducted by Isotron Corporation at the Oak Ridge K-25 facility in Tennessee. The developed process, called electrosorb process, uses a patented cylinder to control buffering conditions *in situ* and an ion exchange polymer matrix called Isolock to trap metal ions. The tested soil was polluted with mercury, lead, and chrome. Preliminary results indicate that at the 5-ppm mercury concentrations found at the site, mercury was difficult to remediate; however, the process showed good results on lead and chrome.

29.3 ELECTROKINETIC REMEDIATION SETUP IN FIELD APPLICATIONS

All the above-mentioned processes followed the fundamental electrokinetic techniques, but with some modifications in the process either by adding additional chemicals or in the configuration or construction of electrodes. From the cited literatures, an electrokinetic remediation setup essentially consists of the following elements.

29.3.1 Electrode System

The electrode system includes an array of electrodes, wells for electrodes, and other site- and process-specific equipments for the effective installation of electrodes. The electrodes can be installed vertically or horizontally. In order to cover large areas, rows of anodes and cathodes are installed one in front of each other, although the disposition with one cathode surrounded by anodes is also common. The latter disposition permits the concentration of the heavy metals in a small area around the cathode. Considerations are to be given in the selection of electrode material, choosing the optimum electrode spacing, layout, and configuration, and the right choice of these factors results in achieving uniform voltage distribution and minimum effective power loss. These aspects are explained in detail by Alshawabkeh, Yeung, and Bricka (1999a) and by Alshawabkeh *et al.* (1999b).

29.3.2 Power Distribution System

Electric power is applied to the electrode system by a power supply. Several units can be connected either in series or in parallel to give the desired power output. The power units can be operated under constant voltage or current conditions in pulsed or in continuous mode. The most common operation is under constant voltage. Thus, the driving force (i.e. the electric field) for the transportation of ions by electromigration remains constant, and the electric current intensity evolutes along the treatment time as a function of the electrical resistance of the system, which includes the resistance of soil, the resistance of the electrolytic solutions, and the resistance upon the electrode surface. The major resistance corresponds to the soil. The electric resistance in the system can be interpreted as a measure of the available ions in solution that migrate, transporting the electric charges from one electrode to the other. The operation at electric current intensity implies the continuous variation of the voltage to assure a constant transport of charge. In any case, the resistivity of the system varies along the treatment time, and therefore, efficient control of the power system is necessary for the operation.

29.3.3 Circulation System

Pollutants present in the soil are recovered from the electrode solutions and also, the chemicals to enhance the separation and transportation of pollutants are supplied to the electrode solutions. Therefore, a circulation system is important for the management of the electrode solution and to maintain the parameters like temperature and pH of the anolyte as well as the catholyte. A proper circulation control system enables complete mixing of electrode solutions and sample withdrawal also if necessary. It might also include a process piping to distribute any chemical amendments to electrode wells and to extract the contaminants from the electrode solutions (e.g. precipitation and ionic exchange).

29.3.4 Contaminant Recovery and Disposal

The contaminant recovery and disposal system forms an integral part of the whole electrokinetic setup. Generally, the contaminants that are concentrated in the electrode wells are recovered by processing the electrolyte. After regeneration, the electrolytes can be reused, thereby improving the economic balance of the process and reducing the amount of wastes. Contaminants can be recovered by other methods also, such as electroplating, adsorption to electrodes, the use of ion exchange resins, and so on (Gent *et al.*, 2004).

29.3.5 Others

Apart from the above-mentioned systems, an electrokinetic setup might consist of other ancillary systems depending upon the site-specific characteristics, such as off-gas extraction and treatment systems, by-product monitoring systems, and so on.

29.4 OUTCOME OF FIELD-SCALE EXPERIMENTS

29.4.1 Geokinetics Projects

Despite the fact that there were some major hindrances during the operation, some of the projects in literature were able to meet their remediation goals at least partially. This was true especially in the case of Geokinetics, which has pioneered in this field in Europe. The results of its field projects as published show that electrokinetics is a promising technology for the effective remediation of soil (Lageman, 1993). The field trials held in the Netherlands were for the removal of metals like lead, copper, zinc, arsenic, cadmium, nickel, chromium, and so on from different types of soil like peat and sand.

The first demonstration in Groningen in 1987, which aimed at removing lead and copper from peat soil, was a great success. The contaminants were leached down from the waste sludge from a paint production factory. The sludge was heavily polluted with metals in the form of solid particles. The Pb concentration was in the order between 300 and more than 5000 ppm, and Cu concentrations were in the order of 500–1000 ppm. The test area was 70-m long and 3-m wide. The electrode setup consisted of one horizontal cathode and a row of vertical anodes at a depth of 1 m. The process resulted in a removal of 70% lead and 80% copper. The main factor that contributed to the success of the process was the soil type in the treatment area. The relatively low pH of the peat soil (pH 4) facilitated the mobilization of the metals, necessitating a low-energy inducement for the extraction and dissolution of the heavy metals. However, some difficulty was encountered due to the presence of soil sludge layer containing paint particles which acted as new sources of pollution. This shows that a careful site examination prior to the field trial would have led this project to a complete success. The energy consumption for the removal of Cu and Pb was estimated to be 85 kWh/m^3.

In another project held in Delft in a site polluted by a galvanizing plant, the buffering capacity of soil made the remediation effort fail to meet the goal, which was to reduce the zinc concentration in the area below the regulatory limits. Therefore, the soil type made the project successful in the former case and was the cause of failure in this case. This leads to a logical conclusion that it is impossible to follow a "one-for-all" procedure in electrokinetic remediation. However, it was concluded that remediation would be possible by an additional energy inducement to bring the soil pH to 3–4, which in this case means to extend the treatment duration from 8 to 24 weeks, and to increase the electric consumption up to 320 kWh/ton of soil.

This project was followed by an actual cleanup of an arsenic contaminated site in Loppersum in 1989, which was a partial success with three-fourths of the total soil volume remediated. Here, the presence of uncovered metallic objects inhibited the process. These metallic objects served as a path for the electric current and delayed the movement of metallic contaminants in the area.

A similar kind of problem due to the presence of uncovered concretions of CdS acted as the bottleneck in the removal of cadmium during the cleanup of a temporary landfill in Stadskanal in 1990. Though initially the process was progressing well, there was an unexpected increase in the cadmium concentrations, and further analysis revealed the presence of these concretions, which were not identified during the earlier tests. The soil in this area was then pretreated to remove these, and further experiments yielded the desired results.

29.4.2 ISEE System

In the USA, the first ever attempt to evaluate the technical feasibility of electrokinetic remediation was the field study at a Superfund site in 1988 (USEPA, 1998). The technology was studied in combination with pump and treat. The process was adopted for the remediation of a site heavily contaminated with chromium. This was carried out when the knowledge in electrokinetics was not sufficiently developed. However, the results were much encouraging, indicating the potential of electrokinetics as an effective means for the removal of metals from soil by desorbing the ions attached to the soil surface.

The SITE (SITE Program) demonstration of ISEE technology at Albuquerque, New Mexico, was specifically aimed for treating unsaturated soil contaminated with hexavalent chromium at a Superfund site. The system was evaluated from May 15 to November 24, 1996. Several tests were carried out, and the test areas ranged from 3.3- to 6.7-m^2 area and the depth ranged from 2.5 to 4.5 m. The technology involves the *in situ* application of DC to soil. Chromate ions are extracted in the anode effluent. The initial concentrations of Cr(VI) ranged from 0.4 to 6890 mg/kg and total chromium ranged from 7.7 to 26,800 mg/kg. These differences were interpreted as Cr(III). During the electrokinetic treatment, changes in the concentrations of Cr(III) in the specimen were observed in addition to the inherent variability of chromium concentration in the test areas, but no conclusion was drawn regarding the ISEE system's ability to remove trivalent chromium.

The demonstration results indicate that approximately 200 g of hexavalent chromium was removed during about 700 h of system operation, and the average removal rate for the entire system was approximately 0.29 g/h. Nevertheless, a reasonable estimate of the chromium removal based on the pre- and posttreatment soil analysis was not possible in this case due to the nonhomogenous distribution of chromate concentration in pretreated soil and because the demonstration was not carried out until the completion.

The total treatment costs for the ISEE system, which removed 200 g of hexavalent chromium from 16 yd^3 of soil, are estimated to be $1800 per cubic meter. The estimate will vary depending on cleanup goals, soil type, treatment volume, and system design changes. Economic data indicate that soil remediation costs are very high, perhaps because the system demonstrated still requires significant improvements.

29.4.3 NAWS, Point Mugu, California

Another field application was conducted at NAWS, Point Mugu, California (USAEC, 2000) to remediate a soil polluted with Cr and Cd in the presence of brackish water. The total volume treated in the field demonstration was 64 m^3. A new disposition was used in which three cathodes were centered between six anodes. The objective of this was to concentrate the contaminants in the center of the treated area around the cathodes. The anode–cathode spacing was 4.5 m and they were inserted into the ground 3-m deep. A constant voltage of 60 V (~13 V/m) was applied for 20 days and then was reduced to 45 V (10 V/m) for 6 months. Considering the distance between the electrodes, the voltage drop applied in this field test is very high when compared to the standard value of 1 V/cm. Citric acid was used to maintain the cathode pH at 4. After 6 months of treatment, 78% of the soil volume was cleared of chromium (the initial concentration was between 180 and 1100 mg/kg) or was treated to below natural levels. The results also indicated that 70% of the soil between the electrodes was cleared of cadmium contamination (initial concentration was 5–20 mg/kg).

Understanding the lack of necessary field-scale implementation data, during the field demonstration at NAWS, Point Mugu, an extensive laboratory testing was conducted to assess the potential effectiveness of electrokinetic remediation at the site. The laboratory investigations indicated that electrokinetics could be successfully applied to the site. However, the field performance resulted in poor performance and contaminant reduction goals were not met. The initial laboratory tests failed to assess the effects of site water and conditions on the process performance. The chloride content of the surrounding soil and site water resulted in continuous production of chlorine gas, which inhibited the formation of pH front between the electrodes, and hence the movement of metal contaminants was retarded. Furthermore, by-products were generated due to the chemical amendments used. The study was then transitioned to a pilot-scale project and the electrokinetic system was restarted with a reduced number of electrodes, thus inducing a higher current density, making the contaminant removal possible in the brackish conditions (Gent et al., 2004).

29.4.4 Pearl Harbor Naval Shipyard and Intermediate Maintenance Facility, Honolulu, Hawaii

This study evaluated the technology from EDA to remediate a site polluted with Pb. The study was funded by the EPA through the SITE Program (USEPA, 2003). The study was carried out in two stages in 2001 and 2002, which lasted 1.5 and 2.5 months, respectively. The tests were carried out in a tank of 3-m^3 capacity, and a solution of EDTA was injected to dissolve and complex the Pb as Pb–EDTA. The complex can migrate horizontally and vertically in the cell. The solution is removed and treated for the extraction of lead, which is then reused in the process in order to limit the comsumption of EDTA. Three different tests were run at different initial Pb concentrations that ranged from 8 to 82 g/kg. The primary objective of the technology was to reduce the concentration of Pb below 2000 mg/kg, but unfortunately, this goal was only achieved in 6% of the samples (1 of 18 posttreatment samples). The average lead removal efficiency was 59% for all the tests. This was also due to the retention of the Pb–EDTA complex on the surface of the soil particles. However, the extensive flushing effectively removed sorbed solution, yielding a removal of 81%.

In conclusion, the EDA technology effectively removes lead with a high level of treatment efficiency and it does not destroy or degrade the physical properties of the affected media. It also presents minimal short-term risks to workers and nearby community, including noise and exposure to airborne contaminants.

The technology is a mobile system that can be transported anywhere and can be applied to any amount of material. There is no maximum or minimum volume. System is not labor intensive and requires minimal physical effort, although the equipment cannot be left to run unattended.

Based on the results of this study, the estimated cost for the application of this technology is $45,063 per ton for treating 5 tons of soil and $28,252 per ton for treating 25 tons of soil. These costs include personnel, equipment and supplies, chemicals, and technical support. But the authors stressed that the real cost is site specific and depends on the media to be treated, severity of contamination, contaminant, and amount of material to be treated.

29.5 FACTORS THAT LIMIT THE APPLICABILITY OF ELECTROKINETIC TECHNOLOGY

Electrokinetics is applicable to soils of low hydraulic permeability and is efficient when the cation exchange capacity and salinity are low; if not, the large amount of ions present in the soil competes with the pollutants for the charge transportation, requiring much more charge transport (more electric power consumption) for the same removal. The technology is especially useful for remediating inaccessible sites (e.g. under buildings, factories) with a minimum disruption on the surface, where other technologies fail; in fact, the electrokinetic treatment is able to induce hydraulic flow (electroosmosis) in low-permeability soils where the hydraulic gradient is ineffective. Nevertheless, there are several factors that limit the technology performance in the field. They are, but not limited to, the following list.

29.5.1 Soil Heterogeneities

The soil structure and its heterogeneities are one of the major factors that limit the applicability of a treatment process to a specific site. Heterogeneities of soil in the form of rocks or other uncovered objects can cause a discontinuity in the current flow path, which results in poorly remediated zones. On the other hand, the presence of conducting materials in the soil would also adversely affect the technology performance by acting as preferential flow paths for the current, leaving the surrounding soil unaffected by the treatment, or even the dissolution of metallic objects into the soil acts as a new source of ions that generates more pollution and more electric power consumption.

29.5.2 Remediation Time

It is impossible/difficult to predict the time required for complete remediation. Laboratory tests could give an approximate time for remediation but may not always give a true indication due to the different conditions of the test in the laboratory and in the actual field. In the laboratory test, electroosmotic flux and electromigration are defined by the placement of the electrodes and the shape and configuration of the experimental setup. Moreover, the soil sample is commonly homogenized before testing. In the open field, the expected electroosmotic flow and electromigration is affected by soil heterogeneity, flow of groundwater, and even weather conditions. Furthermore, the stagnant zones present in the field would also affect the migration rate of metal species.

29.5.3 Formation of By-Products

The electric current generates at the electrodes several products that can dramatically affect the electrokinetic treatment. The effect of water electrolysis is well known, but any type of enhancements by the addition of other chemicals should consider its potential for the generation of harmful/hazardous by-products. By-products can also be formed by certain undesirable reactions of the metal species with naturally occurring ions. Therefore, the selection of any chemical for pH control or for heavy metal complexing must consider the possible reactions upon the electrodes and into the soil with the pollutant or other components of the soil.

29.5.4 Species Development

A number of factors affect the development of metal species, especially in the field in the presence of naturally occurring ions, which in turn affect the solubility of the contaminants and their desorption from the soil. Enhancement of the electrokinetic treatment with the chemical conditioning of the electrode solutions and/or the addition of chemicals to the soil aids in the extractability of heavy metals and in the formation of metal complexes to stabilize the metal in solution even in adverse pH conditions. The process was found to be less efficient when the nontarget ion concentration is higher than the target ion concentration due to the competition of the different metallic species (GWRTAC, 1997).

29.5.5 Choice of Electrodes

There is little understanding on the basis for choosing the right electrodes. The electrode material should stand the deteriorating conditions that could develop during the process due to electrolytic or other reactions. Furthermore, the electrode materials should not introduce any residues to the soil. The positioning of electrodes is also necessary to extract contaminants from the entire treatment area, and the electric field developed should not cause heating and potential losses. Despite the importance of the electrodes in the process, there are few studies about the spatial disposition of the electrodes and its influence in the remediation (Alshawabkeh *et al.*, 1999b). Usually, noble materials such as Pt, platinized titanium, or coated titanium are used as the electrode material when the interest is the stability of the electrode. However, in order to reduce the cost of the process, materials such us graphite, iron, or stainless steel are used for their lower cost and availability.

29.5.6 Soil Saturation

Electrokinetics is applicable to both saturated and unsaturated soils. However, the soil moisture content should be high enough for electromigration to take place and at the same time less than saturation to avoid the competing effects of tortuosity and pore water content (GWRTAC, 1997).

29.6 PREREQUISITES AND SITE INFORMATION NEEDED

Before implementing electrokinetics, it is necessary to find if the site conditions match the requirements necessary for the successful performance of the technology. This can be done by a number of field and laboratory screening tests:

- *site screening*—to identify any stagnant zones or buried metallic objects in the site;
- *electrical conductivity surveys*—to determine the soil as well as pore water electrical conductivities;
- *pH*—to determine both soil and pore water pH; the pore water pH affects the contaminant characteristics such as valence, solubility and sorption;
- *chemical analysis of pore water*—to identify the presence of competing ions; and
- *chemical analysis of the soil*—to determine the chemical characteristics of the soil such as its buffering capacity, salinity, cation exchange capacity, organic content, and so on, and to evaluate the potential for by-product formation.

Based on the above information, the following is a list of data needed prior to the implementation of the technology (ITRC, 1997; Kim, Han, and Cho, 2002):

1. General data
    i. Contaminated area size and depth
    ii. Depth to water table
    iii. Soil type and grain size
    iv. Moisture content profile
2. Chemical data
    i. Contaminant type and its solubility
    ii. Contaminant concentration and distribution
    iii. Soil permeability
    iv. Total ionic concentration
    v. Cation exchange capacity
3. Other data
    i. Maximum amount of electrical current
    ii. Surface area of electrode plates
    iii. Electroosmotic permeability
    iv. Salinity
    v. Contaminant transference number
    vi. Identification of half-cell potentials

29.7 ADVANTAGES AND DISADVANTAGES OF THE TECHNOLOGY

The major advantages of electrokinetic technology are that it can be used

- to treat low-permeability soil,
- as an *in situ* treatment,
- to treat inaccessible areas that cannot be excavated,
- to control the flow of water and contaminants,
- to treat the entire soil mass between the electrodes,
- in conjunction with other techniques,
- to treat both saturated and unsaturated soil, and
- to treat multiple contaminants in one operation.

Though electrokinetics is being advocated as an effective treatment mechanism, a lot of process-related issues, such as its ability to achieve the remediation goals, are yet to be addressed. Some of the major disadvantages of the technology are

- precipitation of metal species near the cathode,
- problems due to electrode corrosion, and
- excess soil heating.

29.8 SUMMARY

In most of the site demonstrations, the major hindrance was the site characteristics rather than any shortcomings in the technology. In spite of the fact that the projects undertaken have been tested or performed after detailed investigations and laboratory treatability tests, there were some missing factors that ultimately inhibited the field performance. This could be due to inexperience in field trials and the lack of experimental data. The number of completed field projects is very few till date. Moreover, there were discrepancies between the laboratory studies and performance in the field.

In summary, an analysis of these field demonstrations reveals that a detailed study of contaminated sites is necessary for the successful application of the technology. Since no sites are alike, a case-to-case study is relevant in each case. Pretreatment of soil is also to be undertaken if necessary. Methods should be further developed to be more specific and selective as the success of the field remediation is largely operationally defined.

REFERENCES

Acar YB, Alshawabkeh A. (1993). Principles of electrokinetic remediation. *Environmental Science and Technology* **27**(13):2638–2647.

Alshawabkeh AN, Yeung A, Bricka RM. (1999a). Practical aspects of in situ electrokinetic remediation. *Journal of Environmental Engineering* **125**(1):27–35.

Alshawabkeh AN, Gale RJ, Ozsu-Acar E, Bricka RM. (1999b). Optimization of 2-D electrode configuration for electrokinetic remediation. *Journal of Soil Contamination* **8**(6): 617–635.

Banerjee S, Horng JJ, Ferguson JF, Nelson PO. (1989). *Field Scale Feasibility Study of Electrokinetic Remediation*. Cincinnati, OH: USEPA.

Chung HI, Kang BH. (1999). Lead removal from contaminated marine clay by electrokinetic soil decontamination. *Engineering Geology* **53**:139–150.

Gent DB, Bricka RM, Alshawabkeh AN, Larson SL, Fabian G, Granade S. (2004). Bench and field-scale evaluation of chromium and cadmium extraction by electrokinetics. *Journal of Hazardous Materials* **110**:53–62.

GWRTAC. (1997). Electrokinetics. *Technology Overview Report*. Groundwater Remediation Technologies Analysis Center, Pittsburgh, PA.

Hansen HK, Ottosen LM, Kliem BK, Villumsen A. (1997). Electrodialytic remediation of soils polluted with Cu, Cr, Hg, Pb and Zn. *Journal of Chemical Technology and Biotechnology* **70**(1):67–73.

ITRC. (1997). Emerging Technologies for the Remediation of Metals in Soils. *Electrokinetics*. Final Report. Interstate Technology and Regulatory Cooperation. http://www.itrcweb.org/Documents/MIS-4.pdf

Kim SS, Han SJ, Cho YS. (2002). Electrokinetic remediation strategy considering ground strata: A review. *Geosciences Journal* **6**:57–75.

Kim S-O, Kim W-S, Kim K-W. (2005). Evaluation of electrokinetic remediation of arsenic-contaminated soils. *Environmental Geochemistry and Health* **27**(5–6):443–453.

Lageman R. (1993). Electroreclamation: Applications in the Netherlands. *Environmental Science and Technology* **27**(13):2648–2650.

Lageman R, Clarke RL, Pool W. (2005). Electro-reclamation, a versatile soil remediation solution. *Engineering Geology* **77**:191–201.

Le-Hécho I, Tellier S, Astruc M. (1998). Industrial site soils contaminated with arsenic or chromium: Evaluation of the electrokinetic method. *Environmental Technology* **19**(11): 1095–1102.

Ottosen LM, Eriksson T, Hansen HK, Ribeiro AB. (2002). Effects from different types of construction refuse in the soil on electrodialytic remediation. *Journal of Hazardous Materials* **91**(1–3):205–219.

Ottosen LM, Hansen HK, Ribeiro AB, Villumsen A. (2001). Removal of Cu, Pb and Zn in an applied electric field in calcareous and non-calcareous soils. *Journal of Hazardous Materials* **85**(3):291–299.

Pomes V, Fernandez A, Costarramone N, Grano B, Houi D. (1999). Fluorine migration in a soil bed submitted to an electric field: Influence of electric potential on fluorine removal. *Colloids and Surfaces A: Physicochemical and Engineering Aspects* **159**(2–3):481–490.

Reddy KR, Chinthamreddy S. (2003). Effects of initial form of chromium on electrokinetic remediation in clays. *Advances in Environmental Research* **7**:353–365.

Reddy KR, Parupudi US, Devulapalli SN, Xu CY. (1997). Effects of soil composition on the removal of chromium by electrokinetics. *Journal of Hazardous Materials* **55**:135–158.

Reddy KR, Xu CY, Chinthamreddy S. (2001). Assessment of electrokinetic removal of heavy metals from soils by sequential extraction analysis. *Journal of Hazardous Materials* **84**(2–3):279–296.

Ricart MT, Hansen HK, Cameselle C, Lema JM. (2004). Electrochemical treatment of a polluted sludge: Different methods and conditions for manganese removal. *Separation Science and Technology* **39**(15):3679–3689.

USAEC. (2000). In Situ Electrokinetic Remediation of Metal Contaminated Soils Technology Status Report. US Army Environmental Center. *Report Number SFIM-AEC-ET-CR-99022*.

USEPA. (1997). *Recent Developments for In Situ Treatment of Metal Contaminated Soils*. Washington, DC: Office of Solid Waste and Emergency Response, Technology Innovation Office. 20460.

USEPA. (1998). In Situ Electrokinetic Extraction System. SITE Technology Capsule. United States Environmental Protection Agency, Office of Research and Development, Cincinnati, OH. *Report Number EPA/540/R-97/509a*.

USEPA. (2003). Electrochemical Design Associates, Lead Recovery Technology Evaluation. United States Environmental Protection Agency, Office of Research and Development, National Risk Management Research Laboratory, Cincinnati, Ohio. *Report Number EPA/540/R-04/506*.

Virkutyte J, Sillanpää M, Latostenmaa P. (2002). Electrokinetic soil remediation—Critical overview. *Science of the Total Environment* **289**(1–3):97–121.

Yang GCC, Hung C-H, Tu H-C. (2008). Electrokinetically enhanced removal and degradation of nitrate in the subsurface using nanosized Pd/Fe slurry. *Journal of Environmental Science and Health Part A* **43**(8):945–951.

Zhou DM, Deng CF, Cang L, Alshawabkeh AN. (2005). Electrokinetic remediation of a Cu–Zn contaminated red soil by controlling the voltage and conditioning catholyte pH. *Chemosphere* **61**:519–527.

30

FIELD STUDIES: ORGANIC-CONTAMINATED SOIL REMEDIATION WITH LASAGNA TECHNOLOGY

CHRISTOPHER J. ATHMER AND SA V. HO

30.1 INTRODUCTION

The use of electrokinetics as an *in situ* method for soil remediation, the technology focus of this book, has received much attention due to its unique applicability to low-permeability soils (Casagrande, 1952; Shapiro, Renaud, and Probstein, 1989; Hamed, Acar, and Gale, 1991; Acar, Li, and Gale, 1992; Bruell, Segall, and Walsh, 1992; Segall and Bruell, 1992; Acar and Alshawabkeh, 1993; Lageman, 1993; Shapiro and Probstein, 1993). Electrokinetics includes transport of water (electroosmosis) and ions (electromigration) as a result of an applied electric field. Electromigration is the more common application of electrokinetics for treating metal-contaminated soils in which the charged metal species are driven by an applied direct current (DC) toward the electrodes where they are removed. For uncharged organic contaminants, the apparent primary transport mechanism is electroosmotic flow, even though heating of the treated soil usually occurs during remediation, which could introduce additional transport routes and even *in situ* degradation, such as vapor transport and thermal or (bio)catalytic degradation (Truex, Powell, and Lynch, 2007). Inherent advantages with electroosmosis include relatively uniform water flow through heterogeneous/low-permeability soils (e.g. glacial tills, lacustrine clays and silts, alluvial deposits, saprolitic formations, and loess), high degree of control of the flow direction, and very low power required to drive the water flow as compared to hydraulic means.

Lasagna™ technology, so called for its layered configuration, is a novel, *in situ* remediation technology developed for cleaning up contamination in heterogeneous or low-permeability soils. The technology combines electrokinetic transport of

---

Electrochemical Remediation Technologies for Polluted Soils, Sediments and Groundwater,
Edited by Krishna R. Reddy and Claudio Cameselle
Copyright © 2009 John Wiley & Sons, Inc.

contaminants in pore water with *in situ* treatment zones installed in the contaminated soil between electrodes (Brodsky and Ho, 1995; Ho and Brodsky, 1995; Ho et al., 1995). With its *in situ* treatment approach, especially for degrading organic contaminants, the Lasagna technology offers many advantages that include highly flexible treatment configurations and degradation methods, no additional aboveground waste disposal, highly contained system with little excavation, and potential cost-effectiveness.

Various configurations of the Lasagna technology have been studied at both bench and pilot scales, including solid (graphite or steel plate) and granular carbon electrodes; adsorption and degradation treatment zones; and various soil matrices such as homogeneous clay, sand mixed with clay, and a number of actual soils (Ho et al., 1995, 1997, 1998). The original intent was to install the electrodes and treatment zones horizontally using hydrofracturing, in a layered manner, like lasagna. The name "Lasagna" was trademarked by Monsanto for this process. Since then, vertical emplacements have been found to be more practical. Typical field configurations of the Lasagna process are shown in Figure 30.1. The complete system is installed and operated below grade without the need for digging or for any aboveground equipment except for a rectifier.

For treating soils contaminated with chlorinated solvents such as trichloroethylene (TCE), vertical curtains typically comprising iron filings and kaolin clay have been found quite effective in the Lasagna process. The kaolin clay is used as a filler/slurry for iron filing conveyance. The clay can also be used to minimize effects of local horizontal hydraulic gradients by effectively adding a series of slurry walls within the treatment area. Bracketed by treatment zones, contaminated chlorinated solvents in the soil are believed to eventually break down into nonhazardous compounds as they come in contact with the iron particles in the treatment zones. The reduction of TCE by zero-valent iron is a well-documented process (Senzaki and Kumagai, 1989; Gillham and O'Hannesin, 1994; Burris, Campbell, and Manoranjan, 1995; Puls, Powell, and Paul, 1995; Johnson, Scherer, and Tratnyek, 1996; Orth and Gillham, 1996; Roberts et al., 1996; Campbell et al., 1997).

For sites contaminated with dense nonaqueous phase liquid (DNAPL), Lasagna takes advantage of the soil heating as a result of the applied electrical energy. Soil temperatures have been found to readily reach 90–100 °C even with modest levels of applied currents and voltages. As with all thermal technologies, the elevated temperature can cause vapor transport and potentially near-vapor transport within the treated soil mass. Vapor-phase transport provides relatively fast migration of the solvent directly to treatment zones or to cooler areas where the vapors dissolve into pore water moving toward treatment zones by electroosmosis. Near-vapor transport describes the situation when viscosity of the DNAPL solvent drops to the point at which the electroosmotic pressure is sufficient to push the solvent through the soil pores to the treatment zones.

The Lasagna technology was originally developed by a consortium of Monsanto, DuPont, and General Electric companies along with the United States Environmental Protection Agency (USEPA) to treat organic contamination in low-permeability soils. The goal was to develop a technology to treat chlorinated solvent contamination in low-permeability and heterogeneous soils at a cost of $100 per ton or less. In 1992, the consortium members, USEPA, and the Department of Energy (DOE) constituted a working group under the Research Technology Development Forum

FIELD IMPLEMENTATION CONSIDERATIONS **627**

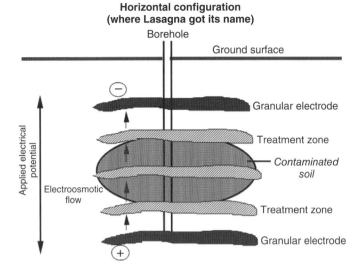

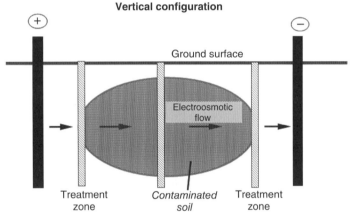

Figure 30.1. Typical configuration of the Lasagna process.

(RTDF) to accelerate the development and deployment of this technology. Funding for this development effort was, in part, supported by grants from the DOE Research Opportunity Announcement (ROA) and by the individual consortium members. The industry–government collaborative effort culminated in the successful cleanup of the Cylinder Drop Test Area at the DOE Gaseous Diffusion plant in Paducah, Kentucky, one of the case studies covered in this chapter.

30.2 FIELD IMPLEMENTATION CONSIDERATIONS

Electrokinetic implementation for full-scale site remediation requires consideration of many practical parameters not typically encountered in laboratory settings.

Operating complications include temperature buildup at electrodes, large amounts of acid and base generated at the electrodes, differences in soil buffering and exchange capacities, aged contaminants binding tightly to the soil, and possible corrosion of anode materials. There are also economic considerations with respect to electrode materials and neutralization schemes. Lasagna with its *in situ* treatment zone approach has been found very effective in field applications for treating low-permeability soils contaminated chlorinated solvents, primarily TCE. Key considerations in the field implementation of the Lasagna process are discussed below.

30.2.1 Site Characteristics

A major factor in applying electrokinetic processes in the field is the site conditions. Careful considerations need to be made regarding soil conductivity. If the soil electrical conductivity is high, the current-to-voltage ratio will be too high, causing excess heating without much water flow. Even though clayey soils tend to be more electrically conductive than sandy soils, the Lasagna process is still applicable in glacial till and lacustrine clay environments. High-salinity soils such as those in coastal areas will not work well as the soil electrical conductivity prevents the application of appropriate voltage gradients due to current limitations.

Presence of underground utilities or metal objects buried within the treatment area may preclude the use of Lasagna or any other DC electrically based processes. Metal objects can "short" the electrical path and corrode very quickly, causing hot spots. A survey of the area during the design phase is important. It is also best to have all building and footers removed from the site to allow complete access to the subsurface.

Groundwater conditions also need to be evaluated since high groundwater velocities may impact the effectiveness of in situ treatment partly due to potentially shorter residence time in the treatment zones and partly due to overall lack of control of water flow in the treated area. The Lasagna process is most effective in low-permeability source areas where groundwater movement is too slow to render more common remediation methods such as pump and treat or extraction effective. However, Lasagna can be adapted for handling narrow zones of preferential flow by using the kaolin-based treatment zones to isolate the permeable areas. This happened at the Paducah site where a 30-cm-thick zone of sandy gravel containing DNAPL levels of TCE (1200 mg/kg) was successfully treated.

30.2.2 Water Management

When a DC potential is applied to a medium containing water and ions, such as soil, acid is generated at the anode and base at the cathode due to the electrolysis of water. The highly acidic environment near the anode could be detrimental to the electroosmosis process if the zeta potential in the soil falls too low or reverses. Two clever approaches are utilized in the Lasagna process to both minimize the acid effect on soil and keep the anode pH between 5 and 7. First, steel plates are used as the anode material to promote iron oxidation as the main anodic reaction instead of water oxidation, which forms acid (H^+). Second, the basic pore water accumulated at the cathode (pH $\geq$ 12) is recycled by gravity back to the anode as makeup water, which is required for continuing the operation of electroosmosis and at the same time for neutralizing acid formed at the anode. Since metals are not being removed in the Lasagna process, the base front and associated metal precipitation is of no consequence to the operation. Steel of sufficient thickness to allow for 2–4 years of operation is typically used as anodes. Steel plates are also used as cathodes, but they do not corrode.

30.2.3 Electrode Configuration

So far, the most effective electrode configuration found for the Lasagna process is the planar one with vertically emplaced steel plate electrodes. This provides an advantage over cylindrical well-based electrodes by significantly lowering the current density. Planar electrodes enable more uniform current distribution and therefore uniform soil resistance and heating. In contrast, cylindrical well-based electrodes focus the current density down to a 10- to 20-cm-diameter well. This constriction of current density can cause excessive heating in local areas and can cause drying of the soil around the electrode well, ultimately resulting in failure.

30.2.4 Soil Temperature Management

When electrical current is passed through the soil, heat is generated by resistive heating. As the soil heats up, it becomes more electrically conductive. This in turn increases the current flow at a fixed voltage causing more heating. Many laboratory experiments are conducted at approximately 1 V/cm of path length. This level is not practical for sustained field operation as most soils would heat above 100 °C within weeks. Voltage gradients for Lasagna systems are limited to around 0.1–0.3 V/cm, which equates to roughly 2–5 W/m^3 for typical glacial tills.

The limiting factor for determining voltage gradient is soil temperature. If the soil heats up beyond 100 °C, water will boil away, drying the soil. Dry soil ultimately causes the electrokinetic process to shut down. Running at lower voltage gradients reduces the current density at the anodes and provides a far more stable long-term operation. Rectifier costs are also lower if the voltage is less than 600 V, and the current is kept as low as practical. Based on rectifier sizes, line power availability, and temperature limits, electrode spacing is generally in the 10- to 60-m range (30–200 ft).

30.2.5 Treatment Zones

The spacing of treatment zones is an important consideration since it greatly affects the cleanup time. At the field-scale voltage gradients mentioned above, the pore water velocity is reduced to only about 0.5 cm/day. To overcome this issue, more treatment zones, thus more closely spaced, result in a faster cleanup time and less electricity usage but higher installation cost. For most installations, treatment zone spacing and the cost of electricity represent the two biggest variables in the overall cost optimization. Typical cost-effective treatment zone spacing is about 1.5–2.5 m (2–8 ft). The cathode itself has been found effective as a treatment zone due to its reductive nature.

The treatment process should be tailored as much as possible to the target contaminant. Zero-valent iron in the form of iron filings has been effectively used in the Lasagna process to treat chlorinated ethenes and ethanes. Iron filings are relatively inexpensive and are easy to install using a clay-based slurry. Lasagna has not yet been used to treat other contaminants at this point. Mulch or activated carbon-based treatment zones may be viable for other volatile organic compounds (VOCs) and for trapping some metals.

30.2.6 Electrical Management

Another field issue to be concerned about is stray current effects on local buried utilities. Any buried metal object in the vicinity of a DC-based system could be subject to corrosion if the stray current is not contained. A corrosion expert should be consulted when applying DC systems underground and testing must be conducted to show that the impact to utilities is minimal. The best way to minimize stray current is to maintain a constant potential throughout the treated site perimeter. With Lasagna, this is done by using two outer anodes and one central cathode. Additional anodes can be used to enclose the site if needed. With the outer electrodes as anodes, the rectifier practically holds the anodic potential at ground level and the cathodic potential strongly negative. This way, all the current flow will be contained within the treated area, and any buried metal object not connected to the anode would serve as a cathode and would thus be inherently protected from corrosion.

Choosing the most cost-effective rectifier is another important consideration in field implementation. Rectifiers convert alternating current (AC) line power to DC by inverting half the sine wave of the AC using diode circuitry and by keeping one terminal at a constant positive potential (anode) and the other at a constant negative potential (cathode). Unless the rectification is highly filtered, there will still be an AC component to the output. The ripple is typically proportional to the difference in the input and output of the rectifier. The key for electrokinetic applications is to minimize the AC component while not overspending for a highly filtered rectifier. The way to do that is to use relatively inexpensive silicon-controlled rectifiers (SCRs) and to match the output voltage with the apparent input voltage. Rectifiers can be purchased with multiple input and output taps on the primary transformer. This allows "step" changes in voltage without creating a high AC component. Typical SCR rectifiers can have an AC "ripple" as low as 5% when the output and input voltages are matched.

30.2.7 Installation

Many significant advances achieved in the Lasagna installation process over the last few years have made it a cost-effective solution to source area remediation. The equipment has evolved to the point where emplacements take weeks instead of months and the material costs have been significantly minimized. Installation rates have risen from 9 linear meters (30 ft) per day for the Paducah installation to over 90 m (300 ft) per day for the Wisconsin installation. Both systems are described below.

The primary piece of equipment for the installation is a tower and mandrel system developed by Nilex, Inc. and DuPont. The Nilex rig, shown in Figure 30.2, consists of a square metal hollow tube, or mandrel, mounted vertically on a tower

Figure 30.2. Picture of Nilex rig.

system. The mandrel is driven into and raised out of the soil using a cable system and hydraulic motors and a vibratory hammer mounted on top of the mandrel. The length of the mandrel is dependent on the depth of penetration needed to install the materials. The tower is connected to a large track excavator, typically in the size range of a CAT-245 to a CAT-375. An expendable "shoe" made from a section of angle iron is used as the leading edge of the mandrel. The shoe acts as a knife edge to open the hole for the mandrel and prevents soil from packing up inside. The shoe comes off when the mandrel is pulled out of the ground.

Prefabricated electrodes are easier to make and dramatically improve installation rates. The electrodes can be made by combining steel plates and drainboards. Drainboards are simply plastic mesh pads covered with geotextile fabric. This allows water to travel in or out of the electrode zones as needed, yet electricity can pass through. The drainboards allow water management without creating electrical resistance at the electrodes. The wicks need to be covered with a geotextile to prevent silt and clay from penetrating the permeable media. Silt and clay particles are negatively charged and are drawn to the anode by electromigration, and can clog up the media and cause drying of the anodes.

Treatment zone material delivery has been dramatically improved. During the first installation at Paducah, the slowest step of the installation was the delivery of the treatment zone materials to the mandrel. The treatment zone material consists of iron filings suspended in a slurry of kaolin clay and water. The consistency of the treatment zone materials was that of thick cement. The materials were mixed in a cement mixer and were transported by cement bucket to the mandrel with a hopper fastened to the top of it. Cycle times were on the order of 15–20 min per drive. Currently, the materials are thinned slightly to the point when it can be pumped using a standard concrete pump yet can keep the iron filings in suspension. The mandrel is now filled while it is being driven to the appropriate depth (typically 5–15 m), allowing it to be immediately withdrawn leaving the materials in place. Cycle times are now on the order of 2–3 min or less.

Connecting the steel plate electrodes together and to a common buss wire was initially a laborious process that required electricians and plenty of special CADWELD® products. That system has been replaced with a simple steel buss bar that is welded to the tops of all the plates. This can be done with standard welding equipment. The buss wire is then connected to the end plate or buss bar using the CADWELD connectors. Once the welding is completed and the buss wire is connected, the whole buss bar and tops of all the plates are sealed in place to prevent corrosion of the buss bar. This arrangement provides and excellent planar electrode array that is durable and easy to install and to maintain.

30.2.8 Safety Issues

Safety is always the most important considerations in construction activities. During installation, the Lasagna installation team has found that placing a layer of crushed stones that have been sieved through a 3-in. screen over the site improves working conditions. The stone layer isolates workers and equipment from the sticky and contaminated soils below, significantly reducing the decontamination effort required at the end of the day as well as at the completion of the treatment. Safety during operations is also very important. Cathodic protection of buried utilities was mentioned above and can be managed with good design and monitoring. Personnel handling the rectifier and wiring must be trained in electrical safety. If securing the site is desired, the use of wood or other nonconductive security fence should be considered to minimize potential hazard. Metal fencing surrounding the entire site tends to attain an average DC potential (anodic), which may cause local DC voltage differences of 10 s of volts between the fence and the ground in cathodic regions. This could cause corrosion of the fence and possible shock to someone touching the fence.

Overall, the Lasagna process has proven to be a very safe method to treat high concentrations of chlorinated solvent. No contaminated material is removed from the site. All materials are emplaced directly into the contaminated soil by the mandrel system, minimizing exposures to the workers. The contamination is degraded below grade and very little, if any, vapor reaches the site worker or nearby residents during construction and operation.

30.3 CASE STUDIES

To date, the Lasagna process has been successfully implemented at two large sites: the DOE Gaseous Diffusion Plant site in Paducah, Kentucky, and the former Quicfrez site in Fond du Lac, Wisconsin. The contaminant treated at both sites is TCE.

30.3.1 Paducah, Kentucky

As mentioned above, in 1992, a consortium of industry (Monsanto, DuPont, and General Electric) along with USEPA and DOE formed a working group under the RTDF to accelerate the development and deployment of the Lasagna technology. Funding for this development effort was supported by grants from the DOE and by the individual consortium members. For the period of 1994–1998, the group actively combined expertise and resources in developing the technology for field implementation.

30.3.1.1 Site Background After surveying several available sites, the group selected a site at the DOE Gaseous Diffusion Plant in Paducah, Kentucky, for field testing of the technology. This particular site, called the Cylinder Drop Test Area and also known as Solid Waste Management Unit 91, was chosen for its low-permeability soil with TCE as the single major contaminant. The site is located at the south end of the C-745-B Cylinder Storage Yard in the northwest quadrant of the plant and is mostly clay loam with highly compacted gravel and clay overburden that has been used as a truck road. From late 1964 until early 1965 and in February 1979, cylinder drop tests were conducted to test the structural integrity of steel cylinders used to store and to transport uranium hexafluoride. A pit that was lined with plastic and was filled with TCE and dry ice was used as part of the testing process. As a result of these tests, the surrounding shallow soil and groundwater were contaminated with TCE.

The Paducah soil is classified as clay loam with fairly low organic content (0.2%) and much higher cation exchange activity than that of kaolinite (13.4 meq vs. 1.1 meq/100 g). The soil is made up of 22% sand, 46% silt, and 32% clay, with a bulk density of $2 g/cm^3$, 40% porosity, 20% moisture content, and estimated hydraulic conductivity of 10^{-8} to 10^{-6} cm/s. Cone penetrometry tests were conducted at three locations in the vicinity of the test area to approximately 15 m deep. It appears that the water table is not present until about 9–12 m. However, there are saturated zones above this at various depths. Based on lithology logs, transmittable water is probably in the coarser zones, which reside on top of the tight clay zones.

In preparation for the Lasagna initial field test, nine soil borings were made to confirm the northern edge of the plume and to determine the baseline of TCE concentrations in the test area. TCE concentrations ranged from below 1 mg/kg (ppm) to approximately 1500 ppm and appeared to taper off to below detection limits at about 9–11 m deep. With a soil density of 2 g/cc and 40% porosity (~20 wt % moisture at saturation), a TCE soil concentration of 220 mg/kg would result in pore water saturated with TCE (1100 mg/l). Thus, a level of 1500 ppm is a definite indication of the presence of DNAPL. From 1995 through early 2002, a total of three field operations was conducted at the Paducah site, first for proof-of-concept testing, then large-scale demonstration, and finally the full-scale site remediation.

30.3.1.2 Proof-of-Concept Field Test An initial small field test called Phase I was first conducted at the Paducah site to test the electroosmotic water transport of the Lasagna technology under actual field conditions. The treatment covered an area 3.0×4.8 m and reached 4.8 m deep. Activated carbon was used as the treatment material to trap the TCE transported from the contaminated soil for mass balance determination. The test was successfully completed in May of 1995 after 4 months of operation. Very high removal of TCE from the soil (~99%) was accomplished, even in regions of possible DNAPL. The scale-up from the laboratory to this field operation was very successful with respect to key electrical parameters and electroosmotic flow (Ho et al., 1999a).

30.3.1.3 Large-Scale Demonstration Test A second and larger field test, called Phase IIa, was carried out at the same site to demonstrate the effectiveness of the full Lasagna technology for *in situ* cleanup of TCE. Phase IIa was designed to perfect methods for installing treatment and electrode zones to a depth of 13.5 m (45 ft) and to demonstrate the effectiveness of the *in situ* degradation of TCE to this depth. Phase IIa began in August 1996 and originally planned to operate for 6 months. However, with extensive contamination uncovered during the operation, it was extended to 1 year to allow the process sufficient time to demonstrate its effectiveness in meeting the target cleanup level. A more detailed review of the Phase IIa field test has been published elsewhere (Landis *et al.*, 1998; DOE, 1998a; Ho *et al.*, 1999b).

The dimensions chosen for this field test were 6.3 m (21 ft) by 9.0 m (30 ft) and 13.5 m (45 ft) deep, the depth to which TCE had infiltrated after more than two decades since the original spill. The two electrodes were 6.3 m apart with three treatment zones, each 3.75 cm (1.5 in.) thick, inserted in between. The first treatment zone was 2.1 m (7 ft) from the anode, the second one 1.5 m (5 ft) from the first one, and the last one 0.6 m (2 ft) from the second one and 2.1 m (7 ft) from the cathode. These variable spacings were intended to help determine the optimum spacing as well as to provide information on soil conditions at various stages of treatment. Smaller treatment zone spacings would result in shorter treatment times but higher material and installation costs. A passive, underground fluid circulation system was designed to recycle water, which was electroosmotically transported to the cathode, back to the anode. A schematic diagram of the field setup is depicted in Figure 30.3. This modular design was intended to be duplicated over larger areas for further remediation of the site.

Based on extensive data in the literature and from its own studies, the consortium selected zero-valent iron in the form of iron filings as the reagent to degrade TCE. The dechlorination of TCE using zero-valent iron has been studied extensively by many research groups. The effectiveness of this approach has been demonstrated either as permeable reactive walls or as packed bed reactors at many field sites (Gillham, 1995; Puls, Powell, and Paul, 1995). The dechlorination apparently goes through a stepwise process in which one chlorine atom is removed each time:

$$Fe^0 + RCl_n + H^+ \rightarrow Fe(II) + RHCl_{n-1} + Cl^-,$$

where R stands for hydrocarbons. According to this mechanism, TCE dechlorination products would occur in the following order: cis-dichloroethylene (cis-DCE), vinyl chloride (VC), ethylene, and ethane. However, while ethylene and ethane are readily observed as reaction products, very little of the intermediates cis-DCE and VC could be detected (typically less than 2% of the initial TCE level), suggesting that they may remain bound to the iron surface and may undergo further dechlorination.

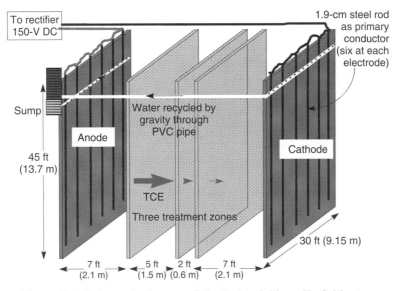

Figure 30.3. Schematic diagram of the Paducah Phase IIa field setup.

Issues related to TCE degradation in the Lasagna configuration at this site included effectiveness of iron filings in an electroosmotic environment, degradation products of TCE, effectiveness of iron for degrading DNAPL TCE, and role of electrodes in TCE degradation. Other important operational issues were installation of electrodes and treatment zones to the required depth through stiff clay soil without generating solid waste, long-term operation of the process, and extent of soil cleanup as a function of treatment time.

A site-specific action level of 5.6 mg TCE per kilogram of soil on the average for the treated area was set by the Commonwealth of Kentucky as the criterion for soil cleanup at this site. This level was based on an assessment model in which the groundwater originating from the site is required to contain less than $5\,\mu g/l$ (the maximum concentration limit for drinking water) by the time it reaches the plant's fence line. This action level can be used for this site as long as groundwater monitoring also confirms the modeling results.

The degradation of DNAPL level contamination was not anticipated but was substantiated by in-process field data and final sampling results (Ho *et al.*, 1999b). Analysis of *in situ* carbon-based samplers indicated that far more TCE had been adsorbed on the carbon than can be accounted for by dissolution and electroosmotic transport and diffusion. Also, gas samples taken at the surface above the operating Phase IIa process using an EPA landfill/flux chamber method (Kienbusch, 1986) found large quantities of acetylene relative to the ethene and ethane gas. These results suggest the continued presence of pure-phase TCE in contact with the zero-valent iron. No TCE or breakdown products were detected in the gas samples as there was a 2-in.-thick layer of treatment zone material spread over the site at the completion of the installation as a preventive measure.

The reason for the apparent transport of DNAPL level concentrations is not fully understood. However, the implications are profound. This observed result implies that only two pore volume exchanges may be needed to reduce the highest concentrations to target levels instead of the 10-s or even the 100-s or pore volumes required by dissolution alone and transport.

The soil temperature did rise above 60 °C during the treatment. Apparently, breakdown of TCE DNAPL at elevated temperatures has been observed at many sites treated with the electrical resistive heating (ERH) method. Truex, Powell, and Lynch (2007) report that TCE degradation at intermediate temperatures occurs primarily via biological reactions, while at elevated temperatures (80 °C), the mechanism is dominated by abiotic reactions as indicated by the detection of acetylene, even though acetylene can also be produced by acetogenesis.

Overall, TCE in the target soil area was reduced by approximately 95%–99%, including regions of pure-phase residual TCE. Based on these field results, the design for the full-scale site cleanup had all the contaminated soil bracketed with treatment zones, with primarily 1.5-m spacing and two to three pore volumes. In regions of suspect high DNAPL, some 0.75-m (2.5 ft)-spaced additional treatment zones would also be used to ensure sufficient flushing between treatment zones for complete TCE destruction.

30.3.1.4 Full-Scale Remediation In July 1998, DOE issued the Record of Decision (ROD) for the waste unit (DOE, 1998b), which designated Lasagna as the selected remedial alternative for reducing the concentration of TCE in Solid Waste Management Unit (SWMU) 91. The basic design for the full-scale installation, called Phase IIb, was patterned from the Phase IIa demonstration with slight modifications to the electrodes. Electrodes and treatment zones were installed to a depth of 13.5 m (45 ft) using a hollow mandrel system. The mandrel was driven to depth using a vibratory hammer and mast tower managed by a track hoe excavator, much like sheet pile driving. An expendable drive shoe is used to open the hole as the mandrel is driven to depth. Upon extraction, the drive shoe is left behind at the bottom of the hole allowing emplacement materials to fill in the void of the retreating mandrel. A 30-cm-wide by 7.5-cm-thick mandrel was used for the electrode emplacement, while a 50-cm (20-in.)-wide by 5-cm (2-in.)-thick mandrel was used for treatment zone material emplacement. A 5-cm-thick mandrel resulted in a treatment zone thickness of about 2.5 cm as witnessed in a soil core sample.

The electrodes for the full-scale system were preassembled using steel plate, thick enough to last for 3 years, and drainboards were wrapped in a geotextile fabric. The drainboards are expanded plastic hydraulic channels that allow water to pass freely across the face of the electrode plates yet let the DC current pass through. Keeping the electrodes hydrated is a key operations issue. The drainboards for each electrode were connected to a horizontal header placed at the top of each electrode row. The headers were then plumbed together, with a sump, to allow complete recycle of the pore water collected at the cathode back to the anodes. The sump was used as a water level control point and access point for sampling. Three rows of electrodes were installed. The outermost rows were anodes while the center row was the cathode. This resulted in electroosmotic flow converging at the center cathode.

The electrode spacing and configurations were determined based on the soil electrical and thermal properties and the available power of the DC rectifier. A spacing of 13.5 m was used for the electrodes. As it turned out, the buildup of heat

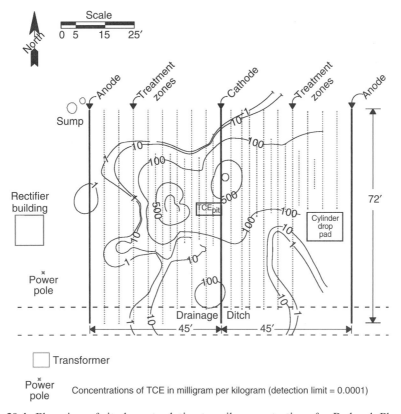

Figure 30.4. Plan view of site layout relative to soil concentrations for Paducah Phase IIb.

in the soil mass was the actual control parameter. The soil temperature should not exceed 90 °C to prevent overdrying of the soil, which results in severe loss of electrical conductivity. At this site, the process was frequently shut down due to high temperatures.

The rows of treatment zones were spaced at 1.5-m intervals between the electrode rows. This was determined to be the most cost-effective spacing based on operational time and installation costs. Extra treatment zone emplacements were made in areas of known or suspected DNAPL levels of TCE. As the contaminated pore water travels by electroosmosis through the treatment zones, the iron filings (zero-valent iron) act to degrade the chlorinated organics. Based on the electroosmotic velocity of 0.5 cm/day, the retention time in the 2.5-cm-thick treatment zones is about 120 h. This is more than sufficient time to degrade nearly all TCE to ethane and chloride, especially at the elevated temperature observed.

The construction of Phase IIb required about 3 months. Figure 30.4 shows a plan view of the site layout relative to soil concentrations. Phase IIb began operation in December 1999 and operated for 2 years. CDM Federal, Paducah, provided installation management as well as operations and management of the full-scale Phase IIb installation.

The full-scale process at this site was operated in steps regarding power input. The system was first powered up at about 415 V, drawing about 700 A. Once the center soil temperature reached 50 °C, the power was reduced to 290 V drawing 650 A then to 220 V and 500 A. After about 6 months, when the center temperature reached 90 °C, the unit was operated in pulse mode, several days on followed by several days off to cool. Simply reducing the voltage further was not practical. As the difference between the input voltage (240-V AC at the time) and output voltage increased, so did the AC ripple effect, causing more soil heating with less electro-osmotic flow.

Near the end of the planned operational period, soil samples were collected from the treatment area to determine if the soil concentrations were reduced to less than 5.6 mg/kg. These samples, taken only in areas of known high-preoperation TCE concentrations, were collected to get a conservative estimate of the concentration reductions. This preliminary soil sampling showed that the process had worked and that the ROD limit would be met. The unit was then operated until the end of the contract period. Based on the laboratory-determined electroosmotic permeability (adjusted for temperature), applied voltage, and operational time, a total of 1.6 pore volumes of treatment occurred. No groundwater (or pore water) samples were collected during the operation of Phase IIb.

Verification soil sampling was conducted in May 2002 and showed that the TCE soil concentrations were reduced to a level well below 5.6 mg/kg. An intensely reviewed sampling and analysis plan (SAP) was approved and implemented. The SAP called for a statistically based grid sampling in and around the treatment area totaling over 80 soil samples. The results of the verification sampling indicate an average TCE concentration of 0.38 mg/kg with a high concentration of 4.5 mg/kg (DOE, 2002). Figure 30.5 shows the post-Lasagna TCE soil concentrations. Only one sample had a detectable amount of cis-1,2-dichloroethylene (0.002 mg/kg). No VC was detected in any samples. The lack of degradation products duplicated earlier field demonstration results, indicating the nearly complete dechlorination of TCE to ethane, ethane, or acetylene.

The above results were better than anticipated based on the amount of TCE contamination and the relatively large amount of downtime due to temperature limits. Additional investigative soil samples above (near the surface), below, and to the sides of the treated area revealed little, if any, outward migration due to vapor-phase transport. The highest concentrations after treatment appeared to be adjacent to and south of the cement drop pad where no treatment zones were installed. The effort required to remove the drop pad, coupled with the results of previous soil samples showing very little contamination in that area, led to leaving the pad in place and terminating the treatment zones at the north edge of the pad.

The remediation effort at Paducah Gaseous Diffusion Plant (PGDP) was determined to be successful in all aspects. The Lasagna system met all remediation targets and budgets. The data quality objectives established before the project were all met or exceeded.

30.3.1.5 Cost The cost for the development of Lasagna through the Phase IIa site demonstration was shared among the consortium members and DOE through the ROA grant of the DOE EM-50 program. The full-scale site remediation was funded by the DOE EM-40 Program through Bechtel Jacobs Company LLC. The full-scale system installed at Paducah cost $3,958,000 or about $475 per cubic meter ($366 per cubic yard). The amount was well above the target cost for the remedia-

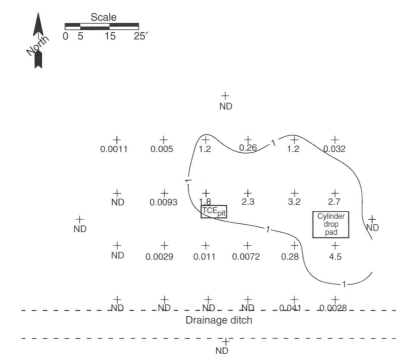

Concentrations of TCE in milligram per kilogram (detection limit = 0.001)

Figure 30.5. Post-Lasagna TCE soil concentrations for Paducah Phase IIb. ND, not detected.

tion system but was encouraging considering the extra costs associated with detailed data collection, rigorous sampling, extensive monitoring, and the usual administration and oversight required at federal DOE facilities. For example, before construction even started, $235,000 was spent on subcontractor environmental liability insurance and the installation of 480-volt power lines to the site. Based on just system installation and operation, the actual cost at a commercial or industrial facility would have been around $325 per cubic meter ($250 per cubic yard).

30.3.2 Fond du Lac, Wisconsin

30.3.2.1 Site Background The Lasagna installation in Fond du Lac, Wisconsin, was very similar to the Paducah installation. The site was an abandoned industrial facility, called Quicfrez, where TCE was used and spilled. The major differences were location and soil characteristics. The old Quicfrez site is located in an urban setting along the East Branch of the Fond du Lac River. The Quicfrez complex was used for refrigerator manufacturing from the 1920s through the early 1960s. After being vacant for some time, three buildings were used for storage under various ownerships. The city of Fond du Lac bought the property in 2001 with an eye toward green space and a bike path that follows the old railroad bed through the site. Evidently, TCE was spilled in the building basements and along the cement retaining wall at the river's edge.

A total of 28 soil borings were completed during the course of the preliminary investigations. These borings, along with the installation of 10 monitoring wells, helped determine the areas of contamination and the areas needed corrective action. The heaviest petroleum hydrocarbon and heavy metal contamination was removed by excavation. A large area of TCE contamination remained and reached a depth of 7.5 m (25 ft). This contamination has the potential to impact the regional aquifer, which is approximately 18 m (60 ft) below the site surface.

The soils at the Quicfrez site consists of an approximately 1.5 m of fill material sitting on top of roughly 7.5-m-thick layer of lean to fat clay deposits with approximately 60% clay and 30% silt and 10% sand. Below the 7.5-m-thick till layer exists approximately 10 m of dense hard clay. Under the hard clay is the sandstone bedrock aquifer that provides drinking water to the city and nearby residents. The area requiring remediation of TCE contamination measured approximately 30 m long along the river by roughly 24 m wide, extending some 3 m under the river, and approximately 7.5 m deep. The soil TCE concentrations ranged up to 2100 mg/kg, indicating a DNAPL condition.

30.3.2.2 Field Operation After the extensive site investigations by the Wisconsin Department of Natural Resources (WDNR) and Miller Engineers and Scientists, a total of five remedial options were compared and presented to the WDNR to clean up the TCE contamination at Quicfrez. The Lasagna process was chosen as the remedy of choice for this site due to its cost, effectiveness in clayey soils, and the ability to remove contaminants from under the river bed. The design called for a "terrace" to be built out into the river by installing a sheet pile wall out in the river bed and by backfilling with clay. Figure 30.6 shows the designed layout for the Quicfrez site.

Site preparations began in the fall of 2005 with site expansion into the river and removal of the remaining building debris and footers. A hydrologic evaluation was undertaken to show the terrace would not increase flooding risks in the Fond du Lac River upstream of the site. The terrace would restrict the river at low flood stages but had a low-enough finish elevation as to allow flood waters to pass over the top. Modeling showed the sheet pile wall and backfill extending 20 ft out in the river would not restrict flow during extreme flooding events. The structure was shown to not change the 100-year flood levels upstream of the sheet pile wall. Larger stone, however, was recommended for the river bottom to avoid scouring and erosion at the river's bottom due to the possible increase in velocity caused by the constricting of flow. It was also determined that the top of the Lasagna site should be covered with riprap to minimize erosion during high-flood stages.

The Lasagna process was installed during the summer of 2006. The site was laid out such that it measured 33.0 × 24.0 × 7.5 m deep (110 × 80 × 25 ft). Approximately 1.5 m of fill and building demolition debris, along with the old building footers, were removed immediately prior to Lasagna installation. This action cleared most of the subsurface obstructions and created a level surface to work on. However, during a test run with the heavy installation equipment, it was determined that the soils, primarily the fill clays, were too soft and the Cat-375 track hoe began to sink in. The sheet pile wall also began to bulge from the weight of the excavator. To solve the wall problem, the sheet pile wall was reinforced using 11 30-cm (12-in.) I-beams driven behind sheet pile and welded to the wall. Additionally, a 30- to 45-cm layer

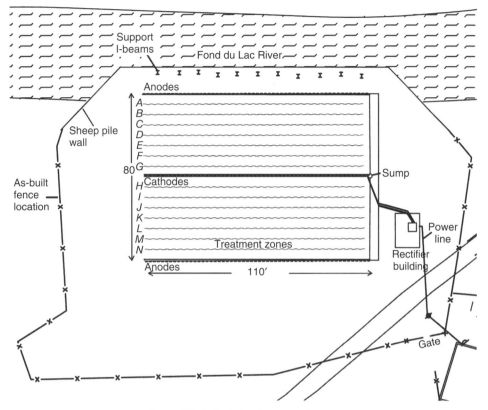

Figure 30.6. Quicfrez Lasagna layout.

of a size 7.5-cm (3-in.) ballast rock was spread, with a geogrid support, over the site. These actions were sufficient to support the excavator. The rock layer was likely an overall cost saving by making the working conditions cleaner and by minimizing the decontamination effort at the end of the installation.

The electrode and treatment rows were installed parallel to the river. The north anode was emplaced in the river bed, roughly 2.4 m (8 ft) from sheet pile wall. The electrode zones were constructed of vertical 1-cm (3/8-in)-thick steel plates, 30 cm wide and 6 m (20 ft) long, and were emplaced to a depth of 7.5 m (25 feet) below grade (5–25 ft below ground surface [bgs]). Each plate is separated by 30 cm. There were 55 electrode plates along each 33-m-long electrode row. The steel plates were connected electrically by welding a steel buss bar long the top of each plate and connecting to the buss wire cable.

The treatment zones consisted of "curtains" of zero-valent iron suspended in a kaolin clay slurry. The clay, water, and iron filings were mixed in a cement mixer and were delivered to the installation equipment using a standard concrete pump. The slurry contained 50% iron by weight. The iron filings remain in place after operations and will continue to treat residual TCE in the soil long after the DC power is turned off. This prevents any rebound issues associated with some extractive remediation systems for low-permeability or heterogeneous sites.

Once the electrodes and treatment zone materials were installed, the electrodes were wired and plumbed. The electrical system consisted of a steel buss bar welded across the tops of the steel plates. The end plate was connected to the buss wire, which was in turn connected to a large DC rectifier. The rectifier, relocated from the Paducah site and reconfigured, converts standard 208-V, three-phase AC power (600-A service) to a maximum of 190-V and 1000-A DC. There were two anodes rated at 500 A each and one central cathode with two buss wires.

Keeping the anodes wet and supplied with water is crucial. As pore water moves away from the anode by electroosmosis, makeup water must be available to prevent soil drying and burned-out anodes. Horizontal wells were installed along the tops of the electrodes along with the steel bus bars. The cathode trench sloped toward a common sump at the east end of the cathode. The sump also had lateral lines running to the north and south anodes. The anode trenches sloped from east to west and were constructed 6-in. deeper than the cathode trench. Hydraulically conductive wick drains were installed with the steel plate electrodes to allow free movement of water in or out of the electrodes as needed. The sump contained a level sensor to monitor the water level. The electroosmotic water flows from the cathode to the sump and then to each anode as needed by gravity.

The DC rectifier (Rapid Power Corporation) was located in a small storage shed built at the site. The shed also housed a Web-based computer acquisition system that monitors the voltage, current, sump level, and soil temperatures throughout the site. The system will automatically shut down the rectifier if any of several key monitoring points exceed limits.

The installation was completed in October 2006. The 1.5 m of removed surface fill was placed back over the site, after removing large debris, and the riprap was placed over the terrace extending into the river. Due to innovations in emplacement equipment and treatment delivery methods, the installation was performed 17% under budget, including the modifications to the sheet pile wall and the ballast rock added to the site.

On November 8, 2006, the system was started up for full-time operation. The voltage was set to 190-V DC and each anode was delivering approximately 250 A. The soil temperatures climbed quickly with the center of the soil mass reaching 80 °C within 3 months and the current reached 400 A per side. The south anode warmed to around 55 °C, and the north anode only reached 40 °C due to the cooling effect of the sheet pile wall and river during the winter months.

To address concerns relating to public safety and protection of nearby underground utilities such as gas and waterlines, stray voltage/current tests were conducted by Cathodic Protection Management in cooperation with Alliant Energy using a copper/copper sulfate reference junction and voltmeter. Initial testing was performed prior to Lasagna start-up and during a test start-up in August showing no impact to local utilities. Additional testing was performed after full-time start-up in November 2006. Originally, there was a 30-V DC potential at several locations at the chain-link security fence surrounding the site, primarily near the cathode ends. This occurrence was due to the fence attaining an anodic potential from being located close to the anodes and the soils near the cathode having a cathodic charge. In response to this issue, the fence was moved 3.6 m further away from the energized cathode areas, and several sacrificial electrodes were installed between the fence and the cathode ends. The voltage measured at the fence was reduced to 12 V.

Although this voltage was not a concern for safety, it did have the potential to corrode the fence quickly in those areas. Monitoring continues but shows no significant corrosion of the fence posts.

At the time of this writing, Lasagna setup continues to perform as designed. The steady-state voltage of 107 V, with 170 A per side, is holding most of the soil temperatures steady except for the season fluctuations at the north anodes. Other than some minor corrosion of the buss bars, there have been no operation issues. It is anticipated the cleanup goals will be met by mid- to late 2008. As of fall 2007, monitoring well data shows a definite downward trend.

The first performance soil sampling event was conducted in August 2007 and showed approximately 60% removals. One of the soil sampling locations increased slightly while the others dropped, some dramatically. Some soil samples can increase due to the local redistribution of DNAPL caused by the heating and vapor transport. This was also seen at the Paducah Lasagna site where the interim sampling concentrations were higher than the initial in some locations. However, the second interim sampling at Paducah showed remarkable decreases at all locations and was verified by a statistically valid confirmation soil sampling program.

It is anticipated that the Lasagna system at the old Quicfrez site will meet the cleanup goal of 1-ppm TCE in the soil sometime mid-2008. Meanwhile, plans for the site continue to develop. Designs are being drawn up to add a stormwater retention pond and bikeway through the site. The terrace will be removed once the remediation is complete. The Lasagna treatment zones will remain in place to allow for continued treatment of any residual TCE for many years.

30.3.2.3 Cost The cost to install Lasagna at this site was approximately $1,030,000. An additional $250,000 was spent on site preparations to extend the remediation process into the river. Operations are estimated at an additional $250,000 for electricity, rectifier rental, site oversight, and monitoring for 2 years. The complete site cleanup will cost an estimated $2,000,000, of which $1,280,000 is directly for Lasagna. Based on the 6000-m^3 (8000-yd^3) site volume, the Lasagna unit costs equal $213 per cubic meter ($160 per cubic yard or about $100 per ton).

30.4 SUMMARY AND FUTURE ACTIVITIES

Currently, the Lasagna process is a commercially viable soil remediation process that has been proven especially effective in treating chlorinated solvent contamination in low-permeability soils. The use of DC electrical current to heat the soil and to mobilize pore water and contaminants directionally into degradative treatment zones makes *in situ* remediation possible. The process is a truly treatment in-place method since no contamination is brought aboveground. Transport and destruction take place within the soil mass.

Lasagna can be installed at many sites with diverse conditions and surrounding environments. The system works well in saturated or vadose zones. Installations have been installed and operated safely in semirural and urban settings. With the process being contained underground, the only pieces of equipment required are the rectifier and a remote data monitoring system. The only structure required is a shed on site to house the rectifier and the data system. A telephone line and typical 400- to 600-A electrical service are all the required utilities.

The cost of the Lasagna process continues to drop with increasing field experience and technology adaptation. Current installation methods and equipment allow the costs to approach $100 per ton. The materials required, iron filings, kaolin clay, and steel plates, are as natural and innocuous as they can be. The cost of materials and rectifiers is determined primarily by steel prices. Electricity costs are only about 10% of the total and may be further managed by negotiation with the electric company.

Two more Lasagna systems will be installed at an industrial facility in eastern Ohio during the spring of 2008. These units will treat TCE contamination in saturated lacustrine clay. The location is an active industrial plant surrounded by residential and commercial properties. The contamination is from degreasing operations in the 1950s and 1960s and approaches DNAPL level in some areas. It is anticipated that the cost of the two Lasagna installations will be significantly less than long-term groundwater and indoor air monitoring programs.

Future developments may include alternative treatment zones to treat aromatic hydrocarbons, metals, or radionuclides. The basic installation procedures and equipment can easily be modified to utilize other treatment zone materials. Any solid that can be handled as granular (fluid or semifluid) or introduced as a slurry can be used as treatment zone material with the current equipment.

REFERENCES

Acar YB, Alshawabkeh AN. (1993). Principles of electrokinetic remediation. *Environmental Science and Technology* **27**(13):2638–2647.

Acar YB, Li H, Gale RJ. (1992). Phenol removal from kaolinite by electrokinetics. *Journal of Geotechnical Engineering* **118**(11):1837–1852.

Brodsky PH, Ho SV. (1995). Monsanto Company. US Patent 5,398,756, issued March 21, 1995.

Bruell CJ, Segall BA, Walsh TM. (1992). Electroosmotic removal of gasoline hydrocarbons and from clay. *Journal of Environmental Engineering* **118**(1):68–83.

Burris DR, Campbell TJ, Manoranjan VS. (1995). Sorption of trichloroethylene and tetrachloroethylene in a batch reactive metallic iron-water system. *Environmental Science and Technology* **29**(11):2850–2855.

Campbell TJ, Burris DR, Roberts AL, Wells JR. (1997). Trichloroethylene and tetrachloroethylene reduction in a metallic iron–water–vapor batch system. *Environmental Toxicology and Chemistry* **16**(4):625–630.

Casagrande L. (1952). Electro-osmotic stabilization of soils. *Journal of Boston Society of Civil Engineers* **39**:51–83.

DOE. (1998a). Rapid Commercialization Initiative (RCI) Final Report for an Integrated In-Situ Remediation Technology (Lasagna™). *Report Number DOE/OR/22459-1*.

DOE. (1998b). Record of Decision for Remedial Action at Solid Waste Management Unit 91 of Waste Area Group 27 at the Paducah Gaseous Diffusion Plant, Paducah, Kentucky. United States Department of Energy Office of Environmental Management, Paducah, KY. *DOE/OR/06-1527&D2*.

DOE. (2002). Final Remedial Action Report for Lasagna™ Phase IIb In Situ Remediation of Solid Waste Management Unit 91 at the Paducah Gaseous Diffusion Plant, Paducah,

Kentucky. United States Department of Energy Office of Environmental Management, Paducah, KY. *DOE/OR/07-2037&D0*.

Gillham RW. (1995). Resurgence of research concerning organic transformations enhanced by zero valent metals and potential application in remediation of contaminated ground water. National Meeting of the American Chemical Society, Div. Environs. *Chemistry* **35**:691–694.

Gillham RW, O'Hannesin SF. (1994). Enhanced degradation of halogenated aliphatics by zero-valent iron. *Ground Water* **32**:958–967.

Hamed J, Acar YB, Gale RJ. (1991). Pb (II) Removal from kaolinite by electrokinetics. *Journal of Geotechnical Engineering* **117**(2):241–270.

Ho SV, Athmer CJ, Brackin JM, Heitkamp MA, Sheridan PW, Weber D, Brodsky PH. (1998). The Lasagna™ process for in-situ bioremediation of low permeability soils. In *Bioremediation: Principles and Practices*, Vol. **3** (eds. SK Sikdar, RL Irvine). Lancaster, PA: Technomic Publishing, pp. 393–417.

Ho SV, Athmer CJ, Sheridan PW, Hughes BM, Brodsky PH, Shapiro A, Thornton R, Salvo J, Schultz D, Landis R, Griffith R, Shoemaker S. (1999a). The Lasagna technology for in situ soil remediation: 1. Small field test. *Environmental Science and Technology* **22**:1086–1091.

Ho SV, Athmer CJ, Sheridan PW, Hughes BM, Brodsky PH, Shapiro A, Thornton R, Salvo J, Schultz D, Landis R, Griffith R, Shoemaker S. (1999b). The Lasagna technology for in situ soil remediation: 2. Large field test. *Environmental Science and Technology* **22**: 1092–1099.

Ho SV, Athmer CJ, Sheridan PW, Shapiro AP. (1997). Scale-up aspects of the Lasagna process for in situ soil decontamination. *Journal of Hazardous Materials* **55**:39–60.

Ho SV, Brodsky PH. (1995). Monsanto Company. US Patent 5,476,992, issued December 19, 1995.

Ho SV, Sheridan PW, Athmer CJ, Heitkamp MA, Brackin JM, Weber D, Brodsky PH. (1995). Integrated in situ soil remediation technology: The Lasagna process. *Environmental Science and Technology* **29**(10):2528–2534.

Johnson TL, Scherer MM, Tratnyek PG. (1996). Kinetics of halogenated organic compound degradation by iron metal. *Environmental Science and Technology* **30**(8):2634–2640.

Kienbusch MK. (1986). *Measurement of Gaseous Emission Rates from Land Surfaces using an Emission Isolation Flux Chamber*. EPA/600/8-86/008.

Lageman R. (1993). Electroreclamation (applications in the Netherlands). *Environmental Science and Technology* **27**(13):2648–2650.

Landis RC, Griffith RJ, Shoemaker SH, Schultz DS, Quinton GC. (1998). Emplacement Technology - an Evaluation of Phase IIa and Alternative Lasagna Emplacement Methods, DOE Phase IIA Report. E.I. du Pont Nemours and Co., Wilmington, DE. *Contract Number DE-AC05-96OR22459*.

Orth WS, Gillham RW. (1996). Dechlorination of trichloroethene in aqueous solutions using Fe^0. *Environmental Science and Technology* **30**(1):66–71.

Puls RW, Powell MR, Paul CJ. (1995). In situ remediation of ground water contaminated with chromate and chlorinated solvents using zero valent iron: A field study. 209th American Chemical Society National Meeting, Anaheim, CA. Division of Environmental Chemistry Preprints of Papers, vol. 35, no. 1, pp. 788–791.

Roberts AL, Totten A, Arnold WA, Burris DR, Campbell TJ. (1996). Reductive elimination of chlorinated ethylenes by zero-valent metals. *Environmental Science and Technology* **30**(8):2654–2659.

Segall BA, Bruell CJ. (1992). Electroosmotic contaminant-removal processes. *Journal of Environmental Engineering* **118**(1):84–100.

Senzaki T, Kumagai Y. (1989). Removal of chlorinated organic compounds from wastewater by reduction process: II–treatment of trichloroethylene with iron powder. *Kogyo Yosui (Japan)* **369**:19–25.

Shapiro AP, Probstein RF. (1993). Removal of contaminants from saturated clay by electroosmosis. *Environmental Science and Technology* **27**:283–291.

Shapiro AP, Renaud P, Probstein RF. (1989). Preliminary studies on the removal of chemical species from saturated porous media by electroosmosis. *Physicochemical Hydrodynamics* **11**(5/6):785–802.

Truex M, Powell T, Lynch K. (2007). In situ dechlorination of TCE during aquifer heating. *Ground Water Monitoring and Remediation* **27**(2):96–105.

31

COUPLED ELECTROKINETIC PRB FOR REMEDIATION OF METALS IN GROUNDWATER

HA IK CHUNG AND MYUNGHO LEE

31.1 INTRODUCTION

Since the industrial revolution, pollution of groundwater resources with toxic metals and metalloids, such as cadmium (Cd) as well as volatile organic compounds including trichloroethylene (TCE), has been a significant threat and damage to human health and the environment (Vangronsveld and Cunningham, 1998). For example, cadmium causes the notorious *itai-itai* disease resulting in serious skeleton deformation and kidney damage (Yeung and Hsu, 2005). Also, TCE is harmful to the respiratory system, the circulatory system, and the central nervous system of human bodies (Yang and Liu, 2001). A need thus exists for an effective and economical technique to remediate these constituents and others that impact the subsurface. Many techniques exist for the remedial treatment of contaminated land including isolation, immobilization, toxicity reduction, physical separation, and extraction.

31.2 ELECTROKINETIC (EK) EXTRACTION SYSTEM

Reuss discovered the EK phenomena in 1809, and L. Casagrande first employed the technique in 1939 for stabilizing a long railroad cutting. Since then, the EK technique became an established technique for stabilizing slopes, dams, embankments, and excavation sites for construction (Perry, 1963; Fetzer, 1967; Chappell and Burton, 1975). Recently, the EK remediation technology has been proven to be a very effective tool to clean up heavy metal-contaminated fine-grained soils. A schematic diagram of EK soil processing is illustrated in

Electrochemical Remediation Technologies for Polluted Soils, Sediments and Groundwater,
Edited by Krishna R. Reddy and Claudio Cameselle
Copyright © 2009 John Wiley & Sons, Inc.

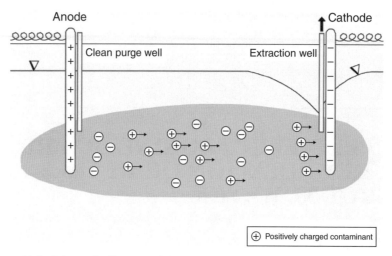

Figure 31.1. Schematic diagram of electrokinetic soil processing (US EPA, 1995b).

Figure 31.1. The EK remediation is a relatively new method that involves passing a low electrical current between electrode pairs imbedded in the ground for the removal of subsurface contaminants via electrophoresis (EP), electroosmosis (EO), and electromigration (EM) (Acar and Alshawabkeh, 1993).

EM rates in the subsurface depend on electric current, soil pore fluid, grain size, ionic mobility, and contamination level. The direction and quantity of contaminant movement are influenced by soil type, pore fluid chemistry, contamination level, and electric current (Yeung, 1994). EK remediation can be used for both saturated and unsaturated soils, but for better efficiency, the soil moisture content should be high enough to allow EM. Nonionic species would be transported along with the electroosmotically induced fluid flow. The efficiency of extraction relies on several factors such as species type, solubility, electrical charge, and concentration relative to other species (Mitchell, 1993).

This technology is particularly effective in low-permeability soils where hydrodynamic techniques would be unsuitable (Yeung, Hsu, and Menon, 1996; Page and Page, 2002; Reddy and Saichek, 2003). Contaminants that can be treated by EK processes include (Yeung, 1994)

- heavy metals (cadmium, zinc, lead, mercury, copper, chromium, etc.);
- toxic anions (nitrates, sulfates);
- radioactive species (Cs, Sr, Co, Ur);
- dense non-aqueous-phase liquids (DNAPLs);
- petroleum hydrocarbons (gasoline, diesel fuel, oils); and
- TCE; benzene, toluene, ethylbenzene, and xylenes (BTEX); polycyclic aromatic hydrocarbon (PAH).

31.3 PERMEABLE REACTIVE BARRIER (PRB) SYSTEM

Various remediation techniques have been developed and adopted to clean up contaminated groundwater. Among them, the PRB system has been considered as

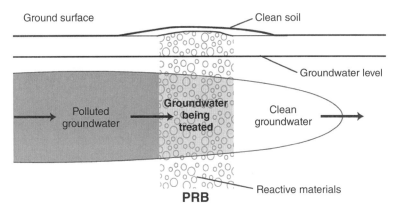

Figure 31.2. Schematic diagram of a permeable reactive barrier (US EPA, 2001).

an innovative technology because the PRB system is usually quicker and cheaper than other common methods. PRB technology was developed in the early 1990s and was first used economically in 1994 (US EPA, 1995a).

A schematic diagram of the PRB system is presented in Figure 31.2. As seen in the figure, the PRB system is a kind of permeable wall built by digging a long, narrow trench in the path of contaminated groundwater so that groundwater flows through tiny holes. The trench is filled with a reactive material that can change the chemicals into harmless ones or can trap harmful chemicals such as heavy metals. Thus, the clean groundwater can flow out of the wall (Simon and Meggyes, 2000; Morrison, 2003).

Common types of reactive materials that can be used are carbon, limestone, and iron. The wall is part of a funnel that directs the polluted groundwater to the reactive part of the wall, and the funnel is usually covered with soil as illustrated in Figure 31.3. The material used to fill the trench depends on the types of harmful chemicals in the groundwater (US EPA, 1998). Different materials clean up pollution through different methods by

- trapping or sorbing chemicals on their surface (e.g. carbon),
- precipitating chemicals that are dissolved in water (e.g. limestone), and
- changing the chemicals into harmless ones (e.g. iron).

TCE has been successfully treated by the PRB using zero-valent iron (ZVI) (Pulsa, Blowesb, and Gillhamb, 1999; Su and Puls, 2001a,b). The PRB using ZVI has also been used for the remediation of arsenic (As)-contaminated ground by the effect of sorption (Farrell *et al.*, 2001; Melitas, Conklin, and Farrell, 2002a; Melitas *et al.*, 2002b). As groundwater usually moves a few inches to hundreds of feet per year so that cleaning groundwater using the PRB system may take a few years. The processing time for the PRB system depends on the following two major factors that vary from site to site:

- type and amount of pollution present in the groundwater and
- how fast the groundwater moves through the PRB.

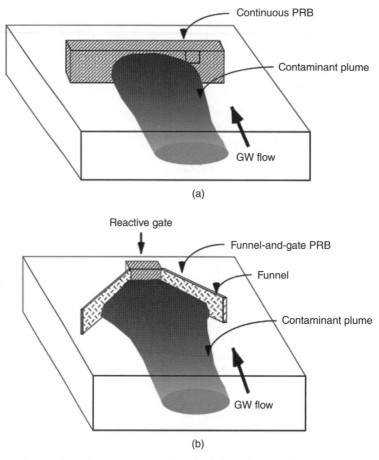

Figure 31.3. Contaminated plume capture by the (a) continuous PRB trenched system and by the (b) funnel-and-gate PRB system (US EPA, 1998). GW, groundwater.

The PRB system works best at sites with loose, sandy soil and with a steady flow of groundwater. The depth of pollution that can be treated by the PRB system should be within 50 ft. The remediation by the PRB system can be effective (i.e. faster) and economic (i.e. cheaper) compared to other methods as there is no need to pump contaminated groundwater (US EPA, 2001).

31.4 COMBINED SYSTEM OF ELECTROKINETICS AND PERMEABLE REACTIVE BARRIER

The reactive materials placed in the PRB trench are not harmful to the groundwater as well as to people. The contaminated groundwater can be cleaned underground, and therefore people can avoid contact with pollutants. The remediation using

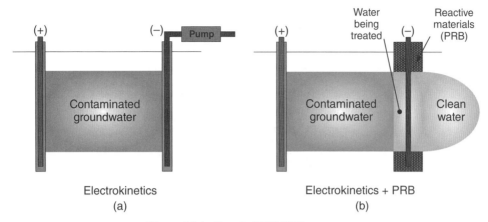

Figure 31.4. Coupled EK-PRB system.

PRB usually needs to remove some soil, which may be contaminated, when digging the trench.

Figure 31.4a shows a schematic diagram of the *in situ* EK remediation, which requires a pumping system at the cathode in order to extract the contaminant flux migrated from the anode toward the cathode due to the combined effects of EO and EM. However, as shown in Figure 31.4b, if the cathode electrode was installed with reactive materials, the migrating pollutants from the anode toward the cathode could be trapped or destroyed by passing through the PRB system (Chung and Lee, 2007).

If the soil is polluted, another cleanup method may be required to remediate contaminated ground. Thus, the coupled electrokinetic permeable reactive barrier system (EK-PRB) system should be effective to clean up polluted groundwater by the PRB system as well as to remediate contaminated ground by the EK system. The following section describes the investigations on a feasible field application of the EK remediation coupled with the PRB system to remediate contaminated ground of low-permeability soils.

31.5 FIELD APPLICATION

31.5.1 Site Description

The coupled EK-PRB system was applied to a vicinity of landfill site located in GyeongGi province, Korea (see Fig. 31.5). Such landfill site is one of the simple, unregulated, and old-fashioned landfills in Korea, and there may have been possibilities of groundwater contamination due to the leachates from the landfill. Thus, the *in situ* remediation of the coupled EK-PRB system was applied to the surroundings of the landfill site to remediate contaminated soils and groundwater as well as to investigate the applicability of the combined system of EK-PRB in the field. The treatment zone was determined by the groundwater level: that is, between 3 and 5 m below the ground surface as shown in Figure 31.6.

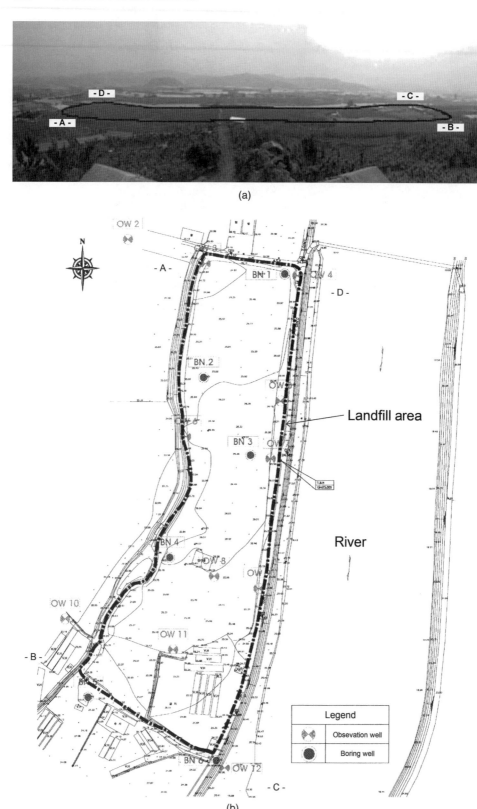

Figure 31.5. Landfill site: (a) aerial photo and (b) plane view.

FIELD APPLICATION 653

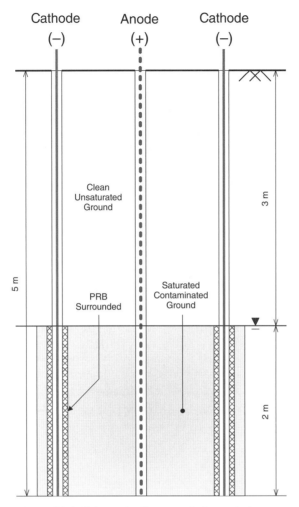

Figure 31.6. Schematic diagram of electrode layout.

31.5.2 System Installation

The *in situ* remediation of the EK-PRB system basically consists of an array of electrodes, electrode wells, a direct current (DC) power supply, a vacuum pump, a water reservoir, and a process tubing to inject water to the anode electrode well and to extract contaminants from the cathode electrode wells as shown in Figure 31.7.

The electrode array consists of a series of anode and cathodes housed in wells. Figure 31.8 illustrates the array of anode and cathodes that were installed within the testing area. It can be seen that the anode well is located in the center, and the cathode wells are located in radial shape at the distance of approximately 1 m apart from the central anode well. Various types of PRB materials (i.e. ZVI, zeolite, slag, and sand) were used. The applied electrical potential difference between the electrodes was approximately 100 V (i.e. voltage gradient of 1 V/cm). Two different

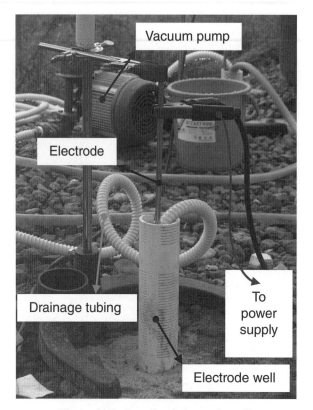

Figure 31.7. Details of electrode well.

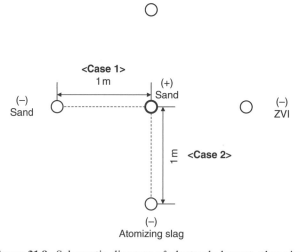

Figure 31.8. Schematic diagram of electrode layout, plan view.

Figure 31.9. Experimental setup for case 1.

conditions, the EK extraction system (both anode and cathode wells filled with sand) and the EK-PRB extraction system (cathode well filled with atomizing slag), were considered. Slotted polyvinyl chloride (PVC) columns (0.08-m inner diameter (i.d.) and 5 m long) were used as the electrode chambers.

Figure 31.9 presents the experimental setup of the EK-PRB system. After the installation of PVC chambers into bore holes (0.16-m i.d.), the surroundings were filled with various reactive materials for various purposes. Thereafter, electrodes (stainless steel rod of 0.01-m outer diameter (o.d.) and 5.5 m long) and drainage tubes were installed into the electrode wells. The electrodes were connected to a DC power supply (Agilent E3612A, Agilent Technologies, Inc., Santa Clara, CA), and the inlet and outlet tubes were connected to a water reservoir and vacuum pump, respectively. A constant electric field, rather than a constant electrical current, was applied to the ground because significant power loss may occur due to the increase of soil resistance by the depletion of ions, resulting in the premature termination of the experiment (Yeung et al., 1997).

31.5.3 Materials

Atomizing slag (commercial name: PS Ball, worldwide patent) is a product manufactured by the atomizing method. Conventional slag is cooling slowly in the open yard, having a composition that contains CaO, FeO, Fe_2O_3, and so on. The atomizing slag is almost the same in its composition as the conventional converter slag, but its compounds become different due to the different method of slag treatment: it has much better strength and does not release toxic materials. The atomizing slag has a higher Fe_2O_3 than the converter slag and contains little metallic Fe due to the atomizing process.

The atomizing slag is also relatively inexpensive compared to other PRB materials, such as ZVI PRB or zeolite PRB (US EPA, 1995a). Therefore, the atomizing

TABLE 31.1. Material Properties of Atomizing Slag and Sand as Filter Media

Parameter	Atomizing Slag	Sand
Specific gravity	3.54 or 3.42	2.57
Effective size (mm)	0.5 ~ 1.2	0.5 ~ 1.2
Uniformity coefficient	1.22	1.64
Porosity	0.41	0.52
Filtration rate (m^3/m^2-min)	0.08	0.08
Strength (kg/m^2)	223	135
Bed expansion (%)	15 ~ 20	20 ~ 40
Absorption (%)	0.1	—
Moisture content (%)	0.79 ~ 0.98	—

TABLE 31.2. Chemical Composition of Atomizing Slag and Sand

Parameter	Atomizing Slag (%)	Sand (%)
SiO_2	14.6	92.83
Al_2O_3	4.6	1.34
Fe_2O_3	29.8	0.95
CaO	38.7	1.97
MgO	11.8	2.23
Na_2O	0.04	0.11
K_2O	0.11	0.14
Ignition loss	0.35	0.43
Total	100	100

slag was chosen to be used for the PRB material. Tables 31.1 and 31.2 present the material properties and chemical compositions of the atomizing slag by comparison to standard sand.

31.5.4 Results and Discussion

The effluent was continuously collected from the cathode well, and the effluent pH was measured during EK soil processing. It was found that there were no big differences in the effluent pH values under the EK and the EK-PRB system.

The initial chlorine concentrations for both the EK system and the EK-PRB system were approximately 16 mg/l, and the chlorine concentration under the EK system rather maintained the initial value as shown in Figure 31.10. However, in the case of the EK-PRB system, the chloride concentration was continuously increasing during the EK treatment. It appears that the migrating organic pollutants from the anode toward the cathode, due to the electroosmotic advective flow passing through the PRB material, were destroyed by the reactions with the atomizing slag.

Toxic pollutants were found to be extracted from the contaminated ground by both the EK system and the EK-PRB system as shown in Figures 31.11 and 31.12.

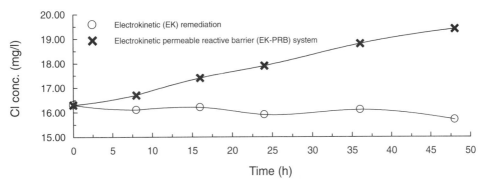

Figure 31.10. Chlorine concentration against time results.

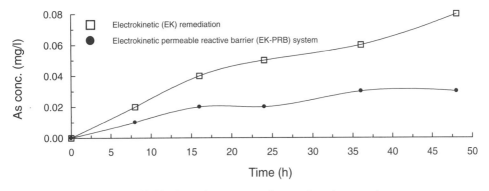

Figure 31.11. Arsenic concentration against time results.

There is an increasing tendency for the migrating pollutants toward the cathode well due to the EK treatment. In the case of the EK-PRB system, the migrating contaminants toward the cathode well appear to adsorb onto the reactive material, and those of concentrations were not high. The removal rate of heavy metal contaminants was slightly higher than that of organic contaminants due to the effects of EO and the additional effect of EM for their positive charges.

31.6 SUMMARY

Field testing results showed that the effluent pH under the EK-PRB system was maintained constant during the treatment, which may be due to the reactions between the hydroxyl ions and the reactive materials. The chloride concentrations were higher under the EK-PRB system than those under the EK system, which seemed to be caused by the depletion of organic contaminants. The heavy metal contaminants were removed by the effects of EO and EM under the EK system. In the case of the EK-PRB system, the majority of migrating heavy metals were adsorbed onto the atomizing slag, and hence the heavy metal concentrations were lower than those under the EK system. From the preliminary field investigations,

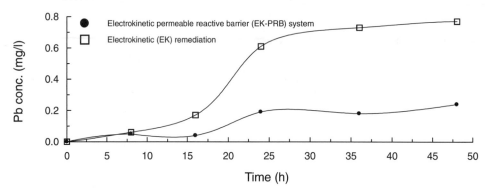

Figure 31.12. Lead concentration against time results.

the coupled technology of the EK-PRB system would be effective to remediate contaminated grounds without the extraction of pollutants from the subsurface due to the reactions between the reactive materials and the contaminants.

REFERENCES

Acar YB, Alshawabkeh AN. (1993). Principles of electrokinetic remediation. *Environmental Science and Technology* **27**(13):2638–2647.

Chappell BA, Burton PL. (1975). Electro-osmosis applied to unstable embankment. *Journal of Geotechnical Engineering, ASCE* **101**(8):733–739.

Chung HI, Lee MH. (2007). A new method for remedial treatment of contaminated clayey soils by electrokinetics coupled with permeable reactive barriers. *Electrochimica Acta* **52**:3427–3431.

Farrell J, Wang J, O'Day P, Conklin M. (2001). Electrochemical and spectroscopic study of arsenate removal from water using zero-valent iron media. *Environmental Science and Technology* **35**(10):2026–2032.

Fetzer CA. (1967). Electroosmotic stabilization of west branch dam. *Journal of the Soil Mechanics and Foundation Division, ASCE* **93**(4):85–106.

Melitas N, Conklin M, Farrell J. (2002a). Electrochemical study of arsenate and water reduction on iron media used for arsenic removal from potable water. *Environmental Science and Technology* **36**(14):3188–3193.

Melitas N, Wang J, Conklin M, O'Day P, Farrell J. (2002b). Understanding soluble arsenate removal kinetics by zero-valent iron media. *Environmental Science and Technology* **36**(9):2074–2081.

Mitchell JK. (1993). *Fundamentals of Soil Behavior*, 2nd ed. New York: Wiley.

Morrison SJ. (2003). Performance evaluation of a permeable reactive barrier using reaction products as tracers. *Environmental Science and Technology* **37**:2302–2309.

Page MM, Page CL. (2002). Electroremediation of contaminated soils. *Journal of Environmental Engineering, ASCE* **128**(3):208–219.

Perry W. (1963). Electro-osmosis dewaters foundation excavation. *Construction Methods and Equipment* **49**(9):116–119.

Pulsa RW, Blowesb DW, Gillhamb RW. (1999). Long-term performance monitoring for a permeable reactive barrier at the US Coast Guard Support Center, Elizabeth City, North Carolina. *Journal of Hazardous Materials* **68**(1–2):109–124.

Reddy KR, Saichek RE. (2003). Effect of soil type on electrokinetic removal of phenanthrene using surfactants and cosolvents. *Journal of Environmental Engineering, ASCE* **129**(4): 336–346.

Simon FG, Meggyes T. (2000). Removal of organic and inorganic pollutants from groundwater using permeable reactive barriers. *Land Contamination & Reclamation* **8**(2): 103–116.

Su C, Puls RW. (2001a). Arsenate and arsenite removal by zerovalent iron: Kinetics, redox transformation, and implications for in situ groundwater remediation. *Environmental Science and Technology* **35**:1487–1492.

Su C, Puls RW. (2001b). Arsenate and arsenite removal by zerovalent iron: Effects of phosphate, silicate, carbonate, borate, sulfate, chromate, molybdate and nitrate, relative to chlorine. *Environmental Science and Technology* **35**:4562–4568.

US EPA. (1995a). In Situ Remediation Technology Status Report: Treatment Walls. *US EPA Research Report EPA/ 542/K-94/004*.

US EPA. (1995b). In Situ Remediation Technology Status Report: Electrokinetics. *US EPA Research Report EPA/542/K-94/007*.

US EPA. (1998). Permeable Reactive Barrier Technologies for Contaminant Remediation. *US EPA Research Report EPA/600/R-98/125*.

US EPA. (2001). A Citizen's Guide to Permeable Reactive Barriers. *EPA/542/F-01/005*.

Vangronsveld J, Cunningham SD. (1998). *Metal-Contaminated Soils: In Situ Inactivation and Phytorestoration*. Berlin: Springer.

Yang GCC, Liu C. (2001). Remediation of TCE contaminated soils by in situ EK-Fenton process. *Journal of Hazardous Materials* **B85**:317–331.

Yeung AT. (1994). Electrokinetic flow processes in porous media and their applications. *Advances in Porous Media* **2**:309–395.

Yeung AT, Hsu C. (2005). Electrokinetic remediation of cadmium-contaminated clay. *Journal of Environmental Engineering, ASCE* **131**(2):298–304.

Yeung AT, Hsu C, Menon RM. (1996). EDTA-enhanced electrokinetic extraction of lead. *Journal of Geotechnical Engineering, ASCE* **122**(8):666–673.

Yeung AT, Scott TB, Gopinath S, Menon RM, Hsu C. (1997). Design, fabrication, and assembly of an apparatus for electrokinetic remediation studies. *Geotechnical Testing Journal, ASTM* **20**(2):199–210.

32

FIELD STUDIES ON SEDIMENT REMEDIATION

J. KENNETH WITTLE, SIBEL PAMUKCU, DAVE BOWMAN,
LAWRENCE M. ZANKO AND FALK DOERING

32.1 INTRODUCTION

In this chapter, field study experiences with an alternative sediment remediation technology known as electrochemical geo-oxidation (ECGO) are presented. The chapter provides conceptual, technological, and operational context for ECGO, focusing on *in situ* and *ex situ* field applications of the technology in Duluth, Minnesota, USA, from 2002 to 2007, and on a 2006 *ex situ* application in Copenhagen, Denmark.

Duluth, Minnesota, is located at the far westerly tip of Lake Superior near the geographic center of the North American continent (Fig. 32.1). The harbor is shared by two states—Minnesota and Wisconsin. In the maritime industry, the port is known as Duluth-Superior, with iron ore docks, coal docks, grain elevators, and specialized cargo facilities lining the industrial waterfronts of Duluth, Minnesota and Superior, Wisconsin. Duluth-Superior is the largest port on the Great Lakes in total cargo volume and is one of the premier bulk cargo ports in North America.

The initial 2002–2003 Duluth, Minnesota experience, which targeted polycyclic aromatic hydrocarbons (PAHs), was a simulated *in situ* sediment treatment demonstration of ECGO, done under the direction of—and with support from—the United States Environmental Protection Agency, Great Lakes National Program Office (USEPA-GLNPO) and the US Army Corps of Engineers (USACE). Additional ECGO investigations in Duluth from 2004 to 2007 focused on modifying field operating conditions (a) by injecting air into the system to change the *in situ* sediment redox (oxidation-reduction potential, or ORP) state (2004) and (b) by heaping the sediments and treating them *ex situ* (2005–2007). While these

---

Electrochemical Remediation Technologies for Polluted Soils, Sediments and Groundwater,
Edited by Krishna R. Reddy and Claudio Cameselle
Copyright © 2009 John Wiley & Sons, Inc.

Figure 32.1. Duluth, Minnesota, location. *Source*: http://www.nrcs.usda.gov/technical/NRI/maps/meta/m4015.html.

supplemental investigations were independently undertaken by ECGO technology developers Electro-Petroleum, Inc. (EPI) and electrochemical process (ecp) llc, they were still observed with interest by the USACE throughout the testing period. All Duluth-based activities were overseen and managed by the Natural Resources Research Institute, University of Minnesota Duluth (NRRI-UMD), while the ECGO system was operated and maintained by Harrison Marine Electronics.

In the Duluth project, average total PAH levels declined by 17.3% overall in the test cell during the 2002–2003 evaluation period. The three-ring compounds (acenaphthene, fluorene, phenanthrene, anthracene, and fluoranthene), which represent about 40% of total PAH by weight, showed an average reduction of 25.2%. These values were lower than what had been anticipated, for example, >50% reduction, but empirical evidence suggested the ECGO technology was indeed working during this period. The *ex situ* (heaped) treatment of the Duluth sediments did not reduce total PAH levels, even though the heaped application of ECGO in Copenhagen was found to effectively reduce contaminant levels there.

This difference in efficiency can be attributed to several complicating factors which contributed to the Duluth outcome, the most significant of which is the unanticipated but overwhelming influence of *solid-phase particulate coal tar on both sampling and analytics*. These factors and additional analytical results are discussed and presented in greater detail in the remainder of this chapter, as are lessons learned and related recommendations and observations.

32.2 BACKGROUND ON THE NEED FOR REMEDIATION AND THE DULUTH PROJECT

Nationwide, the majority of inland and coastal ports and harbors contain sediments contaminated with pollutants ranging from polychlorinated biphenyls (PCBs) to heavy metals. Major projects undertaken to remove contamination such as PCBs in New Bedford Harbor, Massachusetts, the Fox River in Green Bay, Wisconsin, and ongoing contamination removal in the Hudson River will cost tens of millions of dollars. In place of removal, a number of suggestions have been made including the concept of capping in place and natural attenuation. The concept of dredging and soil washing and/or vitrification has also been proposed although not widely used.

The USACE sponsored a number of treatment programs designed to reduce the levels of contamination after the material has been dredged. Alternatives have focused on

1. techniques to separate sand or debris from the dredged material, thereby reducing the quantities placed in confined disposal facilities (CDFs), and
2. techniques to remove contaminants from dredged material in the CDF environment.

Alternative 2 can be further broken down into active decontamination techniques and into more passive techniques, such as composting and phytoremediation. Table 32.1 presents the typical sediment remediation techniques used. *In situ* treatment, however, is another option.

Treatment options for CDFs in the Great Lakes region have received particular attention because, after years of being used for disposal of contaminated dredged materials from navigation dredging projects where open water disposal was not permitted, many CDFs are nearing capacity. It is expensive to site, design, and construct new CDFs, and alternatives to traditional CDF disposal of contaminated dredged materials are needed. To address this need, the US Army Engineer District, Detroit (CELRE) and the Dredging Operations and Environmental Research Program (DOER) at the Engineering Research Development Center (ERDC) partnered with the USEPA-GLNPO to test the feasibility of using various technologies such as composting and phytoremediation to bioremediate dredged materials in CDFs; and with the Natural Resources Research Institute (NRRI) (in 2000) (a) to demonstrate the use of hydrocyclone particle separation technology, first developed for the mining industry, to remove clean sands from contaminated dredged materials in CDFs, and (b) to conduct a literature review of the potential of increased temperatures and pressures combined with oxidizer addition for remediation of PAHs in Lake Superior sediments.

The larger objective of the partnered efforts has been to convert CDFs from perpetual containment facilities to storage, rehandling, treatment, and (ideally) beneficial reuse facilities. The partnered efforts have, thus far, shown that no single technology is fully capable of dealing with the unique chemical and physical properties of contaminated dredged materials in CDFs, and none of the technologies investigated in the partnered efforts were applicable to *in situ* sediment treatment.

TABLE 32.1. Current Cleanup Methods for Contaminated Sediments

Highlight: Cleanup Methods of Contaminated Sediments	
In Situ Methods	Removal *Ex Situ* Methods

Monitored natural recovery
    Physical processes
    Chemical processes
    Biological processes
	Wet dredging (underwater)
	    Hydraulic
	    Mechanical

In situ capping
    Single-layer granular cap
    Multilayer granular cap
    Combination granular/
        geotextile caps
	Dry dredging (not underwater)
	    Dry season work
	    Winter dewatered dredging

In situ treatment
    Currently experimental only
	Containment (disposal of sediment or treatment residuals)
	    Upland
	        On-site landfill
	        Off-site (regional) landfill
	    Upland confined disposal facility
	    In-water disposal
	        Confined disposal facility (above water surface)
	    Confined aquatic disposal (underwater cap)
	    Ex situ treatment
	        Chemical
	        Biological
	        Extraction/washing
	        Thermal

While *in situ* treatment is not typically required for sediments in navigation channels, the areas upstream and outside of the navigation channels are often sites of sediment contamination and are prime candidates for *in situ* remediation. Given the prohibitively high combined cost of removal, transport, and placement of contaminated sediments in landfills, effective *in situ* remediation is a desirable alternative. For example, a recent sediment remediation project in Wisconsin has been projected to have a transport and land filling cost of $308 million, including costs for dredging, transport, offloading, dewatering, and treatment.

High costs like these prompted the federal government to provide funding to both the USEPA-GLNPO and the USACE (also known as the Corps) to evaluate techniques to reduce contaminant concentrations in sediments and in dredged materials (Averett *et al.*, 1990). Programs and authorizations specific to the Great Lakes region include (a) the Assessment and Remediation of Contaminated Sediments (ARCS) program, authorized by the US Congress in 1987; (b) the Alternative Technologies program in the Water Resources Development Act (WRDA), 1996; and (c) the Section 541 Program. The Section 541 program provides authority for the Corps to evaluate emerging remediation technologies at the pilot demonstration scale, specific to sediments from the Duluth Harbor in Minnesota [under Section 541(a)] that fit the section as modified by Section 526 in the WRDA of 2000. Because sediments in the Federal Navigation Channel in Duluth are considered uncontaminated, the Corps has focused on boat slips, embayments, and

Figure 32.2. Great Lakes–St. Lawrence River Basin Areas of Concern (AOC), with arrow denoting St. Louis Bay/River AOC, Duluth/Superior Harbor. *Source*: http://www.ec.gc.ca/raps-pas/D91BD30F-DDC9-4009-ABE0-B3B560EF5324/aoc-e1.jpg.

embankments where contaminated sediments remain. As a result, a project to treat contaminated sediments from one of those Duluth locations, Minnesota Slip, was undertaken in 2002. It is that project which forms the basis for this chapter.

32.2.1 The Duluth Project Sediment Source

In 1987, concerns over environmental quality conditions prompted the International Joint Commission to designate the lower St. Louis River as one of 43 Great Lakes Areas of Concern (AOCs) (Fig. 32.2). The boundaries of the St. Louis River AOC include the river's mouth at Lake Superior, that is, Duluth-Superior Harbor. In recent years, several investigations have taken place in the St. Louis River AOC to assess contaminated sediments. One of these AOC sites, Minnesota Slip, has received particular attention.

Minnesota Slip is located on the northern edge of the Duluth Harbor, as depicted in Figure 32.3. Figure 32.3a shows the location of the treatment site (Erie Pier CDF) for the ECGO demonstration project, which is the main subject of this chapter, while Figure 32.3b shows Minnesota Slip's location between the shipping canal (foreground) and the Duluth Entertainment and Convention Center (background). The slip is also the permanent berth of a retired iron ore carrier and floating maritime museum, the SS William A. Irvin.

Studies by the Minnesota Pollution Control Authority (MPCA) show that slip sediments are contaminated with PAHs, PCBs, mercury, and other miscellaneous

Figure 32.3. Google Earth image of Duluth-Superior Harbor, showing locations of Minnesota Slip (MN Slip) and the Erie Pier Confined Disposal Facility (CDF) (a), and an aerial photo of Minnesota Slip and Duluth Harbor, Duluth, Minnesota (b). *Source*: Jerry Paulson, Duluth, Minnesota, http://www.duluth-mn-usa.com/.

contaminants. Bioaccumulation studies indicate that benthic worms accumulate PAH compounds, but little PCB or mercury, from Minnesota Slip sediments. Thus, PAHs are contaminants of primary concern. Work by Crane, Schubauer-Berigan, and Schmude (1997) and Crane *et al.* (2002) showed that Minnesota Slip sediment contamination was heterogeneous, and total PAH concentrations ranged from 9.8

TABLE 32.2. The List and Properties of the Polycyclic Aromatic Hydrocarbons (PAHs) Analyzed

PAH Compound	PAH Molecular Weight	Number of Rings
Naphthalene	128.17	2
Acenaphthylene	152.2	3
Acenaphthene	154.21	3
Fluorene	166.22	3
Phenanthrene	178.23	3
Anthracene	178.23	3
Fluoranthene	202.26	4
Pyrene	202.0	4
Chrysene	228.0	4
Benzo(a)anthracene	228.0	4
Benzo(b)fluoranthene	252.0	5
Benzo(k)fluoranthene	252.0	5
Benzo(a)pyrene	252.0	5
Indeno(1,2,3-cd)pyrene	276.0	6
Dibenz(a,h)anthracene	278.0	5
Benzo(g,h,l)perylene	276.0	6
2-Methylnaphthalene	242.2	2

to 1370.0 mg/kg (or ppm). The mean total PAH concentration was 142 mg/kg, and the median was 66 mg/kg. The type and properties of the PAHs found at the site are given in Table 32.2. These studies did not, however, highlight the issues of coal tar, a source of PAHs, in the sediment.

32.2.2 The *In Situ* Pilot Project

Based on the issues just discussed and previous investigations, the USACE and Great Lakes National Program Office (GLNPO) partnered on the development of a pilot-scale demonstration project to assess a potential *in situ* sediment treatment technology called ECGO. The demonstration project, Emerging Technology Demonstration at the Erie Pier Confined Disposal Facility, was funded by GLNPO to test the ECGO process, via Contract Number DACW35-02-0019 (Zanko and Oreskovich, unpublished observations). A summary of that project, conducted in 2002 and 2003, but followed up on with modified demonstrations through 2007, is presented in the section The Remediation of Minnesota Slip Sediments at the Erie Pier CDF in Duluth, Minnesota of this chapter. But first, an overview of the technological and scientific concepts that provide the basis for ECGO is presented in the next section.

32.3 WHAT IS ECGO TECHNOLOGY?

ECGO is an emerging European technology developed by Dr. Falk Doering for sediment treatment that has been tested in Austria, France, Germany, and Denmark.

As will be described, the technology has shown promise as a cost-effective and rapid remediation option. ECGO uses low voltage and amperage to induce reduction–oxidation reactions at the microscale. In essence, the method advocates that each sediment particle acts as a microcapacitor that charges and discharges in a cyclic fashion. Even though low voltage and amperage are used, the energy burst on discharge at the microscale is intense, resulting in destruction of organic contaminants, theoretically to carbon dioxide and water.

ECGO has also been shown to offer several technical advantages as follows: noninvasive, *in situ* or *ex situ* application, suitable for a wide variety of grain sizes and mineralogies and multicontaminant sites, no waste streams produced, and no additives or noise pollution. It should be noted that since the technology is currently available at demonstration/documentation scale, no regulatory evaluation for full-scale implementation has been conducted in the USA. Hence, in this chapter, we only present and discuss the long-term results of the demonstration efforts of ECGO undertaken on sediments from the Minnesota Slip site.

32.3.1 The Concept of Induced Polarization (IP)

The patented base technology for the electrochemical remediation of organic and metal–organic pollutants in soils referred to as ECGO is a continuation in part of the research into IP (Vacquier *et al.*, 1957; Marshall and Madden, 1959). IP describes certain phenomena when a direct electric current at high voltages is introduced via *in situ* electrodes into a body of surface conductive porous materials such as soil or rock (Ward and Fraser, 1967), and is switched off after some time in order to measure the apparent resistivity of the subsurface. If an electric field is introduced into the ground, the polarization sites on the surface of mineral grains will become polarized, with the down-gradient side of each site positively charged and the up-gradient side negatively charged, such that each polarization site becomes an electric dipole in opposition to the inducing electric field. The individual site polarizations occur in a cascading manner, much like a row of falling dominoes.

Electric conduction in aqueous solutions in soil pores occurs by ion transport. Cations (positively charged ions) move toward the cathode (negatively polarized electrode), and anions (negatively charged ions) move toward the anode (positively charged electrode). On the particle surfaces, the passage of electricity is electronic conduction based on the model of the Helmholtz double layer comprising the inner Helmholtz layer (IHL) and outer Helmholtz layer (OHL) lining the capillaries and forming so-called cation "selective membranes." An external electrical field drives cations and their associated water molecules through the cation-selective membrane. If the anions and their associated water molecules cannot enter the pore throat filled with Helmholtz double layers, a chemical imbalance results across the electrodes. In this fashion, electrical energy input to the system is stored as chemical energy.

In IP, there exist two paths by which current may pass the interface between the solid particle and the electrolyte: the faradaic and nonfaradaic paths. Current passage in the faradaic path is the result of electrochemical reactions (redox reactions) and the diffusion of charge toward or off the Helmholtz double layer and aqueous solution interface, that is, "Warburg impedance." In the nonfaradaic case, charged particles do not cross the interface. Instead, the current is carried by charging and discharging the double layer, behaving as a capacitor. Empirical evidence indicates that both types simultaneously exist (Ward and Fraser, 1967).

IP defines "soil" as a complex system of mineral particles in an aqueous matrix. This aqueous matrix comprises (a) soil water (or gravitational water), in general flowing through the capillaries, and (b) hygroscopic (colloquially, "captive") water, fixed by and onto the mineral particles. Any soil particle, once being in contact with the groundwater, builds up a "double layer." Driving forces of this phenomenon are electrostatic forces such as van der Waals, London, and Coulomb forces creating the so-called adsorption complex comprising two water layers. The layer near the solid surface is a highly viscous, highly pressurized (water pressure = 10,000–25,000 bars) hygroscopic water hull, whereas the second layer is captive water, which marks the transition area to the bulk solution (pore water) with very little electrostatic forces. When an electrical field is applied across the double-layer structure, a dielectric layer is generated between the hygroscopic hull and the captive water. This dielectric layer is the so-called solvation water layer, in which ions are converted into a solvate. Since natural electrical fields in the range of some millivolts to about 1 V are ubiquitous, the generation of the solvation water layer is a standard phenomenon in soils. By this phenomenon, the double layer is converted into a triple layer having the properties of a capacitor. Under undisturbed field conditions, soils in general have a capacitance in the range of 5–15 μF.

The above nomenclature can, without any difficulty, be made compatible with the electrochemical terminology. The hygroscopic water corresponds to the IHL; the solvation water to the term OHL, and the captive water to the term diffuse layer. In addition, the colloid structure provides for an interface between the OHL and the diffuse layer, called outer Helmholtz plane (OHP), the precondition for any and all electrochemical reactions. The electrical potential between the OHL and the diffuse layer is known as zeta potential, or electrokinetic potential.

The diffuse layer shrinks over time due to the water consumption by electrolysis, reducing the magnitude of the zeta potential (to lower negative values) and the potential gradient across the diffuse layer, hence causing the capacity of this layer to increase. To provide an order of magnitude, in a test project in New York State, the capacitance of a soil cell of 1 m^3 increased within 90 days from 13 to 1020 μF at a voltage of 100-V direct current (DC) and 5 A.

32.3.2 ECGO: A Low-Voltage Phenomena

The first electrical requirement for ECGO is the definition of all soil particles as microelectrodes having the triple layer structure described previously. The observation of the electrical behavior of the OHL and the diffuse layer identifies different phenomena, which are decisive for ECGO. The first phenomenon relates to the conductors in soils. Rahner, Ludwig, and Rohrs (2002) gave preference to the electronic conductor in soils. With exception of deposits of ores, the chemical composition of most soils does not support Rahner's hypothesis of soils being solely an electronic conductor. The second conductor is well known as the ionic conductor, that is, the conductance by the groundwater in capillaries in the soil, as used by IP. The resistance of the groundwater serving as carrier for even high electrical currents required for electrokinetics, that is, electrokinetic transport of ions, complex ions, and colloids, is fairly high (>20 Ω). Due to the lack of interface(s) in the ionic

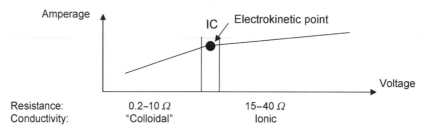

Figure 32.4. Depiction of the electrokinetic point in ECGO (IC = ionic conduction).

conductor, almost no electrochemical reactions take place in it. When applying low voltages, a third conductor has been detected, called the colloid conductor. The colloid conductor is represented by the diffuse double layer (DDL) formation. The DDL comprises high concentrations of ions, complex ions, and colloids, which permit a resistance between 0 and <10 Ω at low voltages (regularly <75-V DC). When increasing the voltage by a rheostat, the transition from the colloid conductivity to the ionic conductivity can be measured by an oscilloscope. The point where this transition happens is named "electrokinetic point." When voltage is increased, the discharge signals caused by the capacitance in oscillograms disappear at this particular potential. Figure 32.4 depicts the typical relation between amperage and voltage around the electrokinetic point.

The second electrical requirement of ECGO is direct electric current application using *in situ* electrodes. In field remediation projects, the electrodes, which create an electric field in the soil, are often placed at distances of 5–30 m, and in special applications, the distance may be as large as 300 m. This means that the electric field applied ranges from 0.0025 to 0.25 V/cm for up to 100 V applied across electrodes. In this electrical field, the soil DDL lining the pore space becomes polarized and electrons begin to migrate from the working anode to the cathode primarily within the diffuse layer of each individual soil particle, with the assumption that the DDLs are connected and continuous. The electrons, as they migrate across the boundaries of the diffuse layers between adjacent particles, essentially flow off on one side (donation) and are added on the other (acceptance). If the rate and quantity of electron donation and acceptance are constant throughout a continuous thread of double layers, then no other effect but ionic conduction and electroosmotic flux should be observed. At high applied electric fields, that is, which surpass the "electrokinetic point," further contraction of the diffuse layer lowers the electrokinetic potential at OHP to such values that electronic conduction no longer can compete with ionic conduction; hence, the ionic transport takes over the process.

Yet at low electric fields (0.0025–0.25 V/cm) and without the idealization of continuous and connected DDLs lining the pore walls, as electrons leave or enter the diffuse layer, physical discontinuities that do not allow easy "hopping" of the electrons from one DDL to another result in permanent chemical transformations in the form of electrochemical reactions in the DDL. This means that each soil particle serves as a "microelectrode" accepting electrons for reduction of the species within, or donating electrons for oxidation of species in the adjacent layer. Except in those events when a direct oxidation occurs by electron transfer, ECGO generates itself, *in situ*, the agents for reduction (H as ion or radical) and oxygen (O elemental, OH

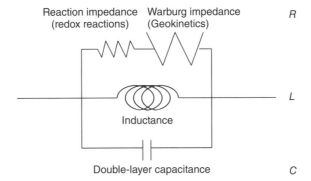

Figure 32.5. ECGO equivalent circuit model. *R*, resistance; *C*, capacitance; *L*, induction.

and its radicals, HO_2 and its radicals) and—when combining the generation of hydrogen peroxide (H_2O_2) with the corrosion products of steel anodes—also Fenton's reagent (Fenton's reagent is defined as a solution of hydrogen peroxide and an iron catalyst used for remediation).

The third electrical requirement for ECGO refers to the component related to the alternating current (AC). Under the conditions of polarization, the triple layer acts as a capacitor that loads and discharges electricity in the frequency of the AC. Looking more closely into this electric behavior, we find that the soil particle may fulfill the requirements of an RCL element which describes a simple electronic amplifier, whereby *R* stands for resistance (soil resistance, Warburg impedance), *C* for capacitance, and *L* for induction, as shown in Figure 32.5. The RCL element amplifies the electric discharges of the double-layer structure by a factor of 8000–10,000. This phenomenon of amplification, on the other hand, is responsible for ECGO working even at low voltages and amperages supplying the required energies (including activation energy) to the redox reactions. The central problem, however, is not the low primary energy delivered to the electrodes, but the control of the amplification of electric potential delivered, which is mainly a function of inductance, capacitance, and conductance of the soil and the frequency of the induced AC.

32.3.3 Comparison of IP to ECGO

In developing ECGO, five distinctions are observed from IP:

1. In ECGO, the subject matter is soils comprising a broad spectrum and mixture of minerals.
2. Whereas IP analyzes the decay of the voltage when shut off, ECGO applies a continuous DC without interruption for a longer period.
3. ECGO distinguishes between two types of electrodes in soils: (a) the working electrodes serving the introduction of electricity into the soil and (b) the soil particles serving as microelectrodes including the solid/electrolyte interface.
4. IP is applied at high voltages, whereas ECGO requires low voltages regularly below 100V, and electric fields of 0.0025–0.25 V/cm.

5. The basic phenomena of ECGO are released only by a coupled AC/DC current having a ripple in the range of >10% of the DC input.

32.4 THE REMEDIATION OF MINNESOTA SLIP SEDIMENTS AT THE ERIE PIER CDF IN DULUTH, MINNESOTA

The Duluth demonstration project described in this section targeted PAHs. The demonstration began in 2002–2003 as a simulated *in situ* sediment treatment pilot project. After 2003, it continued as a project to test ECGO under modified field operating conditions and concluded in 2007 as an *ex situ* (heaped) treatment test.

32.4.1 Demonstration Project I: *In Situ* ECGO Treatment of Erie Pier CDF Cell

The Erie Pier CDF was constructed in the mid 1970s by the USACE to contain dredged material from the Duluth-Superior Harbor. Because the CDF is more than 90% filled (nearing capacity), contaminated and uncontaminated dredged material would eventually have to be transported to a landfill at considerable expense. Given the pressing need to find a more economic solution to this impending problem, the USACE and the Duluth Seaway Port Authority (owner of the CDF) were willing to set aside a small area in the corner opposite the offloading site for testing sediment treatment technologies. Conducting the study within the CDF rather than in Minnesota Slip location also allowed

1. more thorough mixing (and theoretical homogenization) of the dredged sediments and
2. better control of the sediments by avoiding new contaminants being introduced during the course of the study or of contaminants being carried away by currents.

The decision was made to evaluate the treatment of sediments in *in situ* conditions since the candidate areas, that is, boat slips like Minnesota Slip, were not being considered for navigation dredging. Consequently, the ECGO technology was selected as the *in situ* treatment method.

The overall purpose, that is, to investigate the efficacy of the ECGO technology for *in situ* sediment treatment in a demonstration test in the Erie Pier CDF, Duluth, Minnesota, meant the study required an independent evaluation of the technology in a controlled and monitored test situation, but at a sufficient scale to provide realistic information on costs, effectiveness, and ease of implementation. Hence the specific objectives were identified as

- proof of concept that ECGO can treat representative organic compounds in Great Lakes sediments, especially PAHs;
- to simulate *in situ* ECGO sediment treatment;
- bulk sediment monitoring throughout the duration of the ECGO technology demonstration to document expected reductions in chemical concentrations over time; and
- to investigate changes in toxicity before and after treatment.

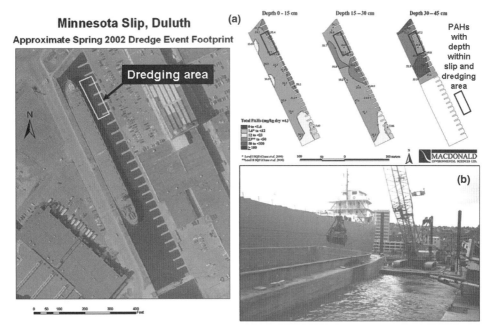

Figure 32.6. Minnesota Slip dredging location and PAH levels (a) and land-based dredging of sediment (b). *Source*: Streitz and Johnson (2005). Detailed Investigation of the Minnesota Slip Featuring Laser Induced Fluorescence. *MPCA Technical Document, tdr-g1-01* (a).

From this, the controlled pilot test was designed, with the work involving the following phases:

Phase 1. Coordination. CELRE coordination with MPCA, United States Environmental Protection Agency (USEPA) Region V, ERDC the technology vendors (EPI and ecp), the NRRI, Harrison Marine Electronics, and others as appropriate was essential for the successful execution of all aspects of the project, from permitting through sampling and analysis.

Phase 2. Test Cell Preparation. Test cell preparation involved raising berms for experimental and control cells, one each, at the Erie Pier CDF. Berms were constructed using dredged material that is already in the CDF. The test cells were anticipated to provide a minimum containment volume of $382\,m^3$ ($500\,yd^3$) each with approximate dimensions of 2-m (6-ft) depth by 27 m (90 ft) by 7 m (24 ft), based on the amount dredged.

Phase 3. Dredging and Dredged Material Placement. Sediment was mechanically dredged from the Minnesota Slip in spring 2002, following MCPA approval of the permit application. Figure 32.6a is an aerial photograph of Minnesota Slip showing the candidate dredging area. MPCA data were reviewed to confirm that PAH concentrations were sufficient to validate ECGO technology (inset Fig. 32.6a), and this area was dredged using a clamshell excavator (Fig. 32.6b). Dredged materials were transported to and placed in the test cells constructed in Phase 2 at the Erie Pier CDF. One cell served as the ECGO treatment (test) cell, and one served as a control. The experimental cells were filled using dump

674 FIELD STUDIES ON SEDIMENT REMEDIATION

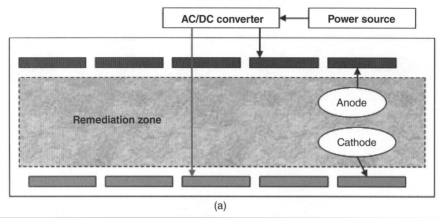

Figure 32.7. ECGO test cell treatment layout (a) and ECGO cell construction at the Erie Pier site (b) [*cell construction photo source: Mr. Curt Anderson, USACE Duluth Area Office (retired)].

trucks, so it was possible to alternate the dumps between the two cells, thereby providing comparable materials and conditions in both cells.

Phase 4. Electrode Installation. Electrode installation took place in June, 2002. Electrodes were to be installed parallel to the long side of the rectangular cell (Fig. 32.7a). Figure 32.7b shows the stages of cell construction including electrode placement. The electrodes (anodes and cathodes) were constructed from steel pipe, although steel sheet pile was another option; it was simply a matter of vendor preference. Typical anode–cathode separations range from 5 to 15 m, but the actual layout depends on the test cell size, which in turn depends on the size of the dredging project. A generator, underground conduit for running electrical wiring, and other project equipment were installed at this time.

Phase 5. Quality Assurance Project Plan (QAPP). A detailed QAPP was developed for review by GLNPO. The QAPP documented all sampling and analytical chemistry procedures and health and safety concerns.

Phase 6. System Start-Up. Activation of the ECGO system included installation of the power supply, monitoring and recording equipment, and control system; and troubleshooting, calibrating, optimizing, and stabilizing system equipment and operating parameters. Typical ECGO applications use voltages and electrical currents that vary from 20 to 100 V and are approximately 0.05–50.0 A, using local utility power. For this demonstration, however, a portable propane generator was needed to supply power.

Phase 7. Operation. Experimental and control cells were maintained in a saturated condition during the demonstration by addition of water. The volume of water delivered to and maintained in each cell was also recorded. The purpose of maintaining a saturated condition in each cell was to simulate *in situ* treatment and to maximize treatment efficiency. By simulating in situ treatment, the results of the test were more generally applicable to AOCs in the Great Lakes. Further, conducting the test under wet conditions did not preempt application of the results to CDFs because CDFs that contain fine-grained dredged material are usually saturated throughout most of their vertical profile.

Phase 8. Monitoring and Sampling. Monitoring included sediment sampling for chemical analyses and ECGO-specific performance metrics. Five sampling events were originally planned for 2002 only, but the system was restarted in 2003 and was sampled twice more. The initial sampling event took place prior to ECGO system start-up. During each sampling event, five cores were collected at random locations from each cell. Each core was split into two sections representing the upper and lower halves of a cell. The subsamples were thoroughly mixed prior to analysis. Initial and final 2002 samples were analyzed for 17 PAHs, PCB aroclors, and 60 PCB congeners. Samples from the intermediate sampling events were analyzed for PAHs only. All samples were analyzed for total organic carbon (TOC) and pH.

The sampling schedule was as follows:

Event	Day
Initial (July 25, 2002)	0
First intermediate	33
Second intermediate	68
Third/fourth intermediate	97
Fifth intermediate	342
Final (November 11, 2003)	474

32.4.1.1 Discussion of Chemical Analysis To a regulator, chemical analysis reveals a number to use in evaluating the contamination at a site provided the correct contaminant is picked for analysis. To the regulated party, it provides the change in numbers to demonstrate that the site is being treated or is clean. As we found in our study, one needs to be sure of all of the constituents in the material being sampled. If the site is contaminated with additional source material that (a) is due to sample preparation protocols, calls for its removal from the sample prior to analysis, or (b) is a precursor of a chemical being analyzed, then the results may be misleading. In our study, we experienced both, the most significant of which is the quantity of coal tar present in the mixture. The presence and impact of coal tar is a recurring theme that is addressed in upcoming sections.

As the contracting authority for the technology vendor, the NRRI became more familiar with the ECGO technology as the project progressed. Likewise, NRRI's on-site sampling responsibilities and in-house laboratory work presented opportunities for observing potential treatment effects that may not have been reflected by the PAH analyses alone. The following observations were made and recorded:

- Persistent gas formation (bubbling) in the ECGO treatment cell throughout the project indicated that electrochemical reactions were occurring. No such gas formation was observed in the control cell. The gas ignited when touched with a flame, and was most likely hydrogen (as a product of electrolysis) or methane (as a product of biological activity perhaps enhanced by the ECGO process).
- Core from the test cell often contained gas pockets; core from the control cell did not.
- Sediment residue on the exterior of test cell sample tubes over time generally appeared to be lighter colored (brown) than residues on sample tubes withdrawn from the control cell, which often appeared very dark brown to almost black. Figure 32.8 illustrates this color change over a few months' period of time in collected samples.
- While empirical field observations suggested evidence of reactions taking place in the presence of the electric field, the numeric analytical data were statistically inconclusive regarding ECGO technology reducing total contaminant (PAH or PCB) levels.
- There was no decrease in TOC and no change in pH.

PAH Results. The graph presented in Figure 32.9 summarizes the compiled PAH data based on the following approach. All control cell analyses and the initial (prestart) test cell sample analyses are combined, by PAH compound, and are averaged to establish a full project (2002 and 2003) "baseline" (or non-ECGO treatment) PAH level. All poststart analyses from the test cell are combined, averaged, and compared to this baseline, and are expressed as a percent "reduction" if negative (or percent "increase," if positive) from the full project baseline levels, compound by compound.

The acenaphthylene column (second compound from the left) is blank because its analytical values were usually too low to be determinate. 2-Methylnaphthalene (rightmost column) is slightly positive, indicating an increase. The analysis shows an overall 17.3% reduction (test vs. control) in total PAHs. The thee-ring com-

THE REMEDIATION OF MINNESOTA SLIP SEDIMENTS AT THE ERIE PIER CDF 677

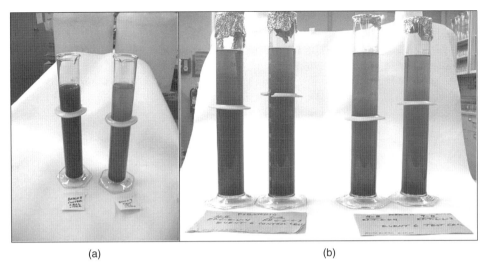

Figure 32.8. Color comparison of sediments in suspension: July 2003 samples (a) and November 2003 samples (b).

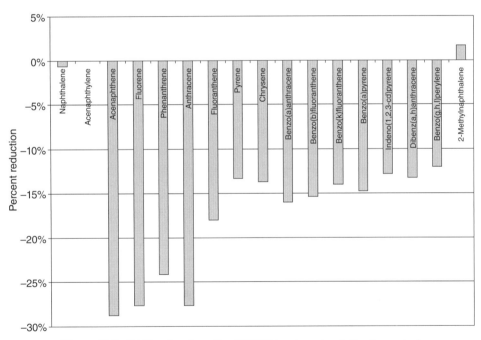

Figure 32.9. PAH reduction after ECGO treatment relative to control.

pounds (acenaphthene, fluorene, phenanthrene, anthracene, and fluoranthene), which represent about 40% of total PAH by weight, show an average reduction of 25.2%. For reference, Table 32.2 presents the molecular weight and number of rings for the 17 PAH compounds analyzed for this project.

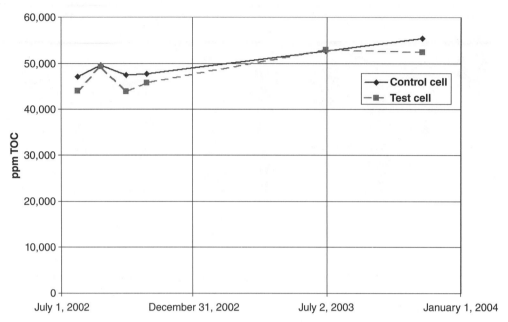

Figure 32.10. Total organic carbon (TOC) variation over treatment time, in parts per million.

TOC Results. Figure 32.10 shows there was no decrease in TOC. The fact that TOC increased by 18–19% in both cells, especially between 2002 and 2003, suggests that the vegetative growth that developed in 2002 and flourished in 2003 may have contributed to the increase. The photographs in Figure 32.11 show how vegetation conditions changed in the test cell from 2002 (a) to 2003 (b). Nearly identical changes in vegetation conditions occurred in the control.

Humic substances (HS) have also been postulated as a contributor to the ECGO treatment inefficiencies. HS are "a series of relatively high-molecular-weight, brown to black colored substances formed by secondary synthesis reactions. The term is used as a generic name to describe colored material or its fractions obtained on the basis of solubility characteristics" (http://www.ar.wroc.pl/~weber/humic.htm).

- humic acids (HAs)
- fulvic acids (FAs)
- humins

HAs. HA is the fraction of HS that is not soluble in water under acidic conditions (pH < 2) but is soluble at higher pH values. HA can be extracted from soil by various reagents and is insoluble in dilute acid. HAs are the major extractable component of soil HS. They are dark brown to black in color.

FAs. FA is the fraction of HS that is soluble in water under all pH conditions. It remains in solution after removal of HA by acidification. FAs are light yellow to yellow brown in color.

THE REMEDIATION OF MINNESOTA SLIP SEDIMENTS AT THE ERIE PIER CDF 679

(a)

(b)

Figure 32.11. 2002 aerial view of test cell vegetation conditions (a) and 2003 ground-level view of test cell vegetation conditions (b).

Humin. Humin is the fraction of HS that is not soluble in water at any pH value and in alkali. Humins are black in color.

The pH of sediment samples was determined to be neutral, with pH levels typically ranging from 7.00 to 7.15. Therefore, any humic and FAs present would have been in a soluble form.

The Duluth-Superior Harbor (St. Louis River) contains HS from upstream peat bogs, which explains its tea-colored waters. The pH of its waters is also in the neutral range. Figure 32.12 (Stevenson, 1982) illustrates the range of colors associated with HS and how these colors are related to molecular weight, carbon content, oxygen content, acidity, and solubility.

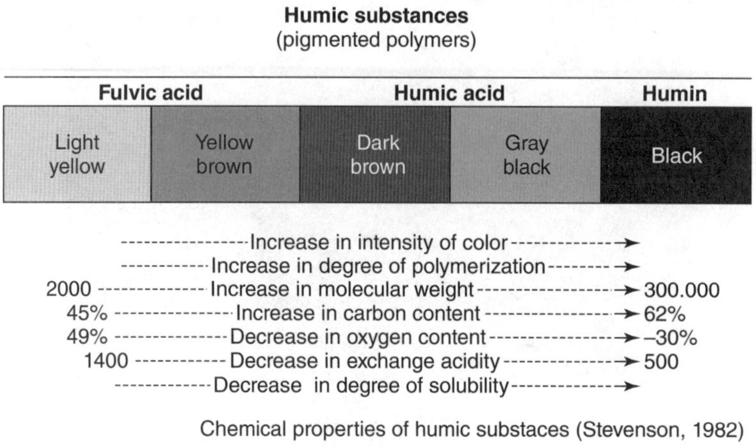

680 FIELD STUDIES ON SEDIMENT REMEDIATION

Humic substances
(pigmented polymers)

Fulvic acid		Humic acid		Humin
Light yellow	Yellow brown	Dark brown	Gray black	Black

```
---------------Increase in intensity of color------------→
------------Increase in degree of polymerization---------→
2000 -------------- Increase in molecular weight-----------------→ 300.000
45% ---------------- Increase in carbon content ----------------→ 62%
49% -------------- Decrease in oxygen content ----------------→ -30%
1400 ------------ Decrease in exchange acidity --------------→ 500
------------Decrease in degree of solubility-------------→
```
Chemical properties of humic substances (Stevenson, 1982)

Figure 32.12. Range of colors associated with humic substances (Stevenson, 1982).

Referring again to Figure 32.8a, the left-hand cylinder contains upper control cell material, and the right-hand cylinder contains an equal amount of upper test cell material from the July 2003 sampling event; in Figure 32.8b, the two cylinders on the left contain upper and lower control cell material, and the two cylinders on the right contain upper and lower test cell material from the November 2003 sampling event. Material in all cylinders was mixed and placed into suspension simultaneously, and was allowed to settle overnight before being photographed. Both photographs show a color difference between the control (darker) and test (lighter) materials, although the difference is subtler in Figure 32.8b. While these comparisons are purely empirical, they do provide a field-based illustration of HS colors summarized by Stevenson (1982).

Did the colors reflect a difference in cell conditions, possibly influenced by the ECGO process, for example, by a change in ORP or by degradation of higher-molecular-weight substances, or were they merely an observational coincidence? Although not possible to state with any certainty without a more rigorous and controlled investigation of the comparison technique, another tool is potentially available in the investigation of sediment treatment technologies.

PCB Results. Lastly, it was concluded that PCBs had decreased, but not in response to ECGO. PCBs were analyzed in 2002, but not in 2003. Analyses from the first (pre-start-up) sample event (July 24–25, 2002) for both cells were combined to establish a PCB "baseline" against which the analyses from the October 29, 2002 sample event were compared. Both the test and control cells showed similar reductions in PCBs from initial baseline levels, although the test cell had slightly lower final values. Given that the PCB levels were very low to begin with (parts per billion vs. the parts per million levels found at PCB-contaminated sites) and that only one set of post-start-up samples was analyzed, the results were difficult to assign significance and are mentioned here primarily for information purposes only.

32.4.1.2 The Significance of ORP in the Treatment Process ORP is related to the concentration of oxidizers or reducers in a system and indicates their activity

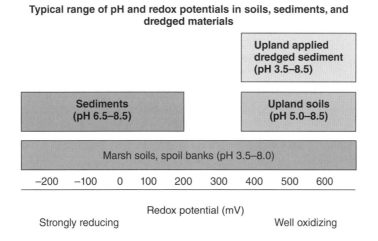

Figure 32.13. Range of expected ORP for various types and sources of soils and sediments (Patrick, Gambrell, and DeLaune, 2003).

or strength. ORP is a measure of the system ability to oxidize (accept electrons) or to reduce (donate electrons) in the system and subsequently to oxidize or to reduce other components in the system. In soil and sediment remediation, if the ORP is positive, less oxidant will be required to oxidize a component. If it is positive and the objective is to reduce a component, more reducing agent will be required. In most soils and sediments, the ORP is generally found to be more positive in systems exposed to the air and to be more negative in systems such as sediments not exposed to air. In Figure 32.13, the range of expected ORP for various types and sources of soils and sediments is shown (Patrick, Gambrell, and DeLaune, 2003).

In addition to the ORP (or E_h), the pH of the system will play a part in the range that oxidation and reduction can take place. Referred to as an E_h–pH relation, a diagram of which normally shows the anticipated stable species that may be available in the sample at a given pH and E_h is illustrated in Figure 32.14 for near-surface sediments (Dragun, 1988).

Following the first year of ECGO testing at Erie Pier, it became apparent that the electrochemical treatment of PAHs was not proceeding as fast as anticipated. When testing was continued into 2003 and similar results were returned, it was speculated that the ORP of the system may have been too negative (reductive). To test this, ORP readings were made of the same core taken for chemical analysis. The results are shown in Figure 32.15 and indicate a large drop in the ORP at the water–sediment interface. A wide fluctuation in the ORP with depth is also observed, with swings for the test samples T, T1, T2, and T3. It should be noted that in sample T1-32, a large drop is associated with a tar fragment at 3 ft of depth, which possibly had a large negative impact on the ORP. Subsequent chemical analysis of tar in the sediment indicated high concentrations of PAHs, a finding that will be discussed in greater detail in upcoming sections.

In order to mitigate the influence of *in situ* low ORP on the process, two changes in operation were implemented. First, remove vegetation from the site and

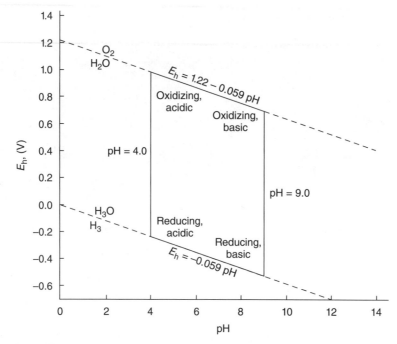

Figure 32.14. pH and E_h relation for near-surface sediments (Dragun, 1988).

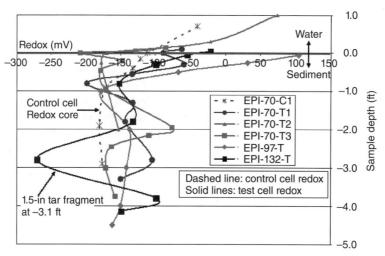

Figure 32.15. Measured ORP distribution in test and control cells in year 2 of treatment.

cover to prevent further plant growth and to minimize potential humate-producing plant decay. Second, develop a method to increase the ORP. Both changes were accomplished by pulling all plants, covering the cell with black plastic membrane, and injecting air into the cell at a depth of approximately 3 ft through 0.5-in. tubing. The tubes were inserted on an approximately 3 × 3 foot grid. A picture of the modified test cell is shown in Figure 32.16. The rate of air injection was set at about

Figure 32.16. Aerated test cell configuration with plastic membrane cover.

12 cfm at 4–5 psi, using an oil-less air compressor. ORP measurements were made on cored samples during each sampling period. The overall results indicate that the ORP was increased slightly (by +25 mV) using the air injection process.

32.4.2 Demonstration Project II: *Ex Situ* ECGO Treatment of the Erie Pier CDF Cell

Following approval by the Corps and the Duluth Seaway Port Authority to continue testing ECGO at the Erie Pier site, EPI and ecp converted the Erie Pier test from an *in situ* to an *ex situ* (heaped) condition in 2005. The context and basis for evaluating ECGO as a potential *ex situ* (heaped) sediment remediation technology in the Duluth, Minnesota, project is explained in the following European field case study overview.

32.4.2.1 A European Field Case Study of **Ex** *Situ (Heaped) ECGO* In 1990, a Danish Soil Recycling Company started operation of seven soil recycling centers in Denmark and Sweden, applying primarily microbiological processes for the off-site remediation of heaped soils polluted by total petroleum hydrocarbons (TPHs) and PAH. The soil masses were delivered by truck to the sites, where the masses were weighed and sieved. The sieved (<20-mm grain size) fraction was heaped up regularly in heaps of about 2000–3000 tons, to be used for landscaping after remedial action. A coarser grain size (>20 but <50 mm) was heaped up separately for later use as gravel, whereas the coarsest (>50 mm) debris and stones were separated for use in road construction and coast protection. Organic substances, grass, paper, and so on were treated separately.

These soil remediation activities were governed by the regulatory cleanup levels for the total hydrocarbons (C6–C 35) (TPH): 200 mg/kg dry mass (dm); for gasoline (C6–C10) including aromatic substances: 35 mg/kg dm; for diesel fuel fractions (C15–C25): 75 mg/kg dm; and for heavy oils (C25–C35): 200 mg/kg dm. For benzene, toluene, ethylbenzene, and xylenes (BTEX) chemicals, the cleanup level was 10 mg/kg dm, thereof for benzene: 1.5 mg/kg dm. The official cleanup level for seven PAHs [fluoranthene, benzo(b)fluoranthene, benzo(j)-fluoranthene, benzo(k)fluoranthene, benzo(a)pyrene, dibenzo(a,h)anthracene, and indeno(1,2,3-cd)pyrene] totaled 15 mg/kg dm, and that for special individual species like benzo(a)pyrene, dibenzo(a,h) anthracene, and naphthalene totaled 1 mg/kg each.

In order to assess the efficiency of the different technologies, including the experimental ones, a performance rate, k, has been calculated as follows:

$$C_t = C_0 \cdot e^{-kt}, \tag{32.1}$$

where C_t is the concentration at time t; C_0 is the initial concentration; k is the rate of decrease of the concentration (% decrease per day), and t is the time.

In 2006, the Soil Recycling Company was dissatisfied with the technological/commercial performance of the microbiological processes, and 12 heaps (or about 40,000 tons) of soil did not reach the cleanup levels at all and had to be disposed of in landfills. The average k for TPH was calculated at $k_{microbio} = 0.04$, which means that 0.04% of the concentration is eliminated each day. This k factor was considered as the soil's "natural attenuation capability." A total of 61 heaps or about 171,800 tons of soils were successfully cleaned up at an average $k = 0.35$, requiring treatment periods of about 1025 days on the average.

In mid-June 2006, the company started examining the performance of the ECGO treatment of 16,200 tons of soil, piled up in 11 heaps. Figure 32.17 shows a picture of a typical heap of treated soil, where hollow steel pipes were used as the sacrificial electrodes. On the average, each heap comprises 15 increments between 50 and 400 tons. The soils in general were sandy silts, with origin in Greater Copenhagen. About 6600 tons were fresh material, that is, being on the site for less than 100 days, whereas about 9600 tons had already been treated on the site without success by microbiology for a 500- to 1750-day period (on the average for 1025 days), or about 2.8 years. The electrochemical treatment of all the heaps was completed, on the average, in 177 days.

The remedial action was directed by a quality control system. Prior to heaping, the soil to be analyzed had undergone a sieving procedure, sorting out stones, debris, and waste >2 cm of diameter. By sieving, the soil was homogenized in portions of about 15 tons (= 1 truck load). By heaping up, the soil was homogenized again. Heaps, each containing about 1500 tons of soil, were sampled by a handheld auger at 20 sampling points per heap, with each sample comprising four increments. Sampling events included baseline sampling and periodic sampling, which continued every 2 months until the final sampling, for a total of seven rounds of sampling. In the laboratory, the incremental samples were homogenized per sampling point and thereafter were subcored for analysis. Sampling followed the standards provided by the International Organization for Standardization (ISO) ISO 10381-1 to ISO 10381-5, whereas soil pretreatment was governed by ISO 14507. Chemical analyses were based on analytical protocols issued by the Danish Government. As

Figure 32.17. A picture of ECGO treatment heaps with vertical steel electrodes, Copenhagen, Denmark.

to TPH, the analyses were performed by gas chromatography/flame ionization detector (GC/FID) using pentane as solvent for substances C6 to <C35 at the analytical temperature between 50 and 300 °C. PAHs were analyzed according to EPA 3546 resp. 8270. All chemical analyses were performed by accredited external laboratories.

Table 32.3 summarizes the chemical analysis results for nine heaps at the end of 177 days. Two heaps were excluded from the assessment due to the presence of a mixed organic/inorganic pollution. Of special interest was TPH in C25–C35, which could not be remedied by microbiological methods. In all heaps, the different types of petroleum product responded to ECGO. TPH after 177 days of treatment was substantially below regulatory cleanup levels. As to PAH, the concentrations prior to the test were already below the cleanup levels and are reproduced only for purposes of completeness.

The average performance rate for TPH when using ECGO was $k_{ECGO} = 1.08$. In other words, approximately 1% of the soil contaminants had been eliminated each day. The comparative k factors determined for different treatment methods of the recycled soil are given in Table 32.4. As observed, the k factor for ECGO treatment was determined to be 26 times higher than that of natural attenuation and was three times higher than the biological methods.

32.4.2.2 The Duluth, Minnesota, Field Case Study of Ex Situ (Heaped) ECGO

Air injection into the sediment at Erie Pier was marginally successful, so a second approach was tried, starting in year 4 (2005). This included heaping the test cell material as shown in Figure 32.18 and incorporating horizontal perforated pipes

TABLE 32.3. Performance of Selected Heaps in Milligram per Kilogram Dry Mass in 177 Days (Danish Soil Recycling Treatment)

		Hydrocarbons	C6-C10	C10-C25	C25-C35	7 PAH
Cleanup levels (mg/kg dm)		200	35	75	200	15
Heap number						
1	Baseline	2085	145	1624	141	0.3
	Final	44	<2	37	36	0.25
	Reduction (%)	98	100	98	75	17
2	Baseline	988	37	894	93	0.19
	Final	30	<2	25	23	0.3
	Reduction (%)	97	100	97	75	−58
3	Baseline	510	0.29	126	384	3.345
	Final	49	<2	17	32	3.79
	Reduction (%)	75	100	73	75	−13
4	Baseline	959	81	843	104	3.33
	Final	72	<2	31	41	2.68
	Reduction (%)	93	100	96	60	20
5	Baseline	129	3.4	112	35	4.76
	Final	64	<2	32	33	2.89
	Reduction (%)	50	100	72	6	39
6	Baseline	462	1	133	321	5.8
	Final	150	<2	30	122	7.9
	Reduction (%)	68	100	78	62	−36
7	Baseline	455	0.8	135	321	4.08
	Final	145	<2	40	106	2.96
	Reduction (%)	68	100	70	67	27
8	Baseline	1542	36	1264	228	0.065
	Final	83	<2	34	49	0.157
	Reduction (%)	95	100	97	79	−142
9	Baseline	516	0.91	171	365	6.72
	Final	98	<2	24	73	3.7
	Reduction (%)	81	100	86	80	45

TABLE 32.4. Comparison of Decay Rates, k (Danish Soil Recycling Treatment)

Elimination Rates	k—average	k—maximum	k—minimum	Number of Cases
Untreated soil and soil unsuccessfully treated by biology ("natural attenuation")	0.04	—	—	12
Successful biological treatment	0.35	1.98	0.02	61
Geo-oxidation (ECGO)	1.08	1.90	0.36	11

located approximately 3 and 6 ft from the surface and stretched horizontally through the heap. In this configuration, the heap, drained of free water, and the ECGO testing continued during years 4–6 (through 2007). During year 4, only one core for redox measurement was retrieved, and in year 5 (2006), a number of cores were taken. Figure 32.19 shows that the heap is in an unsaturated state at the end of

Figure 32.18. Heap formation of the contaminated sediments for ECGO treatment at the Erie Pier site.

Figure 32.19. Coring locations in the ECGO heap of contaminated sediments at the Erie Pier site.

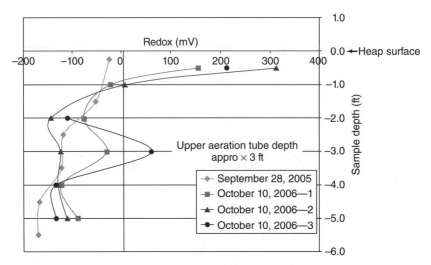

Figure 32.20. The distribution and change of ORP in the Erie Pier heap over 2 years.

year 5 and is also perforated by numerous holes where vertical core samples were collected.

The results of the redox measurements shown in Figure 32.20 indicate that the average ORP increased to approximately −100 mV from the previous −125 mV at depths below 3 ft and reached higher positives in the top 1 ft of the heap. Of note, the positive ORP of the top section may be related to the drier/dewatered nature of the material closer to the surface. Also, the trend to a less negative/more positive ORP in the coring at a depth of 3 ft may indicate the influence of the upper cross-heap perforated ventilation pipes, which were positioned at an approximate depth of 3 ft. Overall, the measurements suggested that venting and heaping the material improved (made less negative) the redox conditions. Also, material taken from the upper 1–2 ft of the heap was (a) drier and was (b) more oxidized in appearance, that is, brownish, in comparison to the dark gray and moist material at depth.

32.4.3 PAH Analytical Results Summary for Original (Simulated *In Situ*), ORP-Modified (Simulated *In Situ*), and Heaped (*Ex Situ*) Demonstrations in Duluth, Minnesota (2002–2007)

Total PAH results spanning the entire Duluth, Minnesota, field demonstration period are presented in Figure 32.21. Note the wide range of results.

32.4.4 Comments on Sediment Variability, Sampling, and Analytics, and the Impact of Particulate Coal Tar

In 2002, approximately 750 yd^3 (570 m^3) of sediment was dredged from the back half of Minnesota Slip for the alternative technology investigation, and the dredged test material was transported to the Erie Pier CDF and was placed into two identically constructed control and test cells.

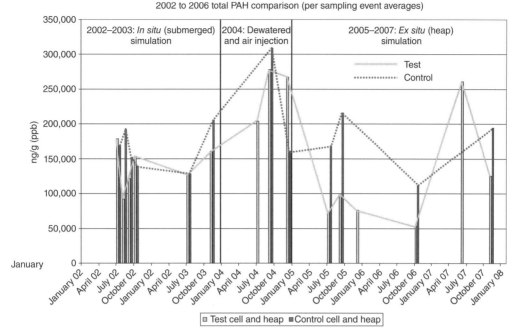

Figure 32.21. Comparison of control and test cell total PAH results spanning the entire Duluth, Minnesota, field demonstration period.

During the 2002–2003 phase of the Erie Pier project, 35 core samples were collected from 1600 sq ft (150 m²) sampling areas established in both cells (Fig. 32.22). This translated to an average sample spacing of 6.76 ft (or just over 2 m). Further, an upper and lower sample split was taken for each core sample. Therefore, 70 analyses were conducted within each cell during the project for a total of 140 analyses.

According to the recommended protocol, large objects such as rocks, metal, glass, wood, coal fragments, and other miscellaneous debris were noted in the field notebook and were removed from the core samples prior to homogenization. Only one sample out of 140 was "free" of miscellaneous debris, and coal tar was most likely present in all samples. Material heterogeneity complicates sampling and analysis. Furthermore, extreme contaminant concentration levels can occur at the micro and macro physical scales, for example, with coal tar.

Coal tar is produced by the carbonization, or coking, of coal. Its color is almost black, occurs as a thick liquid or semisolid, and is only slightly soluble in water. It has a specific gravity of 1.18–1.23. A specific gravity of 1.20 has been determined for the Minnesota Slip coal tar. Streitz and Johnson (2005) described how their use of a laser-induced fluorescence (LIF) tool indicated a thick sequence of material identified as coal tar constituents in the back half of Minnesota Slip. Significantly, this location is near the 2002 project dredging site. PAH analysis of a Minnesota Slip coal tar sample taken from the Erie Pier site in 2005 is summarized in Table 32.5. Note the extreme values, that is, percent total PAH.

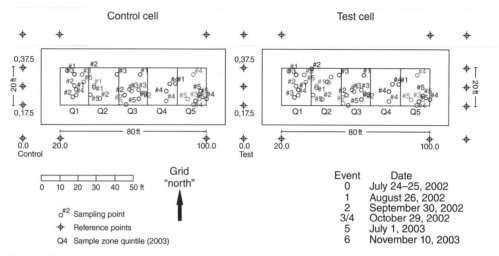

Figure 32.22. Schematic drawing of sampling locations in the adjacent test and control cells at the Erie Pier site.

TABLE 32.5. Minnesota Slip Coal Tar Analysis (Analytical Results Courtesy of USACE-ERDC, Vicksburg, MS, 2005)

	mg/kg (ppm)	ng/g (ppb)	Percent (%)
Naphthalene	1,200	1,200,000	0.12
Acenaphthylene	—	—	—
Acenaphthene	3,900	3,900,000	0.39
Fluorene	4,900	4,900,000	0.49
Phenanthrene	16,900	16,900,000	1.69
Anthracene	6,700	6,700,000	0.67
Fluoranthene	11,600	11,600,000	1.16
Pyrene	8,000	8,000,000	0.80
Benzanthracene	4,800	4,800,000	0.48
Chrysene	4,600	4,600,000	0.46
Benzo(b)fluoranthene	5,600	5,600,000	0.56
Benzo(k)fluoranthene	—	—	—
Benzo(a)pyrene	3,400	3,400,000	0.34
Benzo(g,h,i)perylene	1,100	1,100,000	0.11
Indenopyrene	1,400	1,400,000	0.14
Dibenz(a,h)anthracene	—	—	—
2-Methylnaphthalene	900	900,000	0.09
Total	75,000	75,000,000	7.50

32.4.4.1 Coal Tar Assessment To assess the potential influence of coal tar particles on the field study's analytical results, a simple test was begun in 2006. In that test, crushed (−1 mm) coal tar retrieved from the Erie Pier site was blended with a PAH-free inorganic silt substrate (Fig. 32.23) at a ratio of 10 g of coal tar per 10,000 g of silt and was placed in two plastic bins. Assuming the coal tar sample was pure product, that is, 100% PAH, and that 100% extraction and recovery were achiev-

Figure 32.23. Experimental setup for ECGO coal tar tests.

able, perfect mixing would theoretically result in a sample returning 1000-ppm total PAH, or 0.1%, by weight.

The first bin acted as an untreated control, and the second bin was treated with ECGO. The bins were sampled three times during a 9-month period. During each sampling, six cores were removed from each bin and were composited, and the samples were submitted for gas chromatography/mass spectrometry (GC/MS) PAH analysis. While the results showed no discernible ECGO effect on total PAH, the coal tar test clearly illustrated that even a carefully controlled and blended ("homogenized") sampling medium containing only one contaminant (particulate coal tar) produced highly variable analytical results (range of 55- to 386-ppm total PAH, in the control bin; the ECGO bin samples exhibited similar variability; Table 32.6).

A regression analysis, which compared relative concentrations of individual PAH compounds in 2002–2003 sediment samples having the highest total PAH levels, correlated more closely to coal tar than did samples having the lowest total PAH levels, showing that the dominant PAH source at Erie Pier was indeed coal tar (Fig. 32.24):

Tar versus high PAH content sediments: R^2 of 0.90

Tar versus low PAH content sediments: R^2 of 0.59

Based, in part, on the coal tar test results just described, we conclude that coal tar and coal tar particles present within Minnesota Slip sediment treated at Erie

TABLE 32.6. Coal Tar Test Results: Total PAH

Sample Type	Sampling Date	Total PAH (ppm)
Coal tar control	September 14, 2006	101
Coal tar control	November 8, 2006	386
Coal tar control	June 11, 2007	55
Coal tar test	September 14, 2006	138
Coal tar test	November 8, 2006	72
Coal tar test	June 11, 2007	173

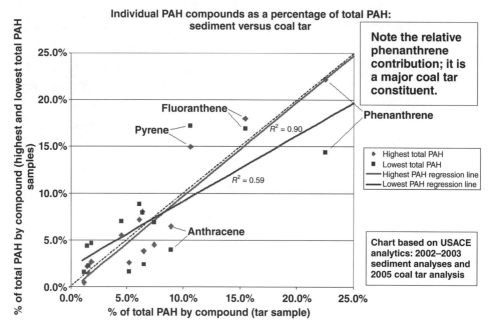

Figure 32.24. Regression comparison of coal tar PAH analyses with Minnesota Slip/Erie Pier sediment PAH analyses.

Pier, as shown in Figure 32.25, contributed to a significant PAH "nugget effect." For example, the 2-mm tar particle pictured in Figure 32.25b weighs approximately 0.005 g (or 0.5, 000th of 10 g). Therefore, if a 10-g sediment sample is analyzed and 100% extraction is achieved, this single particle alone could contribute 32.5-ppm total PAH to the overall analysis. If the cleanup goal is <50-ppm total PAH, sand- and silt-sized coal tar particles within a 10-g sample split could make that goal virtually impossible to achieve. Consequently, this nugget effect—when applied not only to the various stages of the Duluth project described in this chapter but also to projects that face similar conditions in other locales—shows how it can impact sampling strategies, complicate the interpretation of analytical results, and call into question how meaningful the analytical results are from a remediation evaluation and regulatory perspective.

As described previously, the 2002–2003 period of the Erie Pier project collected 35 core samples from each cell, giving an average sample spacing of 6.76 ft (or just

Figure 32.25. Coal tar at the Erie Pier site (a) and microscopic image of sediment showing particulate coal tar "nugget" and the heterogeneous nature of the samples collected for analysis (b).

over 2m), and a total of 140 analyses were performed. Even with such a high sampling density and frequency, it remained a challenge to interpret the highly variable analytical results. Regulators must find it similarly challenging to make decisions based on field samples that are frequently much more widely spaced and that also encounter heterogeneous conditions.

Contaminated sediments are inherently complex and variable. However, it appears that unless the applied treatment technology generates analytical results that are discernible from (a) the inherent heterogeneity of contaminant levels within the sediment in question and b) the allowable variability of the analytical technique applied to the compound(s) being analyzed within the sediment, for example, ±30% to 40% for PAHs according to USEPA, then no confident conclusions can be drawn about a given technology's effectiveness, positive or negative. As a result, potentially viable technologies run the risk of being dismissed outright if they do not overcome this built-in statistical hurdle. As the project QAPP has acknowledged, "... The treatment process must be effective enough to overcome dredged material heterogeneities; otherwise, monitoring can never demonstrate a treatment effect, although one may be present."

The evaluation undertaken at the Erie Pier site suggests that the ECGO technology imparted a treatment effect. However, it was unable to destroy the primary PAH source: particulate coal tar. Unless physical separation and removal or thermal/ vitrification methods are used, sediments or soils that contain solid-phase contaminants like coal tar will continue to be a challenge to remediate with low-energy nondestructive technologies. If alternative remedies for sediment cleanup remain elusive, then the options are limited to the following: (a) do nothing; (b) leave the contaminated sediments in place, but cap them; or (c) remove/dredge the contaminated sediments and store them in CDFs. None of these options is scientifically attractive or provides a permanent solution, so continued evaluation of alternatives that show promise, like ECGO, is recommended.

32.5 SUMMARY

A large-scale field demonstration of ECGO was undertaken using pooled cells of dredged contaminated sediments from Minnesota Slip in the Duluth-Superior Harbor, Duluth, Minnesota. The main objective of the demonstration was to determine if application of the ECGO technology led to a clear-cut reduction in the test cell's PAH levels relative to an untreated control cell. The average total PAH levels declined by 17.3% overall in the test cell during the 2002–2003 evaluation period, which was significantly lower than what had been anticipated, for example, >50% reduction. The combination of reducing conditions (negative ORP), sediment heterogeneity, presence of HS, abundant vegetative growth, and the relatively high organic carbon content of the sediment all contributed to the treatment inefficiencies that were experienced. However, the presence of essentially pure PAH in the form of a solid-phase contaminant (particulate coal tar) likely had the biggest impact on the project's treatment and analytical outcomes. Therefore, ECGO technology may need to be tailored and/or augmented to meet the specific characteristics of each treatment site for optimal performance. Alternatively, significant reduction of PAH contaminant levels within complex and organic-rich sediments using ECGO (or other low-energy and physically nondestructive technologies) may simply be too difficult to achieve efficiently *in situ* without concurrent modification of the *in situ* conditions themselves, for example, by changing the sediment's initial ORP state. In summary, the following conclusions were made based on the long-term observations and analytical results:

- While PAHs were not being destroyed to the degree anticipated, analysis of samples taken from one *confined area* of the test cell and analyzed for 60 PAH

compounds showed a significant reduction of 46.4%; hence, smaller-size treatment cells or volumes may be viable.
- PCBs in low concentrations were being reduced in concentration.
- System pH did not change by the ECGO process.
- Plants and animals continued to live in both the test and control cells.
- Sediments in Erie Pier were very reductive (negative ORP) and hence required enhancement of the process by the introduction of oxidants; aerated cell showed increase in system ORP.
- Pyrolysis GC/MS indicated the influence of HS on redox activity in the sediment.
- The more TOC in the system, the more "oxidant" is required to destroy the organic matter, including the targeted contaminates but also other materials such as humates. Hence, in sediments, such as the sediments from Minnesota Slip, this oxygen demand may be considerable.
- The sediment contained coal tar, coal, wood and vegetation fragments, slag, naturally occurring HS, oil, and a variety of foreign objects, which compromised reliability of sampling and analysis, even after a statistically sound number of samples were collected and sound homogenization techniques were used.
- The presence of pure PAH in the form of particulate coal tar likely had the most significant impact on the project's outcome.

These points and this project have shown that evaluations and interpretation of alternative remediation technologies like ECGO (a) should take into account empirical data and overall data trends; (b) can benefit from analytical techniques that provide more detail about the fate and levels of individual PAH compounds, especially those that degrade into, or are degradation products of, the PAH compounds that are—and are not—included in the EPA "Standard 16"; and (c) benefit from the scales of project operation like that used at Erie Pier, that is, much larger then bench top. This latter point is important, because while bench-scale, for example, milliliter (beaker-sized) treatment studies might be statistically satisfying to the lab practitioner, they can also be of limited practical value for providing realistic (scaled-up) remediation solutions to regulatory agencies and private contractors who encounter physically heterogeneous and chemically complex contaminated sediments like those described in this chapter.

REFERENCES

Averett DE, Perry BD, Torrey EJ, Miller JA. (1990). Review of Removal, Containment and Treatment Technologies for Remediation of Contaminated Sediment in the Great Lakes. Department of the Army, Corps of Engineers, prepared for U.S. Environmental Protection Agency, Great Lakes National Program Office, December 1990.

Crane JL, Schubauer-Berigan M, Schmude K. (1997). Sediment Assessment of Hotspot Areas in the Duluth/Superior Harbor. U.S. Environmental Protection Agency, Great Lakes National Program Office, Chicago, IL. *EPA-905-R97-020.*

Crane JL, Smorong DE, Pillard DA, MacDonald DD. (2002). Sediment Remediation Scoping Project in Minnesota Slip, Duluth Harbor. U.S. Environmental Protection Agency, Great Lakes National Program Office, Chicago, IL. *EPA-905-R02-002*.

Doering F. *EP0578 925 B1*, August 12, 1998; EP0729796B1, February 2, 2000; EP1123755B1, May 11, 2005; US5738778, April 14, 1998; US6280601B1, August 28, 2001; US6984306B2, January 10, 2006.

Dragun J. (1988). *The Soil Chemistry of Hazardous Materials*. Silver Springs, MD: HMCRI.

Marshall D, Madden T. (1959). Induced polarization, a study of its cause. *Geophysics* **24**(4): 790–817.

Patrick WH Jr., Gambrell RP, DeLaune RD. (2003). Physicochemical factors controlling stability of toxic heavy metals in sediments. *Workshop on Environmental Stability of Chemicals in Sediments*, April 8–10, 2003, San Diego, CA.

Rahner D, Ludwig G, Rohrs J. (2002). Electrochemically induced reactions in soils-a new approach to the in-situ remediation of contaminated soils? Part 1: The microconductor principle. *Electrochimica Acta* **47**:1395–1403.

Stevenson FJ. (1982). *Humus Chemistry. Genesis, Composition and Reactions*. New York: Wiley Interscience, pp. 443–465.

Streitz A, Johnson S. (2005). Detailed Investigation of the Minnesota Slip Featuring Laser Induced Fluorescence. Minnesota Pollution Control Agency, St. Paul, MN, *MPCA Technical Document, tdr-g1-01*.

Vacquier V, Holmes CR, Kintzinger PR. (1957). Prospecting for groundwater by induced electrical polarization. *Geophysics* **22**(3):660.

Ward SH, Fraser DC. (1967). Conduction of electricity in rocks. In *Mining Geophysics*, Part B, Vol. II (eds. DA Hanson *et al.*). Tulsa, OK, pp. 197–223.

33

EXPERIENCES WITH FIELD APPLICATIONS OF ELECTROKINETIC REMEDIATION

REINOUT LAGEMAN AND WIEBE POOL

33.1 INTRODUCTION

Electroreclamation (ER) is a soil remediation technology that uses electrokinetic effects to remove inorganic contamination. It can, for example, be used to remove heavy metals, all types of cyanides, arsenic, and other ionic or polar compounds. The basic principle involves applying a difference in potential, thus causing charged particles to migrate to the cathode or the anode. A special electrolyte system is used to both condition physical parameters around the electrodes and in the soil, and to remove the contaminants that have collected around the electrodes.

Applying this soil remediation technology requires special expertise; this chapter describes the technology and its applications, indicating the materials and decontamination methods that can be used and how the system should be initiated and controlled. It also specifies the samples that need to be taken in order to monitor the decontamination process. Before remediation can commence, however, it is first necessary to clarify where contamination is present in the area of soil concerned and in what form. It is furthermore necessary to perform electrokinetic laboratory tests with one or preferably more representative soil samples. Finally, it is explained how the data are analyzed and used for carrying out the design of the remediation system.

33.2 ER

33.2.1 Principle and Fundamentals

ER belongs to the group of so-called physicochemical remediation technologies. Under the influence of an electric field, a number of electrokinetic phenomena occur:

Electrochemical Remediation Technologies for Polluted Soils, Sediments and Groundwater,
Edited by Krishna R. Reddy and Claudio Cameselle
Copyright © 2009 John Wiley & Sons, Inc.

1. electromigration or the movement of ionic species in pore water or groundwater,
2. electroosmosis or movement of water from anode to cathode, and
3. electrophoresis or movement of charged particles.

In the following sections, these phenomena will be described in more detail.

When inert electrodes are placed in water and a direct current (DC) is passed, changes occur at the anode and cathode according to the scheme presented in Equation 33.1:

$$2H_2O - 4e^- \rightarrow O_2 + 4H^+. \tag{33.1}$$

At the anode or positive electrode, electrons are being stripped from water molecules. Oxygen is evolved and protons, H^+, are formed and travel through the electrolyte toward the cathode. Meanwhile, the cathode is donating electrons to water molecules creating hydroxyl ions and liberating hydrogen gas (Eq. 33.2):

$$4H_2O + 4e^- \rightarrow 2H_2 + 4OH^-. \tag{33.2}$$

Note that the two reactions, generation of protons at the anode and hydroxyl ions at the cathode, are in balance and no net change in pH of the electrolyte occurs. In a liquid electrolyte, sufficient mixing occurs such that the local pH changes around the anode and cathode are difficult to detect unless the electrodes are well separated. In soil, sludge, concrete, and gels, however, mixing is reduced or eliminated so that the area around the anode becomes acidic and the area around the cathode becomes alkaline due to inhibition of the remixing of the electrolyte. This has a significant impact on the process in soil.

33.2.1.1 Soil as an Electrolyte Most soils are conductive due to the presence of dissolved ions, such as calcium, magnesium, sodium, potassium, (bi) carbonate, some soluble fatty acids, nitrate, phosphate, sulfate, and chloride ions. Most seemingly, dry soils have more than 5% moisture, sufficient to provide a continuous path for these ions to move. This is essential for plants as the roots need access to these nutrients and ion transport across membranes is the means with which they extract them. The most significant feature of natural soils, with respect to their contamination and subsequent remediation, is the high ion exchange capacity (Table 33.1).

Heavy metals such as cadmium, lead, iron, and zinc, in their metallic state, corrode and form salts and bases, which take up cationic sites on soil particles. In some cases, land is often contaminated from the spillage of heavy metal ions directly from aqueous plating shop wastes or airborne pollution from metal smelters. Soil has the capacity to immobilize significant quantities of heavy metal ions, to the 2%–3% level in some cases, such as the top soil around lead smelters.

Soil is made up of several components derived from the weathering of rocks and the addition of organic materials from growth and decay of plants and organisms. These materials have a very high ion exchange capacity and are usually closely bound to clay particles. Clays have ion exchange capacity in their own right; they derive their ion exchange capacity from the basic silicate structure, which acts like a Lewis acid. Note that the basic unit contains excess oxygen atoms, which are able

TABLE 33.1. Cation Exchange Capacity (CEC) of Some Clay Minerals

Clay Mineral	CEC (meq/100 g)
Kaolinite	3–15
Illite	10–40
Chlorite	10–40
Montmorillonite	60–120
Vermiculite	100–160

Kaolinite is widely used for electrokinetic laboratory experiments, but because of the low CEC or buffering capacity, the results are of no great practical value as kaolinite occurs in abundance only in soils that have been formed in hot moist climates, for example, in tropical rainforest areas.

Note: There are no data available in literature for anion exchange capacity.

to form chains, sheets, or three-dimensional networks. End groups in such networks provide cation vacancies. Such complex silicates behave as ionic exchangers for metal cations. Multivalent cations attached to silicates can also provide sites for the attachment of anions such as arsenites, sulfates, cyanide, and carbonates and hydroxide ions.

At pH 7, the concentration of H^+ is too low to significantly affect interactions between ion sites on the soil and protons in the water. This explains why washing soil with water as a remediation technique for contaminated soil and clay is ineffective.

33.2.1.2 The Interaction of Electrochemical and Electrokinetic Phenomena

When anodes are placed in soil and a DC is passed between them, the amount of available H^+ is raised significantly. The ion travels much faster than other ions and carries a disproportionate share of the total current. The understanding of ion transport and ion mobilities (i.e., classical electrochemical phenomena) applies to the movement of ions in soil as it does in liquid electrolytes, except for one major difference. Ions interact with ion exchange sites on the soil and their progress is modified by this. Although the presence of particles and the difference in path length due to the tortuous route taken by ions will affect the values of ion mobilities in water, the differences in mobilities between the various ions still hold. For ions, the average mobility amounts to $5 \cdot 10^{-8}\,m^2/U \cdot s$, where U = drop in potential in volts (Table 33.2).

Along the gradient in potential between the electrode pairs, a hydrogen ion moves almost twice as fast ($33 \cdot 10^{-8}\,m^2/U \cdot s$) as hydroxyl ion ($18 \cdot 10^{-8}\,m^2/U \cdot s$) and five to six times faster than metal ions. In soil however, the hydrogen ion is quickly mopped up by the ion exchange sites on the soil particles; as a result, the metal ions are displaced into the electrolyte. Even if the soil picks up the metal ion on a new ion exchange site, there is a net drift toward the cathode as the ion exchange sites fill up by protons emanating from the anode. Metal ions will be in equilibrium with other available sites on the soil in front of the proton sweep. The effective ion mobility in soils is therefore much lower than in liquid solutions (Table 33.3).

TABLE 33.2. Average Ion Mobilities in Aqueous Environment Found in Literature

Cations	Mobility $\times 10^{-8}$ m^2/V·s	Anions	Mobility $\times 10^{-8}$ m^2/V·s
H$^+$	32.7	OH$^-$	18
Li$^+$	3.4	F$^-$	4.9
Na$^+$	4.5	Cl$^-$	6.8
K$^+$	6.6	Br$^-$	7.0
Cs$^+$	7.0	I$^-$	6.4
NH$^+$	6.6	NO$_3^-$	3.6
Ag$^+$	5.6	CH$_4$COO$^-$	2.9
½Mg^{2+}	4.6	C$_3$H$_7$COO$^-$	7.0
½Ca^{2+}	5.3	½SO$_4^{2-}$	6.2
½Ba^{2+}	5.7	½CO$_3^{2-}$	3.6
½Pb^{2+}	6.4	⅓Fe(CN)$_6^{3-}$	10.5
		¼Fe(CN)$_6^{4-}$	9.8

TABLE 33.3. Effective Ion Mobilities in the Soil during Electroreclamation

Cations	Average Mobility $\times 10^{-8}$ m^2/V·s	Range $\times 10^{-8}$ m^2/V·s	Anions	Average Mobility $\times 10^{-8}$ m^2/V·s	Range $\times 10^{-8}$ m^2/V·s
½Cd^{2+}	0.28	0.16–0.41	⅓As^{3-}	0.14	0.05–0.5
½Cr^{2+}	0.25	0.07–0.55	—	—	—
½Cu^{2+}	0.26	0.05–0.54	—	—	—
½Pb^{2+}	0.30	0.15–0.56	—	—	—
½Ni^{2+}	0.16	0.09–0.18	—	—	—
½Zn^{2+}	0.27	0.04–0.55	—	—	—

A similar but slower drift of hydroxyl ions moves from the cathode displacing anions adsorbed onto the soil particles. Free anions in the water electrolyte will be in equilibrium with available sites on the soil. Note, however, that these sites compete with OH$^-$ created at the cathode, and so there will be a net drift of displaced anions to the anode due to electromigration.

As a result, the soil acidifies from the anode into the direction of the cathode and if no measures are taken, practically all the energy will be used for the transport of H$^+$ ions. Likewise there is increased alkalinity into the direction of the anode, ultimately leading to precipitation of metal hydroxides in the soil (Fig. 33.1).

Electromigration will occur with any species that will form ions in aqueous environments. The process can therefore be applied to contaminants such as

- inorganic anions and cations,
- organic carboxylic acids and phenols,
- sulfonated aliphatic and aromatic compounds like some dyestuffs, and
- detergents and some pesticides like paraquat and diquat.

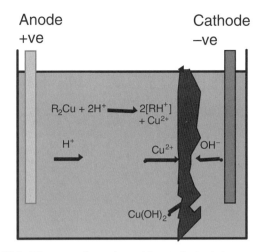

Figure 33.1. Electroreclamation without electrolyte management.

33.2.1.3 Electroosmosis Electroosmosis is the movement of pore water or groundwater under the influence of a DC field. With electroosmosis, the direction of flow is from anode to cathode. In some cases, one can also observe a flow direction from cathode to anode. This phenomenon is known as electroendosmosis. Electroosmosis is determined by the following factors:

- the mobility of the ions and charged particles present in the pore fluid, including those ions and particles entering into the pore fluid via ion exchange;
- the hydration of ions and charged particles present in the pore fluid;
- the electrical charge and direction of movement of ions and particles, resulting in a net water transport;
- the ionic strength or ion concentration;
- the viscosity of the pore fluid;
- the temperature; and
- the soil porosity.

Note that the volume of water removed per unit of time is directly proportional to the electrical power used per unit of volume of removed water. Thus, the faster the water transport, the more power is necessary to remove the same volume of water.

33.2.1.4 Electrophoresis Electrophoresis, or cataphoresis, is the movement of particles under the influence of a DC field. The term "particles" includes all charged particles like colloidal, clay, and organic matter particles suspended in the pore fluid. The movement of these particles is similar to the movement of ions. In the pore fluid of clay soils, the particles participate in the transfer of electrical charges and influence the electrical conductivity and the electroosmotic flow.

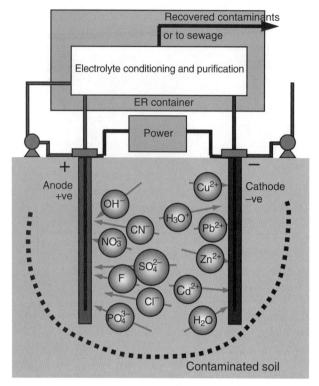

Figure 33.2. Schematic view of electrokinetic installation.

Clay minerals can polarize in two ways. The first is the permanent dipole moment, which results from the structure and depends on the atomic masses. Its is oriented parallel to the long axis of the clay particles. The second polarity is perpendicular to the first one and is a result of the external electrical field. It depends on the polarization capacity of the electrical double layer. Thus, the mobility of clay particles depends on the combined action of these two moments and is therefore low, varying between 1.10^{-10} and $3.10^{-9}\,m^2/U\cdot s$.

33.2.2 Practical Realization

33.2.2.1 Electrodes and Electrolyte Management System (EMS) The key elements of an electrokinetic installation are as follows (Fig. 33.2):

- Ion-permeable electrolyte wells are placed in the contaminated medium and are connected to a centralized EMS. Each well has an electrode inside. The result is alternating rows of anodes and cathodes. Electrolytes are circulated in a closed loop between the electrode wells and the EMS. Via these electrolytes, pH is maintained at a predetermined value. The electrodes are then energized. The water in the electrolytes is electrolyzed, forming H^+ ions and $O_2\uparrow$ at the anodes, and OH^- ions and $H_2\uparrow$ at the cathodes. These ions are then made to migrate through the casing into the soil to generate a temporary and localized

pH shift, which desorbs the contaminating ions. No acids are pumped directly into the soil.
- Once desorbed, the contaminating ions migrate to their respective electrodes, under the influence of the potential applied (electromigration). The anions migrate to the anodes, the cations to the cathodes. Here they pass through the electrode well screen and are taken up by the circulating electrolytes.
- Critical for the control of system performance is the careful management of the pH and other electrolyte conditions within the electrode casings.
- Contaminants are recovered from the circulating electrolytes by both precipitation and filtration, by electrochemical ion exchange (EIX®), or by other ion-collecting systems.

33.2.2.2 Electrolyte Purification When the electrical conductivity of the electrolytes has reached approximately 20 S/m, the electrolytes are pumped into a connected treatment installation. Depending on the type of contaminants, they are treated:

- with lye (NaOH), to precipitate metal hydroxides, which are then removed by filter press; the amount of filter cake produced depends not only on the concentrations of the heavy metals but also on concentrations of iron, alkaline earth and alkaline ions, carbonate, bicarbonate, magnesium, and so on; as a rule of thumb, 0.05%–0.1% of the total volume of treated soil is ultimately collected as waste;
- by ion exchange through resins or EIX, a process developed by AEA Technology (Harwell) that uses electrochemically driven adsorption and desorption to remove ions from dilute electrolytes; and
- by using a so-called EnViroCell® for specific heavy metals and cyanide.

The electrolyte management and purification systems are housed in containers, together with the electrical power supply. If necessary, electricity cables and circulation ducts and pipes can be installed underground.

33.2.3 Application Area

Electrokinetic remediation can be deployed in four separate ways:

- *In Situ (Fig. 33.2).* A network of electrode casings is placed directly in the ground, covering part—if the total area is too large to be remediated in one go—or the whole of the contaminated area. The distance between electrodes of both equal and opposite charges depends on site-specific conditions but, in general, amounts to 1.5 or 2.0 m. Contamination is recovered with minimal disturbance to the site. Electrodes can also be placed inside or underneath buildings, either vertically or horizontally.
- *Batch.* In this configuration, contaminated soil is transported to a mobile batch facility or a temporary lagoon and is treated *ex situ* (Fig. 33.3). This process has been operated at 2 and 7400 m^3 in size.

Figure 33.3. Electroreclamation in temporary depot.

- *Electrokinetic Fence (EKF)*. This uses a chain of electrode pairs deployed in the ground to halt the migration of contaminated groundwater from a point source (see Chapter 22).
- *Electrokinetic Biofence (EBF)*. Besides the chain of electrodes, a row of filters with nutrients is placed upstream of the electrodes. Groundwater transports the dissolved nutrients toward the electrodes. Under the influence of the electrical field, the electrically charged nutrients are dispersed homogeneously between the electrodes, enhancing biodegradation (see Chapter 22).

The technology is applicable for diffusely dispersed pollutants both in the non-saturated and saturated zone and in clay, sand, and peat soils. Contaminants that can be recovered by ER can be heavy metals, arsenic, nitrates, phosphates, halogenides, and polar and/or water-soluble organic compounds such as cyanides, phenols, and nitro aromatics (such as trinitrotoluene [TNT]). Minimal moisture content should be 15%–20%. The technology is not economically applicable to heavy metals in metallic form, such as metal grindings, slag and cinder, concretions, and paint particles (putty).

33.2.4 Combination with Other Technologies

The technology can be combined with electrobioreclamation for removing cocktail of inorganic and organic compounds.

33.2.5 Supplementary Provisions

The availability of an electric power supply is a prerequisite. If no connection via the grid is possible, a generator can be used. Because of the high noise levels, a generator can only be deployed outside residential areas. Temperatures >40 °C can have an adverse effect on some coatings of subsurface cables. The presence and location of these cables must be known so that they may be insulated. Cathodic protection of subsurface metal cables, pipes, and objects such as tanks may be necessary when they are directly in the electric current zone.

33.2.6 Remediation Time

Remediation time is dependent on the nature, concentration, and extent of the pollution and the capacity of the remediation equipment. More extensive polluted sites are cleaned up in sections. Cleanup duration varies from a few months to several years.

33.3 INVESTIGATION AND DESIGN OF ER

33.3.1 Laboratory and Field Tests

The design and dimensions of the ER system to be deployed at a site are based on and derived from the data collected during the preceding investigations and from one or more electrokinetic laboratory tests. If possible, some electrical resistivity soundings should be performed on the site in order to measure the electrical resistance of the ground.

33.3.1.1 Laboratory Tests Electrokinetic laboratory tests are done on soil samples from the site, preferably taken at the contamination hot spots, that is, the locations with the highest concentrations of contaminants. The client can choose between two separate or combined test configurations:

1. turbo test, which requires a soil sample of 4–6 kg or
2. standard test, which requires a soil sample of 10- to 15-kg soil.

For both tests, the soil samples are homogenized and analyzed for their heavy metal content, pH, iron, Ca, Mg, Na, Cl, and clay and organic matter contents. Before the tests start, the cation exchange capacity (CEC) of the soil is determined using a buffer test.

33.3.1.2 CEC and Anion Exchange Capacity (AEC) Test The method used to find the CEC or AEC indicates the total amount of exchangeable cations or anions. Cations could be alkaline earth and alkaline, metals, and organic complexes. Anions comprise cyanides, arsenic, and so on. The purpose of the CEC test is

- to get information on the buffer capacity of the contaminated soil and
- to get information on the amount of energy or acid needed to lower pH to an optimum value to mobilize the contaminants. The pH change is effectuated by protons that are generated at the anodes and are needed to lower pH to an optimum level. It can be calculated in faraday (A·s).

The purpose of the AEC test is

- to get information on the buffer capacity of the contaminated soil and
- to get information on the amount of energy or lye needed to increase pH to an optimum value to mobilize the contaminants. The pH change is effectuated by hydroxyl ions that are generated at the cathodes and are needed to increase pH to an optimum level. It can be calculated in faraday (A·s).

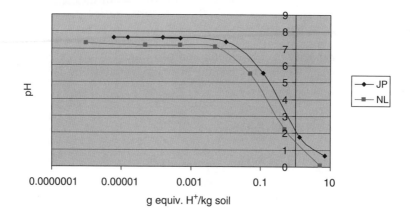

pH solution	gequiv	gequiv/kg dm	pH solution + 10-g dm
4.01	4,886E−06	0,000488619	7.62
5.5	1,581E−07	1,58114E−05	7.64
5.9	6,295E−08	6,29463E−06	7.63
4.5	1,581E−06	0,000158114	7.63
2.7	9,976E−05	0,009976312	7.41
1.62	0,0011994	0,119941646	5.56
0.58	0,0131513	1,315133996	1.78
−0.15	0,0706269	7,062687723	0.67
6.7	9,976E−09	9,97631E−07	7.3
5	0,0000005	0,00005	7.2
4	0,000005	0,0005	7.2
3	0,00005	0,005	7.1
2	0,0005	0,05	5.5
1	0,005	0,5	2.2
0	0,05	5	0.1

Figure 33.4. Example of a CEC test. dm, dry matter; JP, Japan; NL, Netherlands.

The method is as follows: several 50 ml volumes of demineralized water are acidified with hydrochloric acid or with another acid, depending on the contamination, to different pH levels, and are then mixed with 10 g (dry matter) of the sample. After thorough mixing for 24 h, the mixture is left for a time, to settle out, and then the pH of the solution on top of the sample material is determined. The results are generally presented in a diagram, where the pH of the solution is plotted as a function of the milliequivalent H^+ added per kilogram dry matter (Fig. 33.4).

33.3.1.3 Turbo Test In the turbo test (Fig. 33.5), the soil material is divided equally between several small containers, each of which is given a different energy supply and/or electrolyte solution (different inorganic or organic acids). This kind of test is used if data about the origin and type of contaminants are scanty or not available, or the client wants to have a quick scan of the possibilities of deploying ER. It gives less accurate data on energy requirements and remediation time but

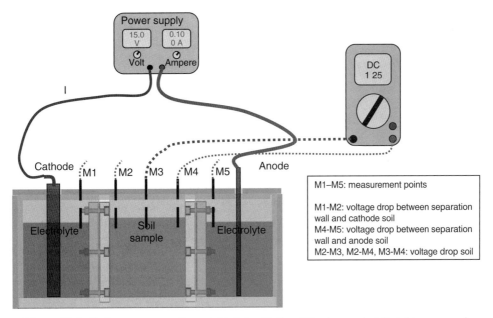

Figure 33.5. Turbo test setup. Copyright © by Holland Enviroment. All rights reserved.

indicates whether the contaminants can be mobilized at all, using a specific kind of electrolyte solution.

A turbo test lasts 2–4 weeks. The time can be shortened to 1 week if high current densities are used. Each container is sampled once at the end of the test. Five samples for analysis are taken from each soil compartment: anode side, anode middle, middle, cathode middle, and cathode side. Precipitates are prepared from both the cathode and anode compartments and are analyzed.

33.3.1.4 Standard Test In the standard test, all soil material is treated in one laboratory setup. This setup is used when the client has supplied ample data on the origin of the sample and its contaminants. This kind of test gives accurate data on energy requirements and remediation time. The most reliable results are obtained when the two tests are combined: a turbo test followed by a standard test run using the electrolyte solutions that came out best during the turbo test.

A standard test lasts about 8–10 weeks. Sampling frequency is 1 week. At the time of analysis, five samples are taken and analyzed: anode side, middle and cathode side, and precipitates in the cathode and anode compartments.

33.3.1.5 Test Results The electrokinetic laboratory tests are finalized with a report. This presents relevant data, such as sample preparation, analysis results, concentration decrease, type of electrolyte solutions, electrolyte conditioning, and electrical and electrokinetic parameters. The electrical parameters are voltage (V), drop in potential (V/m), current (A), current density (A/m^2), electrical power (kW/m^3), and resistivity (Ohm·m). Conditioning parameters comprise type of acid

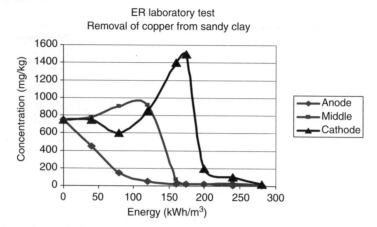

Figure 33.6. Standard test with Cu contamination.

or lye and amount of faraday (A·s) necessary to reach and maintain an optimum pH level. Electrokinetic parameters pertain to electrokinetic mobility (EKM in $m^2/V·s$), electrokinetic velocity (EKV in m/s = $m^2/V·s$ × V/m), effective electrokinetic mobility [EEKM in $m^2/V·s$ = $A/m^2/kWh/m^3$ × distance (m)], energy (kWh/m^3), electroosmotic mobility (EOM in $m^2/V·s$), electroosmotic velocity (EOV in m/s = $m^2/V·s$ × V/m), and electroosmotic transport (EOT in m^3/s). Particular detail is given on the decrease in concentration of the contaminants in relation to the amount of energy used (kWh/ton of per m^3) and the time involved. Examples of standard laboratory tests with soil contaminated with copper, zinc, and arsenic are given in Figures 33.6–33.8, respectively.

Figure 33.6 is a good example of the movement of metal ions (copper in this case) under the influence of an electric field. Copper ions are first disappearing at the anode side and are moving into the direction of the cathode. After an energy inducement of ca. $180 kWh/m^3$, the bulk of the copper ions has reached the cathode side and is then collected in the catholyte. After ca. $270 kWh/m^3$ of energy inducement, all copper has disappeared from the sample.

The diagram of Figure 33.7 shows that zinc ions are removed quickly and with low energy input. This is caused by the low buffering capacity of the soil and the initially low pH of the sample (pH = 4). The zinc ions are desorbed easily and move into the direction of the cathode. There is no buildup of Zn concentration in the middle and at the cathode side.

In Figure 33.8, results are shown from a standard test with soil with a high buffering capacity, contaminated with arsenic. After ca. $300 kWh/m^3$, the arscnic concentration at the anode side has reached the target level, but in the middle and at the cathode side, concentrations are still too high. Even after an energy inducement of $450 kWh/m^3$, target level has not been reached. In this case, the laboratory test indicated that it would take too much energy too clean this soil by ER.

33.3.2 Field Test

A geoelectrical survey enables the measurement of formation resistivities of the subsoil. These are necessary so that the distance between electrodes and also the

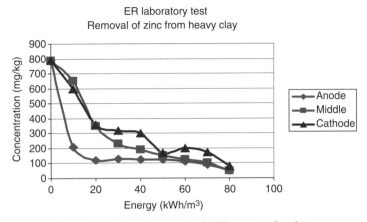

Figure 33.7. Standard test with Zn contamination.

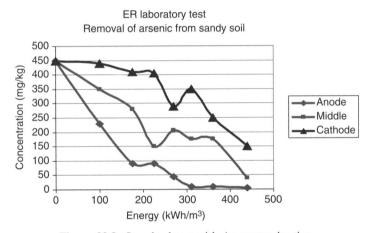

Figure 33.8. Standard test with As contamination.

amount of energy necessary to migrate ions or polar constituents through the soil can be calculated in the most efficient way.

The electrical resistivity method is a nonintrusive exploration method for determining electrical resistance. It does so by applying an electric current (I) to metal stakes (outer electrodes) driven into the ground. The apparent potential difference (ΔV) is then measured between two inner electrodes (nonpolarizing DC type, i.e. porous pots filled with $CuSO_4$ solution, or metal stakes in AC type) buried or driven into the ground. Figure 33.9 gives an indication of the field setup of the resistivity method.

Increasing the spacing of electrodes A and B increases the depth of penetration of the current. The apparent electrical resistivity ρa, obtained at different spacing (m) by measuring the resistance R ($=\Delta V/I$), is plotted on a log–log paper against the spacing between the outer electrodes. The depth at which the current enters a formation of higher or lower resistivity is signaled by a change in the resistivities

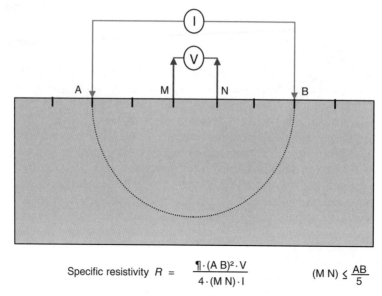

Specific resistivity $R = \dfrac{\P \cdot (A\,B)^2 \cdot V}{4 \cdot (M\,N) \cdot I}$ $\qquad (M\,N) \leq \dfrac{AB}{5}$

Figure 33.9. Field setup of geoelectrical sounding (Schlumberger method).

recorded at the ground surface. The field curves are interpreted using indirect or direct interpretation computer programs. This makes it possible to identify clay and sand layers and also changes in water quality (e.g. fresh, saline, or contaminated).

33.3.3 Remediation Data and Parameters

In order to successfully electroremediate a site contaminated with inorganic species like heavy metals, cyanide, arsenic, or other ionic or polar species, an inventory of the site has to be made by collecting basic data. These data include

- *General Data.* Present owner, type of company and production processes, type of chemicals used, type of waste products, possible causes of pollution
- *Historical Data.* Date on which the site came into use, date on which production processes started, location of production processes, calamities, earlier investigations, or remediation actions.
- *Site-Specific Data.* Location, topography, underground tanks, ducts and cables, neighboring facilities, electricity supplies and their availability, sewage and main water connections.
- *Hydrogeological and Lithological Data.* Groundwater levels, thickness of aquifers and aquitards, hydraulic conductivity, groundwater flow velocity, recharge and/or discharge areas, soil types, silt content, organic matter content, porosity.
- *Chemical and Physicochemical Data and Parameters.* Type and concentration of chemical constituents, pH, groundwater conductivity, soil resistivity.

33.3.4 Sampling Strategy

Site investigations are relatively expensive because they consist of many borings, samples, and analyses. It is therefore essential to base them on a practical and well-designed sampling plan. Sampling strategy is of prime importance, and it should be based on expert assessment of soil type, soil stratification, hydrogeological situation, type of contamination, and company data. It is also advisable to investigate the site in phases. This means choosing the location of every new sampling point or monitoring well on the basis of the contaminant situation established during the foregoing investigations.

Some of the data mentioned above may have already been collected during earlier investigations. It is not always possible to collect all the relevant basic data, but the essential data are the lithological and hydrogeological data and the chemical and physicochemical parameters. Additional soil investigations should focus on

- demarcating the contaminants, both vertically and laterally, by drilling boreholes and installing groundwater monitoring wells and by taking soil and groundwater samples. The density of the network of boreholes and monitoring wells depends on the size of the site and the number of potentially hazardous locations within the site (hot spots). At these hot spots, sampling density is higher than in the surrounding less-contaminated areas.
- taking soil and groundwater samples for laboratory analyses. Soil samples should be taken from different depths.
- Soil samples for laboratory tests should be representative of the contaminant situation. They should preferably be taken from the known hot spots where the highest concentrations of contaminants are found.

Note: Often, groundwater flow has carried the contaminants beyond the site boundaries. If this is the case, it may be necessary to consult with competent authorities and the "neighbors."

33.3.5 Remediation Strategy

The aim of any soil and groundwater remediation is to clean up a contaminated site or area within a certain period of time and for a certain price. These basic principles also apply when deploying ER. That is why basic data are important. Only if there is abundant and reliable data can realistic estimates be made of the cost and duration of the remediation.

As soon as it has been decided to remediate a site, a remediation plan has to be drawn up. The first step is to work out possible remediation options in relation to the basic assumptions. This phase is called the remediation investigation. When assessing the remediation options, it is necessary to consider the technical as well as the environmental and financial aspects. Sometimes, ER is not the sole remediation option; in certain cases, it should be combined with other methods in order to maximize the remediation efficiency.

Briefly, a remediation plan includes

- a description of basic environmental assumptions;*
- the reason(s) for choosing the particular remediation method or methods;

- a detailed description of the chosen remediation technique or techniques;
- the design and dimensions of the remediation method(s);
- a timetable and the phases envisaged;
- a description of the remediation target and of the procedures to establish that this target has been reached;*
- a monitoring plan and sampling procedures;
- the proposed treatment of excavated soil, if applicable, or of soil from borings;
- the proposed treatment of groundwater, if applicable;
- the proposed treatment of electrolytes and filter cake;
- the environmental supervision and reporting;
- a schedule for applying for the necessary permits;
- the electricity and main water and sewage connections required;
- a safety plan;
- a calculation of the costs; and
- the payment schedule.

Depending on the size and complexity of the contaminant situation, the remediation plan may vary from a short description to an extensive report.

Note: The most important aspect of any soil or groundwater remediation is that the competent authorities, the client, and the remediation contractor have agreed about the remediation target. The target values laid down by the government are generally very low and in certain cases are unrealistic or hard to achieve. From the outset, it should therefore be completely clear what the remediation target will be and how it will be verified.

33.3.6 Design and Dimensions

The design and dimensions of the ER system to be deployed at a site are based on and derived from the data collected during the foregoing investigations and from one or more ER laboratory tests.

33.3.6.1 Field Setup Anodes and cathodes are installed in separate rows parallel to each other. The design of the field setup is based on the following data and parameters:

- size (area and depth) and shape of the area or areas to be remediated;
- length of the anode and cathode rows of electrodes;
- distance between the electrodes of equal charge, that is, anode–anode and cathode–cathode;
- distance between the electrodes of opposite charge, that is, anode–cathode;
- number of anodes connected up to one anode electrical cable;†
- depth to water table;
- distance between the remediation area(s) and the remediation equipment installed in containers;

Figure 33.10. Electroreclamation in a temporary lagoon (Woensdrecht).

- distance to the grid connection;
- length of electrode cables; and
- length of pipes for circulating anode and cathode electrolytes.

†*Note*: In certain cases, the number of anodes needed in the field exceeds the number of anode connections available on the rectifying units (see following section). In such cases, up to a maximum of three anodes can be connected to one connection point, provided the total current per connection does not exceed 20 A.

In Figure 33.10, an example of a field layout is presented.

33.3.6.2 Remediation Equipment An installation for ER consists of an electrical power unit, an electrolyte management unit, and, optionally, an electrolyte treatment unit and an EnViroCell. All the equipment is built into separate containers that are each 20 ft long.

The electrolyte management unit consists of several buffer tanks for anode and cathode electrolytes, plus acid and base dosing units. The number of buffer tanks corresponds to the size of the electrical power units.

The electrolyte treatment unit consists of a mixing tank, buffer tanks, and a filter press.

The EnViroCell is a special device for treating certain heavy metal solutions and electrolytes containing cyanides.

Which power and electrolyte management units are deployed depend on the size of the remediation area, the number of electrodes, how much electricity is needed, and the regulations concerning the use of electricity. For practical and safety purposes, the current should not exceed 6000 A. The electrolyte treatment and the EnViroCell units are not needed if existing on-site water purification facilities can be used.

33.3.6.3 Electrode Materials

A variety of electrolytic processes take place at the electrodes. At the anode, a wide range of peroxides occur, in addition to hydrogen ions and oxygen, meaning that the anode has to be made of a precious metal. A relatively inexpensive alternative is to construct the electrodes of titanium with a precious metal coating. Titanium has the advantage of being light; it can also be welded electrically. It has the disadvantage of a relatively high specific resistance for a metal. Various models of titanium electrode are available as wires, mesh, sheets, and so on. If ions such as fluoride are present in the anode electrolyte, the coated titanium electrode will dissolve. In the case of such contaminants, the anode can be made of Ebonex, a ceramic titanium oxide material. The latter is also available in a wide variety of forms, but the delivery time is longer and the price higher than that of titanium electrodes.

At the cathode, electrolysis produces hydroxyl ions and hydrogen, and various reduction processes take place. The cathode material can basically consist of such materials as iron gas pipes, concrete reinforcement steel, cables, and so on. When the ER process is running well, the hydroxyl ions are compensated for by supplying acid to the cathode electrolyte. The average acidity of the cathode electrolyte is decided depending on the contaminants, the current strength, and the flow of electrolyte. Although the cathode is "cathode protected," corrosion of iron will occur around the water level in the cathode well if the pH of the cathode electrolyte is <1.7. Suitable cathode materials under such conditions are graphite pipes and a special type of steel also used as the anode material when cathodic protection is applied. The use of coated titanium as the cathode material cannot be recommended because the development of hydrogen will cause the coating of precious metal to come away from the surface of the titanium. As a result, the titanium surface will corrode and will consequently cease to conduct electricity effectively.

The ER system makes use of a one-sided rectified current. This has the advantage that the neutral point of the transformer can be used as the cathode. In the case of a two-sided rectified current, there are problems if the minuses of the diode bridges are connected to one another.

33.3.7 Risks and Uncertainties

33.3.7.1 Risks

The general definition of risk is the danger of injury, harm, or loss. With respect to soil and groundwater remediation, this definition can be extended to the danger of failure or the likelihood of an unwanted situation occurring. The probability that one or more of these calamities will actually happen is the risk factor.

During remediation operations, one can furthermore distinguish between direct and indirect risks, which in turn can be divided into foreseeable and unforeseeable risks:

- *Direct* risks pertain to calamities occurring immediately during the execution of an activity, for example, as a consequence of neglect or inattentive behavior and/or actions.
- *Indirect* risks pertain to calamities occurring at a later time because of negligent actions.
- *Foreseeable* risks are the unwanted situations that may be expected but can be averted by taking the right countermeasures beforehand.

TABLE 33.4. Definition of Risk Factors

Risk Factor	Indication	Risk Profile
0	No risk	Little or no chance that remediation will be delayed or unsuccessful
1	Low risk	Potential problems can be anticipated and/or prevented, or fixed with little effort
2	Medium risk	There is a fair chance that remediation will be slowed down, but problems can be dealt with and there are alternatives.
3	High risk	A high chance that remediation will be slowed down or targets will be met only partly, if at all; alternative options are expensive and entail much effort.
4	Extreme risk	A very high chance of failure to meet remediation targets; no remediation alternatives are available. Project acceptance and execution only under very stringent conditions.

- *Unforeseeable* risks are those calamities or unwanted situations that occur completely unexpectedly.

When establishing the risk factor for remediation projects, the following classification and definitions are used (Table 33.4):

Situations that might influence the risk factor are the following:

- Contaminant situation differs from start-up situation;
- influence of other contaminants;
- power failure;
- malfunctioning of EMS;
- insufficient remediation progress;
- excessive rainfall;
- high outdoor temperature (in summer); and
- low outdoor temperature (in winter).

33.3.7.2 Uncertainties However, thorough the preparations preceding an ER project, some uncertainty will always remain, especially with respect to the calculated, estimated, or predicted decrease in contaminant concentrations during a certain period of time. It should always be kept in mind that

- Field investigations never give a total picture of the contaminant situation. Usually, the number of samples in relation to the size of the contaminated site will merely indicate the existing situation. Only by using an experienced team of remediation specialists (hydrogeologists, chemists, soil specialists, etc.) will it be possible to minimize the uncertainties and thus the risks.
- The laboratory experiments generally consist of one standard test or turbo test. Again, the amount of soil subjected to the laboratory tests is but a tiny fraction of the total volume of soil that has to be remediated. When taking one or more

TABLE 33.5. Results of Some Electroreclamation Projects and Large Field Tests

Project	Volume (m³)	Start (mg/kg)	End (mg/kg)	Notes
Former paint factory, Groningen	300 peat/clay soil	Cu: >5000 Pb: >500–1000	Cu: <200 Pb: <280	First electroreclamation pilot project, 1987
Galvanizing plant, Delft	250 clay soil	Zn: >1400	Zn: 600	*In situ* electroreclamation of area outside of operational plant
Former timber impregnation plant, Loppersum	300 heavy clay soil	As: >250	As: <30	*In situ* electroreclamation at the site of a former timber impregnation plant; after remediation site became residential area
Temporary landfill, Stadskanaal	2500 agrillaceous sand	Cd: >180 (also Pb, Zn, CN)	Cd average: 11	*Ex situ* electroreclamation in two batches. Goal was to reduce Cd to <10 ppm. Total remediation time: 2 years (1990–1992)
Military airbase, Woensdrecht	3500 clay soil	Cr: 7300 Ni: 860 Cu: 770 Zn: 2600 Pb: 730 Cd: 660	Cr: 755 Ni: 80 Cu: 98 Zn: 289 Pb: 108 Cd: 47	*Ex situ* electroreclamation in temporary landfill. Goal was to reduce Cd to <50 mg/kg. Total remediation time: 2 years (1992–1994)
Batch test, Hallschlag	40 loamy soil	TNT 49 DNT 188 DNB 553 PAH 40 Org. As 11	TNT 10 DNT 3.3 DNB 6.8 PAH nd Org. As 0.1	*Ex situ* electroreclamation test with soil from an old World War I munitions factory in Germany (which exploded in 1918). During the test, nutrients were added to enhance biodegradation.
Kapelle	130 clay soil	CN 120 PAH 45	CN 18 PAH 2	Pilot project *ex situ* batch electroreclamation during 12 weeks.
Former gaswork at Oostburg	120 clay and sandy clay soil	CN 930	28	Pilot project *in situ* electroreclamation, partly underneath building. Depth up to 4-m bgs. Duration: 3 months. Removal percentages varied from 83% to 97%.
Operational galvanizing plant at 's-Heerenberg	4300 sandy loam and silty sand	Ni: 1350 Zn: 1300 Groundwater Ni: 3500 µg/l	Ni: 15 Zn: 75 Groundwater Ni: 15 µg/l	*In situ* electroreclamation project, inside production plant, started at the beginning of May 2003. Depth up to 6-m bgs
The Hague, former galvanizing plant	5800 medium-fine sand and silty clay layer	Groundwater Zn: 2000 µg/l	Groundwater Zn average: <600 µg/l after 2 years	*In situ* remediation of groundwater contaminated with zinc using electrostimulated groundwater extraction (ESGE). Running project. Target value Zn <400 µ/l. Project ends in 2009.

bgs, below ground surface; CN, cyanide; DNB, dinitrobenzene; DNT, dinitrotoluene.

soil samples for the laboratory experiments, those site locations where the highest concentrations are found (hot spots) should be chosen. At least the results of the laboratory experiments will then indicate the required energy and period of time necessary to clean up those hot spots.

33.4 SOME PROJECT RESULTS

In Table 33.5, results of a number of ER projects and large-scale field tests are shown.

33.5 SUMMARY

In this chapter, we have outlined and discussed what it takes to start and to finish a "real" commercial ER project, where one has to deal with clients and authorities and to reach a target value within a preset period. What parameters should be known to calculate necessary energy and time to reach a certain remediation value? How are these parameters obtained and what other field information is needed and what field equipment should be used, and finally how is an electrokinetic project designed and operated? Many parts of the technology are proprietary information, so we could not go into much detail with regard to field equipment.

INDEX

Abiotically/biotically dehaloelimination 523
Acenaphthene 677
Acetic acid 154
Acetone 202, 209
Acid front 70, 494
Acidic conditioning 168
Acidic front 105, 112
Activated carbon 20, 180, 400, 484, 633
Adenosine triphosphate (ATP) 376
Adsorption
  capacity 288, 290
  of heavy metals 99
Adsorption–desorption of NACs 267
Advanced oxidation technology 442
Advection–dispersion 570
Advective flux 540
Advective transport 6, 145
Advective velocity 39
Aerobic degradation 392
Agricultural soils 236
Air Quality Standards (NAAQS) 593
Alkaline conditions 114, 144, 307
Alkaline front 99
Alkaline pH conditions 475
Alkyl polyglucosides 204
Alternating current (AC) 671

Ammonia 86, 114
Amphoteric metals 161
Anaerobic microbial processes 525
Anaerobic oxidation 392
Anion exchange capacity (AEC) 705
Anionic metals 12
Anionic pollutants 141
Anionic reactive membranes 322, 326
Anionic species 12
Anionic surfactants 200, 244, 245, 269
Anionic-nonionic mixed surfactants 208
Anode-generated radicals 473
Anodic oxidation 478, 479
Anodophilic bacteria 375
Anthracene 677
Anthropogenic processes 198
Application area 703
Applications
  of electrokinetic technologies 590
  of electrokinetics in Europe 583
Aqueous solubility 515
Arrhenius equation 508
Arsenic 13, 105, 112, 114, 307, 352, 616, 617
Arsenic-contaminated soil 497
Atomic Energy Act (AEA) 604
Atomizing slag 458, 499, 655

Electrochemical Remediation Technologies for Polluted Soils, Sediments and Groundwater,
Edited by Krishna R. Reddy and Claudio Cameselle
Copyright © 2009 John Wiley & Sons, Inc.

Atrazine 250, 261
Availability of nutrients 22, 369, 393
Average tortuosity 289

Bacterial adhesion 375, 377
Bacterial mobility 396, 377
Bacteriological decomposition 362
Barrier filled with FeOOH 497
Based treatment zones 628, 630
Bench-scale electrobioremediation 409
Bench-scale test 609
Bentazone 252, 261
Bimetallic nanoparticles 443
Bioaccessibility 373
Bioaccumulation 223, 666
Bioaugmentation 395
Bioavailability 372, 531
    of contaminants 21
    of electron acceptors 393
    of pollution 370
Biocatalytic reactions 530
Biodegradability and biocompatibility 201
Biodegradable contaminants 390
Biodegradation 197
    capacity 529
    of chlorinated compounds 403
    constraints 393
    of diesel-contaminated soil 379
    of explosives 272
    pathways 390
    reaction 525
    of VOCs 366, 363
Biological activity 88
Biological immobilization 21
Biological processes in soil 423
Bioreduction 492
Bioremediation
    of chromium-contaminated soil 403
    limitations 392
    of organic contaminants 317
Biostimulation 13, 395
Biosurfactant-enhanced EK 210
Biosurfactants 14, 201, 203
Biotransformation 370
    of contaminants 391
    rates 373
Bipolar electrodes 115
Boiling point of the mixture 509
Boron-doped diamond electrodes 479, 481
Brackish conditions 619
Brij-30 204
Brij-35 202

BTEX 208, 212, 390, 684
Buffer capacity 65, 82
Buffering capacity 10, 187, 190
Bulk conductivity 43
By-product formation 620

CAA Requirements 601
Cadmium 105, 112, 296, 299, 340, 458, 616, 618
Calcareous soils 113
Capacitance 53
Carbon dioxide (CO_2) 509
Carbon electrodes 345
Carbonate complexes 170
Carcinogenic effects 198
Catabolically active microorganism 373
Catalytic iron wall 451
Catalyze direct reactions 480
Cathodic production of hydrogen peroxide 476
Cathodic protection 632, 704, 714
Cation exchange capacity (CEC) 46, 145, 266, 297, 699, 705
Cation transport 571
Cationic heavy metals 12
Cationic surfactants 243, 269
Cavitation phenomenon 442
Cesium 128, 130, 131, 134
Cetyltrimethylammonium bromide 208, 244, 269
Chain reactions 452
Change of polarity 479
Charge density 31
Charge flux 40, 570
Charged complex species 10
Chelating agents 16
Chemical adsorbents 484
Chemical energy 668
Chemical equilibrium 547
Chemical forms of metals 293
Chemical gradients 67
Chemical oxidation 197, 440
Chemical potential gradient 568, 570
Chemical precipitators 484
Chemical reactions 41
Chemical reduction of nitrate 442
Chemical speciation
    changes 10
    models 24
Chemical washing of sediments 151
Chemiosmosis 391
Chemistry of the bulk fluid 37

Chloride concentration 480, 512, 656
Chloride index 365
Chloride ions 141
Chlorinated benzenes 227
Chlorinated compounds 512
　aliphatic 526
　aliphatic hydrocarbon 14, 219
　aromatic 315
　organic 219, 498
Chlorinated pesticides 14, 235
Chlorinated solvents 390, 632
Chlorinated-hydrocarbon contaminated site 450
Chlorine dioxide 480
Chlorine 479
　production 480
Chlorobenzenes 14, 219
Chloroethene biodegradation 404
Chlorophenols 14, 219, 223, 448
Chromate 353
Chromium 13, 87, 105, 113, 299, 306, 458, 616, 618
　electromigration 187
　speciation 10
Citric acid 86, 165, 173
Clean Air Act (CAA) 590, 593
Clean Water Act (CWA) 590, 594
Cleanup methods for contaminated sediments 664
Clogging in the PRB 485
Coal tar 689
　particles 690
Coarse-grained soils 98
Coated titanium 714
Coefficient of electroosmotic conductivity 32
Colloid conductivity 670
Combination of electric and hydraulic gradients 339
Combined barrier 352
Cometabolic processes 526
Competitive ions 6
Complex charge 72
Complex microbial habitat 422
Complexation 11, 65, 83
Compost 425, 431
Comprehensive Environmental Response, Compensation, and Liability Act (CERCLA) 590, 591, 596
Conditioning liquids 322
Conduction and convection 23
Conductivity 42

Confined aquatic disposal 149, 150
Confined disposal facilities 150
Conservation of charge 571
Consolidation 17
Contaminant
　disposal 616
　mixtures 15, 22
　movement 610
　plume 633
　recovery 616
　spread 568
Contaminated ground 651
Contamination level 496, 648
Copper 111, 113, 296, 343, 616
Copper ions 708
　velocity 350
Corrosion of the ZVI 494
Cosolvents 14, 15, 200, 455
　for chlorinated pesticides 246
　-enhanced EK 209
　-enhanced systems 209
　extracting solution 320
Cost
　breakdown 586
　of electrochemical remediation 584
　of electrokinetic remediation 583
　of installations 584, 586
　of the Lasagna process 644
　of mixed contamination treatment 328
　treatment 155, 618
　variables 585
Cost-effective rectifier 630
Counterions 72
Coupled bio-electro processes 406
Coupled EK–chemical redox 440
Coupled EK–ultrasonic process 459
Coupled electrokinetic permeable reactive barrier system (EK-PRB) 27, 651
Coupled technologies 18
Coupled transport equations 67
Coupling electrokinetics-bioremediation 369
Coupling electrolysis-microbial degradation 404
Cr(VI)
　reduction 190, 191, 487, 489, 490
　-contaminated clay 492
　-contaminated soils 181
Creosote
　-contaminated clay 452
　-contaminated soil 378

Critical micelle concentration (CMC) 200, 239
Cross-contamination of water 249
Cumulative flow 38
Current density 40
CWA Requirements 601
Cyclic electric field 115
Cyclodextrins 14, 15, 87, 154, 201, 203, 210, 212, 229, 232, 299, 320, 453
   chemistry 269
   -enhanced EK 209

Dalton's law 509
Darcy's law 7
Darcy's velocity 237
Data on energy requirements 706
Data on remediation time 706
DC rectifier 636, 642
DDT 240, 242, 243, 245
Debye screening length 31
Dechlorinate VOCs 363
Dechlorinating microorganisms 528
Dechlorination reaction agents 498
Decomposition
   of water 6
   of VOCs 363
Degradation
   of contaminants 19
   of nitrate 464
   of organic compounds 152, 440
   rate 270
   of soils 419
Dehydrohalogenation 523
Demarcating the contaminants 711
Demonstration sites 583
Dense nonaqueous phase liquids (DNAPLs) 626, 497, 505
Density and temperature 516
Dental/skeletal fluorosis 143
Depletion of organic contaminants 657
Depolarization of cathode 166
Desorb/dissolve the heavy metals 97
Desorption
   /dissolution process 98
   of HOCs 228
   kinetics 228
   of PAHs 202
   phenomena 290
Detergents 315
Dewatering 17
   sediments 149, 151
Dichloroethene 390, 359

Dichloroethylene 499
Dichlorophenol 225
Dichloropropane 521
Dielectric constants 8, 244
   of pore fluid 8
Dielectric layer 669
Diesel-contaminated soil 208, 379
Diffuse double layer (DDL) 7, 51
Diffusion 9, 30
Diffusion potential 31
Diffusion coefficient 288, 289, 337
Diffusion of nanoiron 461
Diffusional flux of ions 40
Diffusive flux 540
Dimension stable anodes (DSAs) 473
Direct anodic oxidation 445, 446
Direct oxidation 473
Direct risks 714
Dissolved gases in groundwater 514
Dissolved ions 698
Distribution of contaminants 496
DNAPL 635
Dodecyl sulfate 269
Donated/accepted electrons 11
Double layer 8, 11
Double-layer structure 669
Dredged material/sediment 153, 663, 673
Dynamic chemical equilibrium 378
Dynamic geochemistry 10

Ebonex 714
Ebullition in saturated porous media 509
ECGO 671, 673
   ex situ treatment 683
   field demonstration 26
   in situ treatment 672
   technology 672, 694
Economic considerations 24
   of electrokinetics 584
Ecotoxicological assessment criteria 304
EDTA 85, 154, 172, 173, 299, 310, 324, 619
   -enhanced test 165
   -enhanced treatment 80, 162
   metal complexes 163
Effective diffusion coefficient 570
Effective diffusivity 373
Effective electrokinetic mobility 708
Effective ion mobilities 289, 700
Effective porosity 8
Efficient biotransformation 376
Eh–pH diagram 88

Einstein–Townsend relation 541
EK. *See* electrokinetics
Electric conductivity 161
Electric dipole 668
Electric double layer 30
Electric field strength 7
Electric fields 32, 59
Electric potential applied 6
Electric power supply 704
Electrical conductivity 621, 637
Electrical conductor 67
Electrical dipoles 11
Electrical energy 668
Electrical gradients 67
Electrical management 630
Electrical power requirements 365
Electrical resistance 543, 553
  heating (ERH) 22, 505, 636
  method 709
Electrobiofence 358, 366
Electrobioremediation 370, 403
Electrochemical biotransformation 377
Electrochemical degradation 224
Electrochemical Design Associates (EDA) 25, 614
Electrochemical geo-oxidation (ECGO) 21, 661
Electrochemical ion exchange 703
Electrochemical leaching 135
Electrochemical microreaction sites 447
Electrochemical model 52
Electrochemical oxidation 152, 445, 473, 540
Electrochemical oxidation/reduction 445
Electrochemical phenomena 699
Electrochemical production of electron donors/acceptors 396, 403
Electrochemical reactions 66, 88, 377
Electrochemical reduction 540
Electrochemical remediation costs 24
Electrochemical solid bed reactor 447
Electrochemical stabilization 22
Electrochemical transformations 11, 55
Electrochemical transport 29, 50
Electrochemistry and geochemistry 23
Electrodes 585
  array 653
  conditioning 299
  configuration 629
  filters 357
  installation 674
  materials 445, 479, 714
  placement 584
  polarity 449
  potentials 68
  reactions 46, 407
  rows 359, 361, 636
  selection 621
  system 615
Electrodialysis 117
Electrodialytic remediation 116, 303
Electroendosmosis 8
Electrokinetic Ltd. 484
Electrokinetic Biofence (EBF) 19, 357, 360, 362, 403, 704
Electrokinetic Fence (EKF) 339, 704
Electrokinetics (EK) 5, 379
  advantages 622
  assisted surface transformations 56
  barriers 17, 18, 24, 336, 352, 572
  biobarrier 18, 19
  bioremediation 18, 21
  chemical oxidation/reduction 18, 20
  coupled chemical redox processes 465
  delivery of nanoiron 462
  disadvantages 622
  dispersal of bacteria 371
  enhanced bioremediation 403
  enhanced flushing 18
  enhanced remediation 273
  enhanced by ultrasonic 321
  extraction 18
  facilitating agents 198
  Fenton process 447, 220
  gradient 568
  induced mass transfer 395
  induced mass transport 395
  laboratory tests 697, 705, 707
  limitations 619
  mechanisms 440
  mobility 708
  nanoparticle treatment 460
  optimization by soil type 279
  and permeable reactive barriers (PRBs) 18, 20, 499, 653
  phenomena 697, 699
  phytoremediation 18, 21
  point 670
  recovery of heavy metals 291
  remediation for mixed contamination 318
  remediation of sediments 5, 152, 173, 609
  removal of chlorobenzenes 229
  removal of toxic metabolites 377

Electrokinetics (EK) (*cont'd*)
  screens 129
  schematic diagram 647
  stabilization 18, 22
  technology application 422
  thermal treatment 18
  transport 71, 440
  transport of nutrients 400
  velocity 482, 708
Electro-Klean™ 484
Electrolysis
  reactions 8, 9, 23
  of water 290
Electrolytes 479
  circulation 361, 616
  decomposition of water 68
  effect of 211
  management 362, 702
  processes 714
  purification 703
  reactive barriers 18, 19, 26
  solution circulation 357
  stimulation 405
Electromigration 6, 30, 98, 337
  flux 540
  of ionic species 698
  of ions 40, 568
  nitrate transport 402
Electromobility of anionic surfactants 245
Electron acceptors 21, 317, 391, 400
Electron donors 21, 363, 391, 528
Electroneutrality 30, 70, 557
Electronic conductor in soils 669
Electro-osmosis 32, 337, 698, 701
  advection 7, 15, 492
  conductivity 23, 567
  efficiency coefficient 35
  flow 8, 14, 65, 73, 336, 481
  flow direction 76, 567
  flow rate 86
  flow velocity 7, 32, 237, 708
  flux 540
  mobility 38, 708
  permeability 489
  permeability coefficient 8, 450
  transport 32, 708
Electro-oxidation 453
Electro-oxidation–electroreduction 454
Electrophoresis 9, 244, 698, 701
Electrophoretic mobility 74
Electroreclamation (ER) 5, 611, 697

Electroremediation 5
Electrosorb process 25, 614
Electrosynthesis methods 21
Elevated temperatures 530
Energetic compounds 13, 15, 266
Energy
  consumption 171, 327, 492, 708
  requirements 540
  transport 506, 522
Engineer Research and Development
    Center (ERDC) 614
Engineering of bioremediation 389
Enhanced biodegradation 23, 395
Enhanced electrochemical remediation 5
Enhanced reduction 55
Enhanced reductive dechlorinization 359
Enhancement
  agents 71
  of contaminant bioavailability 395
  methods 111
  by reagent addition 112
  solutions 11
  strategies 12
EnViroCell® 703, 713
Environmental biotechnology 369
Environmental laws 25, 589
Environmental properties of cosolvents 201
Environmental regulations 25, 590
Equation for the mass transport 570
Equilibrium equations 540, 548
Equivalent circuit 54
ER laboratory tests 712
Escherichia coli 375
Estimated costs 24
Ethyl benzene 208
Evolution of oxygen 480
Ex situ field applications 661
Ex situ/in situ method 5
Excavation methods 265
Exoelectrogens 375
Exposure to pollutants 421
Extracting agents 151

Facilitating agents 232
Faradaic current 11, 52, 55
Faradaic paths 668
Faraday constant 31, 30, 68, 288, 568
Faraday's law 68, 405, 544
Federal Insecticide, Fungicide, and
    Rodenticide Act 590
Fenton process 440

Fenton's reagents 20, 21, 441, 476
Fermentation reactions 391
Fick's second law 40
Field applications 25
    in Europe 26
    in Korea 27
    setup 615
Field biofence 362
Field investigations 715
Field Setup 712
Field-scale application 609
Field-scale applications of electrokinetics 610
Fine-grained soils 65, 75, 98
Finite element model 569, 571
Flow of the current 30
Fluid viscosity 8
Fluoranthene 677
Fluorene 677
Fluoride 12
    complexes 146
    contamination 143
    mobility 146
Fluorite solubility 146
Fluvic acids 678
Fly ashes 118
Food production 250
Foreseeable risks 714
Formation of H^+ and OH^- ions 578
Freshwater sediments 118
Fuel hydrocarbons 315
Fugacity of CO_2 514
Full-Scale test 636
Full-time operation 642
Fungicides 223
Future directions 27

Gasoline hydrocarbons 197
Gene transfer between soil bacteria 377
Generation
    of hydroxyl ions 558, 698
    of protons 558, 698
Geochemical characteristics 26
Geochemical model 551
Geochemical processes 11, 66, 79
Geochemical reactions 22, 431
Geokinetics International, Inc. 611
Geotechnical stabilization 179
Geotextile fabric 631, 636
Gibbs free energy 391, 525, 527
Granular carbon electrodes 626
Granular Fe^0 492

Graphite rod 492
Groundwater 3
    conditions 628
    contaminant plume 19
    dissolved gases 514
    flow direction 359
    in situ remediation 484
    movement 628
    natural flow 336
    pollution 18, 358, 651
    redox potential 525
    samples 711
    and soil contamination 420
    steady flow 650
    temperature 512
Growth rate of thermophiles 530

Halogenated alkanes 519
Harbor sediments 118, 155
Harmful/hazardous by-products 620
Hazardous and Solid Waste Amendments (HSWA) 589, 590
Hazardous under RCRA 600
Heat capacity 508
Heat of vaporization 508
Heat transport 23
Heavily polluted sludge 616
Heavy metals 3, 5, 11, 15, 315, 698
    complexes 112
    contamination 287, 541, 657
    and organics 115
    remediation 440
    retention 99
    simultaneous removal of, with organic pollutants 322, 329
    and soil type 99
    -spiked kaolinite 609
HEDPA 86
Helmholtz double layer 668
Helmholtz–Smoluchowski
    equation 33, 305
    model 72, 73
    theory 7, 204
Henry's law constant 518
Henry's law 509
Herbicides 13, 14, 223
Heterogeneity of contaminant levels 693
Hexachlorobenzene 227, 240, 453
Hexavalent chromium 180
High-pH environment 9, 13
High-tech electrode materials 474

HOC 14, 80, 197, 228, 369
  aqueous solubility 210
  bioavailability 210, 379
  -bioremediation efficiency 379
Human health, impact on 250
Humic acids 678
Humic substances 35
Humin 679
Hydration shells 267
Hydraulic conductivity 7, 521, 568
Hydraulic flow 6, 7, 8
Hydraulic gradient 7, 67, 339, 567
Hydrocarbon contaminants 459
Hydrodechlorination 444
Hydrofracturing techniques 460
Hydrogen and hydroxyl ions 46
Hydrogen ions (H^+) 6, 9
Hydrogen peroxide 21, 321, 450, 474, 476
Hydrogenolysis 444
Hydrolysis 519, 532
Hydrophobic chemicals 372
Hydrophobic organic compounds (HOCs) 14, 80, 197, 228, 369
  aqueous solubility 210
  bioavailability 210, 379
  -bioremediation efficiency 379
Hydrophobic organic contaminations 474
Hydroxyl (OH^-) ions 6, 9, 546
Hydroxyl radicals 21, 440, 441, 450, 473, 481

Igepal 207, 209
Immobility of contaminants 79
Impact of electric fields on microorganisms 380
Implementing electrokinetic remediation 597
Implementing electrokinetics 621
In situ biodegradation 530
In situ electrodes 670
In Situ Electrokinetic Extraction (ISEE) 611
In situ field applications 661
In situ heating 517
In situ pilot project 667
In situ remediation
  bioremediation 279
  groundwater 484
  sediment remediation 154, 664
  thermal remediation 505, 526, 530
Inaccessibility of contaminants 393
Indirect risks 714

Induced mass transfer/transport 400
Induced polarization 668
Industrial sludge 118
Inhibited microbial activity 402
Inhibitory effect of multiple metals 297
Inhibitory effect 288
Initial ion concentration 6
Injection nanoscale zero-valent iron 463
Injection of nutrients 21
Inner Helmholtz layer (IHL) 668
Inner Helmholtz plane (IHP) 51
Innovative removal of contaminants 420
Inorganic pollutants 11, 17
Insecticides 223
Integration with conventional technologies 5
Interactions of geochemical processes 89
Interfacial tension 200
Intermittent current 153
Iodide/iodine 85
Ion charge 7
Ion exchange
  capacity 698
  membranes 16
  reactions 550
Ion mobility 7
Ion transport 43, 39
Ion velocity 41
Ionic conductivity 670
Ionic conductor 669
Ionic current 539, 543
Ionic migration 6, 492, 539
Ionic mobility 7, 70, 288, 289
Ionic surfactants 14
Ionic valence 7, 288
Iron
  -chromium speciation 191
  electrodes as anodes 477
  filing conveyance 626
  filings 632
  nanoparticles for nitrate reduction 443
  oxide/oxyhydroxides 184
  reactive wall 226
  -reducing bacteria 485
  -rich barrier 89
  -rich electrodes 180, 181
  -rich environment 187
  -rich mineral phase 191
  -rich sacrificial electrodes 22
  wall 456, 458
Irreversible permeabilization of cell membranes 408

Irreversible sorption 268
ISEE technology 617
Isoelectric point 567
Isomorphic substitution 72
Isopropyl alcohol 202
Isotron Corporation 614
Iterative numerical integration 544

Joule heating 23

Kinetics of biodegradation 529
Kirchhoff's laws 554

Laboratory experiments 715
Laboratory-scale studies 610
Land disposal restrictions 589
Large-scale test 634
Lasagna™ Process 26, 409, 455, 484, 583, 625
  configuration 626
  costs 638, 643
  implementation 627
  installation 630, 640
  systems 586
  treatment zones 643
Leaching technique 304
Lead Recovery Technology 614
Lead 111, 113, 130, 296, 352, 616
  -EDTA complexes 326
  -oxide electrodes 481
Light nonaqueous phase liquids (LNAPLs) 517
Liquid-soil heating 522
Low aqueous solubility 199, 370
Low buffering capacity 401
Low electrical fields 59
Low volatility 519
Low-permeability soils 5, 625
Low-pH conditions 9, 12, 22, 75, 76
Low-polarity organic molecules 268

Major cost variables 585
Management
  of nonhazardous solid wastes 592
  of pH 703
Mass conservation equation 540, 541, 552, 558
Mass transport equation 47
Materials requirements 584
Mathematical models 23, 540
Mechanisms
  in electrokinetic removal 288
  of metal removal 160
Membranes, selective 668
Mercury 111, 113, 114, 154, 309, 665
  -iodide complex 86
Metabolic activity 23, 530
Metabolic dechlorination 526
Metal complexes 161
Metal hydroxides 13
Metal hyperaccumulators
  aquatic species 428
  plants 427
Metal partitioning 158
Metal species 620
Metal transport competition 86
Metal–EDTA complexes 17, 84
Metal-tolerant plant 428
Methane production 514
Methanogenic conditions 528
Microbes 392, 431
  activity 408
  constraints 392
  degradation 394, 531
  metabolism 391
  nutrients 400
  pollution degradation 145
  temperature tolerance 529
  transport in porous media 373
  uptake rate 373
Microconductors 447
Microelectrodes 11, 50
Microorganisms 462, 484. *See also* microbes
Mineralogical and geochemical data 496
Mineralogy of soil 37
Mines
  drainage 335
  tailings 116
  waste 118
Mixed surfactants 202
Mixtures of contaminants 316
Mobility
  of contaminants 316
  of phenanthrene 326
  of surfactants 244
Mobilization
  of HOCs 238
  of hydrocarbons 396
  of phenol 225
Model
  of the electrokinetic flow phenomena 569
  of the electrokinetic remediation 539
  for ionic species 45

Model (cont'd)
   of the migration 569
   simulations 575
   validation 571
Modified nanoparticles 460
Moisture content 704
Molecular diffusion coefficient 7
Molinate 251, 261
Monitoring wells 365
Monodentate ligands 83
Movement
   of heavy metals 97, 338
   of pore water 539
Multidentate ligands 83
Multiphase flow 521
Multiple organic compounds 15
Municipal drinking water 223

NaCl purging solution 278
Nanoiron 57, 443
Nanomaterials 459
Nanoscale iron particles (NIPs) 461
Nanoscale ZVI 460
Nanosized iron slurry 461
Nanosized palladium/iron 463
Nanotechnology 279
Naphthalene 203, 515
NAPLs. *See* Nonaqueous phase liquids
National Ambient 593
National Contingency Plan (NCP) 591
National Pollutant Discharge Elimination System (NPDES) 594
National Priorities List (NPL) 591
n-butylamine 87, 209
Negative redox potentials 525
Nernst equation 543
Nernst–Einstein relation 39, 568
Nernst–Einstein–Townsend equation 7
Nernst–Planck equations 30
Net proton charge 72
Neumann boundary 45
Newton–Raphson method 550
Nickel 111, 113, 299, 318, 616
Ni–EDTA complexes 326
Nitrates 12
   barrier 351
   contamination 141, 142
   reduction 524
   transformation 145
Nitric acid 154, 168
Nitroaromatic compounds 15, 267

Nitrobenzene 275
Nonaqueous phase liquids (NAPLs) 81, 238, 515
   azeotropic systems 509
   density 516
   volatilization 532
Nonfaradaic paths 668
Noninsulated electrodes 153
Nonionic polyoxyethylenes 202
Nonionic surfactant 207, 245, 462
Nonlinear equation system 548
Nonuniform electric field 225
NPO Radon organization 136
Nuclear wastes 127
Numerical integration 542, 552
Numerical model 49
Numerical solution 541
Nutrients 358, 361
   dispersion 403
   restoration of 431
   transfer of charged 379
   transport 317, 378

Objects, buried 628
One-dimensional mass transport 47
One-dimensional model 24, 541
Operation on the field 640
Organic carbon demand 524
Organic compounds 3, 5
Organic contaminants 22, 625, 657
Organic cosolvents 232
Organic matter 212
Organic pollutants 13, 17, 440
ORP 681
OSHA requirements 603
Outer Helmholtz layer (OHL) 668
Outer Helmholtz plane (OHP) 51
Oxalic silica-alumina gel 131
Oxidation
   of hydrocarbons 477
   mediators 474
   of nanoiron 462
   of phenanthrene 321
   state 306
   –reduction potential 23, 55, 523
   –reduction reactions 10
Oxidative biodegradation 21
Oxidative degradation 391
Oxidizing agents 524
Oxygen 478
Ozone 478

PAHs. *See* Polycyclic aromatic hydrocarbons
Particle separation technology 663
Partitioning coefficient 241
Pb–EDTA complex 619
PCBs 665, 675
    treatment 680
Pentachlorophenol (PCP) 224, 450, 461
Perchloroethene 359, 390
Perchloroethylene (PCE) 220, 456, 498
Periodic electric potential 14, 15
Periodic voltage application 320
Permanent chemical transformations 670
Permanent structural charge 72
Permanganate 474, 477, 478
Permeable electrolyte wells 702
Permeable reactive barrier (PRB) 226, 455, 484, 648
    schematic diagram 649
Permittivity
    of the fluid 73
    of the solvent 32
Peroxodisulfate 480
Persistence of CPs 223
Persulfates 477
    production 481
Pesticides 13, 315, 519
    contamination 249
Petroleum hydrocarbons 683
pH
    adjustment 14
    buffering 211, 291
    change 46
    control 76, 211, 359
    gradient 6, 11, 65, 70, 79, 401
    high-pH environment 9, 13
    jump 9
    low-pH conditions 9, 12, 22, 75, 76
Phenanthrene 202, 203, 204, 209, 212, 321, 452, 677, 318
    removal 325
Phenols 197
    -contaminated soils 449
    phenolic compounds 155
Phosphates 145
Photolytic degradation 227
Physical–chemical interactions 540
Physicochemical cell surface properties 374
Physicochemical data 710
Physicochemical processes 324
Physicochemical remediation technologies 697

Phytoaccumulation 426
Phytoelectrokinetic 429
Phytoextraction 426
Phytofiltration 426
Phytoremediation 115, 317, 424, 430, 433
    EK 18, 21
    limitations 425
    wetlands 426
Phytostabilization 425, 426
Planar electrodes 629, 632
Plant toxicity 22
Platinum
    -coated titanium 137
    electrodes 479
Point of zero charge (PZC) 8, 73, 227
Poisson's equations 30
Polar organic solvents 200
Polarity, change of 479
Polarization 52
Polonium 130
Polychlorinated biphenyls (PCBs) 13, 315, 393, 604
Polychlorinated organic compounds 13
Polycyclic aromatic hydrocarbons (PAHs) 13, 14, 81, 87, 155, 197, 315, 390, 452, 515, 661, 665, 675
    -degrading soil bacteria 376
    properties 198
    results 688
    solubility 200, 202, 530
    source 694
    thermophilic degradation 531
Pore fluid chemistry 648
Porosity 8
Porous media, high 489
Positive zeta potential 460
Potential for emissions 600
Power
    consumption and requirements 584
    system 615
Precious metal 714
Precipitate/adsorb 99
Precipitation of heavy metal 540
Precipitation reactions 550
Precipitation–dissolution reactions 10, 297
Production/consumption rates 547
Protection
    clay screens 132
    electrokinetic barriers 129
    screens 129, 130, 132
Pseudonomas species 210, 203, 399
Public safety and protection 642

Pulsed electric fields 115
Pulsed high electrical field gradients 377
Pulsed voltage 346
Pump and treat remediation 265, 575
PVC chambers 655

Quantity of organic matter 316

Radical generation 475
Radioactive concentration 136
Radioactive contamination 127
Radioactive nuclides (radionuclides) 3, 5,
    11, 13, 127, 132, 315
  cleaning 132
  diffusion 129
  leaching 129
  pollutants 134
  removal 129
Radium 130
Raoult's law 509
RCRA regulations 600
RDX 265
Reactive iron barrier 353
Reactive materials 400, 649
Reactive media 20
Reactive membranes 326
Reactive zone 484, 486
Recalcitrant NAPL 530
Recultivation 419
Redox biodegradation 524
Redox chemistry 10
Redox potential of groundwater 525
Redox reactions 11, 52, 65, 87, 112
  at the microscale 668
Reducing agents 88
Reducing/oxidizing agents 10
Reduction
  of Cr(VI) 180, 184
  of TCE 626
  of ZVI 493
Reduction potential 88
Reductive anaerobic transformation
    processes 392
Reductive dechlorination 392, 443, 461,
    524
Reductive dehalogenation reactions
    528
Regulation
  considerations 25
  limits 157, 617, 585
  for RD&D 601
  for wastes and residuals 595

Remediation
  of CAHs 220
  data 710
  equipment 713
  limits 419
  options 711
  time 620, 705
Removal
  of CPs 223
  of heavy metals 97
  mechanisms of mixed metals 305
Resource Conservation and Recovery Act
    (RCRA) 589, 590, 592
Reverse EOF 243, 449, 493
Rhizofiltration 426
Rhizosphere 429
Rubidium 130
Runge–Kuta method 545

Sacrificial aluminum electrodes 179
Sacrificial electrodes 187, 642
Sacrificial iron electrodes 89, 353
Safe Drinking Water Act (SDWA) 590, 594,
    603
Safety issues 632
Salt removal 17
Sandia National Laboratories (SNL) 25,
    611
Saturated or vadose zones 643
Scanning electron microscope (SEM)
    182
SDWA. *See* Safe Drinking Water Act
Secondary reactions 480
Sediments 117
  characterization 150
  consolidation 153
  contaminated 3, 17, 149, 229, 693
  pH 160, 163
  treatment 151
Selective membranes 668
Selective oxidation 476
Sequential biodegradation pathway 404
Sequential electrochemical treatment 16
Sequential extracting solutions 319
Sequential extraction 304
Sequential oxidation and reduction 19
Settling velocity 153
Simultaneous removal of heavy metals and
    organic pollutants 322, 329
Site characteristics 628
Site preparation 585, 586
Sodium dodecyl sulfate 202

Sodium 269
  transport 340
Soil
  acidification 105
  aged polluted 97
  agricultural 236
  arsenic-contaminated 497
  biological processes 423
  biology 22
  bioremediation 389
  buffering capacity 11, 12
  calcareous 113
  changes in structure 423
  chemical analysis 621
  chromium-contaminated 403
  coarse-granied 98
  conductivity 6, 292
  -contaminant interactions 65
  controlling alkalization 191
  Cr(VI)-contaminated 181
  creosote-contaminated 378
  degradation 418, 419
  decontaminated materials 419
  diesel-contaminated 208, 379
  electronic conductor in 669
  fertility 427
  fine-grained 65, 75, 98
  gene transfer between bacteria 377
  and heavy metals 99
  heterogeneities 5, 620
  high buffering capacity 112, 113
  high organic content 207
  legislation 417
  low-permeability 5, 625
  mineralogy 37
  natural functions 417
  organic matter 268
  PAH-degrading bacteria 376
  permanent-charge 73
  permeability of granular 521
  pH 12, 16, 144
  phenol-contaminated 449
  pollution problem 418, 420
  porosity 6, 7
  properties 212, 232, 318, 431, 496
  quality 22
  remediation technology 697
  saturation 5, 621
  secondary pollution 475
  /solution chemistry 207
  structure changes 423, 432
  substance transformation 419
  surfactant-enhanced remediation 202
  TCEs in 222
  temperature 629, 638
  tortuosity 7
  types 417
  vapor extraction 572
  variable-charge 73
  washing 151
  -water sorption coefficients 517
Solar power 346, 365
Solidification/stabilization 151
Solid–liquid distribution coefficient 241
Solid-phase particulate coal tar 662
Solubility
  -enhancing agents 14, 197, 204, 232, 274
  of organics 239
  of PAHs 202, 530
  solubilization/desorption of PAHs 200
Soluble complexes 16
Soluble organic pollutants 197
Soluble-phase redox couples 88
Solvation water layer 669
Solvent mixtures 199
Solvent-flushing processes 200, 203
Sonochemical degradation 441
Sorption
  coefficient 238
  -desorption 10, 65, 79
  of hydrophobic organic compounds 237
  isotherm 238
Speciation, of heavy metals 294, 304
Specific conductance 38
Stability constants 85
Stainless steel 137
Standard reduction potentials 528
Steam stripping 514
Stern layer 52
Stimulated biodegradation 370, 400
Stimulation of bioremediation 145
Strategy of remediation 429
Streaming potential 569
Stripped contaminants 514
Strontium 130, 131
Sulfate-reducing conditions 528
Superfund sites 25
  cleanup process 592
Superfund Amendments and Reauthorization Act (SARA) 591
Surface charge of the solid matrix 8
Surface complexion reactions 550
Surface functional groups 72

Surfactants 14, 15, 81, 87, 199, 200, 238, 320,
   characteristics 200
   classification 200
   -coupled electrokinetic methods 244
   -enhanced remediation 202, 204, 238
   micelles 208, 243

TCE 633
   contamination 220, 400, 640
   dechlorination 634
   in low-permeability soils 222
   residual 641
Tergitol 204, 207
Terminal electron acceptors 369, 523
Test cell preparation 673
Tetrachloroethane 451
Tetrahydrofuran 209
Thermal conductivity 506, 507
Thermal diffusivity 507
Thermal oxidation and desorption 151
Thermal properties of materials 507
Thermal treatment 419
Thermally enhanced hydrolysis 519
Thermophilic degradation of PAHs 531
Thorium 127, 128, 130
Titanium 479, 714
Tortuosity 8
Total organic carbon (TOC) 453, 676, 678
Toxic radioactive elements 127, 128
Toxic Substances Control Act (TSCA) 590
Transference number 42, 298
Transient ORP 57
Transport distance 112
Transport number 556
Transport
   of charge 30
   of nutrients 317
   of oxidants 473, 474
   processes 6, 9, 23
Treatment costs 155, 618
Treatment zone spacing 629
Trichloroethane (TCA) 220
Trichloroethene 359, 390, 511
Trichloroethylene (TCE) 14, 197, 220, 463, 448
Trichloropropane 521
Trinitrotoluene 265
Triton X-100 202

Turbo test 706
Tween 80 202, 204, 207, 240, 242
Two-dimensional model 24, 552

Ultrasonic technology 21, 115, 459
Uncertainties in field applications 715
Unenhanced electrochemical remediation 5
Unenhanced sediment treatment 159
Unforeseeable risks 715
United States Army Environmental Center (USAEC) 25, 614
United States Environmental Protection Agency (USEPA) 25, 604, 611
Unpredictable performance 610
Uranium 130

Validation of the model 578
Vapor pressure 518
Vinyl chloride (VC) 363, 390, 499
Viscosity
   of organic liquids 517
   of pore fluid 8
   of solvent 32
Volatile organic compounds (VOCs) 360, 517, 596
Voltage gradient 8
   effect 213
   small 567
Volumetric flow rate 8

Waste disposal 3
Waste-water sludge 118
Water management 628
Water quality standards 594
Wetlands phytoremediation 426

Zeolite 499, 653
Zero-valent iron (ZVI) 20, 22, 13, 115, 145, 456, 497, 626, 629, 634, 653, 485, 443
   aggregate 499
   thickness of barrier 489
Zero-valent metal 222, 456
Zero-valent zinc 498
Zeta potential 8, 11, 23, 33, 65, 73, 208, 244, 306, 338, 628
Zinc 111, 113, 296, 343, 616
ZVI. *See* zero-valent iron
Zwitterionic surfactants 17, 322